STUDENT'S SOLUTIONS MANUAL

DAVID ATWOOD
Rochester Community and Technical College

A GRAPHICAL APPROACH TO COLLEGE ALGEBRA

FIFTH EDITION

John Hornsby
University of New Orleans

Margaret L. Lial
American River College

Gary Rockswold
Minnesota State University, Mankato

Addison-Wesley
is an imprint of

Reproduced by Pearson Addison-Wesley from electronic files supplied by the author.

Publishing as Pearson Addison-Wesley, 75 Arlington Street, Boston, MA 02116.

ISBN-13: 978-0-321-66461-7
ISBN-10: 0-321-66461-2

1 2 3 4 5 6 BRR 14 13 12 11 10

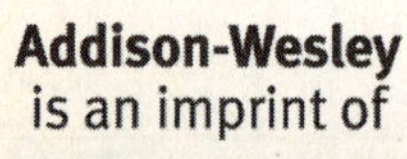

www.pearsonhighered.com

Contents

Preface

This solutions manual is written as an aid for students studying the text *A Graphical Approach to College Algebra*, 5th edition, by Hornsby, Lial, and Rockswold. It contains answers and solutions to the odd-numbered exercises as well as all solutions to the Relating Concepts, Reviewing Basic Concepts, Chapter Review, and Chapter Test exercises.

This manual is consistent with the text and gives solutions that are accessible to college algebra students. The solutions are written in a clear format including hundreds of traditional graphs and TI calculator screens to help with the learning process. Solutions frequently include both symbolic and graphical solutions to help students understand the problem more completely. It is recommended that you make a genuine attempt to solve a problem before looking in this solutions manual for help. Regular class attendance is also recommended. Following these suggestions will promote insight and understanding of college algebra.

It is this author's hope that you will find this manual helpful when studying the text *A Graphical Approach to College Algebra.* Please email any comments or questions to David.Atwood@roch.edu. Your opinion is important. Best wishes for an enjoyable and successful college algebra course.

David Atwood
Rochester Community and Technical College

Chapter 1: Linear Functions, Equations, and Inequalities

1.1: Real Numbers and the Rectangular Coordinate System

1. (a) The only natural number is 10.

   (b) The whole numbers are 0 and 10.

   (c) The integers are $-6, -\frac{12}{4}\text{ (or } -3), 0, 10.$

   (d) The rational numbers are $-6, -\frac{12}{4}\text{(or } -3), -\frac{5}{8}, 0, .31, .\overline{3},$ and 10.

   (e) The irrational numbers are $-\sqrt{3},\ 2\pi$ and $\sqrt{17}$.

   (f) All of the numbers listed are real numbers.

3. (a) There are no natural numbers listed.

   (b) There are no whole numbers listed.

   (c) The integers are $-\sqrt{100}\text{(or } -10)\text{ and } -1.$

   (d) The rational numbers are $-\sqrt{100}\text{ (or } -10), -\frac{13}{6}, -1, 5.23, 9.\overline{14}, 3.14,$ and $\frac{22}{7}$.

   (e) There are no irrational numbers listed.

   (f) All of the numbers listed are real numbers.

5. The number 10,600,000,000,000 is a natural number, integer, rational number, and real number.

7. The number -17 is an integer, rational, and real number.

9. The number $\frac{1}{5}$ is a rational and real number.

11. The number $5\sqrt{2}$ is a real number.

13. Natural numbers would be appropriate because population is only measured in positive whole numbers.

15. Rational numbers would be appropriate because shoes come in fraction sizes.

17. Integers would be appropriate because temperature is given in positive and negative whole numbers.

19. −4 −3 −2 −1 0 1

21. 0 $\frac{5}{3}$ −.5 .75 3.5

23. A rational number can be written as a fraction, $\frac{p}{q}$, $q \neq 0$, where p and q are integers. An irrational number cannot be written in this way.

25. The point $\left(2, \frac{5}{7}\right)$ is in Quadrant I. See Figure 25-34.

27. The point $(-3,-2)$ is in Quadrant III. See Figure 25-34.

29. The point $(0,5)$ is located on the y-axis, therefore is not in a quadrant. See Figure 25-34.

31. The point $(-2,4)$ is in Quadrant II. See Figure 25-34.

33. The point $(-2,0)$ is located on the x-axis, therefore is not in a quadrant. See Figure 25-34.

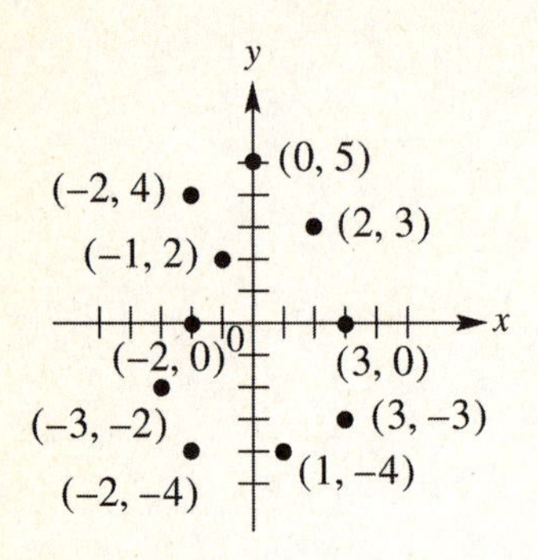

Figure 25-33

35. If $xy>0$, then either $x>0$ and $y>0 \Rightarrow$ Quadrant I, or $x<0$ and $y<0 \Rightarrow$ Quadrant III.

37. If $\frac{x}{y}<0$, then either $x>0$ and $y<0 \Rightarrow$ Quadrant IV, or $x<0$ and $y>0 \Rightarrow$ Quadrant II.

39. Any point of the form $(0,b)$ is located on the y-axis.

41. $[-5,5]$ by $[-25,25]$

43. $[-60,60]$ by $[-100,100]$

45. $[-500,300]$ by $[-300,500]$

47. See Figure 47.

49. See Figure 49.

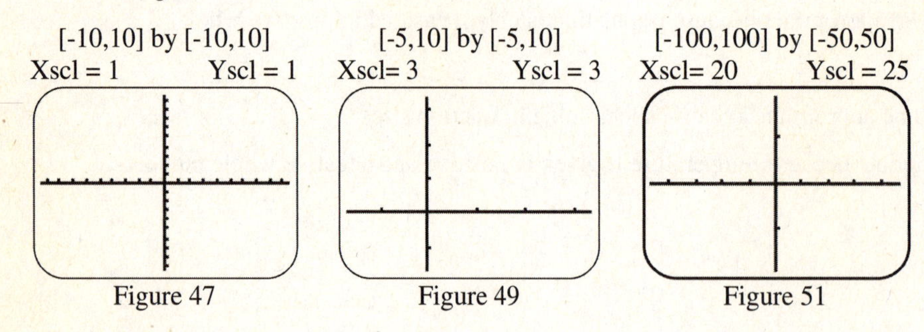

Figure 47 Figure 49 Figure 51

51. See Figure 51.

53. There are no tick marks, which is a result of setting Xscl and Yscl to 0.

55. $\sqrt{58} \approx 7.615773106 \approx 7.616$

57. $\sqrt[3]{33} \approx 3.20753433 \approx 3.208$

59. $\sqrt[4]{86} \approx 3.045261646 \approx 3.045$

61. $19^{1/2} \approx 4.35889844 \approx 4.359$

63. $46^{1.5} \approx 311.9871792 \approx 311.987$

65. $(5.6-3.1)/(8.9+1.3) \approx .25$

67. $\sqrt{(\pi \wedge 3+1)} \approx 5.66$

69. $3(5.9)^2 - 2(5.9) + 6 = 98.63$

71. $\sqrt{(4-6)^2 + (7+1)^2)} \approx 8.25$

73. $\sqrt{(\pi-1)} / \sqrt{(1+\pi)} \approx .72$

75. $2/(1-\sqrt[3]{5}) \approx -2.82$

77. $a^2 + b^2 = c^2 \Rightarrow 8^2 + 15^2 = c^2 \Rightarrow 64 + 225 = c^2 \Rightarrow 289 = c^2 \Rightarrow c = 17$

79. $a^2 + b^2 = c^2 \Rightarrow 13^2 + b^2 = 85^2 \Rightarrow 169 + b^2 = 7225 \Rightarrow b^2 = 7056 \Rightarrow b = 84$

81. $a^2 + b^2 = c^2 \Rightarrow 5^2 + 8^2 = c^2 \Rightarrow 25 + 64 = c^2 \Rightarrow 89 = c^2 \Rightarrow c = \sqrt{89}$

83. $a^2 + b^2 = c^2 \Rightarrow a^2 + (\sqrt{13})^2 = (\sqrt{29})^2 \Rightarrow a^2 + 13 = 29 \Rightarrow a^2 = 16 \Rightarrow a = 4$

85. (a) $d = \sqrt{(2-(-4))^2 + (5-3)^2} = \sqrt{(6)^2 + (2)^2} = \sqrt{36+4} = \sqrt{40} = 2\sqrt{10}$

(b) $M = \left(\frac{-4+2}{2}, \frac{3+5}{2}\right) = \left(\frac{-2}{2}, \frac{8}{2}\right) = (-1, 4)$

87. (a) $d = \sqrt{(6-(-7))^2 + (-2-4)^2} = \sqrt{(13)^2 + (-6)^2} = \sqrt{169+36} = \sqrt{205}$

(b) $M = \left(\frac{-7+6}{2}, \frac{4+(-2)}{2}\right) = \left(\frac{-1}{2}, \frac{2}{2}\right) = \left(-\frac{1}{2}, 1\right)$

89. (a) $d = \sqrt{(2-5)^2 + (11-7)^2} = \sqrt{(-3)^2 + (4)^2} = \sqrt{9+16} = \sqrt{25} = 5$

(b) $M = \left(\frac{5+2}{2}, \frac{7+11}{2}\right) = \left(\frac{7}{2}, \frac{18}{2}\right) = \left(\frac{7}{2}, 9\right)$

91. (a) $d = \sqrt{(-3-(-8))^2 + ((-5)-(-2))^2} = \sqrt{(5)^2 + (-3)^2} = \sqrt{25+9} = \sqrt{34}$

(b) $M = \left(\frac{-8+(-3)}{2}, \frac{-2+(-5)}{2}\right) = \left(\frac{-11}{2}, \frac{-7}{2}\right) = \left(-\frac{11}{2}, -\frac{7}{2}\right)$

93. (a) $d = \sqrt{(6.2-9.2)^2 + (7.4-3.4)^2} = \sqrt{(-3)^2 + (4)^2} = \sqrt{9+16} = \sqrt{25} = 5$

(b) $M = \left(\frac{9.2+6.2}{2}, \frac{3.4+7.4}{2}\right) = \left(\frac{15.4}{2}, \frac{10.8}{2}\right) = (7.7, 5.4)$

95. (a) $d = \sqrt{(6x-13x)^2 + (x-(-23x))^2} = \sqrt{(-7x)^2 + (24x)^2} = \sqrt{49x^2 + 576x^2} = \sqrt{625x^2} = 25x$

(b) $M = \left(\frac{13x+6x}{2}, \frac{-23x+x}{2}\right) = \left(\frac{19x}{2}, \frac{-22x}{2}\right) = \left(\frac{19}{2}x, -11x\right)$

97. Using the midpoint formula we get: $\left(\frac{7+x_2}{2},\frac{-4+y_2}{2}\right)=(8,5)\Rightarrow\left(\frac{7+x_2}{2}\right)=8\Rightarrow 7+x_2=16\Rightarrow x_2=9$ and

$\frac{-4+y_2}{2}=5\Rightarrow -4+y_2=10\Rightarrow y_2=14$. Therefore the coordinates are: $Q(19,14)$.

99. Using the midpoint formula we get: $\left(\frac{5.64+x_2}{2},\frac{8.21+y_2}{2}\right)=(-4.04,1.60)\Rightarrow\frac{5.64+x_2}{2}=-4.04\Rightarrow$

$5.64+x_2=-8.08\Rightarrow x_2=-13.72$ and $\frac{8.21+y_2}{2}=1.60\Rightarrow 8.21+y_2=3.20\Rightarrow y_2=-5.01$. Therefore the

coordinates are: $Q(-13.72,-5.01)$.

101. $M=\left(\frac{2008+2004}{2},\frac{25143+20082}{2}\right)=\left(\frac{4012}{2},\frac{45225}{2}\right)=(2006,22612.5)$; the cost was $\approx$ \$22612.5

103. In 2001, $M=\left(\frac{1999+2003}{2},\frac{17029+18810}{2}\right)=\left(\frac{4002}{2},\frac{35839}{2}\right)=(2001,17919.5)$; poverty level

was $\approx$\$17,920. In 2005, $y_1=kx_1$ it was $\approx$\$20,007.

105. (a) See Figure 105.

(b) $d=\sqrt{(50-0)^2+(0-40)^2}=\sqrt{(50)^2+(-40)^2}=\sqrt{2500+1600}=\sqrt{4100}\approx 64.0$ miles.

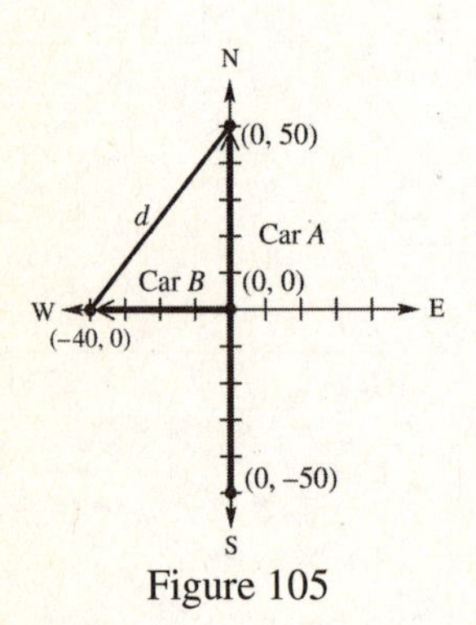

Figure 105

107. Using the area of a square produces: $(a+b)^2=a^2+2ab+b^2$. Now, using the sum of the small square and the four right triangles produces $c^2+4\left(\frac{1}{2}ab\right)=c^2+2ab$. Therefore $a^2+2ab+b^2=c^2+2ab$, and subtracting $2ab$ from both sides produces $a^2+b^2=c^2$.

1.2: Introduction to Relations and Functions

1. See Figure 1
3. See Figure 3

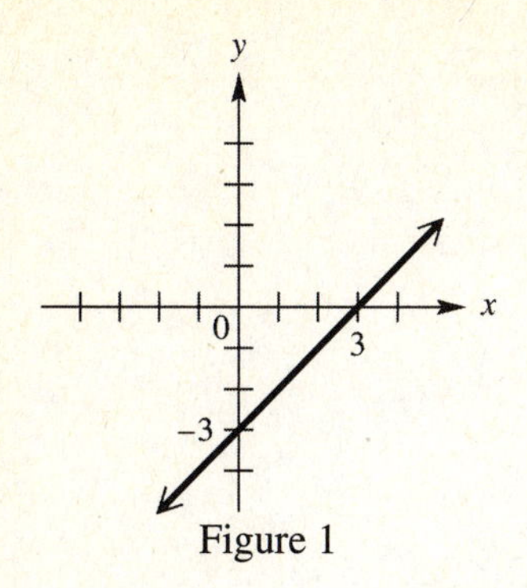

Figure 1

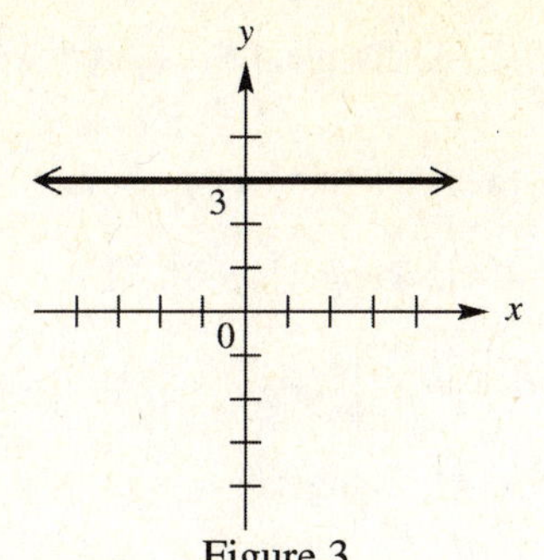

Figure 3

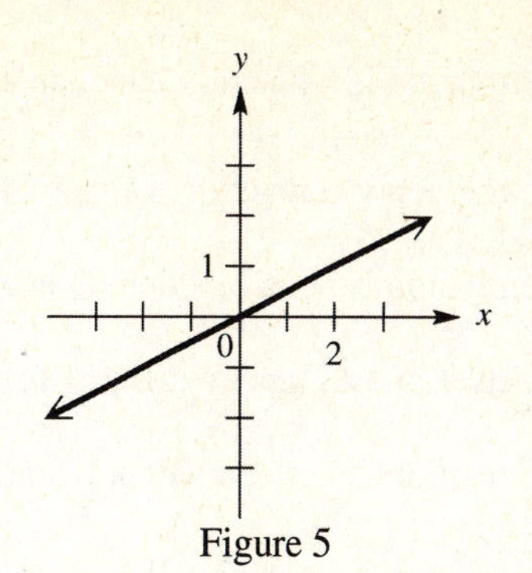

Figure 5

5. See Figure 5

7. See Figure 7

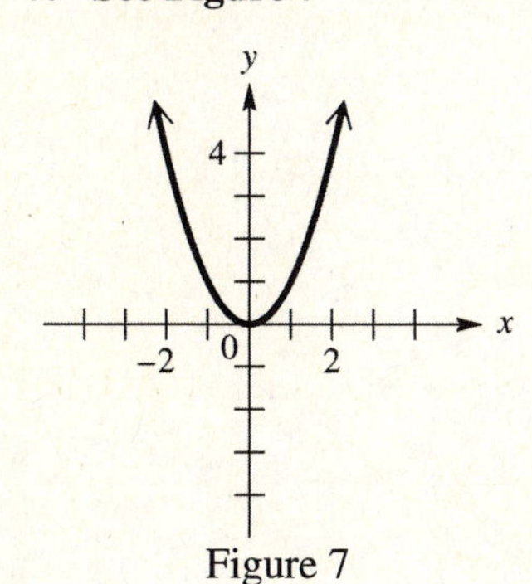

Figure 7

9. The interval is $(-1,4)$.

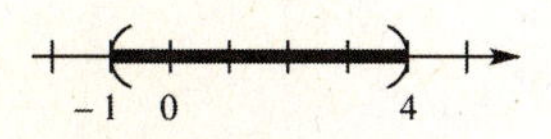

11. The interval is $(-\infty,0)$.

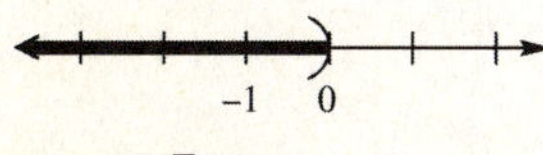

13. The interval is $[1,2)$.

15. $(-4,3) \Rightarrow \{x \mid -4 < x < 3\}$

17. $(-\infty,-1] \Rightarrow \{x \mid x \le -1\}$

19. $\{x \mid -2 \le x < 6\}$

21. $\{x \mid x \le -4\}$

23. A parenthesis is used if the symbol is $<$ or $>$. A square bracket is used if the symbol is $\le$ or $\ge$.

25. The relation is a function. Domain: $\{5,3,4,7\}$ Range: $\{1,2,9,6\}$.

27. The relation is a function. Domain: $\{1,2,3\}$, Range: $\{6\}$.

29. The relation is not a function. Domain: $\{4,3,-2\}$, Range: $\{1,-5,3,7\}$.

31. The relation is a function. Domain: $\{11,12,13,14\}$, Range: $\{-6,-7\}$.

33. The relation is a function. Domain: $\{0,1,2,3,4\}$, Range: $\{\sqrt{2},\sqrt{3},\sqrt{5},\sqrt{6},\sqrt{7}\}$.

35. The relation is a function. Domain: $(-\infty,\infty)$, Range: $(-\infty,\infty)$.

37. The relation is not a function. Domain: $[-4,4]$, Range: $[-3,3]$.

39. The relation is a function. Domain: $[2,\infty)$, Range: $[0,\infty)$.

41. The relation is not a function. Domain: $[-9,\infty)$, Range: $(-\infty,\infty)$.

43. The relation is a function. Domain: $\{-5,-2,-1,-.5,0,1.75,3.5\}$, Range: $\{-1,2,3,3.5,4,5.75,7.5\}$.

45. The relation is a function. Domain: $\{2,3,5,11,17\}$ Range: $\{1,7,20\}$.

47. From the diagram, $f(-2)=2$.

49. From the diagram, $f(11)=7$.

51. f(1) is undefined since 1 is not in the domain of the function.

53. $f(-2)=3(-2)-4=-6-4=-10$

55. $f(1)=2(1)^2-(1)+3=2-1+3=4$

57. $f(4)=-(4)^2+(4)+2=-16+4+2=-10$

59. $f(9)=5$

61. $f(-2)=\sqrt{(-2)^3+12}=\sqrt{-8+12}=\sqrt{4}=2$

63. $f(8)=|5-2(8)|=|-11|=11$

65. Since $f(-2)=3$, the point $(-2,3)$ lies on the graph of f.

67. Since the point $(7,8)$ lies on the graph of f, $f(7)=8$.

69. From the graph: (a) $f(-2)=0$, (b) $f(0)=4$, (c) $f(1)=2$, and (d) $f(4)=4$.

71. From the graph: (a) $f(-2)$ is undefined, (b) $f(0)=-2$, (c) $f(1)=0$, and (d) $f(4)=2$.

73. (a) – (f) Answers will vary. Refer to the definitions in the text.

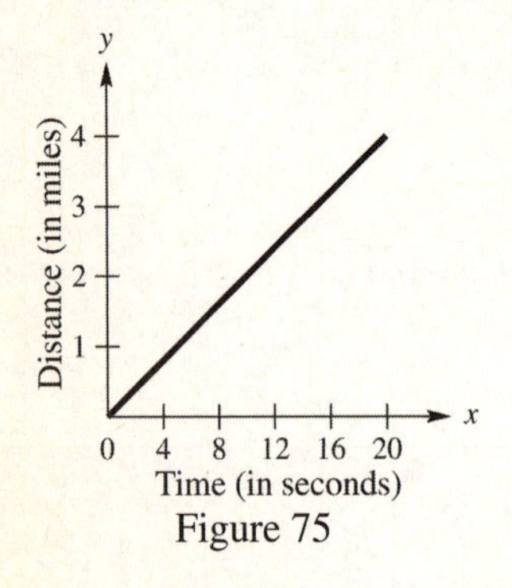

Figure 75

75. (a) $f(15)=\dfrac{15}{5}=3$. When the delay is 15 seconds, there are 3 miles between the lightning and the observer.

(b) See Figure 75.

77. The function that calculates a 7.5% tax on a sale of x dollar is $f(x) = 0.075x$. The tax on a purchase of \$86 is $f(86) = .075(86) = 6.45$ or \$6.45.

79. The function that calculates tuition and fees for taking x credits is $f(x) = 192x + 275$. The cost of taking 11 credits is $f(11) = 192(11) + 275 = 2387$ or \$2387.

Reviewing Basic Concepts (Sections 1.1 and 1.2)

1. See Figure 1.
2. The distance is $d = \sqrt{(6-(-4))^2 + (-2-5)^2} = \sqrt{100+49} = \sqrt{149}$.

   The midpoint is $M = \left(\frac{-4+6}{2}, \frac{5-2}{2}\right) = \left(1, \frac{3}{2}\right)$.
3. $\sqrt{5+\pi} / (\sqrt[3]{3}+1) \approx 1.168$
4. Using Pythagorean Theorem, $11^2 + b^2 = 61^2 \Rightarrow b^2 = 61^2 - 11^2 \Rightarrow b^2 = 3600 \Rightarrow b = 60$ inches.
5. The set $\{x \mid -2 < x \leq 5\}$ is the interval $(-2, 5]$. The set $\{x \mid x \geq 4\}$ is the interval $[4, \infty)$.
6. The relation is not a function because it does not pass the vertical line test. Domain: $[-2, 2]$, Range: $[-3, 3]$.
7. $f(-5) = 3 - 4(-5) = 23$
8. From the graph, $f(2) = 3$ and $f(-1) = -3$.
9. See Figure 9.
10. $d = \sqrt{(27-(-3))^2 + ((-4)-12)^2} = \sqrt{30^2 + (-16^2)} = \sqrt{900+256} = \sqrt{1156} = 34$

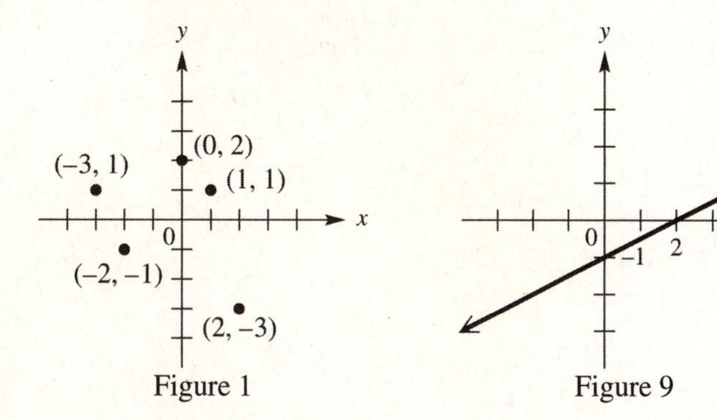

Figure 1

Figure 9

1.3: Linear Functions

1. The graph is shown in Figure 1.

   (a) x-intercept: 4 (b) y-intercept: -4 (c) Domain: $(-\infty, \infty)$ (d) Range: $(-\infty, \infty)$

   (e) The equation is in slope-intercept form, therefore $m = 1$.
3. The graph is shown in Figure 3.

(a) x-intercept: 2 (b) y-intercept: –6 (c) Domain: $(-\infty,\infty)$ (d) Range: $(-\infty,\infty)$

(e) The equation is in slope-intercept form, therefore $m = 3$.

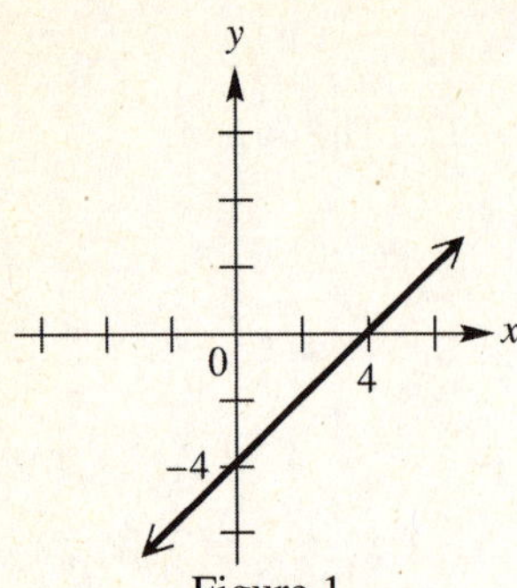

Figure 1

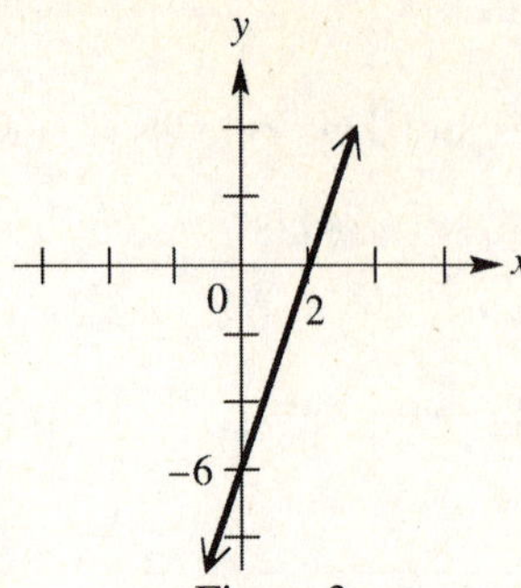

Figure 3

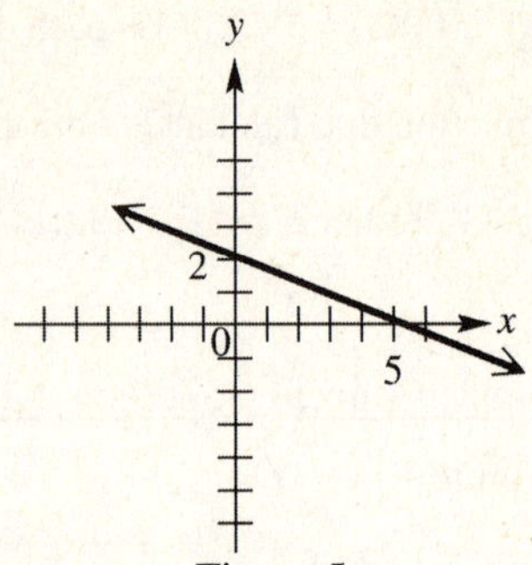

Figure 5

5. The graph is shown in Figure 5.

(a) x-intercept: 5 (b) y-intercept: 2 (c) Domain: $(-\infty,\infty)$ (d) Range: $(-\infty,\infty)$

(e) The equation is in slope-intercept form, therefore $m = -\frac{2}{5}$.

7. The graph is shown in Figure 7.

(a) x-intercept: 0 (b) y-intercept: 0 (c) Domain: $(-\infty,\infty)$ (d) Range: $(-\infty,\infty)$

(e) The equation is in slope-intercept form, therefore $m = 3$.

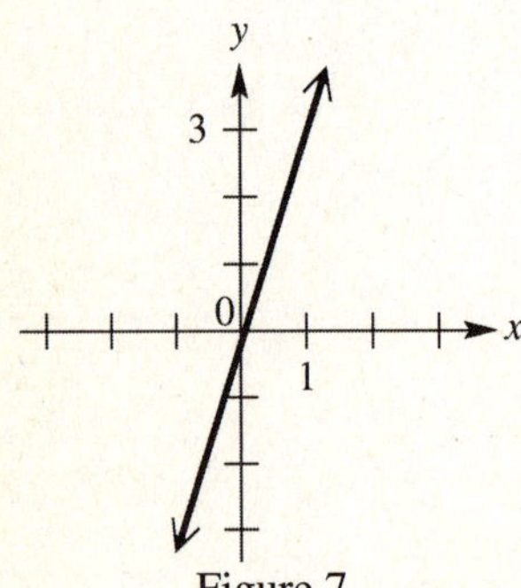

Figure 7

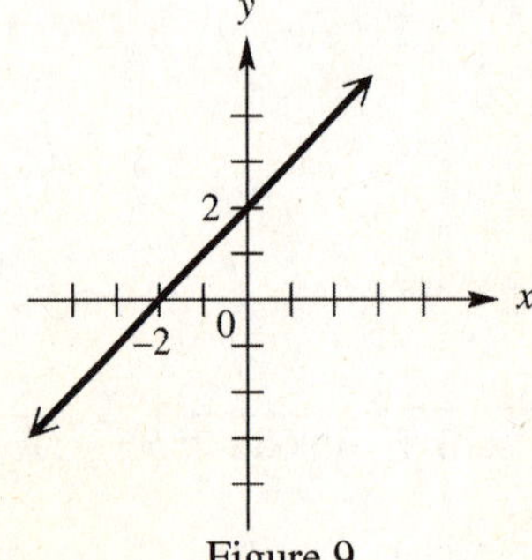

Figure 9

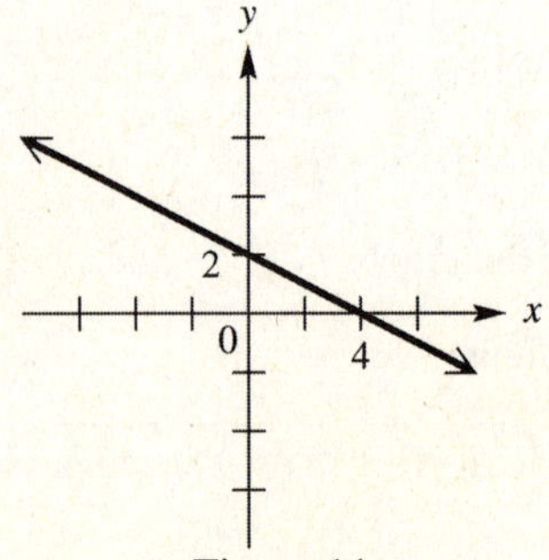

Figure 11

9. (a) $f(-2) = (-2) + 2 = 0$ and $f(4) = (4) + 2 = 6$

(b) $x + 2 = 0 \Rightarrow x = -2$

(c) The x-intercept is –2 and corresponds to the zero of f. See Figure 9.

11. (a) $f(-2) = 2 - \frac{1}{2}(-2) = 3$ and $f(4) = 2 - \frac{1}{2}(4) = 0$

(b) $2 - \frac{1}{2}x = 0 \Rightarrow \frac{1}{2}x = 2 \Rightarrow x = 4$

(c) The x-intercept is 4 and corresponds to the zero of f. See Figure 11.

13. (a) $f(-2) = \frac{1}{3}(-2) = -\frac{2}{3}$ and $f(4) = \frac{1}{3}(4) = \frac{4}{3}$

(b) $\frac{1}{3}x = 0 \Rightarrow x = 0$

(c) The x-intercept is 0 and corresponds to the zero of f. See Figure 13..

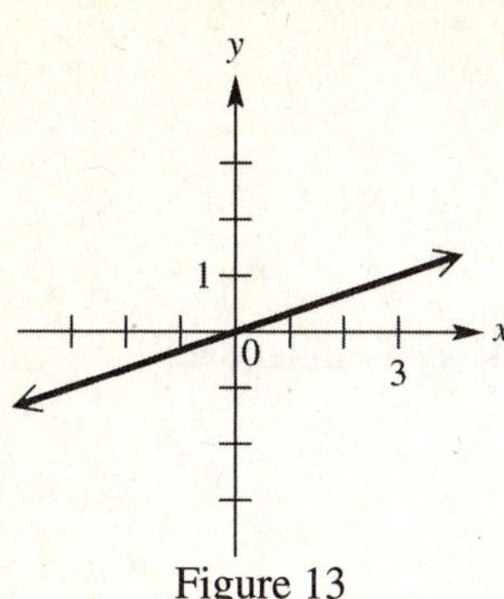

Figure 13

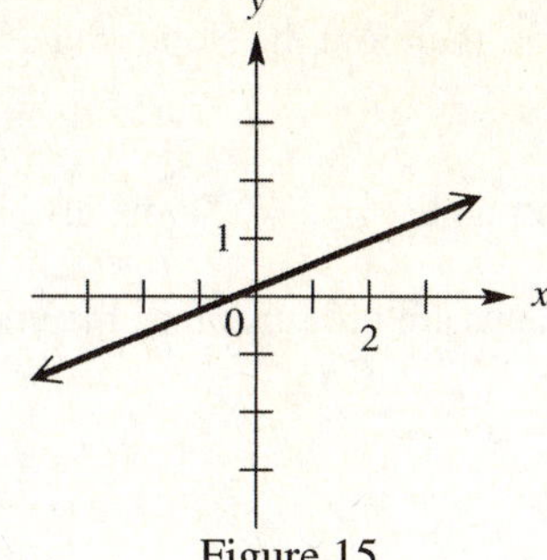

Figure 15

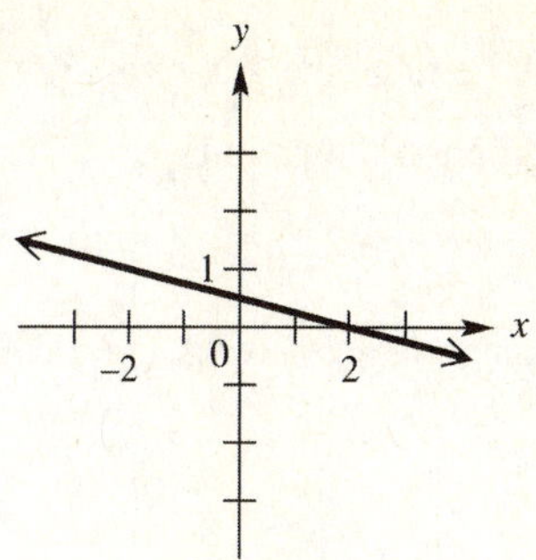

Figure 17

15. (a) $f(-2)=.4(-2)+.15=-.65$ and $f(4)=.4(4)+.15=1.75$

(b) $.4x+.15=0 \Rightarrow .4x=-.15 \Rightarrow x=-.375$

(c) The x-intercept is $-.375$ and corresponds to the zero of f. See Figure 15.

17. (a) $f(-2)=\dfrac{2-(-2)}{4}=1$ and $f(4)=\dfrac{2-(4)}{4}=-\dfrac{1}{2}$

(b) $\dfrac{2-x}{4}=0 \Rightarrow 2-x=0 \Rightarrow x=2$

(c) The x-intercept is 2 and corresponds to the zero of f. See Figure 17.

19. (a) $f(-2)=\dfrac{3(-2)+\pi}{2}=\dfrac{\pi-6}{2}$ and $f(4)=\dfrac{3(4)+\pi}{2}=\dfrac{12+\pi}{2}$

(b) $\dfrac{3x+\pi}{2}=0 \Rightarrow 3x+\pi=0 \Rightarrow x=-\dfrac{\pi}{3}$

(c) The x-intercept is $-\dfrac{\pi}{3}\approx -1.047$ and corresponds to the zero of f. See Figure 19.

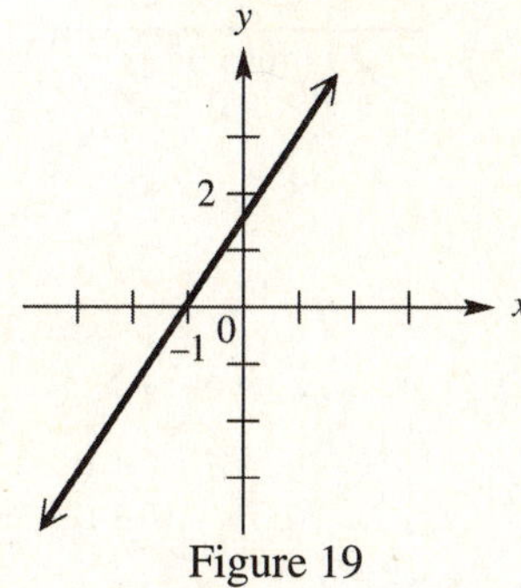

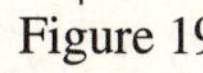
Figure 19

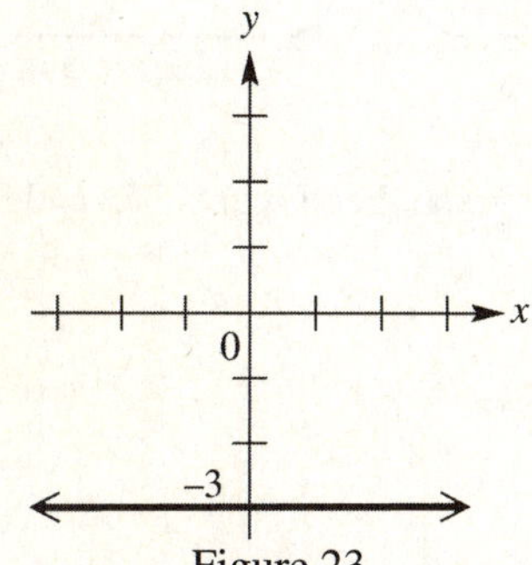

Figure 23

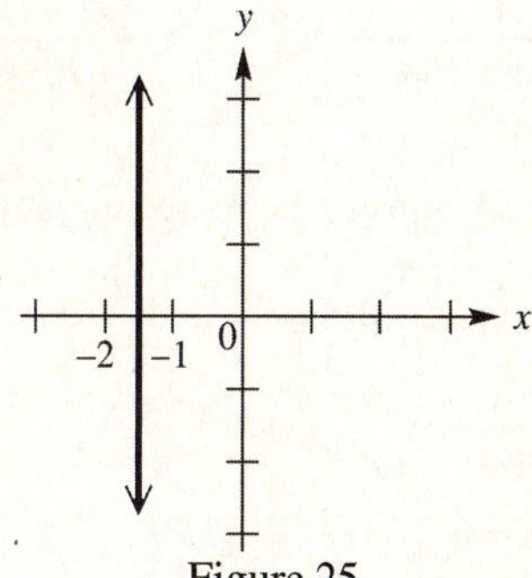

Figure 25

21. The graph of $y=ax$ always passes through (0, 0).

23. The graph is shown in Figure 23.

(a) x-intercept: none (b) y-intercept: -3 (c) Domain: $(-\infty,\infty)$ (d) Range: $\{-3\}$

(e) The slope of all horizontal line graphs or constant functions is $m=0$.

25. The graph is shown in Figure 25.

(a) x-intercept: -1.5 (b) y-intercept: none (c) Domain: $\{-1.5\}$ (d) Range: $(-\infty,\infty)$

(e) All vertical line graphs are not functions, therefore the slope is undefined.

27. The graph is shown in Figure 27.

(a) x-intercept: 2 (b) y-intercept: none (c) Domain: $\{2\}$

(d) Range: $(-\infty,\infty)$ (e) All vertical line graphs are not functions, therefore the slope is undefined.

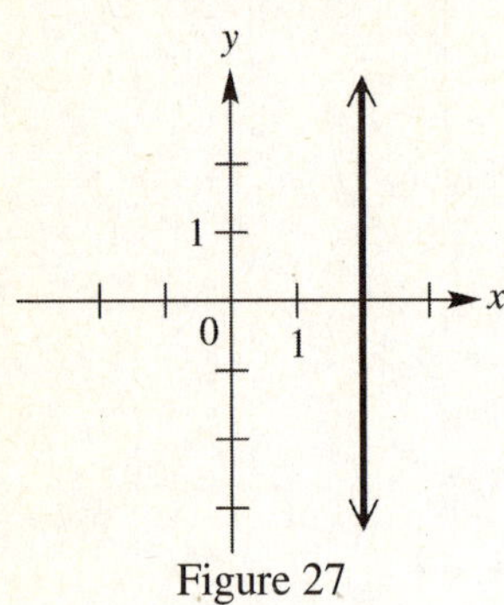

Figure 27

29. All functions in the form $f(x) = a$ are constant functions.

31. This is a horizontal line graph, therefore $y = 3$.

33. This is a vertical line graph on the y-axis, therefore $x = 0$.

35. Window B gives the more comprehensive graph. See Figures 35a and 35b.

[-10,10] by [-10,10]
Xscl = 1 Yscl = 1

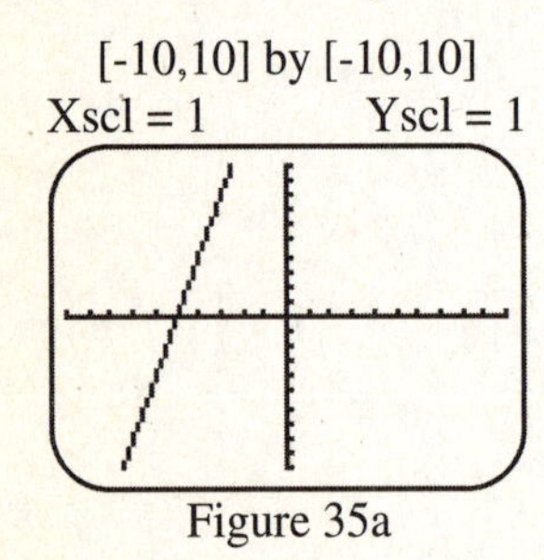

Figure 35a

[-10,10] by [-5,25]
Xscl = 1 Yscl = 5

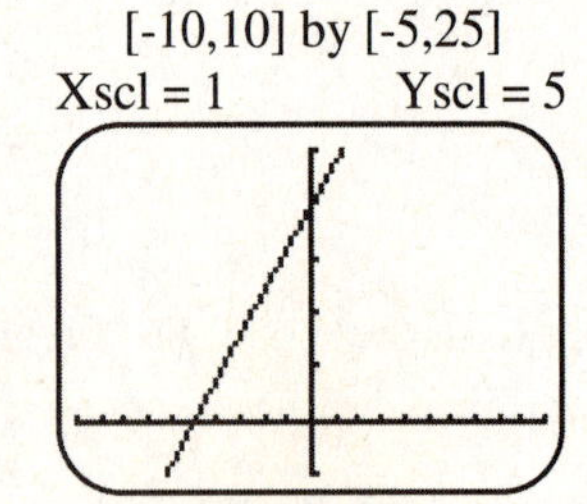

Figure 35b

[-3,3] by [-5,5]
Xscl= 1 Yscl = 1

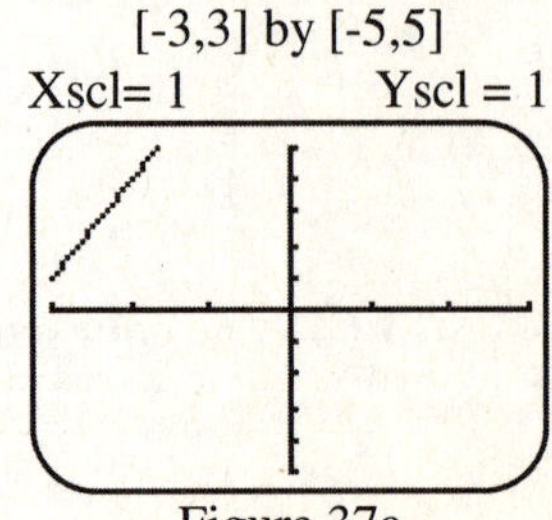

Figure 37a

[-5,5] by [-10,14]
Xscl = 1 Yscl= 2

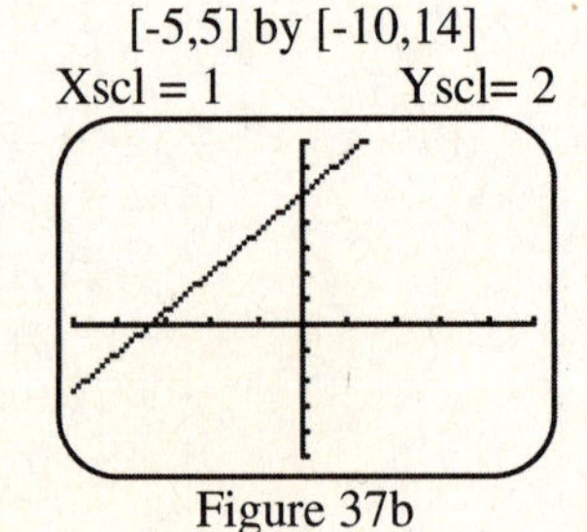

Figure 37b

37. Window B gives the more comprehensive graph. See Figures 37a and 37b..

39. $m = \dfrac{6-1}{3-(-2)} = \dfrac{5}{5} = 1$

41. $m = \dfrac{4-(-3)}{8-(-1)} = \dfrac{7}{9}$

43. $m = \dfrac{5-3}{-11-(-11)} = \dfrac{2}{0} \Rightarrow$ undefined slope

45. $m = \dfrac{9-9}{\frac{1}{2}-\frac{2}{3}} = \dfrac{0}{-\frac{1}{6}} \Rightarrow 0$

47. $m = \dfrac{-\frac{2}{3} - \frac{1}{6}}{\frac{1}{2} - \left(-\frac{3}{4}\right)} = \dfrac{-\frac{5}{6}}{\frac{5}{4}} \Rightarrow -\frac{2}{3}$

49. Since $m = 3$ and $b = 6$, graph A most closely resembles the equation.

51. Since $m = -3$ and $b = -6$, graph C most closely resembles the equation.

53. Since $m = 3$ and $b = 0$, graph H most closely resembles the equation.

55. Since $m = 0$ and $b = 3$, graph B most closely resembles the equation.

57. (a) The graph passes through (0,1) and (1,-1) $\Rightarrow m = \dfrac{-1-1}{1-0} = \dfrac{-2}{1} = -2$. The y-intercept is 1 and the x-intercept is $\dfrac{1}{2}$.

(b) Using the slope and y-intercept, the formula is $f(x) = -2x + 1$.

(c) The x-intercept is the zero of $f \Rightarrow \dfrac{1}{2}$.

59. (a) The graph passes through (0, 2) and (3,1) $\Rightarrow m = \dfrac{1-2}{3-0} = \dfrac{-1}{3} = -\dfrac{1}{3}$. The y-intercept is 2 and the x-intercept is 6.

(b) Using the slope and y-intercept, the formula is $f(x) = -\dfrac{1}{3}x + 2$.

(c) The x-intercept is the zero of $f \Rightarrow 6$.

61. (a) The graph passes through (0, 300) and (2, −100) $\Rightarrow m = \dfrac{-100-300}{2-0} = \dfrac{-400}{2} = -200$.

The y-intercept is 300 and the x-intercept is $\dfrac{3}{2}$.

(b) Using the slope and y-intercept, the formula is $f(x) = -200x + 300$.

(c) The x-intercept is the zero of $f \Rightarrow \dfrac{3}{2}$.

63. Using (0, 2) and (1, 6), $m = \dfrac{6-2}{1-0} = \dfrac{4}{1} = 4$. From the table, the y-intercept is 2. Using these two answers and slope-intercept form, the equation is $f(x) = 4x + 2$.

65. Using (0, −3.1) and (.2, −3.38), $m = \dfrac{-3.38-(-3.1)}{.2-0} = \dfrac{-.28}{.2} = -1.4$. From the table, the y-intercept is −3.1. Using these two answers and slope-intercept form, the equation is $f(x) = -1.4x - 3.1$.

67. The graph of a constant function with positive k is a horizontal graph above the x-axis. Graph A

69. The graph of an equation of the form $x = k$ with $k > 0$ is a vertical line right of the y-axis. Graph D

71. Using (−1, 3) with a rise of 3 and a run of 2, the graph also passes through (1, 6). See Figure 71.

73. Using (3, –4) with a rise of –1 and a run of 3, the graph also passes through (6, –5). See Figure 73.

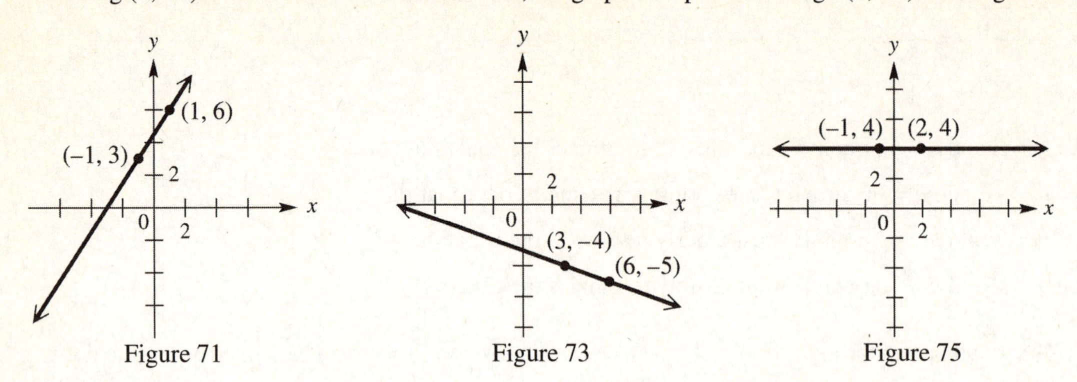

Figure 71 Figure 73 Figure 75

75. Using (–1, 4) with slope of 0, the graph is a horizontal line which also passes through (2, 4). See Figure 75.

77. Using (0, –4) with a rise of 3 and a run of 4, the graph also passes through (4, –1). See Figure 77

79. Using (–3, 0) with undefined slope, the graph is a vertical line which also passes through (–3, 2). See Figure 79.

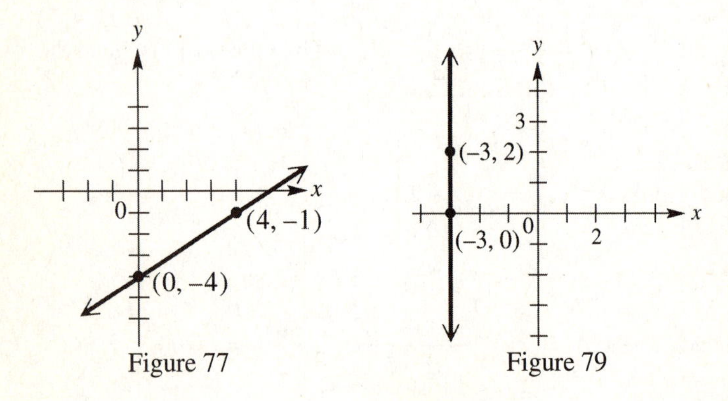

Figure 77 Figure 79

81. (a) Using the points (0, 2000) and (4, 4000), $m = \dfrac{4000-2000}{4-0} = \dfrac{2000}{4} = 500$. The y-intercept is $b = 2000$. The formula is $f(x) = 500x + 2000$.

(b) Water is entering the pool at a rate of 500 gallons per hour. The pool contains 2000 gallons initially.

(c) From the graph $f(7) = 5500$ gallons. By evaluating, $f(7) = 500(7) + 2000 = 5500$ gallons.

83. (a) The rain fell at a rate of $\frac{1}{4}$ inches per hour, so $m = \frac{1}{4}$. The initial amount of rain at noon was 3 inches, so $b = 3$. The equation $f(x) = \frac{1}{4}x + 3$.

(b) By 2:30 P.M. ($x = 2.5$), the total rainfall was $f(2.5) = \frac{1}{4}(2.5) + 3 = 3.625$ in.

85. (a) The birth rate is decreasing at a rate of .096 births per year, so $m = -0.096$. The initial birth rate was 14.8 births per 1000 people in 1995, so $b = 14.8$. The equation is $f(x) = -0.096x + 14.8$ (per thousand).

(b) Since $x = 8$ for 2001, $f(8) = -.096(8) + 14.8 = 14.032$ births per thousand people. This favorably compares to the actual number of 14.1 births per thousand people

1.4: Equations of Lines and Linear Models

1. Using Point-Slope Form yields $y-3 = -2(x-1) \Rightarrow y-3 = -2x+2 \Rightarrow y = -2x+5.$

3. Using Point-Slope Form yields $y-4 = 1.5(x-(-5)) \Rightarrow y-4 = 1.5x+7.5 \Rightarrow y = 1.5x+11.5.$

5. Using Point-Slope Form yields $y-1 = -.5(x-(-8)) \Rightarrow y-1 = -.5x-4 \Rightarrow y = -.5x-3.$

7. Using Point-Slope Form yields $y-(-4) = 2\left(x-\frac{1}{2}\right) \Rightarrow y+4 = 2x-1 \Rightarrow y = 2x-5.$

9. Using Point-Slope Form yields $y-\frac{2}{3} = \frac{1}{2}\left(x-\frac{1}{4}\right) \Rightarrow y-\frac{2}{3} = \frac{1}{2}x-\frac{1}{8} \Rightarrow y = \frac{1}{2}x+\frac{13}{24}.$

11. Use the points to (–4, –6) and (6, 2) find the slope: $m = \frac{2-(-6)}{6-(-4)} \Rightarrow m = \frac{4}{5}$. Now using Point-Slope Form yields $y - 2 = \frac{4}{5}(x-6) \Rightarrow y-2 = \frac{4}{5}x-\frac{24}{5} \Rightarrow y = \frac{4}{5}x-\frac{14}{5}.$

13 Use the points (–12, 8) and (8, –12) to find the slope: $m = \frac{-12-8}{8-(-12)} \Rightarrow m = \frac{-20}{20} \Rightarrow m = -1$. Now using Point-Slope Form yields $y - 8 = -1(x + 12) \Rightarrow y - 8 = -x - 12 \Rightarrow y = -x - 4.$

15. Use the points (4, 8) and (0, 4) to find the slope: $m = \frac{4-8}{0-4} \Rightarrow m = \frac{-4}{-4} \Rightarrow m = 1$. Now using Slope-Intercept Form yields $b = 4 \Rightarrow y = x+4.$

17. Use the points (3, –8) and (5, –3) to find the slope: $m = \frac{-3-(-8)}{5-3} \Rightarrow m = \frac{5}{2}$. Now using Point-Slope Form yields $y-(-8) = \frac{5}{2}(x-3) \Rightarrow y+8 = \frac{5}{2}x-\frac{15}{2} \Rightarrow y = \frac{5}{2}x-\frac{31}{2}.$

19. Use the points (2, 3.5) and (6, –2.5) to find the slope: $m = \frac{-2.5-3.5}{6-2} \Rightarrow m = \frac{-6}{4} \Rightarrow m = -1.5$. Now using Point-Slope Form yields $y-3.5 = -1.5(x-2) \Rightarrow y-3.5 = -1.5x+3 \Rightarrow y = -1.5x+6.5.$

21. Use the points (0, 5) and (10, 0) to find the slope: $m = \frac{0-5}{10-0} \Rightarrow m = \frac{-5}{10} \Rightarrow m = -\frac{1}{2}$. Now using Point-Slope Form yields $y-5 = -\frac{1}{2}(x-0) \Rightarrow y-5 = -\frac{1}{2}x \Rightarrow y-\frac{1}{2}x+5.$

23. Use the points (–5, –28) and (–4, –20) to find the slope: $m = \frac{-20-(-28)}{-4-(-5)} \Rightarrow m = \frac{8}{1} \Rightarrow m = 8$. Now using Point-Slope Form yields $y-(-20) = 8(x-(-4)) \Rightarrow y+20 = 8x+32 \Rightarrow y = 8x+12.$

25. Use the points (2, - 5) and (4, – 11) to find the slope: $m = \frac{-11-(-5)}{4-2} \Rightarrow m = \frac{-6}{2} \Rightarrow m = -3.$ Now using Point-Slope Form yields $y-(-5) = -3(x-2) \Rightarrow y+5 = -3x+6 \Rightarrow y = -3x+1.$

27. To find the x-intercept set $y = 0,$ then $x - 0 = 4 \Rightarrow x = 4.$ Therefore (4, 0) is the x-intercept. To find the y-intercept set $x = 0,$ then $0 - y = 4 \Rightarrow y = -4.$ Therefore (0, –4) is the y-intercept. See Figure 27.

29. To find the x-intercept set $y = 0,$ then $3x - 0 = 6 \Rightarrow 3x = 6 \Rightarrow x = 2.$ Therefore (2, 0) is the x-intercept. To find the y-intercept set $x = 0,$ then $3(0) - y = 6 \Rightarrow y = -6.$ Therefore (0, –6) is the y-intercept. See Figure 29.

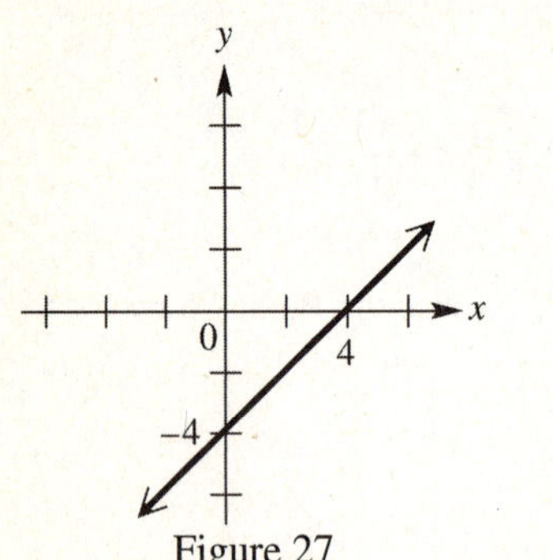

Figure 27

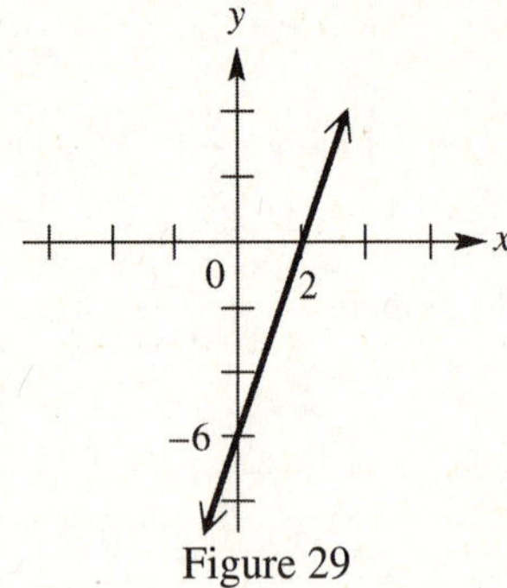

Figure 29

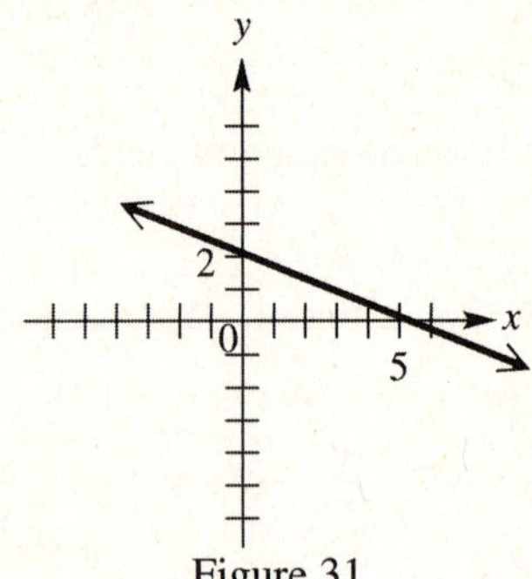

Figure 31

31. To find the x-intercept set $y = 0,$ then $2x + 5(0) = 10 \Rightarrow 2x = 10 \Rightarrow x = 5.$ Therefore $(5,0)$ is the x-intercept. To- find the y-intercept set $x = 0,$ then $2(0) + 5y = 10 \Rightarrow 5y = 10 \Rightarrow y = 2.$ Therefore $(0,2)$ is the y-intercept. See Figure 31.

33. To find a second point set $x = 1,$ then $y = 3(1) \Rightarrow y = 3.$ A second point is $(1,3).$ See Figure 33.

35. To find a second point set $x = 4,$ then $y = -.75(4) \Rightarrow y = -3.$ A second point is $(4,-3)$. See Figure 35.

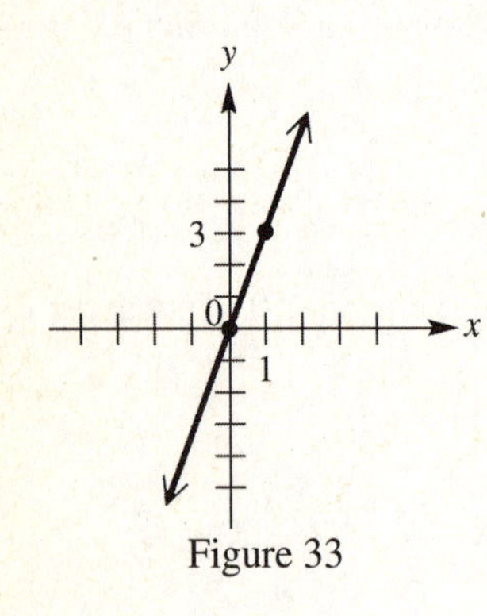

Figure 33

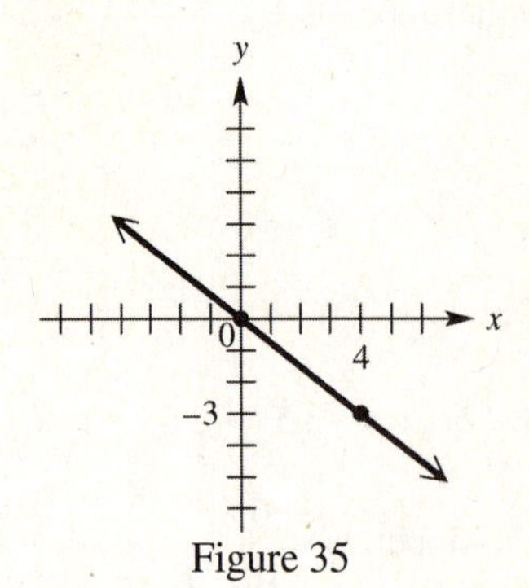

Figure 35

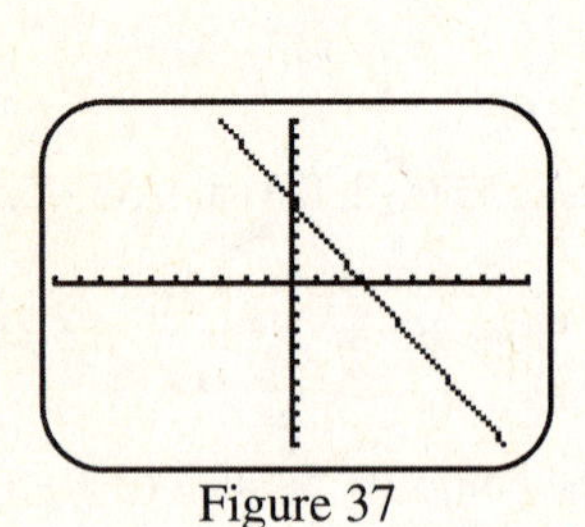
Figure 37

37. $5x + 3y = 15 \Rightarrow 3y = -5x + 15 \Rightarrow y = -\frac{5}{3}x + 5.$ See Figure 37.

39. $-2x + 7y = 4 \Rightarrow 7y = 2x + 4 \Rightarrow y = \frac{2}{7}x + \frac{4}{7}.$ See Figure 39.

41. $1.2x + 1.6y = 5.0 \Rightarrow 12x + 16y = 50 \Rightarrow 16y = -12x + 50 \Rightarrow y = -\frac{12}{16}x + \frac{50}{16} \Rightarrow y = -\frac{3}{4}x + \frac{25}{8}.$ See Figure 41.

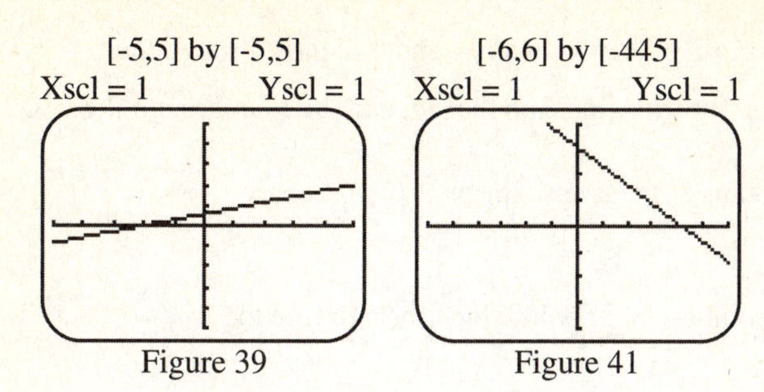

Figure 39 Figure 41

43. Put into slope-intercept form to find slope: $x+3y=5 \Rightarrow 3y=-x+5 \Rightarrow y=-\frac{1}{3}x+\frac{5}{3} \Rightarrow m=-\frac{1}{3}$.

Since parallel lines have equal slopes, use $m=-\frac{1}{3}$ and $(-1, 4)$ in point-slope form to find the equation:

$y-4=-\frac{1}{3}(x-(-1)) \Rightarrow y-4=-\frac{1}{3}x-\frac{1}{3} \Rightarrow y=-\frac{1}{3}x+\frac{11}{3}$.

45. Put into slope-intercept form to find slope: $3x+5y=1 \Rightarrow 5y=-3x+1 \Rightarrow y=-\frac{3}{5}x+\frac{1}{5} \Rightarrow m=-\frac{3}{5}$. Since perpendicular lines have negative reciprocal slopes, use $m=\frac{5}{3}$ and $(1, 6)$ in point-slope form to find the equation: $y-6=\frac{5}{3}(x-1) \Rightarrow y-6=\frac{5}{3}x-\frac{5}{3} \Rightarrow y=\frac{5}{3}x+\frac{13}{3}$.

47. The equation $y=-2$ has a slope $m=0$. A line perpendicular to this would have an undefined slope which would have an equation in the form $x=a$. An equation in the form $x=a$ through $(-5, 7)$ is $x=-5$.

49. The equation $y=-.2x+6$ has a slope $m=-0.2$. Since parallel lines have equal slopes, use $m=-2$ and $(-5, 8)$ in point-slope form to find the equation

$y-8=-0.2(x-(-5)) \Rightarrow y-8=-0.2x-1 \Rightarrow y=-0.2x+7$.

$y-(-7)=-1(x-(-4)) \Rightarrow y+7=-x-4 \Rightarrow y=-x-11$.

51. Put into slope-intercept form to find slope: $2x+y=6 \Rightarrow y=-2x+6 \Rightarrow m=-2$. Since perpendicular lines have negative reciprocal slopes, use $m=\frac{1}{2}$ and the origin $(0, 0)$ in point-slope form to find the equation

$y-0=\frac{1}{2}(x-0) \Rightarrow y=\frac{1}{2}x$.

53. The equation $x=3$ has an undefined slope. A line perpendicular to this would have a slope $m=0$, which would have an equation in the form $y=b$. An equation in the form $y=b$ through $(1, 2)$ is $y=2$.

55. We will first find the slope of the line through the given points: $m=\dfrac{\frac{2}{3}-\frac{1}{2}}{-3-(-5)}=\dfrac{\frac{1}{6}}{2} \Rightarrow m=\frac{1}{12}$. Since perpendicular lines have negative reciprocal slopes, use $m=-12$ and the point (-2,4) in point-slope form to find the equation $y-4=-12(x-(-2)) \Rightarrow y=-12x-20$.

57. The slope of the perpendicular bisector will have a negative reciprocal slope and will pass through the midpoint of the line segment joined by the two points. We will first find the slope of the line through the given points: $m=\frac{10-2}{2-(-4)}=\frac{8}{6}\Rightarrow m=\frac{4}{3}$. The midpoint of the line segment is $\left(\frac{-4+2}{2},\frac{2+10}{2}\right)=(-1,6)$. Use $m=-\frac{3}{4}$ and the point (-1,6) in point-slope form to find the equation $y-6=-\frac{3}{4}(x-(-1))\Rightarrow y=-\frac{3}{4}x+\frac{21}{4}$.

59. (a) The Pythagorean Theorem and its converse.

(b) Using the distance formula from (0, 0) to (x_1,m_1x_1) yields: $d(0,P)=\sqrt{(x_1)^2+(m_1x_1)^2}$.

(c) Using the distance formula from (0, 0) to (x_2,m_2x_2) yields: $d(0,Q)=\sqrt{(x_2)^2+(m_2x_2)^2}$.

(d) Using the distance formula from (x_1,m_1x_1) to (x_2,m_2x_2) yields:

$d(P,Q)=\sqrt{(x_2-x_1)^2+(m_2x_2-m_1x_1)^2}$.

(e) Using Pythagorean Theorem yields: $[d(0,P)]^2+[d(0,Q)]^2=[d(P,Q)]^2\Rightarrow$

$(x_1)^2+(m_1x_1)^2+(x_2)^2+(m_2x_2)^2=(x_1-x_2)^2+(m_1x_1-m_2x_2)^2\Rightarrow (x_1)^2+(m_1x_1)^2+(x_2)^2+(m_2x_2)^2=$

$(x_1)^2-2x_1x_2+(x_2)^2+(m_1x_1)^2-2m_1m_2x_1x_2+(m_2x_2)^2\Rightarrow 0=-2m_1m_2x_1x_2-2x_1x_2$.

(f) $0=-2x_1x_2-2m_1m_2x_1x_2\Rightarrow 0=-2x_1x_2(1+m_1m_2)$

(g) By the zero-product property, for $-2x_1x_2(1+m_1m_2)=0$ either $-2x_1x_2=0$ or $1+m_1m_2=0$. Since $x_1\neq 0$ and $x_2\neq 0$, $-2x_1x_2\neq 0$, and it follows that $1+m_1m_2=0\Rightarrow m_1m_2=-1$.

(h) The product of the slopes of two perpendicular lines, neither of which is parallel to an axis, is -1.

61. (a) Use the given points to find slope, then $m=\frac{161-128}{4-1}=\frac{33}{3}\Rightarrow m=11$. Now use point-slope form to find the equation: $y-128=11(x-1)\Rightarrow y-128=11x-11\Rightarrow y=11x+117$.

(b) From the slope the biker is traveling 11 mph.

(c) At $x=0$, $y=11(0)+117\Rightarrow y=117$, therefore 117 miles from the highway.

(d) Since at 1 hour and 15 minutes $x=1.25$, then $y=11(1.25)+117\Rightarrow y=130.75$, so 130.75 miles away.

63. (a) Use the points (2002, 4), (2005, 10) to find slope, then $m=\frac{10-4}{2005-2002}=\frac{6}{3}\Rightarrow m=2$. Now use point-slope form to find the equation: $y-4=2(x-2002)\Rightarrow y-4=2x-4004\Rightarrow y=2x-4000$.

(b) y = 2(2008) – 4000 =16. There was approximately $16 billion in betting losses in 2008.

65. (a) Since the plotted points form a line, it is a linear relation. See Figure 65.

(b) Using the first two points find the slope: $m=\frac{0-(-40)}{32-(-40)}=\frac{40}{72}=\frac{5}{9}$, now use slope-intercept form to find the function: $C(x)-0=\frac{5}{9}(x-32)\Rightarrow C(x)=\frac{5}{9}(x-32)$. The slope of $\frac{5}{9}$ means that the Celsius temperature changes 5° for every 9° change in Fahrenheit temperature.

(c) $C(83)=\frac{5}{9}(83-32)=28\frac{1}{3}°C$

67. (a) The slope is $m=\frac{59.3-51.5}{2007-1980}=\frac{7.8}{27}=\frac{13}{45}$. Using point-slope form produces the equation: $y-51.5=\frac{13}{45}(x-1980)$.

(b) At $x=1990$, $y-51.5=\frac{13}{45}(1990-1980)\Rightarrow y=\frac{13}{45}(10)+51.5\Rightarrow y\approx 54.4\%$.

At $x=1995$, $y-51.5=\frac{13}{45}(1995-1980)\Rightarrow y=\frac{13}{45}(15)+51.5\Rightarrow y\approx 55.8\%$.

At $x=2000$, $y-51.5=\frac{13}{45}(2000-1980)\Rightarrow y=\frac{13}{45}(20)+51.5\Rightarrow y\approx 57.3\%$.

The values for all years are too low.

69. (a) Enter the years in L_1 and enter tuition and fees in L_2. The regression equation is: $y\approx 837.701x-1{,}657{,}993$.

(b) See Figure 69.

(c) At $x=2005$, $y\approx 837.701(2005)-1{,}657{,}993\Rightarrow \$21{,}598$. This is close to the actual value of \$21,235.

[-50,250] by [-50,110]
Xscl = 50 Yscl = 50

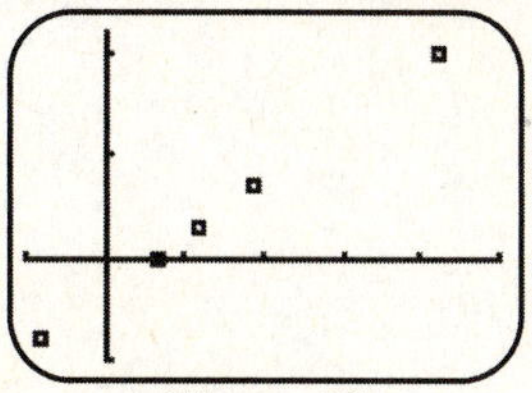

Figure 65

[1980,2010] by [5000,26000]
Xscl = 700 Yscl = 700

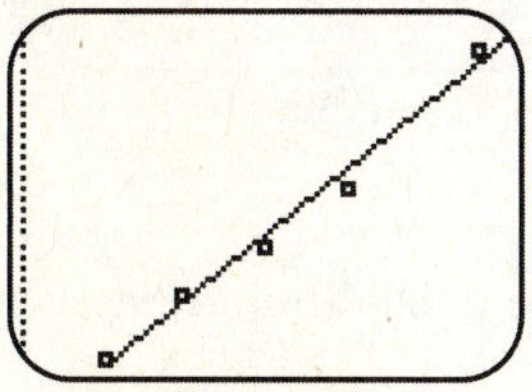

Figure 69

71. (a) Enter the distance in L_1 and enter velocity in L_2. The regression equation is: $y\approx 14.68x+277.82$.

(b) At $y=37{,}000$, $37{,}000\approx 14.68x+277.82\Rightarrow 14.68x\approx 36{,}722.18\Rightarrow x\Rightarrow 2501.5$, or approximately 2500 light-years.

73. Enter the Gestation Period in L_1 and enter Life Span in L_2. The regression equation is: $y\approx .101x+11.6$ and the correlation coefficient is: $r\approx .909$. There is a strong positive correlation, because .909 is close to 1.

Reviewing Basic Concepts (Sections 1.3 and 1.4)

1. Since $m = 1.4$ and $b = -3.1$, slope-intercept form gives the function: $f(x) = 1.4x - 3.1$.

   $f(1.3) = 1.4(1.3) - 3.1 \Rightarrow f(1.3) = -1.28$

2. See Figure 2. x-intercept: $\frac{1}{2}$, y-intercept: 1, slope: -2, domain: $(-\infty, \infty)$, range: $(-\infty, \infty)$

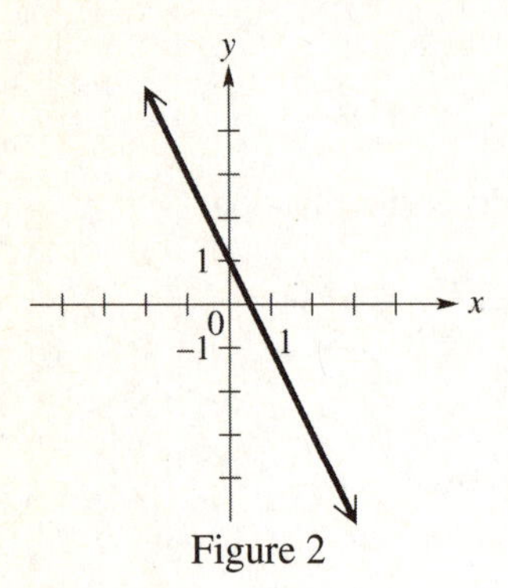

Figure 2

3. $m = \frac{6-4}{5-(-2)} = \frac{2}{7}$

4. Vertical line graphs are in the form $x = a$; through point $(-2, 10)$ would be $x = -2$.

   Horizontal line graphs are in the form $y = b$; through point $(-2, 10)$ would be $y = 10$.

5. See Figures 5a and 5b.

[-10,10] by [-10,10]
Xscl = 1 Yscl = 1

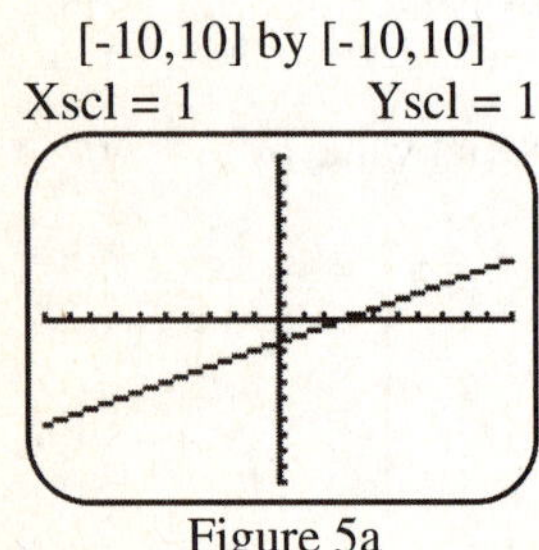
Figure 5a

[-10,10] by [-10,10]
Xscl = 1 Yscl = 1

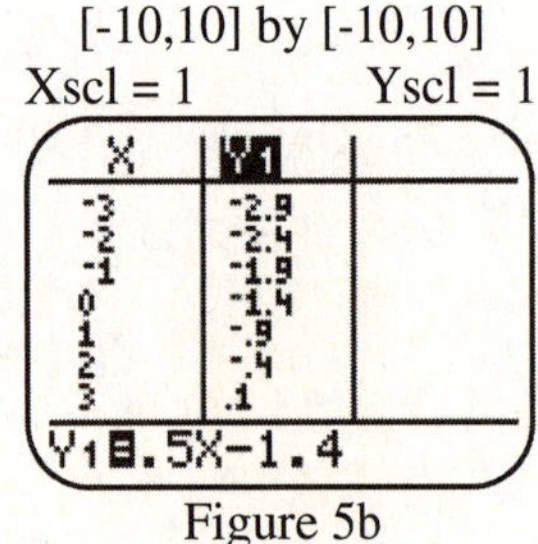

Figure 5b

[1950,2010] by [2,4]
Xscl= 10 Yscl = .5

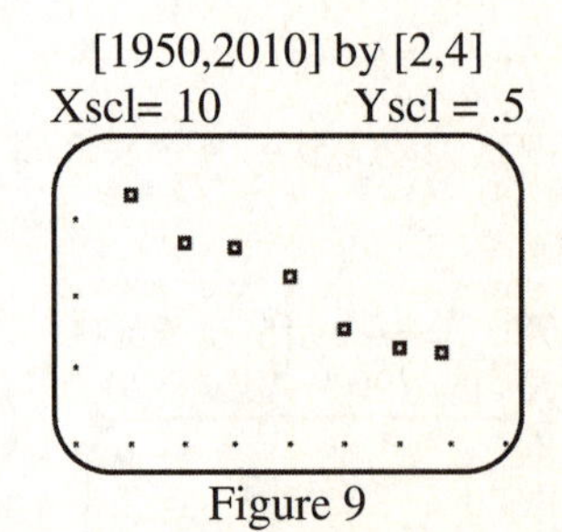
Figure 9

6. The line of the graph rises 2 units for each 1 unit to the right, therefore the slope is: $m = \frac{2}{1} = 2$.

   The y-intercept is: $b = -3$. The slope-intercept form of the equation is: $y = 2x - 3$.

7. The slope is: $m = \frac{4-2}{(-2)-5} = \frac{2}{-7} = -\frac{2}{7}$; now using point-slope form the equation is:

   $$y - 4 = -\frac{2}{7}(x+2) \Rightarrow y - 4 = -\frac{2}{7}x - \frac{4}{7} \Rightarrow y = -\frac{2}{7}x + \frac{24}{7}.$$

8. Find the given equation in slope-intercept form: $3x - 2y = 5 \Rightarrow -2y = -3x + 5 \Rightarrow y = \frac{3}{2}x - \frac{5}{2}$.

   The slope of this equation is $m = \frac{3}{2}$, therefore the slope of a perpendicular line will be the negative reciprocal: $m=-\frac{2}{3}$. Using point-slope form yields the equation:

$$y-3=-\frac{2}{3}(x+1)\Rightarrow y-3=-\frac{2}{3}x-\frac{2}{3}\Rightarrow y=-\frac{2}{3}x+\frac{7}{3}.$$

9. (a) See Figure 9.

   (b) As x increases, y decreases, therefore a negative correlation coefficient.

   (c) Enter the years in L_1 and enter people per household in L_2. The regression equation is:

   $y\approx -0.0171716x+36.9175$ and the correlation coefficient is: $r=-0.9727$.

   (d) The regression equation is: $y\approx -0.0171716(1975)+36.9175\Rightarrow y\approx 3.00$, which is close to the actual value 2.94.

1.5: Linear Equations and Inequalities

1. $-3x-12=0\Rightarrow -3x=12\Rightarrow x=-4$

3. $5x=0\Rightarrow x=0$

5. $2(3x-5)+8(4x+7)=0\Rightarrow 6x-10+32x+56=0\Rightarrow 38x=-46\Rightarrow x=-\frac{46}{38}\Rightarrow x=-\frac{23}{19}$

7. $3x+6(x-4)=0\Rightarrow 3x+6x-24=0\Rightarrow 9x=24\Rightarrow x=\frac{24}{9}\Rightarrow x=\frac{8}{3}$

9. $1.5x+2(x-3)+5.5(x+9)=0\Rightarrow 1.5x+2x-6+5.5x+49.5=0\Rightarrow 9x=-43.5\Rightarrow$

   $x=\frac{-43.5}{9}\Rightarrow x=-\frac{29}{6}$

11. The solution to $y_1=y_2$ is the intersection of the lines or $x=\{10\}$.

13. The solution to $y_1=y_2$ is the intersection of the lines or $x=\{1\}$.

15. When $y_1=y_2$, $y=0$. $y=0$ when the graph crosses the x-axis or at the zero $x=\{3\}$.

17. When $x=10$ is substituted into each function the result is 20.

19. There is no real solution if y_1-y_2 yields a contradiction, $y=b$, where $b\neq 0$. This equation is called a contradiction and the solution set is: $\varnothing$.

21. $2x-5=x+7\Rightarrow x-5=7\Rightarrow x=12$ **Check:** $2(12)-5=12+7\Rightarrow 19=19$ The graphs of the left and right sides of the equation intersect when $x=12$. The solution set is $\{12\}$.

23. $0.01x+3.1=2.03x-2.96\Rightarrow 3.1=2.02x-2.96\Rightarrow 6.06=2.02x\Rightarrow x=3$

    Check: $0.01(3)+3.1=2.03(3)-2.96\Rightarrow .03+3.1=6.09-2.96\Rightarrow 3.13=3.13$

    The graphs of the left and right sides of the equation intersect when $x=3$. The solution set is $\{3\}$.

25. $-(x+5)-(2+5x)+8x=3x-5\Rightarrow -x-5-2-5x+8x=3x-5\Rightarrow 2x-7=3x-5\Rightarrow -2=x$

    Check: $-(-2+5)-(2+5(-2))+8(-2)=3(-2)-5\Rightarrow -2-5-2+10-16=-6-5\Rightarrow -11=-11$. The graphs of the left and right sides of the equation intersect when $x=-2$. The solution set is $\{-2\}$.

27. $\frac{2x+1}{3}+\frac{x-1}{4}=\frac{13}{2} \Rightarrow 12\left(\frac{2x+1}{3}+\frac{x-1}{4}\right)=12\left(\frac{13}{2}\right) \Rightarrow 8x+4+3x-3=78 \Rightarrow 11x+1$

$=78 \Rightarrow 11x=77 \Rightarrow x=7$ **Check:** $\frac{2(7)+1}{3}+\frac{7-1}{4}=\frac{13}{2} \Rightarrow 5+\frac{6}{4}=\frac{13}{2} \Rightarrow \frac{13}{2}=\frac{13}{2}$

The graphs of the left and right sides of the equation intersect when $x=7$. The solution set is $\{7\}$.

29. $\frac{1}{2}(x-3)=\frac{5}{12}+\frac{2}{3}(2x-5) \Rightarrow 12\left[\frac{1}{2}(x-3)=\frac{5}{12}+\frac{2}{3}(2x-5)\right] \Rightarrow 6x-18=5+16x-40 \Rightarrow$

$-10x=-17 \Rightarrow x=\frac{17}{10}$ **Check:** $\frac{1}{2}\left(\frac{17}{10}-3\right)=\frac{5}{12}+\frac{2}{3}\left(2\left(\frac{17}{10}\right)-5\right) \Rightarrow \frac{1}{2}\left(-\frac{13}{10}\right)=\frac{5}{12}+\frac{2}{3}\left(-\frac{16}{10}\right)$

$\Rightarrow \frac{13}{20}=\frac{5}{12}+\left(\frac{32}{30}\right) \Rightarrow \frac{78}{120}=\frac{50}{120}+\left(-\frac{128}{120}\right) \Rightarrow \frac{78}{120}=-\frac{78}{120}$. The graphs of the left and right sides of the equation intersect when $x=\frac{17}{10}$. The solution set is $\left\{\frac{17}{10}\right\}$.

31. $0.1x - 0.05 = -0.07x \Rightarrow 0.17x=0.05 \Rightarrow 17x=5 \Rightarrow x=\frac{5}{17}$

Check: $1\left(\frac{5}{17}\right)-0.05=-0.07\left(\frac{5}{17}\right) \Rightarrow 10\left(\frac{5}{17}\right)-5=-7\left(\frac{5}{17}\right) \Rightarrow \frac{50}{17}-5=-\frac{35}{17}=-\frac{35}{17}$

The graphs of the left and right sides of the equation intersect when $x=\frac{5}{17}$. The solution set is $\left\{\frac{5}{17}\right\}$.

33. $0.40x+0.60(100-x)=0.45(100) \Rightarrow 0.40x+60-0.60x=45 \Rightarrow -0.20x=-15 \Rightarrow 20x=-1500 \Rightarrow x=75$

Check: $0.40(75)+0.60(100-75)=0.45(100) \Rightarrow 30+15=45 \Rightarrow 45=45$. The graphs of the left and right sides of the equation intersect when $x=75$. The solution set is $\{75\}$.

35. $2[x-(4+2x)+3]=2x+2 \Rightarrow 2[x-4-2x+3]=2x+2 \Rightarrow 2[-x-1]=2x+2 \Rightarrow$

$-2x-2=2x+2 \Rightarrow -4x=4 \Rightarrow x=-1$

Check: $2[-1-(4+2(-1))+3]=2(-1)+2 \Rightarrow 2[-1-2+3]=0 \Rightarrow 2[0]=0 \Rightarrow 0=0$

The graphs of the left and right sides of the equation intersect when $x=-1$. The solution set is $\{-1\}$.

37. $\frac{5}{6}x-2x+\frac{1}{3}=\frac{1}{3} \Rightarrow 6\left(\frac{5}{6}x-2x+\frac{1}{3}=\frac{1}{3}\right) \Rightarrow 5x-12x+2=2 \Rightarrow -7x=0 \Rightarrow x=0$

Check: $\frac{5}{6}(0)-2(0)+\frac{1}{3}=\frac{1}{3} \Rightarrow \frac{1}{3}=\frac{1}{3}$

The graphs of the left and right sides of the equation intersect when $x=0$. The solution set is $\{0\}$.

39. $5x-(8-x)=2[-4-(3+5x-13)] \Rightarrow 6x-8=2[-5x+6] \Rightarrow 6x-8=-10x+12 \Rightarrow$

$16x=20 \Rightarrow x=\frac{20}{16}=\frac{5}{4}$

Check: $5\left(\frac{5}{4}\right)-\left(8-\frac{5}{4}\right)=2\left[-4-\left(3+5\left(\frac{5}{4}\right)-13\right)\right]\Rightarrow\frac{25}{4}-\frac{27}{4}=2\left[-4-\left(\frac{25}{4}\right)-10\right]\Rightarrow$

$-\frac{2}{4}=2\left[6-\frac{25}{4}\right]\Rightarrow-\frac{1}{2}=2\left[-\frac{1}{4}\right]\Rightarrow-\frac{1}{2}=-\frac{1}{2}$. The graphs of the left and right sides of the equation

intersect when $x=\frac{5}{4}$.The solution set is $\left\{\frac{5}{4}\right\}$.

41. When $x=4$, both Y_1 and Y_2 have a value of 8. Therefore the solution set is $\{4\}$.

43. Graph $Y_1=4\left(0.23+\sqrt{5}\right)$ and $Y_2=\sqrt{2}x+1$ as shown in Figure 43. The graphs intersect when $x\approx16.07$.

Therefore the solution set is $\{16.07\}$.

45. Graph $Y_1=2\pi x+\sqrt[3]{4}$ and $Y_2=0.5\pi x-\sqrt{28}$ as shown in Figure 45. The graphs intersect when $x\approx-1.46$.

Therefore the solution set is $\{-1.46\}$.

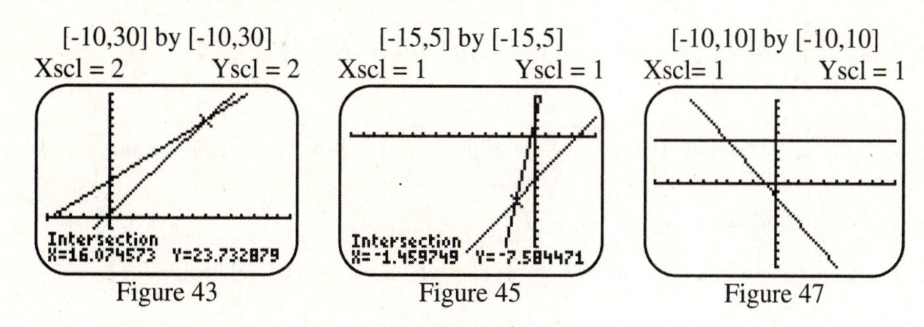

Figure 43 Figure 45 Figure 47

47. Graph $Y_1=0.23\left(\sqrt{3}+4x\right)-0.82(\pi x+2.3)$ and $Y_2=5$ as shown in Figure 47. The graphs intersect when

$x\approx-3.92$. Therefore the solution set is $\{-3.92\}$.

49. $5x+5=5(x+3)-3\Rightarrow5x+5=5x+15-3\Rightarrow5x+5=5x+12\Rightarrow5=12\Rightarrow$ Contradiction. The

solution set is $\varnothing$ The table of $Y_1=5x+5$ and $Y_2=5(x+3)-3$ never produces the same answers,

therefore supports the Contradiction.

51. $6(2x+1)=4x+8\left(x+\frac{3}{4}\right)\Rightarrow12x+6=4x+8x+6\Rightarrow12x+6=12x+6\Rightarrow6=6\Rightarrow$ Identity. The solution

set is $(-\infty,\infty)$ The table of $Y_1=6(2x+1)$ and $Y_2=4x+8\left(x+\frac{3}{4}\right)$ produces all the same answers,

therefore supports the Identity.

53. $7x-3[5x-(5+x)]=1-4x\Rightarrow7x-3[4x-5]=1-4x\Rightarrow7x-12x+15=1-4x\Rightarrow$
$-5x+15=1-4x\Rightarrow-x=-14\Rightarrow x=14\Rightarrow$ Conditional.

The solution set is 14. The table of $Y_1=7x-3[5x-(5+x)]$ and $Y_2=1-4x$ shows that the answers are the

same when x = 14.

55. $0.2(5x-4)-0.1(6-3x)=0.4\Rightarrow x-0.8-0.6+0.3x=0.4\Rightarrow1.3x-1.4=0.4\Rightarrow1.3x=1.8\Rightarrow$

$x=\frac{18}{13}\Rightarrow$ Conditional. The solution set is $\frac{18}{13}$ The table of $Y_1 = 0.2(5x-4)-0.1(6-3x)$ and $Y_2 =$ 0.4 shows that the answers are the same when $x = \frac{18}{13}$.

57. $-4[6-(-2+3x)]=21+12x\Rightarrow -4[8-3x]=21+2x\Rightarrow -32+12x=21+2x\Rightarrow -32=21\Rightarrow$

Contradiction. The solution set is $\varnothing$. The table of $Y_1=-4[6-(-2+3x)]$ and $Y_2=21+12x$ never produces the same answers, therefore supports the Contradiction.

59. $\frac{1}{2}x-2(x-1)=2-\frac{3}{2}x\Rightarrow\frac{1}{2}x-2x+2=2-\frac{3}{2}x\Rightarrow-\frac{3}{2}+2=-\frac{3}{2}+2\Rightarrow 2=2\Rightarrow$ Identity.

The solution set is $(-\infty,\infty)$. The table of $Y_1=\frac{1}{2}x-2(x-1)$ and $Y_2=2-\frac{3}{2}x$ produces all the same answers, therefore supports the Identity.

61. $\frac{x-1}{2}=\frac{3x-2}{6}\Rightarrow 6\left[\frac{x-1}{2}=\frac{3x-2}{6}\right]\Rightarrow 3(x-2)=3x-2\Rightarrow 3x-6=3x-3\Rightarrow -6=-3\Rightarrow$

Contradiction. The solution set is $\varnothing$. The table of $Y_1=\frac{x-1}{2}$ and $Y_2=\frac{3x-2}{6}$ never produces the same answers, therefore supports the Contradiction.

63. For the given functions, $f(x)=g(x)$ when the graphs intersect or when $x=3$. The solution is $\{3\}$.

65. For the given functions, $f(x)<g(x)$ when the graph of $f(x)$ is below the graph of $g(x)$ or when $x>3$. The solution is $(3,\infty)$.

67. For the given inequality, $y_1-y_2\geq 0\Rightarrow f(x)-g(x)\geq 0\Rightarrow f(x)\geq g(x)$ when the graph of $f(x)$ is above or intersects the graph of $g(x)$ or when $x\leq 3$. The solution is $(-\infty,3]$.

69. For the given functions, $f(x)\leq f(x)$ when the graph of is $f(x)$ below or intersects the graph $g(x)$ of or when $x\geq 3$. The solution is $[3,\infty)$.

71. For the given functions, $f(x)\leq 2$ when the graph of $f(x)$ is below or equal to 2 or when $x\geq 3$. The solution is $[3,\infty)$.

73. (a) The function $f(x)>0$ when the graph is above the x-axis for the interval $(20,\ \infty)$.

(b) The function $f(x)<0$ when the graph is below the x-axis for the interval $(-\infty,\ 20)$.

(c) The function $f(x)\geq 0$ when the graph intersects or is above the x-axis for the interval $[20,\ \infty)$.

(d) The function $f(x)\leq 0$ when the graph intersects or is below the x-axis for the interval $(-\infty,\ 20]$.

75. (a) If the solution set of $f(x)\geq g(x)$ is $[4,\ \infty)$, then $f(x)=g(x)$ at the intersection of the graphs, $x=4$ or $\{4\}$.

(b) If the solution set of $f(x) \geq g(x)$ is $[4, \infty)$, then $f(x) > g(x)$ is the same, but does not include the intersection of the graphs for the interval $(4, \infty)$.

(c) If the solution set of $f(x) \geq g(x)$ is $[4, \infty)$, then $f(x) < g(x)$ is left of the intersection of the graphs for the interval: $(-\infty, 4)$.

77. (a) $9-(x+1)<0 \Rightarrow -x+8<0 \Rightarrow -x<-8 \Rightarrow x>8 \Rightarrow$ the interval is $(8, \infty)$. The graph of $y_1 = 9-(x+1)$ is below the x-axis for the interval $(8, \infty)$.

(b) If $9-(x+1)<0$ for $(8, \infty)$, then $9-(x+1) \geq 0$ for the interval $(-\infty, 8]$. The graph of $y_1 = 9-(x+1)$ intersects or is above the x-axis for the interval $(-\infty, 8]$.

79. (a) $2x-3>x+2 \Rightarrow x-3>2 \Rightarrow x>5 \Rightarrow$ the interval is $(5, \infty)$. The graph of $y_1 = 2x-3$ is above the graph of $y_2 = x+2$ for the interval $(5, \infty)$.

(b) If $2x-3>x+2$ for $(5, \infty)$, then $2x-3 \leq x+2$ for the interval $(-\infty, 5]$. The graph $y_1 = 2x-3$ intersects or is below the graph $y_2 = x+2$ for the interval $(-\infty, 5]$.

81. (a) $10x+5-7x \geq 8(x+2)+4 \Rightarrow 3x+5 \geq 8x+20 \Rightarrow -5x \geq 15 \Rightarrow x \leq -3 \Rightarrow$ the interval is $(-\infty, -3]$. The graph of $y_1 = 10x+5-7x$ intersects or is above the graph of $y_2 = 8(x+2)+4$ for the interval $(-\infty, -3]$.

(b) If $10x+5-7x \geq 8(x+2)+4$ for $(-\infty, -3)$, then $10x+5-7x < 8(x+2)+4$ for the interval $(-3, \infty)$. The graph of $y_1 = 10x+5-7x$ is below the graph of $y_2 = 8(x+2)+4$ for the interval $(-3, \infty)$.

83. (a) $x+2(-x+4)-3(x+5)<-4 \Rightarrow x-2x+8-3x-15<-4 \Rightarrow -4x<3 \Rightarrow x>-\frac{3}{4} \Rightarrow$ the interval is $\left(-\frac{3}{4}, \infty\right)$. The graph of $y_1 = x+2(-x+4)-3(x+5)$ is below the graph of $y_2 = -4$ for the interval $\left(-\frac{3}{4}, \infty\right)$.

(b) If $x+2(-x+4)-3(x+5)<-4$ for $\left(-\frac{3}{4}, \infty\right)$, then $x+2(-x+4)-3(x+5) \geq -4$ for the interval $\left(-\infty, -\frac{3}{4}\right]$. The graph of $y_1 = x+2(-x+4)-3(x+5)$ intersects or is above the graph $y_2 = -4$ for the interval $\left(-\infty, -\frac{3}{4}\right]$.

85. $\frac{1}{3}x-\frac{1}{5}x\le 2\Rightarrow\left[\frac{1}{3}x-\frac{1}{5}x\le 2\right]\Rightarrow 5x-3x\le 30\Rightarrow 2x\le 30\Rightarrow x\le 15\Rightarrow(-\infty,15]$. The graph of

$y_1=\frac{1}{3}x-\frac{1}{5}x$ intersects or is below the graph of $y_2=2$ for the interval $(-\infty,15]$.

87. $\frac{x-2}{2}-\frac{x+6}{3}>-4\Rightarrow 6\left[\frac{x-2}{2}-\frac{x+6}{3}>-4\right]\Rightarrow 3x-6-(2x+12)>-24\Rightarrow$

$x-18>-24\Rightarrow x>-6\Rightarrow(-6,\infty)$. The graph of $y_1=\frac{x-2}{2}-\frac{x+6}{3}$ is above the graph of

$y_2=5$ for the interval: $(-6,\infty)$.

89. $0.6x-2(0.5x+.2)\le 0.4-0.3x\Rightarrow .6x-1x-0.4\le 0.4-0.3x\Rightarrow 10[0.6x-1x-0.4\le 0.4-0.3x]\Rightarrow$

$6x-10x-4\le 4-3x\Rightarrow -4x-4\le 4-3x\Rightarrow -x\le 8\Rightarrow x\ge -8\Rightarrow[-8,\infty)$. The graph of

$y_1=0.6x-2(0.5x+.2)$ intersects or is below the graph of $y_2=0.4-0.3x$ for the interval $[-8,\infty)$.

91. $-\frac{1}{2}x+0.7x-5>0\Rightarrow 10\left[-\frac{1}{2}x+0.7x-5>0\right]\Rightarrow -5x+7x-50>0\Rightarrow 2x>50\Rightarrow$

$x>25\Rightarrow(25,\infty)$. The graph of $y_1=-\frac{1}{2}x+.7x-5$ is above the graph of $y_2=0$ for the interval $(25,\infty)$.

93. $-4(3x+2)\ge -2(6x+1)\Rightarrow -12x-8\ge -12x-2\Rightarrow -8\ge -2$; since this is false the solution is $\varnothing$.

The graph of $y_1=-4(3x+2)$ never intersects or is above the graph of $y_2=-2(6x+1)$, therefore the solution is $\varnothing$.

95. (a) As time increases, distance increases, therefore the car is moving away from Omaha.

(b) The distance function $f(x)$ intersects the 100 mile line at 1 hour and the 200 mile line at 3 hours.

(c) Using the answers from (b) the interval is [1, 3].

(d) Because x hours is $0\le x\le 6$, the interval is (1, 6].

97. $4\le 2x+2\le 10\Rightarrow 2\le 2x\le 8\Rightarrow 1\le x\le 4\Rightarrow[1,4]$ The graph of $y_2=2x+2$ is between the graphs of $y_1=4$ and $y_3=10$ for the interval [1, 4].

99. $-10>3x+2>-16\Rightarrow -12>3x>-18\Rightarrow -4>x>-6\Rightarrow -6<x<-4\Rightarrow(-6,-4)$

The graph of $y_2=3x+2$ is between the graphs of $y_1=-10$ and $y_3=-16$ for the interval $(-6,-4)$.

101. $-3\le\frac{x-4}{-5}<4\Rightarrow 15\ge x-4>-20\Rightarrow 19\ge x>-16\Rightarrow -16<x\,19\Rightarrow(-16,19]$ The graph of

$y_2=\frac{x-4}{-5}$ is between the graphs of $y_1=-3$ and $y_3=4$ for the interval $(-16,19]$.

103. $\sqrt{2}\le\frac{2x+1}{3}\le\sqrt{5}\Rightarrow 3\sqrt{2}\le 2x+1\le 3\sqrt{5}\Rightarrow 3\sqrt{2}-1\le 2x\le 3\sqrt{5}-1\Rightarrow$

$\frac{3\sqrt{2}-1}{2} \le x \le \frac{3\sqrt{5}-1}{2} \Rightarrow \left[\frac{3\sqrt{2}-1}{2}, \frac{3\sqrt{5}-1}{2}\right]$; The graph of $y_2 = \frac{2x+1}{3}$ is between the graphs of

$y_1 = \sqrt{2}$ and $y = \sqrt{5}$. for the interval $\left[\frac{3\sqrt{2}-1}{2}, \frac{3\sqrt{5}-1}{2}\right]$.

105. The graph of $y_1 = 3.7x - 11.1$ crosses the x-axis at $x = 3$. There is one solution to this equation. Because a linear equation can only cross the x-axis in one location, there is only one solution to any linear equation.

106. $3.7x - 11.1 < 0 \Rightarrow 3.7x < 11.1 \Rightarrow x < 3 \Rightarrow (-\infty, 3)$ $3.7x - 11.1 > 0 \Rightarrow 3.7x > 11.1 \Rightarrow x > 3 \Rightarrow (3, \infty)$

The value of $x = 3$ given by the equation represents the boundary between the sets of real numbers given by the inequality solutions $(-\infty, 3)$ and $(3, \infty)$.

107. The graph of $y_1 = -4x + 6$ crosses the x-axis at $x = 1.5$.

$-4x + 6 < 0 \Rightarrow -4x < -6 \Rightarrow x > \frac{-6}{-4} \Rightarrow x > 1.5 \Rightarrow (1.5, \infty)$

$-4x + 6 > 0 \Rightarrow -4x > -6 \Rightarrow x < \frac{-6}{-4} \Rightarrow x > 1.5 \Rightarrow (-\infty, 1.5)$

108. (a) If $a \ne 0$, then $ax + b = 0 \Rightarrow ax = -b \Rightarrow x = -\frac{b}{a}$

(b) If $a > 0$, a positive slope, then $ax + b < 0 \Rightarrow ax < -b \Rightarrow x < \frac{-b}{a} \Rightarrow \left(-\infty, \frac{-b}{a}\right)$

If $a > 0$, a positive slope, then $ax + b < 0 \Rightarrow ax > -b \Rightarrow x > \frac{-b}{a} \Rightarrow \left(\frac{-b}{a}, \infty\right)$

(c) If $a < 0$, a negative slope, then $ax + b < 0 \Rightarrow ax < -b \Rightarrow x > \frac{-b}{a} \Rightarrow \left(\frac{-b}{a}, \infty\right)$

If $a < 0$, a positive slope, then $ax + b > 0 \Rightarrow ax > -b \Rightarrow x < \frac{-b}{a} \Rightarrow \left(-\infty, \frac{-b}{a}\right)$

109. (a) The graph of $T(x) = 65 - 19x$ intersects the graph of $D(x) = 50 - 5.8x$ at $\approx (1.136, 43.41)$. Since the x-coordinate is altitude, the clouds will not form below 1.14 miles or for the interval: [0, 1.14).

(b) Clouds will not form when air temperature is above dew point temperature or $T(x) > D(x)$.

Then $65 - 19x > 50 - 5.8x \Rightarrow -13.2x > -15 \Rightarrow x < \frac{15}{13.2}$ for the interval $\left[0, \frac{15}{13.2}\right)$.

111. Since $C = 2\pi r$ and radius is in the range $1.99 \le r \le 2.01$, circumference is in the range $2\pi(1.99) \le 2\pi r \le 2\pi(2.01) \Rightarrow 3.98\pi \le C \le 4.02\pi$.

1.6: Applications of Linear Functions

1. $.75(40) = 30L$

3. When combining a 26% acid solution to a 32% acid solution the result will be a solution with between 26% and 32% acid. (A) 36% is not in between these percent values and not a possible concentration.

5. If x is the second number, then $6x-3$ is the first number. The equation with the sum of these two numbers equal to 32 is: (D) $(6x-3)+x=32$.

7. If $P=2L+2W$ then $P=2L+2(19) \Rightarrow 98=2L+38 \Rightarrow 60=2L \Rightarrow L=30\text{cm}$.

9. Let x = width and $2x-2.5$ = length. If $P=2W+2L$, then

$40.6=2x+2(2x-2.5) \Rightarrow 40.6=6x-5 \Rightarrow 45.6=6x \Rightarrow x=7.6$. The width is 7.6 cm.

11. Let x = the original square side length and $2x - 3$ = the new square side length. If $P = 4s$, then

$4(x+3)=2x+40 \Rightarrow 4x+12=2x+40 \Rightarrow 2x=28 \Rightarrow x=14$. The original side length is 14 cm.

13. With an aspect ratio of 4:3, let x = width and $\frac{4}{3}x$ = length. If $P=2W+2L$, then

$98=2x+2\left(\frac{4}{3}x\right) \Rightarrow 98=2x+\frac{8}{3}x \Rightarrow 98=\frac{14}{3}x \Rightarrow 294=14x \Rightarrow x=21$. The width is 21 inches and the length is $\frac{4}{3}(21)=28$ inches. Use the Pythagorean theorem to find the diagonal

$c^2=(21)^2+(28)^2 \Rightarrow c^2=441+784 \Rightarrow c^2=1225 \Rightarrow c=35$. The television is advertised as a 35 inch screen.

15. Let x = the short side length and $2x$ = the longer two side lengths. If $P=s+s+s$, then

$30=x+2x+2x \Rightarrow 30=5x \Rightarrow x=6$. The shortest side is 6 cm long.

17. Let x = the number of hours traveled at 70 mph, and (6-x) = the number of hours traveled at 55 mph. Since D = RT and the total distance traveled by the car was 372 miles, then 55(6-x) + 70x = $372 \Rightarrow 330-55x+70x=372 \Rightarrow 15x=42 \Rightarrow x=2.8$. The car traveled for 2.8 hours at 70 mph and (6-2.8)=3.2 hours at 55 mph.

19. Let x = gallons of 5% acid solution. Then

$5(.10)+x(.05)=(x+5)(.07) \Rightarrow .50+.05x=.07x+.35 \Rightarrow .15=.02x \Rightarrow 15=2x \Rightarrow x=7.5$.

Mix in 7.5 gallons of 5% acid solution.

21. Let x = gallons of pure alcohol. Then

$20(.15)+x(1.00)=(x+20)(.25) \Rightarrow 3+x=.25x+5 \Rightarrow .75x=2 \Rightarrow x=2.67$ or $2\frac{2}{3}$.

Mix in $2\frac{2}{3}$ gallons of pure alcohol.

23. Let x = milliliters of water. Then $8(.06)+x(0)=(x+8)(.04) \Rightarrow .48=.04x+.32 \Rightarrow$

$0.16=0.04 \Rightarrow 4x=16 \Rightarrow x=4$. Mix in 4 milliliters of water.

25. Let x = liters of fluid to be drained and pure antifreeze added. Then

$16(0.80)-x(0.80)+x(1.00)=16(0.90) \Rightarrow 12.8-0.80x+x=14.4 \Rightarrow 0.20x=1.6$

$\Rightarrow 20x = 160 \Rightarrow x = 8$. Drain and add in 8 liters of antifreeze.

27. Let $x =$ gallons of 94-octane gasoline. Then

$400(0.99) + x(0.94) = (x+400)(0.97) \Rightarrow 396 + 0.94x = 0.97x + 388$

$8 = 0.03x \Rightarrow 3x = 800 \Rightarrow x = 266.67$ or $266\frac{2}{3}$. Mix in $266\frac{2}{3}$ gallons of 94-octane gasoline.

29. (a) Since each student needs 15 ft^3/min and there are 60 min/hr, the ventilation required by x students per hour is $V(x) = 60(15x) \Rightarrow V(x) = 900x$.

(b) The number of air exchanges needed per hour is $A(x) = \dfrac{V(x)}{\text{ft}^3 \text{ in room}} = \dfrac{900x}{15000} \Rightarrow A(x) = \dfrac{3}{50}x$.

(c) $A(x) = \dfrac{3}{50}(40) \Rightarrow A(x) = 2.4$ ach are needed.

(d) The increase is $\dfrac{\text{smoking}}{\text{nonsmoking}} = \dfrac{50\,\text{ft}^3}{15\,\text{ft}^3} = 3\frac{1}{3}$ times. Smoking areas require more than triple the ventilation.

31. (a) Use the points (2003, 17) and (2006,26) to find the slope, $m = \dfrac{26-17}{2006-2003} = \dfrac{9}{3} = 3$ and the point (2003,17) to find the equation of the line.

$S(x) - 17 = 3(x - 2003) \Rightarrow S(x) - 17 = 3x - 6009 \Rightarrow S(x) = 3x - 5992$.

(b) Sales increased , on average by \$3 billion per year.

(c) $S(x) = 3x - 5992 \Rightarrow 41 = 3x - 5992 \Rightarrow 6033 = 3x \Rightarrow 2011 = x$. The sales will reach \$41 billion in 2011.

33. (a) If the fixed cost $= \$200$ and the variable cost $= \$.02$ the cost function is: $C(x) = 0.02x + 200$.

(b) If she gets paid \$.04 per envelope stuffed and $x =$ number of envelopes, the revenue function is $R(x) = 0.04x$.

(c) $R(x) = C(x)$ when $0.02x + 200 = 0.04x \Rightarrow 200 = 0.02x \Rightarrow x = 10{,}000$.

(d) Graph $C(x)$ and $R(x)$, see Figure 33. Rebecca takes a loss when stuffing less than 10,000 envelopes and makes a profit when stuffing over 10,000 envelopes.

35. (a) If the fixed cost $= \$2300$ and the variable cost $= \$3.00$ the cost function is: $C(x) = 3.00x + 2300$.

(b) If he gets paid \$5.50 per delivery and $x =$ number of deliveries, the revenue function is: $R(x) = 5.50x$.

(c) $R(x) = C(x)$ when $3.00x + 2300 = 5.50x \Rightarrow 2300 = 2.50x \Rightarrow x = 920$.

(d) Graph $C(x)$ and $R(x)$. See Figure 35. Tom takes a loss when making fewer than 920 deliveries and makes a profit when making over 920 deliveries.

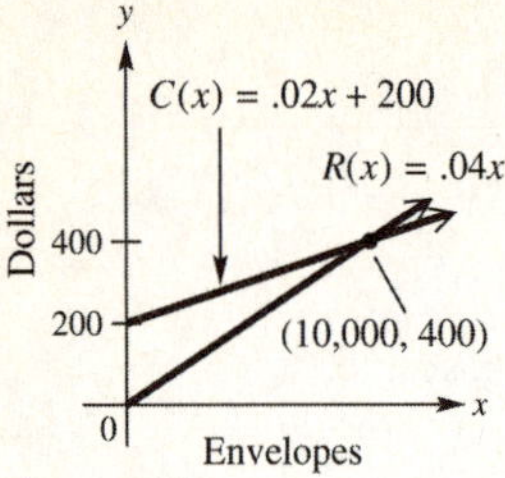

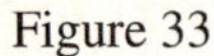
Figure 33

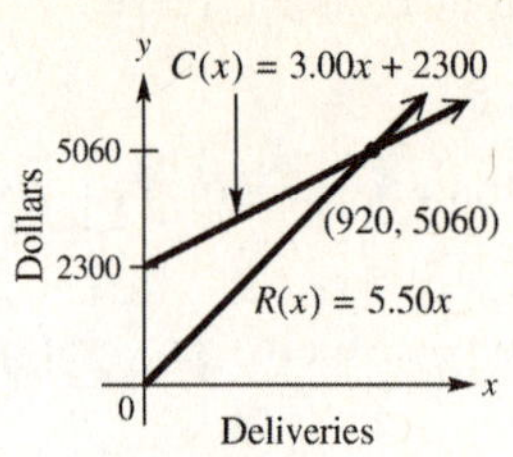

Figure 35

37. If $y = kx$, $x = 3$, and $y = 7.5$, then $7.5 = k(3) \Rightarrow k = 2.5$. Now, with $k = 2.5$ and $x = 8$, $y = 2.5(8) \Rightarrow y = 20$ when $x = 8$.

39. If $y = kx$, $x = 25$, and $y = 1.5$, then $1.50 = k(25) \Rightarrow k = 0.06$. Now, with $k = 0.06$ and $y = 5.10$, $5.10 = 0.06x \Rightarrow x = \$85$ when $y = \$5.10$

41. Let $y =$ pressure and $x =$ depth for the direct proportion: $y = kx$. Then $13 = k(30) \Rightarrow k = \frac{13}{30}$.

Now use $k = \frac{13}{30}$ and a depth of 70 feet to find the pressure: $y = \frac{13}{30}(70) \Rightarrow y = \frac{91}{3} \Rightarrow y = 30\frac{1}{3}\ \text{lb/in}^2$.

43. Let $t =$ tuition and $c =$ credits taken for the direct proportion $t = kc$. Then $720.50 = k(11) \Rightarrow k = 65.5$. Now use the constant of variation $k = 65.5$ and 16 credits to find the tuition: $y = 65.5(16) \Rightarrow y = \$1048$.

45. First use proportion to find the radius of the water at a depth of 44 feet $\frac{5}{11} = \frac{x}{6} \Rightarrow 11x = 30 \Rightarrow x \approx 2.727$.

Now use the cone volume formula to find the water's volume:

$V = \frac{1}{3}\pi r^2 h \Rightarrow V = \frac{1}{3}\pi(2.727)^2(6) \approx 46.7\ \text{ft}^3$.

47. Since the triangles are similar, use a proportion to solve: $\frac{1.75}{2} = \frac{45}{x} \Rightarrow 1.75x = 90 \Rightarrow x \approx 51.43$ or

$x = 51\frac{3}{7}$ feet tall.

49. Let $w =$ weight, $d =$ distance, and use the direct proportion $w = kd$ to find k: $3 = k(2.5) \Rightarrow k = 1.2$. Now use $k = 1.2$ and a weight of 17 pounds to find the stretch length: $17 = 1.2(d) \Rightarrow d = 14.17$ or $14\frac{1}{6}$ in.

51. With direct proportion $y_1 = kx_1$ and $y_2 = kx_2$, then $k = \frac{y_1}{x_1} = \frac{y_2}{x_2}$. Now let $y_1 = 250$ tagged fish, $y_2 = 7$ tagged fish, $x_2 = 350$ sample fish, and $x_1 =$ total fish. Therefore

$\frac{250}{x_1} = \frac{7}{350} \Rightarrow 7x_1 = 87{,}500 \Rightarrow x_1 = 12{,}500$ is the estimate for total population of fish in the lake.

53. (a) Let $x =$ number of heaters produced and $y =$ cost. Then (10, 7500) and (20, 13900) are two points on the graph of the linear function. Find the slope: $m = \dfrac{13900-7500}{20-10} = \dfrac{6400}{10} = 640$.

Now use point-slope form to find the linear function:

$y-7500 = 640(x-10) \Rightarrow y-7500 = 640x-640 \Rightarrow y = 640x+1100.$

(b) $y = 640(25)+1100 \Rightarrow y = 16,000+1100 \Rightarrow y = \$17,100.$

(c) Graph $y = 640x+1100$ and locate the point (25, 17,100) on the graph.

55. (a) Let $x =$ number of years after 1998 and $y =$ value, then (0, 120,000) and (10, 146,000) are two points on the graph of the linear function. Find the slope: $m = \dfrac{146,000-120,000}{10-0} = \dfrac{26,000}{10} = 2,600.$

The y-intercept is: 120,000. Therefore the linear function is: $y = 2,600x+120,000.$

(b) $y = 2,600(7)+120,000 \Rightarrow y = 18200+120,000 \Rightarrow y = \$138,200$ value of the house.

(c) The value of the house increased, on average, \$2,600 per year.

57. (a) Surface area = $4\pi r^2 \Rightarrow 4\pi(3960)^2 \approx 197,000,000 \text{ mi}^2$.

(b) $(197,000,000)(0.71) \approx 140,000,000 \text{ mi}^2$

(c) $\dfrac{680,000}{140,000,000} \approx 0.00486$ miles. Converted to feet is $(0.00486)(5280) \approx 25.7$ feet.

(d) Since this height is greater than the heights of both Boston and San Diego these cities would be flooded.

(e) We know from above that oceans cover approximately 140,000,000 square miles of the earth. The Antarctic ice cap contains 6,300,000 cubic miles of water.

$\dfrac{6,300,000}{140,000,000} = 0.045$ miles. Converted to feet is $(0.045)(5280) \approx 238$ feet..

59. (a) $y = \dfrac{5}{3}(27)+455 \Rightarrow y = 45+455 \Rightarrow y = 500\text{cm}^3.$

(b) $605 = \dfrac{5}{3}x+455 \Rightarrow 150 = \dfrac{5}{3}x \Rightarrow x = 90°\text{ C}.$

(c) $0 = \dfrac{5}{3}x+455 \Rightarrow -455 = \dfrac{5}{3}x \Rightarrow x = -273°\text{C}.$

61. $I = PRT \Rightarrow \dfrac{I}{RT} = P$ or $P = \dfrac{I}{RT}$

63. $P = 2L+2W \Rightarrow P-2L = 2W \Rightarrow W = \dfrac{P-2L}{2}$ or $W = \dfrac{P}{2}-L$

65. $A = \dfrac{1}{2}h(b_1+b_2) \Rightarrow 2A = h(b_1+b_2) \Rightarrow h = \dfrac{2A}{b_1+b_2}$

67. $S = 2LW + 2WH + 2HL \Rightarrow S - 2LW = H(2W + 2L) \Rightarrow \dfrac{S-2LW}{2W+2L}$

69. $V = \dfrac{1}{3}\pi r^2 h \Rightarrow 3V = \pi r^2 h \Rightarrow h = \dfrac{3V}{\pi r^2}$

71. $S = \dfrac{n}{2}(a_1 + a_n) \Rightarrow 2S = n(a_1 + a_n) \Rightarrow n = \dfrac{2S}{a_1 + a_n}$

73. $s = \dfrac{1}{2}gt^2 \Rightarrow 2s = gt^2 \Rightarrow g = \dfrac{2s}{t^2}$

75. Let P = the amount put into the short-term note, then $240,000 - P$ = the amount put into the long-term note. With \$13,000 one year interest income, solve:
$P(.06)(1) + (240,000 - P)(0.05)(1) = 13,000 \Rightarrow 0.06P + 120,000 - 0.05P = 13,000 \Rightarrow$
$0.01P = 1,000 \Rightarrow P \Rightarrow 100,000.$ The short-term note was \$100,000 and the long-term was \$140,000.

77. Let P = the amount deposited at 2.5% interest rate, then $2P$ = the amount deposited at 3% interest rate. With a one year interest income of \$850, solve:
$0.025P(1) + 0.03(2P)(1) = 850 \Rightarrow 0.025P + 0.06P = 850 \Rightarrow 0.085P = 850 \Rightarrow P = 10,000.$
Therefore, \$10,000 was deposited at 2.5% and \$20,000 was deposited at 3%.

79. After taxes, Marietta was able to invest 70% of the original winnings. This is $0.70(200,000) = \$140,000$. Now let P = amount invested at 1.5%, then $140,000 - P$ = the amount invested at 4%. With a one year interest income of \$4350, solve: $0.015P + 0.04(140,000 - P) = 4350 \Rightarrow 0.015P + 5600 - 0.04P = 4350 \Rightarrow$
$-0.025P = -1250 \Rightarrow P = 50,000.$ Therefore, \$50,000 was invested at 1.5% and \$90,000 was invested at 4%.

Reviewing Basic Concepts (Sections 1.5 and 1.6)

1. $3(x-5) + 2 = 1 - (4 + 2x) \Rightarrow 3x - 15 + 2 = 1 - 4 - 2x \Rightarrow 3x - 13 = -2x - 3 \Rightarrow 5x = 10 \Rightarrow x = 2.$ The graphs of the left and right sides of the equation intersect at $x = 2$; this supports the solution set: $\{2\}$.

2. Graph $y_1 = \pi(1 - x)$ and $y_2 = .6(3x - 1)$. See Figure 2. The graphs intersect at $x = .757$.

3. $0 = \dfrac{1}{3}(4x-2) + 1 \Rightarrow 0 = \dfrac{4}{3}x - \dfrac{2}{3} + 1 \Rightarrow -\dfrac{1}{3} = \dfrac{4}{3}x \Rightarrow -\dfrac{1}{4}.$ Graph $y_1 = \dfrac{1}{3}(4x-2) + 1$. See Figure 3.

The graph intersects the x-axis at $x = -\dfrac{1}{4}$.

[-10,10] by [-10,10]
Xscl = 1 Yscl = 1

[-10,10] by [-10,10]
Xscl = 1 Yscl = 1

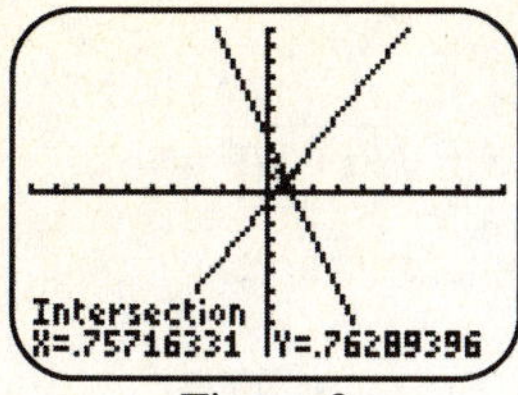

Figure 2

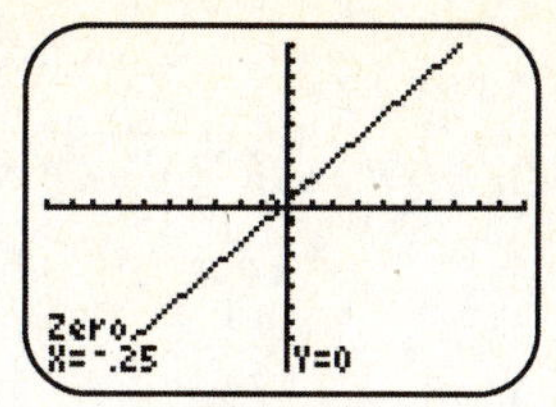

Figure 3

4. (a) $4x-5=-2(3-2x)+3 \Rightarrow 4x-5=-6+4x+3 \Rightarrow -5=-3$. Since this is false, the equation is a contradiction and the solution set is: $\varnothing$.

   (b) $5x-9=5(-2+x)+1 \Rightarrow 4x-5=-6+4x+3 \Rightarrow -5=-3$. Since this is true the equation is an identity and the solution set is: $(-\infty,\infty)$.

   (c) $5x-4=3(6-x) \Rightarrow 5x-4=18-3x \Rightarrow 8x=22 \Rightarrow x=\frac{11}{4}$. The equation is a conditional equation and the solution set is: $\left\{\frac{11}{4}\right\}$.

5. $2x+3(x+2)<1-2x \Rightarrow 2x+3x+6<1-2x \Rightarrow 7x<-5 \Rightarrow x<-\frac{5}{7}$. The solution set is: $\left(-\infty,\frac{5}{7}\right)$.

   Graph $y_1=2x+3(x+2)$ and $y_2=1-2x$; the graph of y_1 is below the graph of y_2 when $x<-\frac{5}{7}$, which supports the original solution.

6. $-5 \le 1-2x<6 \Rightarrow -6 \Rightarrow -2x<5 \Rightarrow 3 \ge x>-\frac{5}{2} \Rightarrow -\frac{5}{2}<x \le 3 \Rightarrow \left(-\frac{5}{2},3\right]$

7. (a) The graphs intersect at $x=2 \Rightarrow \{2\}$

   (b) The graph of $f(x)$ intersects or is below the graph of $g(x)$ for $x \ge 2$, on $[2,\infty)$.

8. Since the triangles formed by the shadows are similar, use proportion to solve.

   $\frac{x}{27}=\frac{6}{4} \Rightarrow 4x=162 \Rightarrow x=40.5$ ft.

9. (a) Since the income from each disc is \$5.50, a function R for revenue from selling x discs is: $R(x)=5.50x$.

   (b) Since the cost of producing each disc is \$1.50 and there is a one-time equipment cost of \$2500, a function C for cost of recording x discs is: $C(x)=1.50x+2500$.

   (c) Solve $R(x)=C(x)$: $5.5x=1.5x+2500 \Rightarrow 4x=2500 \Rightarrow x=625$ discs.

10. $V=\pi r^2 h \Rightarrow h=\frac{V}{\pi r^2}$

Chapter 1 Review Exercises

1. Use the distance formula: $d=\sqrt{(-1-5)^2+(16-(-8))^2} \Rightarrow d \Rightarrow \sqrt{(-6)^2+24^2} \Rightarrow d=\sqrt{36+576} \Rightarrow$

   $d=\sqrt{612}=6\sqrt{17}.$

2. Use the midpoint formula: Midpoint $=\left(\frac{-1+5}{2},\frac{16-8}{2}\right)=(2,4).$

3. Use the slope formula: $m=\frac{16-(-8)}{-1-5} \Rightarrow m=\frac{24}{-6}=-4.$

4. Use point-slope form and slope from ex. 3: $y-16=-4(x-(-1)) \Rightarrow y-16=-4x-4 \Rightarrow y=-4x+12.$

5. Change to slope-intercept form: $3x+4y=144 \Rightarrow 4y=-3x+144 \Rightarrow y=-\frac{3}{4}x+36 \Rightarrow m=-\frac{3}{4}.$

6. For the x-intercept, $y=0$. Therefore, $3x+4(0)=144 \Rightarrow 3x=144 \Rightarrow x=48.$

7. For the y-intercept, $x=0$. Therefore, $3(0)+4y=144 \Rightarrow 4y=144 \Rightarrow y=36.$

8. One possible window is: $[-10, 50]$ by $[-40, 40]$.

9. Since $f(3)=6$ and $f(-2)=1$, $(3,6)$ and $(-2,1)$ are points on the graph of the line. Using these points, find

   the slope: $m=\frac{6-1}{3-(-2)}=\frac{5}{5}=1.$ Use point-slope form to find the function:

   $f(x)-6=1(x-3) \Rightarrow f(x)-6=x-3 \Rightarrow f(x)=x+3.$ Now solve for $f(8)$: $f(8)=8+3=11.$

10. The slope of the given equation is -4. A line perpendicular to this will have a slope of $\frac{1}{4}$. Using this and

    point-slope form produces: $y-4=\frac{1}{4}(x-(-2)) \Rightarrow y-4=\frac{1}{4}x+\frac{1}{2} \Rightarrow y=\frac{1}{4}x+\frac{9}{2}.$

11. (a) $m=\frac{-4-5}{2-(-1)} \Rightarrow m=\frac{-9}{3}=-3$

    (b) Use point-slope form: $y-5=-3(x-(-1)) \Rightarrow y-5=-3x-3 \Rightarrow y=-3x+2$

    (c) Midpoint $=\left(\frac{-1+2}{2},\frac{5+(-4)}{2}\right)=\left(\frac{1}{2},\frac{1}{2}\right)$ or $(0.5,0.5)$

    (d) $d=\sqrt{(2-(-1))^2+(-4-5)^2} \Rightarrow d\sqrt{3^2+(-9)^2} \Rightarrow d=\sqrt{9+81}=\sqrt{90}=3\sqrt{10}$

12. (a) $m=\frac{1.5-(-3.5)}{-1-(-3)} \Rightarrow m=\frac{5}{2}=2.5$

    (b) Use point-slope form: $y-1.5=2.5(x-(-1) \Rightarrow y-1.5=2.5x+2.5 \Rightarrow y=2.5x+4$

    (c) Midpoint $=\left(\frac{-1+(-3)}{2},\frac{1.5+(-3.5)}{2}\right)=\left(-\frac{4}{2},-\frac{2}{2}\right)=(-2,-1)$

    (d) $d=\sqrt{(-1-(-3))^2+(1.5-(-3.5))^2} \Rightarrow d=\sqrt{2^2+(5)^2} \Rightarrow d=\sqrt{4+25}=\sqrt{29}$

13. C most closely represents: $m < 0, b < 0$.

14. F most closely represents: $m > 0, b < 0$.

15. A most closely represents: $m < 0, b > 0$.

16. B most closely represents: $m > 0, b > 0$.

17. E most closely represents: $m = 0$.

18. D most closely represents: $b = 0$.

19. The Domain is: $[-6,6]$ and the Range is: $[-6,6]$.

20. False, the slopes are different. Although the difference is small, the lines are not parallel and will intersect.

21. $f(x) = g(x)$ when the graphs intersect or when $x = -3$. I $\{-3\}$ is the best match.

22. $f(x) > g(x)$ when the graph of $f(x)$ is above the graph of $g(x)$ or when $x > -3$. K $(-3, \infty)$ is the best match.

23. $f(x) < g(x)$ when the graph of $f(x)$ is below the graph of $g(x)$ or when $x < -3$. B $(-\infty, 3)$ is the best match.

24. $g(x) \geq f(x)$ when the graph of $g(x)$ intersects or is above the graph of $f(x)$ or when $x \leq -3$. A $(-\infty, -3]$ is the best match.

25. $y_2 - y_1 = 0 \Rightarrow y_2 = y_1 \Rightarrow g(x) = f(x)$ when the graphs intersect or when $x = -3.1$. I $\{-3\}$ is the best match.

26. $f(x) < 0$ when the graph of $f(x)$ is below the x-axis or when $x < -5$. M $(-\infty, -5)$ is the best match.

27. $g(x) > 0$ when the graph of $g(x)$ is above the x-axis or when $x < -2$. O $(-\infty, -2)$ is the best match.

28. $y_2 - y_1 < 0 \Rightarrow y_2 < y_1 \Rightarrow g(x) < f(x)$ when the graph of $g(x)$ is below the graph of $f(x)$ or when $x > -3$. K $(-3, \infty)$ is the best match.

29. $5[3+2(x-6)] = 3x+1 \Rightarrow 5[2x-9] = 3x+1 \Rightarrow 10x-45 = 3x+1 \Rightarrow 7x = 46 \Rightarrow x = \frac{46}{7}$.

The graphs of $y_1 = 5[3+2(x-6)]$ and $y_2 = 3x+1$ intersect at: $\left\{\frac{46}{7}\right\}$, which supports the result.

30. $\frac{x}{4} - \frac{x+4}{3} = -2 \Rightarrow 12\left[\frac{x}{4} - \frac{x+4}{3} = -2\right] \Rightarrow 3x - 4(x+4) = -24 \Rightarrow -x-16 = -24 \Rightarrow -x = -8 \Rightarrow x = 8$.

The graphs of $y_1 = \frac{x}{4} - \frac{x+4}{3}$ and $y_2 = -2$ intersect at: $\{8\}$, which supports the result.

31. $-3x-(4x+2) = 3 \Rightarrow -7x-2 = 3 \Rightarrow -7x = 5 \Rightarrow x = -\frac{5}{7}$. The graphs of $y_1 = -3x-(4x+2)$ and $y_2 = 3$ intersect at: $\left\{-\frac{5}{7}\right\}$, which supports the result.

32. $-2x+9+4x = 2(x-5)-3 \Rightarrow 2x+9 = 2x-10 \Rightarrow 9 = -10$. This is false, therefore this is a

contradiction and the solution is: $\varnothing$. The graphs of $y_1 = -2x + 9 + 4x$ and $y_2 = 2(x-5) - 3$ are parallel and do not intersect, therefore no solution, $\varnothing$, which supports the result.

33. $0.5x + 0.7(4-3x) = 0.4x \Rightarrow 0.5x + 2.8 - 2.1x = 0.4x \Rightarrow 5x + 28 - 21x = 4x \Rightarrow -20x = -28 \Rightarrow$

$x = \frac{7}{5}$. The graphs of $y_1 = .5x + .7(4-3x)$ and $y_2 = .4x$ intersect at: $\left\{\frac{7}{5}\right\}$, which supports the result.

34. $\frac{x}{4} - \frac{5x-3}{6} = 2 - \frac{7x+18}{12} \Rightarrow 12\left[\frac{x}{4} - \frac{5x-3}{6} = 2 - \frac{7x+18}{12}\right] \Rightarrow 3x - 2(5x-3) =$

$24 - (7x+18) \Rightarrow -7x + 6 = -7x + 6 \Rightarrow 6 = 6$. This is true, therefore an identity and the solution is:

$(-\infty, \infty)$. The graphs of $y_1 = \frac{x}{4} - \frac{5x-3}{6}$ and $y_2 = 2 - \frac{7x+18}{12}$ are the same line; the solution is

$(-\infty, \infty)$, which supports the result.

35. $x - 8 < 1 - 2x \Rightarrow 3x < 9 \Rightarrow x < 3$. The graph of $y_1 = x - 8$ is below the graph of $y_2 = 1 - 2x$ for the interval: $(-\infty, 3)$, which supports the result.

36. $\frac{4x-1}{3} \geq \frac{x}{5} - 1 \Rightarrow 15\left[\frac{4x-1}{3} \geq \frac{x}{5} - 1\right] \Rightarrow 5(4x-1) \geq 3x - 15 \Rightarrow 20x - 5 \geq 3x - 15 \Rightarrow$

$17x \geq -10 \Rightarrow x \geq -\frac{10}{17}$. The graph of $y_1 = \frac{4x-1}{3}$ intersects or is above the graph of $y_2 = \frac{x}{5} - 1$

for the interval: $\left[-\frac{10}{17}, \infty\right)$, which supports the result.

37. $-6 \leq \frac{4-3x}{7} < 2 \Rightarrow -42 \leq 4 - 3x < 14 \Rightarrow -46 \leq -3x < 10 \Rightarrow \frac{46}{3} \geq x > -\frac{10}{3} \Rightarrow -\frac{10}{3} < x \leq \frac{46}{3}$ or

for the interval: $\left(-\frac{10}{3}, \frac{46}{3}\right]$.

38. (a) Graph $y_1 = 5\pi + (\sqrt{3})x - 6.24(x - 8.1) + (\sqrt[3]{9})x$. See Figure 38. Find the x- intercept: $x \approx \{-3.81\}$.

(b) $f(x) < 0$ when the graph of $f(x)$ is below the x-axis. This happens for the interval: $(-\infty, -3.81)$.

(c) $f(x) \geq 0$ when the graph of $f(x)$ intersects or is above the x-axis. This happens for the interval: $[-3.81, \infty)$.

[-10,10] by [-10,10]
Xscl = 1 Yscl = 1

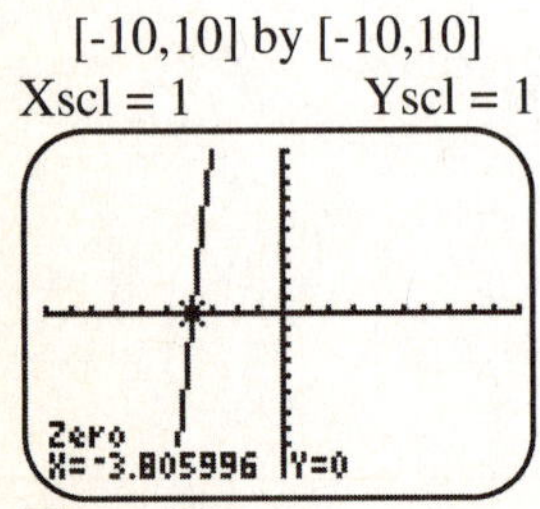

Figure 38

39. It costs \$30 to produce each CD and there is a one-time advertisement cost, therefore: $C(x) = 30x + 150$.

40. Each tape is sold for \$37.50, therefore: $R(x) = 37.50x$.

41. $C(x) = R(x)$ when $30x + 150 = 37.50x \Rightarrow 150 = 7.50x \Rightarrow x = 20$.

42. When the graph of $R(x)$ is below $C(x)$ the company is losing money, when $R(x)$ intersects $C(x)$ the company breaks even, and when $R(x)$ is above $C(x)$ the company makes money. This happens as follows: losing money when $x < 20$, breaking even when $x = 20$, and making money when $x > 20$.

43. $A = \dfrac{24f}{B(p+1)} \Rightarrow AB(p+1) = 24f \Rightarrow f = \dfrac{AB(p+1)}{24}$.

44. $A = \dfrac{24f}{B(p+1)} \Rightarrow AB(p+1) = 24f \Rightarrow B = \dfrac{24f}{A(p+1)}$.

45. (a) $f(x) = -3.52(5) + 58.6 \Rightarrow f(x) = -17.6 + 58.6 = 41^\circ$ F.

(b) $-15 = -3.52x + 58.6 \Rightarrow -73.6 = -3.52x \Rightarrow x \approx 20.9$ or about 21,000 feet.

(c) Graph $y_1 = -3.52x + 58.6$. Find the coordinates of the point where $x = 5$ to support the answer in (a). Find the coordinates of the point where $y = -15$ to support the answer in (b).

46. (a) Linear regression gives the model: $f(x) \approx 0.10677x - 211.69$. Answers may vary.

(b) $f(1987) = 0.10677(1987) - 211.69 \Rightarrow f(1987) \approx 0.462$ The cost of a 30 second Super Bowl ad in 1987 was approximately \$0.5 million.

(c) $f(x) = 3.2 \Rightarrow 3.2 = 0.10677x - 211.69 \Rightarrow 214.89 = 0.10677x \Rightarrow x \approx 2012.64$ Thus, the cost for a 30 second Super Bowl ad could reach \$3.2 million2013.

47. Let x = bat speed and y = ball travel distance; then (50, 320) and (80, 440) are two points on the graph of the function. Use the points to find slope: $m = \dfrac{440 - 320}{80 - 50} \Rightarrow m = \dfrac{120}{30} = 4$. Now use point-slope form to find the equation for the model: $y - 320 = 4(x - 50) \Rightarrow y - 320 = 4x - 200 \Rightarrow y = 4x + 120$. Because the slope is 4, the ball will travel 4 feet further for each additional 1 mph in bat speed.

48. Since surface area is: $A = 2(lw) + 2(lh) + 2(wh)$ we can solve $496 = 2(18)(8) + 2(18)h + 2(8)h \Rightarrow 496 = 288 + 36h + 16h \Rightarrow 208 = 52h \Rightarrow h = 4$. The height of the box is 4 feet.

49. Since there are 5280 feet in a mile, there are $5280 \times 26 = 137,280$ feet in a marathon. Since there are 3.281 feet in a meter, there are $100 \times 3.281 = 328.1$ feet in a 100 meter dash. Now use a proportion to solve: $\dfrac{9.69}{328.1} = \dfrac{x}{137,280} \Rightarrow 328.1x = 1340475.84 \Rightarrow x = 4085.571$ seconds to run a marathon. Divide by 60 to get minutes run: $4085.571 \div 60 = 68.09$ minutes run or 1 hour 8 minutes and 6 seconds. This time is 55 minutes and 53 seconds faster.

50. $C = \dfrac{5}{9}(864 - 32) \Rightarrow C = \dfrac{5}{9}(832) \Rightarrow C = \dfrac{4160}{9} \Rightarrow 462\dfrac{2}{9}\,^\circ$C.

51. Find the constant of variation: $4k = 3000 \Rightarrow k = 750$. Use this to find the pressure: $10(750) = 7500 \text{ kg/m}^2$.

52. (a) Use any two points to find slope: $m = \frac{1.8-3}{1-0} \Rightarrow m = \frac{-1.2}{1} = -1.2$ Now use point-slope form to find the equation: $y-3=-1.2(x-0) \Rightarrow y=-1.2x+3$.

(b) $y=-1.2(-1.5)+3 \Rightarrow y=1.8+3 \Rightarrow y=4.8$ when $x=-1.5$.

$y=-1.2(3.5)+3 \Rightarrow y=-4.2+3 \Rightarrow y=-1.2$ when $x=3.5$.

53. (a) Enter the years in L_1 and enter test scores in L_2. The regression equation is: $f(x)=1.2x-1886.4$

(b) $y=1.2(2003)-1886.4 \Rightarrow y=2403.6-1886.4 \Rightarrow y \approx 517$.

(c) $y=1.2(2008)-1886.4 \Rightarrow y=2409.6-1886.4 \Rightarrow y \approx 523$.

54. Let $x =$ the amount of 5% solution to be added, then solve:

$120(.20)+x(.05)=(120+x)(.10) \Rightarrow 24+.05x=12+.10x \Rightarrow 12=.05x \Rightarrow x=240$ mL of 5% solution needs to be added.

55. The company will at least break even when $R(x) \geq C(x)$, therefore solve:

$8x \geq 3x+1500 \Rightarrow 5x \geq 1500 \Rightarrow x \geq 300$ or for the interval: $[300,\infty)$. 300 or more DVD's need to be sold to at least break even.

56. Let m = mental age and c = chronological age, then $IQ = \frac{100m}{c}$.

(a) $130 = \frac{100m}{7} \Rightarrow 910 = 100m \Rightarrow m = 9.1$ years.

(b) $IQ = \frac{100(20)}{16} \Rightarrow IQ = \frac{2000}{16} \Rightarrow IQ = 125$ years.

57. (a) Enter the heights in L_1 and enter weights in L_2. The regression equation is: $y \approx 4.512x-154.4$

(b) $y=4.51(75)-154.4 \Rightarrow y=338.25-154.4 \Rightarrow y \approx 184$.

58. (a) Enter the heights in L_1 and enter weights in L_2. The regression equation is: $y \approx 4.465x-133.3$

(b) $y=4.465(80)-133.3 \Rightarrow y=357.2-133.3 \Rightarrow y \approx 224$.

Chapter 1 Test

1. (a) Domain: $(-\infty,\infty)$, Range: $[2,\infty)$, x-intercept: none, y-intercept: 3

(b) Domain: $(-\infty,\infty)$, Range: $(-\infty,0]$, x-intercept: 3, y-intercept: -3

(c) Domain: $[-4,\infty)$, Range: $[0,\infty)$, x-intercept: -4, y-intercept: 2

2. (a) $f(x) = g(x)$ when the graph of $f(x)$ intersects the graph of $g(x)$ therefore $\{-4\}$.

(b) $f(x) < g(x)$ when the graph of $f(x)$is below the graph of $g(x)$, for the interval: $(-\infty,-4)$.

(c) $f(x) \geq g(x)$ when the graph of $f(x)$intersects or is above the graph of $g(x)$, for the interval: $[-4,\infty)$.

(d) $f(\;y_2 - y_1 = 0 \Rightarrow y_2 = y_1 \Rightarrow g(x) = f(x)$ when the graph of $g(x)$ intersects the graph of $f(x)$, therefore $\{-4\}$.

3. (a) $y_1 = 0$ when the graph of $f(x)$ intersects the x-axis, therefore $\{5.5\}$.

(b) $y_1 < 0$ when the graph of $f(x)$ is below the x-axis, for the interval $(-\infty, 5.5]$.

(c) $y_1 > 0$ when the graph of $f(x)$ is above the x-axis, for the interval $(5.5, \infty)$.

(d) $y_1 \le 0$ when the graph of $f(x)$ intersects or is below the x-axis, for the interval $(-\infty, 5.5]$.

4. (a) $3(x-4)-2(x-5) = -\;2(x+1)-3 \Rightarrow 3x-12-2x+10 = -2x-2-3 \Rightarrow x-2 = -2x-5$

$\Rightarrow 3x = -3 \Rightarrow x = -1$. Check: $3(-1-4)-2(-1-5) = -2(-1+1)-3 \Rightarrow$

$3(-5)-2(-6) = -2(0)-3 \Rightarrow -15+12 = 0-3 \Rightarrow -3 = -3$

(b) Graph $y_1 = 3(x-4)-2(x-5)$ and $y_2 = -2(x+1)-3$. See Figure 4. $f(x) > g(x)$ for the interval: $(-1, \infty)$ because the graph of $y_1 = f(x)$ is above the graph of $y_2 = g(x)$ for domain values greater than -1.

(c) See Figure 4. $f(x) < g(x)$ for the interval $(-\infty, -1)$ because the graph of $y_1 = f(x)$ is below the graph of $y_2 = g(x)$ for domain values less than -1.

5. (a) $-\frac{1}{2}(8x+4)+3(x-2) = 0 \Rightarrow -4x-2+3x-6 = 0 \Rightarrow -x-8 = 0 \Rightarrow x = -8$ or $\{-8\}$.

(b) $-\frac{1}{2}(8x+4)+3(x-2) \le 0 \Rightarrow -4x-2+3x-6 \le 0 \Rightarrow -x-8 \le 0 \Rightarrow x \ge -8$ or for the interval: $[-8, \infty)$.

(c) Graph $y_1 - \frac{1}{2}(8x+4)+(3(x-2)$. See Figure 5. The x-intercept is -8 supporting the result in part (a). The graph of the linear function lies below or on the x-axis for domain values greater than or equal to -8, supporting the results in part (b).

[-10,10] by [-10,10]
Xscl = 1 Yscl = 1

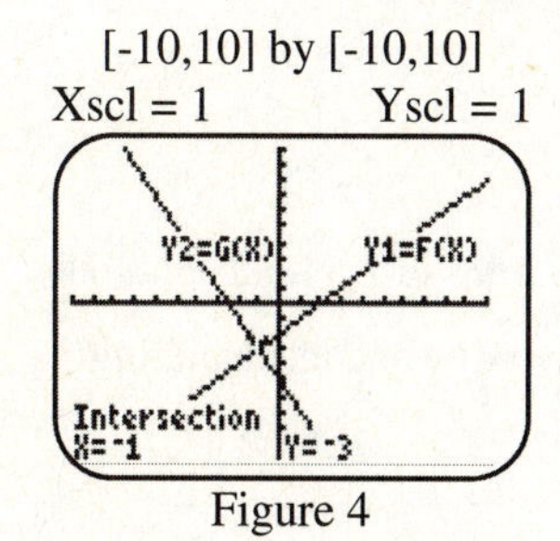

Figure 4

[-10,10] by [-10,10]
Xscl = 1 Yscl = 1

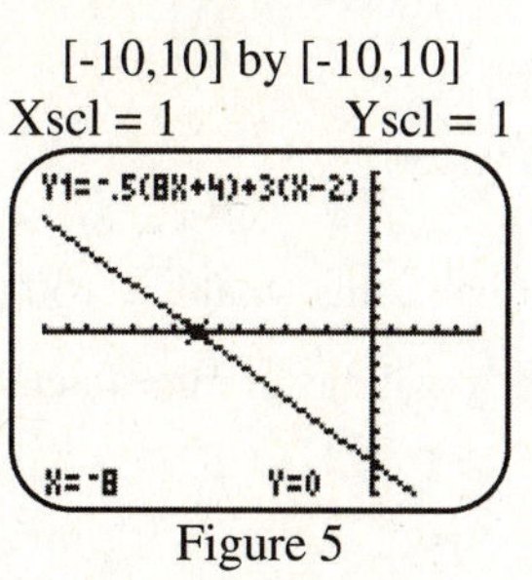

Figure 5

6. (a) Using the points $(1982, 8.30)$ and $(2008, 30.53)$ find the midpoint:

$$M.P. = \left(\frac{1982+2008}{2}, \frac{8.30+30.53}{2}\right) \Rightarrow M.P. = (1995, 19.42).$$ \$19.42 is the approximate cost in 1995.

(b) Using the points $(1982, 8.30)$ and $(2008, 30.53)$ find the slope:

$m=\dfrac{30.53-8.30}{2008-1982}\Rightarrow m=\dfrac{22.23}{26}\Rightarrow m=0.855$. This means that during the years 1982 to 2008, the monthly rate increased, on the average, by about \$0.855 per year.

7. (a) Since the given line has a slope of -2 and parallel lines have equal slopes, our new line has a slope of -2. Now use point-slope form: $y-5=-2(x-(-3))\Rightarrow y-5=-2x-6\Rightarrow y=-2x-1$.

   (b) The equation: $-2x+y=0\Rightarrow y=2x$ has a slope of 2. Since perpendicular lines have slopes whose product equals -1, our new line has a slope of $-\frac{1}{2}$. Now use point-slope form:

   $y-5=-\frac{1}{2}(x-(-3))\Rightarrow y-5=-\frac{1}{2}x-\frac{3}{2}\Rightarrow y=-\frac{1}{2}x+\frac{7}{2}$.

8. For the x-intercept $y=0$, therefore: $3x-4(0)=6\Rightarrow 3x=6\Rightarrow x=2$. The x-intercept is 2. For the y-intercept $x=0$, therefore: $3(0)-4y=6\Rightarrow -4y=6\Rightarrow y=-\frac{6}{4}=-\frac{3}{2}$. The y-intercept is: $-\frac{3}{2}$. Using the intercepts: $\left(0,-\frac{3}{2}\right)$ and $(2,0)$, the slope is $m=\dfrac{0-\left(-\frac{3}{2}\right)}{2-0}=\dfrac{\frac{3}{2}}{2}\Rightarrow m=\frac{3}{4}$.

9. The equation of the horizontal line passing through $(-3,7)$ is $y=7$. The equation of the vertical line passing through $(-3,7)$ is $x=-3$.

10. (a) Enter the wind speed in L_1 and enter degrees in L_2. The regression equation is: $Y\approx -.246x+35.7$ and the correlation coefficient is: $r\approx -.96$.

    (b) $y\approx -.246(40)+35.7\Rightarrow y\approx -9.84+35.7\Rightarrow y\approx 25.9\ ^\circ F$. This is $3.3\ ^\circ F$ lower than the actual value.

11. (a) Use the volume of a cylinder: $V=\pi r^2h$ and the fact that 20 feet $=240$ inches.

    $V=\pi(240)^2(1)\Rightarrow V=57{,}600\pi\Rightarrow V\approx 180{,}956\text{ in}^3$.

    (b) Since 1 gallon = 231 cubic inches, the equation is: $g(x)=\dfrac{180{,}956}{231}x$.

    (c) $g(x)=\dfrac{180{,}956}{231}(2.5)\Rightarrow g(x)\approx 1958$ gallons of water.

    (d) Since the downspout can handle 400 gallons per hour, this would be 1000 gallons in 2.5 hours and one downspout would not be sufficient handle the 1958 gallons of rain water. Two downspouts handling 2000 gallons would be sufficient.

Chapter 2: Analysis of Graphs and Functions

2.1: Graphs of Basic Functions and Relations; Symmetry

1. $(-\infty,\infty)$.

3. $(0,0)$

5. increases

7. x-axis

9. odd

11. The domain can be all real numbers; therefore, the function is continuous for the interval $(-\infty,\infty)$.

13. The domain can only be values where $x \geq 0$; therefore, the function is continuous for the interval $[0,\infty)$.

15. The domain can be all real numbers except -3; therefore, the function is continuous for the interval $(-\infty,-3)\cup(-3,\infty)$.

17. (a) The function is increasing for the interval $[3,\infty)$.

    (b) The function is decreasing for the interval $(-\infty,3]$.

    (c) The function is never constant; therefore, none.

    (d) The domain can be all real numbers; therefore, the interval $(-\infty,\infty)$.

    (e) The range can only be values where $y \geq 0$; therefore, the interval $[0,\infty)$.

19. (a) The function is increasing for the interval $(-\infty,1]$.

    (b) The function is decreasing for the interval $[4,\infty)$.

    (c) The function is constant for the interval $[1,4]$.

    (d) The domain can be all real numbers; therefore, the interval $(-\infty,\infty)$.

    (e) The range can only be values where $y \leq 3$; therefore, the interval $(-\infty,3]$.

21. (a) The function is never increasing; therefore, none

    (b) The function is decreasing for the intervals $(-\infty,-2]$ and $[3,\infty)$.

    (c) The function is constant for the interval $(-2,3)$.

    (d) The domain can be all real numbers; therefore, the interval $(-\infty,\infty)$.

    (e) The range can only be values where $y \leq 1.5$ or $y \geq 2$; therefore, the interval $(-\infty,1.5]\cup[2,\infty)$.

23. Graph $f(x)=x^5$. See Figure 23. As x increases for the interval $(-\infty,\infty)$, y increases; therefore, the function is increasing.

25. Graph $f(x)=x^4$. See Figure 25. As x increases for the interval $(-\infty,0]$, y decreases; therefore, the function is decreasing on $(-\infty,0]$.

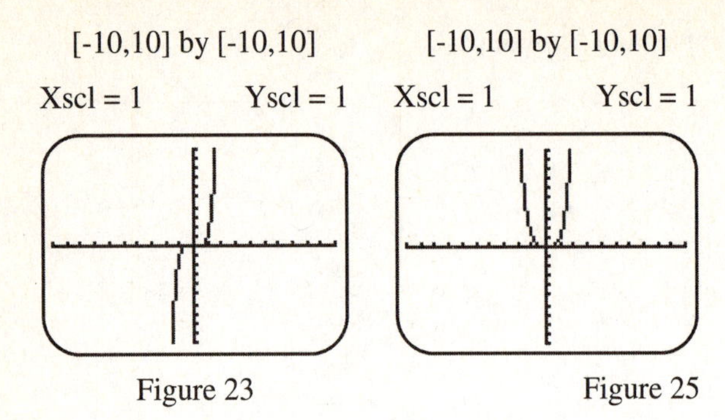

Figure 23 Figure 25

27. Graph $f(x) = -|x|$. See Figure 27. As x increases for the interval $(-\infty, 0]$, y increases; therefore, the function is increasing on $(-\infty, 0]$.

29. Graph $f(x) = -\sqrt[3]{x}$. See Figure 29. As x increases for the interval $(-\infty, \infty)$, y decreases; therefore, the function is decreasing.

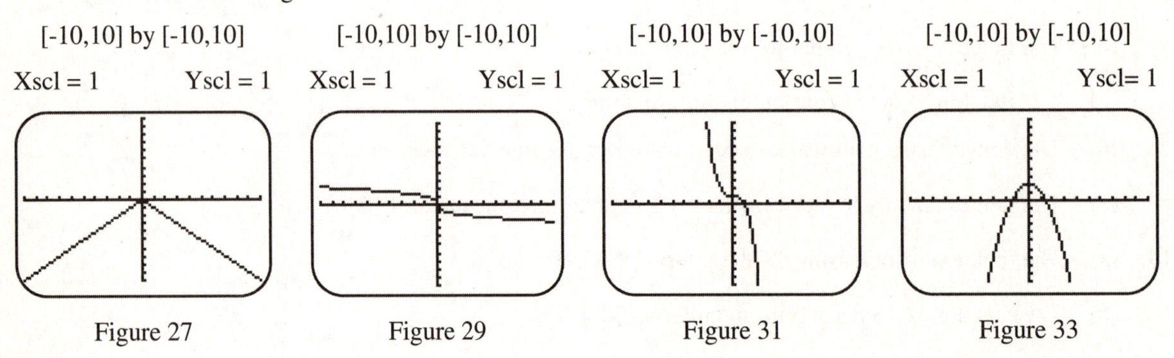

Figure 27 Figure 29 Figure 31 Figure 33

31. Graph $f(x) = 1 - x^3$. See Figure 31. As x increases for the interval $(-\infty, \infty)$, y decreases; therefore, the function is decreasing.

33. Graph $f(x) = 2 - x^2$. See Figure 33. As x increases for the interval $(-\infty, 0]$, y increases; therefore, the function is increasing on $(-\infty, 0]$.

35. (a) No (b) Yes (c) No
37. (a) Yes (b) No (c) No
39. (a) Yes (b) Yes (c) Yes
41. (a) No (b) No (c) Yes
43. (a) Since $f(-x) = f(x)$, this is an even function and is symmetric with respect to the y-axis. See Figure 43a.
    (b) Since $f(-x) = -f(x)$, this is an odd function and is symmetric with respect to the origin. See Figure 43b.

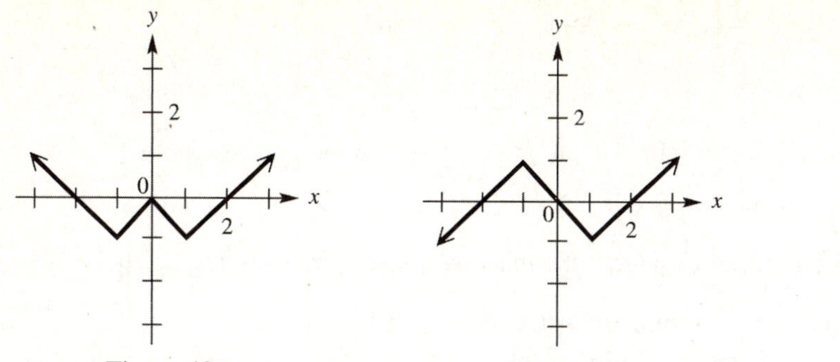

Figure 43a Figure 43b

x	$f(x)$
–3	21
–2	–12
–1	–25
1	–25
2	–12
3	21

Figure 45

45. If f is an even function then $f(-x)=f(x)$ or opposite domains have the same range. See Figure 45

47. This is an even function since opposite domains have the same range.

49. This is an odd function since opposite domains have the opposite range.

51. This is neither even nor odd since the opposite domains are neither the opposite or same range.

53. If $f(x)=x^4-7x^2+6$, then $f(-x)=(-x)^4-7(-x)^2+6 \Rightarrow f(-x)=x^4-7x^2+6$. Since $f(-x)=f(x)$, the function is even.

55. If $f(x)=3x^3-x$, then $f(-x)=3(-x)^3-(-x) \Rightarrow f(-x)=-3x^3+x$ and $-f(x)=-(3x^3-x) \Rightarrow -f(x)=-3x^3+x$. Since $f(-x)=-f(x)$, the function is odd.

57. If $f(x)=x^6-4x^4+5$ then $f(-x)=(-x)^6-4(-x)^4+5 \Rightarrow f(-x)=x^6-4x^4+5$. Since $f(-x)=f(x)$, the function is even.

59. If $f(x)=3x^5-x^3+7x$, then $f(-x)=3(-x)^5-(-x)^3+7(-x) \Rightarrow f(-x)=-3x^5+x^3-7x$ and $-f(x)=-(3x^5-x^3+7x) \Rightarrow -f(x)=-3x^5+x^3-7x$. Since $f(-x)=-f(x)$, the function is odd.

61. If $f(x)=|5x|$, then $f(-x)=|5(-x)| \Rightarrow f(-x)=|5x|$. Since $f(-x)=f(x)$, the function is even.

63. If $(-3,11)$ and $(2,9)$ then $f(-x)=\dfrac{1}{2(-x)} \Rightarrow f(-x)=-\dfrac{1}{2x}$ and $-f(x)=-\left(\dfrac{1}{2x}\right) \Rightarrow -f(x)=-\dfrac{1}{2x}$ Since $f(-x)=-f(x)$, the function is odd.

65. If $f(x)=-x^3+2x$, then $f(-x)=-(-x)^3+2(-x) \Rightarrow f(-x)=x^3-2x$ and $-f(x)=-(-x^3+2x) \Rightarrow -f(x)=x^3-2x$. Since $f(-x)=-f(x)$, the function is symmetric with respect to the origin. Graph $f(x)=-x^3+2x$; the graph supports symmetry with respect to the origin.

67. If $f(x)=0.5x^4-2x^2+1$, then $f(-x)=0.5(-x)^4-2(-x)^2+1 \Rightarrow f(-x)=0.5x^4-2x^2+1$. Since $f(-x)=f(x)$, the function is symmetric with respect to the y-axis. Graph $f(x)=0.5x^4-2x^2+1$; the graph supports symmetry with respect to the y-axis.

69. If $f(x)=x^3-x+3$, then $f(-x)=(-x)^3-(-x)+3 \Rightarrow f(-x)=-x^3+x+3$ and $-f(x)=-(x^3-x+3) \Rightarrow -f(x)=-x^3+x-3$. Since $f(x) \neq f(-x) \neq -f(x)$, the function is not symmetric with respect to the y-axis or the origin.

71. If $f(x)=x^6-4x^3$, then $f(-x)=(-x)^6-4(-x)^3 \Rightarrow f(-x)=x^6+4x^3$ and $-f(x)=-(x^6-4x^3) \Rightarrow -f(x)=-x^6+4x^3$. Since $f(x)\neq f(-x)\neq -f(x)$, the function is not symmetric with respect to the y-axis or the origin. Graph $f(x)=x^6-4x^3$; the graph supports no symmetry with respect to the y-axis or the origin.

73. If $f(x)=-6$, then $f(-x)=-6$, Since $f(-x)=f(x)$, the function is symmetric with respect to the y-axis. Graph $f(x)=-6$; the graph supports symmetry with respect to the y-axis.

75. If $f(x)=\frac{1}{4x^3}$, then $f(-x)=\frac{1}{4(-x)^3} \Rightarrow f(-x)=-\frac{1}{4x^3}$ and $-f(x)=-\left(\frac{1}{4x^3}\right) \Rightarrow -f(x)=-\frac{1}{4x^3}$. Since $f(-x)=-f(x)$, the function is symmetric with respect to the origin. Graph $f(x)=\frac{1}{4x^3}$; the graph supports symmetry with respect to the origin.

2.2: Vertical and Horizontal Shifts of Graphs

1. The equation $y=x^2$ shifted 3 units upward is $y=x^2+3$.

3. The equation $y=\sqrt{x}$ shifted 4 units downward is $y=\sqrt{x}-4$.

5. The equation $y=|x|$ shifted 4 units to the right is $y=|x-4|$.

7. The equation $y=x^3$ shifted 7 units to the left is $y=(x+7)^3$.

9. Shift the graph of f 4 units upward to obtain the graph of g.

11. The equation $y=x^2-3$ is $y=x^2$ shifted 3 units downward; therefore, graph B.

13. The equation $y=(x+3)^2$ is $y=x^2$ shifted 3 units to the left; therefore, graph A.

15. The equation $y=|x+4|-3$ is $y=|x|$ shifted 4 units to the left and 3 units downward; therefore, graph B.

17. The equation $y=(x-3)^3$ is $y=x^3$ shifted 3 units to the right; therefore, graph C.

19. The equation $y=(x+2)^3-4$ is $[-a,-b]$. shifted 2 units to the left and 4 units downward; therefore, graph B.

21. Using $Y_2=Y_1+k$ and $x=0$, we get $-5=-3+k \Rightarrow k=-2$.

23. From the graphs, $(6,2)$ is a point on Y_1 and $(6,-1)$ a point on Y_2. Using $Y_2=Y_1+k$ and $x=6$, we get $-1=2+k \Rightarrow k=-3$.

25. For the equation $y=x^2$, the Domain is $(-\infty,\infty)$ and the Range is $[0,\infty)$. Shifting this 3 units downward gives us: (a) Domain: $(-\infty,\infty)$ (b) Range: $[-3,\infty)$.

27. For the equation $y=|x|$, the Domain is $(-\infty,\infty)$ and the Range is $[0,\infty)$. Shifting this 4 units to the left and 3 units downward gives us: (a) Domain: $(-\infty,\infty)$ (b) Range: $[-3,\infty)$.

29. For the equation $y = x^3$, the Domain is $(-\infty, \infty)$ and the Range is $(-\infty, \infty)$. Shifting this 3 units to the right gives us: (a) Domain: $(-\infty, \infty)$ (b) Range: $(-\infty, \infty)$

31. The graph of $y = f(x)$ is the graph of the equation $y = x^2$ shifted 1 unit to the right. See Figure 31.

33. The graph of $y = x^3 + 1$ is the graph of the equation $y = x^3$ shifted 1 unit upward. See Figure 33.

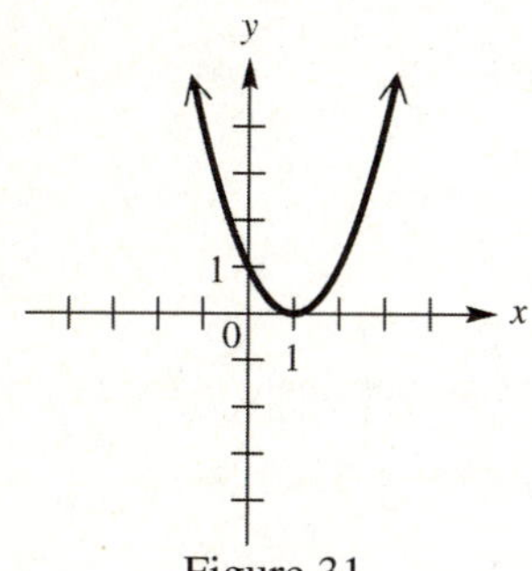

Figure 31

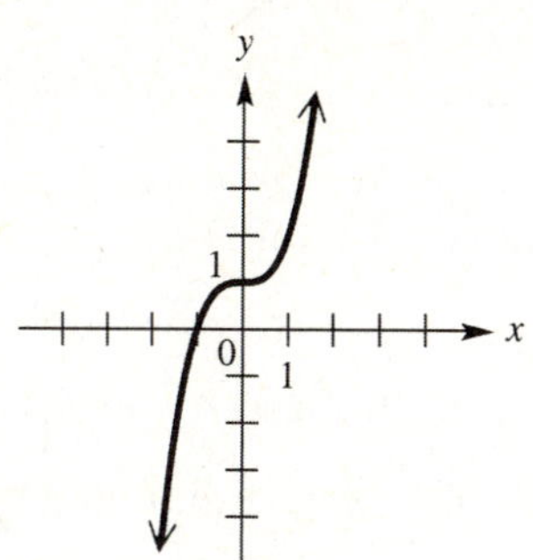

Figure 33

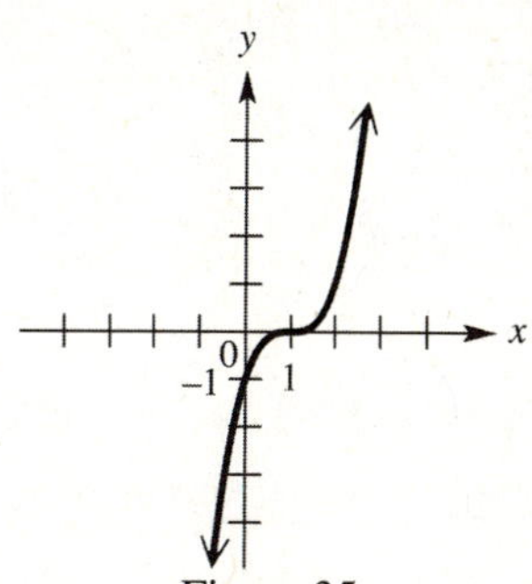

Figure 35

35. The graph of $y = (x-1)^3$ is the graph of the equation $y = x^3$ shifted 1 unit to the right. See Figure 35.

37. The graph of $y = \sqrt{x-2} - 1$ is the graph of the equation $y = \sqrt{x}$ shifted 2 units to the right and 1 unit downward. See Figure 37.

39. The graph of $f(x)$ is the graph of the equation $y = x^2$ shifted 2 units to the left and 3 units upward. See Figure 39.

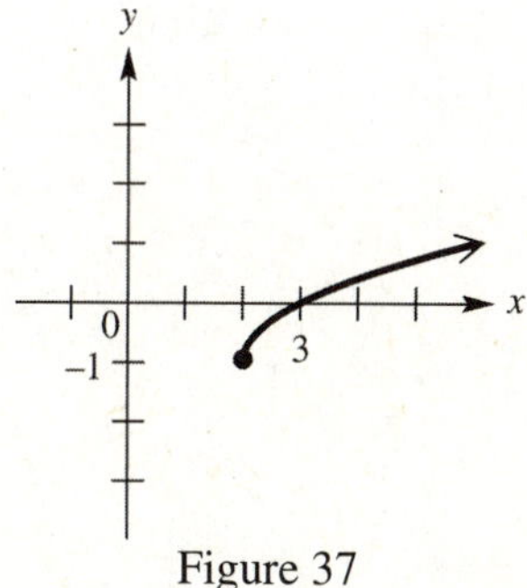

Figure 37

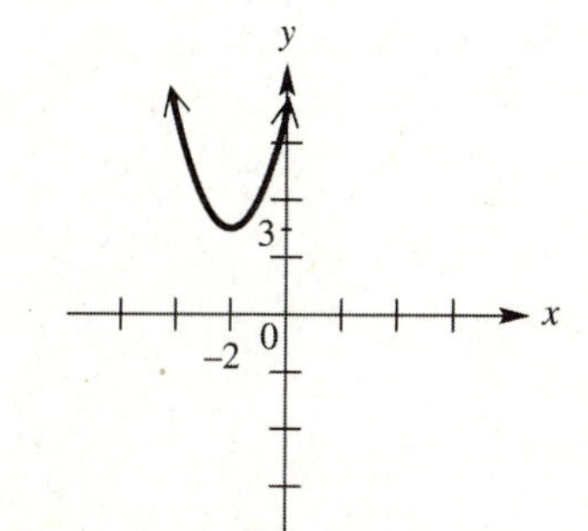

Figure 39

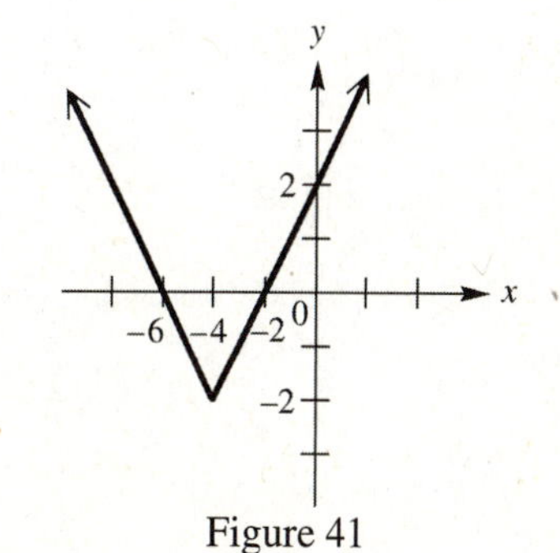

Figure 41

41. The graph of $y = |x+4| - 2$ is the graph of the equation $y = |x|$ shifted 4 units to the left and 2 units downward. See Figure 41.

43. Since h and k are positive, the equation is $y = x^2$ shifted to the right and down; therefore, B.

45. Since h and k are positive, the equation is $y = x^2$ shifted to the left and up; therefore, A.

47. The equation $y = f(x) + 2$ is $y = f(x)$ shifted up 2 units or add 2 to the y-coordinate of each point as follows: $(-3, 2) \Rightarrow (-3, 0); (-1, 4) \Rightarrow (-1, 6); (5, 0) \Rightarrow (5, 2)$. See Figure 47.

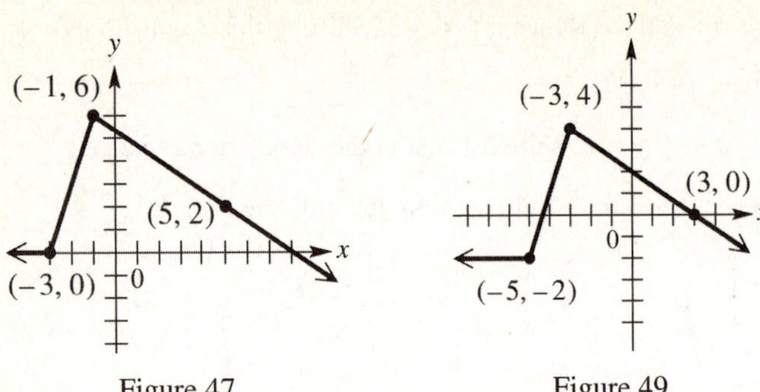

Figure 47 Figure 49

49. The equation $y = f(x+2)$ is $y = f(x)$ shifted left 2 units or subtract 2 from the x-coordinate of each point as follows: $(-3,-2) \Rightarrow (-5,-2); (-1,4) \Rightarrow (-3,4); (5,0) \Rightarrow (3,0)$. See Figure 49.

51. The graph is the basic function $y = x^2$ translated 4 units to the left and 3 units up; therefore, the new equation is $y = (x+4)^2 + 3$. The equation is now increasing for the interval: (a) $[-4,\infty)$ and decreasing for the interval: (b) $(-\infty,-4]$.

53. The graph is the basic function $y = x^3$ translated 5 units down; therefore, the new equation is $y = x^3 - 5$. The equation is now increasing for the interval: (a) $(-\infty,\infty)$ and does not decrease; therefore: (b) none.

55. The graph is the basic function $y = \sqrt{x}$ translated 2 units to the right and 1 unit up; therefore, the new equation is $y = \sqrt{x-2} + 1$. The equation is now increasing for the interval: (a) $[2,\infty)$ and does not decrease; therefore: (b) none.

57. (a) $f(x) = 0$: $\{3,4\}$

(b) $f(x) > 0$: for the intervals $(-\infty,3) \cup (4,\infty)$.

(c) $f(x) < 0$: for the interval $(3,4)$.

58. (a) $f(x) = 0$: $\{\sqrt{2}\}$

(b) $f(x) > 0$: for the interval $\left(\sqrt{2},\infty\right)$.

(c) $f(x) < 0$: for the interval $\left(-\infty,\sqrt{2}\right)$.

59. (a) $f(x) = 0$: $\{-4,5\}$

(b) $f(x) \geq 0$: for the intervals $\left(-\infty,-4\right] \cup \left[5,\infty\right)$

(c) $f(x) \leq 0$: for the interval $[-4,5]$.

60. (a) $f(x) = 0$: never; therefore: $\varnothing$.

(b) $f(x) \geq 0$: for the interval $[1,\infty)$.

(c) $f(x) \leq 0$: never; therefore: $\varnothing$.

61. The translation is 3 units to the left and 1 unit up; therefore, the new equation is $y = |x+3| + 1$. The form $y = |x-h| + k$ will equal $y = |x+3| + 1$ when: $h = -3$ and $k = 1$.

63. (a) Since 0 corresponds to 1978, our equation using exact years would be $y = 1.311(x - 2000) + 68.28$.

(b) $y = 1.311(2003 - 2000) + 68.28 \Rightarrow y = 72.2$ million.

65. (a) Enter the year in L_1 and enter tuition and fees in L_2. The year 1993 corresponds to $x = 0$ and so on. The regression equation is $y \approx 235.4x + 2278.2$.

(b) Since $x = 0$ corresponds to 1993, the equation when the exact year is entered is $y = 235.4(x - 1993) + 2278.2$

(c) $y \approx 235.4(2009 - 1993) + 2278.2 \Rightarrow y \approx \$6000$

67. See Figure 67.

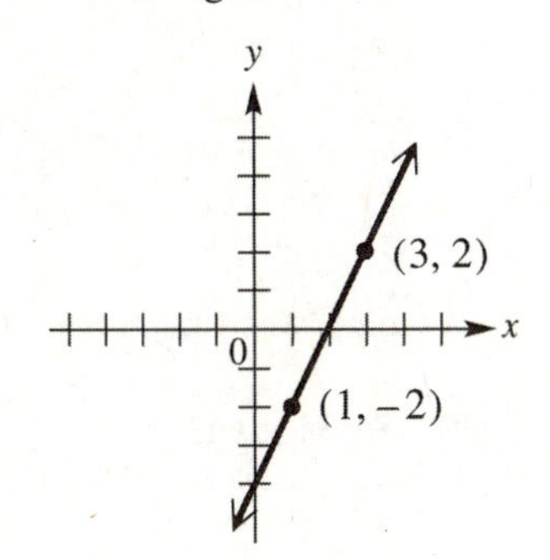

Figure 67

68. $m = \dfrac{2-(-2)}{3-1} \Rightarrow m = \dfrac{4}{2} = 2$

69. Using slope-intercept form yields: $y_1 - 2 = 2(x-3) \Rightarrow y_1 - 2 = 2x - 6 \Rightarrow y_1 = 2x - 4$

70. $(1, -2+6)$ and $(3, 2+6) \Rightarrow (1,4)$ and $(3,8)$

71. $m = \dfrac{8-4}{3-1} \Rightarrow m = \dfrac{4}{2} = 2$

72. Using slope-intercept form yields: $y_2 - 4 = 2(x-1) \Rightarrow y_2 - 4 = 2x - 2 \Rightarrow y_2 = 2x + 2$.

73. Graph $y_1 = 2x - 4$ and $y_2 = 2x + 2$ See Figure 73. The graph y_2 can be obtained by shifting the graph of y_1 upward 6 units. The constant 6, comes from the 6 we added to each y-value in Exercise 70.

[-10,10] by [-10,10]
Xscl = 1 Yscl = 1

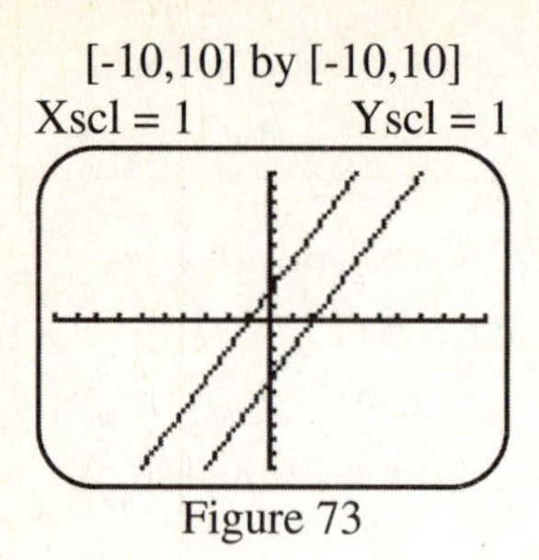

Figure 73

74. c; c; the same as; c; upward (or positive vertical)

2.3: Stretching, Shrinking, and Reflecting Graphs

1. The function $y = x^2$ vertically stretched by a factor of 2 is $y = 2x^2$.
3. The function $y = \sqrt{x}$ reflected across the y-axis is $y = \sqrt{-x}$.
5. The function $y = |x|$ vertically stretched by a factor of 3 and reflected across the x-axis is $y = -3|x|$.
7. The function $y = x^3$ vertically shrunk by a factor of 0.25 and reflected across the y-axis is $y = 0.25(-x^3)$ or $y = -0.25x^3$.
9. Graph $y_1 = x$, $y_2 = x+3$ (y_1 shifted up 3 units), and $y_3 = x-3$ (y_1 shifted down 3 units). See Figure 9.
11. Graph $y_1 = |x|$, $y_2 = |x-3|$ (y_1 shifted right 3 units), and $y_3 = |x+3|$ (y_1 shifted left 3 units). See Figure 11.

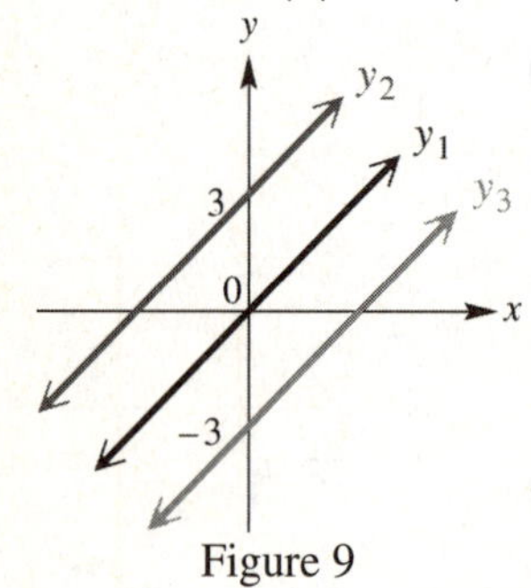

Figure 9

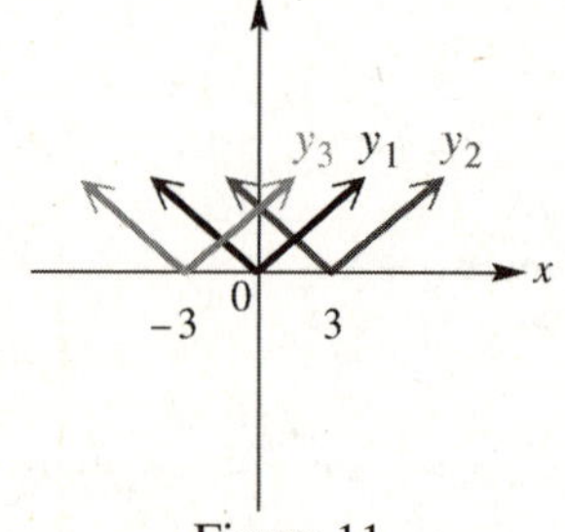

Figure 11

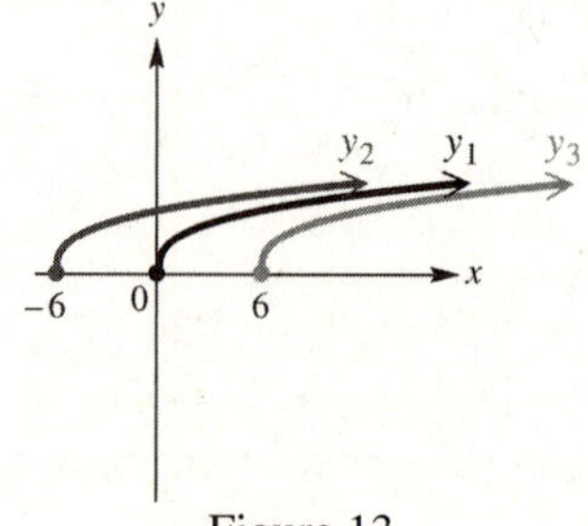

Figure 13

13. Graph $y_1 = \sqrt{x}$, $y_2 = \sqrt{x+6}$ (y_1 shifted left 6 units), and $y_3 = \sqrt{x-6}$ (y_1 shifted right 6 units). See Figure 13.
15. Graph $y_1 = \sqrt[3]{x}$, $y_2 = -\sqrt[3]{x}$ (y_1 reflected across the x-axis), and $y_3 = -2\sqrt[3]{x}$ (y_1 reflected across the x-axis and stretched vertically by a factor of 2). See Figure 15.
17. Graph $y_1 = |x|$, $y_2 = -2|x-1|+1$ (y_1 reflected across the x-axis, stretched vertically by a factor of 2, shifted right 1 unit, and shifted up 1 unit), and $y_3 = -\frac{1}{2}|x|-4$ (y_1 reflected across the x-axis, shrunk by factor of $\frac{1}{2}$, and shifted down 4 units). See Figure 17

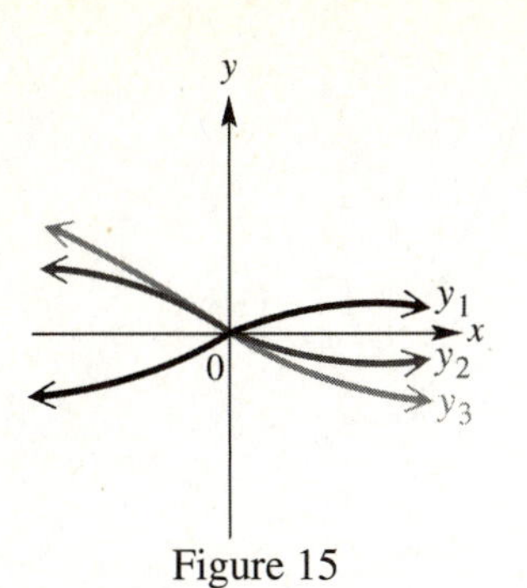

Figure 15

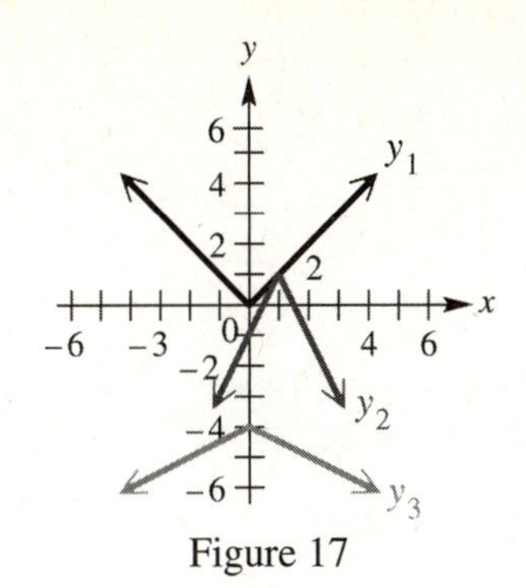

Figure 17

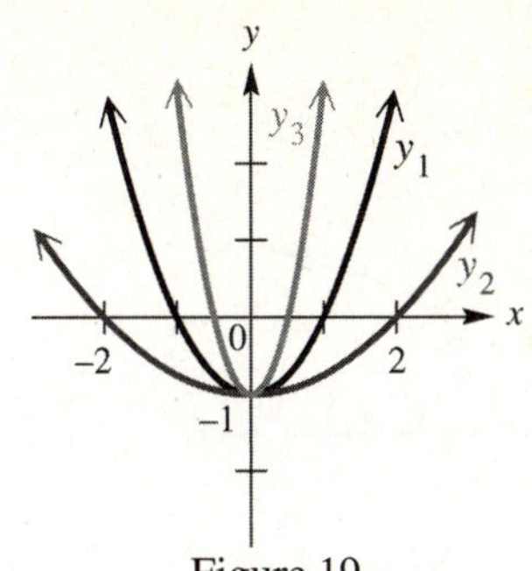

Figure 19

19. Graph $y_1 = x^2 - 1$ (which is $y = x^2$ shifted down 1 unit), $y_2 = \left(\frac{1}{2}x\right)^2 - 1$ (y_1 shrunk vertically by A factor of $\frac{1}{2}$), and $y_3 = (2x)^2 - 1$ (y_1 stretched vertically by a factor of 2^2 or 4). See Figure 19.

21. Graph $y_1 = \sqrt[3]{x}$, $y_2 = \sqrt[3]{-x}$ (y_1 reflected across the y-axis), and $y_3 = \sqrt[3]{-(x-1)}$ (y_1 reflected across the y-axis and shifted right 1 unit). See Figure 21.

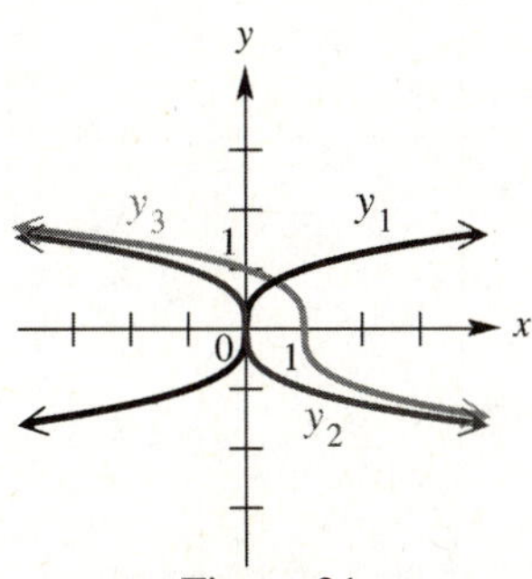

Figure 21

23. 4; x

25. 2; left; $\frac{1}{4}$; x; 3; downward (or negative)

27. 3; right; 6

29. The function $y = x^2$ is vertically shrunk by a factor of $\frac{1}{2}$ and shifted 7 units down; therefore, $y = \frac{1}{2}x^2 - 7$.

31. The function $y = \sqrt{x}$ is shifted 3 units right, vertically stretched by a factor of 4.5, and shifted 6 units down; therefore, $y = 4.5\sqrt{x-3} - 6$.

33. The graph $y = f(x) = x^2$ has been reflected across the x-axis, shifted 5 units to the right, and shifted 2 units downward; therefore, the equation of $g(x)$ is $g(x) = -(x-5)^2 - 2$.

35. The function $f(x) = \sqrt{x-3} + 2$ is $f(x) = \sqrt{x}$ shifted 3 units right and 2 units upward. See Figure 35.

37. The function $f(x) = \sqrt{2x} = \sqrt{2}\sqrt{x}$ is $f(x) = \sqrt{x}$ stretched vertically by a factor of $\sqrt{2}$. See Figure 37

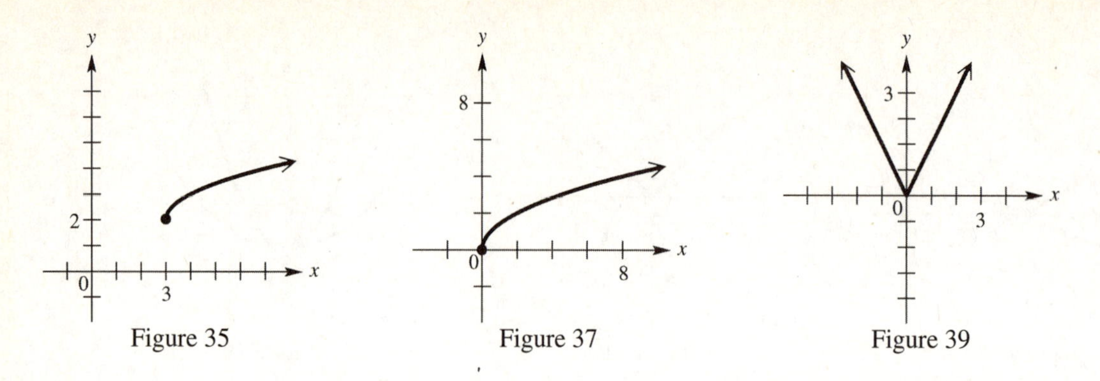

Figure 35 Figure 37 Figure 39

39. The function $f(x)=|2x|=2|x|$ is $f(x)=|x|$ stretched vertically by a factor of 2. See Figure 39.

41. The function $f(x)=1-\sqrt{x}$ is $f(x)=\sqrt{x}$ reflected across the x-axis and shifted 1 unit upward. See Figure 41.

43. The function $f(x)=-\sqrt{1-x}=-\sqrt{-(x-1)}$ is $f(x)=\sqrt{x}$ reflected across both the x-axis, the y-axis and shifted 1 unit right. See Figure 43.

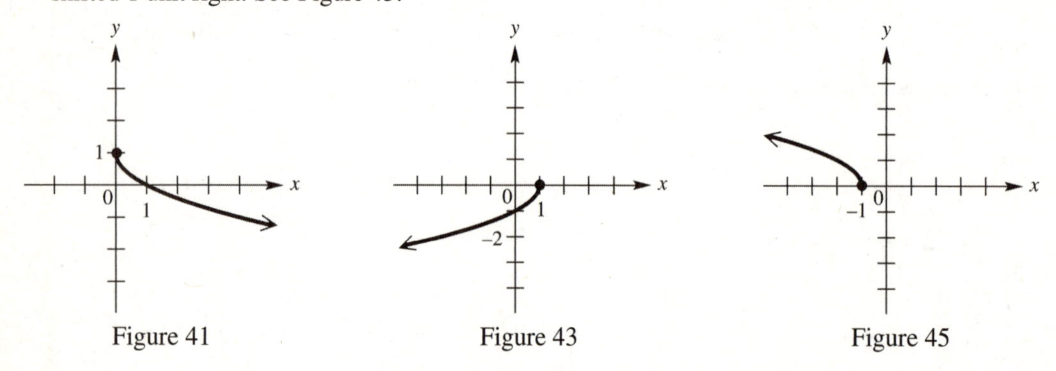

Figure 41 Figure 43 Figure 45

45. The function $f(x)=\sqrt{-(x+1)}$ is $f(x)=\sqrt{x}$ reflected across the y-axis and shifted 1 unit left. See Figure 45.

47. The function $f(x)=(x-1)^3$ is $f(x)=x^3$ shifted 1 unit right. See Figure 47.

49. The function $f(x)=-x^3$ is $f(x)=x^3$ reflected across the x-axis. See Figure 49

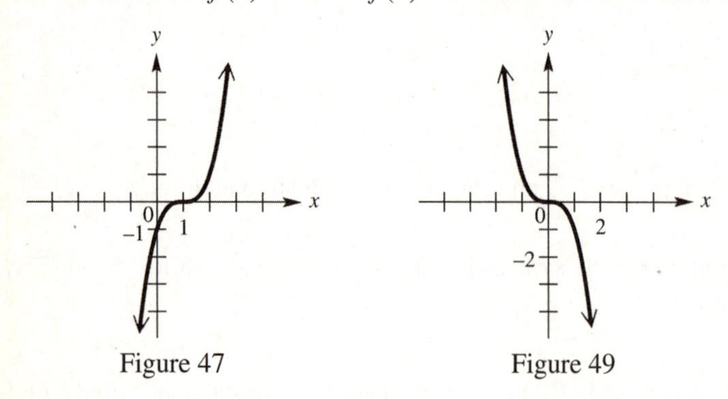

Figure 47 Figure 49

51. (a) The equation $y=-f(x)$ is $y=f(x)$ reflected across the x-axis. See Figure 51a.

(b) The equation $y=f(-x)$ is $y=f(x)$ reflected across the y-axis. See Figure 51b.

(c) The equation $y=2f(x)$ is $y=f(x)$ stretched vertically by a factor of 2. See Figure 51c.

(d) From the graph $f(0)=1$.

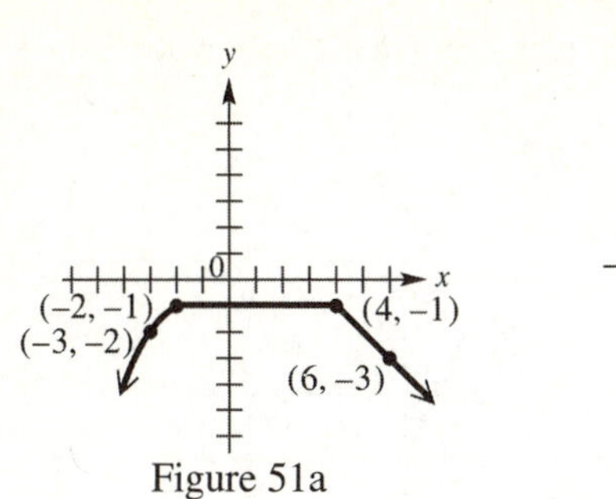

Figure 51a

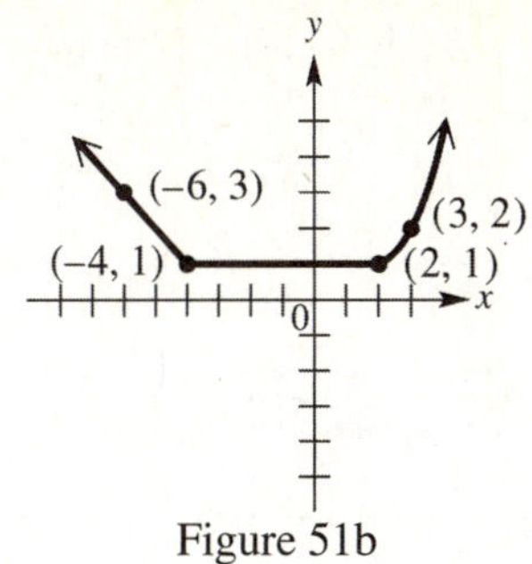

Figure 51b

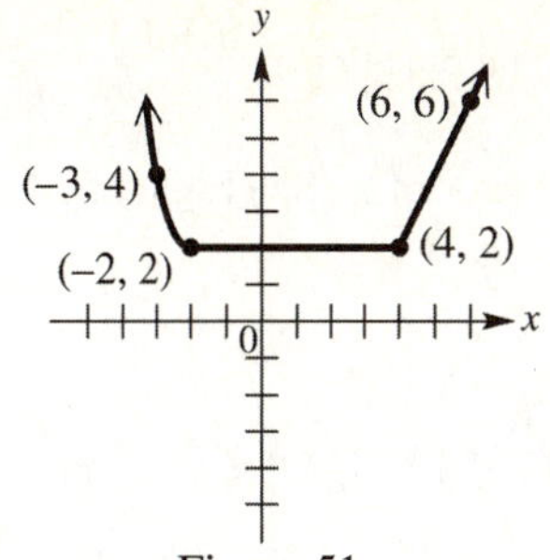

Figure 51c

53. (a) The equation $f(x)$ is $y = f(x)$ reflected across the x-axis. See Figure 53a.

(b) The equation $y = f(-x)$ is $y = f(x)$ reflected across the y-axis. See Figure 53b.

(c) The equation $y = f(x+1)$ is $y = f(x)$ shifted 1 unit to the left. See Figure 53c.

(d) From the graph, there are two x-intercepts, -1 and 4.

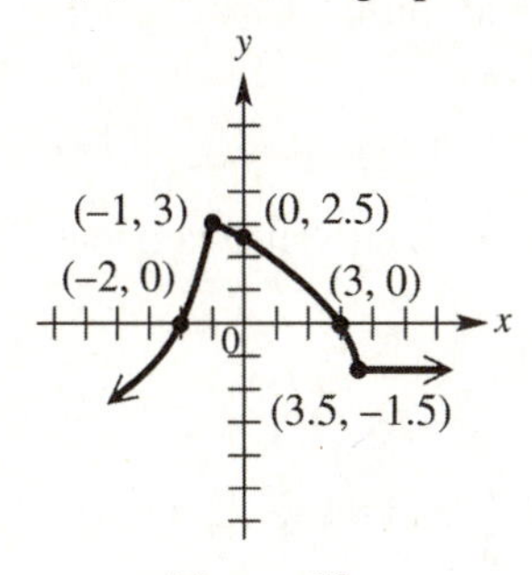

Figure 53a

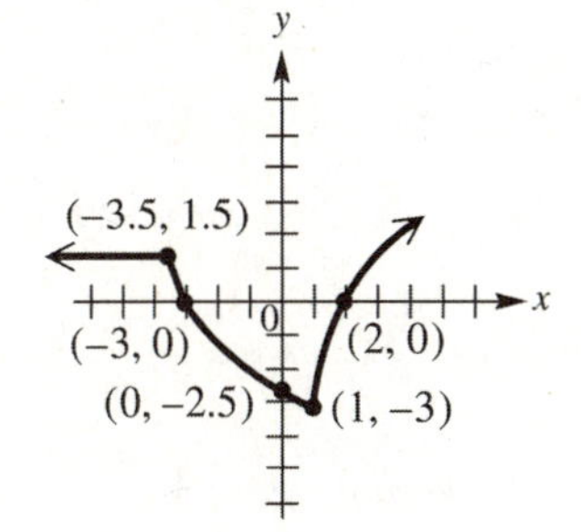

Figure 53b

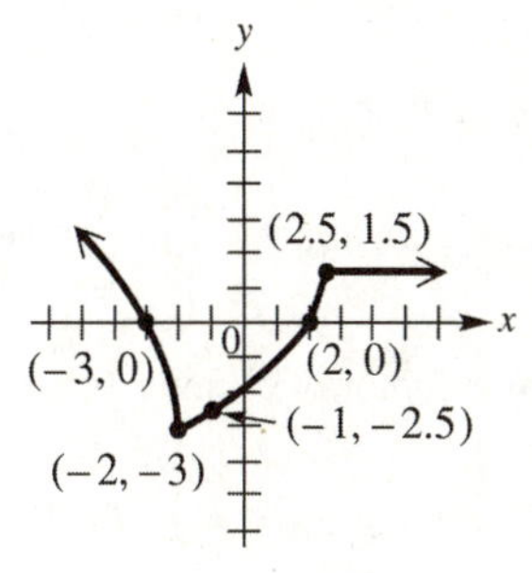

Figure 53c

55. (a) The equation $y = -f(x)$ is $y = f(x)$ reflected across the x-axis. See Figure 55a.

(b) The equation $y = f\left(\frac{1}{3}x\right)$ is $y = f(x)$ stretched horizontally by a factor of 3. See Figure 55b.

(c) The equation $y = 0.5f(x)$ is $y = f(x)$ shrunk vertically by a factor of 0.5. See Figure 55c.

(d) From the graph, symmetry with respect to the origin.

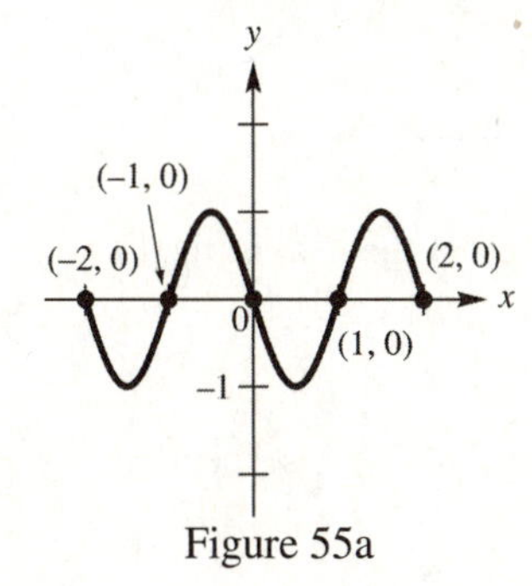

Figure 55a

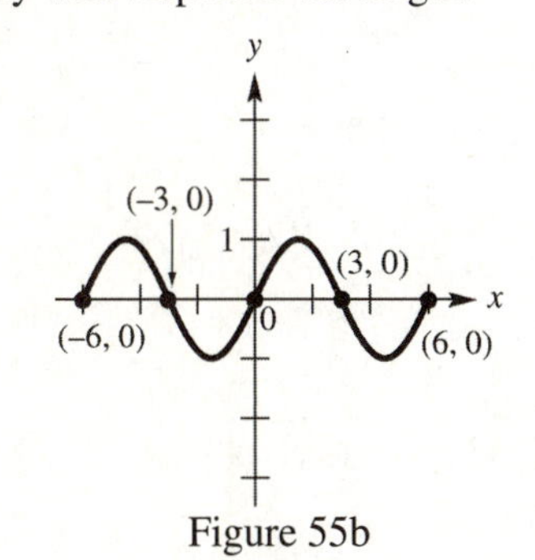

Figure 55b

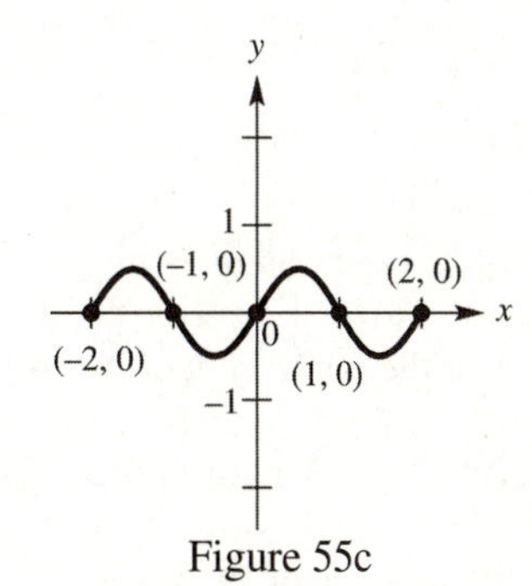

Figure 55c

57. (a) The equation $y = f(x)+1$ is $y = f(x)$ shifted 1 unit upward. See Figure 57a.

(b) The equation -30° is $y = f(x)$ reflected across the x-axis and shifted 1 unit down. See Figure 57b.

(c) The equation $y = 2f\left(\frac{1}{2}x\right)$ is $y = f(x)$ stretched vertically by a factor of 2 and horizontally by a factor of 2. See Figure 57c.

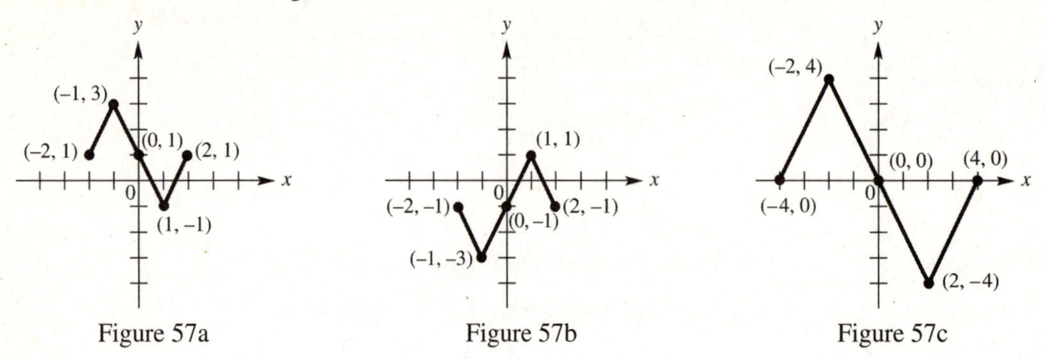

Figure 57a Figure 57b Figure 57c

59. (a) The equation $y = f(2x)+1$ is x shrunk horizontally by a factor of $\{2\}$ and shifted 1 unit upward. See Figure 59a.

(b) The equation $y = 2f\left(\frac{1}{2}x\right)+1$ is $y = f(x)$ stretched vertically by a factor of 2, stretched horizontally by a factor of 2, and shifted 1 unit upward. See Figure 59b.

(c) The equation $y = \frac{1}{2}f(x-2)$ is $y = f(x)$ shrunk vertically by a factor of $\frac{1}{2}$ and shifted 2 units to the right. See Figure 59c.

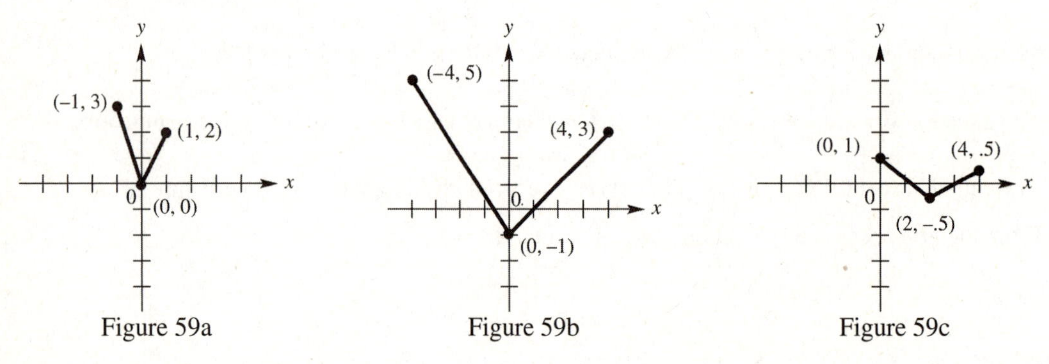

Figure 59a Figure 59b Figure 59c

61. (a) If r is the x-intercept of $y = f(x)$ and $y = -f(x)$ is $y = f(x)$ reflected across the x-axis, then r is also the x-intercept of $y = -f(x)$.

(b) If r is the x-intercept of $y = f(x)$ and $y = f(-x)$ is $f(x)$ reflected across the y-axis, then $-r$ is the x-intercept of $y = -f(x)$

(c) If r is the x-intercept of $y = f(x)$ and $y = -f(-x)$ is $y = f(x)$ reflected across both the x-axis and y-axis, then $(5,0)$, is the x-intercept of $y = -f(-x)$.

63. Since $y = f(x-2)$ is $y = f(x)$ shifted 2 units to the right, the domain of $f(x-2)$ is $[-1+2, 2+2]$ or $[1,4]$, and the range is the same: $[0,3]$.

65. Since $-f(x)$ is $f(x)$ reflected across the x-axis, the domain of $-f(x)$ is the same: $[-1,2]$, and the range is $[-3,0]$. .

67. Since $f(2x)$ is $f(x)$ shrunk horizontally by a factor of $\frac{1}{2}$, the domain of $f(2x)$ is $\left[\frac{1}{2}(-1), \frac{1}{2}(2)\right]$ or $\left[-\frac{1}{2}, 1\right]$, and the range is the same: $[0,3]$.

69. Since $3f\left(\frac{1}{4}x\right)$ is $f(x)$ stretched horizontally by a factor of 4, the domain of $3f\left(\frac{1}{4}x\right)$ is $[4(-1), 4(2)]$ or $[-4,8]$, and stretched vertically by a factor of 3, the range is $[3(0), 3(3)]$ or $[0,9]$.

71. Since $f(-x)$ is $f(x)$ reflected across the y-axis, the domain of $f(-x)$ is $[-(-1), -(2)] = [1,-2]$ or $[-2,1]$; and the range is the same: $[0,3]$.

73. Since $f(-3x)$ is $f(x)$ reflected across the y-axis and shrunk horizontally by a factor of $\frac{1}{3}$, the domain of $f(-3x)$ is $\left[-\frac{1}{3}(-1), -\frac{1}{3}(2)\right] = \left[\frac{1}{3}, -\frac{2}{3}\right]$ or $\left[-\frac{2}{3}, \frac{1}{3}\right]$, and the range is the same: $[0,3]$.

75. Since $y = \sqrt{x}$ has an endpoint of (0, 0), and the graph of $y = 10\sqrt{x-20}+5$ is the graph of $y = \sqrt{x}$ shifted 20 units right, stretched vertically by a factor of 10, and shifted 5 units upward, the endpoint of $y = 10\sqrt{x-20}+5$ is $(0+20, 10(0)+5)$ or $(20,5)$. Therefore, the domain is $[20,\infty)$, and the range is $[5,\infty)$.

77. Since $y = \sqrt{x}$ has an endpoint of (0, 0), and the graph of $y = -.5\sqrt{x+10}+5$ is the graph of $y = \sqrt{x}$ shifted 10 units left, reflected across the x-axis, shrunk vertically by a factor of .5, and shifted 5 units upward, the endpoint of $y = -.5\sqrt{x+10}+5$ is $(0-10, -.5(0)+5)$ or $(-10,5)$. Therefore, the domain is $[-10,\infty)$, and the range, because of the reflection across the x-axis, is $(-\infty,5]$.

79. The graph of $y = -f(x)$ is $y = f(x)$ reflected across the x-axis; therefore, $y = -f(x)$ is decreasing for the interval $[a,b]$.

81. The graph of $y = -f(-x)$ is $y = f(x)$ reflected across both the x-axis and y-axis; therefore, $y = -f(-x)$ is increasing for the interval $[-b,-a]$.

83. (a) the function is increasing for the interval: $[-1,2]$.

    (b) the function is decreasing for the interval: $(-\infty,-1]$.

(c) the function is constant for the interval: $[2,\infty)$.

85. (a) the function is increasing for the interval: $[1,\infty)$.

(b) the function is decreasing for the interval: $[-2,1]$.

(c) the function is constant for the interval: $(-\infty,-2]$..

87. From the graph, the point on y_2 is approximately $(8,10)$.

89. Use two points on the graph to find the slope. Two points are $(-2,-1)$ and $(-1,1)$; therefore, the slope is

$m=\dfrac{1-(-1)}{-1-(-2)}=\dfrac{2}{1}\Rightarrow m=2$. The stretch factor is 2 and the graph has been shifted 2 units to the left and 1

unit down; therefore, the equation is $y=2|x+2|-1$.

91. Use two points on the graph to find the slope. Two points are $(0,2)$ and $(1,-1)$; therefore, the slope is

$m=\dfrac{-1-2}{1-0}=\dfrac{-3}{1}\Rightarrow m=-3$. The stretch factor is 3, the graph has been reflected across the x-axis, and

shifted 2 units upward; therefore, the equation is $y=-3|x|+2$.

93. Use two points on the graph to find the slope. Two points are (0,-4) and (3,0); therefore, the slope is

$m=\dfrac{-4-0}{0-3}=\dfrac{4}{3}\Rightarrow m=\dfrac{4}{3}$. The stretch factor is $\dfrac{4}{3}$ and the graph has been shifted 4 units down.; therefore,

the equation is $y=\dfrac{4}{3}|x|-4$.

95. Since $y=f(x)$ is symmetric with respect to the y-axis, for every (x,y) on the graph, $(-x,y)$ is also on the graph. Reflection across the y-axis reflect onto itself and will not change the graph. It will be the same.

Reviewing Basic Concepts (Sections 2.1—2.3)

1. (a) If $y=f(x)$ is symmetric with respect to the origin, then another function value is $f(-3)=-6$.

(b) If $y=f(x)$ is symmetric with respect to the y-axis, then another function value is $f(-3)=6$.

(c) If $f(-x)=-f(x), y=f(x)$ is symmetric with respect to both the x-axis and y-axis, then another function value is $f(-3)=-6$.

(d) If $y=f(-x)$, $y=f(x)$ is symmetric with respect to the y-axis, then another function value is $f(-3)=6$.

2. (a) The equation $y=(x-7)^2$ is $y=x^2$ shifted 7 units to the right: B.

(b) The equation $y=x^2-7$ is $y=x^2$ shifted 7 units downward: D.

(c) The equation $y=7x^2$ is $y=x^2$ stretches vertically by a factor of 7: E.

(d) The equation $y=(x+7)^2$ is $y=x^2$ shifted 7 units to the left: A.

(e) The equation $y = \left(\frac{1}{3}x\right)^2$ is $y = x^2$ stretches horizontally by a factor of 3: C.

3. (a) The equation $y = x^2 + 2$ is $y = x^2$ shifted 2 units upward: B.

   (b) The equation $y = x^2 - 2$ is $y = x^2$ shifted 2 units downward: A.

   (c) The equation $y = (x+2)^2$ is $y = x^2$ shifted 2 units to the left: G.

   (d) The equation $y = (x-2)^2$ is $y = x^2$ shifted 2 units to the right: C.

   (e) The equation $y = 2x^2$ is $y = x^2$ stretched vertically by a factor of 2: F.

   (f) The equation $y = -x^2$ is $y = x^2$ reflected across the x-axis D.

   (g) The equation $y = (x-2)^2 + 1$ is $y = x^2$ shifted 2 units to the right and 1 unit upward: H.

   (h) The equation $y = (x+2)^2 + 1$ is $y = x^2$ shifted 2 units to the left and 1 unit upward: E.

4. (a) The equation $y = |x| + 4$ is $y = |x|$ shifted 4 units upward. See Figure 4a.

   (b) The equation $y = |x+4|$ is $y = |x|$ shifted 4 units to the left. See Figure 4b.

   (c) The equation $y = |x-4|$ is $y = |x|$ shifted 4 units to the right. See Figure 4c.

   (d) The equation $y = |x+2| - 4$ is $y = |x|$ shifted 2 units to the left and 4 units down. See Figure 4d.

   (e) The equation $y = -|x-2| + 4$ is $y = |x|$ reflected across the x-axis, shifted 2 units to the right, and 4 units upward. See Figure 4e.

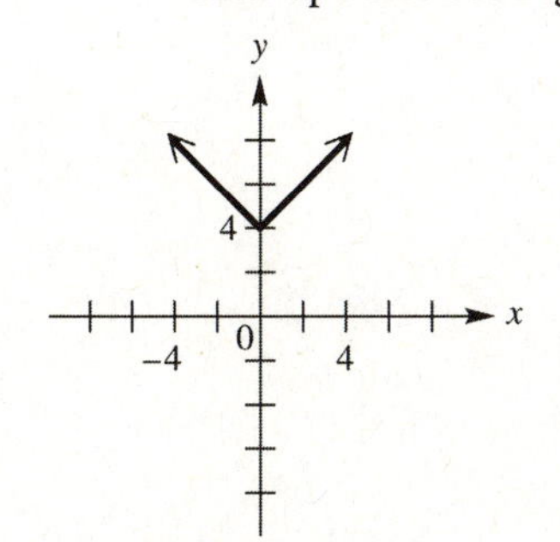

Figure 4a

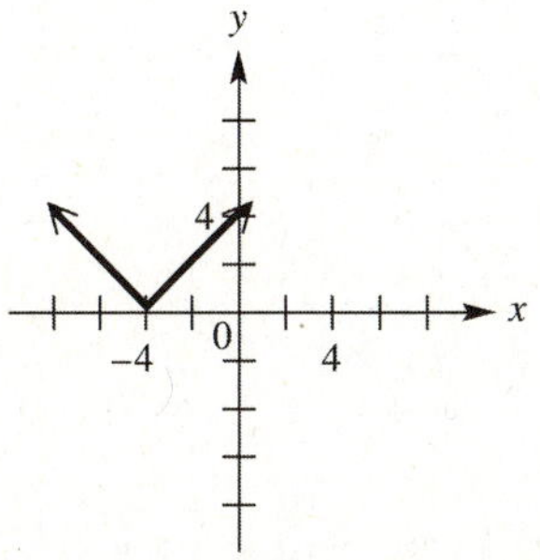

Figure 4b

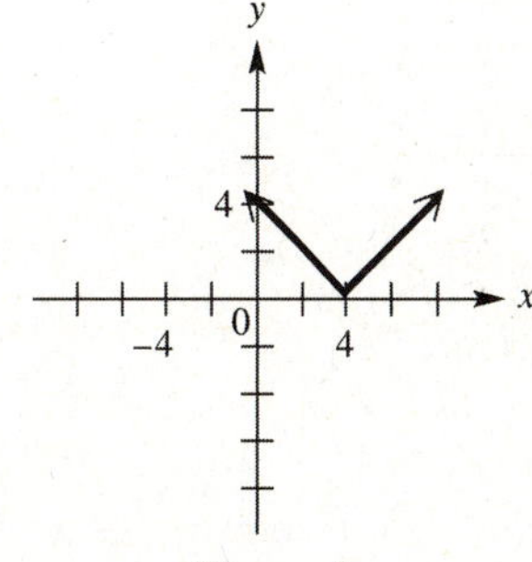

Figure 4c

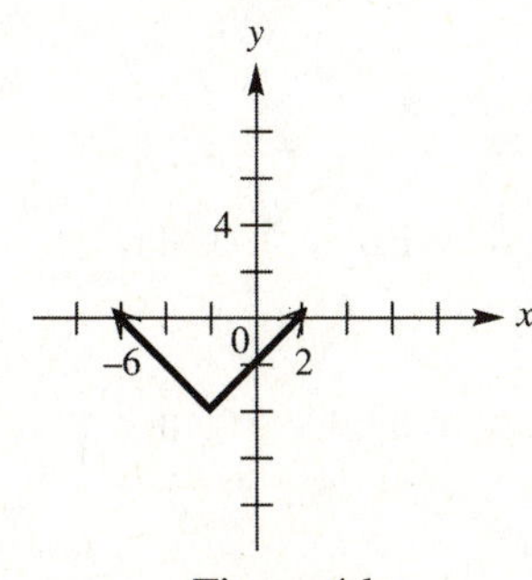

Figure 4d

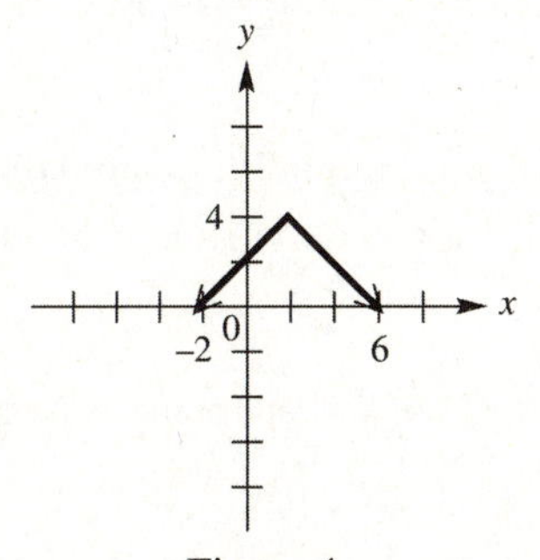

Figure 4e

5. (a) The graph is the function $f(x) = |x|$ reflected across the x-axis, shifted 1 unit left and 3 units upward

   Therefore, the equation is $y = -|x+1| + 3$.

(b) The graph is the function $g(x)=\sqrt{x}$ reflected across the x-axis, shifted 4 units left and 2 units upward. Therefore, the equation is $y=-\sqrt{x+4}+2$.

(c) The graph is the function $g(x)=\sqrt{x}$ stretches vertically by a factor of 2, shifted 4 units left and 4 units downward. Therefore, the equation is $y=2\sqrt{x+4}-4$.

(d) The graph is the function $f(x)=|x|$ shrunk vertically by a factor of $\frac{1}{2}$, shifted 2 units right and 1 unit downward. Therefore, the equation is $y=\frac{1}{2}|x-2|-1$.

6. (a) The graph of $g(x)$ is the graph $f(x)$ shifted 2 units upward. Therefore, $c=2$.

   (b) The graph of $g(x)$ is the graph $f(x)$ shifted 4 units to the left. Therefore, $c=4$.

7. The graph of $y=F(x+h)$ is a horizontal translation of the graph of $y=F(x)$. The graph of $y=F(x)+h$ is not the same as the graph of $y=F(x+h)$, because the graph of $y=F(x)+h$ is a vertical translation of the graph of $y=F(x)$.

8. The effect is either a stretch or a shrink, and perhaps a reflection across the x-axis. If $c>0$, there is a stretch or shrink by a factor of c. If $c<0$, there is a stretch or shrink by a factor of $|c|$, and a reflection across the x-axis. If $|c|>1$, a stretch occurs; if $|c|<1$, a shrink occurs.

9. (a) If f is even, then $f(x)=f(-x)$. See Figure 9a.

   (b) If f is odd, then $f(-x)=-f(-x)$. See Figure 9b.

x	$f(x)$
-3	4
-2	-6
-1	5
1	5
2	-6
3	4

Figure 9a

x	$f(x)$
-3	4
-2	-6
-1	5
1	-5
2	6
3	-4

Figure 9b

10. (a) Since $x=1$ corresponds to 1993, the equation using the actual year is $g(x)=0.223(x-1993)+20.6$.

    (b) $g(2011)=-0.223(2011-1993)+20.6 \Rightarrow g(2011)\approx 24.6\%$

2.4: Absolute Value Functions: Graphs, Equations, Inequalities, and Applications

1. We reflect the graph of $y=f(x)$ across the x-axis for all points for which $y<0$. Where $y\geq 0$, the graph remains unchanged. See Figure 1.

3. We reflect the graph of $y=f(x)$ across the x-axis for all points for which $y<0$. Where $y\geq 0$, the graph remains unchanged. See Figure 3.

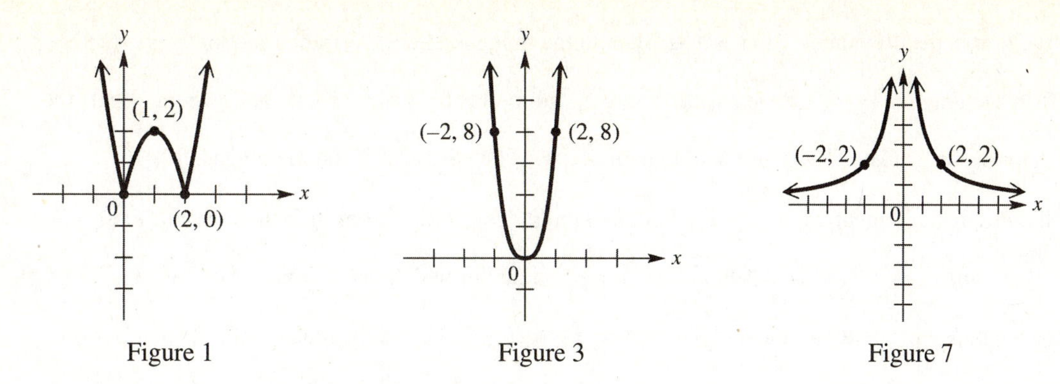

Figure 1 Figure 3 Figure 7

5. Since for all y, $y \geq 0$, the graph remains unchanged. That is, $y = |f(x)|$ has the same graph as $y = f(x)$.
7. We reflect the graph of $y = f(x)$ across the x-axis for all points for which $y < 0$. Where $y \geq 0$, the graph remains unchanged. See Figure 7.
9. We reflect the graph of $y = f(x)$ across the x-axis for all points for which $y < 0$. Where $y \geq 0$, the graph remains unchanged. See Figure 9.

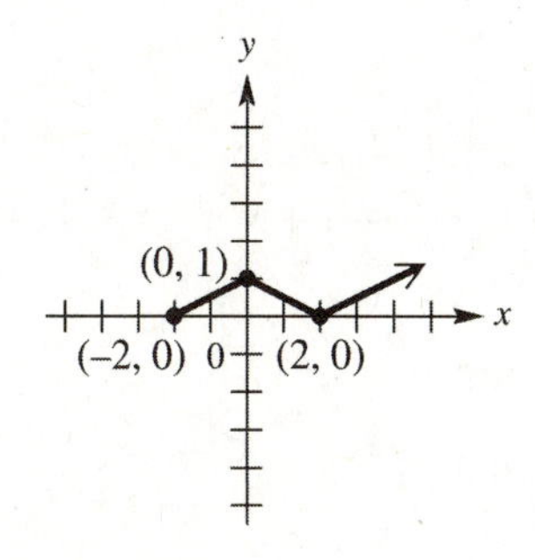

Figure 9

11. If $f(a) = -5$, then $|f(a)| = |-5| = 5$.
13. If $f(x) = -x^2$, then $y = |f(x)| \Rightarrow y = |-x^2| \Rightarrow y = x^2$. Therefore, the range of $y = |f(x)|$ is $[0, \infty)$.
15. If the range of $y = f(x)$ is $(-\infty, -2]$, the range of $y = |f(x)|$ is $[2, \infty)$ since all negative values of y are reflected across the x-axis.
17. From the graph of $y = (x+1)^2 - 2$ the domain of $f(x)$ is $(-\infty, \infty)$, and the range is $[-2, \infty)$. From the graph of $y = |(x+1)^2 - 2|$ the domain of $|f(x)|$ is $(-\infty, \infty)$, and the range is $[0, \infty)$.
19. From the graph of $y = -1 - (x-2)^2$ the domain of $f(x)$ is $(-\infty, \infty)$, and the range is $(-\infty, -1]$. From the graph of $y = |-1 - (x-2)^2|$ the domain of $|f(x)|$ is $(-\infty, \infty)$, and the range is $[1, \infty)$.

21. From the graph, the domain of $f(x)$ is $[-2,3]$, and the range is $[-2,3]$. For the function $y=|f(x)|$, we reflect the graph of $y=f(x)$ across the x-axis for all points for which $y<0$, and, where $y \geq 0$, the graph remains unchanged. Therefore, the domain of $y=|f(x)|$ is [–2, 3], and the range is [0, 3].

23. From the graph, the domain of $f(x)$ is [–2, 3], and the range is [–3, 1]. For the function $y=|f(x)|$, we reflect the graph of $y=f(x)$ across the x-axis for all points for which $y<0$, and, where $y \geq 0$, the graph remains unchanged. Therefore, the domain of $y=|f(x)|$ is [–2, 3], and the range is [0, 3].

25. (a) The function $y=f(-x)$ is the function $y=f(x)$ reflected across the y-axis. See Figure 25a.

    (b) The function $y=-f(-x)$ is the function $y=f(x)$ reflected across both the x-axis and y-axis. See Figure 25b.

    (c) For the function $y=|-f(-x)|$ we reflect the graph of $y=-f(-x)$ (ex. b) across the x-axis for all points for which $y<0$, and where $y \geq 0$, the graph remains unchanged. See Figure 25c.

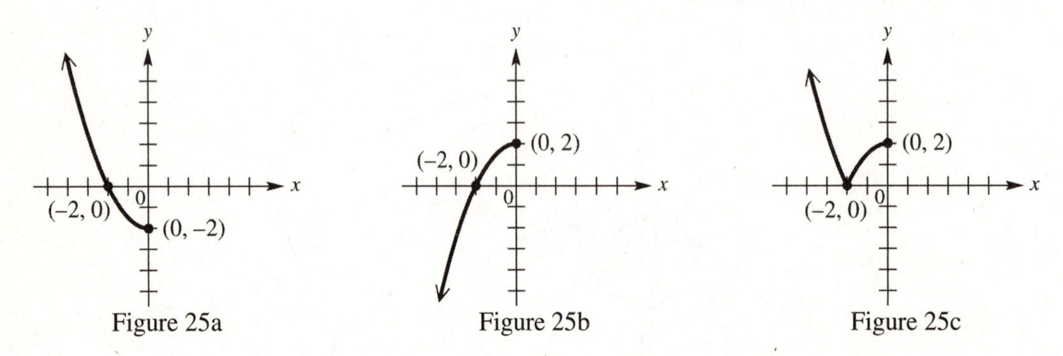

Figure 25a Figure 25b Figure 25c

27. The graph of $y=|f(x)|$ can not be above the x-axis; therefore, Figure A shows the graph of $y=f(x)$, while Figure B shows the graph of $y=|f(x)|$.

29. (a) From the graph, $y_1 = y_2$ at the coordinates $(-1,5)$ and $(6,5)$; therefore, the solution set is $\{-1,6\}$.

    (b) From the graph, $y_1 < y_2$ for the interval $(-1,6)$.

    (c) From the graph, $y_1 > y_2$ for the intervals $(-\infty,-1)\cup(6,\infty)$.

31 (a) From the graph, $y_1 = y_2$ at the coordinate $(4,1)$; therefore, the solution set is $\{4\}$.

    (b) From the graph, $y_1 < y_2$ never occurs; therefore, the solution set is $\varnothing$.

    (c) From the graph, $y_1 > y_2$ for all values for x except 4; therefore, the solution set is the intervals $(-\infty,4)\cup(4,\infty)$.

33. The V-shaped graph is that of $f(x)=|.5x+6|$, since this is typical of the graphs of absolute value functions of the form $f(x)=|ax+b|$

34. The straight line graph is that of $g(x)=3x-14$, which is a linear function.

35. The graphs intersect at $(8,10)$, so the solution set is $\{8\}$.

36. From the graph, $f(x) > g(x)$ for the interval $(-\infty, 8)$.

37. From the graph, $f(x) < g(x)$ for the interval $(8, \infty)$.

38. If $|.5x+6| - (3x-14) = 0$ then $|.5x+6| = 3x-14$. Therefore, the solution is the intersection of the graphs, or $\{8\}$.

39. (a) $|x+4| = 9 \Rightarrow x+4 = 9$ or $x+4 = -9 \Rightarrow x = 5$ or $x = -13$. The solution set is $\{-13, 5\}$, which is supported by the graphs of $y_1 = |x+4|$ and $y_2 = 9$.

(b) $|x+4| > 9 \Rightarrow x+4 > 9$ or $x+4 < -9 \Rightarrow x > 5$ or $x < -13$. The solution is $(-\infty, -13) \cup (5, \infty)$, which is supported by the graphs of $y_1 = |x+4|$ and $y_2 = 9$.

(c) $|x+4| < 9 \Rightarrow -9 < x+4 < 9 \Rightarrow -13 < x < 5$. The solution is $(-13, 5)$, which is supported by the graphs of $|x+4|$ and $y_2 = 9$.

41. (a) $|7-2x| = 3 \Rightarrow 7-2x = 3$ or $7-2x = -3 \Rightarrow -2x = -4$ or $-2x = -10 \Rightarrow x = 2$ or $x = 5$. The solution set is $\{2, 5\}$, which is supported by the graphs of $y_1 = |7-2x|$ and $y_2 = 3$.

(b) $|7-2x| \geq 3 \Rightarrow 7-2x \geq 3$ or $7-2x \leq -3 \Rightarrow -2x \geq -4$ or $-2x \leq -10 \Rightarrow x \leq 2$ or $x \geq 5$. The solution set is $(-\infty, 2] \cup [5, \infty)$, which is supported by the graphs of $y_1 = |7-2x|$ and $y_2 = 3$.

(c) $|7-2x| \leq 3 \Rightarrow -3 \leq 7-2x \leq 3 \Rightarrow -10 \leq -2x \leq -4 \Rightarrow 5 \geq x \geq 2$ or $2 \leq x \leq 5$. The solution is $[2, 5]$, which is supported by the graphs of $y_1 = |7-2x|$ and $y_2 = 3$.

43. (a) $|2x+1| + 3 = 5 \Rightarrow 2x+1 = 2$ or $2x+1 = -2 \Rightarrow 2x = 1$ or $2x = -3 \Rightarrow x = \frac{1}{2}$ or $x = -\frac{3}{2}$. The solution set is $\left\{-\frac{3}{2}, \frac{1}{2}\right\}$, which is supported by the graphs of $y_1 = |2x+1| + 3$ and $y_2 = 5$.

(b) $|2x+1| + 3 \leq 5 \Rightarrow -2 \leq 2x+1 \leq 2 \Rightarrow -3 \leq 2x \leq 1 \Rightarrow -\frac{3}{2} \leq x \leq \frac{1}{2}$. The solution is $\left[-\frac{3}{2}, \frac{1}{2}\right]$, which is supported by the graphs of $y_1 = |2x+1| + 3$ and $y_2 = 5$.

(c) $|2x+1| + 3 \geq 5 \Rightarrow 2x+1 \geq 2$ or $2x+1 \leq -2 \Rightarrow 2x \geq 1$ or $2x \leq -3 \Rightarrow x \geq \frac{1}{2}$ or $x \leq -\frac{3}{2}$. The solution is $\left(-\infty, -\frac{3}{2}\right] \cup \left[\frac{1}{2}, \infty\right)$, which is supported by the graphs of $y_1 = |2x+1| + 3$ and $y_2 = 5$.

45. (a) $|5-7x| = 0 \Rightarrow 5-7x = 0 \Rightarrow 7x = 5 \Rightarrow x = \frac{5}{7}$. The solution set is $\left\{\frac{5}{7}\right\}$, which is supported by the graphs of $y_1 = |5-7x|$ and $y_2 = 0$.

(b) $|5-7x|\ge 0 \Rightarrow 5-7x\ge 0$ or $5-7x\le 0 \Rightarrow 7x\ge 5$ or $7x\le 5 \Rightarrow x\ge \frac{5}{7}$ or $x\le \frac{5}{7}$. The solution is $(-\infty,\infty)$, which is supported by the graphs of $y_1=|5-7x|$ and $y_2=0$.

(c) $|5-7x|\le 0 \Rightarrow 0\le 5-7x\le 0 \Rightarrow 5\ge 7x\ge 5 \Rightarrow \frac{5}{7}\ge x\ge \frac{5}{7}$. The solution set is $\left\{\frac{5}{7}\right\}$, which is supported by the graphs of $y_1=|5-7x|$ and $y_2=0$.

47. (a) Absolute value is always positive; therefore, the solution set is $\varnothing$, which is supported by the graphs of $y_1=\left|\sqrt{2}x-3.6\right|$ and $y_2=-1$.

(b) Absolute value is always positive, and so cannot be less than or equal to -1; therefore, the solution set is $\varnothing$, which is supported by the graphs of $y_1=\left|\sqrt{2}x-3.6\right|$ and $y_2=-1$.

(c) Absolute value is always positive, and so is always greater than -1; therefore, the solution is $(-\infty,\infty)$, which is supported by the graphs of $y_1=\left|\sqrt{2}x-3.6\right|$ and $y_2=-1$.

49. $3|4-3x|-4=8 \Rightarrow 3|4-3x|=12 \Rightarrow |4-3x|=4 \Rightarrow 4-3x=4$ or $4-3x=-4 \Rightarrow -3x=0$ or $-3x=-8 \Rightarrow x=0$ or $x=\frac{8}{3}$. Therefore, the solution set is $\left\{0,\frac{8}{3}\right\}$.

51. $\frac{1}{2}\left|-2x+\frac{1}{2}\right|=\frac{3}{4} \Rightarrow \left|-2x+\frac{1}{2}\right|=\frac{3}{2} \Rightarrow -2x+\frac{1}{2}=\frac{3}{2}$ or $-2x+\frac{1}{2}=-\frac{3}{2} \Rightarrow -2x=1$ or $-2x=-2 \Rightarrow x=-\frac{1}{2}$ or $x=1$. Therefore, the solution set is $\left\{-\frac{1}{2},1\right\}$.

53. $4.2|.5-x|+1=3.1 \Rightarrow 4.2|.5-x|=2.1 \Rightarrow |.5-x|=.5 \Rightarrow .5-x=.5$ or $.5-x=-.5 \Rightarrow -x=0$ or $-x=-1 \Rightarrow x=0$ or $x=1$. Therefore, the solution set is $\{0,1\}$.

55. $|15-x|<7 \Rightarrow -7<15-x<7 \Rightarrow 22>x>8$ or $8<x<22$. Therefore, the solution is $(8,22)$.

57. $|2x-3|>1 \Rightarrow 2x-3>1$ or $2x-3<-1 \Rightarrow 2x>4$ or $\left(\frac{f}{g}\right)(-3)=\frac{-3(-3)-4}{(-3)^2}=\frac{5}{9}$. or $x<1$. Therefore, the solution is $(-\infty,1)\cup(2,\infty)$.

59. $|-3x+8|\ge 3 \Rightarrow -3x+8\ge 3$ or $-3x+8\le -3 \Rightarrow -3x\ge -5$ or $-3x\le -11 \Rightarrow x\le \frac{5}{3}$ or $x\ge \frac{11}{3}$. Therefore, the solution is $\left(-\infty,\frac{5}{3}\right]\cup\left[\frac{11}{3},\infty\right)$.

61. $\left|6-\frac{1}{3}x\right|>0 \Rightarrow 6-\frac{1}{3}x>0$ or $6-\frac{1}{3}x<0 \Rightarrow -\frac{1}{3}x>-6$ or $-\frac{1}{3}x<-6 \Rightarrow x<18$ or $x>18$. Therefore, the solution is every real number except 18: $(-\infty,18)\cup(18,\infty)$.

63. Absolute value is always positive, and so cannot be less than or equal to -6; therefore, the solution set is $\varnothing$.

65. Absolute value is always positive, and so is always greater than −5; therefore, the solution is $(-\infty, \infty)$.

67. (a) $3x+1=2x-7 \Rightarrow x+1=-7 \Rightarrow x=-8$ $3x+1=-(2x-7) \Rightarrow 3x+1=-2x+7 \Rightarrow 5x=6 \Rightarrow x=\frac{6}{5}$.

Therefore, the solution set is $\left\{-8, \frac{6}{5}\right\}$.

(b) Graph $y_1=|3x+1|$ and $y_2=|2x-7|$. See Figure 67. From the graph, $|f(x)|>|g(x)|$ when $y_1>y_2$ which is for the interval $(-\infty,-8)\cup\left(\frac{6}{5},\infty\right)$.

(c) Graph $y_1=|3x+1|$ and $y_2=|2x-7|$ See Figure 67. From the graph, $|f(x)|<|g(x)|$ when $y_1<y_2$ which is for the interval $\left(-8,\frac{6}{5}\right)$.

[-20,20] by [-10,50]
Xscl = 2 Yscl = 5

Figure 67

69. (a) $-2x+5=x+3 \Rightarrow -3x=-2 \Rightarrow x=\frac{2}{3}$ or $-2x+5=-(x+3) \Rightarrow -2x+5=-x-3 \Rightarrow$

$-x=-8 \Rightarrow x=8$. Therefore, the solution set is $\left\{\frac{2}{3}, 8\right\}$.

(b) Graph $y_1=|-2x+5|$ and $y_2=|x+3|$. See Figure 69. From the graph, $|f(x)|>|g(x)|$ when $y_1>y_2$, which it is for the interval $\left(-\infty,\frac{2}{3}\right)\cup(8,\infty)$.

(c) Graph $y_1=|-2x+5|$ and $y_2=|x+3|$. See Figure 69. From the graph, $|f(x)|<|g(x)|$ when $y_1<y_2$, which it is for the interval $\left(\frac{2}{3},8\right)$.

[-10,10] by [-4,16]
Xscl = 1 Yscl = 1

Figure 69

[-6,6] by [-2,10]
Xscl = 1 Yscl = 1

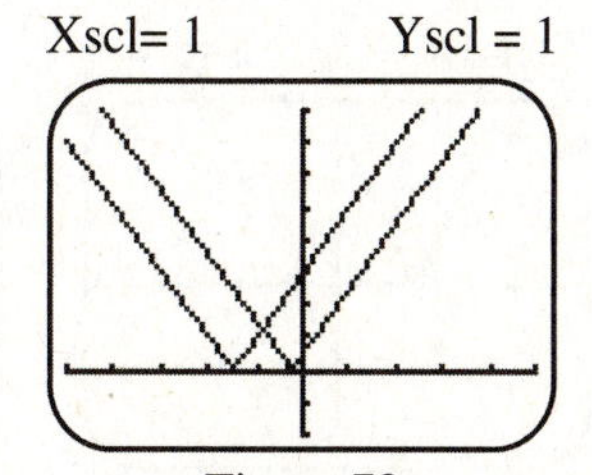

Figure 71

[-10,10] by [-10,10]
Xscl= 1 Yscl = 1

Figure 73

71. (a) $x-\frac{1}{2}=\frac{1}{2}x-2 \Rightarrow \frac{1}{2}x=-\frac{3}{2} \Rightarrow x=-3$ or $x-\frac{1}{2}=-\left(\frac{1}{2}x-2\right) \Rightarrow x-\frac{1}{2}=-\frac{1}{2}x+2 \Rightarrow$

$\frac{3}{2}x=\frac{5}{2}\Rightarrow x=\frac{5}{3}$. Therefore, the solution set is $\left\{-3,\frac{5}{3}\right\}$.

(b) Graph $y_1=\left|x-\frac{1}{2}\right|$ and $y_2=\left|\frac{1}{2}x-2\right|$. From the graph, $|f(x)|>|g(x)|$ when $y_1>y_2$, which it is for the interval $(-\infty,-3)\cup\left(\frac{5}{3},\infty\right)$.

(c) Graph $y_1=\left|x-\frac{1}{2}\right|$ and $y_2=\left|\frac{1}{2}x-2\right|$ .See Figure 71. From the graph $|f(x)|<|g(x)|$ when $y_1<y_2$, which it is for the interval $\left(-3,\frac{5}{3}\right)$.

73. (a) $4x+1=4x+6\Rightarrow 1=6\Rightarrow\varnothing$ or $4x+1=-(4x+6)\Rightarrow 4x=1=-4x-6\Rightarrow 8x=-7\Rightarrow -\frac{7}{8}$.

Therefore, the solution set is $\left\{-\frac{7}{8}\right\}$.

(b). Graph $y_1=|4x+1|$ or $y_2=|4x+6|$. See Figure 73. From the graph $|f(x)|>|g(x)|$ when $y_1>y_2$, which it is for the interval $\left(-\infty,\frac{7}{8}\right)$.

(c) Graph $y_1=|4x+1|$ or $y_2=|4x+6|$. See Figure 73. From the graph, $|f(x)|<|g(x)|$ when $y_1>y_2$, which is for the interval $\left(-\frac{7}{8},\infty\right)$.

75. (a) $0.25x+1=0.75x-3\Rightarrow -0.50x=-4\Rightarrow x=8$ or $0.25x+1=-(0.75x-3)\Rightarrow 0.25x+1=-0.75x+3\Rightarrow$ $x=2$. Therefore, the solution set is $\{2,8\}$.

(b) Graph $y_1=|.25x+1|$ and $y_2=|.75x-3|$. See Figure 75. From the graph, $|f(x)|>|g(x)|$ when $y_1<y_2$, which it is for the interval $(2,8)$.

(c) Graph $y_1=|.25x+1|$ and $y_2=|.75x-3|$. See Figure 75. From the graph, $|f(x)|<|g(x)|$ when $y_1<y_2$, which it is for the interval $(-\infty,2)\cup(8,\infty)$.

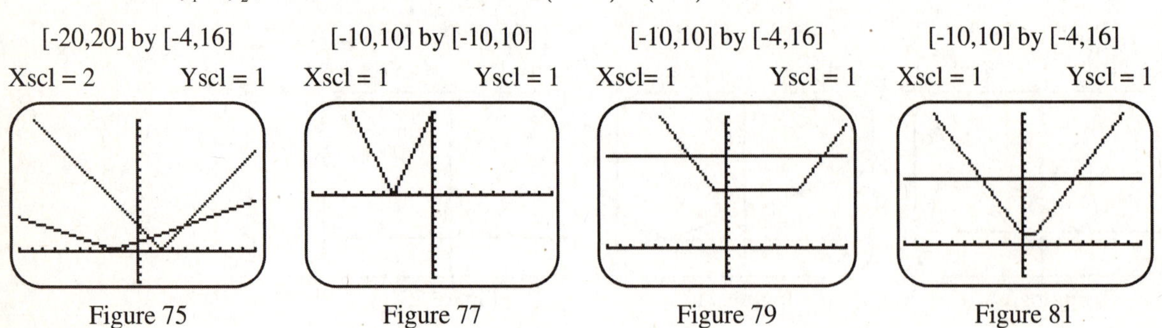

Figure 75 Figure 77 Figure 79 Figure 81

77. (a) $3x+10=-(-3x-10) \Rightarrow 3x+10=3x+10 \Rightarrow$ there are an infinite number of solutions. Therefore, the solution set is $(-\infty,\infty)$.

(b) Graph $y_1=|3x+10|$ and $y_2=|-3x-10|$. See Figure 77. From the graph, $|f(x)|>|g(x)|$ when $y_1>y_2$, for which there is no solution.

(c) Graph $y_1=|3x+10|$ and $y_2=|-3x-10|$. See Figure 77. From the graph, $|f(x)|<|g(x)|$ when $y_1<y_2$, for which there is no solution.

79. Graph $y_1=|x+1|+|x-6|$ and $y_2=11$. See Figure 79. From the graph, the lines intersect at $(-3,11)$ and $(2,9)$. Therefore, the solution set is $\{-3,8\}$.

81. Graph $y_1=|x|+|x-4|$ and $y_2=8$. See Figure 81. From the graph, the lines intersect at $(-2,8)$ and $(6,8)$. Therefore, the solution set is $\{-2,6\}$.

83. (a) $|T-50|\le 22 \Rightarrow -22\le T-50\le 22 \Rightarrow 28\le T\le 72$.

(b) The average monthly temperatures in Boston vary between a low of 28° F and a high of 72° F. The monthly averages are always within 22° of 50° F.

85. (a) $|T-61.5|\le 12.5 \Rightarrow -12.5\le T-61.5\le 12.5 \Rightarrow 49\le T\le 74$..

(b) The average monthly temperatures in Buenos Aires vary between a low of 49° F (possibly in July) and a high of 74° F (possibly in January). The monthly averages are always within 12.5° of 61.5° F.

87. $|x-8.0|\le 1.5 \Rightarrow -1.5\le x-8.0\le 1.5 \Rightarrow 6.5\le x\le 9.5$; therefore, the range is the interval $[6.5,9.5]$.

89. (a) $P_d=|116-125| \Rightarrow P_d=|-9|=9$.

(b) $17=|P-130| \Rightarrow P-130=17 \text{ or } P-130=-17 \Rightarrow P=147 \text{ or } P=113$.

91. If the difference between y and 1 is less than .1, then $|y-1|<.1 \Rightarrow |2x+1-1|<.1 \Rightarrow$ $|2x|<.1 \Rightarrow -.1<2x<.1 \Rightarrow -.05<x<.05$. The open interval of x is $(-.05,.05)$.

93. If the difference between y and 3 is less than .001, then $|y-3|<.001 \Rightarrow |4x-8-3|<.001 \Rightarrow$ $|4x-11|<.001 \Rightarrow -.001<4x-11<.001 \Rightarrow 10.999<4x<11.001 \Rightarrow 2.74975<x<2.75025$. The open interval of x is $(2.74975, 2.75025)$.

95. If $|2x+7|=6x-1$ then $|2x+7|-(6x-1)=0$. Graph $y_1=|2x+7|-(6x-1)$, See Figure 95. The x-intercept is 2; therefore, the solution set is $\{2\}$.

97. If $|x-4|>.5x-6$ then $|x-4|-(.5x-6)>0$. Graph $y_1=|x-4|-(.5x-6)$, See Figure 97. The equation is >0, or the graph is above the x-axis, for the interval: $(-\infty,\infty)$.

[-10,10] by [-10,10]
Xscl = 1 Yscl = 1

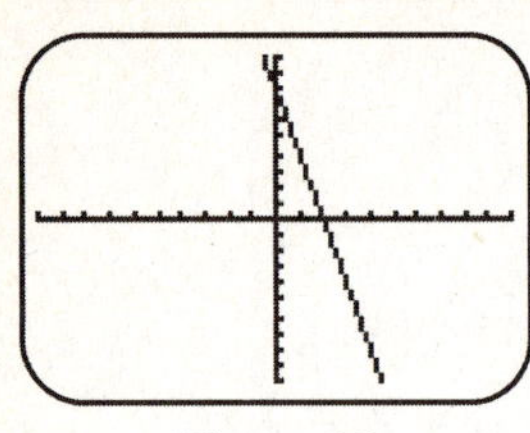

Figure 95

[-10,10] by [-10,10]
Xscl = 1 Yscl = 1

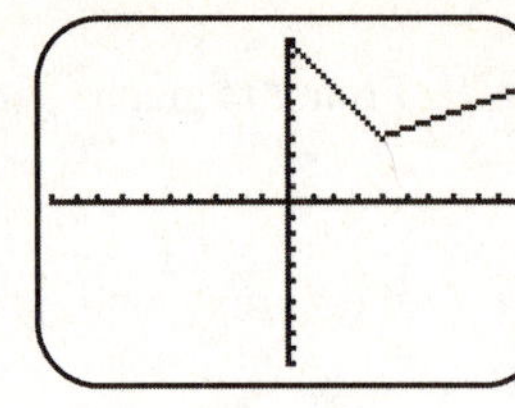

Figure 97

[-10,10] by [-10,10]
Xscl= 1 Yscl = 1

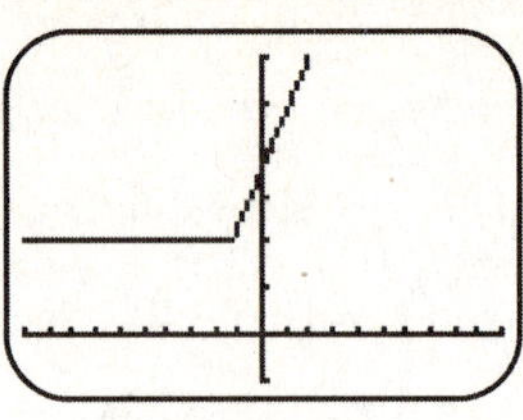

Figure 99

99. If $|3x+4|<-3x-14$ then $|3x+4|-(-3x-14)<0$. Graph $y_1=|3x+4|-(-3x-14)$, See Figure 99. The equation is <0, or the graph is below the x-axis, never or for the solution set: $\varnothing$.

2.5: Piecewise-Defined Functions

1. (a) From the graph, the speed limit is 40 mph.
   (b) From the graph, the speed limit is 30 mph for 6 miles.
   (c) From the graph, $f(5)=40\,\text{mph}$; $f(13)=30\,\text{mph}$; and $f(19)=55\,\text{mph}$.
   (d) From the graph, the graph is discontinuous at $x=4,6,8,12,\text{and}\,16$. The speed limit changes at each discontinuity.
3. (a) From the graph, the Initial amount was: 50,000 gal.; and the final amount was: 30,000 gal.
   (b) From the graph, during the first and fourth days.
   (c) From the graph, $f(2)=45,000\,\text{gal}$; $f(4)=40,000\,\text{gal}$.
   (d) From the graph, between days 1 and 3 the water dropped: $\dfrac{50,000-40,000}{2}=\dfrac{10,000}{2}=5,000$ gal./day.
5. (a) $f(-5)=2(-5)=-10$ (b) $f(-1)=2(-1)=-2$
   (c) $f(0)=0-1=-1$ (d) $f(3)=3-1=2$
7. (a) $f(-5)=2+(-5)=-3$ (b) $f(-1)=-(-1)=1$
   (c) $f(0)=-(0)=0$ (d) $f(3)=3(3)=9$
9. Yes, continuous. See Figure 9.
11. Not continuous. See Figure 11

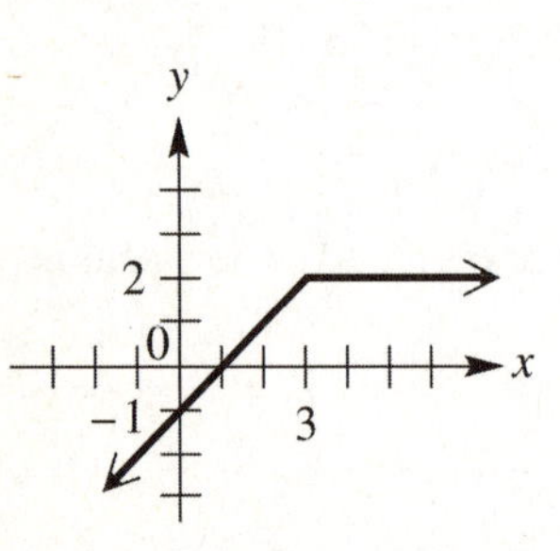

Figure 9

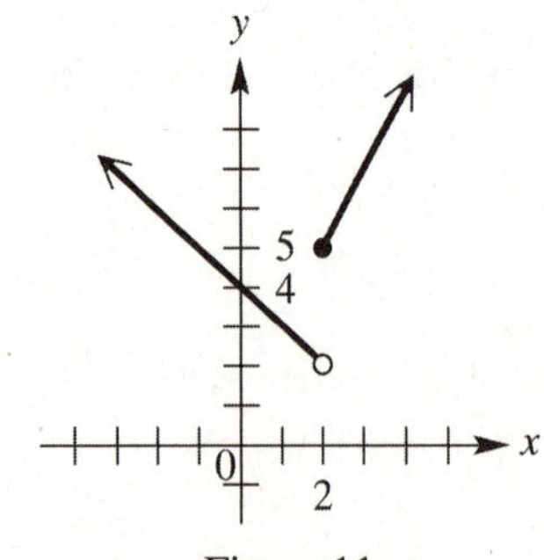

Figure 11

13. Not continuous. See Figure 13.

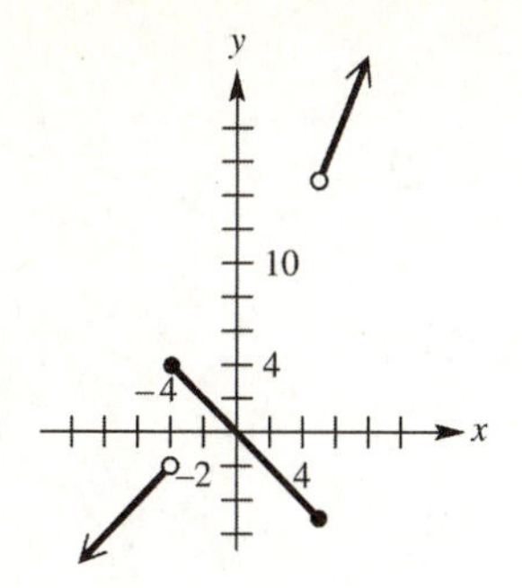

Figure 13

15. Not continuous. See Figure 15.

17. Yes, continuous. See Figure 17.

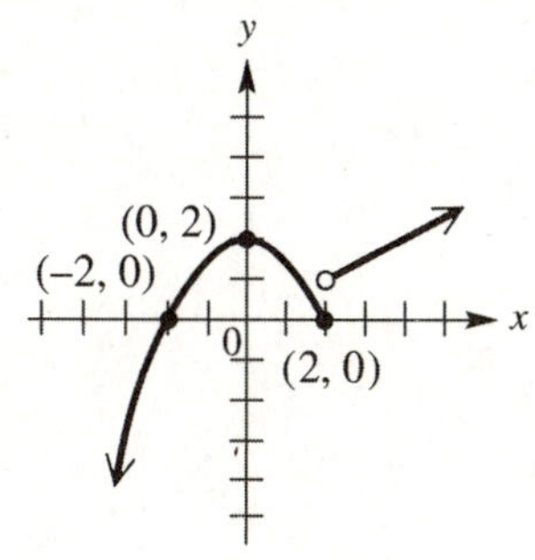

Figure 15

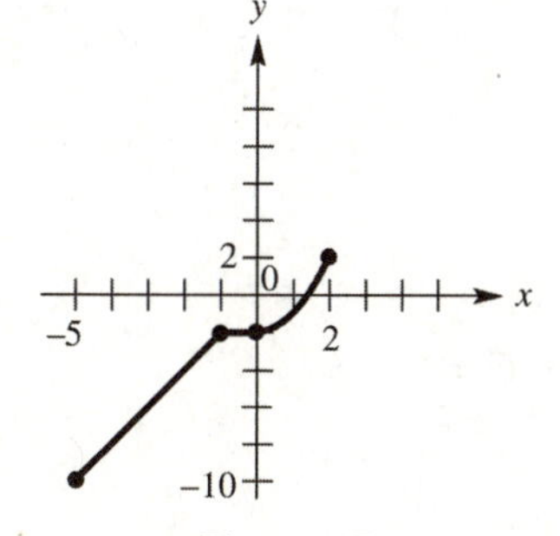

Figure 17

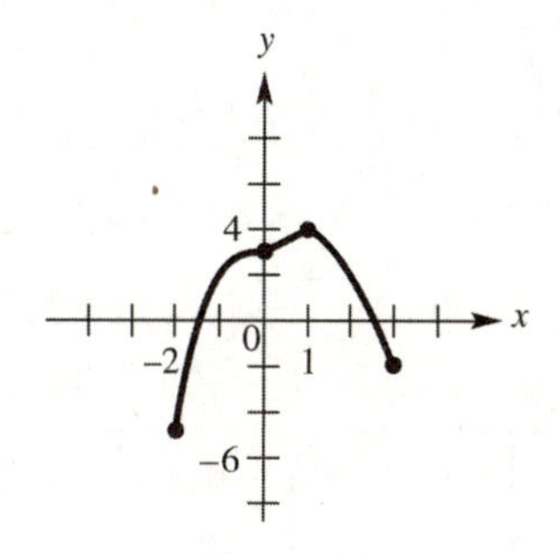

Figure 19

19. Yes, continuous. See Figure 19.

21. Look for a $y = x^2$ graph if $x \geq 0$; and a linear graph if $x < 0$. Therefore: B.

23. Look for a horizontal graph above the x-axis if $x \geq 0$; and a horizontal graph below the x-axis if $x < 0$. Therefore: D.

25. Graph $y_1 = (x-1)(x \leq 3) + (2)(x > 3)$, see Figure 25.

27. Graph $y_1 = (4-x)(x < 2) + (1+2x)(x \geq 2)$, see Figure 27.

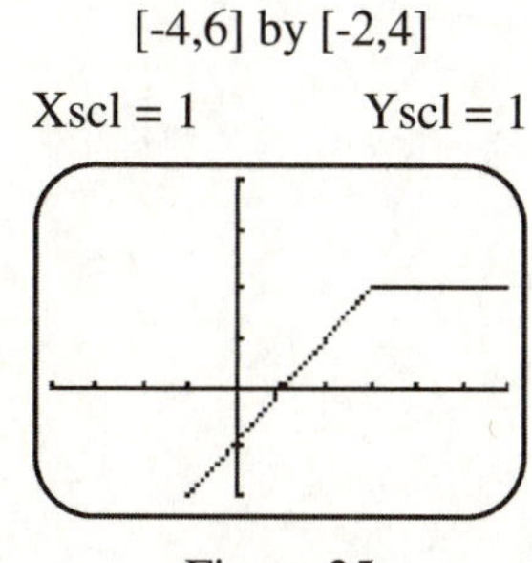

Figure 25

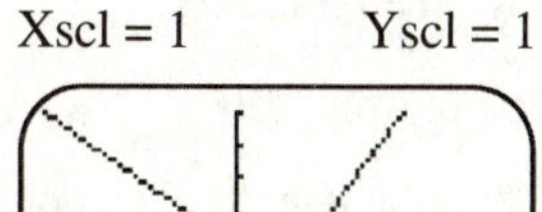

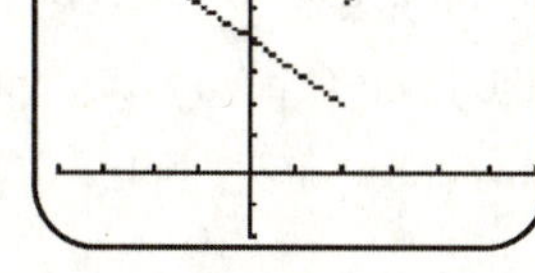

Figure 27

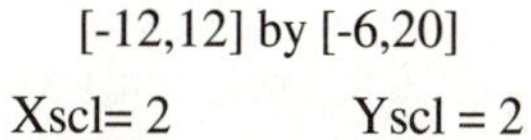

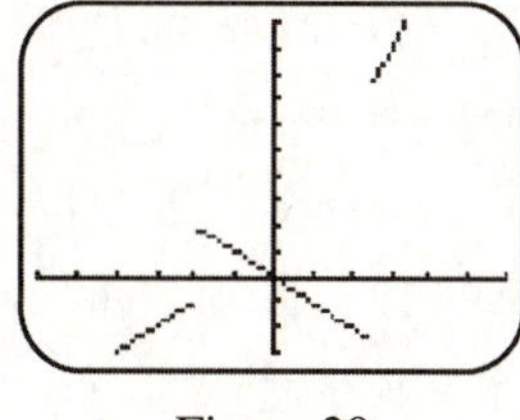

Figure 29

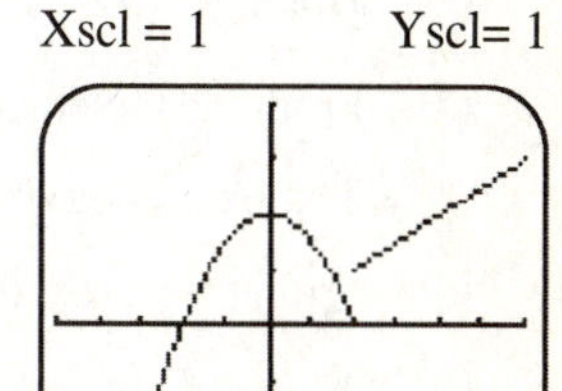

Figure 31

29. Graph $y_1 = (2+x)(x < -4) + (-x)(-4 \leq x \text{ and } x \leq 5) + (3x)(x > 5)$, See Figure 29.

31. Graph $y_1 = \left(-\frac{1}{2}x+2\right)(x \le 2)+\left(\frac{1}{2}x\right)(x>2)$, See Figure 31.

33 From the graph, the function is $f(x)=\begin{cases} 2 & \text{if } x \le 0 \\ -1 & \text{if } x > 1 \end{cases}$; domain : $(-\infty,0]\cup(1,\infty)$; range: $\{-1,2\}$.

35. From the graph, the function is $f(x)=\begin{cases} x & \text{if } x \le 0 \\ 2 & \text{if } x > 0 \end{cases}$; domain : $(-\infty,\infty)$; range: $(-\infty,0]\cup\{2\}$.

37. From the graph, the function is $f(x)=\begin{cases} \sqrt[3]{x} & \text{if } x < 1 \\ x+1 & \text{if } x \ge 1 \end{cases}$; domain : $(-\infty,\infty)$; range : $(-\infty,1)\cup[2,\infty)$.

39. There is an overlap of intervals since the number 4 satisfies both conditions. To be a function, every x-value is used only once.

41. The graph of $y=[\![x]\!]$ is shifted 1.5 units downward.

43. The graph of $y=[\![x]\!]$ is reflected across the x-axis.

45. Graph $y=[\![x]\!]-1.5$, See Figure 45.

47. Graph $y=-[\![x]\!]$, See Figure 47.

[-5,5] by [-3,3]

Xscl = 1 Yscl = 1

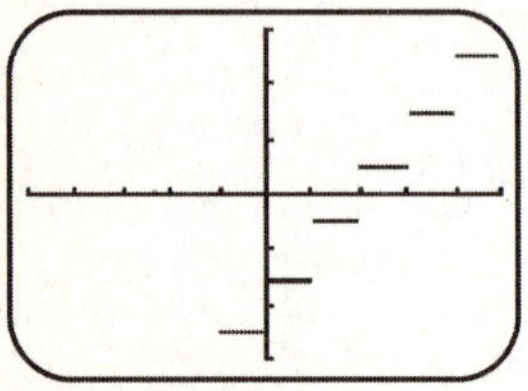

Figure 45

[-5,5] by [-3,3]

Xscl = 1 Yscl = 1

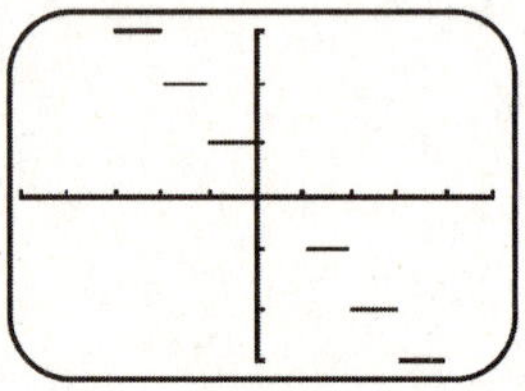

Figure 47

49. When $0 \le x \le 3$ the slope is 5, which means the inlet pipe is open and the outlet pipe is closed; when $3 < x \le 5$ the slope is 2, which means both pipes are open; when $5 < x \le 8$ the slope is 0, which means both pipes are closed; when $8 < x \le 10$ the slope is -3, which means the inlet pipe is closed and the outlet pipe is open.

51. (a) $f(1.5)=0.97, f(3)=1.14$, It costs \$0.97 to mail 1.5 oz and \$1.14 to mail 3 oz.

(b) Domain: $(0,5]$; Range: $\{0.80, 0.97, 1.14, 1.31, 1.48\}$. See Figure 51.

53. (a) From the table, graph the piecewise function. See Figure 53.

(b) The likelihood of being a victim peaks from age 16 up to age 20, then decreases.

55. (a) From the graph, the highest speed is 55 mph and the lowest speed is 30 mph.

(b) There are approximately 12 miles of highway with a speed of 55 mph.

(c) Fromt the graph, $f(4)=40; f(12)=30; f(18)=55$.

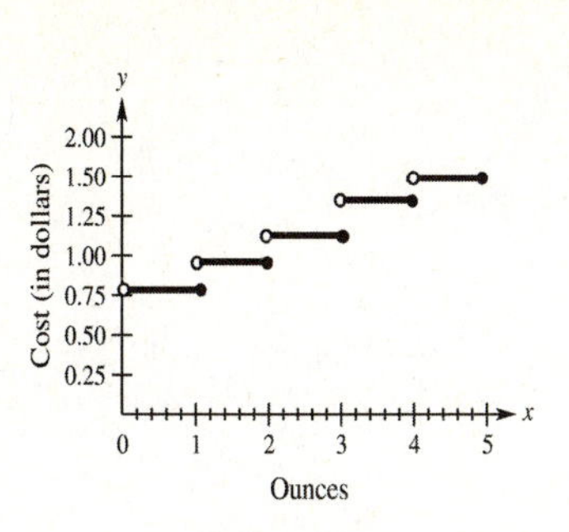

Figure 51

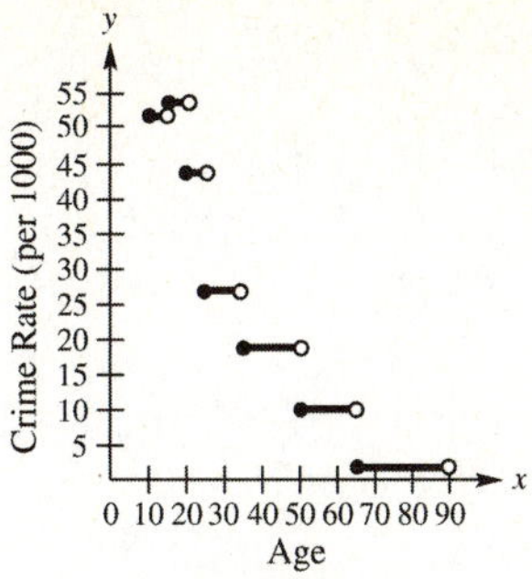

Figure 53

57. (a) A 3.5 minute call would round up to 4 minutes. A 4 minute call would cost: $.50+3(.25)=\$1.25$.

(b) We use a piecewise defined function where the cost increases after each whole number as follows:

$$f(x)=\begin{cases} .50 & \text{if } 0<x\le 1 \\ .75 & \text{if } 1<x\le 2 \\ 1.00 & \text{if } 2<x\le 3 \\ 1.25 & \text{if } 3<x\le 4 \\ 1.50 & \text{if } 4<x\le 5 \end{cases}.$$

Another possibility is $f(x)=\begin{cases} .50 & \text{if } 0<x\le 1 \\ .50-.25[\![1-x]\!] & \text{if } 1<x\le 5 \end{cases}$.

59. For x in the interval $(0,2]$, $y=25$. For x in the interval $(2,3]$, $y=25+3=28$. For x in the interval $(3,4]$, $y=28+3=31$ and so on. The graph is a step function. In this case, the first step has a different width. See Figure 59.

61. Sketch a piecewise function that fills a tank at a rate of 5 gallons a minute for the first 20 minutes (the time it takes to fill the 100 gallon tank) and then drains the tank at a rate of 2 gallons per minute for 50 minutes (the time it takes to drain the 100 gallon tank). See Figure 61.

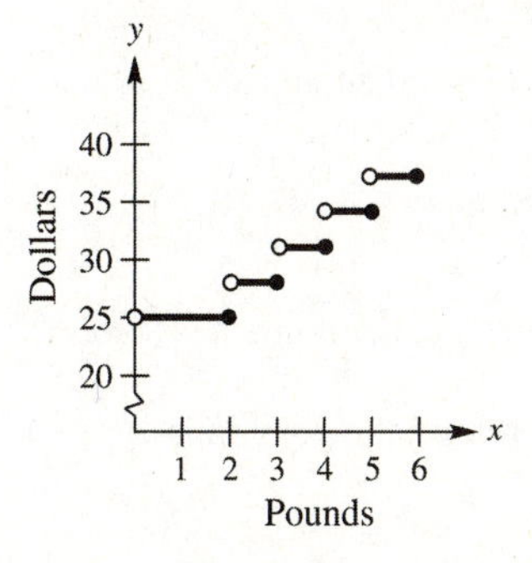

Figure 59

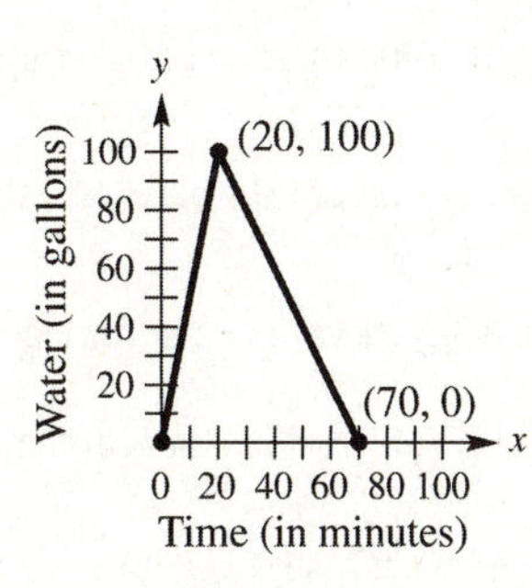

Figure 61

2.6: Operations and Composition

1. $x^2+(2x-5)=x^2+2x-5 \Rightarrow E$.

3. $x^2(2x-5)=2x^3-5x^2 \Rightarrow F$.

5. $(2x-5)^2=4x^2-20x+25 \Rightarrow A$.

7. $(f\circ g)(3)=f(g(3))=f(2(3)-1)=f(5)=5^2+3(5)=40$

9. $(f\circ g)(x)=f(g(x))=f(2x-1)=(2x-1)^2+3(2x-1)=4x^2-4x+1+6x-3=4x^2+2x-2$

11. $(f+g)(3) = f(3) + g(3) = ((3)^2 + 3(3)) + (2(3) - 1) = 23$

13. $(f \cdot g)(4) = f(4) \cdot g(4) = \left((4)^2 + 3(4)\right) \cdot (2(4) - 1) = 196$

15. $\left(\dfrac{f}{g}\right)(-1) = \dfrac{f(-1)}{g(-1)} = \dfrac{(-1)^2 + 3(-1)}{2(-1) - 1} = \dfrac{2}{3}$

17. $(f-g)(2) = f(2) - g(2) = \left((2)^2 + 3(2)\right) - (2(2) - 1) = 7$

19. $(g-f)(-2) = g(-2) - f(-2) = (2(-2) - 1) - \left((-2)^2 + 3(-2)\right) = -3$.

21. $\left(\dfrac{g}{f}\right)(0) = \dfrac{f(0)}{g(0)} = \dfrac{2(0) - 1}{(0)^2 + 3(0)} = \dfrac{-1}{0} \Rightarrow$ undefined.

23. (a) $(f+g)(x) = (4x-1) + (6x+3) = 10x + 2$, $(f-g)(x) = (4x-1) - (6x+3) = -2x - 4$

$(fg)(x) = (4x-1)(6x+3) = 24x^2 + 12x - 6x - 3 = 24x^2 + 6x - 3$

(b) All values can replace x in all three equations; therefore, the domain is $(-\infty, \infty)$ in all cases.

(c) $\left(\dfrac{f}{g}\right)(x) = \dfrac{4x-1}{6x+3}$; all values can replace x, except $-\dfrac{1}{2}$; therefore, the domain is $\left(-\infty, -\dfrac{1}{2}\right) \cup \left(-\dfrac{1}{2}, \infty\right)$.

(d) $(f \circ g)(x) = f[g(x)] = 4(6x+3) - 1 = 24x + 12 - 1 = 24x + 11$; all values can be input for x; therefore, the domain is $(-\infty, \infty)$

(e) $(g \circ f)(x) = g[f(x)] = 6(4x-1) + 3 = 24x - 6 + 3 = 24x - 3$; all values can replace x; therefore, the domain is $(-\infty, \infty)$.

25. (a) $(f+g)(x) = |x+3| + 2x$, $(f-g)(x) = |x+3| - 2x$, $(fg)(x) = |x+3|(2x)$

(b) All values can replace x in all three equations; therefore, the domain is $(-\infty, \infty)$ in all cases.

(c) $\left(\dfrac{f}{g}\right)(x) = \dfrac{|x+3|}{2x}$; all values can replace x, except 0; therefore, the domain is $(-\infty, 0) \cup (0, \infty)$

(d) $(f \circ g)(x) = f[g(x)] = |(2x) + 3| = |2x + 3|$; all values can replace x; therefore, the domain is $(-\infty, \infty)$.

(e) $(g \circ f)(x) = g[f(x)] = 2(|x+3|) = 2|x+3|$; all values can replace x; therefore, the domain is $(-\infty, \infty)$

27. (a) $(f+g)(x) = \sqrt[3]{x+4} + (x^3 + 5) = \sqrt[3]{x+4} + x^3 + 5$, $(f-g)(x) = \sqrt[3]{x+4} - (x^3 + 5) = \sqrt[3]{x+4} - x^3 - 5$

$(fg)(x) = \left(\sqrt[3]{x+4}\right)(x^3 + 5)$

(b) All values can replace x in all three equations; therefore, the domain is $(-\infty, \infty)$ in all cases.

(c) $\left(\frac{f}{g}\right)(x)=\frac{\sqrt[3]{x+4}}{(x^3+5)}$ all values can replace x, except $\sqrt[3]{-5}$, so the domain is $(-\infty,\sqrt[3]{-5})\cup(\sqrt[3]{-5},\infty)$.

(d) $(f\circ g)(x)=f[g(x)]=\sqrt[3]{(x^3+5)+4}=\sqrt[3]{x^3+9}$; all values can replace x, so the domain is $(-\infty,\infty)$.

(e) $(g\circ f)(x)=g[f(x)]=\left(\sqrt[3]{x+4}\right)^3+5=x+4+5=x+9$; all values can replace x, so the domain is $(-\infty,\infty)$.

29. (a) $(f+g)(x)=\sqrt{x^2+3}+(x+1)=\sqrt{x^2+3}+x+1$, $(f-g)(x)=\sqrt{x^2+3}-(x+1)=\sqrt{x^2+3}-x-1$

$(fg)(x)=\left(\sqrt{x^2+3}\right)(x+1)$

(b) All values can replace x in all three equations; therefore, the domain is $(-\infty,\infty)$ in all cases.

(c) $\left(\frac{f}{g}\right)(x)=\frac{\sqrt{x^2+3}}{(x+1)}$; all values can replace x, except -1; therefore, the domain is $(-\infty,-1)\cup(-1,\infty)$.

(d) $(f\circ g)(x)=f[g(x)]=\sqrt{(x+1)^2+3}=\sqrt{x^2+2x+1+3}=\sqrt{x^2+2x+4}$; all values can replace x; therefore, the domain is $(-\infty,\infty)$.

(e) $(g\circ f)(x)=g[f(x)]=\left(\sqrt{x^2+3}\right)+1=\sqrt{x^2+3}+1$; all values can replace x; therefore, the domain is, $(-\infty,\infty)$.

31. (a) From the graph, $4+(-2)=2$.

(b) From the graph, $1-(-3)=4$.

(c) From the graph, $(0)(-4)=0$.

(d) From the graph, $\frac{1}{-3}=-\frac{1}{3}$.

33. (a) From the graph, $0+3=3$.

(b) From the graph, $-1-4=-5$.

(c) From the graph, $(2)(1)=2$.

(d) From the graph, $\frac{3}{0}\Rightarrow$ undefined.

35. (a) From the table, $7+(-2)=5$.

(b) From the table, $10-5=5$.

(c) From the table, $(0)(6)=0$.

(d) From the table, $\frac{5}{0}=$ undefined.

37. See Table 37.

x	$(f+g)(x)$	$(f-g)(x)$	$(fg)(x)$	$\left(\frac{f}{g}\right)(x)$
-2	6	-6	0	0
0	5	5	0	undefined
2	5	9	-14	-3.5
4	15	5	50	2

Table 37

39. From the graph $G(1996) \approx 7.7, B(1996) \approx 11.8$. Since $T(1996) = G(1996) + B(1996)$, $T(1996) = 19.5$.

41. Use the approximate points (1991,14.5) and (1996,19.5) to find the slope for T(x) from 1991-1996. $m = \dfrac{19.5 - 14.5}{1996 - 1991} = 1$. Use the approximate points (1996, 19.5) and (2006,39.5) to find the slope for T(x) from 1996 -2006. $m = \dfrac{39.5 - 19.5}{2006 - 1996} = 2$. Therefore, the number of sodas per week increased more rapidly from 1996-2006.

43. $(T - S)(2000) \approx 19 - 13 \approx 6$; It represents the dollars in billions spent for general science in 2000.

45. From the graph, the only level part of the graph is space and other technologies from 1995-2000.

47. (a) $(f \circ g)(4) = f[g(4)]$, so from the graph find $g(4) = 0$. Now find $f(0) = -4$; therefore, $(f \circ g)(4) = -4$.

(b) $(g \circ f)(3) = g[f(3)]$, so from the graph find $f(3) = 2$. Now find $g(2) = 2$; therefore, $(g \circ f)(3) = 2$.

(c) $(f \circ f)(2) = f[f(2)]$, so from the graph find $f(2) = 0$. Now find $f(0) = -4$; therefore, $(f \circ f)(2) = -4$.

49. (a) $(f \circ g)(1) = f[g(1)]$, so from the graph find $g(1) = 2$. Now find $f(2) = -3$; therefore, $(f \circ g)(1) = -3$.

(b) $(g \circ f)(-2) = g[f(-2)]$, so from the graph find $f(-2) = -3$. Now find $g(-3) = -2$; therefore, $(g \circ f)(-2) = -2$.

(c) $(g \circ g)(-2) = g[g(-2)]$, so from the graph find $g(-2) = -1$. Now find $g(-1) = 0$; therefore, $(g \circ g)(-1) = 0$.

51. (a) $(g \circ f)(1) = g[f(1)]$, so from the table find $f(1) = 4$. Now find $g(4) = 5$; therefore, $(g \circ f)(1) = 5$.

(b) $(f \circ g)(4) = f[g(4)]$, so from the table find $g(4) = 5$. $f(5)$ is undefined; therefore, $(f \circ g)(4)$ is undefined.

(c) $(f \circ f)(3) = f[f(3)]$, so from the table find $f(3) = 1$. Now find $f(1) = 4$; therefore, $(f \circ f)(3) = 4$.

53. From the table, $g(3) = 4$ and $f(4) = 2$.

55. Since $Y_3 = Y_1 \circ Y_2$ and $X = -1$, we solve $Y_1[Y_2(-1)]$. First solve $Y_2 = (-1)^2 = 1$, now solve $Y_1 = 2(1) - 5 = -3$; therefore, $Y_3 = -3$.

57. Since $Y_3 = Y_1 \circ Y_2$ and $X = 7$, we solve $Y_1[Y_2(7)]$. First solve $Y_2 = (7)^2 = 49$, now solve $Y_1 = 2(49) - 5 = 93$; therefore, $Y_3 = 93$.

59. (a) $(f \circ g)(x) = f\left[g(x)\right] = (x^2 + 3x - 1)^3$; all values can be input for x; therefore, the domain is $(-\infty, \infty)$.

(b) $(g \circ f)(x) = g\left[f(x)\right] = (x^3)^2 + 3(x^3) - 1 = x^6 + 3x^3 - 1$; all values can be input for x; therefore, the domain is $(-\infty, \infty)$.

(c) $(f \circ f)(x) = f\left[f(x)\right] = (x^3)^3 = x^9$; all values can be input for x; therefore, the domain is $(-\infty, \infty)$.

61. (a) $(f \circ g)(x) = f\left[g(x)\right] = \left(\sqrt{1-x}\right)^2 = 1 - x$; all values can be input for x; therefore, the domain is $(-\infty, \infty)$.

(b) $(g \circ f)(x) = g\left[f(x)\right] = \left(\sqrt{1-(x^2)}\right) = \sqrt{1-x^2}$; only values where $x^2 \le 1$ can be input for x; therefore, the domain is $\left[-1, 1\right]$.

(c) $(f \circ f)(x) = f\left[f(x)\right] = (x^2)^2 = x^4$; all values can be input for x; therefore, the domain is $(-\infty, \infty)$.

63. (a) $(f \circ g)(x) = f[g(x)] = \dfrac{1}{(5x)+1} = \dfrac{1}{5x+1}$; all values can be input for x, except $-\dfrac{1}{5}$; therefore, the domain is $\left(-\infty, -\dfrac{1}{5}\right) \cup \left(-\dfrac{1}{5}, \infty\right)$.

(b) $(g \circ f)(x) = g[f(x)] = 5\left(\dfrac{1}{x+1}\right) = \dfrac{5}{x+1}$; all values can be input for x, except, -1; therefore, the domain is $(-\infty. -1) \cup (-1, \infty)$.

(c) $(f \circ f)(x) = f[f(x)] = \dfrac{1}{\left(\dfrac{1}{x+1}\right)+1} = \dfrac{1}{\dfrac{1}{x+1} + \dfrac{x+1}{x+1}} = \dfrac{1}{\dfrac{x+2}{x+1}} = \dfrac{x+1}{x+2}$; all values can be input for x, except those that make $\dfrac{x+1}{x+2} = 0$ or undefined. That would be -1 and -2; therefore, the domain is $(-\infty, -2) \cup (-2, -1) \cup (-1, \infty)$.

65. (a) $(f \circ g)(x) = f[g(x)] = 2(4x^3 - 5x^2) + 1 = 8x^3 - 10x^2 + 1$; all values can be input for x; therefore, the domain is $(-\infty, \infty)$.

(b) $(g \circ f)(x) = g[f(x)] = 4(2x+1)^3 - 5(2x+1)^2 = 4(8x^3 + 12x^2 + 6x + 1) - 5(4x^2 + 4x + 1) =$ $32x^3 + 48x^2 + 24x + 4 - (20x^2 + 20x + 5) = 32x^3 + 28x^2 + 4x - 1$; all values can be input for x; therefore, the domain is $(-\infty, \infty)$.

(c) $(f \circ f)(x) = f[f(x)] = 2(2x+1) + 1 = 4x + 3$; all values can be input for x; therefore, the domain is $(-\infty, \infty)$.

67. (a) $(f \circ g)(x) = f(g(x)) = f(5) = 5$, all values can be input for x, therefore the domain is $(-\infty, \infty)$.

(b) $(g \circ f)(x) = g(f(x)) = g(5) = 5$, all values can be input for x, therefore the domain is $(-\infty, \infty)$.

(c) $(f \circ f)(x) = f(f(x)) = f(5) = 5$, all values can be input for x, therefore the domain is $(-\infty, \infty)$.

69. $(f \circ g)(x) = f[g(x)] = 4\left(\dfrac{1}{4}(x-2)\right) + 2 = x - 2 + 2 = x$, $(g \circ f)(x) = g[f(x)] = \dfrac{1}{4}((4x+2)-2) = \dfrac{1}{4}(4x) = x$

71. $(f \circ g)(x) = f\left[g(x)\right] = \sqrt[3]{5\left(\frac{1}{5}x^3 - \frac{4}{5}\right) + 4} = \sqrt[3]{(x^3 - 4) + 4} = \sqrt[3]{x^3} = x$

$(g \circ f)(x) = g\left[f(x)\right] = \frac{1}{5}\left(\sqrt[3]{5x+4}\right)^3 - \frac{4}{5} = \frac{1}{5}(5x+4) - \frac{4}{5} = x + \frac{4}{5} - \frac{4}{5} = x$

73. Graph $y_1 = \sqrt[3]{x-6}$, $y_2 = x^3 + 6$, and $y_3 = x$ in the same viewing window. See Figure 73. The graph of y_2 can be obtained by *reflecting* the graph of y_1 across the line $y_3 = x$.

[-10,10] by [-10,10]

Xscl = 1 Yscl = 1

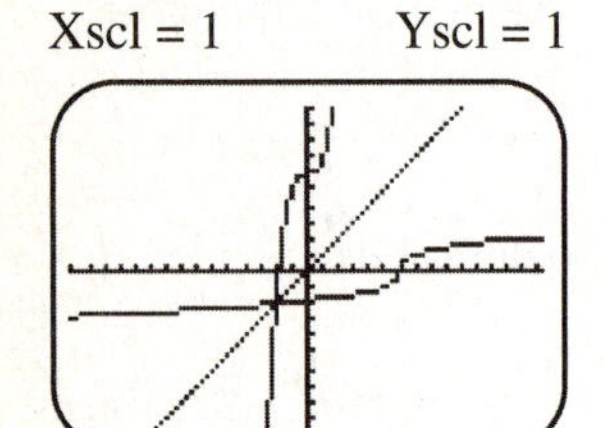

Figure 73

75. Using $\frac{f(x+h)-f(x)}{h}$ gives: $\frac{4(x+h)+3-(4x+3)}{h} = \frac{4x+4h+3-4x-3}{h} = \frac{4h}{h} = 4.$

77. Using $\frac{f(x+h)-f(x)}{h}$ gives: $\frac{-6(x+h)^2-(x+h)+4-(-6x^2-x+4)}{h}$

$= \frac{-6(x^2+2xh+h^2)-x-h+4+6x^2+x-4}{h} = \frac{-6x^2-12xh-6h^2-x-h+4+6x^2+x-4}{h}$

$= \frac{-12xh-6h^2-h}{h} = -12x-6h-1.$

79. Using $\frac{f(x+h)-f(x)}{h}$ gives: $\frac{(x+h)^3-x^3}{h} = \frac{x^3+3x^2h+3xh^2+h^3-x^3}{h} = \frac{3x^2h+3xh^2+h^3}{h} = 3x^2+3xh+h^2.$

81. One possible solution is $f(x) = x^2$ and $g(x) = 6x-2$. Then $(f \circ g)(x) = f[g(x)] = (6x-2)^2$.

83. One possible solution is $f(x) = \sqrt{x}$ and $g(x) = x^2 - 1$. Then $(f \circ g)(x) = f[g(x)] = \sqrt{x^2-1}$.

85. One possible solution is $f(x) = \sqrt{x} + 12$ and $g(x) = 6x$. Then $(f \circ g)(x) = f[g(x)] = \sqrt{6x} + 12$.

87. (a) With a cost of \$10 to produce each item and a fixed cost of \$500, the cost function is $C(x) = 10x + 500$.

(b) With a selling price of \$35 for each item, the revenue function is $R(x) = 35x$.

(c) The profit function is $P(x) = R(x) - C(x) \Rightarrow P(x) = 35x - (10x + 500) \Rightarrow P(x) = 25x - 500$.

(d) A profit is shown when $P(x) > 0 \Rightarrow 25x - 500 > 0 \Rightarrow 25x > 500 \Rightarrow x > 20$. Therefore, 21 items must be produced and sold to realize a profit.

(e) Graph $y_1 = 25x - 500$. The smallest whole number for which $P(x) > 0$ is 21. Use a window of $[0,30]$ by $[-1000,500]$, for example.

89. (a) With a cost of \$100 to produce each item and a fixed cost of \$2700, the cost function is $C(x) = 100x + 2700$.

(b) With a selling price of \$280 for each item, the revenue function is $R(x) = 280x$.

(c) The profit function is $P(x) = R(x) - C(x) \Rightarrow P(x) = 280x - (100x + 2700) \Rightarrow P(x) = 180x - 2700.$

(d) A profit is shown when $P(x) > 0 \Rightarrow 180x - 2700 > 0 \Rightarrow 180x > 2700 \Rightarrow x > 15.$ Therefore, 16 items must be produced and sold to realize a profit.

(e) Graph $y_1 = 180x - 2700,$ the smallest whole number for which $P(x) > 0$ is 16. Use a window of $[0,30]$ by $[-3000,500]$, for example.

91. (a) If $V(r) = \frac{4}{3}\pi r^3$, then a 3 inch increase would be $V(r) = \frac{4}{3}\pi(r+3)^3$, and the volume gained would be

$$V(r) = \frac{4}{3}\pi(r+3)^3 - \frac{4}{3}\pi r^3.$$

(b) Graph $y_1 = \frac{4}{3}\pi(x+3)^3 - \frac{4}{3}\pi x^3$ in the window $[0,10]$ by $[0,1500]$. See Figure 91. Although this appears to be a portion of a parabola, it is actually a cubic function.

(c) From the graph in exercise 91b, an input value of $x = 4$ results in a gain of $y \approx 1168.67$.

(d) $V(4) = \frac{4}{3}\pi(4+3)^3 - \frac{4}{3}\pi(4)^3 = \frac{4}{3}\pi(343) - \frac{4}{3}\pi(64) = \frac{1372}{3}\pi - \frac{256}{3}\pi = \frac{1116}{3}\pi = 372\pi \approx 1168.67.$

93. (a) If $x =$ width, then $2x =$ length. Since the perimeter formula is $P = 2W + 2L$ our perimeter function is $P(x) = 2(x) + 2(2x) = 2x + 4x \Rightarrow P(x) = 6x.$ This is a linear function.

(b) Graph $P(x) = 6x$ in the window $[0,10]$ by $[1,100]$. See Figure 93b. From the graph when $x = 4, y = 24.$ The 4 represents the width of a rectangle and 24 represents the perimeter.

(c) If $x = 4$ is the width of a rectangle then $2x = 8$ is the length. See Figure 93c. Using the standard perimeter formula yields $P = 2(4) + 2(8) = 24.$ This compares favorably with the graph result in part b.

(d) (Answers may vary.) If the perimeter y of a rectangle satisfying the given conditions is 36, then the width x is 6. See Figure 93d.

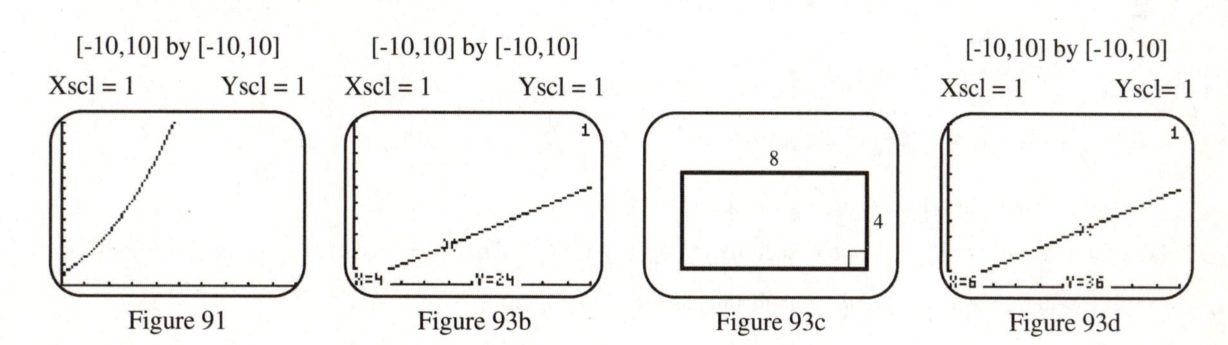

Figure 91 Figure 93b Figure 93c Figure 93d

95. (a) $A(2x) = \frac{\sqrt{3}}{4}(2x)^2 = \frac{\sqrt{3}}{4}(4x^2) \Rightarrow A(2x) = \sqrt{3}x^2$

(b) $A(x) = \frac{\sqrt{3}}{4}(16)^2 = \frac{\sqrt{3}}{4}(256) \Rightarrow A(x) = 64\sqrt{3}$ square units.

(c) On the graph of $y = \frac{\sqrt{3}}{4}x^2$, locate the point where $x = 16$ to find $y \approx 110.85$, an approximation for $64\sqrt{3}$.

97. (a) The function h is the addition of functions f and g.

x	2000	2001	2002	2003
h(x)	82	79	89	103

(b) The function h is the addition of functions f and g. Therefore, $h(x) = f(x) + g(x)$..

99. (a) $(f+g)(1970) = 32.4 + 17.6 = 50.0$

(b) The function $(f+g)(x)$ computes the total SO_2 emissions from burning coal and oil during year x.

(c) Add functions f and g.

x	1860	1900	1940	1970	2000
$(f+g)(x)$	2.4	12.8	26.5	50.0	78.0

101. (a) The function h is the subtraction of function f from g. Therefore, $h(x) = g(x) - f(x)$.

(b) $h(1996) = g(1996) - f(1996) = 841 - 694 = 147$, $h(2006) = g(2006) = f(2006) = 1165 - 1012 = 153$

(c) Using the points (1996, 147) and (2006, 153) from part b, the slope is $m = \frac{153-147}{2006-1996} = \frac{6}{10} = .6$. Now using point slope form: $y - 147 = .6(x - 1996) \Rightarrow y = .6(x - 1996) + 147$.

Reviewing Basic Concepts (Sections 2.4—2.6)

1. (a) $\left|\frac{1}{2}x+2\right| = 4 \Rightarrow \frac{1}{2}x+2=4 \Rightarrow \frac{1}{2}x = 2 \Rightarrow x = 4$ or $\frac{1}{2}x+2=-4 \Rightarrow \frac{1}{2}x=-6 \Rightarrow x=-12$.

Therefore, the solution set is {-12,4}.

(b) $\left|\frac{1}{2}x+2\right| > 4 \Rightarrow \frac{1}{2}x+2>4 \Rightarrow \frac{1}{2}x > 2 \Rightarrow x > 4$ or $\frac{1}{2}x+2<-4 \Rightarrow \frac{1}{2}x<-6 \Rightarrow x<-12$. Therefore, the solution interval is $(-\infty,-12) \cup (4,\infty)$.

(c) $\left|\frac{1}{2}x+2\right| \le 4 \Rightarrow -4 \le \frac{1}{2}x+2 \le 4 \Rightarrow -6 \le \frac{1}{2}x \le 2 \Rightarrow -12 \le x \le 4$.

Therefore, the solution interval is [-12,4].

2. For the graph of $y = |f(x)|$, we reflect the graph of $y = f(x)$ across the x-axis for all points for which $y < 0$. Where $y \ge 0$, the graph remains unchanged. See Figure 2.

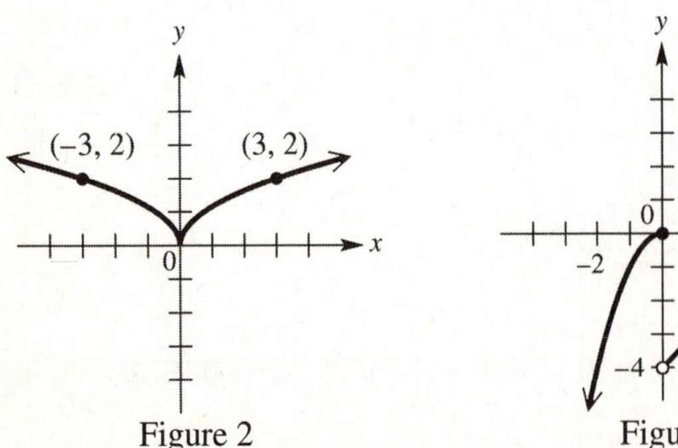

Figure 2

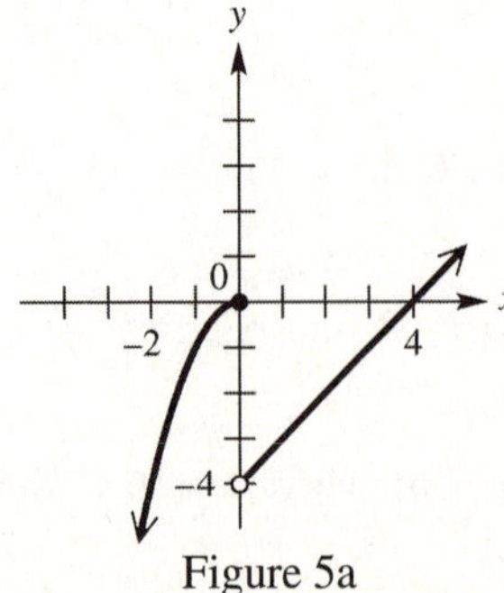

Figure 5a

3. (a) The range of values for $|R_L - 26.75| \le 1.42$ is $-1.42 \le R_L - 26.75 \le 1.42 \Rightarrow 25.33 \le R_L \le 28.17$. The range of values for $|R_E - 38.75| \le 2.17$ is $-2.17 \le R_E - 38.75 \le 2.17 \Rightarrow 36.58 \le R_E \le 40.92$.

(b) If $T_L = 225(R_L)$ then the range for T_L is $225(25.33 \le T_L \le 28.17) = 5699.25 \le T_L \le 6338.25$. If $T_E = 225(R_E)$ then the range for T_E is $225(36.58 \le T_E \le 40.92) = 8230.5 \le T_E \le 9207$.

4. (a) $f(-3) = 2(-3) + 3 = -3$ (b) $f(0) = (0)^2 + 4 = 4$ (c) $f(2) = (2)^2 + 4 = 8$

5. (a) See Figure 5a.

(b) Graph $y_1 = (-x^2)*(x \le 0) + (x-4)*(x > 0)$ in the window [-10,10] by [-10,10]. See Figure 5b.

[-10,10] by [-10,10]
Xscl = 1 Yscl = 1

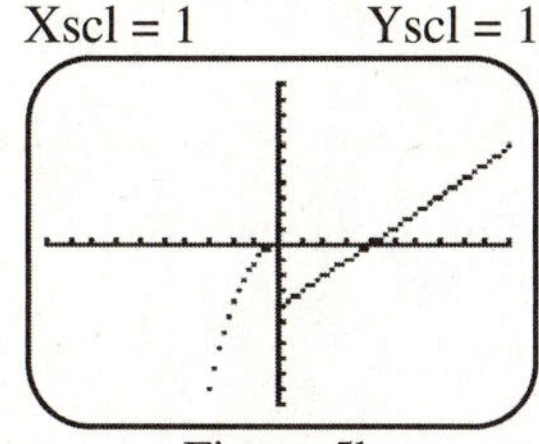

Figure 5b

6. (a) $(f+g)(x) = (-3x-4) + (x^2) = x^2 - 3x - 4$. Therefore, $(f+g)(1) = (1)^2 - 3(1) - 4 = -6$.

(b) $(f-g)(x) = (-3x-4) - (x^2) = -x^2 - 3x - 4$. Therefore, $(f-g)(3) = -(3)^2 - 3(3) - 4 = -22$.

(c) $(fg)(x) = (-3x-4)(x^2) = -3x^3 - 4x^2$. Therefore, $(fg)(1) = -3(-2)^3 - 4(-2)^2 = 24 - 16 = 8$.

(d) $\left(\dfrac{f}{g}\right)(x) = \dfrac{-3x-4}{x^2}$. Therefore, $\left(\dfrac{f}{g}\right)(-3) = \dfrac{-3(-3)-4}{(-3)^2} = \dfrac{5}{9}$.

(e) $(f \circ g)(x) = f[g(x)] = -3(x)^2 - 4 \Rightarrow (f \circ g)(x) = -3x^2 - 4$

(f) $(g \circ f)(x) = g[f(x)] = (-3x-4)^2 \Rightarrow (g \circ f)(x) = 9x^2 + 24x + 16$

7. One of many possible solutions for $(f \circ g)(x) = h(x)$ is $f(x) = x^4$ and $g(x) = x + 2$. Then $(f \circ g)(x) = f[g(x)] = (x+2)^4$.

8. $$\frac{-2(x+h)^2 + 3(x+h) - 5 - (-2x^2 + 3x - 5)}{h} = \frac{-2(x^2 + 2xh + h^2) + 3x + 3h - 5 + 2x^2 - 3x + 5}{h} =$$

$$\frac{-2x^2 - 4xh - 2h^2 + 3x + 3h - 5 + 2x^2 - 3x + 5}{h} = \frac{-4xh - 2h^2 + 3h}{h} = -4x - 2h + 3.$$

9. (a) At 4% simple interest the equation for interest earned is $y_1 = .04x$.

(b) If he invested x dollars in the first account, then he invested x+500 in the second account. The equation for the amount of interest earned on this account is $y_2 = .025(x+500) \Rightarrow y_2 = .025x + 12.5$.

(c) It represents the total interest earned in both accounts for 1 year.

(d) Graph $y_1 + y_2 = .04x + (.025x + 12.5) \Rightarrow y_1 + y_2 = .04x + .025x + 12.5$ in the window [0,1000] by [0,100]. See Figure 9. An input value of $x = 250$, results in \$28.75 earned interest.

(e) At $x = 250$, $y_1 + y_2 = .04(250) + .025(250) + 12.5 = 10 + 6.25 + 12.5 = \$28.75$.

[0,1000] by [-20,100]
Xscl = 100 Yscl = 10

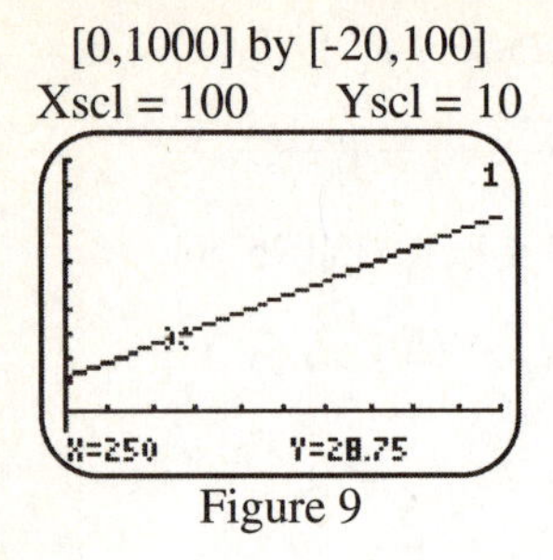

Figure 9

10. If the radius is r, then the height is $2r$ and the equation is

$$S = \pi r\sqrt{r^2+(2r)^2} = \pi r\sqrt{r^2+4r^2} = \pi r\sqrt{5r^2} \Rightarrow S = \pi r^2\sqrt{5}.$$

Chapter 2 Review Exercises

The graphs for exercises 1–10 can be found in the "Function Capsule" boxes located in section 2.1 in the text.

1. True Both $f(x)=x^2$ and $f(x)=|x|$ have the interval: $[0,\infty)$ as the range.
2. True Both $f(x)=x^2$ and $f(x)=|x|$ increase on the interval: $[0,\infty)$.
3. False The function $f(x)=\sqrt{x}$ has the domain: $[0,\infty)$ and $f(x)=\sqrt[3]{x}$ the domain: $(-\infty,\infty)$.
4. False The function $f(x)=\sqrt[3]{x}$ increases on its entire domain.
5. True The function $f(x)=x$ has a domain and range of: $(-\infty,\infty)$
6. False The function $f(x)=\sqrt{x}$ is not defined on $(-\infty,0)$, so certainly cannot be continuous.
7. True All of the functions show increases on the interval: $[0,\infty)$
8. True Both $f(x)=x$ and $f(x)=x^3$ have graphs that are symmetric with respect to the origin.
9. True Both $f(x)=x^2$ and $f(x)=|x|$ have graphs that are symmetric with respect to the y-axis.
10. True No graphs are symmetric with respect to the x-axis.
11. Only values where $x\ge 0$ can be input for x, therefore the domain of $f(x)=\sqrt{x}$ is: $[0,\infty)$
12. Only positive solution are possible in absolute value functions, therefore the range of $f(x)=\sqrt{x}$ is: $[0,\infty)$
13. All solution are possible in cube root functions, therefore the range of $f(x)=\sqrt[3]{x}$ is: $(-\infty,\infty)$.
14. All values can be input for x, therefore the domain of $f(x)=x^2$ is: $(-\infty,\infty)$.
15. The function $f(x)=\sqrt[3]{x}$ increases for all inputs for x, therefore the interval is: $(-\infty,\infty)$.
16. The function $f(x)=|x|$ increases for all inputs where $x\ge 0$, therefore the interval is: $[0,\infty)$
17. The equation is the equation $y=\sqrt{x}$. Only values where $x\ge 0$ can be input for x, therefore the domain of $y=\sqrt{x}$ is: $[0,\infty)$
18. The equation $y^2=x$ is the equation $y=\sqrt{x}$ Square root functions have both positive and negative solutions and all solution are possible, therefore the range of $y=\sqrt{x}$ is: $(-\infty,\infty)$.
19. The graph of $f(x)=(x+3)-1$ is the graph $y=x$ shifted 3 units to the left and 1 unit downward.

See Figure 19.

20. The graph of $f(x)=-\frac{1}{2}x+1$ is the graph reflected $y=x$ across the x-axis, vertically shrunk by a factor of $\frac{1}{2}$, and shifted 1 unit upward. See Figure 20.

21. The graph of $f(x)=(x+1)^2-2$ is the graph $y=x^2$ shifted 1 unit to the left and 2 units downward. See Figure 21.

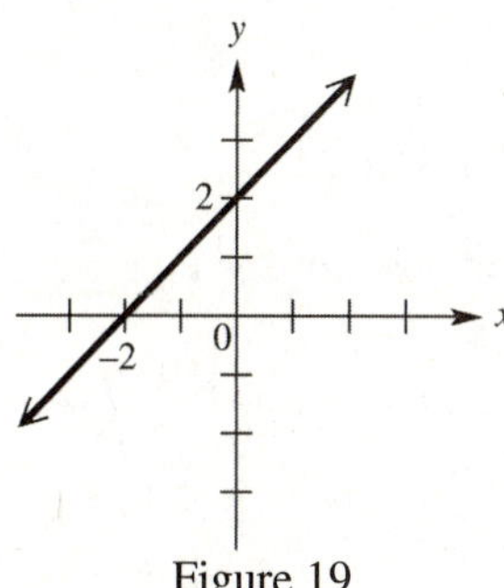

Figure 19

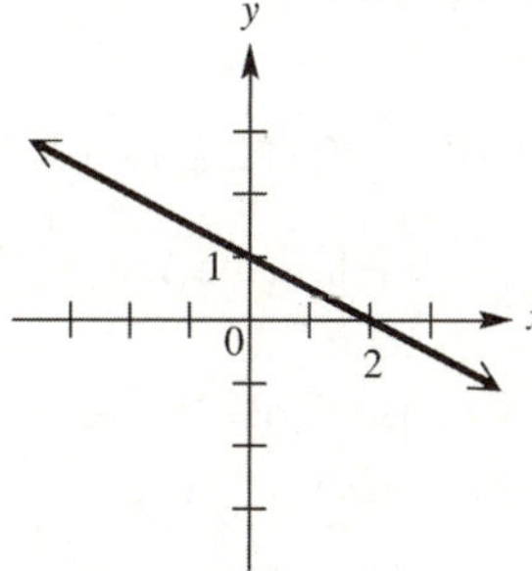

Figure 20

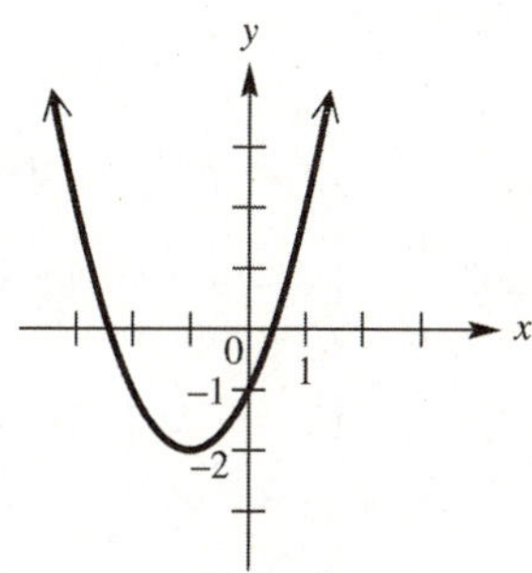

Figure 21

22. The graph of $f(x)=-2x^2+3$ is the graph $y=x^2$ reflected across the x-axis, vertically stretched by a factor of 2, and shifted 3 units upward. See Figure 22.

23. The graph of $f(x)=-x^3+2$ is the graph $y=x^3$ reflected across the x-axis and shifted 2 units upward. See Figure 23.

24. The graph of $f(x)=(x-3)^3$ is the graph $y=x^3$ shifted 3 units to the right. See Figure 24.

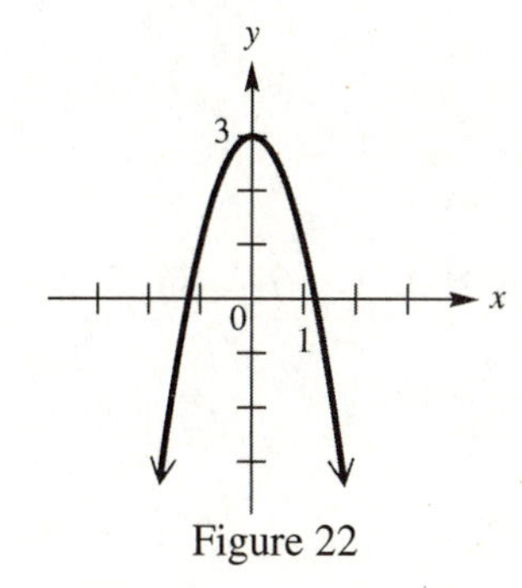

Figure 22

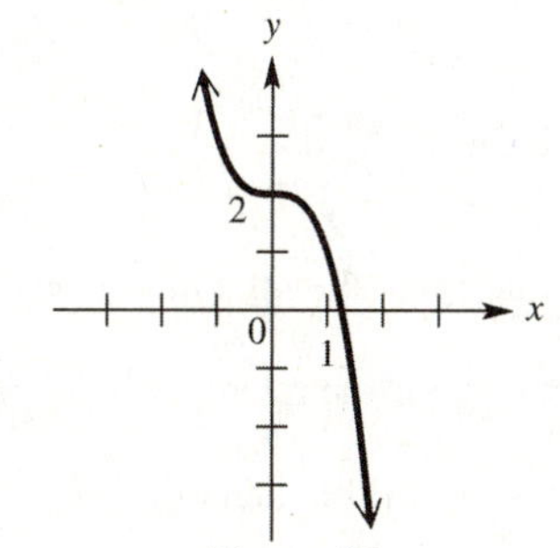

Figure 23

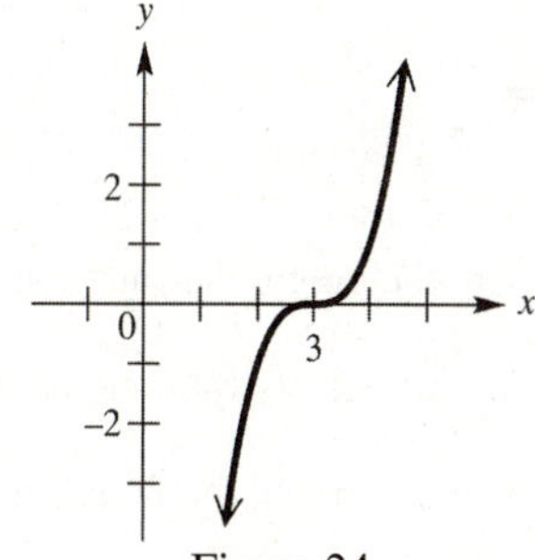

Figure 24

25. The graph of $f(x)=\sqrt{\frac{1}{2}x}$ is the graph $y=\sqrt{x}$ horizontally stretched by a factor of 2. See Figure 25.

26. The graph of $f(x)=\sqrt{x-2}+1$ is the graph $y=\sqrt{x}$ shifted 2 units to the right and 1 unit upward. See Figure 26.

27. The graph of $f(x)=2\sqrt[3]{x}$ is the graph $y=\sqrt[3]{x}$ vertically stretched by a factor of 2. See Figure 27.

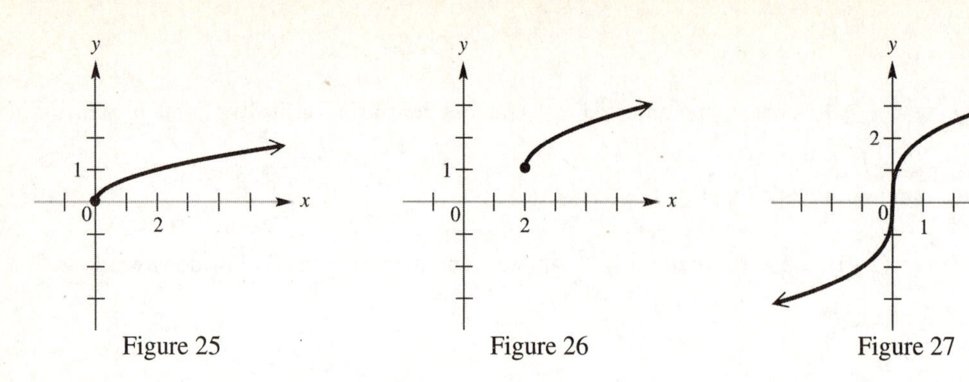

Figure 25 Figure 26 Figure 27

28. The graph of $f(x)=\sqrt[3]{x}-2$ is the graph $y=\sqrt[3]{x}$ shifted 2 units downward. See Figure 28.

29. The graph of $f(x)=|x-2|+1$ is the graph $y=|x|$ shifted 2 units right and 1 unit upward. See Figure 29.

30. The graph of $f(x)=|-2x+3|$ is the graph $y=|x|$ horizontally shrunk by a factor of $\frac{1}{2}$, shifted $\left(\frac{1}{2}\right)(3)$ or $\frac{3}{2}$ units to the left, and reflected across the y-axis. See Figure 30.

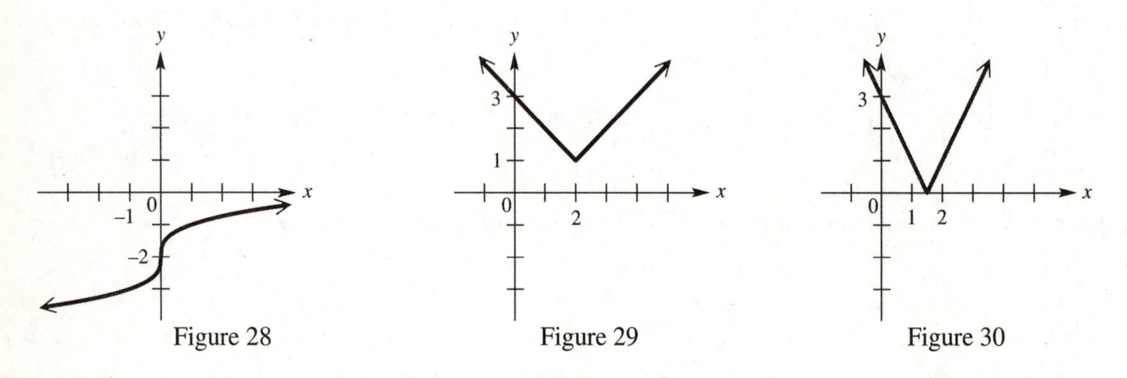

Figure 28 Figure 29 Figure 30

31. (a) From the graph, the function is continuous for the intervals: $(-\infty,-2),[-2,1]$ and $(1,\infty)$

    (b) From the graph, the function is increasing for the interval: $[-2,1]$

    (c) From the graph, the function is decreasing for the interval: $(-\infty,-2)$

    (d) From the graph, the function is constant for the interval: $(1,\infty)$

    (e) From the graph, all values can be input for x, therefore the domain is: $(-\infty,\infty)$

    (f) From the graph, the possible values of y or the range is: $\{-2\}\cup[-1,1]\cup(2,\infty)$

32. $x=y^2-4 \Rightarrow y^2=x+4 \Rightarrow y=\sqrt{x+4}$ and $y=-\sqrt{x+4}$.

33. From the graph, the relation is symmetric with respect to the x-axis, y-axis, and origin. The relation is not a function since some inputs x have two outputs y.

34. If $F(x)=x^3-6$,then $F(-x)=(-x)^3-6 \Rightarrow F(-x)=-x^{-3}-6$ and $-F(x)=-(x^3-6) \Rightarrow -F(x)=-x^3+6$.

    Since $F(x)\neq F(-x)\neq -F(x)$ the function has no symmetry and is neither an ever nor an odd function.

35. If $f(x)=|x|+4$,then $f(-x)=|(-x)|+4 \Rightarrow f(-x)=|x|+4$ and $-f(x)=-|x|-4$ Since $f(-x)=f(x)$ the function is symmetric with respect to the y-axis and is an even function.

36. If $f(x)=\sqrt{x-5}$ then $f(-x)=\sqrt{(-x)-5}$ and $-f(x)=-\sqrt{x-5}$ .Since $f(x)\neq f(-x)\neq -f(x)$, the function has no symmetry and is neither an even nor an odd function.

37. If $y^2=x-5$ then $y=\pm\sqrt{x-5}$. Since $f(x)=-\sqrt{x-5}$ is the reflection of $f(x)=-\sqrt{x-5}$ across the x-axis, the relation has symmetry with respect to the x-axis. Also, one x inputs can produce two y outputs the relation is not a function.

38. If $f(x)=3x^4+2x^2+1$ then $f(-x)=3(-x)^4+2(-x)^2+1 \Rightarrow f(-x)=3x^4+2x^2+1$ and $-f(x)=-3x^4-2x^2-1$. Since $f(-x)=f(x)$ the function is symmetric with respect to the y-axis and is an even function.

39. True, a graph that is symmetrical with respect to the x-axis means that for every (x, y) there is also $(x,-y)$.

40. True, since an even function and one that is symmetric with respect to the y-axis both contain the points (x, y) and $(-x, y)$.

41. True, since an odd function and one that is symmetric with respect to the origin both contain the points (x, y) and $(-x,-y)$.

42. False, for an even function, if (a,b) is on the graph, then $(-a,b)$ is on the graph and not $(a,-b)$. For example, $f(x)=x^2$ is even, and (2, 4) is on the graph, but (2, -4) is not.

43. False, for an odd function, if (a,b) is on the graph, then $(-a,-b)$ is on the graph and not $(-a,b)$ For example, $f(x)=x^3$ is odd, and $(2,8)$ is on the graph, but $(-2,8)$ is not.

44. True, if $(x,0)$ is on the graph of $f(x)=0$ then $(-x,0)$ is on the graph.

45. The graph of $y=-3(x+4)^2-8$ is the graph of $y=x^2$ shifted 4 units to the left, vertically stretched by a factor of 3, reflected across the x-axis, and shifted 8 units downward.

46. The equation $y=\sqrt{x}$ reflected across the y-axis is: $f(x)=\sqrt{-x}$ then reflected across the x-axis is: $y=-\sqrt{-x}$ now vertically shrunk by a factor of $\frac{2}{3}$ is: $y=-\frac{2}{3}\sqrt{-x}$, and finally shifted 4 units upward is: $y=-\frac{2}{3}\sqrt{-x}+4$.

47. Shift the function *f* upward 3 units. See Figure 47.

48. Shift the function *f* to the right 2 units. See Figure 48.

49. Shift the function *f* to the left 3 units and downward 2 units. See Figure 49.

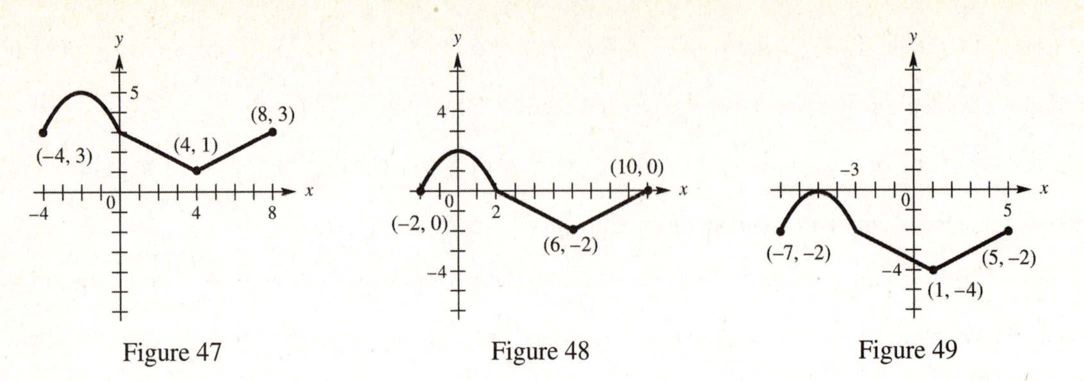

Figure 47 Figure 48 Figure 49

50. For values where $f(x) > 0$ the graph remains the same. For values where $f(x) < 0$ reflect the graph across the x-axis. See Figure 50.

51. Horizontally shrink the function f by a factor of $\frac{1}{4}$. See Figure 51.

52. Horizontally stretch the function f by a factor of 2. See Figure 52.

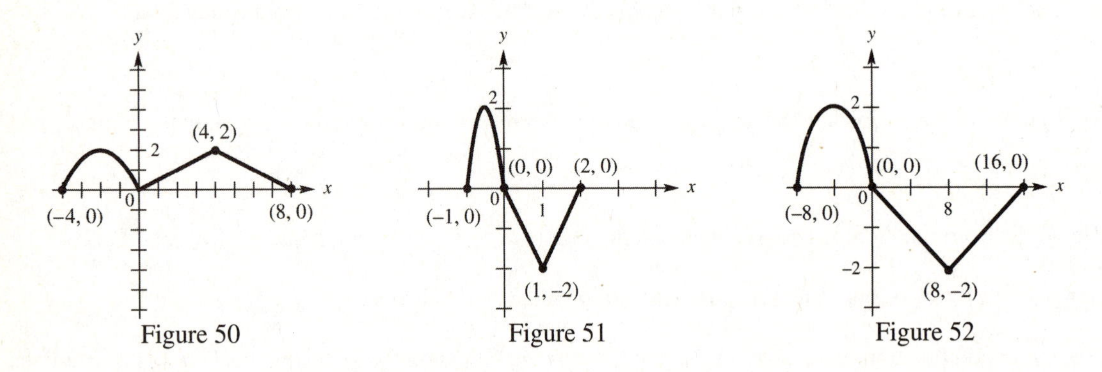

Figure 50 Figure 51 Figure 52

53. The function is shifted upward 4 units, therefore the domain remains the same: $[-3,4]$ and the range is increased by 4 and is: $[2,9]$.

54. The function is shifted left 10 units, therefore the domain is decreased by 10 and is $[-13,-6]$; and the function is stretched vertically by a factor of 5, therefore the range is multiplied by 5 and is: $[-10,25]$.

55. The function is horizontally shrunk by a factor of $\frac{1}{2}$, therefore the domain is divided by 2 and is: $\left[-\frac{3}{2},2\right]$; and the function is reflected across the x-axis, therefore the range is opposites of the original and is: $[-5,2]$.

56. The function is shifted right 1 unit, therefore the domain is increased by 1 and is: $[-2,5]$; and the function is also shifted upward 3 units, therefore the range is increased by 3 and is: $[1,8]$.

57. We reflect the graph of $y = f(x)$ across the x-axis for all points for which $y < 0$. Where $y \geq 0$, the graph remains unchanged. See Figure 57.

58. We reflect the graph of $y = f(x)$ across the x-axis for all points for which < 0. Where $y \geq 0$, the graph remains unchanged. See Figure 58.

59. Since the range is $\{2\}, y \geq 0,$ so the graph remains unchanged.

60. Since the range is $\{-2\}, y < 0,$ so we reflect the graph across the x-axis. See Figure 60.

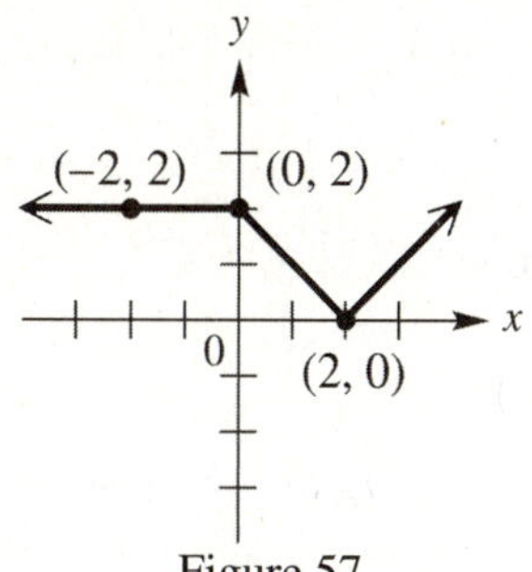

Figure 57

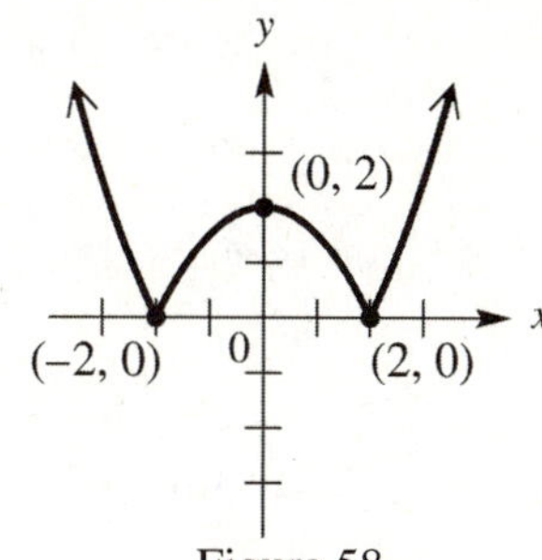

Figure 58

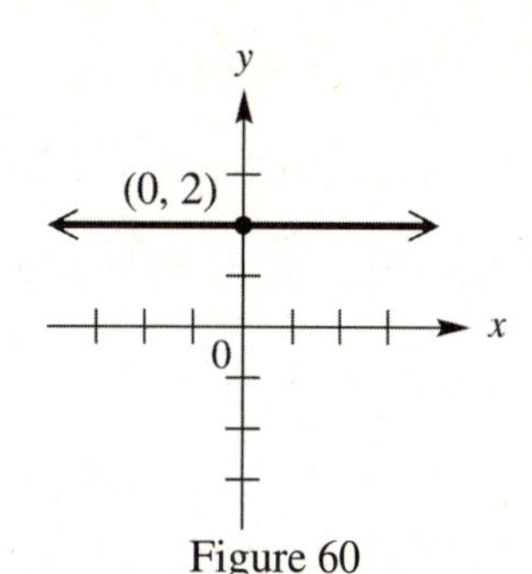

Figure 60

61. $|4x+3| = 12 \Rightarrow 4x+3 = 12 \Rightarrow 4x = 9 \Rightarrow x = \frac{9}{4}$ or $4x+3 = -12 \Rightarrow 4x = -15 \Rightarrow x = -\frac{15}{4}$, therefore the solution set is: $\left\{-\frac{15}{4}, \frac{9}{4}\right\}$.

62. $|-2x-6|+4 = 1 \Rightarrow |-2x-6| = -3$. Since an absolute value equation can not have a solution less than zero, the solution set is: $\varnothing$

63. $|5x+3| = |x+11| \Rightarrow 5x+3 = x+11 \Rightarrow 4x = 8 \Rightarrow x = 2$ or $5x+3 = -(x+11) \Rightarrow 6x = -14 \Rightarrow x = -\frac{14}{6} = -\frac{7}{3}$, therefore the solution set is: $\left\{-\frac{7}{3}, 2\right\}$.

64. $|2x+5| = 7 \Rightarrow 2x+5 = 7 \Rightarrow 2x = 2 \Rightarrow x = 1$ or $2x+5 = -7 \Rightarrow 2x = -12 \Rightarrow x = -6,$ therefore the solution set is: $\{-6, 1\}$.

65. $|2x+5| \leq 7 \Rightarrow -7 \leq 2x+5 \leq 7 \Rightarrow -12 \leq 2x \leq 2 \Rightarrow -6 \leq x \leq 1,$ therefore the interval is: $\{-6, 1\}$.

66. $|2x+5| \geq 7 \Rightarrow 2x+5 \geq 7 \Rightarrow 2x \geq 2 \Rightarrow x \geq 1$ or $2x+5 \leq -7 \Rightarrow 2x \leq -12 \Rightarrow x \leq -6,$ therefore the solution is the interval: $(-\infty, -6] \cup [1, \infty)$.

67. $|5x-12| 0 \Rightarrow 5x-12 > 0 \Rightarrow 5x > 12 \Rightarrow x > \frac{12}{5}$ or $5x-12 < 0 \Rightarrow 5x = 12 \Rightarrow x < \frac{12}{5}$, therefore the solution is the interval: $\left(-\infty, \frac{12}{5}\right) \cup \left(\frac{12}{5}, \infty\right)$ or $\left\{x \mid x \neq \frac{12}{5}\right\}$.

68. Since an absolute value equation can not have a solution less than zero, the solution set is: $\varnothing$

69. $2|3x-1|+1 = 21 \Rightarrow 2|3x-1| = 20 \Rightarrow |3x-1| = 10 \Rightarrow 3x-1 = 10 \Rightarrow 3x = 11 \Rightarrow x = \frac{11}{3}$ or $3x-1 = -10 \Rightarrow 3x = -9 \Rightarrow x = -3,$ therefore the solution set is: $\left\{-3, \frac{11}{3}\right\}$.

70. $|2x+1|=|-3x+1| \Rightarrow 2x+1=-3x+1 \Rightarrow 5x=0 \Rightarrow x=0$ or $2x+1=-(-3x+1) \Rightarrow -x=-2 \Rightarrow x=2$, therefore the solution set is: $\{0,2\}$.

71. The x-coordinates of the points of intersection of the graphs are -6 and 1.Thus, $\{-6,1\}$ is the solution set of $y_1 = y_2$ The graph of y_1 lies on or below the graph of y_2 between -6 and 1, so the solution set of $y_1 \le y_2$ is $[-6,1]$ The graph of y_1 lies above the graph of y_2 everywhere else, so the solution set of $y_1 \ge y_2$ is $(-\infty,-6]\cup[1,\infty)$.

72. Graph $y_1 = |x+1|+|x-3|$ and $\{-3,-1\}$ See Figure 72. The intersections are $x=-3$ and $x=5$, therefore the solution set is: $\{-3,5\}$. **Check:** $|(-3)+1|+|(-3)-3|=8 \Rightarrow |-2|+|-6|=8 \Rightarrow 2+6=8 \Rightarrow 8=8$ and $|(5)+1|+|(5)-3|=8 \Rightarrow |6|+|2|=8 \Rightarrow 6+2=8 \Rightarrow 8=8$

[-10,10] by [-4,16]

Xscl = 1 Yscl = 1

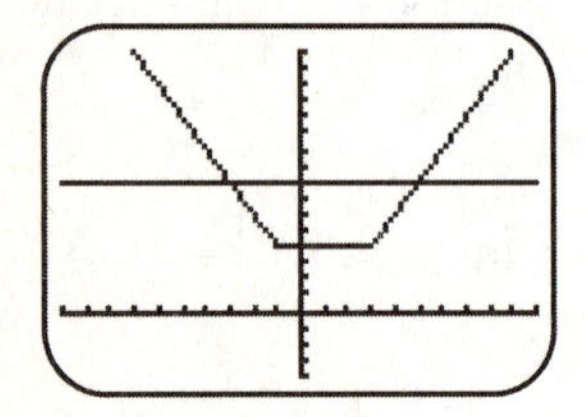

Figure 72

73. Initially, the car is at home. After traveling 30 mph for 1 hr, the car is 30 mi away from home. During the second hour the car travels 20 mph until it is 50 mi away. During the third hour the car travels toward home at 30 mph until it is 20 mi away. During the fourth hour the car travels away from home at 40 mph until it is 60 mi away from home. During the last hour, the car travels 60 mi at 60 mph until it arrives home.

74. See Figure 74

75. See Figure 75

76. See Figure 76

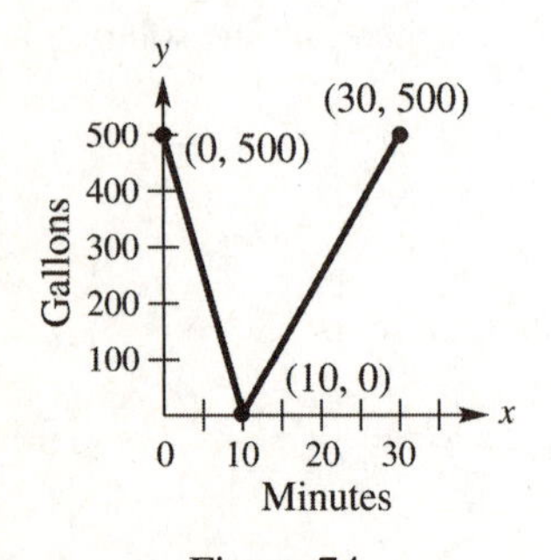

Figure 74

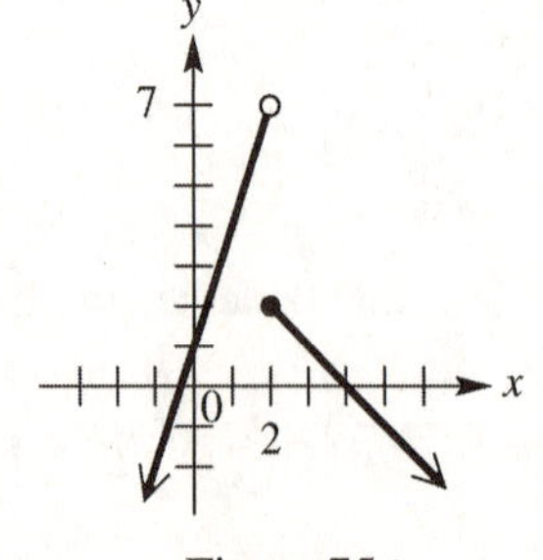

Figure 75

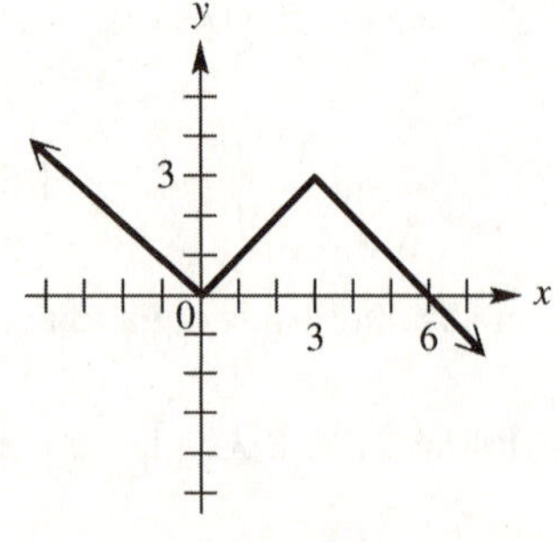

Figure 76

77. Graph $y_1 = (3x+1)*(x<2)+(-x+4)*(x \ge 2)$ in the window $[-10,10]$ by

$|4x+8| > 4 \Rightarrow 4x+8 > 4 \Rightarrow 4x > -4 \Rightarrow -1$ See Figure 77.

78. See Figure 78.

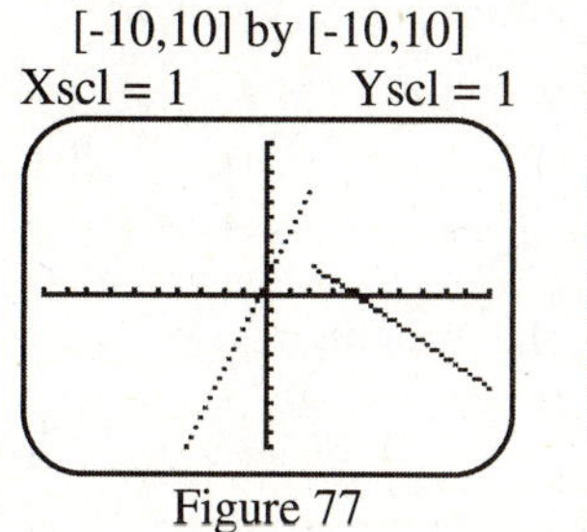

Figure 77

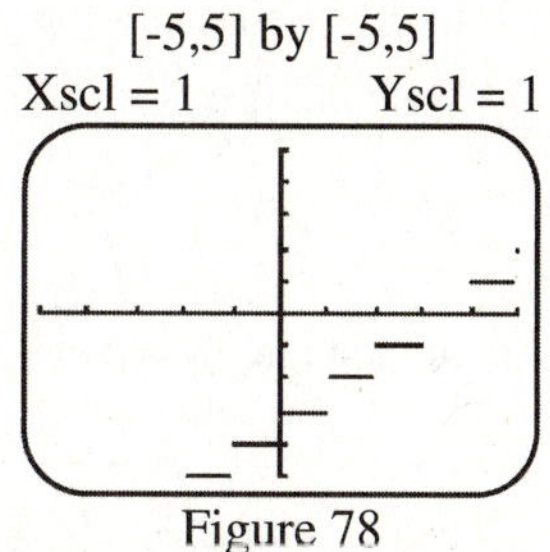

Figure 78

79. From the graphs $(f+g)(1) = 2+3 = 5$

80. From the graphs $(f-g)(0) = 1-4 = -3$

81. From the graphs $(fg)(-1) = (0)(3) = 0$

82. From the graphs $\left(\frac{f}{g}\right)(2) = \frac{3}{2}$

83. From the graphs $(f \circ g)(2) = f[g(2)] = f(2) = 3$

84. From the graphs $(g \circ f)(2) = g[f(2)] = g(3) = 2$

85. From the graphs $(g \circ f)(-4) = g[f(-4)] = g(2) = 2$

86. From the graphs $(f \circ g)(-2) = f[g(-2)] = f(2) = 3$

87. From the table $(f+g)(1) = 7+1 = 8$

88. From the table $(f-g)(3) = 9-1 = 0$

89. From the table $(fg)(-1) = (3)(-2) = -6$

90. From the table $\left(\frac{f}{g}\right)(0) = \frac{5}{0}$, which is undefined.

91. From the tables $(g \circ f)(-2) = g[f(-2)] = g(1) = 2$

92. From the graphs $(f \circ g)(3) = f[g(3)] = f(-2) = 1$

93. $\frac{2(x+h)+9-(2x+9)}{h} = \frac{2x+2h+9-2x-9}{h} = \frac{2h}{h} = 2$

94. $\frac{(x+h)^2-5(x+h)+3-(x^2-5x+3)}{h} = \frac{x^2+2xh+h^2-5x-5h+3-x^2+5x-3}{h} = \frac{2xh+h^2-5h}{h} = 2x+h-5$

95. One of many possible solutions for $(f \circ g)(x) = h(x)$ is: $f(x) = x^2$ and $g(x) = x^3 - 3x$ Then $(f \circ g)(x) = (x^3-3x)^2$.

96. One of many possible solutions for $(f \circ g)(x) = h(x)$ is $f(x) = \frac{1}{x}$ and $g(x) = x - 5$ Then

$(f \circ g)(x) = f[g(x)] = \frac{1}{x-5}$

97. If $V(r) = \frac{4}{3}\pi r^3$, then a 4 inch increase would be: $V(r) = \frac{4}{3}\pi(r+4)^3$, and the volume gained would be:

$V(r) = \frac{4}{3}\pi(r+4)^3 - \frac{4}{3}\pi r^3$.

98. (a) Since $h = d, r = \frac{d}{2}$ and the formula for the volume of a can is: $V = \pi r^2 h$, the function is:

$V(d) = \pi\left(\frac{d}{2}\right)^2 d \Rightarrow V(d) = \frac{\pi d^3}{4}$

(b) Since $h = d, r = \frac{d}{2}, c = 2\pi r$ and the formula for the surface area of a can is: $A = 2\pi rh + 2\pi r^2$, the

function is: $S(d) = 2\pi\left(\frac{d}{2}\right)d + 2\pi\left(\frac{d}{2}\right)^2 \Rightarrow S(d) = \pi d^2 + \frac{\pi d^2}{2} \Rightarrow S(d) = \frac{3\pi d^2}{2}$

99. The function for changing yards to inches is: $f(x) = 36x$ and the function for changing miles to yards is:

$g(x) = 1760x$ The composition of this which would change miles into inches is:

$f[g(x)] = 36[1760(x)] \Rightarrow (f \circ g)(x) = 63{,}360x$.

100. If $x =$ width, then length $= 2x$ A formula for Perimeter can now be written as: $P = x + 2x + x + 2x$ and the function is: $P(x) = 6x$ This is a linear function.

Chapter 2 Test

1. (a) D, only values where $x \geq 0$ can be input into a square root function.

   (b) D, only values where $y \geq 0$ can be the solution to a square root function.

   (c) C, all values can be input for x in a squaring function.

   (d) B, only values where $y \geq 3$ can be the solution to $f(x) = x^2 + 3$

   (e) C, all values can be input for x in a cube root function.

   (f) C, all values can be a solution in a cube root function.

   (g) C, all values can be input for x in an absolute value function.

   (h) D, only values where $y \geq 0$ can be the solution to an absolute value function.

   (i) D, if $x = y^2$ then $y = \sqrt{x}$ and only values where $x \geq 0$ can be input into a square root function.

   (j) C, all values can be a solution in this function.

2. (a) This is $f(x)$ shifted 2 units upward. See Figure 2a.

   (b) This is $f(x)$ shifted 2 units to the left. See Figure 2b.

   (c) This is $f(x)$ reflected across the x-axis. See Figure 2c.

   (d) This is $f(x)$ reflected across the y-axis. See Figure 2d.

   (e) This is $f(x)$ vertically stretched by a factor of 2. See Figure 2e.

(f) We reflect the graph of $y = f(x)$ across the x-axis for all points for which \$2.75 Where $y \geq 0$ the graph remains unchanged. See Figure 2f.

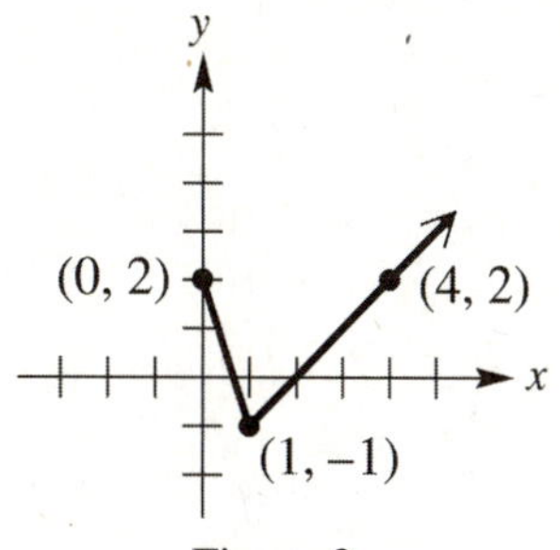

Figure 2a

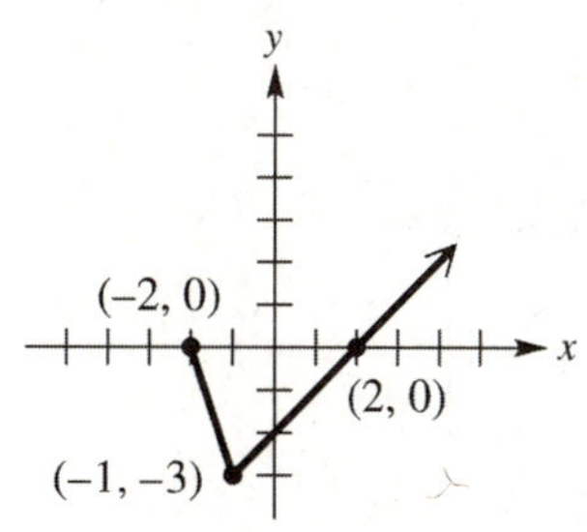

Figure 2b

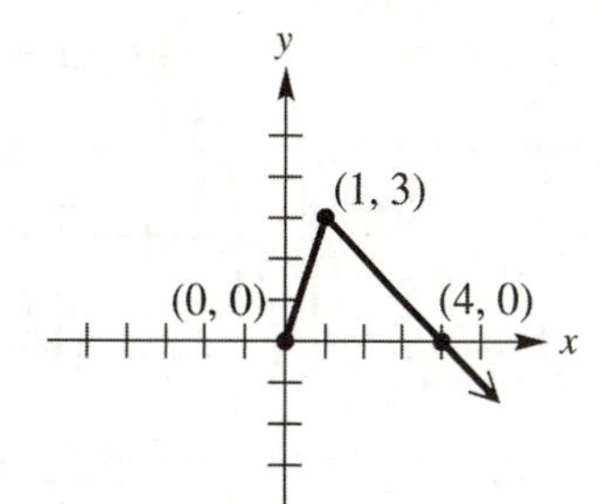

Figure 2c

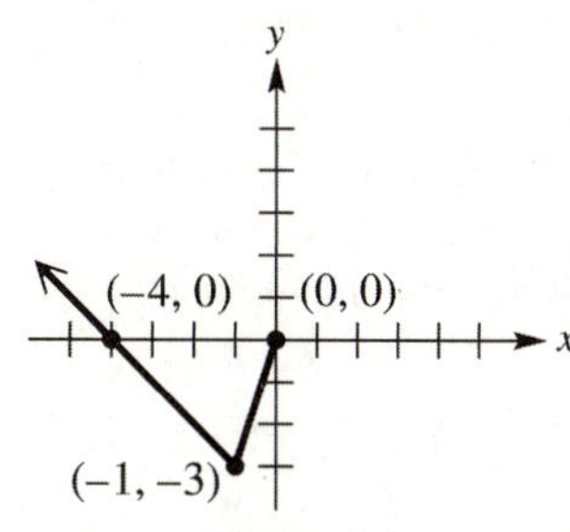

Figure 2d

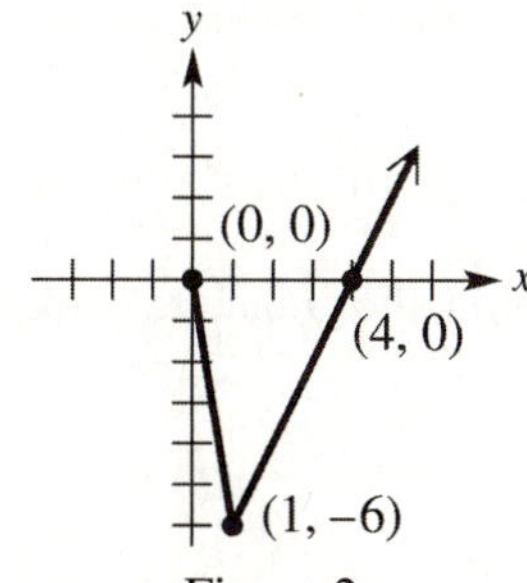

Figure 2e

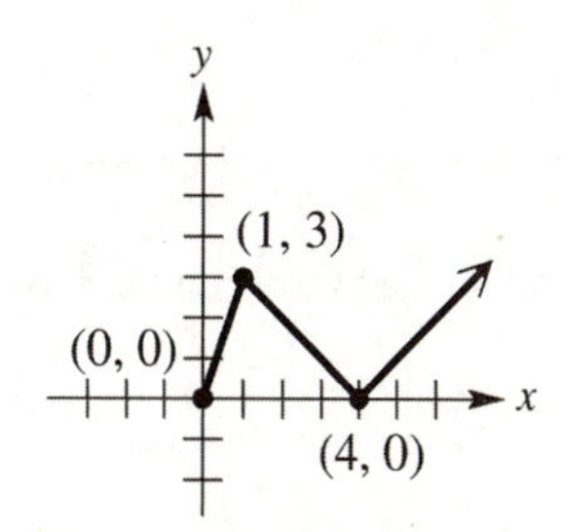

Figure 2f

3. (a) Since $y = f(2x)$ is $y = f(x)$ horizontally shrunk by a factor of $\frac{1}{2}$, the point $(-2, 4)$ on $y = f(x)$ becomes the point $(-1, 4)$ on the graph of $y = f(2x)$.

   (b) Since $y = f\left(\frac{1}{2}x\right)$ is $y = f(x)$ horizontally stretched by a factor of 2, the point $(-2, 4)$ on $y = f(x)$ becomes the point $(-4, 4)$ on the graph of $y = f\left(\frac{1}{2}x\right)$.

4. (a) The graph of $f(x) = -x(x-2)^2 + 4$ is the basic graph $f(x) = x^2$ reflected across the x-axis, shifted 2 units to the right, and shifted 4 units upward. See Figure 4a.

   (b) The graph of $f(x) = -2\sqrt{-x}$ is the basic graph $f(x) = \sqrt{x}$ reflected across the y-axis and vertically stretched by a factor of 2. See Figure 4b.

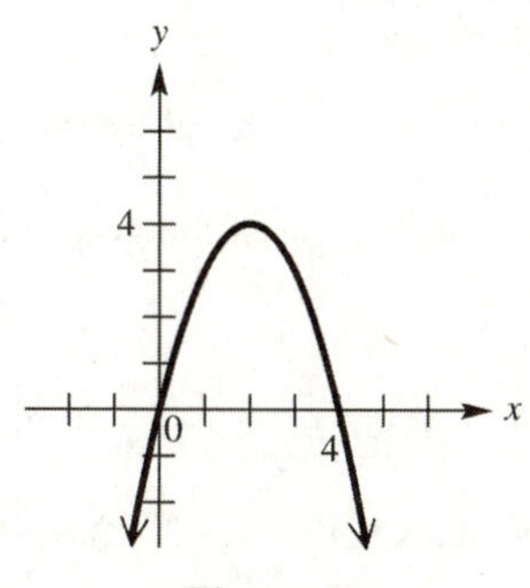

Figure 4a

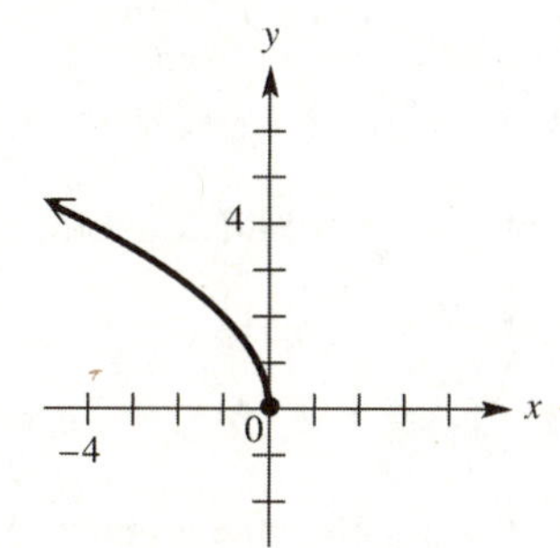

Figure 4b

5. (a) If the graph is symmetric with respect to the y-axis, then $(x, y) \Rightarrow (-x, y)$ therefore $(3, 6) \Rightarrow (-3, 6)$.

(b) If the graph is symmetric with respect to the x-axis, then $(x, y) \Rightarrow (-x, -y)$ therefore $(3,6) \Rightarrow (-3,-6)$.

(c) See Figure 5. We give an actual screen here. The drawing should resemble it.

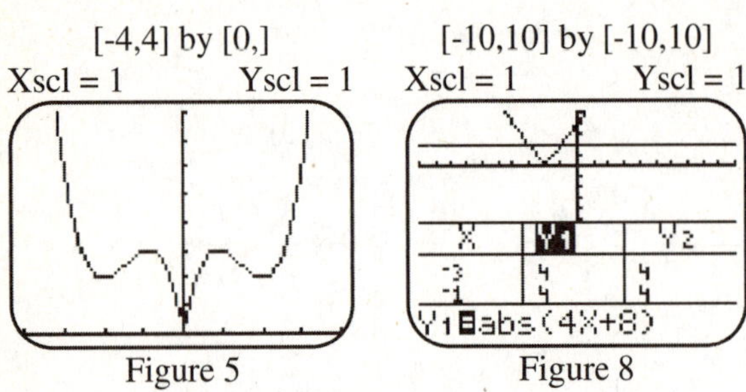

Figure 5 Figure 8

6. (a) Shift the graph of $y = \sqrt[3]{x}$ to the left 2 units, vertically stretch by a factor of 4, and shift 5 units downward.

(b) Graph $y = |x|$ reflected across the x-axis, vertically shrunk by a factor of $\frac{1}{2}$ shifted 3 units to the right, and shifted up 2 units. See Figure 6. From the graph the domain is: $(-\infty, \infty)$; and the range is: $(-\infty, 2]$.

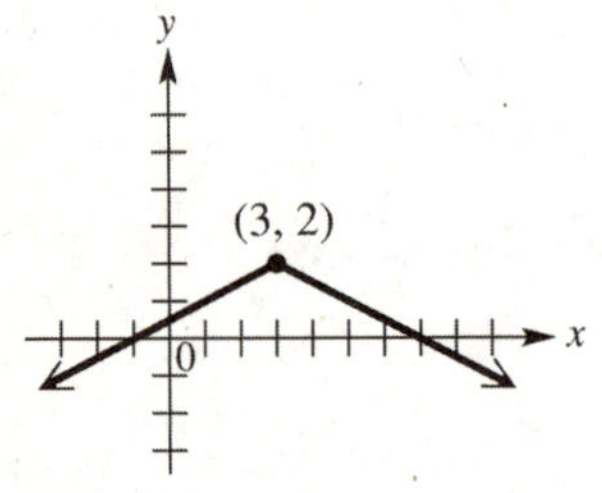

Figure 6

7. (a) From the graph, the function is increasing for the interval: $(-\infty, -3)$

(b) From the graph, the function is decreasing for the interval: $(4, \infty)$

(c) From the graph, the function is constant for the interval: $[-3, 4]$

(d) From the graph, the function is continuous for the intervals: $(-\infty, -3), [-3, 4].(4, \infty)$.

(e) From the graph, the domain is: $(-\infty, \infty)$

(f) From the graph, the range is: $(-\infty, 2)$

8. (a) $|4x+8| = 4 \Rightarrow 4x+8 = 4 \Rightarrow 4x = -4 \Rightarrow x = -1$ or $4x+8 = -4 \Rightarrow 4x = -12 \Rightarrow x = -3$, therefore the solution set is: $\{-3, -1\}$ From the graph, the x-coordinates of the points of intersection of the graphs of Y_1 and Y_2 are -3 and -1 See Figure 8.

(b) $|4x+8| < 4 \Rightarrow -4 < 4x+8 < 4 \Rightarrow -12 < 4x < -4 \Rightarrow -3 < x < -1$, therefore the solution is: $(-3, -1)$ .From the graph, See Figure 8, the graphs of Y_1 lies below the graph of Y_2 for x-values between -3 and -1.

(c) $|4x+8|>4 \Rightarrow 4x+8>4 \Rightarrow 4x>-4 \Rightarrow -1$ or $4x+8<-4 \Rightarrow 4x<-12 \Rightarrow x<-3$ therefore the solution is: $(-\infty,-3)\cup(-1,\infty)$. From the graph, See Figure 8, the graphs of Y_1 lies above the graph of Y_2 for x-values less than -3 or for x-values greater than -1

9. (a) $(f-g)(x)=2x^2-3x+2-(-2x+1) \Rightarrow (f-g)(x)=2x^2-x+1$

(b) $\left(\dfrac{f}{g}\right)(x)=\dfrac{2x^2-3x+2}{-2x+1}$

(c) The domain can be all values for x, except any that make $g(x)=0$. Therefore

$-2x+1\neq 0 \Rightarrow -2x\neq -1 \Rightarrow x\neq \dfrac{1}{2}$ or the interval: $\left(-\infty,\dfrac{1}{2}\right)\cup\left(\dfrac{1}{2},\infty\right)$

(d) $(f\circ g)(x)=f\left[g(x)\right]=2(-2x+1)^2-3(-2x+1)+2=2(4x^2-4x+1)+6x-3+2=$

$8x^2-8x+2+6x-3+2=8x^2-2x+1$

(e) $\dfrac{2(x+h)^2-3(x+h)+2-(2x^2-3x+2)}{h}=\dfrac{2(x^2+2xh+h^2)-3x-3h+2-2x^2+3x-2}{h}=$

$\dfrac{2x^2+4xh+2h^2-3x-3h+2-2x^2+3x-2}{h}=\dfrac{4xh+2h^2-3h}{h}=4x+2h-3$

10. (a) See Figure 10a.

(b) Graph $y_1=(-x^2+3)*(x\leq 1)+\left(\sqrt[3]{x}+2\right)+\left(\sqrt[3]{x}+2\right)*(x>1)$ in the window $[-4.7,4.7]$ by $[-5.1,5.1]$

See Figure 10b.

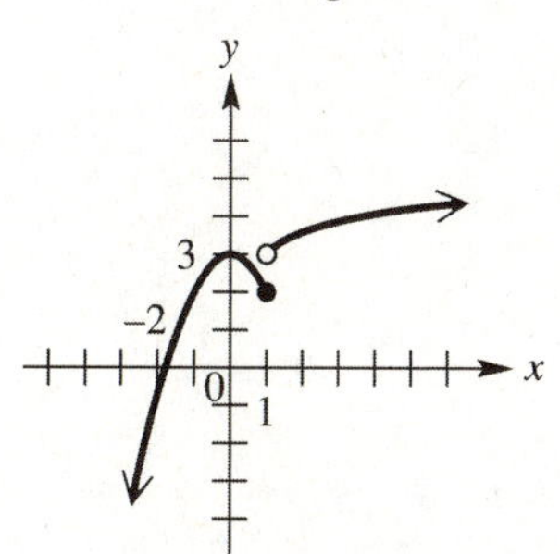

Figure 10a

[-4.7,4.7] by [-5.1,5.1]

Xscl = 1 Yscl = 1

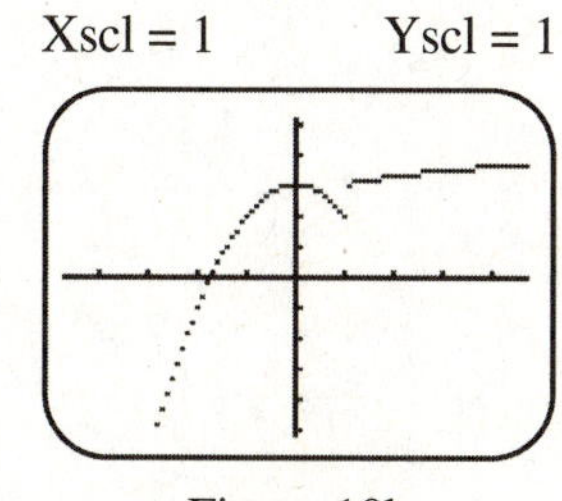

Figure 10b

[0,10] by [0,6]

Xscl = 1 Yscl = 1

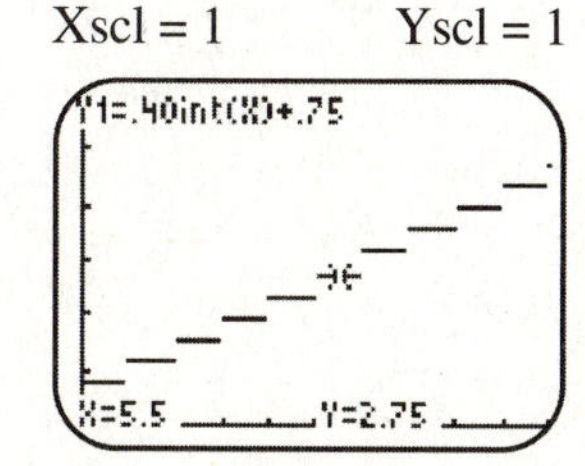

Figure 11

[0,1,000] by [-4,000,4000]

Xscl = 50 Yscl = 500

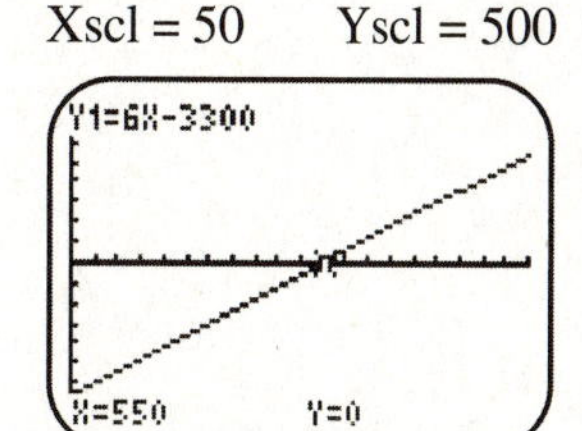

Figure 12

11. (a) See Figure 11.

(b) Set $x=5.5$ then \$2.75 is the cost of a 5.5 minute call. See the display at the bottom of the screen.

12. (a) With an initial set-up cost of \$3300 and a production cost of \$4.50 the function is: $C(x) = 3300 + 4.50x$

(b) With a selling price of \$10.50 the revenue function is: $R(x) = 10.50x$

(c) $P(x) = R(x) - C(x) \Rightarrow P(x) = 10.50x - (3300 + 4.50x) \Rightarrow P(x) = 6x - 3300$

(d) To make a profit $P(x) > 0$, therefore $6x - 3300 > 0 \Rightarrow 6x > 3300 \Rightarrow x > 550$

Tyler needs to sell 551 before he earns a profit.

(e) Graph $y_1 = 6x - 3300$, See Figure 12. The first integer x-value for which $P(x) > 0$ is 551

Chapter 3: Polynomial Functions

3.1: Complex Numbers

1. The complex number $-9i$ can be written $0-9i$.

   (a) The real part is 0. (b) The imaginary part is -9. (c) The number is pure imaginary.

3. The complex number π can be written $\pi+0i$.

   (a) The real part is π. (b) The imaginary part is 0. (c) The number is real.

5. The complex number $3+7i$ is written in standard form.

   (a) The real part is 3. (b) The imaginary part is 7. (c) The number is nonreal complex.

7. The complex number $i\sqrt{7}$ can be written $0+\sqrt{7}i$.

   (a) The real part is 0. (b) The imaginary part is $\sqrt{7}$. (c) The number is pure imaginary.

9. The complex number $\sqrt{-7}$ can be written $0+\sqrt{7}i$.

   (a) The real part is 0. (b) The imaginary part is $\sqrt{7}$. (c) The number is pure imaginary.

11. $3i+5i=(3+5)i=8i$

13. $(-7i)(1+i)=-7i-7i^2=-7i-7(-1)=7-7i$

15. True

17. True

19. False. *Every* real number is a complex number.

21. $\sqrt{-100}=i\sqrt{100}=10i$

23. $-\sqrt{-400}=-i\sqrt{400}=-20i$

25. $-\sqrt{-39}=-i\sqrt{39}$

27. $5+\sqrt{-4}=5+i\sqrt{4}=5+2i$

29. $9-\sqrt{-50}=9-i\sqrt{50}=9-i\sqrt{25.2}=9-5i\sqrt{2}$

31. $i\sqrt{-9}=i^2\sqrt{9}=-3$

33. $\sqrt{-13}\cdot\sqrt{-13}=i\sqrt{13}\cdot i\sqrt{13}=13i^2=-13$

35. $\sqrt{-3}\cdot\sqrt{-8}=i\sqrt{3}\cdot i\sqrt{8}=\sqrt{24}i^2=-2\sqrt{6}$

37. $\dfrac{\sqrt{-30}}{\sqrt{-10}}=\dfrac{i\sqrt{30}}{i\sqrt{10}}=\dfrac{i\sqrt{10\cdot 3}}{i\sqrt{10}}=\dfrac{i\sqrt{10}\cdot\sqrt{3}}{i\sqrt{10}}=\sqrt{3}$

39. $\dfrac{\sqrt{-24}}{\sqrt{8}}=\dfrac{i\sqrt{24}}{\sqrt{8}}=\dfrac{i\sqrt{8\cdot 3}}{\sqrt{8}}=\dfrac{i\sqrt{8}\cdot\sqrt{3}}{\sqrt{8}}=i\sqrt{3}$

41. $\dfrac{\sqrt{-10}}{\sqrt{-40}}=\dfrac{i\sqrt{10}}{i\sqrt{40}}=\dfrac{i\sqrt{10}}{i\sqrt{10\cdot 4}}=\dfrac{i\sqrt{10}}{i\sqrt{10}\cdot\sqrt{4}}=\dfrac{1}{\sqrt{4}}=\dfrac{1}{2}$

43. $\dfrac{\sqrt{-6}\cdot\sqrt{-2}}{\sqrt{3}}=\dfrac{i\sqrt{6}\cdot i\sqrt{2}}{\sqrt{3}}=\dfrac{i^2\sqrt{3}\cdot\sqrt{2}\cdot\sqrt{2}}{\sqrt{3}}=\dfrac{i^2\sqrt{2}\cdot\sqrt{2}}{1}=-\sqrt{4}=-2$

45. $(3+2i)+(4-3i)=(3+4)+(2-3)i=7-i$

47. $(-2+3i)-(-4+3i)=(-2-(-4))+(3-3)i=2+0i=2$

49. $(3-8i)+(2i+4)=(3+4)+(-8+2)i=7-6i$

51. $(2-5i)-(3+4i)-(-2+i)=2-5i-3-4i+2-i=(2-3+2)+(-5i-4i-i)=1-10i$

53. $(-6+5i)+(4-4i)+(2-i)=(-6+4+2)+(5+(-4)+(-1))i=0+0i=0$

55. $(2+i)(3-2i)=6-4i+3i-2i^2=6-i-2(-1)=8-i$

57. $(2+4i)(-1+3i)=-2+6i-4i+12i^2=-2+2i+12(-1)=-14+2i$

59. $(-3+2i)^2=(-3+2i)(-3+2i)=9-6i-6i+4i^2=9-12i+4(-1)=5-12i$

61. $(3+i)(-3-i)=-9-3i-3i-i^2=-9-6i-(-1)=-8-6i$

63. $(2+3i)(2-3i)=4-9i^2=4-9(-1)=13$

65. $\left(\sqrt{6}+i\right)\left(\sqrt{6}-i\right)=6-i^2=6-(-1)=7$

67. $i(3-4i)(3+4i)=i(9-16i^2)=i(9-16(-1))=25i$

69. $3i(2-i)^2=3i(4-4i+i^2)=3i(4-4i+(-1))=3i(3-4i)=9i-12i^2=9i-12(-1)=12+9i$

71. $(2+i)(2-i)(4+3i)=(4-i^2)(4+3i)=(4-(-1))(4+3i)=5(4+3i)=20+15i$

73. $i^5=i^4\cdot i^1=1^1\cdot i=i$

75. $i^{15}=(i^4)^3\cdot i^3=1^3\cdot(-i)=-i$

77. $i^{64}=(i^4)^{16}=1^{16}=1$

79. $i^{-6}=(i^6)^{-1}=[(i^4)^1\cdot i^2]^{-1}=[1^1\cdot(-1)]^{-1}=(-1)^{-1}=-1$

81. $\dfrac{1}{i^9}=i^{-9}=(i^9)^{-1}=[(i^4)^2\cdot i]^{-1}=[1^2\cdot i]^{-1}=(i)^{-1}=\dfrac{1}{i}=\dfrac{1}{i}\cdot\dfrac{i}{i}=\dfrac{i}{i^2}=\dfrac{i}{-1}=-i$

83. $\dfrac{1}{i^{-51}}=i^{51}=(i^4)^{12}\cdot i^3=1^{12}\cdot(-i)=-i$

85. $\dfrac{-1}{-i^{12}}=\dfrac{1}{i^{12}}=\dfrac{1}{(i^4)^3}=\dfrac{1}{1^3}=\dfrac{1}{1}=1$

87. $\left(\dfrac{\sqrt{2}}{2}+\dfrac{\sqrt{2}}{2}i\right)^2=\left(\dfrac{\sqrt{2}}{2}\right)^2+2\left(\dfrac{\sqrt{2}}{2}\right)\left(\dfrac{\sqrt{2}}{2}i\right)+\left(\dfrac{\sqrt{2}}{2}i\right)^2=\dfrac{2}{4}+\dfrac{2}{2}i+\dfrac{2}{4}i^2=\dfrac{1}{2}+i-\dfrac{1}{2}=i$

89. The conjugate of $5-3i$ is $5+3i$.

91. The conjugate of $-18i=0-18i$ is $0+18i=18i$.

93. The conjugate of $-\sqrt{8}=-\sqrt{8}+0i$ is $-\sqrt{8}-0i=-\sqrt{8}$.

95. $\dfrac{3}{-i}=\dfrac{3}{-i}\cdot\dfrac{i}{i}=\dfrac{3i}{-i^2}=\dfrac{3i}{-(-1)}=\dfrac{3i}{1}=3i$

97. $\dfrac{-10}{i}=\dfrac{-10}{i}\cdot\dfrac{-i}{-i}=\dfrac{10i}{-(-1)}=\dfrac{10i}{1}=10i$

99. $\dfrac{1-3i}{1+i}=\dfrac{1-3i}{1+i}\cdot\dfrac{1-i}{1-i}=\dfrac{1-i-3i+3i^2}{1^2-i^2}=\dfrac{-2-4i}{1-(-1)}=\dfrac{-2-4i}{2}=-1-2i$

101. $\dfrac{-3+4i}{2-i}=\dfrac{-3+4i}{2-i}\cdot\dfrac{2+i}{2+i}=\dfrac{-6-3i+8i+4i^2}{2^2-i^2}=\dfrac{-10+5i}{4-(-1)}=\dfrac{-10+5i}{5}=-2+i$

103. $\dfrac{4-3i}{4+3i}=\dfrac{4-3i}{4+3i}\cdot\dfrac{4-3i}{4-3i}=\dfrac{16-12i-12i+9i^2}{4^2-(3i)^2}=\dfrac{7-24i}{16-(-9)}=\dfrac{7-24i}{25}=\dfrac{7}{25}-\dfrac{24}{25}i$

105. The method involves multiplying by 1, the multiplicative identity.

3.2: Quadratic Functions and Graphs

1. Since $a>0$, the parabola opens upward. The vertex is $(4,-3)$. The graph is shown in B.

3. Since $a>0$, the parabola opens upward. The vertex is $(-4,-3)$. The graph is shown in D.

5. (a) $P(x)=x^2-2x-15 \Rightarrow P(x)+15=x^2-2x \Rightarrow P(x)+15+1=x^2-2x+1 \Rightarrow$

   $P(x)+16=(x-1)^2 \Rightarrow P(x)=(x-1)^2-16$

   (b) The vertex is $(1,-16)$.

   (c) See Figure 5a and 5b.

7. (a) $P(x)=-x^2-3x+10 \Rightarrow -P(x)=x^2+3x-10 \Rightarrow -P(x)+10=x^2+3x \Rightarrow$

   $-P(x)+10+\dfrac{9}{4}=x^2+3x+\dfrac{9}{4} \Rightarrow -P(x)+\dfrac{49}{4}=\left(x+\dfrac{3}{2}\right)^2 \Rightarrow -P(x)=\left(x+\dfrac{3}{2}\right)^2-\dfrac{49}{4} \Rightarrow$

   $P(x)=-\left(x+\dfrac{3}{2}\right)^2+\dfrac{49}{4}$

   (b) The vertex is $\left(-\dfrac{3}{2},\dfrac{49}{4}\right)$ or $(-1.5, 12.25)$.

   (c) See Figure 7a and 7b.

9. (a) $P(x)=x^2-6x \Rightarrow P(x)+9=x^2-6x+9 \Rightarrow P(x)+9=(x-3)^2 \Rightarrow P(x)=(x-3)^2-9$

   (b) The vertex is $(3,-9)$.

   (c) See Figure 9a and 9b.

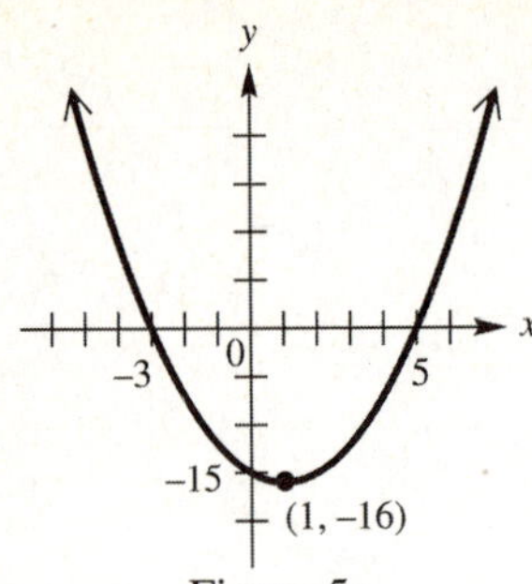

Figure 5a

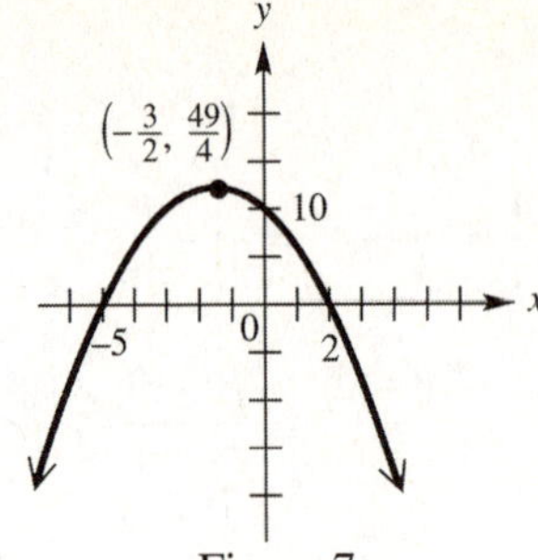

Figure 7a

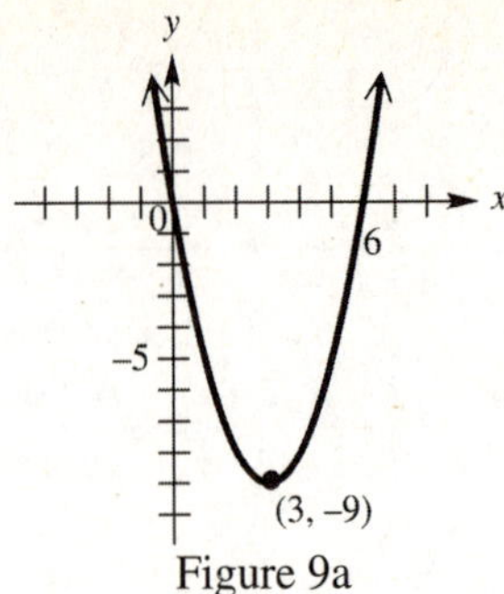

Figure 9a

[-10,10] by [-20,10]
Xscl = 1 Yscl = 5

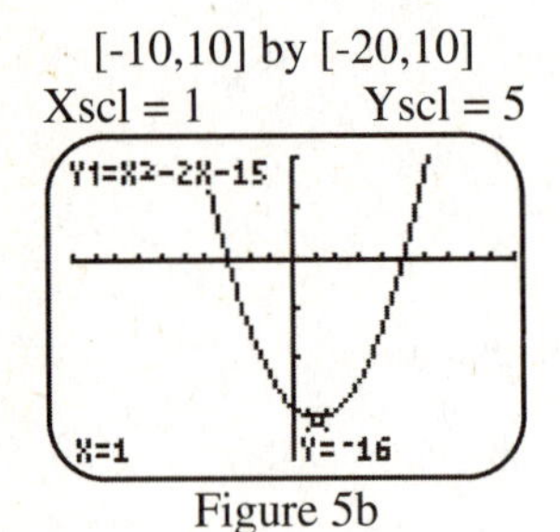

Figure 5b

[-10,10] by [-5,15]
Xscl = 1 Yscl = 1

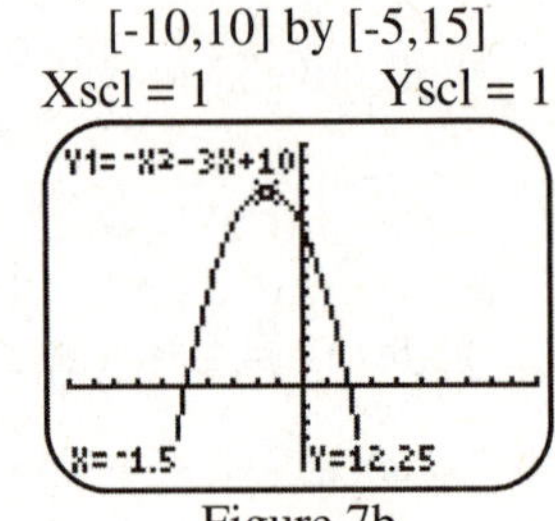

Figure 7b

[-10,10] by [-15,5]
Xscl = 1 Yscl = 1

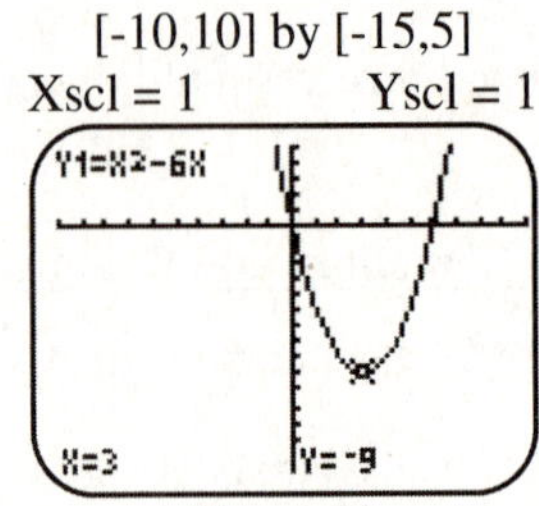

Figure 9b

11. (a). $P(x)=2x^2-2x-24 \Rightarrow \dfrac{P(x)}{2}=x^2-x-12 \Rightarrow \dfrac{P(x)}{2}+12=x^2-x \Rightarrow$

$$\frac{P(x)}{2}+12+\frac{1}{4}=x^2-x+\frac{1}{4} \Rightarrow \frac{P(x)}{2}+\frac{49}{4}=\left(x-\frac{1}{2}\right)^2 \Rightarrow \frac{P(x)}{2}=\left(x-\frac{1}{2}\right)^2-\frac{49}{4} \Rightarrow P(x)=2\left(x-\frac{1}{2}\right)^2-\frac{49}{2}$$

(b) The vertex is $\left(\dfrac{1}{2},-\dfrac{49}{2}\right)$ or $(0.5,-24.5)$.

(c) See Figure 11a and 11b.

13. (a) $P(x)=-2x^2+6x \Rightarrow \dfrac{P(x)}{-2}=x^2-3x \Rightarrow \dfrac{P(x)}{-2}+\dfrac{9}{4}=x^2-3x+\dfrac{9}{4} \Rightarrow$

$$\frac{P(x)}{-2}+\frac{9}{4}=\left(x-\frac{3}{2}\right)^2 \Rightarrow \frac{P(x)}{-2}=\left(x-\frac{3}{2}\right)^2-\frac{9}{4} \Rightarrow P(x)=-2\left(x-\frac{3}{2}\right)^2+\frac{9}{2}$$

(b) The vertex is $\left(\dfrac{3}{2},\dfrac{9}{2}\right)$ or $(1.5, 4.5)$.

(c) See Figure 13a and 13b.

15. (a) $P(x)=3x^2+4x-1 \Rightarrow \dfrac{P(x)}{3}=x^2+\dfrac{4}{3}x-\dfrac{1}{3} \Rightarrow \dfrac{P(x)}{3}+\dfrac{1}{3}=x^2+\dfrac{4}{3}x \Rightarrow$

$$\frac{P(x)}{3}+\frac{1}{3}+\frac{4}{9}=x^2+\frac{4}{3}x+\frac{4}{9} \Rightarrow \frac{P(x)}{3}+\frac{7}{9}=\left(x+\frac{2}{3}\right)^2 \Rightarrow \frac{P(x)}{3}=\left(x+\frac{2}{3}\right)^2-\frac{7}{9} \Rightarrow P(x)=3\left(x+\frac{2}{3}\right)^2-\frac{7}{3}$$

(b). The vertex is $\left(-\dfrac{2}{3},-\dfrac{7}{3}\right)$ or $(-0.67,-2.33)$.

(c) See Figure 15a and 15 b.

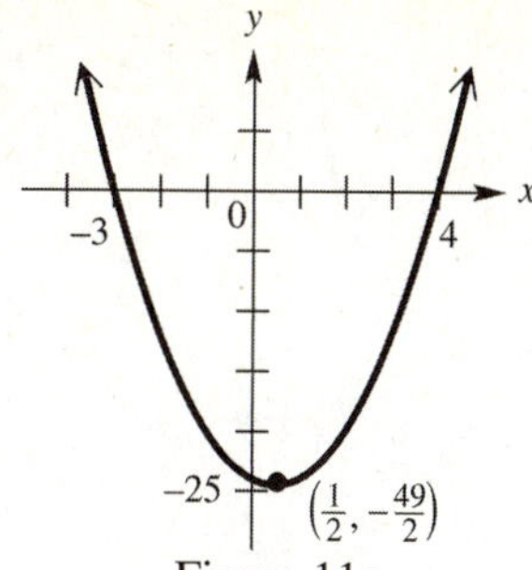

Figure 11a

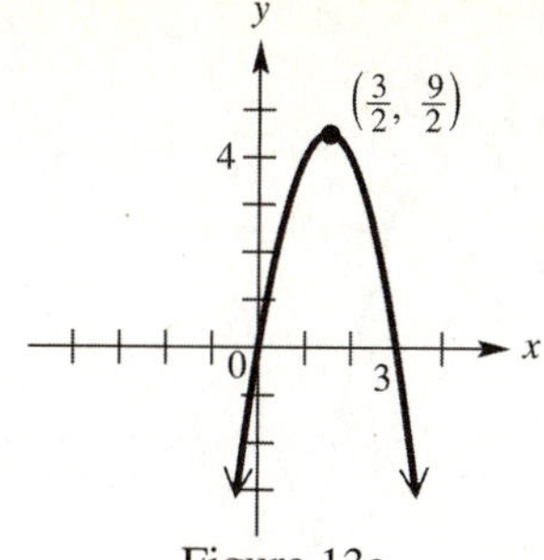

Figure 13a

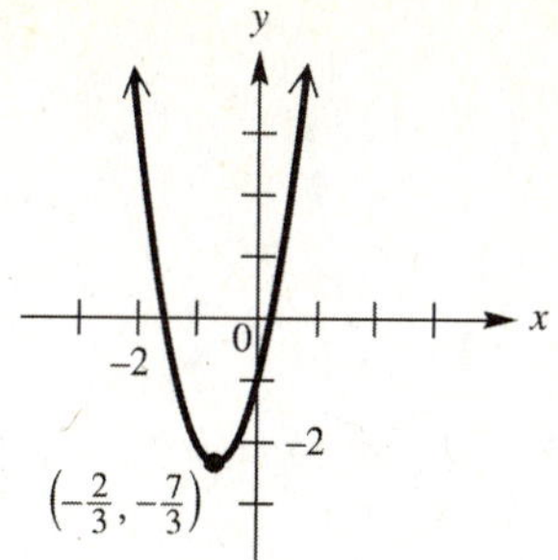

Figure 15a

[-5,5] by [-5,5]
Xscl = 1 Yscl = 1

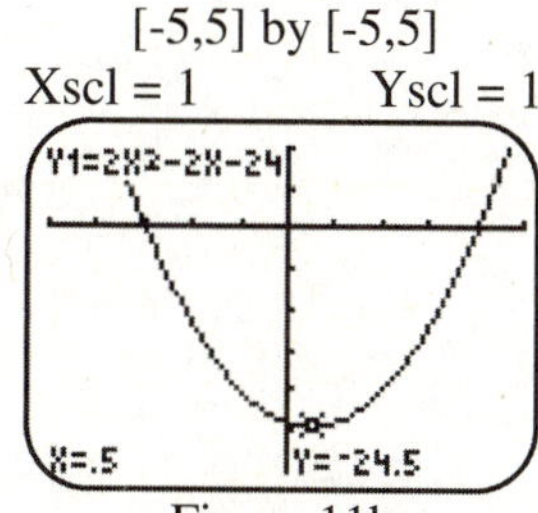

Figure 11b

[-5,5] by [-10,10]
Xscl = 1 Yscl = 1

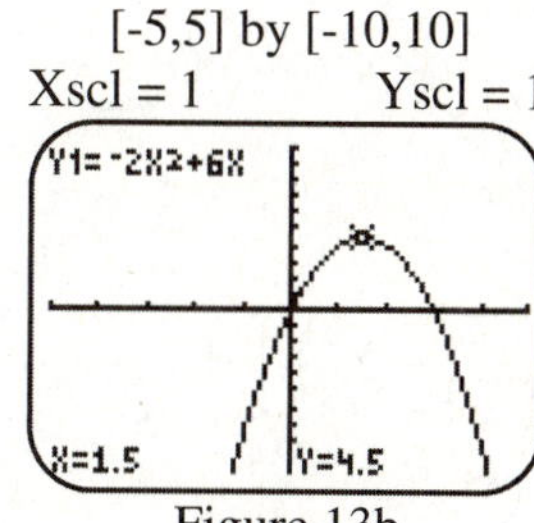

Figure 13b

[-5,5] by [-5,15]
Xscl = 1 Yscl = 1

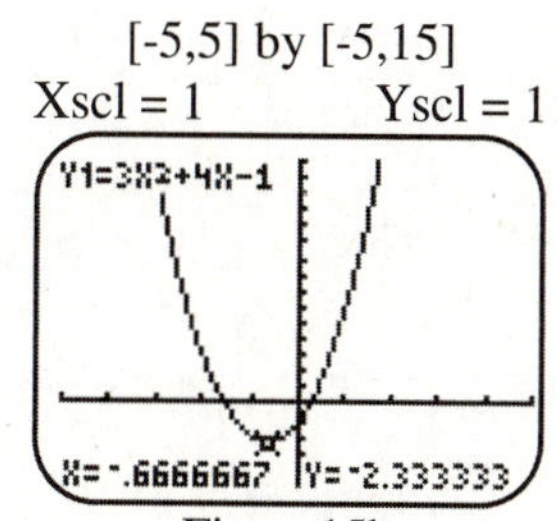

Figure15b

17. (a) Since $a > 0$, the parabola opens upward. The vertex is (4, −2). The graph is given in D.

    (b) Since $a > 0$, the parabola opens upward. The vertex is (2, −4). The graph is given in B.

    (c) Since $a < 0$, the parabola opens downward. The vertex is (4, −2). The graph is given in C.

    (d) Since $a < 0$, the parabola opens downward. The vertex is (2, −4). The graph is given in A.

19. (a) $(2,0)$ (b) $D:(-\infty,\infty)$, $R:[0,\infty)$ (c) $x=2$ (d) $[2,\infty)$ (e) $(-\infty,2]$ (f) Min.: $P(2)=0$

21. (a) $(-3,-4)$ (b) $D:(-\infty,\infty)$, $R:[-4,\infty)$ (c) $x=-3$ (d) $[-3,\infty)$ (e) $(-\infty,-3]$ (f) Min.: $f(-3)=-4$

23. (a) $(-3,2)$ (b) $D:(-\infty,\infty)$, $R:(-\infty,2]$ (c) $x=-3$ (d) $(-\infty,-3]$ (e) $[-3,\infty)$ (f) Max.: $P(-3)=2$

25. (a) $x=-\frac{b}{2a}=-\frac{-10}{2(1)}=5;\ y=P(5)=(5)^2-10(5)+21=-4\Rightarrow$ Vertex: (5, -4)

    (b) See Figure 25.

27. (a) $x=-\frac{b}{2a}=-\frac{4}{2(-1)}=2;\ y=-(2)^2+4(2)-2=2\Rightarrow$ Vertex: (2, 2)

    (b) See Figure 27.

29. (a) $x=-\frac{b}{2a}=-\frac{(-4)}{2(2)}=1;\ y=f(1)=2(1)^2-4(1)+5=3\Rightarrow$ Vertex: (1, 3)

    (b) See Figure 29.

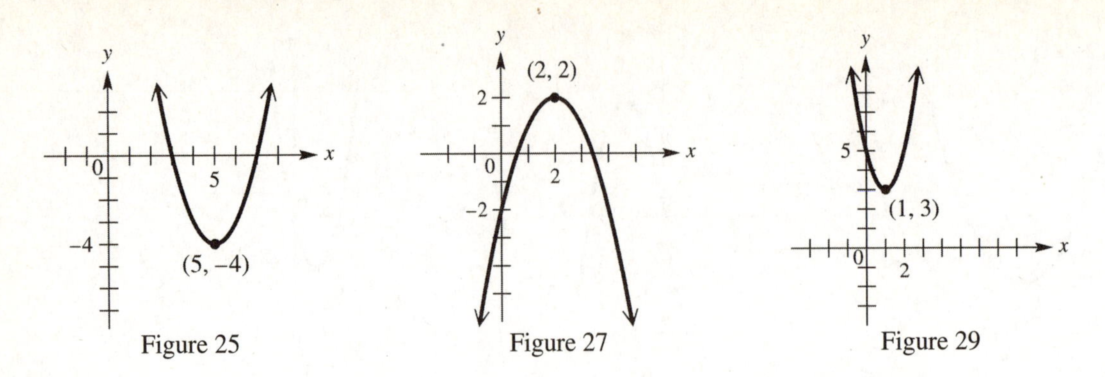

Figure 25 Figure 27 Figure 29

31. (a) $x=-\dfrac{b}{2a}=-\dfrac{24}{2(-3)}=4;\ y=f(4)=-3(4)^2+24(4)-46=2\Rightarrow$ Vertex: (4, 2)

    (b) See Figure 31.

33. (a) $x=-\dfrac{b}{2a}=-\dfrac{-2}{2(2)}=\dfrac{1}{2};\ y=f\left(\dfrac{1}{2}\right)=2\left(\dfrac{1}{2}\right)^2-2\left(\dfrac{1}{2}\right)+1=\dfrac{1}{2}\Rightarrow$ Vertex: $\left(\dfrac{1}{2},\dfrac{1}{2}\right)$.

    (b) See Figure 33.

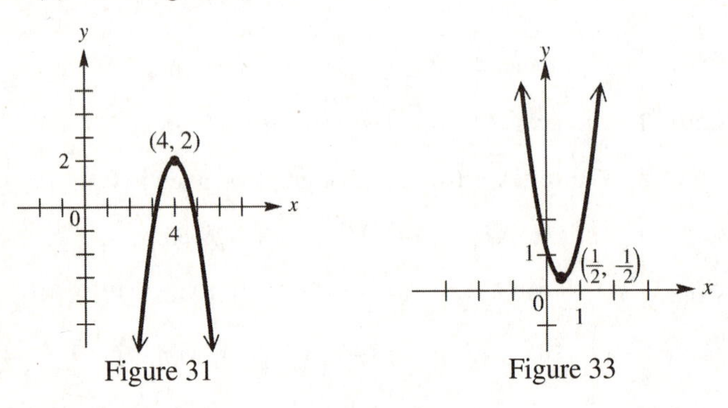

Figure 31 Figure 33

35. The graph is shown in Figure 35.

    (a) The vertex is approximately (2.71, 5.20) (b) The x-intercepts are approximately −1.33 and 6.74.

37. The graph is shown in Figure 37.

    (a) The vertex is approximately (1.12, 0.56). (b) There are no x-intercepts.

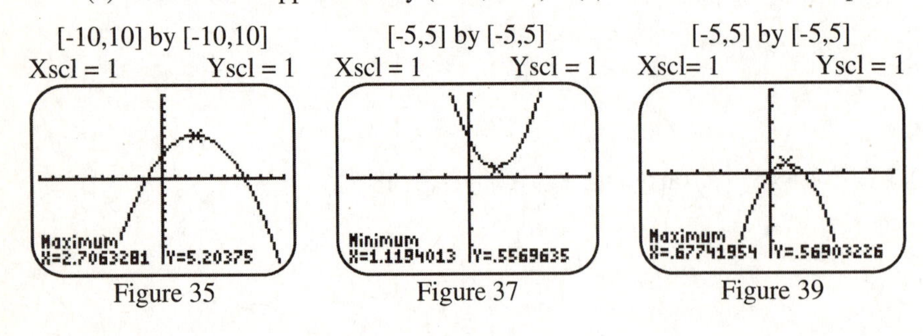

Figure 35 Figure 37 Figure 39

39. The graph is shown in Figure 39.

    (a) The vertex is approximately (0.68, 0.57). (b) The x-intercepts are approximately 0 and 1.35.

41. The minimum value of 3 is found at the vertex.

43. The graph would not intersect the line $y=1$. There are no solutions to $f(x)=1$.

45. (a) From the symmetry of the y-values in the table, the vertex is $(4, -12)$.

(b) Since all other y-values are larger than –12, the vertex is a minimum point.

(c) The minimum value of the function is –12 and is located at the vertex.

(d) Since the function is quadratic with a minimum value of –12, the range is $[-12, \infty)$.

47. (a) From the symmetry of the y-values in the table, the vertex is $(1.5, 2)$.

(b) Since all other y-values are smaller than 2, the vertex is a maximum point.

(c) The maximum value of the function is 2 and is located at the vertex.

(d) Since the function is quadratic with a maximum value of 2, the range is $(-\infty, 2]$.

49. Since the data are in the shape of a parabola that opens downward, a quadratic model with $a < 0$ is appropriate.

51. Since the data are in the shape of a parabola that opens upward, a quadratic model with $a > 0$ is appropriate.

53. Since the data are in the shape of a line that rises from left to right, a linear model with $m > 0$ is appropriate.

55. (a) Since the parabola opens upward and the vertex is (4,90), the function decreases from 0 to 4 and increases from 4 to 8. Therefore, the heart rate decreases during the first 4 minutes and increases during the next 4 minutes.

(b) Since parabola opens upward the minimum occurs at the vertex (4, 90). Therefore, the minimum heart rate is 90 bpm after 4 minutes.

57. (a) The data are not linear since the data increase and then decrease.

(b) Using coordinate (2,120), $f(x) = a(x-2)^2 + 120$. To find , use $(0,84) \Rightarrow$

$84 = a(0-2)^2 + 120 \Rightarrow -36 = a(4) \Rightarrow a = -9$; $f(x) = -9(x-2)^2 + 120$.

(c) $D = \{x \mid 0 \le x \le 4\}$

59. (a) The value of t cannot be negative because t represents elapsed time after the rock is launched.

(b) The original height of the rock is $s_0 = 0$ which represents ground level.

(c) Since $v_0 = 90$ and $s_0 = 0$, $s(t) = -16t^2 + v_0 t + s_0 \Rightarrow s(t) = -16t^2 + 90t$.

(d) $s(1.5) = -16(1.5)^2 + 90(1.5) = 99$ feet

(e) $x = -\dfrac{b}{2a} = -\dfrac{90}{2(-16)} = 2.8125$; $y \approx s(2.8125) = -16(2.8125)^2 + 90(2.8125) = 126.5625$. The vertex is (2.8125, 126.5625). The rock reaches a maximum height of 126.5625 feet after 2.8125 seconds. A graph of $y = -16x^2 + 90x$ (not shown) also gives a vertex of (2.8125, 126.5625).

(f) A graph of $y = -16x^2 + 90x$ (not shown) has x-intercepts at (0, 0) and (5.625, 0). The rock will hit the ground after 5.625 seconds.

61. (a) The graphs of $-16x^2 + 150x$ and $y_2 = 355$ are shown in Figure 61a. The graphs do not intersect, which indicates that the ball does not reach a height of 355 feet.

(b) The graphs of $y_1 = -16x^2 + 250x + 30$ and $y_2 = 355$ are shown in Figure 61b. The graphs intersect when $x \approx 1.43$ and $x \approx 14.19$ which indicates that the ball is 355 feet high at about 1.4 and 14.2 seconds.

[0,10] by [300,400]
Xscl = 1 Yscl = 10

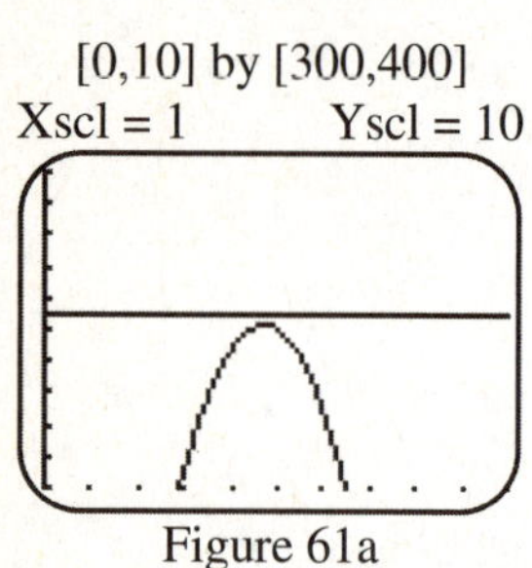

Figure 61a

[0,20] by [0,1200]
Xscl = 5 Yscl = 100

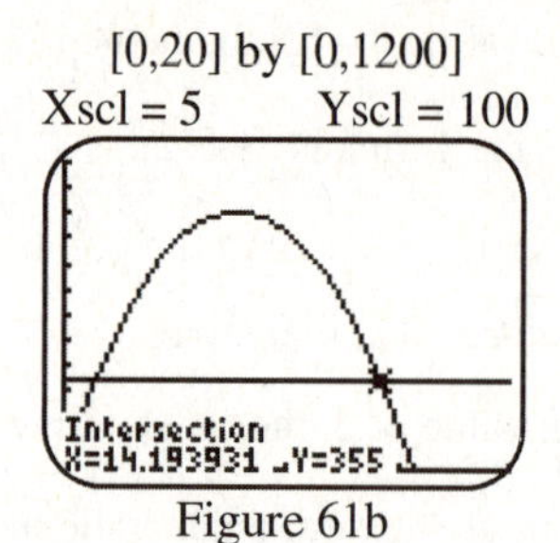

Figure 61b

63. Let $h = -1$, $k = -4$, $x = 5$ and $P(x) = 104$ in $P(x) = a(x-h)^2 + k$ and determine the value of a.

$104 = a(5+1)^2 - 4 \Rightarrow 104 = a(6)^2 - 4 \Rightarrow 108 = 36a \Rightarrow a = 3$

$P(x) = 3(x+1)^2 - 4 \Rightarrow P(x) = 3(x^2 + 2x + 1) - 4 \Rightarrow P(x) = 3x^2 + 6x - 1$

65. Let $h = 8$, $k = 3$, $x = 10$ and $P(x) = 5$ in $P(x) = a(x-h)^2 + k$ and determine the value of a.

$5 = a(10-8)^2 + 3 \Rightarrow 5 = a(2)^2 + 3 \Rightarrow 2 = 4a \Rightarrow a = 0.5$. $P(x) = 0.5(x-8)^2 + 3 \Rightarrow$

$P(x) = 0.5(x^2 - 16x + 64) + 3 \Rightarrow P(x) = 0.5x^2 - 8x + 35$ or $P(x) = \frac{1}{2}x^2 - 8x + 35$

67. Let $h = -4$, $k = -2$, $x = 2$ and $P(x) = -26$ in $P(x) = a(x-h)^2 + k$ and determine the value of a.

$-26 = a(2+4)^2 - 2 \Rightarrow -26 = a(6)^2 - 2 \Rightarrow -24 = 36a \Rightarrow a = -\frac{2}{3}$

$P(x) = -\frac{2}{3}(x+4)^2 - 2 \Rightarrow P(x) = -\frac{2}{3}(x^2 + 8x + 16) - 2 \Rightarrow P(x) = -\frac{2}{3}x^2 - \frac{16}{3}x - \frac{38}{3}$

69. See Figure 69.

71. See Figure 71.

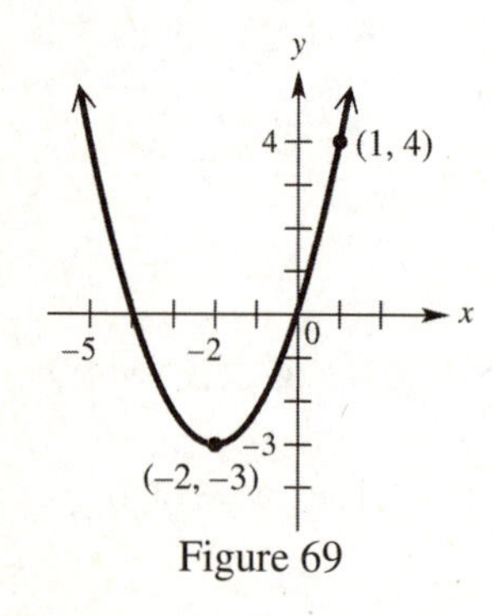

Figure 69

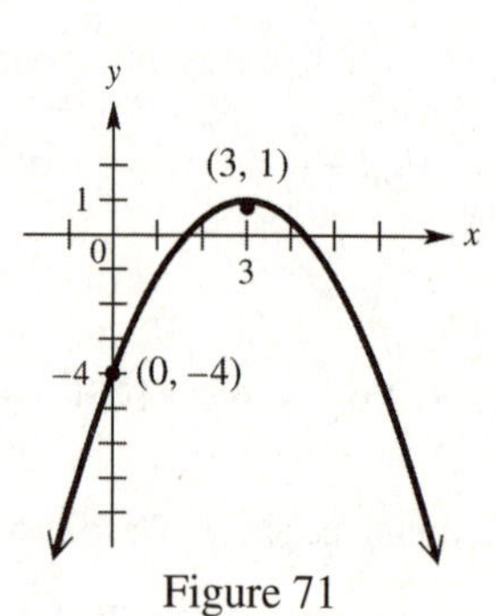

Figure 71

3.3: Quadratic Equations and Inequalities

1. $x^2 = 4 \Rightarrow \sqrt{x^2} = \sqrt{4} \Rightarrow x = \pm 2$; therefore, G.

3. $x^2 + 2 = 0 \Rightarrow x^2 = -2 \Rightarrow \sqrt{x^2} = \sqrt{-2} \Rightarrow x = \pm i\sqrt{2}$; therefore, C.

5. $x^2 = -8 \Rightarrow \sqrt{x^2} = \sqrt{-8} \Rightarrow x = \pm 2i\sqrt{2}$; therefore, H.

7. $x-2=0 \Rightarrow x=2$; therefore, D.

9. Equation D, which is in the form $ab=0$, is set up for the zero product property. Either

$3x+1=0 \Rightarrow x=-\frac{1}{3}$ or $x-7=0 \Rightarrow x=7$. The solution set is $\left\{-\frac{1}{3}, 7\right\}$.

11. Equation C, which has an x^2 coefficient of 1, does not require step 1 of the method of completing the square.

$x^2+x=12 \Rightarrow x^2+x+\frac{1}{4}=12+\frac{1}{4} \Rightarrow \left(x+\frac{1}{2}\right)^2=\frac{49}{4} \Rightarrow x+\frac{1}{2}=\pm\frac{7}{2} \Rightarrow x=-\frac{1}{2}\pm\frac{7}{2}$. The solution set is $\{-4, 3\}$.

13. $x^2=16 \Rightarrow x=\pm\sqrt{16} \Rightarrow x=\{\pm 4\}$; For graphical support, show that the graphs of $y_1=x^2$ and $y_2=16$ intersect when $x=\pm 4$.

15. $2x^2=90 \Rightarrow x^2=45 \Rightarrow x=\pm\sqrt{45} \Rightarrow x=\pm 3\sqrt{5}$; For graphical support, show that the graphs of $y_1=2x^2$ and $y_2=90$ intersect when $x=\pm 3\sqrt{5} \approx \pm 6.7$.

17. $x^2=-16 \Rightarrow x=\pm i\sqrt{16} \Rightarrow x=\pm 4i$; For graphical support, show that the graphs of $y_1=x^2$ and $y_2=-16$ do not intersect; therefore, no real solutions.

19. $x^2=-18 \Rightarrow x=\pm i\sqrt{18} \Rightarrow x=\pm 3i\sqrt{2}$; For graphical support, show that the graphs of $y_1=x^2$ and $y_2=-18$ do not intersect; therefore, no real solutions.

21. $(3x-1)^2=12 \Rightarrow 3x-1=\pm\sqrt{12} \Rightarrow 3x-1=\pm 2\sqrt{3} \Rightarrow 3x=\pm 2\sqrt{3}+1 \Rightarrow x=\frac{1\pm 2\sqrt{3}}{3}$. For graphical support, show that the graphs of $y_1=(3x-1)^2$ and $y_2=12$ intersect when $x=\frac{1\pm 2\sqrt{3}}{3}$; $x \approx -0.8$ or 1.5.

23. $(5x-3)^2=-3 \Rightarrow 5x-3=\pm\sqrt{-3} \Rightarrow 5x-3=\pm i\sqrt{3} \Rightarrow 5x=\pm i\sqrt{3}+3 \Rightarrow x=\frac{3}{5}\pm\frac{\sqrt{3}}{5}i$. For graphical support, show that the graphs of $y_1=(5x-3)^2$ and $y_2=-3$ do not intersect; therefore, there are no real solutions.

25. $x^2=2x+24 \Rightarrow x^2-2x-24=0 \Rightarrow (x-6)(x+4)=0 \Rightarrow x-6=0$ or $x+4=0 \Rightarrow x=-4$ or 6. For graphical support, show that the graph of $y_1=x^2-2x-24=0$ has x-intercepts $x=-4$ and 6.

27. $3x^2-2x=0 \Rightarrow x(3x-2)=0 \Rightarrow x=0$ or $3x-2=0 \Rightarrow x=0$ or $\frac{2}{3}$. For graphical support, show that the graph of $y_1=3x^2-2x$ has x-intercepts $x=0$ and $\frac{2}{3}$.

29. $x(14x+1)=3 \Rightarrow 14x^2+x-3=0 \Rightarrow (2x+1)(7x-3)=0 \Rightarrow 2x+1=0$ or $7x-3=0 \Rightarrow x=-\frac{1}{2}$ or $\frac{3}{7}$. For graphical support, show that the graph of $y_1=14x^2+x-3$ has x-intercepts $x=-\frac{1}{2}$ and $\frac{3}{7}$.

31. $-4+9x-2x^2=0 \Rightarrow (-4+x)(1-2x)=0 \Rightarrow -4+x=0$ or $1-2x=0 \Rightarrow x=\frac{1}{2}$ or 4. For graphical support, show that the graph of $y_1=-4+9x-2x^2$ has x-intercepts $x=\frac{1}{2}$ and 4.

33. $\frac{1}{3}x^2-\frac{1}{6}x=24 \Rightarrow \frac{1}{3}x^2+\frac{1}{3}x-24=0 \Rightarrow x^2-x-72=0 \Rightarrow (x-9)(x+8)=0 \Rightarrow x-9=0$ or $x+8=0 \Rightarrow x=-8$ or 9. For graphical support, show that the graph of $y_1=\frac{1}{3}x^2-\frac{1}{3}x-24$ has x-intercepts $x=-8$ and 9.

35. $(x+2)(x-1)=7x+5 \Rightarrow x^2+x-2=7x+5 \Rightarrow x^2-6x-7=0 \Rightarrow (x-7)(x+1)=0 \Rightarrow x-7=0$ or $x+1=0 \Rightarrow x=-1$ or 7. For graphical support, show that the graph of $y_1=x^2-6x-7$ has x-intercepts $x=-1$ and 7.

37. Use the quadratic formula to solve $x^2-2x-4=0$; therefore, $a=1,\ b=-2$ and $c=-4$

$$x=\frac{-(-2)\pm\sqrt{(-2)^2-4(1)(-4)}}{2(1)}=\frac{2\pm\sqrt{4+16}}{2}=\frac{2\pm\sqrt{20}}{2}=\frac{2\pm2\sqrt{5}}{2}=\frac{2\left(1\pm\sqrt{5}\right)}{2}\Rightarrow x=1\pm\sqrt{5}.$$ For graphical support, show that the graph of $y_1=x^2-2x-4$ has x-intercepts $x=1\pm\sqrt{5}$; $x\approx-1.24$ and 3.24.

39. Use the quadratic formula to solve $y_1=2x^2+2x=-1 \Rightarrow 2x^2+2x+1=0$; therefore, $a=2,\ b=2$ and $c=1$

$$x=\frac{-2\pm\sqrt{(2)^2-4(2)(1)}}{2(2)}=\frac{-8\pm\sqrt{4-8}}{4}=\frac{-2\pm\sqrt{-4}}{4}=\frac{-2\pm2i}{4}=\frac{2(-1\pm i)}{4}\Rightarrow x=-\frac{1}{2}\pm\frac{1}{2}i.$$ For graphical support, show that the graph of $y_1=2x^2+2x+1$ has no x-intercepts; therefore, there are no real solutions.

41. Use the quadratic formula to solve $x(x-1)=1 \Rightarrow x^2-x-1=0$; therefore, $a=1,\ b=-1$ and $c=-1$

$$x=\frac{-(-1)\pm\sqrt{(-1)^2-4(1)(-1)}}{2(1)}=\frac{1\pm\sqrt{1+4}}{2}\Rightarrow x=\frac{1\pm\sqrt{5}}{2}.$$ For graphical support, show that the graph of $y_1=x^2-x-1$ has x-intercepts $x=\frac{1\pm\sqrt{5}}{2}$; $x\approx-0.62$ and 1.62.

43. Use the quadratic formula to solve $x^2-5x=x-7 \Rightarrow x^2-6x+7=0$; therefore, $a=1,\ b=-6$, and $c=7$.

$$x=\frac{-(-6)\pm\sqrt{(-6)^2-4(1)(7)}}{2(1)}=\frac{6\pm\sqrt{36-28}}{2}=\frac{6\pm\sqrt{8}}{2}=\frac{6\pm2\sqrt{2}}{2}=\frac{2\left(3\pm\sqrt{2}\right)}{2}\Rightarrow x=3\pm\sqrt{2}.$$ For graphical support, show that the graph of $y_1=x^2-6x+7$ has x-intercepts $x=3\pm\sqrt{2}$; $x\approx1.59$ and 4.41.

45. Use the quadratic formula to solve $4x^2-12x=-11 \Rightarrow 4x^2-12x+11=0$; therefore, $a=4,\ b=-12$ and $c=11$. $$x=\frac{-(-12)\pm\sqrt{(-12)^2-4(4)(11)}}{2(4)}=\frac{12\pm\sqrt{144-176}}{8}=\frac{12\pm\sqrt{-32}}{8}=\frac{12\pm4i\sqrt{2}}{8}=\frac{4\left(3\pm i\sqrt{2}\right)}{8}\Rightarrow$$

$x=\frac{3}{2}\pm\frac{\sqrt{2}}{2}i$. For graphical support, show that the graph of $y_1=4x^2-12x+11$ has no x-intercepts; therefore, there are no real solutions.

47. Use the quadratic formula to solve $\frac{1}{3}x^2+\frac{1}{4}x-3=0 \Rightarrow 12\left(\frac{1}{3}x^2+\frac{1}{4}x-3=0\right) \Rightarrow 4x^2+3x-36=0$; therefore, $a=4, b=3$ and $c=-36$. $x=\frac{-3\pm\sqrt{(3)^2-4(4)(-36)}}{2(4)}=\frac{-3\pm\sqrt{9+576}}{8}=\frac{-3\pm\sqrt{585}}{8} \Rightarrow x=\frac{-3\pm3\sqrt{65}}{8}$. For graphical support, show that the graph of $y_1=4x^2+3x-36$ has x-intercepts $x=\frac{-3\pm3\sqrt{65}}{8}$; $x\approx-3.4$ and 2.6.

49. Use the quadratic formula to solve $(3-x)^2=25 \Rightarrow 9-6x+x^2=25 \Rightarrow x^2-6x-16=0$; therefore, $a=1, b=-6$ and $c=-16$. $x=\frac{-(-6)\pm\sqrt{(-6)^2-4(1)(-16)}}{2(1)}=\frac{6\pm\sqrt{36+64}}{2}=\frac{6\pm\sqrt{100}}{2} \Rightarrow \frac{6\pm10}{2} \Rightarrow$ $x=-2$ or 8. For graphical support, show that the graph of $y_1=x^2-6x-16$ has x-intercepts $x=-2$ and 8.

51. Use the quadratic formula to solve $2x^2-4x=1 \Rightarrow 2x^2-4x-1=0$; therefore $a=2$, $b=-4$ and $c=-1$. $x=\frac{-(-4)\pm\sqrt{(-4)^2-4(2)(-1)}}{2(2)}=\frac{4\pm\sqrt{16+8}}{4}=\frac{4\pm\sqrt{24}}{4} \Rightarrow \frac{4\pm2\sqrt{6}}{4}=\frac{2\left(2\pm\sqrt{6}\right)}{4} \Rightarrow x=\frac{2\pm\sqrt{6}}{2}$. For graphical support, show that the graph of $y_1=2x^2-4x-1$ has x-intercepts $x=\frac{2\pm\sqrt{6}}{2}$; $x\approx-0.2$ and 2.2.

53. Use the quadratic formula to solve $x^2=-1-x \Rightarrow x^2+x+1=0$; therefore, $a=1$, $b=1$ and $c=1$. $x=\frac{-1\pm\sqrt{(1)^2-4(1)(1)}}{2(1)}=\frac{-1\pm\sqrt{1-4}}{2}=\frac{-1\pm\sqrt{-3}}{2} \Rightarrow x=-\frac{1}{2}\pm\frac{\sqrt{3}}{2}i$. For graphical support, show that the graph of $y_1=x^2+x+1$ has no x-intercepts; therefore, there are no real solutions.

55. Use the quadratic formula to solve $4x^2-20x+25=0$; therefore, $a=4$, $b=-20$ and $c=25$. $x=\frac{-(-20)\pm\sqrt{(-20)^2-4(4)(25)}}{2(4)}=\frac{20\pm\sqrt{400-400}}{8}=\frac{20\pm0}{8} \Rightarrow x=\frac{20}{8}=\frac{5}{2}$. For graphical support, show that the graph of $y_1=4x^2-20x+25$ has x-intercept $x=\frac{5}{2}=2.5$.

57. Use the quadratic formula to solve $-3x^2+4x+4=0$; therefore, $a=-3$, $b=4$ and $c=4$. $x=\frac{-4\pm\sqrt{(4)^2-4(-3)(4)}}{2(-3)}=\frac{-4\pm\sqrt{16+48}}{-6}=\frac{-4\pm\sqrt{64}}{-6}=\frac{-4\pm8}{-6}=\frac{-2(2\pm4)}{-6}=\frac{2\pm4}{3} \Rightarrow x=-\frac{2}{3}$ or 2. For graphical support, show that the graph of $y_1=-3x^2+4x+4$ has x-intercepts $x=-\frac{2}{3}$ or 2; $x\approx-0.7$ and 2.

59. Use the quadratic formula to solve $(x+5)(x-6)=(2x-1)(x-4) \Rightarrow x^2-x-30=2x^2-9x+4 \Rightarrow$

$x^2-8x+34=0$; therefore, $a=1$, $b=-8$, and $c=34$. $x=\dfrac{-(-8)\pm\sqrt{(-8)^2-4(1)(34)}}{2(1)}=\dfrac{8\pm\sqrt{64-136}}{2}=$

$\dfrac{8\pm\sqrt{-72}}{2}=\dfrac{8\pm 6i\sqrt{2}}{2}=\dfrac{2(4\pm 3i\sqrt{2})}{2} \Rightarrow x=4\pm 3i\sqrt{2}$. For graphical support, show that the graph of

$y_1=2x^2-9x+4$ has no x-intercepts; therefore, there are no real solutions.

61. For the equation $x^2+8x+16=0$; $a=1$, $b=8$, and $c=16$; therefore, the discriminant is

$(8)^2-4(1)(16)=64-64=0$. Because the discriminant is 0, there is 1 real solution. Since a, b, and c are nonzero integers and the discriminant is a square of an integer the solution is rational.

63. For the equation $4x^2=6x+3 \Rightarrow 4x^2-6x-3=0$; $a=4$, $b=-6$, and $c=-3$; therefore, the discriminant is

$(-6)^2-4(4)(-3)=36+48=84$. Because the discriminant is positive, there are 2 real solutions. Since a, b, and c are nonzero integers and the discriminant is not the square of an integer, the solutions are irrational.

65. For the equation $9x^2+11x+4=0; a=9$, $b=11$, and $c=4$; therefore, the discriminant is

$(11)^2-4(9)(4)=121-144=-23$. Because the discriminant is negative, there are no real solutions.

67. If $x=4$ or 5, then $(x-4)(x-5)=0 \Rightarrow x^2-9x+20=0$; and $a=1$, $b=-9$, and $c=20$.

69. If $x=1-\sqrt{2}$ or $1+\sqrt{2}$, then $\left(x-\left(1-\sqrt{2}\right)\right)\left(x-\left(1+\sqrt{2}\right)\right)=0 \Rightarrow \left(x-1+\sqrt{2}\right)\left(x-1-\sqrt{2}\right)=0 \Rightarrow$

$x^2-x-x\sqrt{2}-x+1+\sqrt{2}+x\sqrt{2}-\sqrt{2}-2=0 \Rightarrow x^2-2x-1=0$; and $a=1$, $b=-2$, and $c=-1$.

71. If $x=2i$ or $-2i$, then $(x-2i)(x+2i)=0 \Rightarrow x^2-4i^2=0 \Rightarrow x^2+4=0$; and $a=1, b=0, c=4$.

73. If $x=2-\sqrt{5}$ or $2+\sqrt{5}$, then $\left(x-\left(2-\sqrt{5}\right)\right)\left(x-\left(2+\sqrt{5}\right)\right)=0 \Rightarrow \left(x-2+\sqrt{5}\right)\left(x-2-\sqrt{5}\right)=0 \Rightarrow$

$x^2-2x-x\sqrt{5}-2x+4+2\sqrt{5}+x\sqrt{5}-2\sqrt{5}-5=0 \Rightarrow x^2-4x-1=0$; and $a=1, b=-4, c=-1$.

75. The graph of the function $f(x)=ax^2+bx+c$ is a parabola, $a<0$ will make the parabola open downward, and $b^2-4ac=0$ gives us 1 real solution or 1 x-intercept. See Figure 75.

77. The graph of the function $f(x)=ax^2+bx+c$ is a parabola, $a<0$ will make the parabola open downward, and $b^2-4ac<0$ gives us no real solutions or no x-intercepts. See Figure 77

79. The graph of the function $f(x)=ax^2+bx+c$ is a parabola, $a>0$ will make the parabola open upward, and $b^2-4ac>0$ gives us 2 real solutions or 2 x-intercepts. See Figure 79.

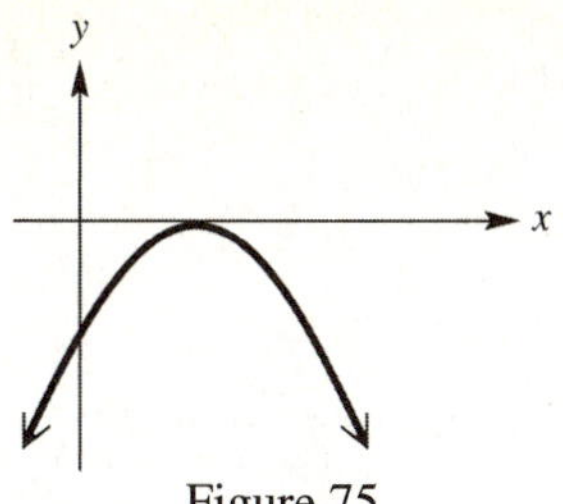

Figure 75

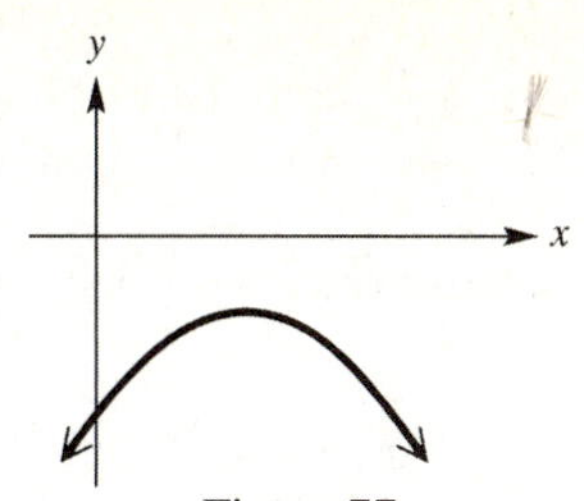

Figure 77

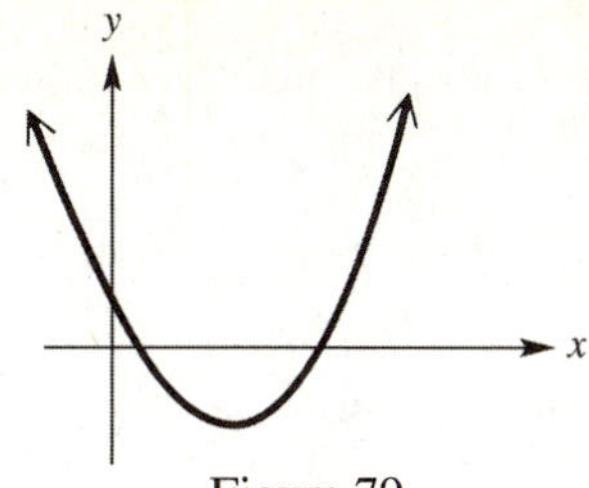

Figure 79

81. From the graph, $f(x)=0$ when 2 or 4.

83. From the graph, $f(x)>0$ for the interval $(-\infty,2)\cup(4,\infty)$.

85. From the graph, $g(x)<0$ for the interval $(-\infty,3)\cup(3,\infty)$.

87. From the graph, $h(x)>0$ always, but no x-intercepts; therefore, the solution is $(-\infty,\infty)$.

89. From the graph, $h(x)=0$ has no real solutions, because $h(x)$ has no x-intercepts. The graph is completely above the x-axis; therefore, there are two nonreal complex solutions.

91. The x-coordinate of the vertex of the graph of $y=f(x)$ is the midpoint of the x-intercepts or $\dfrac{2+4}{2}=3$.

93. From the graph, $y=g(x)$ will have a y-intercept and it will be negative.

95. (a) First, set $x^2+4x+3=0$, then $(x+1)(x+3)=0$ and the endpoints will be $x=\{-3,-1\}$. The numbers -3 and -1 divide the number line into 3 intervals. The interval $(-\infty,-3)$ has a positive product, the interval $(-3,-1)$ has a negative product, and the interval $(-1,\infty)$ has a positive product. Therefore, $x^2+4x+3\geq 0$ for the interval $(-\infty,-3]\cup[-1,\infty)$. The graph of $y_1=x^2+4x+3$ intersects or is above the x-axis for the interval $(-\infty,-3]\cup[-1,\infty)$.

(b) First, set $x^2+4x+3=0$, then $(x+1)(x+3)=0$ and the endpoints will be $x=\{-3,-1\}$. The numbers -3 and -1 divide the number line into 3 intervals. The interval $(-\infty,-3)$ has a positive product, the interval $(-3,-1)$ has a negative product, and the interval $(-1,\infty)$ has a positive product. Therefore, $x^2+4x+3<0$ for the interval $(-3,-1)$. The graph of $y_1=x^2+4x+3$ is below the x-axis for the interval $(-3,-1)$.

97. (a) First, $2x^2-9x>-4\Rightarrow 2x^2-9x+4>0$, now set $2x^2-9x+4=0$ then $(2x-1)(x-4)=0$ and the endpoints will be $x=\dfrac{1}{2}$ and $x=4$. The numbers $\dfrac{1}{2}$ and 4 divide the number line into 3 intervals. The interval $\left(-\infty,\dfrac{1}{2}\right)$ has a positive product, the interval $\left(\dfrac{1}{2},4\right)$ has a negative product, and the interval $(4,\infty)$

has a positive product. Therefore, $2x^2 - 9x > -4$ for the interval $\left(-\infty, \frac{1}{2}\right) \cup (4, \infty)$. The graph of

$y_1 = 2x^2 - 9x + 4$ is above the x-axis for the interval $\left(-\infty, \frac{1}{2}\right) \cup (4, \infty)$.

(b) First, $2x^2 - 9x \le -4 \Rightarrow 2x^2 - 9x + 4 \le 0$, now set $2x^2 - 9x + 4 = 0$ then $(2x-1)(x-4) = 0$ and the endpoints will be $x = \frac{1}{2}$ and $x = 4$. The numbers $\frac{1}{2}$ and 4 divide the number line into 3 intervals. The interval $\left(-\infty, \frac{1}{2}\right)$ has a positive product, the interval $\left(\frac{1}{2}, 4\right)$ has a negative product, and the interval $(4, \infty)$ has a positive product. Therefore, $2x^2 - 9x \le -4$ for the interval $\left(\frac{1}{2}, 4\right)$. The graph of $y_1 = 2x^2 - 9x + 4$ intersects or is below the x-axis for the interval $\left[\frac{1}{2}, 4\right]$.

99. (a) First, set $-x^2 - x = 0$, then $-x(x+1) = 0$ and the endpoints will be $x = -1$ and $x = 0$. The numbers -1 and 0 divide the number line into 3 intervals. The interval $(-\infty, -1)$ has a negative product, the interval $(-1, 0)$ has a positive product, and the interval $(0, \infty)$ has a negative product. Therefore, $-x^2 - x \le 0$ for the interval $(-\infty, -1] \cup [0, \infty)$. The graph of $y_1 = -x^2 - x$ intersects or is below the x-axis for the interval $(-\infty, -1] \cup [0, \infty)$.

(b) First, set $-x^2 - x = 0$, then $-x(x+1) = 0$ and the endpoints will be $x = -1$ and $x = 0$. The numbers -1 and 0 divide the number line into 3 intervals. The interval $(-\infty, -1)$ has a negative product, the interval $(-1, 0)$ has a positive product, and the interval $(0, \infty)$ has a negative product. Therefore, $-x^2 - x > 0$ for the interval $(-1, 0)$. The graph of $y_1 = -x^2 - x$ is above the x-axis for the interval $(-1, 0)$.

101. (a) First, set $x^2 - x + 1 = 0$, then by the quadratic formula $x = \frac{-(-1) \pm \sqrt{(-1)^2 - 4(1)(1)}}{2(1)} =$

$\frac{1 \pm \sqrt{-3}}{2} \Rightarrow x = \frac{1}{2} \pm \frac{\sqrt{3}}{2} i$. The inequality has no real solutions, the graph is a parabola opening upward which does not intersect, but is completely above the x-axis. Therefore, $x^2 - x + 1 < 0$ never happens and the solution is $\varnothing$.

(b) First, set $x^2 - x + 1 = 0$, then by the quadratic formula $x = \frac{-(-1) \pm \sqrt{(-1)^2 - 4(1)(1)}}{2(1)} =$

$\frac{1\pm\sqrt{-3}}{2} \Rightarrow x = \frac{1}{2} \pm \frac{\sqrt{3}}{2}i.$ The graph of this equation is a parabola opening upward, has no real solutions, and does not intersect, but is completely above the x-axis. Therefore, $x^2 - x + 1 \geq 0$ for all values of x and the solution is the interval $(-\infty, \infty)$.

103. (a) First, $2x+1 \geq x^2 \Rightarrow x^2 - 2x - 1 \leq 0,$ now set $x^2 - 2x - 1 = 0,$ then by the quadratic formula

$$x = \frac{-(-2) \pm \sqrt{(-2)^2 - 4(1)(-1)}}{2(1)} = \frac{2 \pm \sqrt{8}}{2} = \frac{2(1 \pm \sqrt{2})}{2} \Rightarrow x = 1 \pm \sqrt{2},$$ and the endpoints will be $x = 1 - \sqrt{2}$ and $x = 1 + \sqrt{2}$. The numbers $1 - \sqrt{2}$ and $1 + \sqrt{2}$ divide the number line into 3 intervals. The interval $(-\infty, 1-\sqrt{2})$ has a positive product, the interval $(1-\sqrt{2}, 1+\sqrt{2})$ has a negative product, and the interval $(1+\sqrt{2}, \infty)$ has a positive product. Therefore, $x^2 - 2x - 1 \leq 0$ for the interval $[1-\sqrt{2}, 1+\sqrt{2}]$. The graph of $y_1 = x^2 - 2x - 1$ intersects or is below the x-axis for the interval $[1-\sqrt{2}, 1+\sqrt{2}]$.

(b) First, $2x+1 < x^2 \Rightarrow x^2 - 2x - 1 > 0,$ now set $x^2 - 2x - 1 = 0,$ then by the quadratic formula

$$x = \frac{-(-2) \pm \sqrt{(-2)^2 - 4(1)(-1)}}{2(1)} = \frac{2 \pm \sqrt{8}}{2} = \frac{2(1 \pm \sqrt{2})}{2} \Rightarrow x = 1 \pm \sqrt{2},$$ and the endpoints will be $x = 1 - \sqrt{2}$ and $x = 1 + \sqrt{2}$. The numbers $1 - \sqrt{2}$ and $1 + \sqrt{2}$ divide the number line into 3 intervals. The interval $(-\infty, 1-\sqrt{2})$ has a positive product, the interval $(1-\sqrt{2}, 1+\sqrt{2})$ has a negative product, and the interval $(1+\sqrt{2}, \infty)$ has a positive product. Therefore, $x^2 - 2x - 1 > 0$ for the interval $(-\infty, 1-\sqrt{2}) \cup (1+\sqrt{2}, \infty)$. The graph of $y_1 = x^2 - 2x - 1$ is above the x-axis for the interval $(-\infty, 1-\sqrt{2}) \cup (1+\sqrt{2}, \infty)$.

105. (a) First, $x - 3x^2 > -1 \Rightarrow 3x^2 - x - 1 < 0,$ now set $3x^2 - x - 1 = 0,$ then by the quadratic formula

$$x = \frac{1 \pm \sqrt{(-1)^2 - 4(3)(-1)}}{2(3)} \Rightarrow x = \frac{1 \pm \sqrt{13}}{6},$$ and the endpoints will be $x = \frac{1-\sqrt{13}}{6}$ and $x = \frac{1+\sqrt{13}}{6}$. The numbers $\frac{1-\sqrt{13}}{6}$ and $\frac{1+\sqrt{13}}{6}$ divide the number line into 3 intervals. The interval $\left(-\infty, \frac{1-\sqrt{13}}{6}\right)$ has a positive product, the interval $\left(\frac{1-\sqrt{13}}{6}, \frac{1+\sqrt{13}}{6}\right)$ has a negative product, and the interval $\left(\frac{1+\sqrt{13}}{6}, \infty\right)$ has positive product. Therefore, $3x^2 - x - 1 < 0$ for the

interval $\left(\frac{1-\sqrt{13}}{6},\frac{1+\sqrt{13}}{6}\right)$. The graph of $y_1 = 3x^2 - x - 1$ is below the x-axis for the interval $\left(\frac{1-\sqrt{13}}{6},\frac{1+\sqrt{13}}{6}\right)$.

(b) First, $x - 3x^2 \le -1 \Rightarrow 3x^2 - x - 1 \ge 0$, now set $3x^2 - x - 1 = 0$, then by the quadratic formula $x = \frac{1 \pm \sqrt{(-1)^2 - 4(3)(-1)}}{2(3)} \Rightarrow x = \frac{1 \pm \sqrt{13}}{6}$, and the endpoints will be $x = \frac{1-\sqrt{13}}{6}$ and $x = \frac{1+\sqrt{13}}{6}$. The numbers $\frac{1-\sqrt{13}}{6}$ and $\frac{1+\sqrt{13}}{6}$ divide the number line into 3 intervals. The interval $\left(-\infty,\frac{1-\sqrt{13}}{6}\right)$ has a positive product, the interval $\left(\frac{1-\sqrt{13}}{6},\frac{1+\sqrt{13}}{6}\right)$ has a negative product, and the interval $\left(\frac{1+\sqrt{13}}{6},\infty\right)$ has positive product. Therefore, $3x^2 - x - 1 \ge 0$ for the intervals $\left(-\infty,\frac{1-\sqrt{13}}{6}\right]$ or $\left[\frac{1+\sqrt{13}}{6},\infty\right)$. The graph of $y_1 = 3x^2 - x - 1$ is above the x-axis for the interval $\left(-\infty,\frac{1-\sqrt{13}}{6}\right] \cup \left[\frac{1+\sqrt{13}}{6},\infty\right)$.

107. $s = \frac{1}{2}gt^2 \Rightarrow \frac{2s}{g} = t^2 \Rightarrow t = \pm\sqrt{\frac{2s}{g}} \Rightarrow t = \pm\frac{\sqrt{2sg}}{g}$

109. $a^2 + b^2 = c^2 \Rightarrow a^2 = c^2 - b^2 \Rightarrow a = \pm\sqrt{c^2 - b^2}$

111. $S = 4\pi r^2 \Rightarrow \frac{S}{4\pi} = r^2 \Rightarrow r = \pm\sqrt{\frac{S}{4\pi}} \Rightarrow r = \pm\frac{\sqrt{S\pi}}{2\pi}$

113. $V = e^3 \Rightarrow e = \sqrt[3]{V}$

115. $F = \frac{kMv^4}{r} \Rightarrow \frac{Fr}{kM} = v^4 \Rightarrow v = \pm\sqrt[4]{\frac{Fr}{kM}} \Rightarrow v = \frac{\pm\sqrt[4]{FrkMkMkM}}{kM} \Rightarrow v = \pm\frac{\sqrt[4]{Frk^3M^3}}{kM}$

117. $P = \frac{E^2R}{(r+R)^2} \Rightarrow P(r+R)^2 = E^2R \Rightarrow P(r^2 + 2rR + R^2) - E^2R = 0 \Rightarrow \Pr^2 + 2\Pr R + \mathrm{P}R^2 - E^2R = 0$

$\Rightarrow PR^2 + (2\Pr - E^2)R + \Pr^2 = 0$. Now use the quadratic formula:

$$\mathrm{R} = \frac{-(2\Pr - E^2) \pm \sqrt{4P^2r^2 - 4E^2\Pr + E^4 - 4(P)(\Pr^2)}}{2(P)} = \frac{E^2 - 2\Pr \pm \sqrt{E^4 - 4E^2\Pr}}{2P} \Rightarrow$$

$$R = \frac{E^2 - 2\Pr \pm E\sqrt{E^2 - 4\Pr}}{2P}$$

119. Use the quadratic formula: $x=\dfrac{-y\pm\sqrt{y^2-4(1)(y^2)}}{2(1)}=\dfrac{-y\pm\sqrt{-3y^2}}{2}\Rightarrow x=-\dfrac{y}{2}\pm\dfrac{\sqrt{3}}{2}yi.$ For $y^2+xy+x^2=0$

use the quadratic formula: $y=\dfrac{-x\pm\sqrt{x^2-4(1)(x^2)}}{2(1)}=\dfrac{-x\pm\sqrt{-3x^2}}{2}\Rightarrow\ y=-\dfrac{x}{2}\pm\dfrac{\sqrt{3}}{2}xi.$

121. $3y^2+4xy-9x^2=-1\Rightarrow -9x^2+4yx+(3y^2+1)=0.$ Now use the quadratic formula:

$$x=\frac{-4y\pm\sqrt{(4y)^2-4(-9)(3y^2+1)}}{2(-9)}=\frac{-4y\pm\sqrt{16y^2+108y^2+36}}{-18}=$$

$$\frac{-4y\pm\sqrt{124y^2+36}}{-18}=\frac{-4y\pm\sqrt{4(31y^2+9)}}{-18}=\frac{-4y\pm 2\sqrt{31y^2+9}}{-18}\Rightarrow x=\frac{2y\pm\sqrt{31y^2+9}}{9}$$

$3y^2+4xy-9x^2=-1\Rightarrow 3y^2+4xy+(-9x^2+1)=0.$ Now use the quadratic formula:

$$x=\frac{-4x\pm\sqrt{(4x)^2-4(3)(-9x^2+1)}}{2(3)}=\frac{-4x\pm\sqrt{16x^2+108x^2-12}}{6}=$$

$$\frac{-4x\pm\sqrt{124x^2-12}}{6}=\frac{-4x\pm\sqrt{4(31x^2-3)}}{6}=\frac{-4x\pm 2\sqrt{31x^2-3}}{6}\Rightarrow y=\frac{-2x\pm\sqrt{31x^2-3}}{3}$$

123. (a) $f(0)=\dfrac{4}{5}(0-10)^2+80=160$ and $f(2)=\dfrac{4}{5}(2-10)^2+80=131.2.$ Initially when the person stops exercising the heart rate is 160 beats per minute, and after 2 minutes the heart rate has dropped to about 131 beats per minute.

(b) Graph $Y_1=100$, $Y_2=0.8(x-10)^2+80$ and $Y_3=120$ (Not shown). The person's heart rate is between 100 and 120 when the graph of Y_2 is between Y_1 and Y_3. This occurs between approximately 2.9 minutes and 5 minutes after the person stops exercising.

Reviewing Basic Concepts (Sections 3.1—3.3)

1. $(5+6i)-(2-4i)-3i=5+6i-2+4i-3i=3+7i$

2. $i(5+i)(5-i)=i(25-i^2)=i(25-(-1))=26i$

3. $\dfrac{-10-10i}{2+3i}\cdot\dfrac{2-3i}{2-3i}=\dfrac{-20+30i-20i+30i^2}{4-9i^2}=\dfrac{-20+10i-30}{4+9}=-\dfrac{50}{13}+\dfrac{10}{13}i$

4. See Figure 4.

[-10,10] by [-10,10]
Xscl = 1 Yscl = 1

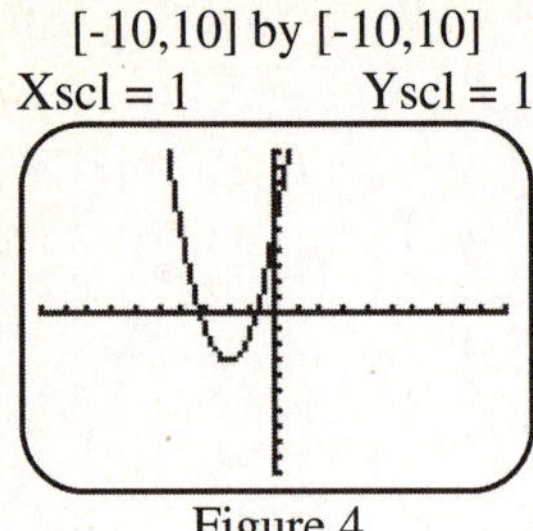

Figure 4

5. Use the vertex formula: $\frac{-b}{2a} = \frac{-8}{2(2)} = -2$, now input $x = -2$ into the equation: $2(-2)^2 + 8(-2) + 5 = -3$. The vertex is $(-2, -3)$ and it is a minimum because $a = 2$, which is positive; therefore, the parabola opens up.

6. Since the vertex is $(-2, -3)$ and the graph is a parabola opening up, the axis of symmetry is $x = -2$.

7. From the graph, the domain is $(-\infty, \infty)$ and the range is $[-3, \infty)$.

8. $9x^2 = 25 \Rightarrow x^2 = \frac{25}{9} \Rightarrow x = \pm\frac{5}{3}$

9. $3x^2 - 5x = 2 \Rightarrow 3x^2 - 5x - 2 = 0 \Rightarrow (3x+1)(x-2) = 0 \Rightarrow 3x + 1 = 0$ or $x - 2 = 0 \Rightarrow x = -\frac{1}{3}$ and $x = 2$.

10. For $-x^2 + x + 3 = 0$ use the quadratic formula: $x = \frac{-1 \pm \sqrt{1^2 - 4(-1)(3)}}{2(-1)} = \frac{-1 \pm \sqrt{13}}{-2} \Rightarrow x = \frac{1 \pm \sqrt{13}}{2}$.

11. First, set $3x^2 - 5x - 2 = 0$, then $g(x)$ and the endpoints will be $x = -\frac{1}{3}$ and $x = 2$. The numbers $-\frac{1}{3}$ and 2 divide the number line into 3 intervals. The interval $\left(-\infty, -\frac{1}{3}\right)$ has a positive product, the interval $\left(-\frac{1}{3}, 2\right)$ has a negative product, and the interval $(2, \infty)$ has a positive product. Therefore, $3x^2 - 5x - 2 \le 0$ for the interval $\left[-\frac{1}{3}, 2\right]$. The graph of $y_1 = 3x^2 - 5x - 2$ intersects or is below the x-axis for the interval $\left[-\frac{1}{3}, 2\right]$.

12. First, set $x^2 - x - 3 = 0$ then by the quadratic formula $x = \frac{-(-1) \pm \sqrt{(-1)^2 - 4(1)(-3)}}{2(1)} \Rightarrow x = \frac{1 \pm \sqrt{13}}{2}$ and the endpoints will be $x = \frac{1 - \sqrt{13}}{2}$ and $x = \frac{1 + \sqrt{13}}{2}$. The numbers $\frac{1 - \sqrt{13}}{2}$ and $\frac{1 + \sqrt{13}}{2}$ divide the number line into 3 intervals. The interval $\left(-\infty, \frac{1 - \sqrt{13}}{2}\right)$ has a positive product, the interval $\left(\frac{1 - \sqrt{13}}{2}, \frac{1 + \sqrt{13}}{2}\right)$ has a negative product, and the interval $\left(\frac{1 + \sqrt{13}}{2}, \infty\right)$ has a positive product. Therefore, $x^2 - x - 3 > 0$ for the

interval $\left(-\infty,\frac{1-\sqrt{13}}{2}\right)\cup\left(\frac{1+\sqrt{13}}{2},\infty\right)$. The graph of $y_1=x^2-x-3$ is above the x-axis for the interval $\left(-\infty,\frac{1-\sqrt{13}}{2}\right)\cup\left(\frac{1+\sqrt{13}}{2},\infty\right)$.

13. First, $x(3x-1)\le -4\Rightarrow 3x^2-x+4\le 0,$ now set $3x^2-x+4=0,$ then by the quadratic formula $x=\frac{-(-1)\pm\sqrt{(-1)^2-4(3)(4)}}{2(3)}=\frac{1\pm\sqrt{-47}}{6}\Rightarrow x=\frac{1}{6}\pm\frac{\sqrt{47}}{6}i.$ The inequality has no real solutions, the graph is a parabola opening upward which does not intersect, but is completely above the x-axis. Therefore, $3x^2-x+4\le 0$ never happens and the solution is $\varnothing$.

Section 3.4 Further Applications of Quadratic Functions and Models

1. The y-coordinate of the vertex is maximum y-value. Using the vertex formula yields:

   $x=-\frac{b}{2a}=-\frac{32}{2(-16)}=1.$ If $x=1,$, then $y=-16(1)^2+32(1)+100=116\Rightarrow$ the maximum y-value is 116.

3. The y-coordinate of the vertex is minimum y-value. Using the vertex formula yields:

   $x=-\frac{b}{2a}=-\frac{-24}{2(3)}=4.$ If $x=4,$, then $y=3(4)^2-24(4)+50=2\Rightarrow$ the minimum y-value is 2.

5. $-4x^2+5x=1\Rightarrow 4x^2-5x+1=0\Rightarrow(4x-1)(x-1)=0\Rightarrow(4x-1)=0\Rightarrow x=\frac{1}{4}$ and $(x-1)=0\Rightarrow x=1$.

7. $\frac{1}{2}x^2+3=6x\Rightarrow\frac{1}{2}x^2-6x+3=0.$ Using the quadratic equation we have the following:

$$\frac{6\pm\sqrt{(-6)^2-4\left(\frac{1}{2}\right)(3)}}{2\left(\frac{1}{2}\right)}=6\pm\sqrt{30}$$

9. A, The area of a rectangle is $L\cdot W=A\Rightarrow x(2x+2)=40{,}000.$

11. (a) $30-x$ would be the other number.

    (b) Since both numbers are positive and their sum is 30, the restrictions are $0<x<30$.

    (c) Multiplying $x(30-x)$ would give the function $P(x)=30x-x^2\Rightarrow P(x)=-x^2+30x.$

    (d) Using the Vertex formula yields $x=\frac{-b}{2a}=\frac{-30}{2(-1)}\Rightarrow x=15.$ If $x=15,$ then $y=30-(15)\Rightarrow y=15.$

    Also, if $x=15$ the maximum product is $P(15)=-(15)^2+30(15)=-225+450=225.$ The graph of $y=-x^2+30x$ has a vertex point of $(15,15)$, which supports the result.

13. Perimeter of fence $= 2l + 2w = 1000 \Rightarrow l = 500 - w$. If $A = l \cdot w$ then $A = (500 - w)w = 500w - w^2$. This is a parabola opening downward and by the vertex formula, the maximum area occurs when $w = -\frac{b}{2a} = -\frac{500}{2(-1)} = 250$. The dimensions that maximize area are 250 ft. by 250 ft.

15. (a) $640 - 2x$ would be the other side.

(b) Since all three numbers are positive, their sum is 640, and two of them are x, the restrictions are $0 < 2x < 640 \Rightarrow 0 < x < 320$.

(c) Multiplying $x(640 - 2x)$ would give the function $A(x) = 640x - 2x^2 \Rightarrow A(x) = -2x^2 + 640x$.

(d) See Figure 15. From the graph, between 57.04 ft and 85.17 ft or between 234.17 ft and 262.96 ft, will give an area between 30,000 and 40,000 feet.

(e) Using the Vertex formula yields $x = \frac{-b}{2a} = \frac{-640}{2(-2)} \Rightarrow x = 160$. If $x = 160$, then $y = 640 - 2(160) \Rightarrow y = 320$. Also, if $x = 160$, the maximum product is $A(160) = -2(160)^2 + 640(160) = -51,200 + 102,400 \Rightarrow A(x) = 51,200$. The graph of $y = -2x^2 + 640x$ has a vertex point of $(160, 320)$, which supports the result.

[0,320] by [0,55,000]
Xscl = 20 Yscl = 10,000

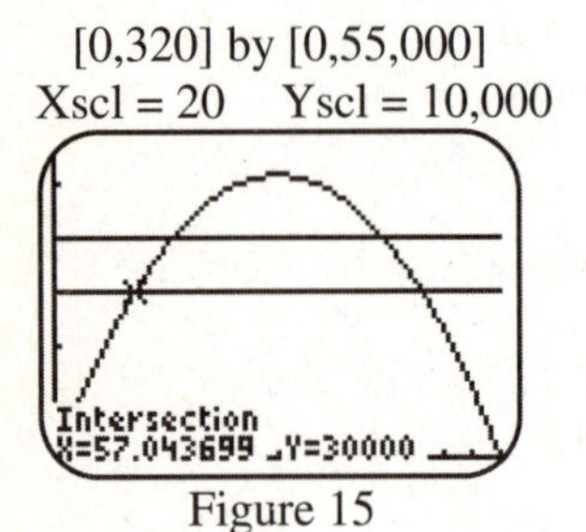

Figure 15

17. (a) $s(1) = -16(1)^2 + 44(1) + 4 = 32$; the baseball is 32 feet high after 1 second.

(b) For $s(t) = -16t^2 + 44t + 4$, the vertex formula gives $t = -\frac{b}{2a} = -\frac{44}{2(-16)} = 1.375$ and $f(1.375) = -16(1.375)^2 + 44(1.375) + 4 = 34.25$; the maximum height is 34.25 feet.

19. The height when it hits the ground will be $0 \Rightarrow 75 - 16t^2 = 0 \Rightarrow 16t^2 = 75 \Rightarrow t^2 = \frac{75}{16} \Rightarrow$

$t = \sqrt{\frac{75}{16}} \Rightarrow t = \frac{\sqrt{75}}{4} \Rightarrow t \approx 2.2$ seconds.

21. (a) If the width is x units long, then the length is twice the width or $2x$ units long.

(b) The width will be $x - 2(2) = x - 4$ units long, and the length will be $2x - 2(2) = 2x - 4$ units long. Since both measurements are positive and 4 inches are removed from each measurement, the restrictions are $x > 4$.

(c) Since volume for the box is $V = L \cdot W \cdot H, V = (2x-4)(x-4)(2)$ and the function is

$V(x) = 4x^2 - 24x + 32$.

(d) $320 = 4x^2 - 24x + 32 \Rightarrow 4x^2 - 24x - 288 = 0 \Rightarrow 4(x^2 - 6x - 72) = 0 \Rightarrow 4(x-12)(x+6) = 0$

$\Rightarrow x = -6$ or 12. Since length cannot be negative, $x = 12$. If $x = 12$, the dimensions are 8 in by 20 in. From the graph of $y = 4x^2 - 24x + 32$, when $x = 12$, $y = 320$. This support our analytical result.

(e) From the graph (not shown) of $y = 4x^2 - 24x + 32$, when $400 < y < 500, 13.0 < x < 14.2$ in.

23. The surface area of the can is $V = \pi r^2 + \pi r^2 + 2\pi r(h) \Rightarrow 54.19 = 2\pi r^2 + 2(4.25)\pi r \Rightarrow$

$6.28x^2 + 26.69x - 54.19 = 0$ now use the quadratic formula: $\dfrac{-26.69 \pm \sqrt{(26.69)^2 - 4(6.28)(-54.19)}}{2(6.28)} =$

$\dfrac{-26.69 \pm \sqrt{2073.6}}{12.56} = \dfrac{-26.69 \pm 45.537}{12.56} \Rightarrow r = -5.751$ or 1.5. Since length cannot be negative, $r = 1.5$ in.

25. If $A = s^2$, then $800 = s^2 \Rightarrow s = \sqrt{800} = 20\sqrt{2}$. Since the lawn is square, the diagonal of the lawn would equal $d^2 = (20\sqrt{2})^2 + (20\sqrt{2})^2 \Rightarrow d^2 = 1600 \Rightarrow d = 40$. The radius of the circular pattern is half the diagonal; therefore, the radius is $r = 20$ ft.

27. If we use Pythagorean Theorem, we get $h^2 + 12^2 = (2h+3)^2 \Rightarrow$

$h^2 + 144 = 4h^2 + 12h + 9 \Rightarrow 3h^2 + 12h - 135 = 0 \Rightarrow (3h+27)(h-5) = 0 \Rightarrow x = -9.5$. Since length cannot be negative, the height of the dock is 5 ft.

29. If we use Pythagorean Theorem, we get $(8+2)^2 + (9+4)^2 = x^2 \Rightarrow 100 + 169 = x^2 \Rightarrow$

$x^2 = 269 \Rightarrow x = \sqrt{269} \Rightarrow x \approx 16.4$. Since a 16 ft ladder will be too short, the ladder must be at least 17 ft.

31. Let $x =$ length of the picture and let $(x-4) =$ the width of the picture. Then $A = l \cdot w \Rightarrow A = x(x-4) \Rightarrow$

$320 = x^2 - 4x \Rightarrow x^2 - 4x - 320 = 0$. $\Rightarrow (x-20)(x+16) = 0 \Rightarrow x = 20$. Thus, the length and width of the picture is 20 and16 inches respectively. The dimensions of the frame will include an additional 4 inches to both the length and width to yield the final dimensions of 20 inches by 24 inches.

33. (a) Since x equals the loss of 1 apartment for each \$20 increase, the number of apartments rented is $80 - x$.

(b) Since each increase is \$20 and x equals the number of increases, the rent per apartment is $400 + 20x$.

(c) Since revenue is the number of apartments rented times rent per apartment, we multiply a and b.

$(80-x)(400+20x) = 32{,}000 + 1200x - 20x^2 \Rightarrow R(x) = -20x^2 + 1200x + 32{,}000$.

(d) $37{,}500 = -20x^2 + 1200x + 32{,}000 \Rightarrow -20x^2 + 1200x - 5{,}500 = 0 \Rightarrow$

$-20(x^2 - 60x + 275) = 0 \Rightarrow -20(x-55)(x-5) = 0 \Rightarrow x = 5$ or 55.

(e) Use the vertex formula: $x = \dfrac{-b}{2a} = \dfrac{-1200}{-40} = 30$. If $x = 30$, then the rent per apartment is

$r(30) = 400 + 20(30) = 400 + 600 = \$1000$.

35. (a) If $f(x)=10$ and $x=15$, then $10=\frac{-16(15)^2}{0.434v^2}+1.15(15)+8 \Rightarrow 10=\frac{-3600}{0.434v^2}+25.25 \Rightarrow$

$-15.25=\frac{-3600}{0.434v^2} \Rightarrow -6.6185v^2=-3600 \Rightarrow v^2=\frac{3600}{6.6185} \Rightarrow v \approx 23.32\text{ft/sec.}$

(b) If $v=23.32$, then $y=\frac{-16x^2}{0.434(23.32)^2}+1.15x+8 \Rightarrow y=-.06779x^2+1.15x+8$. Graph this equation, the graph does pass through points $(0,8)$ and $(15,10)$.

(c) Use the vertex formula: $x=\frac{-b}{2a}=\frac{-1.15}{2(-0.06779)} \approx 8.48$. If $x \approx 8.48$, then the maximum height is $f(8.48)=-0.06779(8.48)^2+1.15(8.48)+8 \Rightarrow f(8.48)=12.88\text{ft}$.

37. (a) If $x=2$, then $h(2)=-0.5(2)^2+1.25(2)+3=-2+2.5+3 \Rightarrow h(2)=3.5\text{ft}$.

(b) If $h(x)=3.25$ .then $3.25=-0.5x^2+1.25x+3 \Rightarrow -0.5x^2+1.25x-0.25=0 \Rightarrow$

$4(-0.5x^2+1.25x+0.25=0) \Rightarrow -2x^2+5x-1=0 \Rightarrow 2x^2-5x+1=0$ Now use the quadratic formula: $x=\frac{5 \pm \sqrt{(-5)^2-4(2)(1)}}{2(2)}=\frac{5 \pm \sqrt{17}}{4} \Rightarrow x=0.219$ or 2.281. Therefore, the frog is 3.25 feet above the ground at approximately 0.2 ft and 2.3 ft.

(c) Using the vertex formula yields $x=\frac{-b}{2a}=\frac{-1.25}{2(-0.5)} \Rightarrow x=1.25$ ft as the horizontal distance.

(d) From part (c) the maximum height was reached when the horizontal distance was $x=1.25$; therefore, $h(1.25)=-0.5(1.25)^2+1.25(1.25)+3=-0.78125+1.5625+3 \Rightarrow h(1.25) \approx 3.78$ ft high.

39. If $f(x)=800$, then $800=\frac{1}{10}x^2-3x+22 \Rightarrow .1x^2-3x-778-0$. Now use the quadratic formula:

$x=\frac{3 \pm \sqrt{9-4(0.1)(-778)}}{2(0.1)}=\frac{3 \pm \sqrt{320.2}}{0.2} \Rightarrow x \approx 104.5$ ft/sec. Converted is $x=\frac{104.5(60)(60)}{5280}=71.25$ mph.

41. (a) See Figure 41a.

(b) Using the defined function and the point (4, 50) yields $f(x)=a(x-4)+50$. Now use (14, 110) to find the function; $110=a(14-4)^2+50 \Rightarrow 60=100a \Rightarrow a=0.6$; therefore, the function is $f(x)=0.6(x-4)^2+50$.

(c) See Figure 41c. From the graph, we see that there is a good fit.

(d) Using the regression feature yields $g(x) \approx 0.402x^2-1.175x+48.343$.

(e) For the year 2006, $x=16$. Therefore, $f(16)=0.6(16-4)^2+50 \approx 136.4$ thousand, and $g(16)=0.402(16)^2-1.175(16)+48.43 \approx 132.5$ thousand. For the year 2010, $x=20$. Therefore,

$f(20) = 0.6(20-4)^2 + 50 \approx 203.6$ thousand, and $g(20) = 0.402(20)^2 - 1.175(20) + 48.43 \approx 185.6$ thousand. For the year 2015, $x = 25$. Therefore, $f(25) = 0.6(25-4)^2 + 50 \approx 314.6$ thousand, and $g(25) = 0.402(25)^2 - 1.175(25) + 48.43 \approx 270.2$ thousand.

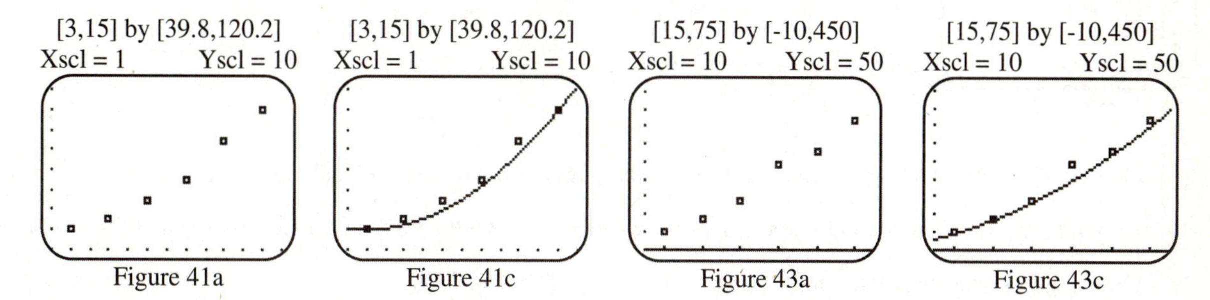

Figure 41a Figure 41c Figure 43a Figure 43c

3.5: Higher-Degree Polynomial Functions and Graphs

1. With three extrema, the minimum degree of *f* is 4.
3. The points (a,b) and (c,d) are local maxima, and (e,t) is a local minimum.
5. The highest point of the graph, *(a, b)* is an absolute maximum.
7. The function *f*, has local maximum values of *b* and *d*; a local minimum value of *t*; and an absolute maximum value of *b*.
9. $P(x)$ is a positive odd-degree polynomial graph; therefore,
11. $P(x)$ is a negative odd-degree polynomial graph; therefore,
13. $P(x)$ is a positive even-degree polynomial graph; therefore,
15. $P(x)$ is a negative even-degree polynomial graph; therefore,
17. $P(x)$ is a positive even-degree polynomial graph; therefore,
19. $P(x)$ is a negative odd-degree polynomial graph; therefore,
21. The graph of $f(x) = x^n$ for $n \in$ {positive odd integers} will take the shape of the graph of $f(x) = x^3$, but gets steeper as *n* and *x* increase.
23. See Figure 23. As the odd exponent *n* gets larger, the graph *flattens out* in the window $[-1,1]$ by $[-1,1]$. The graph of $y = x^7$ will be between $y = x^5$ and the *x*-axis in this window.

[-1,1] by [-1,1]
Xscl = 1 Yscl = 1

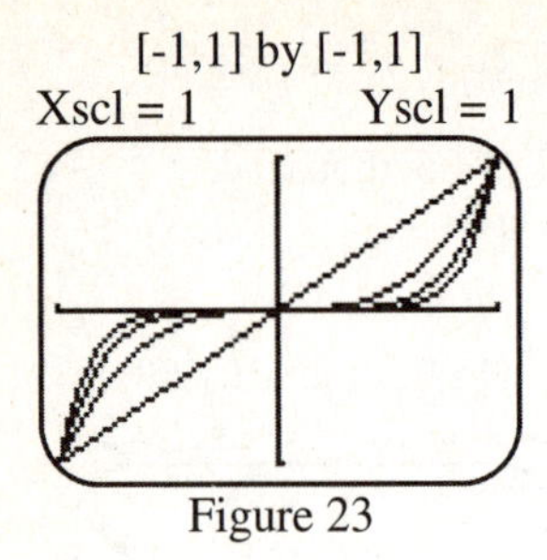

Figure 23

25. Graphing in various windows produces a local maximum of (2, 3.67), and a local minimum of (3, 3.5).

27. Graphing in various windows produces a local maximum of (−3.33, −1.85), and a local minimum of (−4, −2).

29. Graphing in various windows produces two x-intercepts, 2.10 and 2.15.

31. Graphing in various windows produces no x-intercept.

33. $P(x)$ is a positive odd-degree polynomial graph; therefore, D

35. $P(x)$ is a negative even-degree polynomial graph; therefore, B

37. A. The third-degree function can have at most 2 local extrema and a positive lead coefficient will yield a right side end behavior opening up.

39. From the graph, the function graphed in C has 1 real zero.

41. B and D. A third-degree function can have at most 2 local extrema.

43. From the graph, the function graphed in A has 1 *negative* real zero.

45. From the graph, the function graphed in B has a range of approximately $[-100, \infty)$.

47. False. A polynomial function of degree 3 will have at most 3 x-intercepts.

49. True. With a positive y-intercept and negative even-degree end behavior, it must have 2 x-intercepts.

51. True. With a negative even-degree end behavior, which is shifted up 5 units, it will have 2 x-intercepts.

53. False. With odd-degree end behavior, it will have at least 1 and at most 5 x-intercepts. Thus it may have only 1.

55. Shift the graph of $y = x^4$ three units to the left, stretch vertically by a factor of 2, and shift downward 7 units. See Figures 55a and 55b.

56. Shift the graph of $y = x^4$ one unit to the left, stretch vertically by a factor of 3, reflect across the x-axis, and shift upward 12 units. See Figures 56a and 56b.

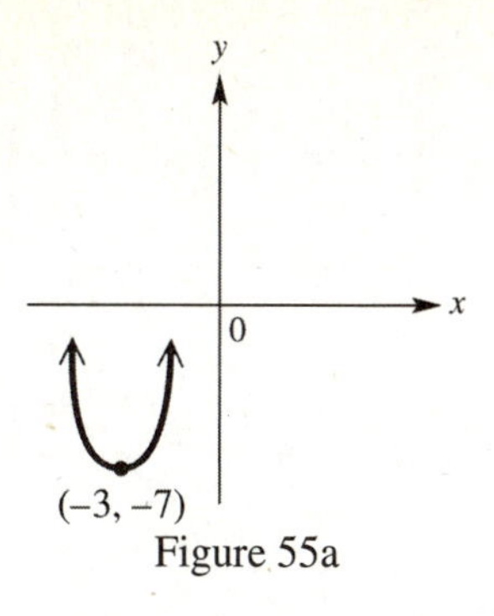

Figure 55a

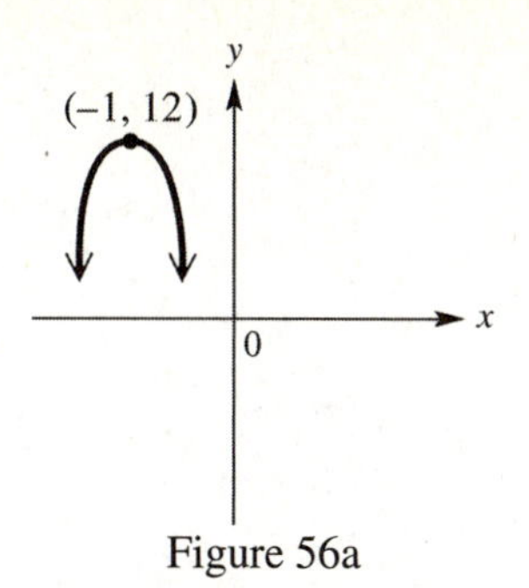

Figure 56a

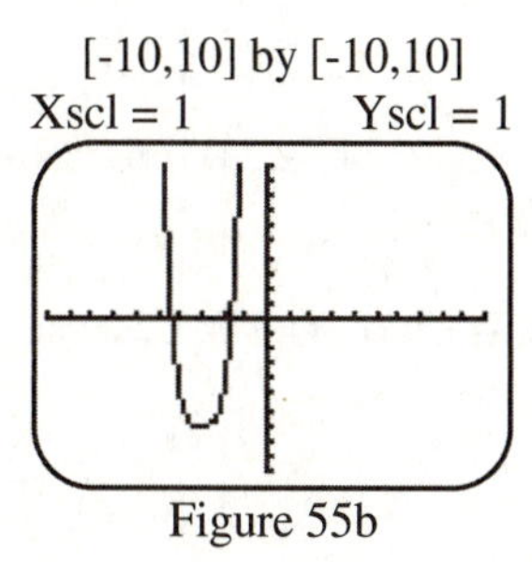

Figure 55b

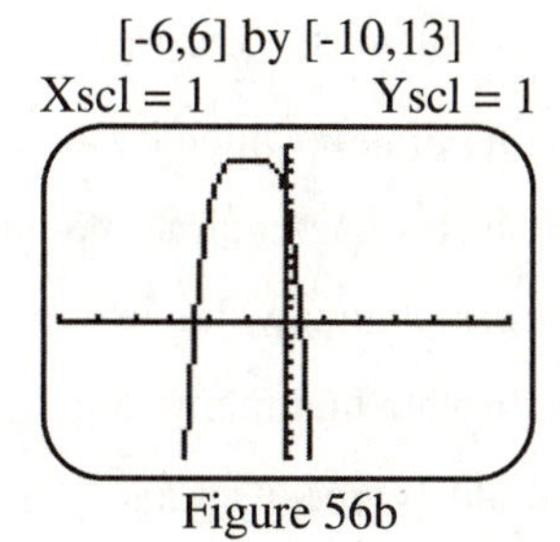

Figure 56b

57. Shift the graph of $y = x^3$ one unit to the right, stretch vertically by a factor of 3, reflect across the x-axis, and shift upward 12 units. See Figures 57a and 57b.

58. Shift the graph of $y = x^5$ one unit to the right, stretch vertically by a factor of .5, and shift upward 13 units. See Figures 58a and 58b.

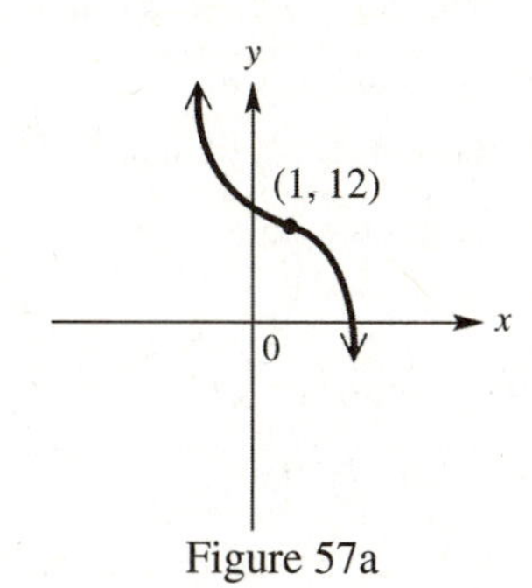

Figure 57a

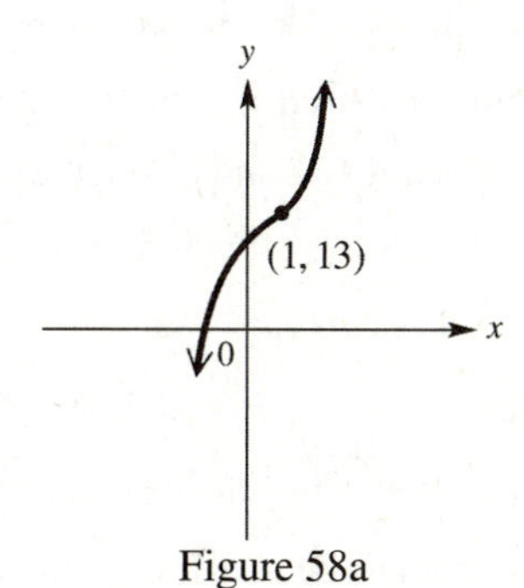

Figure 58a

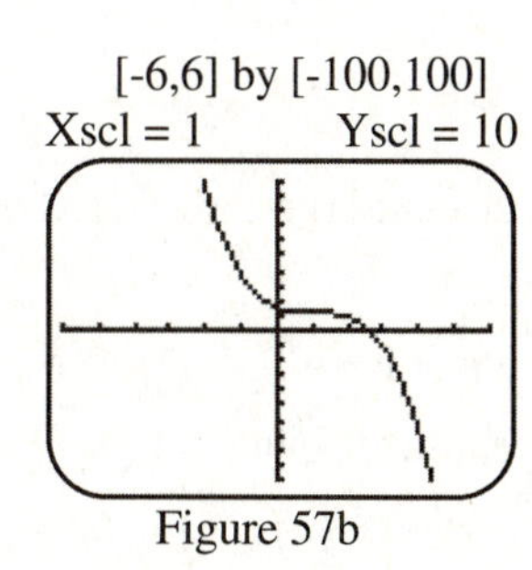

Figure 57b

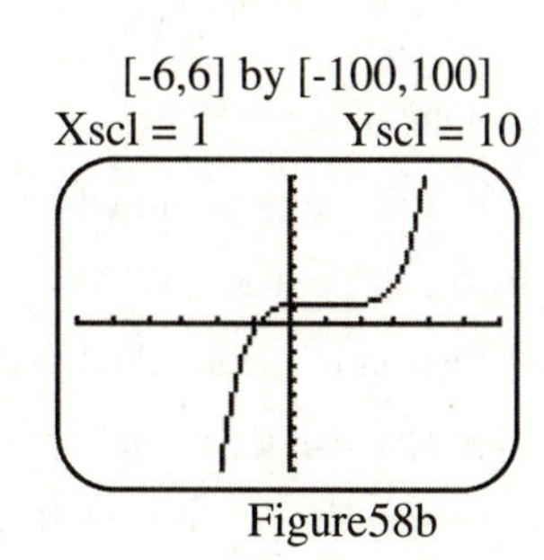

Figure58b

[-10,10] by [-100,100]
Xscl = 1 Yscl = 10

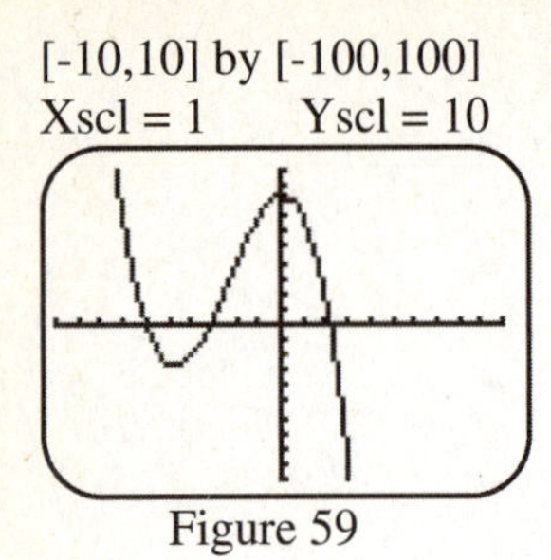

Figure 59

59. See Figure 59 for the graph of this function.

(a) Because it is a polynomial function, its domain is $(-\infty,\infty)$.

(b) From the graph of *P,* there is one local minimum, which by the calculator is (–4.74, –27.03). Because the function is of odd-degree, it is not an absolute minimum point.

(c) From the graph of *P,* there is one local maximum, which by the calculator is (0.07, 84.07). Because the function is of odd-degree, it is not an absolute minimum point.

(d) Because the polynomial function is of odd-degree, its range is $(-\infty,\infty)$.

(e) Using the calculator we find the x-intercepts are –6, –3.19 and 2.19; the y-intercept is 84.

(f) From the graph and our results, the function is increasing for the interval [–4.74, 0.07].

(g) From the graph and our results, the function is decreasing for the intervals $(-\infty,-4.74]$ and $[0.07,\infty)$.

61. See Figure 61 for the graph of this function.

(a) Because it is a polynomial function, its domain is $(-\infty,\infty)$.

(b) From the graph of *P,* there are 2 local minimum points, which by the calculator are (–1.73, –16.39) and (1.35, –3.49). Because the function is of odd-degree, neither is an absolute minimum point.

(c) From the graph of *P,* there are 2 local maximum points, which by the calculator are (–3, 0) and (0.17, 9.52). Because the function is of odd-degree, neither is an absolute maximum point.

(d) Because the polynomial function is of odd-degree, its range is $(-\infty,\infty)$.

(e) Using the calculator, we find the x-intercepts are –3, –0.62, 1, and 1.62; the y-intercept is 9.

(f) From the graph and our results, the function is increasing on $(-\infty,-3],[-1.73,0.17]$ and $[1.35,\infty)$.

(g) From the graph and our results, the function is decreasing for the intervals [–3,–1.73] and [0.17,1.35].

63. See Figure 63 for the graph of this function.

(a) Because it is a polynomial function, its domain is $(-\infty,\infty)$.

(b) From the graph of *P,* there is one absolute minimum point, which by the calculator is (–2.63, –132.69), and one local minimum point, which by the calculator is (1.68, –99.90).

(c) From the graph of *P,* there is one local maximum point, which by the calculator is (–0.17, –71.48). Because the function is of positive even-degree, there is not an absolute maximum point.

(d) Because the positive even-degree function has an absolute minimum, given in (b), its range is $(-132.69,\infty)$.

(e) Using the calculator, we find the x-intercepts are –4, and 3; the y-intercept is –72.

(f) From the graph and our results, the function is increasing for the intervals, [–2.63, –0.17] and $[1.68,\infty)$.

(g) From the graph and our results, the function is decreasing for the intervals, $(-\infty,-2.63]$ and $[-.17,1.68]$.

65. See Figure 65 for the graph of this function.

 (a) Because it is a polynomial function, its domain is $(-\infty, \infty)$.

 (b) From the graph of *P*, there is two local minimum points, which by the calculator are (–2, 0) and (2, 0).

 Because the function is of negative even-degree, there is not an absolute minimum point.

 (c) From the graph of *P*, there are 3 absolute maximum points, which by the calculator are (–3.46, 256), (0,256), and (3.46, 256).

 (d) Because the negative even-degree function has an absolute maximum, its range is $(-\infty, 256]$.

 (e) Using the calculator we find the x-intercepts are –4, –2, 2, and 4; the y-intercept is 256.

 (f) From the graph and our results, the function is increasing on $(-\infty, -3.46]$, $[-2, 0]$ and $[2, 3.46]$.

 (g) From the graph and our results, the function is decreasing on $(-3.46, -2]$, $[0.2]$, and $[3.46, \infty)$.

[-4,4] by [-20,20]
Xscl = 1 Yscl = 5

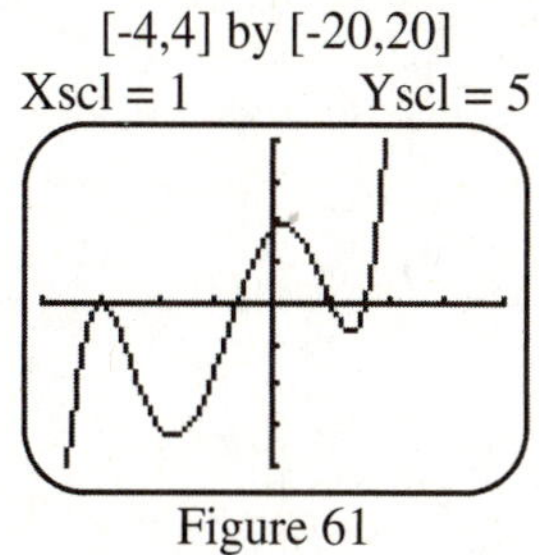

Figure 61

[-10,10] by [-200,100]
Xscl= 1 Yscl = 50

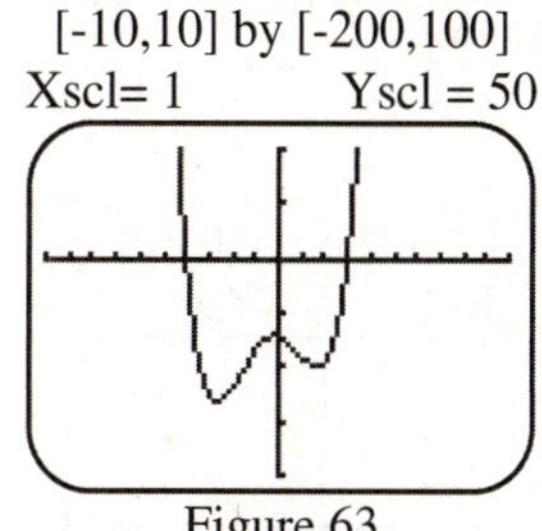

Figure 63

[-6,6] by [-300,300]
Xscl = 1 Yscl = 50

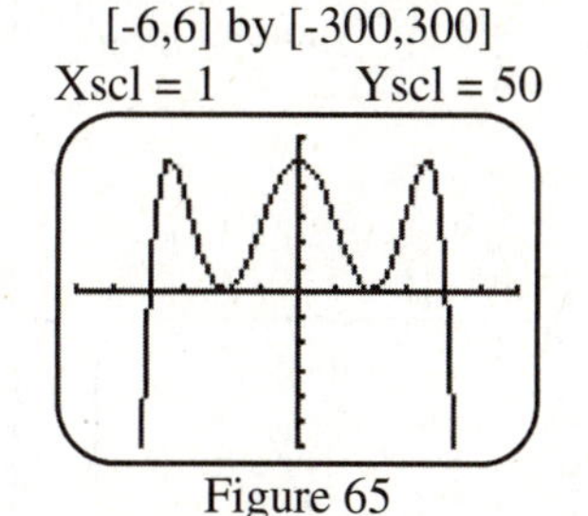

Figure 65

67. There are many possible valid windows, through experimentation one window is [–10, 10] by [–40, 10].

69. There are many possible valid windows, through experimentation one window is [–10, 20] by [–1500, 500].

71. There are many possible valid windows, through experimentation one window is [–10, 10] by [–20, 500].

73. (a) Let x = 3 correspond to March. $f(3) = 0.0145(3)^4 - 0.426(3)^3 + 3.53(3)^2 - 6.23(3) + 72 \approx 74.8$. The average high temperature in March is 74.8^o.

 Let x = 7 correspond to July. $f(7) = 0.0145(7)^4 - 0.426(7)^3 + 3.53(7)^2 - 6.23(7) + 72 \approx 90.1$ The average high temperature in July is 90.1^o.

 (b) From the graph of $f(x)$ we can see that the average temperature is 80 in April and October.

Reviewing Basic Concepts (Sections 3.4 and 3.5)

1. (a) The total length of the fence must equal $2L + 2x = 300 \Rightarrow 2L = 300 - 2x \Rightarrow L = 150 - x$.

 (b) Since width equals x and length equals $150 - x$, the function is $A(x) = x(150 - x)$.

 (c) Since length and width must both be positive, the restrictions are $0 < x < 150$.

 (d) $5000 = x(150 - x) \Rightarrow x^2 - 150x + 5000 = 0 \Rightarrow (x - 50)(x - 100) = 0 \Rightarrow x = 50 \text{ or } 100$. Using either value will yield a garden, which is 50 *m* by 100 *m*.

2. (a) See Figure 2a.

(b) Using the defined function and the point (51, .1) yields $f(x) = a(x-51)+0.1$. Now use (101, 4.7) to find the function: $4.7 = a(101-51)^2 + 0.1 \Rightarrow 4.6 = 2500a \Rightarrow a = 0.0018$; therefore, the function is $f(x) = 0.0018(x-51)^2 + 0.1$.

(c) Using the regression feature yields $g(x) \approx 0.0026x^2 - 0.3139x + 9.426$.

(d) Graph the data and two equations from (b) and (c), See Figure 2d. The regression function fits better because it passes through more data points. Neither function would fit the data for $x < 51$.

3. $P(x)$ is a positive odd-degree polynomial graph; therefore,

4. A polynomial of degree 3 can have at most 2 extrema, and at most 3 x-intercepts.

5. $P(x)$ is a positive even-degree polynomial graph; therefore, .

6. See Figure 6.

[50,120] by [0,5]
Xscl = 10 Yscl = 1

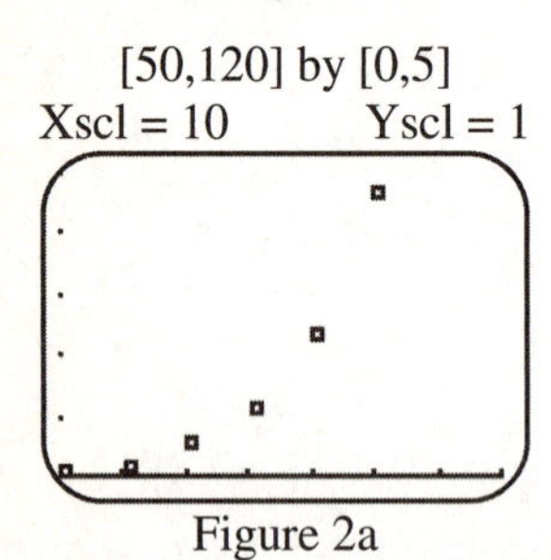

Figure 2a

[50,120] by [0,5]
Xscl = 10 Yscl = 1

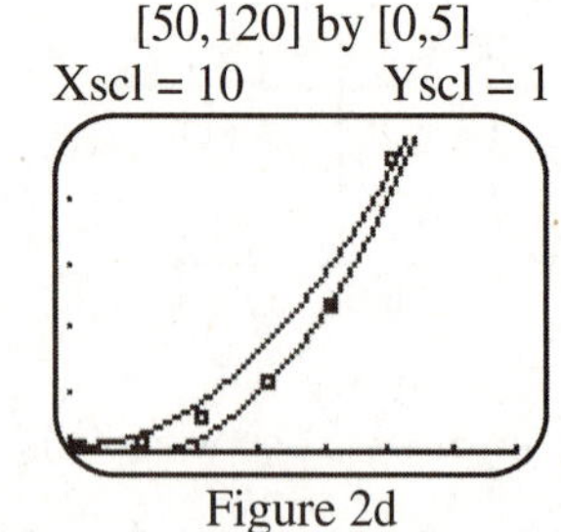

Figure 2d

[-10,10] by [-50,50]
Xscl= 1 Yscl = 10

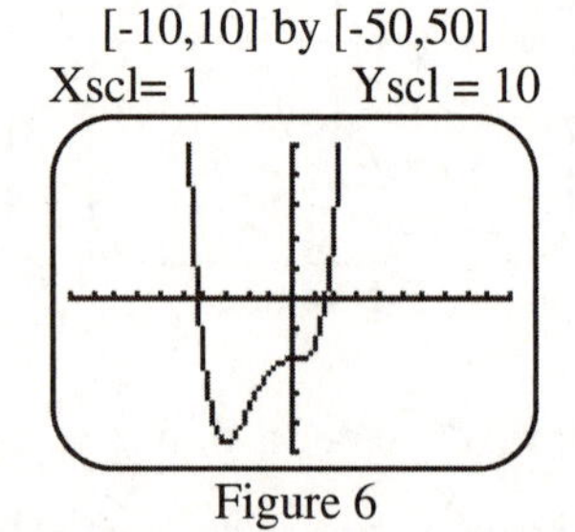

Figure 6

7. From the graph of P, there is one extrema, an absolute minimum point, which by the calculator is (–3, –47).

8. Using the calculator, we find the x-intercepts are –4.26 and 1.53; the y-intercept is –20.

3.6: Topics in the Theory of Polynomial Functions (I)

1. $\frac{10x^6}{5x^3} = 2x^3$

3. $\frac{8x^9}{3x^7} = \frac{8}{3}x^2$

5. $\frac{2x^6 + 3x^3}{2x} = \frac{2x^6}{2x} + \frac{3x^3}{2x} = x^5 + \frac{3}{2}x^2$

7. $\frac{8x^3 - 5x}{2x} = \frac{8x^3}{2x} - \frac{5x}{2x} = 4x^2 - \frac{5}{2}$

9. $P(1) = 3(1)^2 - 2(1) - 6 \Rightarrow P(1) = -5$ and $P(2) = 3(2)^2 - 2(2) - 6 \Rightarrow P(2) = 2$. These answers differ in sign; therefore, there is a real zero between them. Graphed on a calculator, the real zero is approximately 1.79.

11. $P(2)=2(2)^3-8(2)^2+(2)+16 \Rightarrow P(2)=2$ and $P(2.5)=2(2.5)^3-8(2.5)^2+(2.5)+16 \Rightarrow P(2.5)=-0.25$.

These answers differ in sign; therefore, there is a real zero between them. By graphing on the calculator, the real zero is approximately 2.39.

13. $P(1.5)=2(1.5)^4-4(1.5)^2+3(1.5)-6 \Rightarrow P(1.5)=-0.375$ and $P(2)=2(2)^4-4(2)^2+3(2)-6 \Rightarrow P(2)=16$.

These answers differ in sign; therefore, there is a real zero between them. By graphing on the calculator, the real zero is approximately 1.52.

15. $P(2.7)=-(2.7)^4+2(2.7)^3+(2.7)+12 \Rightarrow P(2.7)=0.9219$ and $P(2.8)=-(2.8)^4+2.(2.8)^3+(2.8)+12 \Rightarrow$ $P(2.8)=-2.7616$. These answers differ in sign; therefore, there is a real zero between them. By graphing on the calculator, the real zero is approximately 2.73.

17. $P(-1.6)=(-1.6)^5-2(-1.6)^3+1 \Rightarrow P(-1.6)=-1.29376$ and $P(-1.5)=(-1.5)^5-2(-1.5)^3+1 \Rightarrow$ $P(-1.5)=0.15625$. These answers differ in sign; therefore, there is a real zero between them. By graphing on the calculator, the real zero is approximately -1.51.

19. There is at least one zero between 2 and 2.5.

21.

$$\begin{array}{r|l} & x^2-3x-2 \\ x+5 & x^3+2x^2-17x-10 \\ & \underline{x^3+5x^2} \\ & -3x^2-17x-10 \\ & \underline{-3x^2-15x} \\ & -2x-10 \\ & \underline{-2x-10} \\ & 0 \end{array} \Rightarrow \frac{x^3+2x^2-17x-10}{x+5}=x^2-3x-2$$

23.

$$\begin{array}{r|l} & 3x^2+4x \\ x-5 & 3x^3-11x^2-20x+3 \\ & \underline{3x^3-15x^2} \\ & 4x^2-20x+3 \\ & \underline{4x^2-20x} \\ & 3 \end{array} \Rightarrow \frac{3x^3-11x^2-20x+3}{x-5}=3x^2+4x+\frac{3}{x-5}$$

25.

$$\begin{array}{r} x^3 - x^2 - 6x \\ x-2\overline{)x^4 - 3x^3 - 4x^2 + 12x} \\ \underline{x^4 - 2x^3 } \\ -x^3 - 4x^2 + 12x \\ \underline{-x^3 + 2x^2 } \\ -6x^2 + 12x \\ \underline{-6x^2 + 12x} \\ 0 \end{array} \Rightarrow \frac{x^4 - 3x^3 - 4x^2 + 12x}{x-2} = x^3 - x^2 - 6x$$

27.

$$\begin{array}{r} x^2 + 3x + 3 \\ x-1\overline{)x^3 + 2x^2 + 0x - 3} \\ \underline{x^3 - x^2 } \\ 3x^2 + 0x - 3 \\ \underline{3x^2 - 3x } \\ 3x - 3 \\ \underline{3x - 3} \\ 0 \end{array} \Rightarrow \frac{x^3 + 2x^2 - 3}{x-1} = x^2 + 3x + 3$$

29.

$$\begin{array}{r} -2x^2 + 2x - 3 \\ x+1\overline{)-2x^3 + 0x^2 - x - 2} \\ \underline{-2x^3 - 2x^2 } \\ 2x^2 - x - 2 \\ \underline{2x^2 + 2x } \\ -3x - 2 \\ \underline{-3x - 3} \\ 1 \end{array} \Rightarrow \frac{-2x^3 - x - 2}{x+1} = -2x^2 + 2x - 3 + \frac{1}{x+1}$$

31.

$$
\begin{array}{r|l}
 & x^4 + x^3 + x^2 + x + 1 \\
\hline
x-1 & x^5 + 0x^4 + 0x^3 + 0x^2 + 0x - 1 \\
 & \underline{x^5 - x^4} \\
 & x^4 + 0x^3 + 0x^2 + 0x - 1 \\
 & \underline{x^4 - x^3} \\
 & x^3 + 0x^2 + 0x - 1 \\
 & \underline{x^3 - x^2} \\
 & x^2 + 0x - 1 \\
 & \underline{x^2 - x} \\
 & x - 1 \\
 & \underline{x - 1} \\
 & 0
\end{array}
\Rightarrow \frac{x^5 - 1}{x - 1} = x^4 + x^3 + x^2 + x + 1
$$

33.
$$
\begin{array}{r|rrr}
3 & 1 & -4 & 3 \\
 & & 3 & -3 \\
\hline
 & 1 & -1 & 0
\end{array}
$$
Therefore, $P(3) = 0$.

35.
$$
\begin{array}{r|rrrr}
-2 & 5 & 2 & -1 & 5 \\
 & & -10 & 16 & -30 \\
\hline
 & 5 & -8 & 15 & -25
\end{array}
$$
Therefore, $P(-2) = -25$.

37.
$$
\begin{array}{r|rrr}
2 & 1 & -5 & 1 \\
 & & 2 & -6 \\
\hline
 & 1 & -3 & -5
\end{array}
$$
Therefore, $P(2) = -5$.

39.
$$
\begin{array}{r|rrrr}
0.5 & 1 & 0 & -1 & 4 \\
 & & 0.5 & 0.25 & -.375 \\
\hline
 & 1 & 0.5 & -0.75 & 3.625
\end{array}
$$
Therefore, $P(0.5) = 3.625$.

41.
$$
\begin{array}{r|rrrrr}
\sqrt{2} & 1 & 0 & -1 & 0 & -3 \\
 & & \sqrt{2} & 2 & \sqrt{2} & 2 \\
\hline
 & 1 & \sqrt{2} & 1 & \sqrt{2} & -1
\end{array}
$$
Therefore, $P(\sqrt{2}) = -1$.

43.
$$
\begin{array}{r|rrrr}
\sqrt[3]{4} & -1 & 0 & 1 & 4 \\
 & & -\sqrt[3]{4} & -\sqrt[3]{16} & \sqrt[3]{4} - 4 \\
\hline
 & -1 & -\sqrt[3]{4} & 1 - \sqrt[3]{16} & \sqrt[3]{4}
\end{array}
$$
Therefore, $P(\sqrt[3]{4}) = \sqrt[3]{4}$.

45. $$\begin{array}{r|rrr} 2 & 1 & 2 & -8 \\ & & 2 & 8 \\ \hline & 1 & 4 & 0 \end{array}$$

Yes; since $P(2) = 0$, 2 is a zero.

47. $$\begin{array}{r|rrrr} 4 & 2 & -6 & -9 & 6 \\ & & 8 & 8 & -4 \\ \hline & 2 & 2 & -1 & 2 \end{array}$$

No; since $P(4) \neq 0$, 4 is not a zero.

49. $$\begin{array}{r|rrrr} -0.5 & 4 & 12 & 7 & 1 \\ & & -2 & -5 & -1 \\ \hline & 4 & 10 & 2 & 0 \end{array}$$

Yes; since $P(-0.5) = 0$, -0.5 is a zero.

51. $$\begin{array}{r|rrrr} -5 & 8 & 50 & 47 & 15 \\ & & -40 & -50 & 15 \\ \hline & 8 & 10 & -3 & 30 \end{array}$$

No; since $P(-5) \neq 0$, -5 is not a zero.

53. $$\begin{array}{r|rrrrrrr} \sqrt{6} & -2 & 0 & 5 & 0 & -3 & 0 & 270 \\ & & -2\sqrt{6} & -12 & -7\sqrt{6} & -42 & -45\sqrt{6} & -270 \\ \hline & -2 & -2\sqrt{6} & -7 & -7\sqrt{6} & -45 & -45\sqrt{6} & 0 \end{array}$$

Yes; since $P(\sqrt{6}) = 0, \sqrt{6}$ is a zero.

55. Since the x-intercepts are -3, 1, and 4, the linear factors are $(x-(-3)), (x-1), (x-4)$ or $(x+3), (x-1), (x-4)$.

56. Since the x-intercepts are -3, 1, and 4, the solutions to $P(x) = 0$ are -3, 1, and 4.

57. Since the x-intercepts are -3, 1 and 4; the zeros are -3, 1, and 4

58. Use synthetic division and divide by 2:

$$\begin{array}{r|rrrr} 2 & 1 & -2 & -11 & 12 \\ & & 2 & 0 & -22 \\ \hline & 1 & 0 & -11 & -10 \end{array}$$

The remainder is -10; therefore, $P(2) = -10$.

59. Using the x-intercepts and graph, $P(x) > 0$ for the $(-3,1) \cup (4,\infty)$.

60. Using the x-intercepts and graph, $P(x) < 0$ for the $(-\infty,-3) \cup (1,4)$.

61. First, use synthetic division to factor out the given zero.

$$\begin{array}{r|rrrr} 3 & 1 & -2 & -5 & 6 \\ & & 3 & 3 & -6 \\ \hline & 1 & 1 & -2 & 0 \end{array}$$

Now completely factor the resulting quadratic expression: $x^2 + x - 2 \Rightarrow (x-1)(x+2)$.

The other zeros are -2 and 1.

63. First, use synthetic division to factor out the given zero.

$$\begin{array}{r|rrrr} 1 & 1 & 0 & -2 & 1 \\ & & 1 & 1 & -1 \\ \hline & 1 & 1 & -1 & 0 \end{array}$$

Now solve the quadratic equation, $x^2+x-1=0$, using the quadratic formula:

$x=\dfrac{-1\pm\sqrt{1^2-4(1)(-1)}}{2(1)}=\dfrac{-1\pm\sqrt{5}}{2}$ The other zeros are $\dfrac{-1-\sqrt{5}}{2}$ and $\dfrac{-1+\sqrt{5}}{2}$.

65. First, use synthetic division to factor out the given zero.

$$\begin{array}{r|rrrr} -2 & 3 & 5 & -3 & -2 \\ & & -6 & 2 & 2 \\ \hline & 3 & -1 & -1 & 0 \end{array}$$

Now solve the quadratic equation, $3x^2-x-1=0$, using the quadratic formula:

$x=\dfrac{1\pm\sqrt{(-1)^2-4(3)(-1)}}{2(3)}=\dfrac{1\pm\sqrt{13}}{6}$ The other zeros are $\dfrac{1-\sqrt{13}}{6}$ and $\dfrac{1+\sqrt{13}}{6}$.

67. First, use synthetic division to factor out the first given zero. Then use synthetic division to factor out the second given zero from the resulting expression: $x^3-6x^2-5x+30$.

$$\begin{array}{r|rrrrr} -6 & 1 & 0 & -41 & 0 & 180 \\ & & -6 & 36 & 30 & -180 \\ \hline & 1 & -6 & -5 & 30 & 0 \end{array} \Rightarrow \begin{array}{r|rrrr} 6 & 1 & -6 & -5 & 30 \\ & & 6 & 0 & -30 \\ \hline & 1 & 0 & -5 & 0 \end{array}$$

Finally, solve the resulting equation $x^2-5=0\Rightarrow x^2=5\Rightarrow x=\pm\sqrt{5}$.

Therefore, the other zeros are $-\sqrt{5}$ and $\sqrt{5}$.

69. First, use synthetic division to factor out the given zero.

$$\begin{array}{r|rrrr} 8 & -1 & 8 & 3 & -24 \\ & & -8 & 0 & 24 \\ \hline & -1 & 0 & 3 & 0 \end{array}$$

Now solve the resulting equation: $-x^2+3=0\Rightarrow -x^2=-3\Rightarrow x^2=3\Rightarrow x=\pm\sqrt{3}$

Therefore, the other zeros are $-\sqrt{3}$ and $\sqrt{3}$.

71. First, use synthetic division to factor out the given zero.

$$\begin{array}{r|rrrr} 2 & 2 & -3 & -17 & 30 \\ & & 4 & 2 & -30 \\ \hline & 2 & 1 & -15 & 0 \end{array}$$

Now completely factor the resulting quadratic expression: $2x^2+x-15\Rightarrow(2x-5)(x+3)$.

The linear factors are $P(x)=(x-2)(2x-5)(x+3)$.

73. First, use synthetic division to factor out the given zero.

$$\begin{array}{r|rrrr} -4 & 6 & 25 & 3 & -4 \\ & & -24 & -4 & 4 \\ \hline & 6 & 1 & -1 & 0 \end{array}$$

Now completely factor the resulting quadratic expression: $6x^2+x-1 \Rightarrow (3x-1)(2x+1)$. Therefore, the linear factors are $P(x)=(x+4)(3x-1)(2x+1)$.

75. First, use synthetic division to factor out the given zero.

$$\begin{array}{r|rrrr} -3 & -6 & -13 & 14 & -3 \\ & & 18 & -15 & 3 \\ \hline & -6 & 5 & -1 & 0 \end{array}$$

Now completely factor the resulting quadratic expression: $-6x^2+5x-1 \Rightarrow (-3x+1)(2x-1)$.

The linear factors are $P(x)=(x+3)(-3x+1)(2x-1)$.

77. First, use synthetic division to factor out the given zero.

$$\begin{array}{r|rrrr} -5 & 1 & 5 & -3 & -15 \\ & & -5 & 0 & 15 \\ \hline & 1 & 0 & -3 & 0 \end{array}$$

Now completely factor the resulting quadratic expression: $x^2-3=(x+\sqrt{3})(x-\sqrt{3})$.

The linear factors are $P(x)=(x+5)(x+\sqrt{3})(x-\sqrt{3})$.

79. First, use synthetic division to factor out the given zero.

$$\begin{array}{r|rrrr} -1 & 1 & -2 & -7 & -4 \\ & & -1 & 3 & 4 \\ \hline & 1 & -3 & -4 & 0 \end{array}$$

Now completely factor the resulting quadratic expression: $x^2-3x-4 \Rightarrow (x-4)(x+1)$.

The linear factors are $P(x)=(x+1)(x-4)(x+1)$ or $(x+1)^2(x-4)$.

81.

$$\begin{array}{r|l} & x^3+2 \\ \hline 3x-7 & 3x^4-7x^3+6x-16 \\ & \underline{3x^4-7x^3} \\ & \qquad\quad 6x-16 \\ & \qquad\quad \underline{6x-14} \\ & \qquad\qquad -2 \end{array} \Rightarrow \frac{3x^4-7x^3+6x-16}{3x-7}=x^3+2+\frac{-2}{3x-7}$$

83.

$$\begin{array}{r|l} & 5x^2 - 12 \\ \hline x^2+2 & 5x^4 - 2x^2 + 6 \\ & \underline{5x^4 + 10x^2} \\ & \quad -12x^2 + 6 \\ & \quad \underline{-12x^2 - 24} \\ & \qquad 30 \end{array} \Rightarrow \frac{5x^4 - 2x^2 + 6}{x^2+2} = 5x^2 - 12 + \frac{30}{x^2+2}$$

85.

$$\begin{array}{r|l} & 4x+5 \\ \hline 2x^2-3 & 8x^3 + 10x^2 - 12x - 15 \\ & \underline{8x^3 \qquad\quad - 12x} \\ & \qquad 10x^2 \qquad - 15 \\ & \qquad \underline{10x^2 \qquad - 15} \\ & \qquad\qquad\qquad 0 \end{array} \Rightarrow \frac{8x^3 + 10x^2 - 12x - 15}{2x^2+3} = 4x+5$$

87.

$$\begin{array}{r|l} & x^2 - 2x + 4 \\ \hline 2x^2+3x+2 & 2x^4 - x^3 + 4x^2 + 8x + 7 \\ & \underline{2x^4 + 3x^3 + 2x^2} \\ & \quad -4x^3 + 2x^2 + 8x \\ & \quad \underline{-4x^3 - 6x^2 - 4x} \\ & \qquad\qquad 8x^2 + 12x + 7 \\ & \qquad\qquad \underline{8x^2 + 12x + 8} \\ & \qquad\qquad\qquad\qquad -1 \end{array} \Rightarrow \frac{2x^4 - x^3 + 4x^2 + 8x + 7}{2x^2+3x+2} = x^2 - 2x + 4 + \frac{-1}{2x^2+3x+2}$$

89.

$$\begin{array}{r|l} & \frac{1}{2}x \\ \hline 2x+1 & x^2 + \frac{1}{2}x - 1 \\ & \underline{x^2 + \frac{1}{2}x} \\ & \qquad -1 \end{array} \Rightarrow \frac{x^2 + \frac{1}{2}x - 1}{2x+1} = \frac{1}{2}x + \frac{-1}{2x+1}$$

91.

$$\begin{array}{r} \frac{1}{2}x-\frac{1}{2} \\ 2x^2-1\overline{)x^3-x^2+0x+1} \\ \underline{x^3 \qquad -\frac{1}{2}x} \\ -x^2+\frac{1}{2}x+1 \\ \underline{-x^2 \qquad +\frac{1}{2}} \\ \frac{1}{2}x+\frac{1}{2} \end{array} \Rightarrow \frac{x^3-x^2+1}{2x^2-1}=\frac{1}{2}x-\frac{1}{2}+\frac{\frac{1}{2}x+\frac{1}{2}}{2x^2-1}$$

3.7: Topics in the Theory of Polynomial Functions (II)

1. With the given zeros 4 and $2+i$, the conjugate of $2+i$ or $2-i$ is also a zero. Therefore,

   $P(x)=(x-4)(x-(2+i))(x-(2-i))=(x-4)(x-2-i)(x-2+i)=(x-4)(x^2-4x+5)$

3. With the given zeros 5 and i, the conjugate of i or $-i$ is also a zero. Therefore,

   $P(x)=(x-5)(x-i)(x-(-i))=(x-5)(x^2-i^2)=(x-5)(x^2+1)\Rightarrow P(x)=x^3-5x^2+x-5.$

5. With the given zeros 0 and $3+i$, the conjugate of $3+i$ or $3-i$ is also a zero. Therefore,

   $P(x)=(x-0)(x-(3+i))(x-(3-i))=(x)(x-3-i)(x-3+i)=(x)(x^2-6x+10)$

   $\Rightarrow P(x)=x^3-6x^2+10x.$

7. With the given zeros $-3,-1,$ and 4, the function $P(x)$ is $P(x)=a(x-(-3))(x-(-1))(x-4)=$

   $a(x+3)(x+1)(x-4)=a(x+3)(x^2-3x-4)\Rightarrow P(x)=a(x^3-13x-12)$

   Since $P(2)=5$ we can solve for a:

   $5=a((2)^3-13(2)-12)\Rightarrow 5=a(8-26-12)\Rightarrow -30a=5\Rightarrow a=-\frac{5}{30}=-\frac{1}{6}$

   Therefore, $P(x)=-\frac{1}{6}(x^3-13x-12)\Rightarrow P(x)=-\frac{1}{6}x^3+\frac{16}{3}x+2.$

9. With the given zeros $-2,1$ and 0, the function $P(x)$ is

   $P(x)=a(x-(-2))(x-1)(x-0)=a(x+2)(x-1)(x)\Longrightarrow P(x)=ax(x^2+x-2)$

   Since $P(-1)=-1,$ we can solve for a: $-1=a(-1)((-1)^3+(-1)-2)\Rightarrow -1=-a(1-1-2)$

   $\Rightarrow -a(-2)=-1\Rightarrow 2a=-1\Rightarrow a=-\frac{1}{2}$. Therefore, $P(x)=-\frac{1}{2}x(x^2+x-2)\Rightarrow P(x)=-\frac{1}{2}x^3-\frac{1}{2}x^2+x.$

11. With the given zeros 4, $1+i$, and the conjugate of $1+i$ or $1-i$ also a zero, the function $P(x)$ is

$P(x) = a(x-4)(x-(1+i))(x-(1-i)) = a(x-4)(x-1-i)(x-1+i) =$

$a(x-4)(x^2-2x+2) = a(x^3-2x^2+2x-4x^2+8x-8) \Rightarrow P(x) = a(x^3-6x^2-10x-8)$. Since $P(2)=4$,

we can solve for a: $4 = a((2)^3-6(2)^2+10(2)-8) \Rightarrow 4 = a(8-24+20-8) \Rightarrow -4a = 4 \Rightarrow a = -\frac{1}{1} = -1$.

Therefore, $P(x) = -1(x^3-6x^2+10x-8) \Rightarrow P(x) = -x^3+6x^2-10x+8$.

13. First, use synthetic division to factor out the given zero.

$$\begin{array}{r|rrrr} 3 & 1 & -1 & -4 & -6 \\ & & 3 & 6 & 6 \\ \hline & 1 & 2 & 2 & 0 \end{array}$$

Now solve the resulting quadratic equation, $x^2+2x+2=0$, using the quadratic formula: $x = \frac{-2 \pm \sqrt{(2)^2-4(1)(2)}}{2(1)} = \frac{-2 \pm \sqrt{-4}}{2} = \frac{2 \pm 2i}{2} = -1 \pm i.$

Therefore, the other zeros are $-1-i$ and $-1+i$.

15. First, use synthetic division to factor out the first given zero. Then use synthetic division to factor out the second given zero from the resulting expression x^3-x^2-7x+3

$$\begin{array}{r|rrrrr} -3 & 1 & 2 & -10 & -18 & 9 \\ & & -3 & 3 & 21 & -9 \\ \hline & 1 & -1 & -7 & 3 & 0 \end{array} \Rightarrow \quad \begin{array}{r|rrrr} 3 & 1 & -1 & -7 & 3 \\ & & 3 & 6 & -3 \\ \hline & 1 & 2 & -1 & 0 \end{array}$$

Finally, solve the resulting quadratic equation, $x^2+2x-1=0$, using the quadratic formula:

$$x = \frac{-2 \pm \sqrt{(2)^2-4(1)(-1)}}{2(1)} = \frac{-2 \pm \sqrt{8}}{2} = \frac{-2 \pm 2\sqrt{2}}{2} = -1 \pm \sqrt{2}.$$

Therefore, the other zeros are $-1+\sqrt{2}$ and $-1-\sqrt{2}$.

17. With the given zero $3i$, the conjugate $-3i$ is also a zero. Use synthetic division to factor out these zeros.

$$\begin{array}{r|rrrrr} 3i & 1 & -1 & 10 & -9 & 9 \\ & & 3i & -9-3i & 9+3i & -9 \\ \hline & 1 & -1+3i & 1-3i & 3i & 0 \end{array} \Rightarrow \quad \begin{array}{r|rrrr} -3i & 1 & -1+3i & 1-3i & 3i \\ & & -3i & 3i & -3i \\ \hline & 1 & -1 & 1 & 0 \end{array}$$

Finally, solve the resulting quadratic equation, $x^2-x+1=0$, using the quadratic formula:

$$x = \frac{1 \pm \sqrt{(-1)^2-4(1)(1)}}{2(1)} = \frac{1 \pm \sqrt{-3}}{2} = \frac{1}{2} \pm \frac{\sqrt{3}}{2}i.$$

Therefore, the other zeros are $-3i$, $\frac{1}{2}+\frac{\sqrt{3}}{2}i$, and $\frac{1}{2}-\frac{\sqrt{3}}{2}i$.

19. With the given zeros 5 and –4, $P(x) = a(x-5)(x-(-4)) = a(x-5)(x+4) \Rightarrow P(x) = a(x^2 - x - 20)$.

There are many possible solutions. For example, when $a = 1$, $P(x) = x^2 - x - 20$.

21. With the given zeros –3, 2, *i*, the conjugate of *i* or –*i* is also zero. The function $P(x)$ is

$P(x) = a(x-(-3))(x-2)(x-i)(x-(-i)) = a(x+3)(x-2)(x-i)(x+i) =$

$\Rightarrow P(x) = a(x^2 + x - 6)(x^2 + 1) = a(x^4 + x^3 - 5x^2 + x - 6)$. There are many possible solutions. For example, when $a = 1$, $P(x) = x^4 + x^3 - 5x^2 + x - 6$.

23. With the given zeros $1+\sqrt{3}$, $1-\sqrt{3}$, and 1,

$P(x) = a\left(x-\left(1+\sqrt{3}\right)\right)\left(x-\left(1-\sqrt{3}\right)\right)(x-1) = a\left(x-1-\sqrt{3}\right)\left(x-1+\sqrt{3}\right)(x-1) =$

$a\left(x^2 - x + \sqrt{3}x - x + 1 - \sqrt{3} - \sqrt{3}x + \sqrt{3} - 3\right)(x-1) = a(x^2 - 2x - 2)(x-1) =$

$a(x^3 - x^2 - 2x^2 + 2x - 2x + 2) \Rightarrow P(x) = a(x^3 - 3x^2 + 2)$. There are many possible solutions. For example, when $a = 1$, $P(x) = x^3 - 3x^2 + 2$.

25. With the given zeros −1, 2, and 3 + 2*i*, the conjugate of 3 + 2*i* or 3 – 2*i* is also a zero.

$P(x) = a(x-(3+2i))(x-(3-2i))(x-(-1))(x-2) = a(x-3-2i)(x-3+2i)(x+1)(x-2) =$

$a(x^2 - 3x + 2ix - 3x + 9 - 6i - 2ix + 6i - 4i^2)(x^2 - x - 2) = a(x^2 - 6x + 13)(x^2 - x - 2) =$

$a(x^4 - x^3 - 2x^2 - 6x^3 + 6x^2 + 12x + 13x^2 - 13x - 26) \Rightarrow P(x) = a(x^4 - 7x^3 + 17x^2 - x - 26)$. There are many possible solutions. For example, when $a = 1$, $P(x) = x^4 - 7x^3 + 17x^2 - x - 26$.

27. With the given zeros –1, and $6-3i$, the conjugate of $6-3i$ or $6+3i$ also a zero. The function is

$P(x) = a(x-(6-3i))(x-(6+3i))(x-(-1)) = a(x-6+3i)(x-6-3i)(x+1) =$

$a(x^2 - 6x - 3ix - 6x + 36 + 18i + 3ix - 18i - 9i^2)(x+1) = a(x^2 - 12x + 45)(x+1) =$

$a(x^3 - 12x^2 + 45x + x^2 - 12x + 45) \Rightarrow P(x) = a(x^3 - 11x^2 + 33x + 45)$. There are many possible solutions. For example, when $a = 1$, $P(x) = x^3 - 11x^2 + 33x + 45$.

29. With the given zeros –3 (multiplicity 2) and 2 + *i*, the conjugate of $2+i$ or $2-i$ also a zero.

$P(x) = a(x-(2+i))(x-(2-i))(x-(-3))^2 = a(x-2-i)(x-2+i)(x+3)^2 =$

$a(x^2 - 2x + ix - 2x + 4 - 2i - ix + 2i - i^2)(x^2 + 6x + 9) = a(x^2 - 4x + 5)(x^2 + 6x + 9) =$

$a(x^4 + 6x^3 + 9x^2 - 4x^3 - 24x^2 - 36x + 5x^2 + 30x + 45) \Rightarrow P(x) = a(x^4 + 2x^3 - 10x^2 - 6x + 45)$. There are many possible solutions. For example, when $a = 1$, $P(x) = x^4 + 2x^3 - 10x^2 - 6x + 45$.

31 - 39. See Figures 31-39

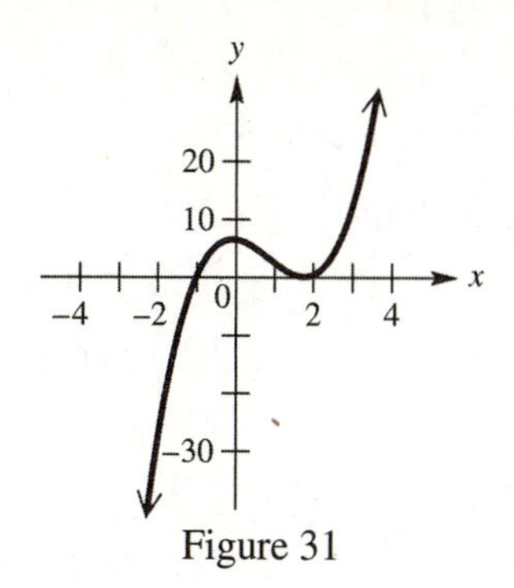

Figure 31

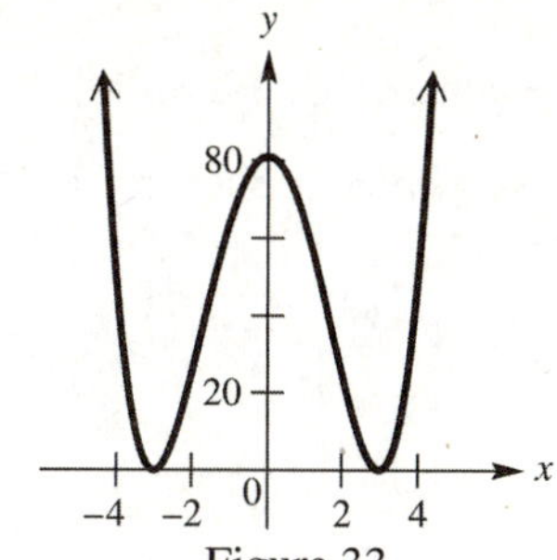

Figure 33

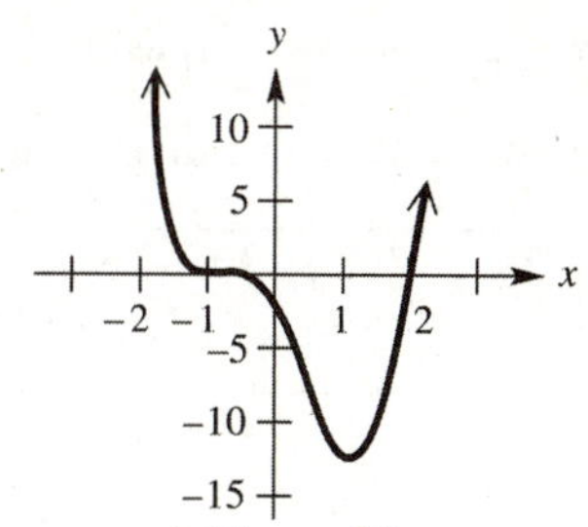

Figure 35

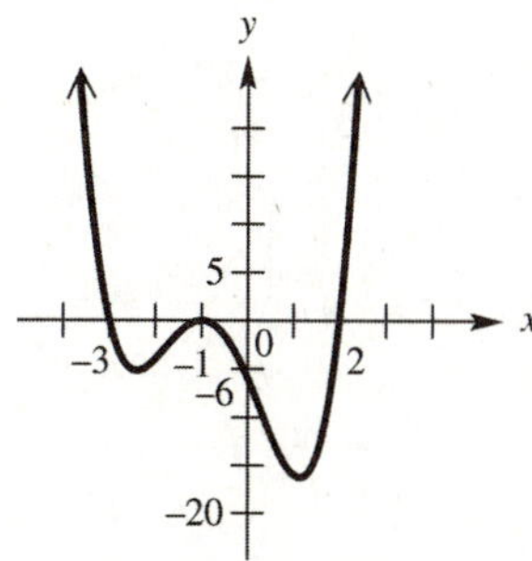

Figure 37

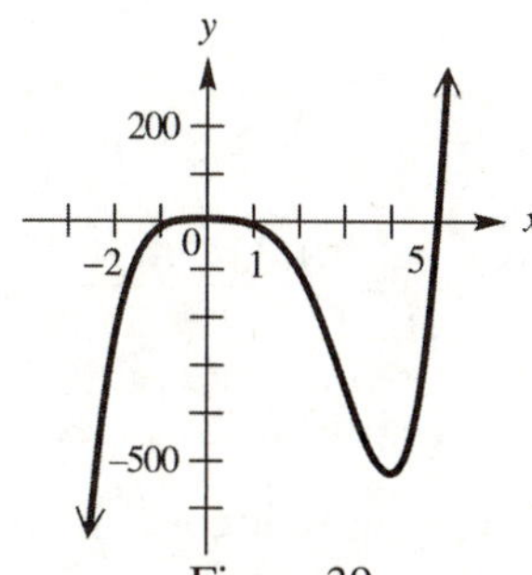

Figure 39

41. First, use synthetic division to factor out the first –2. Then use synthetic division to factor out the second –2 from the resulting expression: x^3+0x^2-7x-6.

$$\begin{array}{r|rrrrr} -2 & 1 & 2 & -7 & -20 & -12 \\ & & -2 & 0 & 14 & 12 \\ \hline & 1 & 0 & -7 & -6 & 0 \end{array} \Rightarrow \qquad \begin{array}{r|rrrr} -2 & 1 & 0 & -7 & -6 \\ & & -2 & 4 & 6 \\ \hline & 1 & -2 & -3 & 0 \end{array}$$

Finally, completely factor the resulting quadratic expression: $x^2-2x-3=(x-3)(x+1)$. Therefore, the zeros are –2, 3, and –1, and the factored form is $P(x)=(x+2)^2(x-3)(x+1)$.

43. Since non real complex zeros occur in pairs, there will be 0, 2, or 4 non real complex zeros, leading to 5, 3, or 1 real zeros.

45. (a) Not Possible. With a zero of $1+i$, its conjugate $1-i$ must also be a zero, and this would give 4 zeros to a degree 3 function, which is not possible.

(b) Possible. Since non real complex zeros occur in pairs, there can be 4 non real complex zeros.

(c) Not Possible. We cannot have a multiplicity of 6 for a degree 5 function.

(d) Possible. With $1+2i$ having a multiplicity of 2, its conjugate $1-2i$ must also have a multiplicity of 2. This means the function has a minimum of 4 zeros which is possible. .

47. (a) Using the Rational Zeros Theorem yields $\frac{p}{q}=\pm\frac{1,2,5,10}{1}=\pm1\pm2\pm5\pm10$.

(b) A graph (not shown) would indicate that there are no zeros less than -2 or greater than 5.

(c) First, use synthetic division to factor using values discovered in steps (a) and (b)

$$\begin{array}{r|rrrr} -2 & 1 & -2 & -13 & -10 \\ & & -2 & 8 & 10 \\ \hline & 1 & -4 & -5 & 0 \end{array}$$

Now completely factor the resulting quadratic expression: $x^2-4x-5 \Rightarrow (x-5)(x+1)$. Therefore, the rational zeros are $-2,-1$ and 5.

(d) The factored form of $P(x)$ is $P(x)=(x+2)(x+1)(x-5)$.

49. (a) Using the Rational Zeros Theorem yields $\frac{p}{q}=\pm\frac{1,2,3,5,6,10,15,30}{1}$

(b) A graph (not shown) would indicate that there are no zeros less than -5 or greater than 2.

(c) First, use synthetic division to factor using values discovered in steps (a) and (b).

$$\begin{array}{r|rrrr} -5 & 1 & 6 & -1 & -30 \\ & & -5 & -5 & 30 \\ \hline & 1 & 1 & -6 & 0 \end{array}$$

Now completely factor the resulting quadratic expression: $x^2+x-6 \Rightarrow (x+3)(x-2)$. Therefore, the rational zeros are $-5,-3$ and 2.

(d) The factored form of $P(x)$ is $P(x)=(x+5)(x+3)(x-2)$.

51. (a) Using the Rational Zeros Theorem yields

$$\frac{p}{q}=\pm\frac{1,2,3,4,6,12}{1,2,3,6}=\pm1,\pm2,\pm3,\pm4,\pm6,\pm12,\pm\frac{1}{2},\pm\frac{3}{2},\pm\frac{1}{3},\pm\frac{2}{3},\pm\frac{4}{3},\pm\frac{1}{6}.$$

(b) A graph (not shown) would indicate that there are no zeros less than -4 or greater than $\frac{3}{2}$.

(c) First, use synthetic division to factor using values discovered in steps (a) and (b).

$$\begin{array}{r|rrrr} -4 & 6 & 17 & -31 & -12 \\ & & -24 & 28 & 12 \\ \hline & 6 & -7 & -3 & 0 \end{array}$$

Now completely factor the resulting quadratic expression: $6x^2-7x-3 \Rightarrow (3x+1)(2x-3)$. Therefore, the rational zeros are $-4,-\frac{1}{3}$, and $\frac{3}{2}$.

(d) The factored form of $P(x)$ is $P(x)=(x+4)(3x+1)(2x-3)$.

53. Using the Rational Zeros Theorem yields

$\frac{p}{q}=\pm\frac{1,2,3,6}{1,2,3,4,6,12}=\pm1,\pm2,\pm3,\pm6,\pm\frac{1}{2},\pm\frac{3}{2},\pm\frac{1}{3},\pm\frac{2}{3},\pm\frac{1}{6},\pm\frac{1}{12},\pm\frac{1}{4},\pm\frac{3}{4}$. Use synthetic division to factor using values from above.

$$\begin{array}{r|rrrr} -\frac{3}{2} & 12 & 20 & -1 & -6 \\ & & -18 & -3 & 6 \\ \hline & 12 & 2 & -4 & 0 \end{array}$$

Now completely factor the resulting quadratic expression: $12x^2+2x-4 \Rightarrow 2(2x-1)(3x+2)$. The factored form of $P(x)$ is $P(x)=(3x+2)(2x+3)(2x-1)$.

55. Using the Rational Zeros Theorem yields $\frac{p}{q}=\pm\frac{1,2,3,4,6,12}{1,2,3,4,6,8,12,24}=$

$\pm1,\pm2,\pm3,\pm4,\pm6,\pm12,\pm\frac{1}{2},\pm\frac{3}{2},\pm\frac{1}{3},\pm\frac{2}{3},\pm\frac{4}{3},\pm\frac{1}{4},\pm\frac{3}{4},\pm\frac{1}{6},\pm\frac{1}{8},\pm\frac{3}{8},\pm\frac{1}{12},\pm\frac{1}{24}$.

Factor out a 2, $P(x)=24x^3+40x^2-2x-12 \Rightarrow P(x)=2(12x^3+20x^2-x-6)$. Now use synthetic division to factor using values from above.

$$\begin{array}{r|rrrr} -\frac{3}{2} & 12 & 20 & -1 & -6 \\ & & -18 & -3 & 6 \\ \hline & 12 & 2 & -4 & 0 \end{array}$$

Now completely factor the resulting quadratic expression: $12x^2+2x-4 \Rightarrow 2(3x+2)(2x-1)$. The factored form of $P(x)$ is $P(x)=2(2x+3)(3x+2)(2x-1)$.

57. To eliminate the fractions, multiply the function by 2; therefore, $P(x)=2\left(x^3+\frac{1}{2}x^2-\frac{11}{2}x-5\right)=$

$2x^3+x^2-11x-10$. Now, use the Rational Zeros Theorem to identify possible zeros:

$\frac{p}{q}=\pm10,\pm5,\pm1,\pm\frac{1}{2},\pm\frac{5}{2}$. Next, from the calculator graph of the equation, we choose to check -1 by synthetic division.

$$\begin{array}{r|rrrr} -1 & 2 & 1 & -11 & -10 \\ & & -2 & 1 & 10 \\ \hline & 2 & -1 & -10 & 0 \end{array}$$

Now, completely factor the resulting quadratic expression: $2x^2-x-10 \Rightarrow (2x-5)(x+2)$. Therefore, the rational zeros are $-2,-1,$ and $\frac{5}{2}$.

59. To eliminate the fractions, multiply the function by 12; therefore,

$P(x)=12\left(\frac{1}{6}x^4-\frac{11}{12}x^3+\frac{7}{6}x^2-\frac{11}{12}x+1\right)=2x^4-11x^3+14x^2-11x+12$. Now, use the Rational Zeros Theorem to identify possible zeros: $\frac{p}{q}=\pm1,\ \pm2,\pm3,\pm4,\pm6,\pm12,\pm\frac{1}{2},\pm\frac{3}{2}$. Next, from the calculator, graph of the equation, we choose to check $\frac{3}{2}$ and 4 by synthetic division. (The second from the result of the first)

$$\begin{array}{r|rrrrr} \frac{3}{2} & 2 & -11 & 14 & -11 & 12 \\ & & 3 & -12 & 3 & -12 \\ \hline & 2 & -8 & 2 & -8 & 0 \end{array} \Rightarrow \begin{array}{r|rrrr} 4 & 2 & -8 & 2 & -8 \\ & & 8 & 0 & 8 \\ \hline & 2 & 0 & 2 & 0 \end{array}$$

Now, completely factor the resulting quadratic expression: $2x^2+2 \Rightarrow 2\left(x^2+1\right) \Rightarrow x^2=-1 \Rightarrow x=\pm i$.

Since $\pm i$ are not real numbers, the rational zeros are $\frac{3}{2}$ and 4.

61. $P(x)=6x^4-5x^3-11x^2+10x-2$. Now use the Rational Zeros Theorem to identify possible zeros: $\frac{p}{q}=\pm1,\pm2,\pm\frac{1}{2},\pm\frac{1}{3},\pm\frac{2}{3},\pm\frac{1}{6}$. Next, from the calculator, graph of the equation, we choose to check $\frac{1}{2}$ and $\frac{1}{3}$ by synthetic division. (The second from the result of the first)

$$\begin{array}{r|rrrrr} \frac{1}{2} & 6 & -5 & -11 & 10 & -2 \\ & & 3 & -1 & -6 & 2 \\ \hline & 6 & -2 & -12 & 4 & 0 \end{array} \Rightarrow \begin{array}{r|rrrr} \frac{1}{3} & 6 & -2 & -12 & 4 \\ & & 2 & 0 & -4 \\ \hline & 6 & 0 & -12 & 0 \end{array}$$

Now completely factor the resulting quadratic expression: $6x^2-12 \Rightarrow 6\left(x^2-2\right) \Rightarrow x^2=2 \Rightarrow x=\pm\sqrt{2}$

The factored form of $P(x)$ is $P(x)=(2x-1)(3x-1)\left(x-\sqrt{2}\right)\left(x+\sqrt{2}\right)$.

63. $P(x)=21x^4+13x^3-103x^2-65x-10$. Now use the Rational Zeros Theorem to identify possible zeros:

$\frac{p}{q}=\pm1,\pm2,\pm5,\pm10,\pm\frac{5}{21},\pm\frac{5}{7},\pm\frac{5}{3},\pm\frac{2}{21},\pm\frac{2}{7},\pm\frac{2}{3},\pm\frac{10}{21},\pm\frac{10}{7},\pm\frac{10}{3},\pm\frac{1}{21},\pm\frac{1}{7},\pm\frac{1}{3}$. Next, from the calculator, graph of the equation, we choose to check $-\frac{2}{7}$ and $-\frac{1}{3}$ by synthetic division. (The second from the result of the first)

$$-\frac{2}{7}\overline{)21 \quad 13 \quad -103 \quad -65 \quad -10} \qquad \Rightarrow \qquad -\frac{1}{3}\overline{)21 \quad 7 \quad -105 \quad -35}$$

$$\begin{array}{rrrrr} & -6 & -2 & 30 & 10 \\ \hline 21 & 7 & -105 & -35 & 0 \end{array} \qquad \begin{array}{rrrr} & -7 & 0 & 35 \\ \hline 21 & 0 & -105 & 0 \end{array}$$

Now completely factor the resulting quadratic expression: $21x^2-105 \Rightarrow 21\left(x^2-5\right) \Rightarrow x^2=5 \Rightarrow x=\pm\sqrt{5}$

The factored form of $P(x)$ is $P(x)=(7x+2)(3x+1)\left(x-\sqrt{5}\right)\left(x+\sqrt{5}\right)$.

65. With the given zero i, the conjugate $-i$ is also a zero. Now use synthetic division to factor out these zeros.

$$i\overline{)1 \quad -1 \quad 5 \quad -5 \quad 4 \quad -4} \Rightarrow -i\overline{)1 \quad -1+i \quad 4-i \quad -4+4i \quad -4i}$$

$$\begin{array}{rrrrrr} & i & -1-i & 1+4i & -4+4i & 4 \\ \hline 1 & -1+i & 4-i & -4+4i & -4i & 0 \end{array} \qquad \begin{array}{rrrrr} & -i & i & -4i & 4i \\ \hline 1 & -1 & 4 & -4 & 0 \end{array}$$

From the graph of the resulting polynomial we can see that 1 is a zero.

$$1\overline{)1 \quad -1 \quad 4 \quad -4}$$

$$\begin{array}{rrrr} & 1 & 0 & 4 \\ \hline 1 & 0 & 4 & 0 \end{array}$$

The resulting polynomial $x^2+4 \Rightarrow x=\pm 2i$. The factors of P(x) are $(x-1)(x-i)(x+i)(x-2i)(x+2i)$.

67. With the given zero $-2i$, the conjugate $2i$ is also a zero. Now use synthetic division to factor out these zeros.

$$-2i\overline{)1 \quad 1 \quad 2 \quad 4 \quad -8} \qquad 2i\overline{)1 \quad 1-2i \quad -2-2i \quad 4i}$$

$$\begin{array}{rrrrr} & -2i & -4-2i & -4+4i & 8 \\ \hline 1 & 1-2i & -2-2i & 4i & 0 \end{array} \Rightarrow \begin{array}{rrrr} & 2i & 2i & -4i \\ \hline 1 & 1 & -2 & 0 \end{array}$$

Factoring the resulting polynomial we have $x^2+x-2=(x+2)(x-1)$. The factors of P(x) are $(x+2)(x-1)(x-2i)(x+2i)$.

69. With the given zero $1+i$, the conjugate $1-i$ is also a zero. Now use synthetic division to factor out these zeros.

$$1+i\overline{)1 \quad -2 \quad 3 \quad -2 \quad 2} \qquad 1-i\overline{)1 \quad -1+i \quad 1 \quad -1+i}$$

$$\begin{array}{rrrrr} & 1+i & -2 & 1+i & -2 \\ \hline 1 & -1+i & 1 & -1+i & 0 \end{array} \Rightarrow \begin{array}{rrrr} & 1-i & 0 & 1-i \\ \hline 1 & 0 & 1 & 0 \end{array}$$

Factoring the resulting polynomial we have $x^2+1 \Rightarrow x=\pm i$. The factors of P(x) are $(x-(1+i))(x-(1-i))(x-i)(x+i)$.

71. Because the function $P(x)$ has coefficient signs: $+\,-\,+\,+$, which is 2 sign changes, the function has 2 or 0 possible positive real zeros. Because $P(-x)=2(-x)^3-4(-x)^2+2(-x)+7=-2x^3-4x^2-2x+7$ has coefficient signs: $-\,-\,-\,+$, which is 1 sign change, the function has 1 negative real zero. From graphing the function, $P(x)$ actually has 0 positive and 1 negative real zeros.

73. Because the function $P(x)$ has coefficient signs: $+\,+\,+\,-$, which is 1 sign change, the function has 1 positive real zero. Because $P(-x)=5(-x)^4+3(-x)^2+2(-x)-9=5x^4+3x^2-2x-9$ has coefficient signs: $+\,+\,-\,-$, which is 1 sign change, the function has 1 negative real zero.

75. Because the function $P(x)$ has coefficient signs: $+\,+\,-\,+\,+$, which is 2 sign changes, the function has 2 or 0 possible positive real zeros. Because $P(-x)=(-x)^5+3(-x)^4-(-x)^3+2(-x)+3=$ $-x^5+3x^4+x^3-2x+3$ has coefficient signs: $-\,+\,+\,-\,+$, which is 3 sign changes, the function has 3 or 1 possible negative real zeros. From graphing the function, $P(x)$ actually has 0 positive and 1 negative real zero.

77. Using the Boundedness Theorem, we divide synthetically by $x-2$.

$$\begin{array}{r|rrrrr} 2 & 1 & -1 & 3 & -8 & 8 \\ & & 2 & 2 & 10 & 4 \\ \hline & 1 & 1 & 5 & 2 & 12 \end{array}$$

The result is all nonnegative; therefore, no real zero greater than 2.

79. Using the Boundedness Theorem, we divide synthetically by $x-(-2)$.

$$\begin{array}{r|rrrrr} -2 & 1 & 1 & -1 & 0 & 3 \\ & & -2 & 2 & -2 & 4 \\ \hline & 1 & -1 & 1 & -2 & 7 \end{array}$$

The result alternates in sign; therefore, no real zero less than -2.

81. Using the Boundedness Theorem, we divide synthetically by $x-1$.

$$\begin{array}{r|rrrrr} 1 & 3 & 2 & -4 & 1 & -1 \\ & & 3 & 5 & 1 & 2 \\ \hline & 3 & 5 & 1 & 2 & 1 \end{array}$$

The result is all nonnegative; therefore, no real zero greater than 1.

83. Using the Boundedness Theorem, we divide synthetically by $x-2$.

$$\begin{array}{r|rrrrrr} 2 & 1 & 0 & -3 & 0 & 1 & 2 \\ & & 2 & 4 & 2 & 4 & 10 \\ \hline & 1 & 2 & 1 & 2 & 5 & 12 \end{array}$$

The result is all nonnegative; therefore, no real zero greater than 2.

$P(x)=\frac{1}{2}(x^3-x^2-32x+60) \Rightarrow P(x)=\frac{1}{2}x^3-\frac{1}{2}x^2-16x+30.$

85. From the graph, the function has zeros $-6, 2$ and 5, the function $P(x)$ is

$P(x) = a(x-(6))(x-2)(x-5) = a(x+6)(x-2)(x-5) = a(x+6)(x^2-7x+10) \Rightarrow$

$P(x) = a(x^3 - x^2 - 32x + 60)$. Since $P(0) = 30$, we can solve for a:

$$30 = a((0)^3 - (0)^2 - 42(0) + 60) \Rightarrow 30 = a(0-0-0+60) \Rightarrow 60a = 30 \Rightarrow a = \frac{30}{60} = \frac{1}{2}.$$

Therefore, $P(x) = \frac{1}{2}(x^3 - x^2 - 32x + 60) \Rightarrow P(x) = \frac{1}{2}x^3 - \frac{1}{2}x^2 - 16x + 30.$

87. (a) Because the function $P(x)$ has coefficient signs: $- - + +$, which is 1 sign change, the function has 1 positive real zero. Because $P(-x) = -2(-x)^4 - (-x)^3 + (-x) + 2 = -2x^4 + x^2 - x + 2$ has coefficient signs: $- + - +$, which is 3 sign changes, the function has 3 or 1 negative real zero.

(b) By the Rational Zero Theorem, the possible rational zeros are $\frac{p}{q} = \pm\frac{1,2}{1,2} = \pm 1, \pm 2, \pm\frac{1}{2}$.

(c) From part (b) and a calculator graph of the equation, we choose 1 and -1 to check by synthetic division. (The second from the result of the first)

$$\begin{array}{r|rrrrr} 1 & -2 & -1 & 0 & 1 & 2 \\ & & -2 & -3 & -3 & -2 \\ \hline & -2 & -3 & -3 & -2 & 0 \end{array} \Rightarrow \begin{array}{r|rrrr} -1 & -2 & -3 & -3 & -2 \\ & & 2 & 1 & 2 \\ \hline & -2 & -1 & -2 & 0 \end{array}$$

The resulting polynomial $-2x^2 - x - 2$ has no real zeros, so the rational zeros are 1 and -1.

(d) No other real zeros, the remaining zeros are imaginary.

(e) Using the quadratic equation yields

$$x = \frac{-(-1) \pm \sqrt{(-1)^2 - 4(-2)(-2)}}{2(-2)} = \frac{1 \pm \sqrt{-15}}{-4} \Rightarrow x = -\frac{1}{4} + i\frac{\sqrt{15}}{4} \text{ and } -\frac{1}{4} - i\frac{\sqrt{15}}{4}.$$

(f) The rational zeros are the x-intercepts; therefore, 1 and -1.

(g) Find $P(0)$: $P(0) = -2(0)^4 - (0)^3 + (0) + 2 \Rightarrow P(0) = 2$. The y-intercept is 2.

(h)
$$\begin{array}{r|rrrrr} 4 & -2 & -1 & 0 & 1 & 2 \\ & & -8 & -36 & -144 & -572 \\ \hline & -2 & -9 & -36 & -143 & -570 \end{array}$$
Therefore, $f(4) = -570$; $(4, -570)$.

(i) $P(x)$ is an negative even-degree polynomial graph; therefore, ∩

(j) See Figure 87.

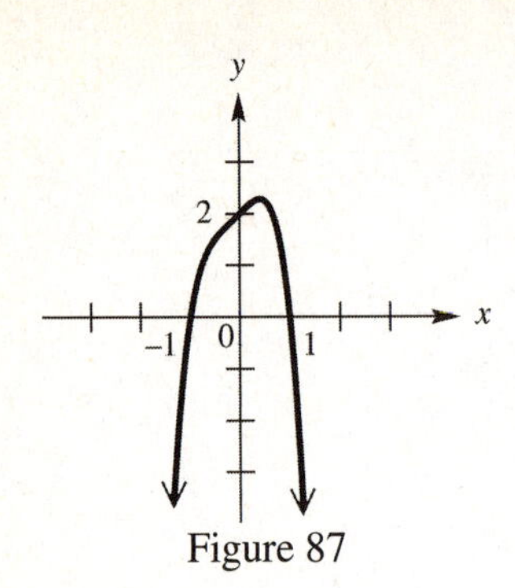

Figure 87

88.(a) Because the function $P(x)$ has coefficient signs: + + + + +, which is 0 sign changes, the function has 0 positive real zeros. Because $P(-x)=4(-x)^5+8(-x)^4+9(-x)^3+27(-x)^2+27(-x)=$ $-4x^5+8x^4-9x^3+27x^2-27x$ has coefficient signs: − + − + −, which is 4 sign changes, the function has 4, 2, or 0 negative real zeros.

(b) By using the Rational Zero Theorem, after factoring out an x, the possible rational zeros are

$$\frac{p}{q}=\pm\frac{1,3,9,27}{1,2,4}=0,\pm1,\pm3,\pm9,\pm27,\pm\frac{1}{2},\pm\frac{3}{2},\pm\frac{9}{2},\pm\frac{27}{2},\pm\frac{1}{4},\pm\frac{3}{4},\pm\frac{9}{4},\pm\frac{27}{4}.$$

(c) Factoring out the x gives 0 as one of the zeros, now from part (b) and a calculator graph of the equation, we choose $-\frac{3}{2}$ (multiplicity 2, because it appears tangent) to check by synthetic division. (The second from the result of the first)

$$\begin{array}{r|rrrrr} -\frac{3}{2} & 4 & 8 & 9 & 27 & 27 \\ & & -6 & -3 & -9 & -27 \\ \hline & 4 & 2 & 6 & 18 & 0 \end{array} \Rightarrow \begin{array}{r|rrrr} -\frac{3}{2} & 4 & 2 & 6 & 18 \\ & & -6 & 6 & -18 \\ \hline & 4 & -4 & 12 & 0 \end{array}$$

The resulting polynomial $4x^2-4x+12$ has no real zeros, so the rational zeros are 0 and $-\frac{3}{2}$ (multiplicity 2).

(d) No other real zeros, the remaining zeros are imaginary.

(e) Using the quadratic formula after factoring $4(x^2-x+3)$ yields

$$x=\frac{-(-1)\pm\sqrt{(-1)^2-4(1)(3)}}{2(1)}=\frac{1\pm\sqrt{-11}}{2}\Rightarrow x=\frac{1}{2}+i\frac{\sqrt{11}}{2} \text{ and } \frac{1}{2}-i\frac{\sqrt{11}}{2}.$$

(f) The rational zeros are the x-intercepts; therefore, 0 and $-\frac{3}{2}$.

(g) Find $P(0)$: $P(0)=4(0)^5+8(0)^4+9(0)^3+27(0)^2+27(0)\Rightarrow P(0)=0$. The y-intercept is 0.

(h)
$$\begin{array}{r|rrrrrr} 4 & 4 & 8 & 9 & 27 & 27 & 0 \\ & & 16 & 96 & 420 & 1788 & 7260 \\ \hline & 4 & 24 & 105 & 447 & 1815 & 7260 \end{array}$$
Therefore, $f(4)=7260$; $(4, 7260)$.

(i) $P(x)$ is an positive odd-degree polynomial graph; therefore, ↗

(j) See Figure 88.

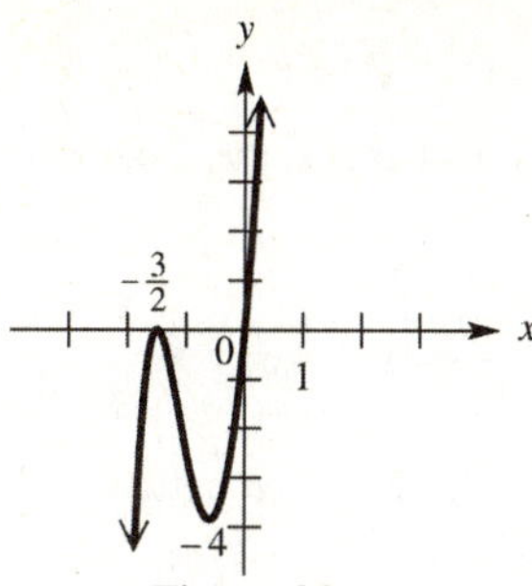

Figure 88

89.(a) Because the function $P(x)$ has coefficient signs: $+--$, which is 1 sign change, the function has 1 positive real zero. Because $P(-x)=3(-x)^4-14(-x)^2-5=3x^4-14x^2-5$ has coefficient signs: $+--$, which is 1 sign change, the function has 1 negative real zero.

(b) By the Rational Zeros Theorem, the possible rational zeros are $\frac{p}{q}=\pm\frac{1,5}{1,3}=\pm1,\pm5,\pm\frac{1}{3},\pm\frac{5}{3}$.

(c) Set equal to 0 and factor the function: $0=3x^4-14x^2-5\Rightarrow 0=(3x^2+1)(x^2-5)$. The factor $3x^2+1$ will yield imaginary zeros and x^2-5 yields irrational zeros; therefore, there are no rational zeros.

(d) From part (c), the factor $\left\{\pm\frac{5}{2},\pm1\right\}=\{-2.5,-1,1,2.5\}$; therefore, $-\sqrt{5}$ and $\sqrt{5}$ are real zeros.

(e) Using the quadratic formula on the factor $3x^2+1=0$ found in part (c), yields

$$x=\frac{-(0)\pm\sqrt{(0)^2-4(3)(1)}}{2(3)}=\frac{0\pm\sqrt{-12}}{6}=\frac{\pm2i\sqrt{3}}{6}\Rightarrow x=-i\frac{\sqrt{3}}{3}\text{ and }i\frac{\sqrt{3}}{3}.$$

(f) The rational zeros are the x-intercepts; therefore, 1 and $-\sqrt{5}$ and $\sqrt{5}$.

(g) Find $P(0)$: $P(0)=3(0)^4-14(0)^2-5\Rightarrow P(0)=-5$. The y-intercept is -5.

(h)

4)3	0	−14	0	−5
	12	48	136	144
3	12	34	136	536

Therefore, $f(4)=539$; $(4, 539)$.

(i) $P(x)$ is a positive even-degree polynomial graph; therefore, $\cup$

(j) See Figure 89.

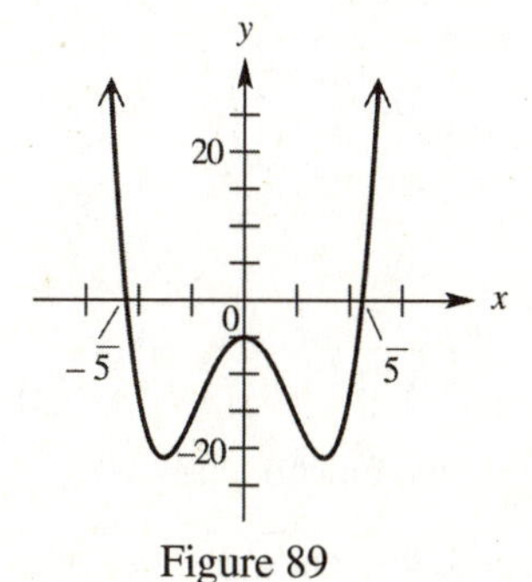

Figure 89

90. (a) Because the function $P(x)$ has coefficient signs: $- - + + - -$, which is 2 sign changes, the function has 2 or 0 positive real zeros. Because $P(-x) = -(-x)^5 - (-x)^4 + 10(-x)^3 + 10(-x)^2 - 9(-x) - 9 =$ $x^5 - x^4 - 10x^3 + 10x^2 + 9x - 9$ has coefficient signs: $+ - - + + -$, which is 3 sign changes, the function has 3 or 1 negative real zeros.

(b) By using the Rational Zeros Theorem, the possible rational zeros are $\dfrac{p}{q} = \pm\dfrac{1,3,9}{1} = \pm 1,\ \pm 3,\ \pm 9$.

(c) From part (b) and a calculator graph of the equation, we choose $-3, 1$ and 3 to check by synthetic division.

$$\begin{array}{r|rrrrrr} -3 & -1 & -1 & 10 & 10 & -9 & -9 \\ & & 3 & -6 & -12 & 6 & 9 \\ \hline & -1 & 2 & 4 & -2 & -3 & 0 \end{array} \Rightarrow \begin{array}{r|rrrrr} 1 & -1 & 2 & 4 & -2 & -3 \\ & & -1 & 1 & 5 & 3 \\ \hline & -1 & 1 & 5 & 3 & 0 \end{array} \Rightarrow$$

$$\begin{array}{r|rrrr} 3 & -1 & 1 & 5 & 3 \\ & 0 & -3 & -6 & -3 \\ \hline & -1 & -2 & -1 & 0 \end{array}$$

Factor the resulting expression to obtain the last 2 zeros. $-x^2 - 2x - 1 = -(x+1)(x+1) = 0 \Rightarrow x = -1$ (multiplicity 2). Therefore, the rational zeros are $-3, -1$ (multiplicity 2), and 3.

(d) No other real zeros, all are rational and given in part (c).

(e) No complex zeros, all are rational.

(f) The rational zeros are the x-intercepts; therefore, $-3, -1$ (multiplicity 2), and 3.

(g) Find $P(0)$: $P(0) = -(0)^5 - (0)^4 + 10(0)^3 + 10(0)^2 - 9(0) - 9 \Rightarrow P(0) = -9$. The y-intercept is -9.

(h)
$$\begin{array}{r|rrrrrr} 4 & -1 & -1 & 10 & 10 & -9 & -9 \\ & & -4 & 20 & -40 & -120 & -516 \\ \hline & -1 & -5 & -10 & -30 & -129 & -525 \end{array}$$
Therefore, $f(4) = -525$; $(4, -525)$.

(i) $P(x)$ is a negative odd-degree polynomial graph; therefore, $\nwarrow \searrow$

(j) See Figure 90.

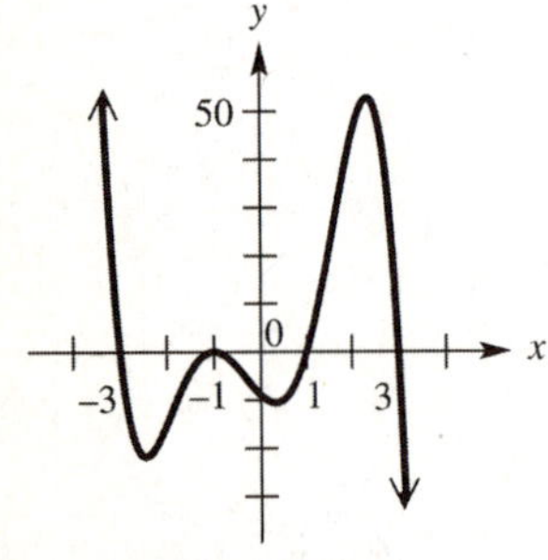

Figure 90

91.(a) Because the function $P(x)$ has coefficient signs, $-+-+-$, which is 4 sign changes, the function has 4, 2 or 0 positive real zeros. Because $P(-x) = -3x^4 - 22x^3 - 55x^2 - 52x - 12$ has coefficient signs, $-----$, which is 0 sign changes, the function has 0 negative real zeros.

(b) By the Rational Zeros Theorem, the possible rational zeros are

$$\frac{p}{q}=\pm\frac{1,\ 2,\ 3,\ 4,\ 6,\ 12}{1,\ 3}=\pm1,\pm2,\ \pm3,\pm4,\pm6,\pm12,\pm\frac{1}{3},\pm\frac{2}{3},\pm\frac{4}{3}$$

(c) From part (b) and a calculator graph of the equation we choose $\frac{1}{3}$ and 3 to check by synthetic division.

$$\begin{array}{r|rrrrr} \frac{1}{3} & -3 & 22 & -55 & 52 & -12 \\ & & -1 & 7 & -16 & 12 \\ \hline & -3 & 21 & -48 & 36 & 0 \end{array} \quad \Rightarrow \quad \begin{array}{r|rrrr} 3 & -3 & 21 & -48 & 36 \\ & & -9 & 36 & -36 \\ \hline & -3 & 12 & -12 & 0 \end{array}$$

Factor the resulting expression to obtain the last 2 zeros. $-3x^2+12x-12=-3(x-2)(x-2)=0\Rightarrow x=2$ (with a multiplicity of 2). Therefore, the rational zeros are $\frac{1}{3}$, 2 (multiplicity 2), and 3.

(d) No other real zeros, all are rational.

(e) No complex zeros, all are rational.

(f) The rational zeros are the x-intercepts; therefore, the answers are $\frac{1}{3}$, 2 (multiplicity 2), and 3.

(g) Find $P(0): P(0)=-3(0)^4+22(0)^3-55(0)^2+52(0)-12\Rightarrow P(0)=-12$. The y-intercept is -12.

(h)

$$\begin{array}{r|rrrrr} 4 & -3 & 22 & -55 & 52 & -12 \\ & & -12 & 40 & -60 & -32 \\ \hline & -3 & 10 & -15 & -8 & -44 \end{array}$$

Therefore, $f(4)=-44$; $(4, -44)$.

(i) $P(x)$ is a negative even-degree polynomial graph; therefore, it is $\frown$

(j) See Figure 91.

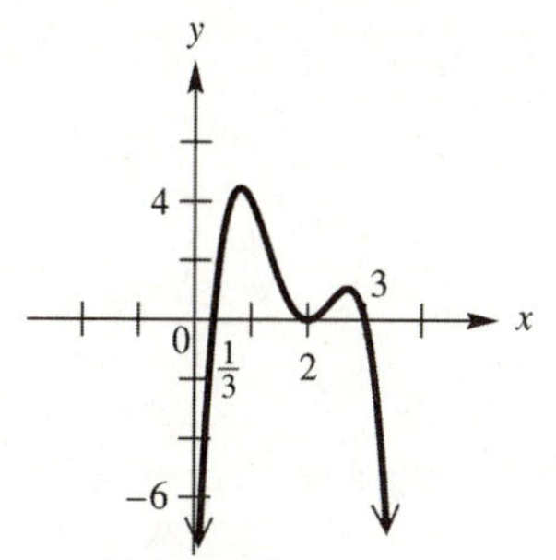

Figure 91

92. The function has 2 irrational zeros of $\pm\sqrt{5}$, which can be approximated to ±2.236.

3.8: Polynomial Equations and Inequalities; Further Applications and Models

1. $x^3-25x=0\Rightarrow x(x^2-25)=0\Rightarrow x(x+5)(x-5)=0\Rightarrow x=0,(x+5)=0,(x-5)=0\Rightarrow x=0,x=-5,x=5.$

3. $x^4-x^2=2x^2+4\Rightarrow x^4-3x^2-4=0\Rightarrow(x^2-4)(x^2+1)=0\Rightarrow(x^2-4)=0,(x^2+1)=0\Rightarrow$

   $x=\pm2$. The solution to x^2+1 is not a real number.

5. $x^3-3x^2-18x=0\Rightarrow x(x^2-3x-18)=0\Rightarrow x(x-6)(x+3)=0\Rightarrow x=0,(x-6)=0,(x+3)=0\Rightarrow$

   $x=0,x=6,x=-3$

7. $2x^3 = 4x^2 - 2x \Rightarrow 2x^3 - 4x^2 + 2x = 0 \Rightarrow 2x(x^2 - 2x + 1) = 0 \Rightarrow 2x(x-1)^2 = 0 \Rightarrow 2x = 0, (x-1)^2 = 0 \Rightarrow x = 0, x = 1.$

9. $12x^3 = 17x^2 + 5x \Rightarrow 12x^3 - 17x^2 - 5x = 0 \Rightarrow x(12x^2 - 17x - 5) = 0 \Rightarrow x(4x+1)(3x-5) = 0 \Rightarrow$

$x = 0, x = -\frac{1}{4}, x = \frac{5}{3}.$

11. $2x^3 + 4 = x(x+8) \Rightarrow 2x^3 + 4 = x^2 + 8x \Rightarrow 2x^3 - x^2 - 8x + 4 = 0 \Rightarrow x^2(2x-1) - 4(2x-1) = 0 \Rightarrow (x^2 - 4)(2x-1) = 0$

$\Rightarrow (x^2 - 4) = 0, (2x-1) = 0 \Rightarrow x = \pm 2, x = \frac{1}{2}.$

13. First, factor the equation: $7x^3 + x = 0 \Rightarrow x(7x^2 + 1) = 0 \Rightarrow x = 0$ and $7x^2 + x = 0$. Now solve:

$7x^2 + 1 = 0 \Rightarrow 7x^2 = -1 \Rightarrow x^2 = -\frac{1}{7} \Rightarrow x = \pm\frac{\sqrt{-1}}{\sqrt{7}} \Rightarrow x = \pm\frac{i}{\sqrt{7}} \Rightarrow x = \pm\frac{i}{\sqrt{7}} \cdot \frac{\sqrt{7}}{\sqrt{7}} \Rightarrow x = \pm\frac{i\sqrt{7}}{7}$. The solution set is $\left\{0, \pm\frac{i\sqrt{7}}{7}\right\}$.

15. First, factor the equation.

$3x^3 + 2x^2 - 3x - 2 = 0 \Rightarrow x^2(3x+2) - 1(3x+2) = 0 \Rightarrow (x^2 - 1)(3x+2) = 0 \Rightarrow (x+1)(x-1)(3x+2) = 0$. Now solve. $x+1 = 0 \Rightarrow x = -1;\ x-1 = 0 \Rightarrow x = 1;\ 3x+2 = 0 \Rightarrow x = -\frac{2}{3}$. The solution set is $\left\{-1, -\frac{2}{3}, 1\right\}$.

17. First, factor the equation: $x^4 - 11x^2 + 10 = 0 \Rightarrow (x^2 - 1)(x^2 - 10) = 0 \Rightarrow (x+1)(x-1)(x^2 - 10) = 0$. Now solve: $x+1 = 0 \Rightarrow x = -1;\ x-1 = 0 \Rightarrow x = 1;\ x^2 - 10 = 0 \Rightarrow x^2 = 10 \Rightarrow x = \pm\sqrt{10}$. The solution set is $\{\pm 1, \pm\sqrt{10}\}$.

19. First, factor the equation: $4x^4 - 25x^2 + 36 = 0 \Rightarrow (4x^2 - 9)(x^2 - 4) = 0$. Now solve:

$4x^2 - 9 = 0 \Rightarrow 4x^2 = 9 \Rightarrow x^2 = \frac{9}{4} \Rightarrow x = \pm\sqrt{\frac{9}{4}} \Rightarrow x = \pm\frac{3}{2};\ x^2 - 4 = 0 \Rightarrow x^2 = 4 \Rightarrow x = \pm\sqrt{4} \Rightarrow x = \pm 2.$ The solution set is $\left\{\pm\frac{3}{2}, \pm 2\right\} = \{-2, -1.5, 1.5, 2\}$. When graphed on the calculator, these values are supported as the x-intercepts of $y = 4x^4 - 29x^2 + 25$.

21. First, factor the equation: $x^4 - 15x^2 - 16 = 0 \Rightarrow (x^2 - 16)(x^2 + 1) = 0$. Now solve:

$x^2 - 16 = 0 \Rightarrow x^2 = 16 \Rightarrow x = \pm\sqrt{16} \Rightarrow x = \pm 4;\ x^2 + 1 = 0 \Rightarrow x^2 = -1 \Rightarrow x = \pm\sqrt{-1} \Rightarrow x = \pm i.$ The solution set is $\{-4, 4, -i, i\}$. When graphed on the calculator, the real values -4 and 4 are supported as the only x-intercepts of $y = x^4 - 15x^2 - 16$.

23. First, factor the equation: $x^3 - x^2 - 64x + 64 = 0 \Rightarrow x^2(x-1) - 64(x-1) = 0 \Rightarrow$

$(x^2 - 64)(x-1) = 0 \Rightarrow (x+8)(x-8)(x-1) = 0.$ Now solve: $x+8 = 0 \Rightarrow x = -8;\ x-8 = 0 \Rightarrow x = 8;$ $x-1 = 0 \Rightarrow x = 1.$ The solution set is $\{-8, 1, 8\}$. When graphed on the calculator, these values are supported as the x-intercepts of $y = x^3 - x^2 - 64x + 64$.

25. First, factor the equation: $-2x^3 - x^2 + 3x = 0 \Rightarrow -x(2x^2 + x - 3) = 0 \Rightarrow -x(2x+3)(x-1) = 0$. Now solve: $-x = 0 \Rightarrow x = 0;\ 2x+3 = 0 \Rightarrow x = -\frac{3}{2};\ x - 1 = 0 \Rightarrow x = 1$. The solution set is $\{-1.5, 0, 1\}$. When graphed on the calculator, these values are supported as the x-intercepts of $-2x^3 - x^2 + 3x = 0$.

27. First, factor the equation: $x^3 + x^2 - 7x - 7 = 0 \Rightarrow x^2(x+1) - 7(x+1) = 0 \Rightarrow (x^2 - 7)(x+1) = 0$. Now solve: $x + 1 = 0 \Rightarrow x = -1;\ x^2 - 7 = 0 \Rightarrow x^2 = 7 \Rightarrow x = \pm\sqrt{7}$. The solution set is $\{-1, \pm\sqrt{7}\}$, When graphed on the calculator, these values are supported as the x-intercepts of $y = x^3 + x^2 - 7x - 7$.

29. First, factor the equation by $-x : -3x^3 - x^2 + 6x = 0 \Rightarrow -x(3x^2 + x - 6) = 0$. Now one solution is $-x = 0 \Rightarrow x = 0$, and we use the quadratic formula on the remaining factor to find the other solutions:

$$x = \frac{-1 \pm \sqrt{1^2 - 4(3)(-6)}}{2(3)} \Rightarrow x = \frac{-1 \pm \sqrt{73}}{6}$$. The solution set is $\left\{0, \frac{-1-\sqrt{73}}{6}, \frac{-1+\sqrt{73}}{6}\right\}$. When graphed on the calculator, these values are supported as the x-intercepts of $y = -3x^3 - x^2 + 6x$.

31. First, factor the equation by $3x : 3x^3 + 3x^2 + 3x = 0 \Rightarrow 3x(x^2 + x + 1) = 0$. Now one solution is $3x = 0 \Rightarrow x = 0$ and we use the quadratic formula on the remaining factor to find the other solutions:

$$x = \frac{-1 \pm \sqrt{1^2 - 4(1)(1)}}{2(1)} \Rightarrow x = \frac{-1 \pm \sqrt{-3}}{2} \Rightarrow x = \frac{-1 \pm i\sqrt{3}}{2}$$. The solution set is $\left\{0,\ -\frac{1}{2} - \frac{\sqrt{3}}{2}i,\ -\frac{1}{2} + \frac{\sqrt{3}}{2}i\right\}$.

When graphed on the calculator, the real value 0 is the only x-intercept of $y = 3x^3 + 3x^2 + 3x$.

33. First, factor the equation: $x^4 + 17x^2 + 16 = 0 \Rightarrow (x^2 + 16)(x^2 + 1) = 0$. Now solve: $x^2 + 16 = 0 \Rightarrow$ $x^2 = -16 \Rightarrow x = \pm\sqrt{-16} \Rightarrow x = \pm 4i;\ x^2 + 1 = 0 \Rightarrow x^2 = -1 \Rightarrow\ x = \pm\sqrt{-1} \Rightarrow x = \pm i$. The solution set is $\{-4i, -i, i, 4i\}$. There are no real solutions, a conclusion which is supported when graphed on the calculator, with no x-intercepts for $y = x^4 + 17x^2 + 16$.

35. First, factor the equation: $x^6 + 19x^3 - 216 = 0 \Rightarrow (x^3 - 8)(x^3 + 27) = 0 \Rightarrow$ $(x-2)(x^2 + 2x + 4)(x+3)(x^2 - 3x + 9) = 0$. Now solve: $x - 2 = 0 \Rightarrow x = 2;\ x + 3 = 0 \Rightarrow x = -3$. We use the quadratic formula on the remaining factors to find the other solutions:

$$x = \frac{-2 \pm \sqrt{2^2 - 4(1)(4)}}{2(1)} = \frac{-2 \pm \sqrt{-12}}{2} = \frac{-2 \pm 2i\sqrt{3}}{2} \Rightarrow -1 \pm i\sqrt{3};$$

$$x = \frac{-(-3) \pm \sqrt{(-3)^2 - 4(1)(9)}}{2(1)} = \frac{3 \pm \sqrt{-27}}{2} \Rightarrow x = \frac{3 \pm 3i\sqrt{3}}{2} = \frac{3}{2} \pm \frac{3\sqrt{3}}{2}i$$. The solution set is

$\left\{-3,\ 2,\ -1 - i\sqrt{3},\ -1 + i\sqrt{3},\ \frac{3}{2} - \frac{3\sqrt{3}}{2}i,\ \frac{3}{2} + \frac{3\sqrt{3}}{2}i\right\}$. When graphed on the calculator, the real values -3

and 2 are supported as the only x-intercepts of $y = x^6 + 19x^3 - 216$.

37. See Figure 37. From the graph the following is found:

(a) $P(x) = 0: \{-2, 1, 4\}$.

(b). $P(x) < 0: (-\infty, -2) \cup (1, 4)$.

(c). $P(x) > 0: (-2, 1) \cup (4, \infty)$.

39. See Figure 39. From the graph the following is found:

(a). $P(x) = 0$: $\{-2.5, 1, 3$ (multiplicity 2)$\}$.

(b). $P(x) < 0: (-2.5, 1)$.

(c). $P(x) > 0: (-\infty, -2.5) \cup (1, 3) \cup (3, \infty)$.

41. See Figure 41. From the graph the following is found:

(a) $P(x) = 0$: $\{-3$ (multiplicity 2), 0, 2$\}$.

(b) $P(x) \geq 0: [-3] \cup [0, 2]$.

(c) $P(x) \leq 0: (-\infty, 0] \cup [2, \infty)$.

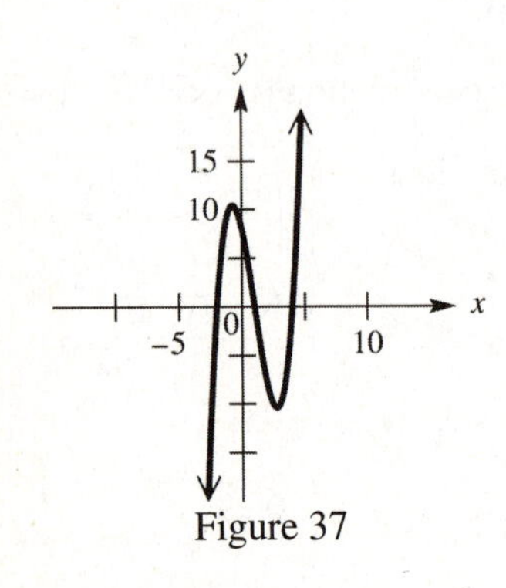

Figure 37

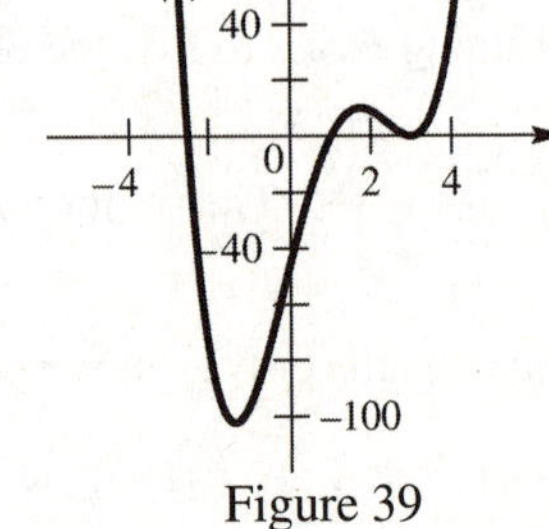

Figure 39

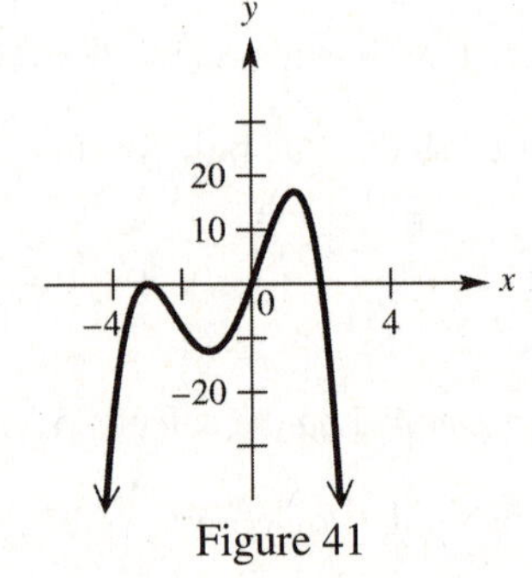

Figure 41

43. From the calculator graph of $y_1 = 0.86x^3 - 5.24x^2 + 3.55x + 7.84$, the solution set is $\{-0.88, 2.12, 4.86\}$.

45. From the calculator graph of $y_1 = -\sqrt{7}x^3 + \sqrt{5}x^2 + \sqrt{17}$, the solution set is $\{1.52\}$.

47. Given the equation $2.45x^4 - 3.22x^3 = -0.47x^2 + 6.54x + 3$, rearrange to get the calculator graph of $y_1 = 2.54x^4 - 3.22x^3 + 0.47x^2 - 6.54x - 3$. The solution set is $\{-0.40, 2.02\}$.

49. $x^2 = -1 \Rightarrow x = \pm\sqrt{-1} \Rightarrow x = \pm i \Rightarrow \{-i, i\}$.

51. First, factor the equation $x^3 = -1 \Rightarrow x^3 + 1 = 0 \Rightarrow (x+1)(x^2 - x + 1) = 0$. Now one solution is $x + 1 = 0 \Rightarrow x = -1$, and we use the quadratic formula on the remaining factor to find the other solutions.

$x = \dfrac{-(-1) \pm \sqrt{(-1)^2 - 4(1)(1)}}{2(1)} \Rightarrow x = \dfrac{1 \pm i\sqrt{3}}{2}$. The solution set is $\left\{-1, \dfrac{1}{2} - \dfrac{\sqrt{3}}{2}i, \dfrac{1}{2} + \dfrac{\sqrt{3}}{2}i\right\}$.

53. First, factor the equation $x^3 = 27 \Rightarrow x^3 - 27 = 0 \Rightarrow (x-3)(x^2 + 3x + 9) = 0$. Now one solution is $x - 3 = 0 \Rightarrow x = 3$. We use the quadratic formula on the remaining factor to find the other solutions.

$x=\frac{-3\pm\sqrt{3^2-4(1)(9)}}{2(1)}\Rightarrow x=\frac{-3\pm\sqrt{-27}}{2}=\frac{-3\pm 3i\sqrt{3}}{2}$. The solution set is $\left\{3,\ -\frac{3}{2}-\frac{3\sqrt{3}}{2}i,\ -\frac{3}{2}+\frac{3\sqrt{3}}{2}i\right\}$.

55. First, factor the equation $x^4=16\Rightarrow x^4-16=0\Rightarrow(x^2-4)(x^2+4)=0\Rightarrow(x-2)(x+2)(x^2+4)=0$. Now two solutions are $x-2=0\Rightarrow x=2$, and $x+2=0\Rightarrow x=-2$. We solve the remaining factor for the last two solutions. $x^2+4=0\Rightarrow x^2=-4\Rightarrow x=\pm\sqrt{-4}\Rightarrow x=\pm 2i$. The solution set is $\{-2, 2, -2i, 2i\}$.

57. First, factor the equation $x^3=-64\Rightarrow x^3+64=0\Rightarrow(x+4)(x^2-4x+16)=0$. Now one solution is $x+4=0\Rightarrow x=-4$. We use the quadratic formula on the remaining factor to find the other solutions.

$x=\frac{-(-4)\pm\sqrt{(-4)^2-4(1)(16)}}{2(1)}=\frac{4\pm\sqrt{-48}}{2}\Rightarrow x=\frac{4\pm 4i\sqrt{3}}{2}=2\pm 2i\sqrt{3}$. The solution set is

$\{-4,\ 2-2i\sqrt{3},\ 2+2i\sqrt{3}\}$.

59. $x^2=-18\Rightarrow x=\sqrt{-18}\Rightarrow x=\pm 3i\sqrt{2}\Rightarrow\{-3i\sqrt{2},\ 3i\sqrt{2}\}$.

61. (a) First, find f(x) for $d=0.8$. $f(x)=\frac{\pi}{3}x^3-5\pi x^2+\frac{500\pi(0.8)}{3}\Rightarrow f(x)=\frac{\pi}{3}x^3-5\pi x^2+\frac{400\pi}{3}$. Now graph this equation on a graphing calculator. We find the smallest positive zero is at $x\approx 7.1286$ cm. or $x\approx 7.13$ cm. The ball floats partly above the surface.

(b) First, find $f(x)$ for $d=2.7$. $f(x)=\frac{\pi}{3}x^3-5\pi x^2+\frac{500\pi(2.7)}{3}\Rightarrow f(x)=\frac{\pi}{3}x^3-5\pi x^2+\frac{1350\pi}{3}$. Now graph this equation on a graphing calculator. We find the graph has no zero; the sphere sinks below the surface because it is more dense than water.

(c) First, find $f(x)$ for $d=1$. $f(x)=\frac{\pi}{3}x^3-5\pi x^2+\frac{500\pi(1)}{3}\Rightarrow f(x)=\frac{\pi}{3}x^3-5\pi x^2+\frac{500\pi}{3}$. Now graph this equation on a graphing calculator. We find the smallest positive zero is at $x=10$ cm; the balloon floats even with the surface.

63. First, find $f(x)$ for $d=.6$. $f(x)=\frac{\pi}{3}x^3-10\pi x^2+\frac{4000\pi(.6)}{3}\Rightarrow f(x)=\frac{\pi}{3}x^3-10\pi x^2+800\pi$. Now graph this equation on a graphing calculator. We find the smallest positive zero is at $x\approx 11.34$ cm.

65. (a) Since the box has sides $x>0$ and since the box must have some width, we have $12-2x>0\Rightarrow -2x>-12\Rightarrow x<6$. Therefore, the restrictions are $0<x<6$.

(b) Since $V=lwh$, length $=18-2x$, and width $=12-2x$, the function is $V(x)=(18-2x)(12-2x)(x)\Rightarrow V(x)=x(4x^2-60x+216)\Rightarrow V(x)=4x^3-60x^2+216x$.

(c) By graphing the function from (b) on a graphing calculator, we can find that the maximum volume is at $x\approx 2.35$ in., which produces a volume of $V(2.35)\approx 4(2.35)^3-60(2.35)^2+216(2.35)\approx 228.16$ in^3.

(d) By graphing the function from (b) on a graphing calculator, we can find the volume to be greater than 80 in^3 for the interval $0.42<x<5$.

67. By graphing the equation on a graphing calculator, we find the smallest positive zero which is less than 10 is

$x \approx 2.61$ in.

69. (a) A length 1 inch less than the hypotenuse (x) is $x-1$.

(b) Using the Pythagorean Theorem yields $l^2+(x-1)^2 = x^2 \Rightarrow l^2 = x^2-(x-1)^2 \Rightarrow$

$l^2 = x^2-(x^2-2x+1) \Rightarrow l^2 = 2x-1 \Rightarrow l = \sqrt{2x-1}$.

(c) Since $A = \frac{1}{2}bh = 84$, the equation is $\frac{1}{2}\left(\sqrt{2x-1}\right)(x-1) = 84 \Rightarrow$

$(\sqrt{2x-1})(x-1) = 168 \Rightarrow [(\sqrt{2x-1})(x-1)]^2 = 168^2 \Rightarrow (2x-1)(x-1)^2 = 28,224 \Rightarrow$

$(2x-1)(x^2-2x+1) = 28,224 \Rightarrow 2x^3-5x^2+4x-28,225 = 0$.

(d) By graphing the equation from (c) on a graphing calculator, we get an x-intercept of 25. Therefore, the hypotenuse is $x = 25$ in., one leg is $x-1 = 24$ in., and the other leg is $\sqrt{2x-1} = \sqrt{49} = 7$ in.

71. (a) Since $l = 11-2x,\ w = 8.5-2x,\ h = x$, and $V = lwh$, we get an equation of $V = (11-2x)(8.5-2x)(x) \Rightarrow V = (93.5-17x-22x+4x^2)(x) \Rightarrow V = 4x^3-39x^2+93.5x$. Because all sides of the box must be positive, the restrictions on this equation will be $8.5-2x > 0 \Rightarrow -2x > -8.5 \Rightarrow x < 4.25 \Rightarrow 0 < x < 4.25$. Now using the table feature on the graphing calculator yields a maximum value, within the restrictions, of $x \approx 1.59$. The volume is $4(1.59)^3-39(1.59)^2+93.5(1.59) \approx 66.15$ in^3.

(b) Using the table feature on the graphing calculator for the equation $V = 4x^3-39x^2+93x$ yields a volume greater than 40 in^3 when x is in the following range: $0.54 \text{ in} < x < 2.92$ in.

73. (a) See Figure 73a.

(b) From the regression feature, the best-fitting quadratic function is $C(x) \approx 0.0035x^2-49x+22$. See Figure 73b.

(c) From the regression feature, the best-fitting cubic function is $C(x) \approx 0.000068x^3+0.00987x^2-0.653x+23$. See Figure 73c.

(d) Using the graphs of each, the cubic function is a slightly better fit.

(e) Using the cubic function and graph, $C(x) > 10$ when $0 \le x < 31.92$.

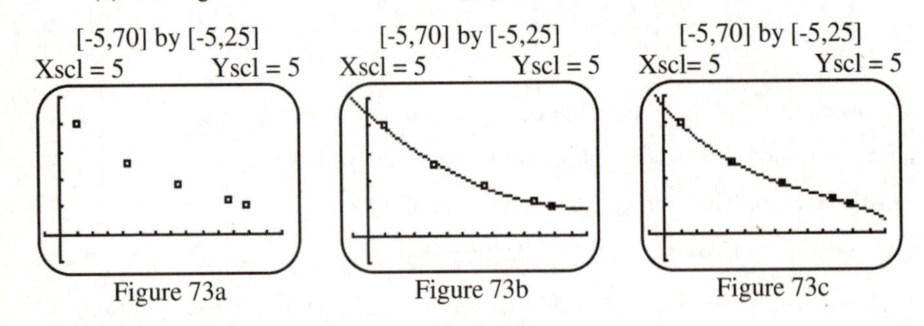

Figure 73a Figure 73b Figure 73c

Reviewing Basic Concepts (Sections 3.6—3.8)

1. $P(3) = 2(3)^4-7(3)^3+29(3)-30 \Rightarrow P(3) = 162-189+87-30 \Rightarrow P(3) = 30$.

2. Since $P(2) = 2(2)^4-7(2)^3+29(2)-30 \Rightarrow P(2) = 32-56+58-30 \Rightarrow P(2) = 4$, $P(2)$ is not a zero.

3. Since $2+i$ is a zero, its conjugate $2-i$ is also a zero. First, use synthetic division to factor out these two zeros from the equation and the resulting equation.

$$\begin{array}{r|rrrr} 2+i & 2 & -7 & 0 & 29 & -30 \\ & & 4+2i & -8+i & -17-6i & 30 \\ \hline & 2 & -3+2i & -8+i & 12-6i & 0 \end{array} \Rightarrow \begin{array}{r|rrrr} 2-i & 2 & -3+2i & -8+i & 12-6i \\ & & 4+2i & 2-i & -12+6i \\ \hline & 2 & 1 & -6 & 0 \end{array}$$

Now completely factor the resulting expression, $2x^2+x-6=(2x-3)(x+2)$. The linear factors are $P(x)=(2x-3)(x+2)(x-2-i)(x-2+i)$.

4. Since the zeros are $\frac{3}{2}$, and i, and the conjugate of i is $-i$, the function is $P(x)=a\left(x-\frac{3}{2}\right)(x-i)(x+i)\Rightarrow$

   $P(x)=a\left(x-\frac{3}{2}\right)(x^2+1)\Rightarrow P(x)=a\left(x^3-\frac{3}{2}x^2+x-\frac{3}{2}\right)$. If $P(x)=15$, then $15=a\left(3^3-\frac{3}{2}(3)^2+3-\frac{3}{2}\right)\Rightarrow$

   $15=a\left(27-\frac{27}{2}+3-\frac{3}{2}\right)\Rightarrow 15=a(30-15)\Rightarrow 15=15a\Rightarrow a=1$. Therefore, the function is

   $P(x)=x^3-\frac{3}{2}x^2+x-\frac{3}{2}$.

5. Since the zeros are $-4,-4,\ 1+2i$, and its conjugate $1-2i$, the function is

   $P(x)=a(x+4)^2(x-1-2i)(x-1+2i)\Rightarrow P(x)=a(x^2+8x+16)(x^2-2x+5)\Rightarrow$

   $P(x)=a(x^4+6x^3+5x^2+8x+80)$. If $a=1$, then $P(x)=x^4+6x^3+5x^2+8x+80$. (Answers may vary).

6. For the function the possible zeros are $x=\pm\frac{1,2,5,10}{1,2}\Rightarrow x=\pm1,\ \pm2,\ \pm5,\ \pm10,\ \pm\frac{1}{2},\ \pm\frac{5}{2}$. From a calculator graph, we can choose $x=-2$ to check by synthetic division.

   $$\begin{array}{r|rrrr} -2 & 2 & 1 & -11 & -10 \\ & & -4 & 6 & 10 \\ \hline & 2 & -3 & -5 & 0 \end{array}$$

   Now solve the resulting quadratic equation: $2x^2-3x-5=0\Rightarrow(2x-5)(x+1)=0\Rightarrow 2x-5=0$ and $x+1=0\Rightarrow$

   $x=\frac{5}{2},\ -1$, and -2.

7. Use the quadratic formula on the equation $x^2=\frac{12\pm\sqrt{(-12)^2-4(3)(1)}}{2(3)}=\frac{12\pm\sqrt{132}}{6}=\frac{12\pm2\sqrt{33}}{6}=\frac{6\pm\sqrt{33}}{3}$.

   Therefore, $x=\pm\sqrt{\frac{6\pm\sqrt{33}}{3}}$.

8. (a) From the regression feature, the best-fitting quadratic function is $P(x)\approx 0.0004815x^3+0.0303x^2-0.1989x+3.1$.

   (b) $P(34)\approx 0.0004815(34)^3+0.0303(34)^2-0.1989(34)+3.1\approx 50.3$ This value is within 300 of the actual value of 50 thousand.

   (c) Using calculator graphs of $y_1=0.0004815x^3+0.0303x^2-0.1989x+3.1.$ and $y_2=100$, the intersection is $x\approx 45$ or approximately the year 2005.

Chapter 3 Review Exercises

1. $(17-i) + (1-3i) = 18-4i.$

2. $(17-i) - (1-3i) = 17-i-1+3i = 16+2i.$

3. $(17-i)(1-3i) = 17-51i-i+3i^2 = 17-52i-3 = 14-52i.$

4. $(17-i)^2 = (17-i)(17-i) = 289-17i+i^2 = 289-34i-1 = 288-34i.$

5. $\dfrac{1}{1-3i}\bullet\dfrac{1+3i}{1+3i} = \dfrac{1+3i}{1-9i^2} = \dfrac{1+3i}{1-9i^2} = \dfrac{1+3i}{10} = \dfrac{1}{10}+\dfrac{3}{10}i.$

6. $\dfrac{17-i}{1-3i}\bullet\dfrac{1+3i}{1+3i} = \dfrac{17+51i-i-3i^2}{1-9i^2} = \dfrac{20+50i}{10} = 2+5i.$

7. Since all real numbers can be input for x, the domain is $(-\infty,\infty)$.

8. By using the vertex formula, we first find the x-coordinate of the vertex. $x=\dfrac{-b}{2a}=\dfrac{-(-6)}{2(2)}=\dfrac{6}{4}=\dfrac{3}{2}$. Now solve for y. $P\left(\dfrac{3}{2}\right)=2\left(\dfrac{3}{2}\right)^2-6\left(\dfrac{3}{2}\right)-8\Rightarrow P\left(\dfrac{3}{2}\right)=\dfrac{9}{2}-\dfrac{18}{2}-\dfrac{16}{2}=-\dfrac{25}{2}$. The vertex is $\left(\dfrac{3}{2},\ -\dfrac{25}{2}\right)$.

9. Since the function is a positive even-degree equation, the end behavior is $\uparrow\!\cup\!\uparrow$

10. At the x-intercept, $P(x)=0$. Therefore, $2x^2-6x-8=0\Rightarrow x^2-3x-4=0\Rightarrow$ $(x-4)(x+1)=0\Rightarrow x=-1,\ 4$. The x-intercepts are –1 and 4.

11. At the y-intercept, $x=0$. Therefore, $P(0)=2(0)^2-6(0)-8\Rightarrow P(0)=-8$. The y-intercept is –8.

12. From a calculator graph of $P(x)=2x^2-6x-8$, the range is $\left[-\dfrac{25}{2},\infty\right)$.

13. From a calculator graph of $P(x)=2x^2-6x-8$, the graph is increasing for the interval $\left[\dfrac{3}{2},\infty\right)$, and decreasing for the interval $\left(-\infty,\dfrac{3}{2}\right]$.

14. (a) $2x^2-6x-8=0\Rightarrow x^2-3x-4=0\Rightarrow(x-4)(x+1)=0\Rightarrow x=-1$ or 4.

(b) First, set $2x^2-6x-8=0$, then $x^2-3x-4=0\Rightarrow(x-4)(x+1)=0$, and the solution set will be $\{-1,4\}$. The numbers –1 and 4 divide the number line into 3 intervals. The interval $(-\infty,-1)$ has a positive product, the interval $(-1,4)$ has a negative product, and the interval $(4,\infty)$ has a positive product. Therefore, $2x^2-6x-8>0$ for the interval $(-\infty,1)\cup(4,\infty)$.

(c) First, set $2x^2-6x-6-8=0$, then $x^2-3x-4=0\Rightarrow(x-4)(x+1)=0$ and the solution set will be $\{-1,4\}$. The numbers -1 and 4 divide the number line into 3 intervals. The interval $(-\infty,-1)$ has a positive product, the interval $(-1,4)$ has a negative product, and the interval $(4,\infty)$ has a positive product.

Therefore, $2x^2-6x-8\le 0$ for the interval $[-1,4]$.

15. The graph intersects the x-axis at –1 and 4, supporting the answer in (a). It lies above the x-axis when $x<-1$ or $x>4$, supporting the answer in (b). It lies below the x-axis when x is between –1 and 4 inclusive, supporting the answer in (c).

16. From problem 8, we know the vertex of $P(x)=2x^2-6x-8$ is $\left(\frac{3}{2},-\frac{25}{2}\right)$. The vertical line of symmetry through this point has the equation $x=\frac{3}{2}$.

17. The discriminate is $b^2-4ac=(5.47)^2-4(-2.64)(3.54)=29.9209+37.3824=67.3033$. Since the discriminate is greater than 0, there are two x-intercepts.

18. From the calculator graph of $P(x)=-2.64x^2+5.47x+3.54$, the x-intercepts are $x=0.52$ and 2.59.

19. From the calculator graph of $P(x)=-2.64x^2+5.47x+3.54$, the vertex is $(1.04,6.37)$.

20. Using the vertex formula yields $x=\frac{-b}{2a}=\frac{-5.47}{2(-2.64)}\approx 1,04$. Now solve for

$P(1.04)=-2.64(1.04)^2+5.47(1.04)+3.54\approx 6.37$. The vertex is $(1.04,6.37)$.

21. (a) From the graph, the maximum value of $f(x)$ is 4.

(b) From the graph, the maximum value of 4 is reached when $x=1$.

(c) From the graph, there would be two intersections with $y=2$; therefore, there are two real solutions.

(d) From the graph, there would be no intersections with $y=6$; therefore, there are no real solutions.

22. At zero seconds, the height is $s(0)=-16(0)^2+800(0)+600=600$. The projectile was fired from 600 feet.

23. From the calculator graph of the equation, the vertex is (25, 10600). The maximum height is reached after 25 sec.

24. From the calculator graph of the equation, the vertex is (25, 10600). The maximum height is 10,600 feet.

25. From the calculator graph of the equation, and $y=5000$, the graph of the equation is above $y=5000$ for the interval (6.3, 43.7). Therefore, the projectile is above 5,000 feet between 6.3 seconds and 43.7 seconds.

26. From the calculator graph of the equation, the x-intercept (when the projectile hits the ground) is $x=50.739$. The projectile will be in the air for approximately 50.7 seconds.

27. (a) Let x = the width and $3x$ = the length of the cardboard. Then the base of the box will be $w=(x-8)$, and $l=(3x-8)$, , and its height is $h=4$. Since $v=lwh$, we get a function

$V(x)=(x-8)(3x-8)(4)\Rightarrow V(x)=(3x^2-32x+64)(4)\Rightarrow V(x)=12x^2-128x+256$.

(b) If the volume is 2496, then $2496=12x^2-128x+256\Rightarrow 0=12x^2-128x-2240\Rightarrow$

$0 = 3x^2 - 32x - 560$. Using the quadratic formula to solve this yields

$x = \frac{-(-32) \pm \sqrt{(-32)^2 - 4(3)(-560)}}{2(3)} = \frac{32 \pm 88}{6} \Rightarrow x = 20, -\frac{28}{3}$. Since we can not have a negative distance we throw out $x = -\frac{28}{3}$. Therefore, the original dimensions are 20 in. by 60 in.

(c) Graphing $y_1 = V(x)$ and $y_2 = 2496$ on a graphing calculator shows that the graphs intersect at $x = 20$.

28. (a) See Figure 28.

(b) Using the form $P(x) = a(x-h)^2 + k$ and a vertex of $(0, 353)$ yields $P(x) = a(x-0)^2 + 353$. Now solve for a using a second point of (285, 2000). $2000 = a(285-0)^2 + 353 \Rightarrow$ $1647 = 81225a \Rightarrow a \approx 0.0203$. Therefore, the function is $P(x) = 0.0203x^2 + 353$.

(c) Using the regression feature of the graphing calculator for the data yields $Q(x) = 0.0167x^2 + 0.968x + 363$.

[-28.5,313.5] by [73,2280]
Xscl = 50 Yscl = 100

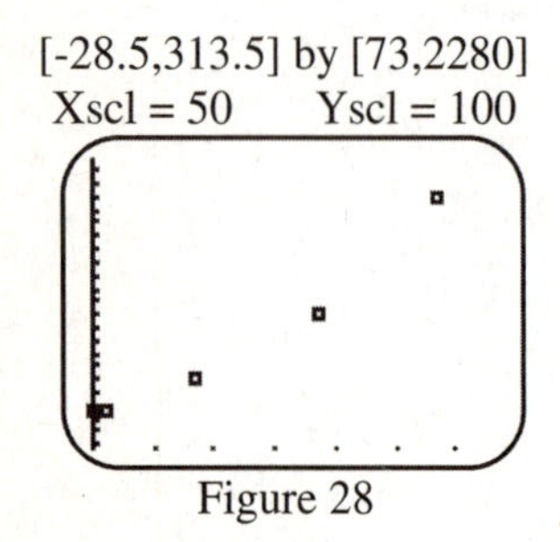

Figure 28

29. First, find $P(-2) = -3(-2)^3 - (-2)^2 + 2(-2) - 4 = 24 - 4 - 4 - 4 \Rightarrow P(-2) = 12$ and $P(-1) = -3(-1)^3 - (-1)^2 + 2(-1) - 4 = 3 - 1 - 2 - 4 \Rightarrow P(-1) = -4$. Since $P(-2) = 12$ and $P(-1) = -4$ differ in sign, the intermediate value theorem assures us that there is a real zero between –2 and –1.

30. (a)

$$\begin{array}{r|rrrr} 3 & 1 & 1 & -11 & -10 \\ & & 3 & 12 & 3 \\ \hline & 1 & 4 & 1 & -7 \end{array}$$

Therefore, $Q(x) = x^2 + 4x + 1$; $R = -7$.

(b)

$$\begin{array}{r|rrrr} -2 & 3 & 8 & 5 & 10 \\ & & -6 & -4 & -2 \\ \hline & 3 & 2 & 1 & 8 \end{array}$$

Therefore, $Q(x) = 3x^2 + 2x + 1$; $R = 8$.

31.

$$\begin{array}{r} 2x^2-2x+2 \\ 3x+1\overline{)6x^3-4x^2+4x+3} \\ \underline{6x^3+2x^2} \qquad\qquad \\ -6x^2+4x \qquad \\ \underline{-6x^2-2x} \qquad \\ 6x+3 \\ \underline{6x+2} \\ 1 \end{array} \Rightarrow \frac{6x^3-4x^2+4x+3}{3x+1}=2x^2-2x+2+\frac{1}{3x+1}$$

32.

$$\begin{array}{r} 2x+1 \\ x^2-3x+1\overline{)2x^3-5x^2+0x+1} \\ \underline{2x^3-6x^2+2x} \qquad \\ x^2-2x+1 \\ \underline{x^2+3x+1} \\ x \end{array} \Rightarrow \frac{2x^3-5x^2+1}{x^2-3x+1}=2x+1+\frac{x}{x^2-3x+1}$$

33. $\begin{array}{r|rrrr} 2 & -1 & 5 & -7 & 1 \\ & & -2 & 6 & -2 \\ \hline & -1 & 3 & -1 & -1 \end{array}$ Therefore, $P(2)=-1$.

34. $\begin{array}{r|rrrr} 2 & 2 & -3 & 7 & -12 \\ & & 4 & 2 & 18 \\ \hline & 2 & 1 & 9 & 6 \end{array}$ Therefore, $P(2)=6$.

35. $\begin{array}{r|rrrrr} 2 & 5 & 0 & -12 & 2 & -8 \\ & & 10 & 20 & 16 & 36 \\ \hline & 5 & 10 & 8 & 18 & 28 \end{array}$ Therefore, $P(2)=28$.

36. $\begin{array}{r|rrrrrr} 2 & 1 & 0 & 0 & 4 & -2 & -4 \\ & & 2 & 4 & 8 & 24 & 44 \\ \hline & 1 & 2 & 4 & 12 & 22 & 40 \end{array}$ Therefore, $P(2)=40$.

37. The conjugate of $7+2i$ must also be a zero; therefore, $7-2i$ is also a zero.

38. With the given zeros of –1, 4, and 7, the function $P(x)$ is

$P(x)=a(x-(-1))(x-4)(x-7)=a(x+1)(x-4)(x-7)=a(x+1)(x^2-11x+28)\Rightarrow$

$P(x)=a(x^3-10x^2+17x+28)$. One of many possible functions, if $a=1$, is $P(x)=x^3-10x^2+17x+28$.

39. With the given zeros of 8, 2, and 3, the function $P(x)$ is

$P(x)=a(x-8)(x-2)(x-3)=a(x-8)(x-2)\ (x-3)=a(x-8)(x^2-5x+6)\Rightarrow$

$P(x)=a(x^3-13x^2+46x-48)$. One of many possible functions, if $a=1$, is $P(x)=x^3-13x^2+46x-48$.

40. With the given zeros of $\sqrt{3},-\sqrt{3},2,$ and 3, the function $P(x)$ is

$P(x)=a(x-\sqrt{3})(x-(-\sqrt{3})(x-2)(x-3)=a(x-\sqrt{3})(x+\sqrt{3})(x-2)(x-3)=$

$a(x^2-3)(x^2-5x+6)\Rightarrow a(x^4-5x^3+3x^2+15x-18)$. One of many possible functions, if $a=1$, is

$P(x)=x^4-5x^3+3x^2+15x-18$.

41. With the given zeros of $-2+\sqrt{5},-2-\sqrt{5}$, -2 and 1, the function $P(x)$ is

$P(x)=a(x-(-2+\sqrt{5}))(x-(-2-\sqrt{5}))(x-(-2))(x-1)=$
$a(x+2-\sqrt{5})(x+2+\sqrt{5})(x+2)(x-1)=a(x^2+4x-1)(x^2+x-2)=a(x^4+5x^3+x^2-9x+2)$. One of many possible functions, if $a=1$, is $P(x)=x^4+5x^3+x^2-9x+2$.

42. Use synthetic division to check if -1 is a zero.

$$\begin{array}{r|rrrrr} -1 & 2 & 1 & -4 & 3 & 1 \\ & & -2 & 1 & 3 & -6 \\ \hline & 2 & -1 & -3 & 6 & -5 \end{array}$$

Since $P(-1)=-5$ and not 0, -1 is not a zero.

43. Use synthetic division to check if $x+1$ or $x=-1$ is a zero.

$$\begin{array}{r|rrrr} -1 & 1 & 2 & 3 & 2 \\ & & -1 & -1 & -2 \\ \hline & 1 & 1 & 2 & 0 \end{array}$$

Since $P(-1)=0$, -1 or $x+1$ is a factor.

44. With the given zeros 3, 1, $-1-3i$, the conjugate of $-1-3i$ is also a zero, so the function $P(x)$ is

$P(x)=a(x-3)(x-1)(x-(-1-3i))(x-(-1+3i))=$

$a(x-3)(x-1)(x+1+3i)(x+1-3i)=a(x^2-4x+3)(x^2+2x+10)=a(x^4-2x^3+5x^2-34x+30)$.

Since $P(2)=-36$, we can solve for a.

$-36=a((2)^4-2(2)^3+5(2)^2-34(2)+30)\Rightarrow -36=a(16-16+20-68+30)\Rightarrow -18a=-36\Rightarrow a=2$.

Therefore, $P(x)=2(x^4-2x^3+5x^2-34x+30)\Rightarrow P(x)=2x^4-4x^3+10x^2-68x+60$.

45. With the given zero $1-i$, the conjugate $1+i$ is also a zero. Use synthetic division to factor out these zeros.

$$\begin{array}{r|rrrrr} 1-i & 1 & -3 & -8 & 22 & -24 \\ & & 1+i & -3+i & -10+12i & 24 \\ \hline & 1 & -2-i & -11+i & 12+12i & 0 \end{array} \Rightarrow \begin{array}{r|rrrr} 1+i & 1 & -2-i & -11+i & 12+12i \\ & & 1+i & -1-i & -12-12i \\ \hline & 1 & -1 & -12 & 0 \end{array}$$

Finally, solve the resulting quadratic equation, $x^2-x-12=0\Rightarrow x^2-x-12=0\Rightarrow$

$(x-4)(x+3)=0\Rightarrow x=-3,\ 4$. The zeros are $x=\{-3,4,1-i,1+i\}$.

46. With the given zeros 1 and $2i$, the conjugate of $2i$, or $-2i$, is also a zero. Use synthetic division to factor

out these zeros.

$$\begin{array}{r|rrrrr} 1 & 2 & -1 & 7 & -4 & -4 \\ & & 2 & 1 & 8 & 4 \\ \hline & 2 & 1 & 8 & 4 & 0 \end{array} \Rightarrow \begin{array}{r|rrrr} 2i & 2 & 1 & 8 & 4 \\ & & 4i & -8+2i & -4 \\ \hline & 2 & 1+4i & 2i & 0 \end{array} \Rightarrow \begin{array}{r|rrr} -2i & 2 & 1+4i & 2i \\ & & -4i & -2i \\ \hline & 2 & 1 & 0 \end{array}$$

Finally, solve the resulting linear equation: $2x+1=0 \Rightarrow 2x=-1 \Rightarrow x=-\frac{1}{2}$. The zeros are

$x=-\frac{1}{2}, 1, 2i, -2i.$

47. Using the Rational Zeros Theorem, the possible zeros are

$x=\pm\frac{1,2,4,8}{1,3}=\pm1, \pm2, \pm4, \pm8, \pm\frac{1}{3}, \pm\frac{2}{3}, \pm\frac{4}{3}, \pm\frac{8}{3}.$ From graphing the equation, we choose to check

possible zeros -2, 4, and $\frac{1}{3}$ by synthetic division.

$$\begin{array}{r|rrrrrr} 2 & 3 & -4 & -26 & -21 & -14 & 8 \\ & & -6 & 20 & 12 & 18 & 8 \\ \hline & 3 & -10 & -6 & -9 & 4 & 0 \end{array} \Rightarrow \begin{array}{r|rrrrr} 4 & 3 & -1 & -6 & -9 & 4 \\ & & 12 & 8 & 8 & -4 \\ \hline & 3 & 2 & 2 & -1 & 0 \end{array} \Rightarrow \begin{array}{r|rrrr} \frac{1}{3} & 3 & 2 & 2 & -1 \\ & & 1 & 1 & 1 \\ \hline & 3 & 3 & 3 & 0 \end{array}$$

Finally, solve the resulting quadratic equation, $x^2+x+1=0$, using the quadratic formula:

$x=\frac{-1\pm\sqrt{1^2-4(1)(1)}}{2(1)}=\frac{-1\pm\sqrt{-3}}{2}.$ This results in non-real solutions; therefore, the rational solutions

are $x=-2, \frac{1}{3}$, and 4.

48. Because the function $P(x)$ has coefficient signs, $++---$, which is 1 sign change, the function has 1 positive real zero. Because $P(-x)=3(-x)^4+(-x)^3-(-x)^2-2(-x)-1=3x^4-x^3-x^2+2x-1$ has coefficient signs, $+--+-$, which is 3 sign changes, the function has 3 or 1 possible negative real zeros.

49. Using the Boundedness Theorem, we divide synthetically by $x-2$.

$$\begin{array}{r|rrrrr} 2 & 2 & 3 & -5 & 8 & -10 \\ & & 4 & 14 & 18 & 52 \\ \hline & 2 & 7 & 9 & 26 & 42 \end{array}$$

The result is entirely non-negative; therefore, there can be no real zero greater than 2. Again using the Boundedness Theorem, we divide synthetically by $x-(-4)$.

$$\begin{array}{r|rrrrr} -4 & 2 & 3 & -5 & 8 & -10 \\ & & -8 & 20 & -60 & 208 \\ \hline & 2 & -5 & 15 & -52 & 198 \end{array}$$

The result alternates in sign; therefore, there is no real zero less than -4.

50. Graphing the equation on a calculator shows x-intercepts of $-1.62, 0.62$, and 3. Therefore, there are 3 real solutions and the integer root is 3.

51. Using synthetic division we can factor out the root 3.

$$\begin{array}{r|rrrr} 3 & 1 & -2 & -4 & 3 \\ & & 3 & 3 & -3 \\ \hline & 1 & 1 & -1 & 0 \end{array}$$

Using the result of the synthetic division the factors are $(x-3)(x^2+x-1)$.

52. The remaining zeros can be found by solving, $x^2+x-1=0$, using the quadratic formula.

$x=\dfrac{-1\pm\sqrt{1^2-4(1)(-1)}}{2(1)}=\dfrac{-1\pm\sqrt{5}}{2}$. The solution set is $\left\{\dfrac{-1-\sqrt{5}}{2},\dfrac{-1+\sqrt{5}}{2}\right\}$.

53. The x-intercepts from the graph are approximately equal to the solutions found in #50. $\dfrac{-1-\sqrt{5}}{2}\approx-1.62$, and $\dfrac{-1+\sqrt{5}}{2}\approx 0.62$.

54. (a) From the graph and the answers of #50, $P(x)>0$ for the interval $\left(\dfrac{-1-\sqrt{5}}{2},\dfrac{-1+\sqrt{5}}{2}\right)\cup(3,\infty)$.

(b) From the graph and the answers of #50, $P(x)\le 0$ for the interval $\left(-\infty,\dfrac{-1-\sqrt{5}}{2}\right]\cup\left[\dfrac{-1+\sqrt{5}}{2},3\right]$.

55. First, factor out an x. $x^3-2x^2+5x=0\Rightarrow x(x^2-2x+5)\Rightarrow x=0$ is a zero. Now solve the quadratic equation using the quadratic formula: $x=\dfrac{-2\pm\sqrt{2^2-4(1)(5)}}{2(1)}=\dfrac{-2\pm\sqrt{-16}}{2}=\dfrac{-2\pm 4i}{2}=-1\pm 2i$. Since two of the zeros are not real, the only x-intercept is $x=0$.

56. The x-intercepts are –2, 1, and 3. The 3 has multiplicity 2, because it appears tangent to the x-axis. The factored form is now: $P(x)=(x+2)(x-1)(x-3)^2$.

57. Since its end behavior has both ends going in the same direction, either up or down, the degree is even.

58. Since its end behavior has one end going up and the other going down, the degree is odd.

59. Since its end behavior has the right end approaching positive infinity, the lead coefficient is positive.

60. Since $g(x)$ has no x-intercepts, it has no real solutions.

61. From the graph, $f(x)<0$ is true for the interval $(-\infty,a)\cup(b,c)$.

62. From the graph, $f(x)>g(x)$ is true for the interval (d,h).

63. If $f(x)-g(x)=0$ then $f(x)=g(x)$. This happens at the intersection of the graphs, which is at $\{d,h\}$.

64. If $r+pi$ is a solution, then its conjugate $r-pi$ is also a solution.

65. Since $f(x)$ has three real zeros, and a polynomial of degree 3 can have at most three zeros, there can be no other zeros, real or non-real complex.

66. False. A 7th degree function can have at most 7 x-intercepts.

67. True. A 7th degree function can have at most 6 local extrema.

68. True. $f(0)=3(0)^7-8(0)^6+9(0)^5+12(0)^4-18(0)^3+26(0)^2-(0)+500 \Rightarrow f(0)=500.$

69. True. The end behavior of an odd-degree polynomial with a positive lead coefficient is down to the left. Because the function has a positive y-intercept, the graph of the equation must cross the negative x-axis at least one time. Therefore, it must have at least one negative x-intercept.

70. True. An even-degree polynomial with a positive lead coefficient will have the end behavior of both ends going upward; therefore, a graph with a negative y-intercept must cross the x-axis at least 2 times and will have at least 2 real zeros.

71. False. The conjugate must also be a zero, but the conjugate of $-\frac{1}{2}+i\frac{\sqrt{3}}{2}$ is $-\frac{1}{2}-i\frac{\sqrt{3}}{2}$, and not $\frac{1}{2}+i\frac{\sqrt{3}}{2}$.

72. From the graph, there are two local maxima.

73. From the graph, there is one local minimum that lies on the x-axis. It has coordinates $(2,0)$.

74. Use synthetic division to factor out $x-5$.

$$\begin{array}{r|rrrrrr} 5 & -2 & 15 & -21 & -32 & 60 & 0 \\ & & -10 & 25 & 20 & -60 & 0 \\ \hline & -2 & 5 & 4 & -12 & 0 & 0 \end{array}$$

From this result, the quotient is the equation $Q(x)=-2x^4+5x^3+4x^2-12x$.

75. Because all real numbers can be input for x, the range is $(-0.97,-54.15)$.

76. Using the calculator functions, this local minimum point has coordinates $(-0.97,-54.15)$.

77. We first factor by grouping and then solve. $3x^3+2x^2-21x-14=0 \Rightarrow x^2(3x+2)-7(3x+2)=0 \Rightarrow$

$(3x+2)(x^2-7)=0 \Rightarrow 3x+2=0 \Rightarrow 3x=-2 \Rightarrow x=-\frac{2}{3}$ or $x^2-7=0 \Rightarrow x^2=7 \Rightarrow x=\pm\sqrt{7}$. Therefore, the solution set is $\left\{-\sqrt{7},-\frac{2}{3},\sqrt{7}\right\}$. Graphing this equation shows x-intercepts of ± 2.65 and -0.67, which are equal to the solution set, and therefore support our analytic solution.

78. An even-degree polynomial with a negative lead coefficient has an end behavior of $\curvearrowright$
The function has zeros $x=-1$ (multiplicity 2,) 2, and 3; therefore, there are x-intercepts at those values of x. The graph will be tangent to the x-axis at $x=-1$ (multiplicity 2) and the y-intercept is $y=-6$.
See Figure 78.

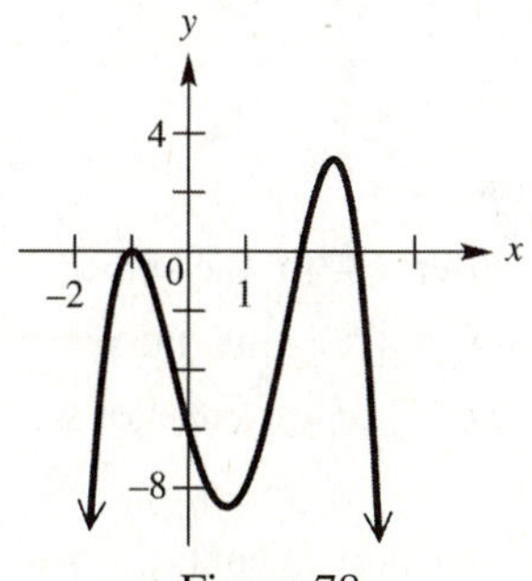

Figure 78

(a) $P(x)=0$ at the x-intercepts; therefore, $x=\{1, 2, 3\}$.

(b) From the graph, $P(x) > 0$ for the interval $(2, 3)$.

(c) From the graph, $P(x) < 0$ for the interval $(-\infty, -1) \cup (-1, 2) \cup (3, \infty)$.

79. If $x =$ cube side length and $V = s^3$, the volume of the cube after the top is cut off is $V = (x)(x)(x-2) = 32 \Rightarrow x^3 - 2x^2 - 32 = 0$. After graphing for possible roots, we use synthetic division to check 4 as a root.

$$\begin{array}{r|rrrr} 4 & 1 & -2 & 0 & -32 \\ & & 4 & 8 & 32 \\ \hline & 1 & 2 & 8 & 0 \end{array}$$

Now solve the resulting equation, $x^2 - 2x + 8 = 0$. $x = \dfrac{-2 \pm \sqrt{2^2 - 4(1)(8)}}{2(1)} = \dfrac{-2 \pm \sqrt{-28}}{2}$. Both of these roots are non-real; therefore, the only real root is 4, and the dimensions of the original cube are 4 in. by 4 in. by 4 in.

80. (a) Since, from a calculator graph of the equation, the zero is at approximately 9.26. Therefore, the restrictions on t are $0 \le t \le 9.26$.

(b) See Figure 80.

(c) From the graph, the object reaches its highest point at approximately $t = 4.08$ seconds.

(d) From the graph, the object reaches its highest point at approximately $s(t) = 131.63$ meters.

(e) The object reaches the ground when $s(t) = 0$; therefore, the equation is $-4.9t^2 + 40t + 50 = 0$. We solve this using the quadratic formula.

$t = \dfrac{-40 \pm \sqrt{40^2 - 4(-4.9)(50)}}{2(-4.9)} = \dfrac{-40 \pm \sqrt{2580}}{-9.8}$; $t \approx -1.10$ or 9.26. Since we cannot have negative time, the object reaches the ground in approximately 9.26 sec.

81. For the year 2001 $x = 11$; therefore, $y = -0.1885(11)^3 + 10.617(11)^2 - 157.075(11) + 2080.8 = 1386.7385$, or approximately 1,387 thousand military personnel were on active duty in 2001.

82. (a) See Figure 82a.

(b) Using the capabilities of the graphing calculator, the quadratic function is $f(x) \approx -0.011x^2 + 0.869x + 11.9$.

(c) Using the capabilities of the calculator, the cubic function is $f(x) \approx -0.00087x^3 + 0.0456x^2 - 0.219x + 17.8$.

(d) See Figure 82b, for the quadratic function, and Figure 82c, for the cubic function.

(e) Both functions approximate the data well. Although the value of R^2 is closer to 1 for the cubic function, the quadratic function is probably better for prediction because it is unlikely that the percent of out-of-pocket spending would decrease after 2025 (as the cubic function shows) unless changes were made in Medicare law.

(f) A linear model would not really be appropriate because the data points lie in a curved nonlinear path.

[0,10] by [0,200]
Xscl = 1 Yscl = 10

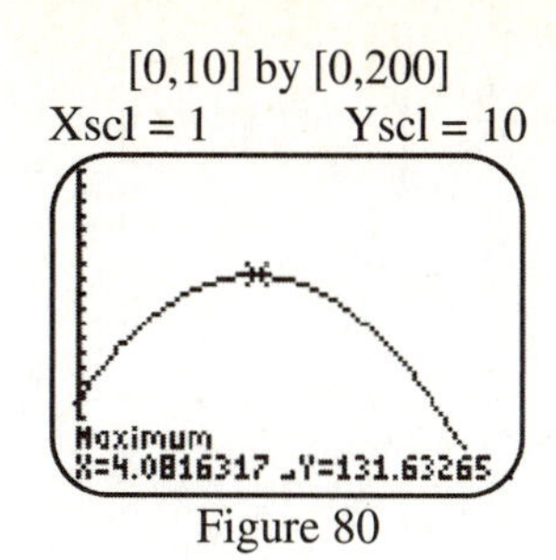

Figure 80

[0,40] by [0,40]
Xscl = 5 Yscl = 5

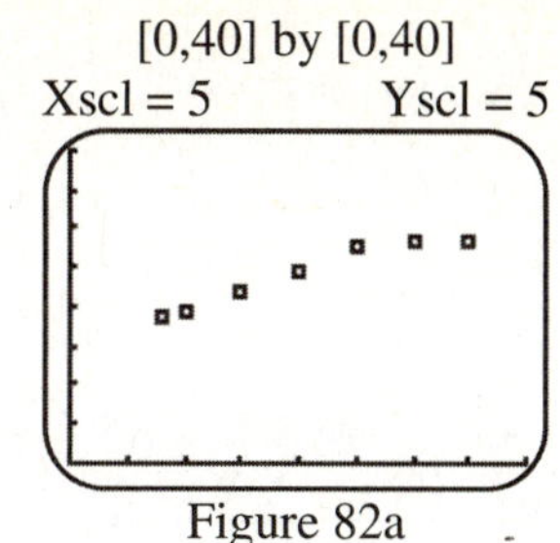

Figure 82a

[0,40] by [0,40]
Xscl= 5 Yscl = 5

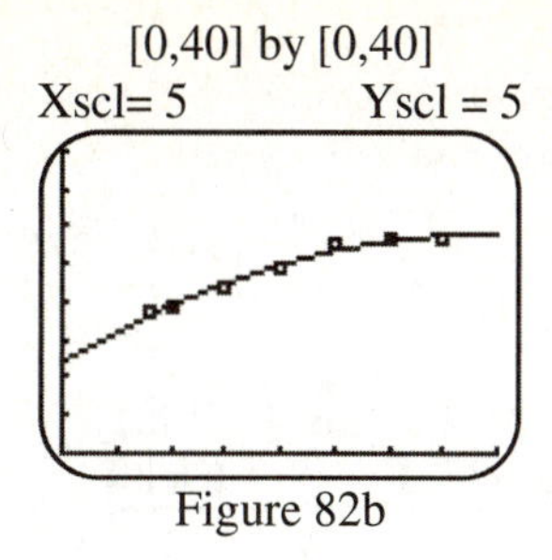

Figure 82b

[0,40] by [0,40]
Xscl = 5 Yscl= 5

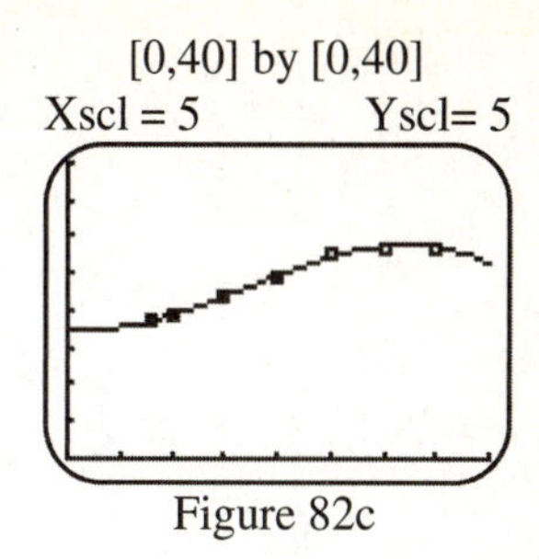

Figure 82c

83. Answers will vary. Example, given 7; x^7

84. Answers will vary.

85. Answers will vary.

86. Answers will vary. Domain: $(-\infty,\infty)$; Range: $(-\infty,\infty)$.

Chapter 3 Test

1. (a) $(8-7i)-(-12+2i)=20-9i.$

(b) $\dfrac{11+10i}{2+3i}\cdot\dfrac{2-3i}{2-3i}=\dfrac{22-33i+20i-30i^2}{4-9i^2}=\dfrac{22-13i+30}{4+9}=\dfrac{52-13i}{13}=4-i.$

(c) $i^{65}=\left(i^4\right)^{16}(i)=1^{16}(i)=i.$

(d) $2i(3-i)^2=2i\left(9-6i+i^2\right)=2i(9-6i-1)=2i(8-6i)=16i-12i^2=12+16i.$

2. (a) Use the vertex formula to find the x-coordinate. $x=-\dfrac{b}{2a}=-\dfrac{-4}{2(-2)}=-1.$ Now solve for $f(-1)$ to find the y-coordinate of the vertex. $f(-1)=-2(-1)^2-4(-1)+6=-2+4+6=8.$ The vertex is $(-1,8)$.

(b) See Figure 2b. Finding the maximum on the graph gives the vertex: $(-1,8)$.

(c) First, we find the zeros by factoring.
$-2x^2-4x+6=0\Rightarrow-2\left(x^2+2x-3\right)=0\Rightarrow-2(x+3)(x-1)=0\Rightarrow x=-3$ or 1. Now support this on the graphing calculator using a graph and a table, See Figure 2c.

(d) For the y-intercept we let $x=0$; therefore, $f(0)=-2(0)-4(0)+6\Rightarrow f(0)=6.$

(e) All values can be input for x. The domain is $(-\infty,\infty)$; since $(-1,8)$ is a maximum, the range is $(-\infty,8]$.

(f) From the graph, the function is increasing for the interval $(-\infty,-1]$, and decreasing for the interval $[-1,\infty)$.

3. (a) We solve the equation using the quadratic formula. $x=\dfrac{-3\pm\sqrt{3^2-4(3)(-2)}}{2(3)}=\dfrac{-3\pm\sqrt{9+24}}{6}\Rightarrow$

$x=\left\{\dfrac{-3-\sqrt{33}}{6},\dfrac{-3+\sqrt{33}}{6}\right\}.$

(b) See Figure 3. From this graph, (i) $f(x)<0$ for the interval $\left(\frac{-3-\sqrt{33}}{6},\frac{-3+\sqrt{33}}{6}\right)$, and (ii) $f(x)\ge 0$ for the interval $\left(-\infty,\frac{-3-\sqrt{33}}{6}\right]\cup\left[\frac{-3+\sqrt{33}}{6},\infty\right)$.

[-9.4,9.4] by [-4.1,8.1]
Xscl = 1 Yscl = 1

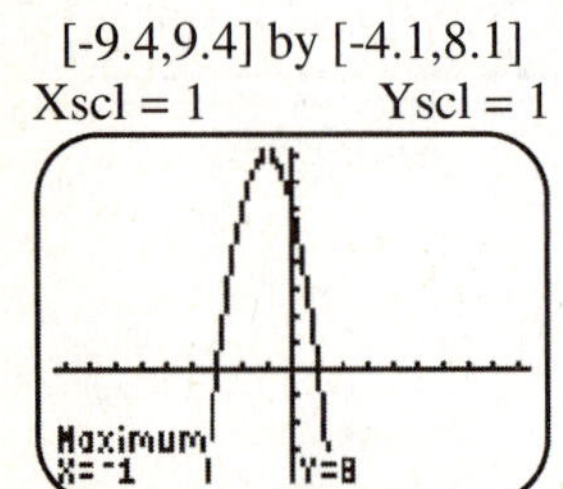
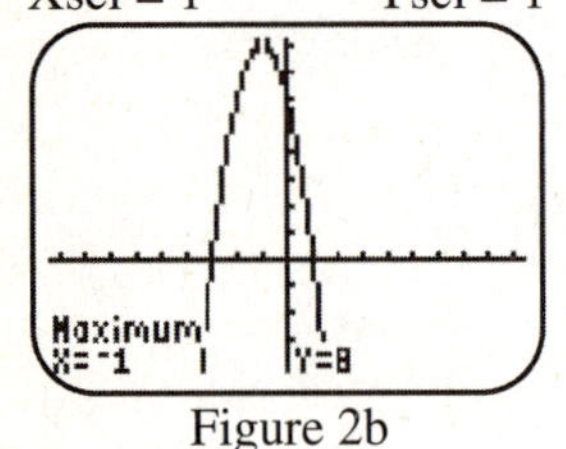

Figure 2b

[-9.4,9.4] by [-4.1,8.1]
Xscl = 1 Yscl = 1

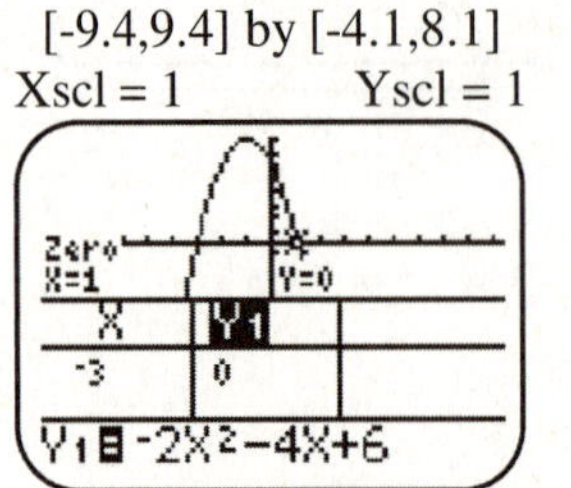
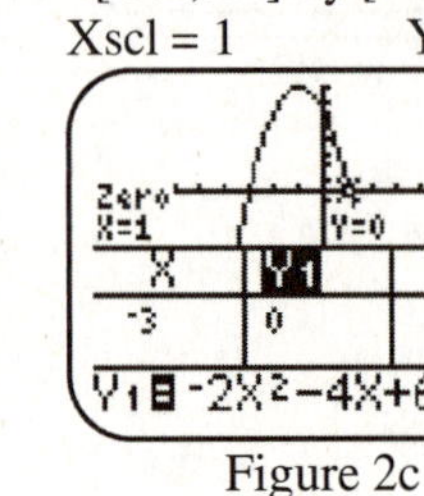

Figure 2c

[-4.7,4.7] by [-3.1,3.1]
Xscl= 1 Yscl = 1

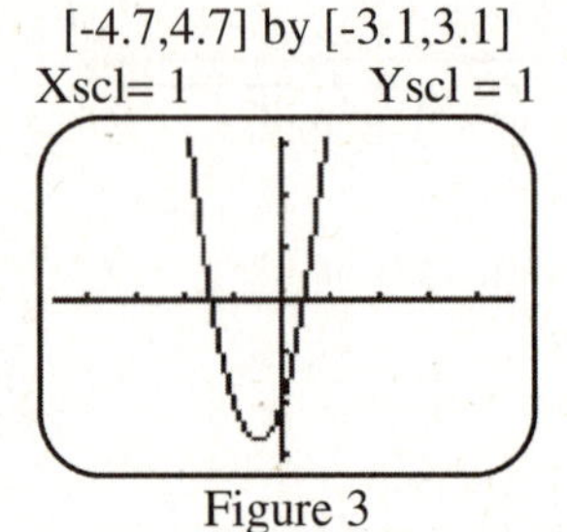

Figure 3

4. Since $V = lwh$, $720=(11+x)(3x)(x)\Rightarrow 720=3x^3+33x^2\Rightarrow 0=3x^3+33x^2-720\Rightarrow 0=x^3+11x^2-240.$

After graphing on a calculator, we use synthetic division to check 4 as a factor.

$$\begin{array}{r|rrrr} 4 & 1 & 11 & 0 & -240 \\ & & 4 & 60 & 240 \\ \hline & 1 & 15 & 60 & 0 \end{array}$$

Now solve the resulting equation, $x^2+15x+60=0$, using the quadratic formula:

$x=\frac{-15\pm\sqrt{15^2-4(1)(60)}}{2(1)}=\frac{-15\pm\sqrt{-15}}{2}$. These solutions will both be nonreal; therefore, the only real solution is 4, and the dimensions are $11+4$, $3(4)$, and 4 or 15 in. by 12 in. by 4 in.

5. (a) Using the capabilities of the graphing calculator, the quadratic function is

$f(x)=0.00019838x^2-0.79153x+791.46$

(b). $f(1975)\approx 1.99$, This value is very close to the actual value of 2.01.

6. (a) Use synthetic division to factor out the given zeros.

$$\begin{array}{r|rrrrrrr} 3 & 1 & -5 & 3 & 1 & 40 & -24 & -72 \\ & & 3 & -6 & -9 & -24 & 48 & 72 \\ \hline & 1 & -2 & -3 & -8 & 16 & 24 & 0 \end{array} \Rightarrow \begin{array}{r|rrrrrr} 3 & 1 & -2 & -3 & -8 & 16 & 24 \\ & & 3 & 3 & 0 & -24 & -24 \\ \hline & 1 & 1 & 0 & -8 & -8 & 0 \end{array} \Rightarrow$$

$$\begin{array}{r|rrrrr} -1 & 1 & 1 & 0 & -8 & -8 \\ & & -1 & 0 & 0 & 8 \\ \hline & 1 & 0 & 0 & -8 & 0 \end{array} \Rightarrow \begin{array}{r|rrrr} 2 & 1 & 0 & 0 & -8 \\ & & 2 & 4 & 8 \\ \hline & 1 & 2 & 4 & 0 \end{array}$$

Now solve the resulting equation, $x^2+2x+4=0$, using the quadratic formula.

$x=\frac{-2\pm\sqrt{2^2-4(1)(4)}}{2(1)}=\frac{-2\pm\sqrt{-12}}{2}=\frac{-2\pm2i\sqrt{3}}{2}=-1\pm i\sqrt{3}$. The other zeros of $f(x)$ are

$x=-1-i\sqrt{3},-1+i\sqrt{3}$.

(b) An even-degree polynomial with a positive lead coefficient has end-behavior $\uparrow\,\uparrow$. See Figure 6.

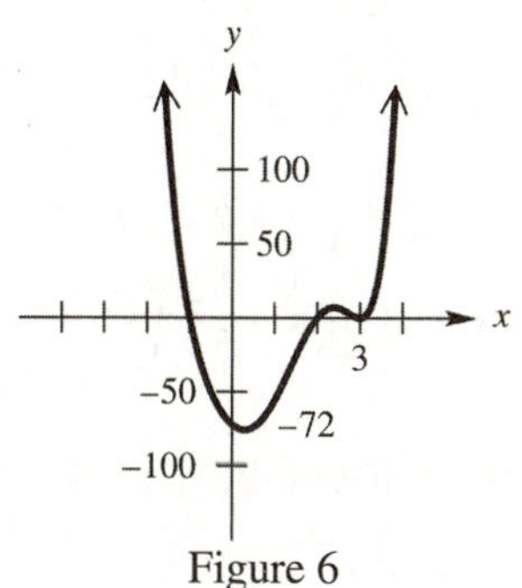

Figure 6

7. (a) We can find the zeros by factoring. $4x^4-21x^2-25=0\Rightarrow(4x^2-25)(x^2+1)=0\Rightarrow$

$4x^2-25=0\Rightarrow4x^2=25\Rightarrow x^2=\frac{25}{4}\Rightarrow x=\pm\sqrt{\frac{25}{4}}=\pm\frac{5}{2}$ or

$x^2+1=0\Rightarrow x^2=-1\Rightarrow x=\pm\sqrt{-1}\Rightarrow x=\pm i$. The zeros are $x=-\frac{5}{2},\frac{5}{2},-i,i$.

(b) Use a graph and table to support the results. See Figure 7.

(c) It is symmetric with respect to the y-axis.

(d) From the graph and the results, (i) $f(x)\geq0$ for the interval $\left(-\infty,-\frac{5}{2}\right]\cup\left[\frac{5}{2},\infty\right)$, and (ii) $f(x)<0$ for the interval $\left(-\frac{5}{2},\frac{5}{2}\right)$.

[-4,4] by [-60,60]
Xscl = 1 Yscl = 10

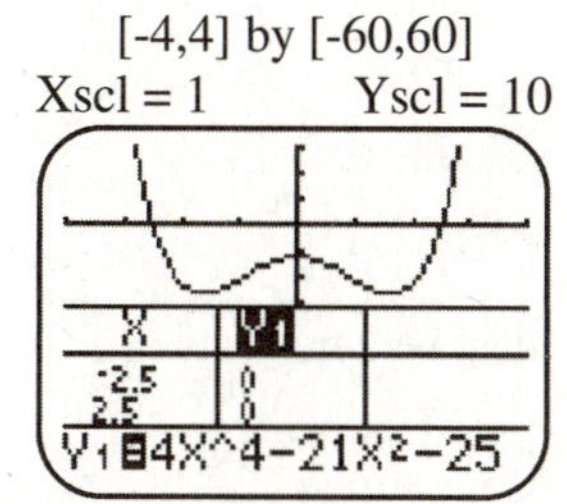

Figure 7

8. (a) Using the Rational Zeros Theorem, the possible zeros are $x=\pm\frac{1,2,11,22}{1,3}=$

$\pm1,\pm2,\pm11,\pm22,\pm\frac{1}{3},\pm\frac{2}{3},\pm\frac{11}{3}$, and $\pm\frac{22}{3}$.

(b) From a calculator graph of the equation, we choose to check -2 and $\frac{1}{3}$ by synthetic division.

$$\begin{array}{r|rrrrr} -2 & 3 & 5 & -35 & -55 & 22 \\ & & -6 & 2 & 66 & -22 \\ \hline & 3 & -1 & -33 & 11 & 0 \end{array} \Rightarrow \begin{array}{r|rrrr} \frac{1}{3} & 3 & -1 & -33 & 11 \\ & & 1 & 0 & -11 \\ \hline & 3 & 0 & -33 & 0 \end{array}$$

Now solve the resulting equation. $3x^2 - 33 = 0 \Rightarrow x^2 - 11 = 0 \Rightarrow x^2 = 11 \Rightarrow x = \pm\sqrt{11}.$ Since these are not rational, the rational zeros are $-2, \frac{1}{3}$.

(c) Since $f(3) = 3(3)^4 + 5(3)^3 + 35(3)^2 - 55(3) + 22 = -80$ and $f(4) = 3(4)^4 + 5(4)^3 + 35(4)^2 - 55(4) + 22 = 330$ differ in sign, the intermediate value theorem assures us that there is a real zero between 3 and 4.

(d) Because the function has coefficient signs $++--+$, which is 2 sign changes, the function has 2 or 0 positive real zeros. Because $f(-x) = 3(-x)^4 + 5(-x)^3 - 35(-x)^2 - 55(-x) + 22 =$ $3x^4 - 5x^3 - 35x^2 + 55x + 22$ has coefficient signs, $+--++$, which is 2 sign changes, the function has 2 or 0 possible negative real zeros.

(e) Using the Boundedness Theorem we divide synthetically by $x - (-5)$.

$$\begin{array}{r|rrrrr} -5 & 3 & 5 & -35 & -55 & 22 \\ & & -15 & 50 & -75 & 650 \\ \hline & 3 & -10 & 15 & -130 & 672 \end{array}$$

The result alternates signs; therefore, there is no real zero less than -5. Also using the Boundedness Theorem, we divide synthetically by $x - 4$.

$$\begin{array}{r|rrrrr} 4 & 3 & 5 & -35 & -55 & 22 \\ & & 12 & 68 & 132 & 308 \\ \hline & 3 & 17 & 33 & 77 & 330 \end{array}$$

The result is all non-negative; therefore, there is no real zero greater than 4.

9. (a) Using the capabilities of the calculator and graphing the equation yields real solutions of $\{0.189, 1, 3.633\}$.

(b) A 5th degree equation has 5 solutions. Since there are 3 real solutions, there must be 2 non-real complex solutions.

10. (a) $\dfrac{8x^3 - 4x^2}{2x} = \dfrac{8x^3}{2x} - \dfrac{4x^2}{2x} = 4x^2 - 2x$

(b)

$$\begin{array}{r|l} & 3x^2-2x-2 \\ x-1 & 3x^3-5x^2+0x+6 \\ & \underline{3x^3-3x^2} \\ & -2x^2+0x \\ & \underline{-2x^2+2x} \\ & -2x+6 \\ & \underline{-2x+2} \\ & 4 \end{array} \Rightarrow \frac{3x^3-5x^2+6}{x-1}=3x^2-2x-2+\frac{4}{x-1}$$

(c)

$$\begin{array}{r|l} & x^3+2x-1 \\ 2x-1 & 2x^4-x^3+4x^2-4x+3 \\ & \underline{2x^4-x^3} \\ & 4x^2-4x \\ & \underline{-4x^2-2x} \\ & -2x+3 \\ & \underline{-2x+1} \\ & 2 \end{array} \Rightarrow \frac{2x^4-x^3+4x^2-4x+3}{2x-1}=x^3+2x-1+\frac{2}{2x-1}$$

(d)

$$\begin{array}{r|l} & -2x+10 \\ x^2+2 & x^4+0x^2-2x+6 \\ & \underline{x^4+2x^2} \\ & -2x^2-2x \\ & \underline{-2x^2\quad -4} \\ & -2x+4+6 \end{array} \Rightarrow \frac{x^4-2x+6}{x^2+2}=x^2-2+\frac{-2x+10}{x^2+2}$$

Chapter 4: Rational, Power, and Root Functions

4.1: Rational Functions and Graphs

1. The only value for x that cannot be used as input is 0. The domain is $(-\infty,0)\cup(0,\infty)$.

   It is not possible for this function to output the value 0. The range is $(-\infty,0)\cup(0,\infty)$.

3. The function decreases everywhere it is defined, $(-\infty,0)\cup(0,\infty)$. It never increases and is never constant.

5. Because the function is undefined when $x=3$, the vertical asymptote has the equation $x=3$. As $|x|$ increases without bound, the graph of the function will move closer and closer to the graph of $y=2$.

7. Because $f(-x)=f(x)$, the function is even. The graph has symmetry with respect to the y-axis.

9. Graphs A, B, and C have domain $(-\infty,3)\cup(3,\infty)$ because each has a vertical asymptote at $x=3$.

11. Graph A has range $(-\infty,3)\cup(3,\infty)$ because it exists above and below the horizontal asymptote at $y=0$.

13. The only graph that would intersect the line $y=3$ exactly one time is graph A.

15. Graphs A, C, and D have the x-axis as a horizontal asymptote.

17. Window C gives the most accurate depiction of the graph. See Figures 17a, 17b and 17c.

[-4.7,4.7] by [-3.1,3.1] Xscl = 1 Yscl = 1

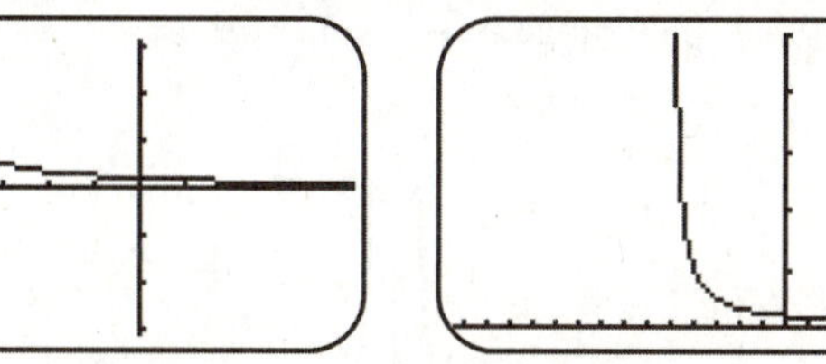

Figure 17a

[-14.4,4.4] by [0,5] Xscl = 1 Yscl = 1

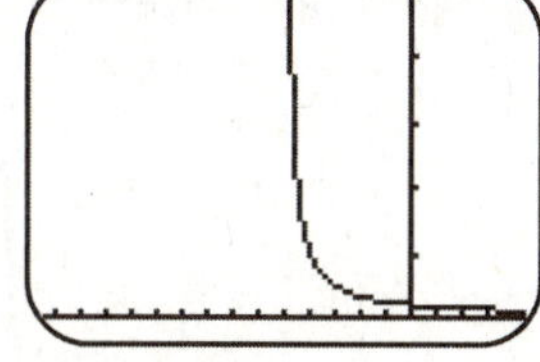

Figure 17b

[-9.4,9.4] by [-3.1,3.1] Xscl= 1 Yscl = 1

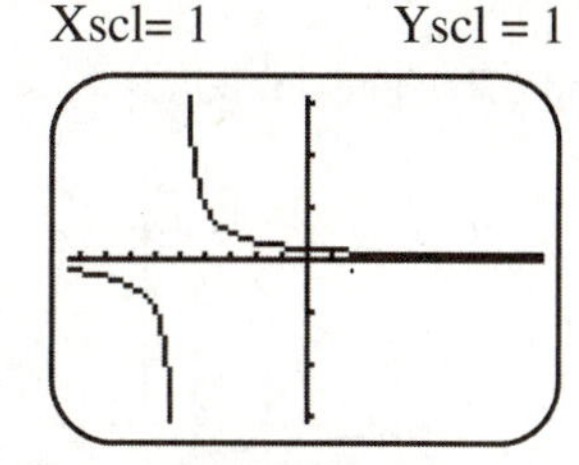

Figure 17c

19. Let $g(x)=y=\frac{1}{x}$, then $f(x)=2g(x)$. To obtain the graph of f, stretch the graph of $y=\frac{1}{x}$ vertically by a factor of 2. See Figures 19a and 19b. The domain is $(-\infty,0)\cup(0,\infty)$. The range is $(-\infty,0)\cup(0,\infty)$.

21. Let $g(x)=y=\frac{1}{x}$, then $f(x)=g(x+2)$. To obtain the graph of f, shift the graph of $y=\frac{1}{x}$ to the left 2 units. See Figures 21a and 21b. The domain is $(-\infty,-2)\cup(-2,\infty)$. The range is $(-\infty,0)\cup(0,\infty)$.

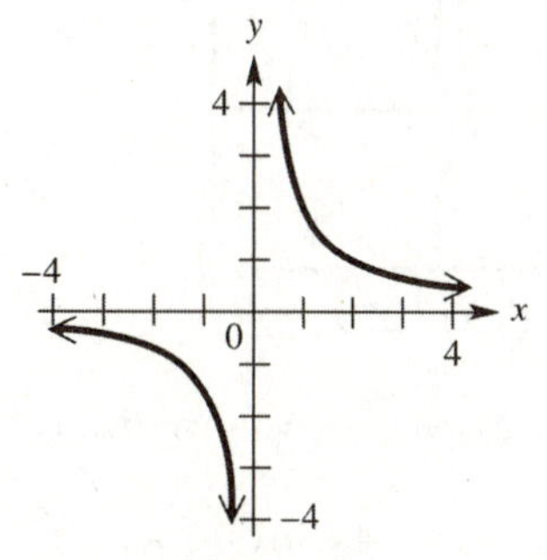

Figure 19a

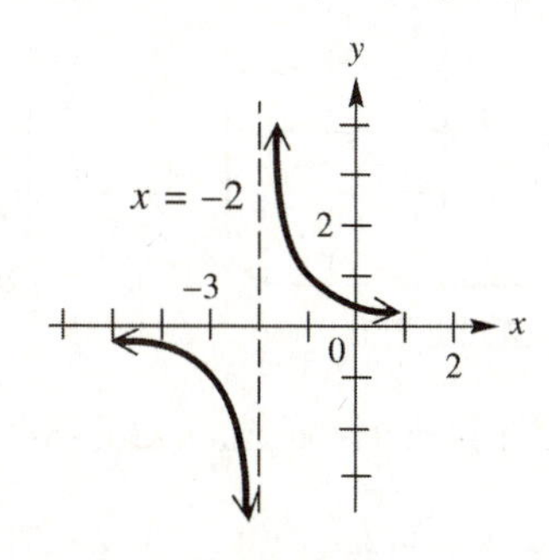

Figure 21a

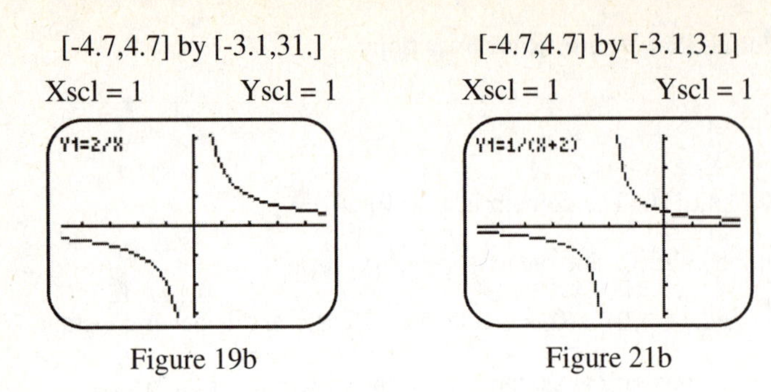

Figure 19b

Figure 21b

23. Let $g(x) = y = \frac{1}{x}$, then $f(x) = g(x) + 1$. To obtain the graph of *f*, shift the graph of $y = \frac{1}{x}$ upward 1 unit. See Figures 23a and 23b. The domain is $(-\infty, 0) \cup (0, \infty)$. The range is $(-\infty, 1) \cup (1, \infty)$.

25. Let $g(x) = y = \frac{1}{x}$, then $f(x) = g(x-1) + 1$. To obtain the graph of *f*, shift the graph of $y = \frac{1}{x}$ to the right 1 unit and upward 1 unit. See Figures 25a and 25b. The domain is $(-\infty, 1) \cup (1, \infty)$. The range is $(-\infty, 1) \cup (1, \infty)$.

27. Let $g(x) = y = \frac{1}{x^2}$, then $f(x) = g(x) - 2$. To obtain the graph of *f*, shift the graph of $y = \frac{1}{x^2}$ downward 2 units. See Figures 27a and 27b. The domain is $(-\infty, 0) \cup (0, \infty)$. The range is $(-2, \infty)$.

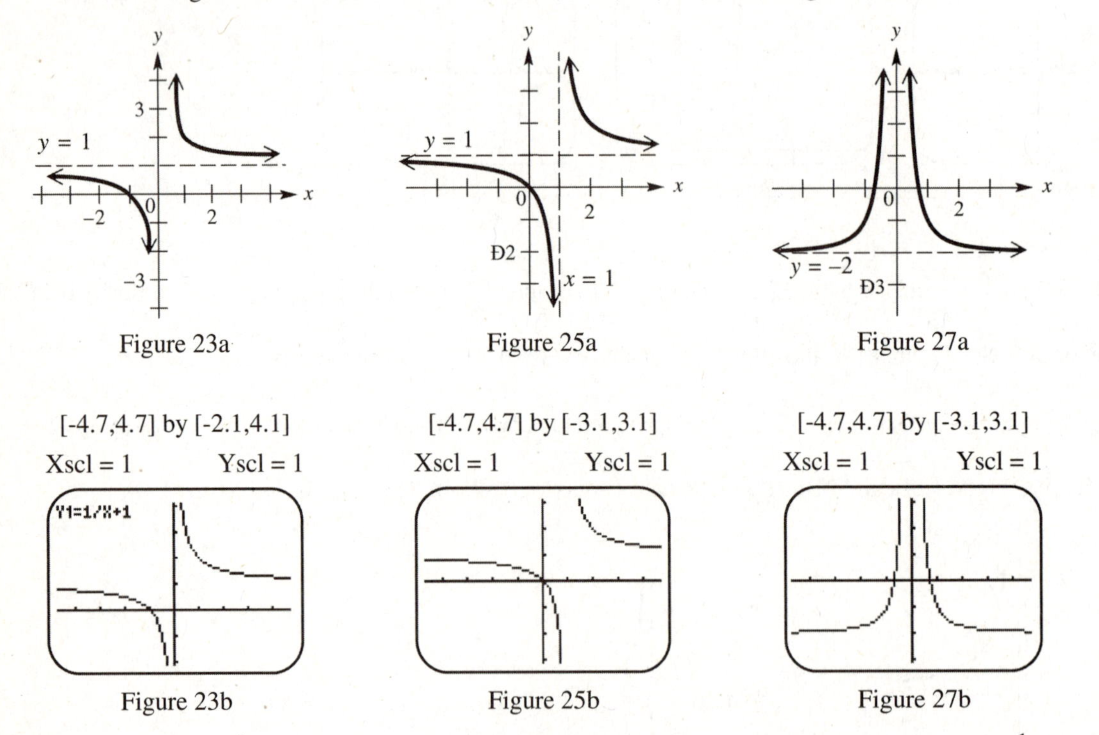

Figure 23a

Figure 25a

Figure 27a

Figure 23b

Figure 25b

Figure 27b

29. Let $g(x) = y = \frac{1}{x^2}$, then $f(x) = -2g(x)$. To obtain the graph of *f*, stretch the graph of $y = \frac{1}{x^2}$ vertically by a factor of 2 and reflect it across the *x*-axis. See Figures 29a and 29b. The domain is $(-\infty, 0) \cup (0, \infty)$. The range is $(-\infty, 0)$

31. Let $g(x) = y = \dfrac{1}{x^2}$, then $f(x) = g(x-3)$. To obtain the graph of *f*, shift the graph of $y = \dfrac{1}{x^2}$ to the right 3 units. See Figures 31a and 31b. The domain is $(-\infty,3)\cup(3,\infty)$. The range is $(0,\infty)$.

$(-\infty,0)$.

33. Let $g(x) = y = \dfrac{1}{x^2}$, then $f(x) = -g(x+2)-3$. To obtain the graph of *f*, shift the graph of $y = \dfrac{1}{x^2}$ to the left 2 units, reflect it across the x-axis and shift it downward 3 units. See Figures 33a and 33b. The domain is $(-\infty,-2)\cup(-2,\infty)$. The range is $(-\infty,-3)$.

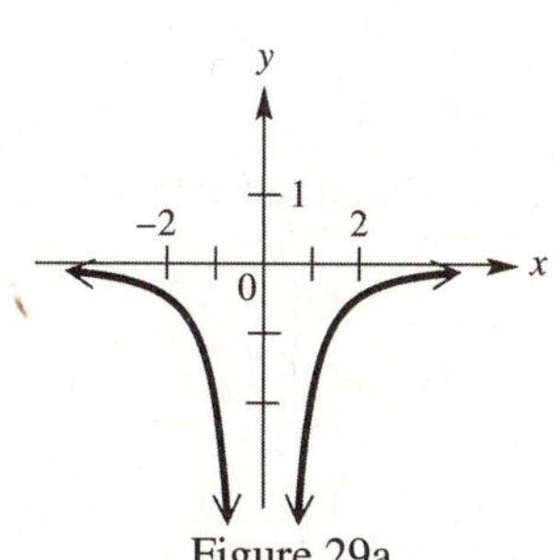

Figure 29a

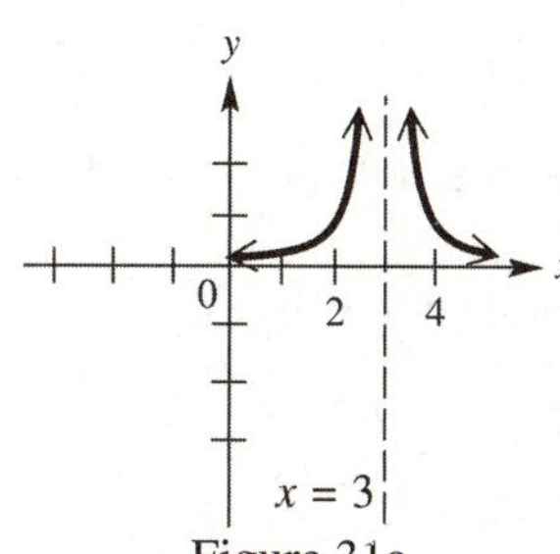

Figure 31a

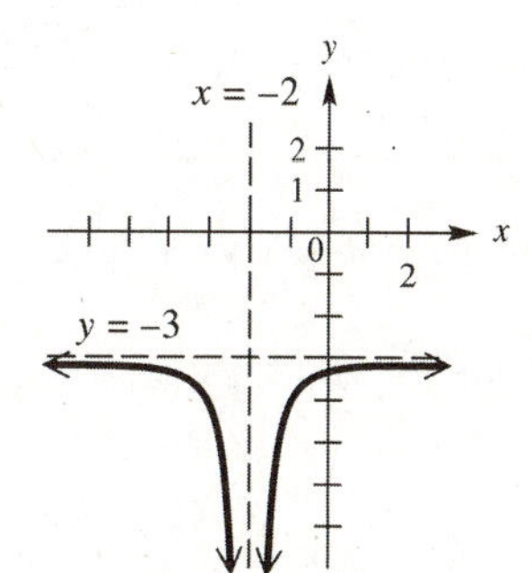

Figure 33a

[-4.7,4.7] by [-3.1,3.1]
Xscl = 1 Yscl = 1

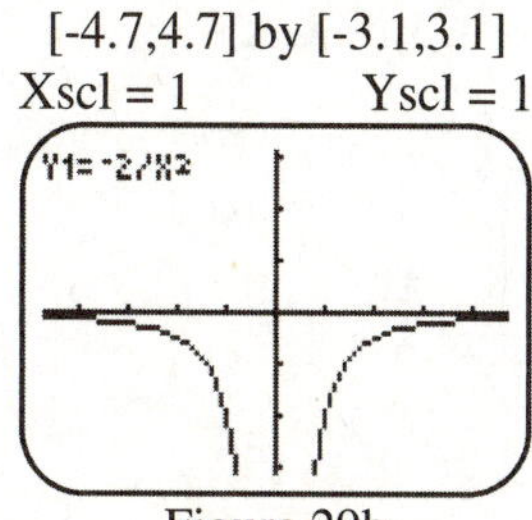

Figure 29b

[-1.7,7.7] by [-5.1,1.1]
Xscl = 1 Yscl = 1

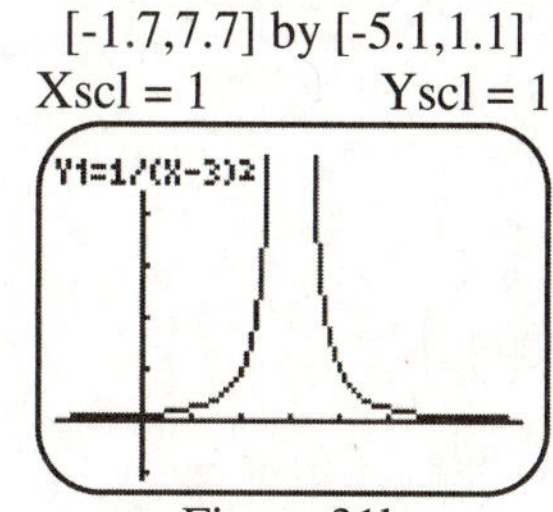

Figure 31b

[-6.7,2.7] by [-5.1,1.1]
Xscl = 1 Yscl = 1

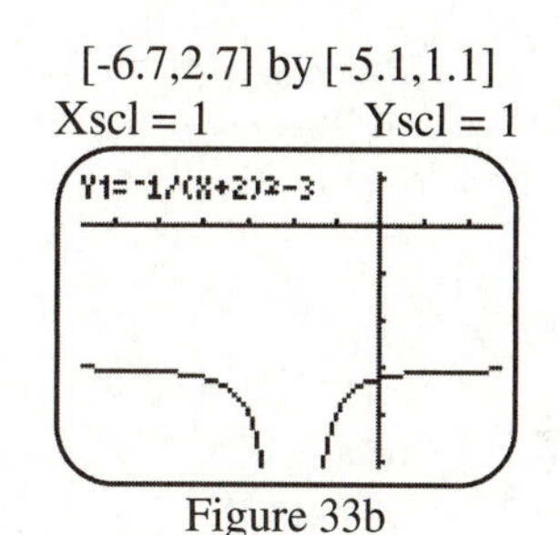

Figure 33b

35. The graph of *f* is obtained by shifting the graph of $y = \dfrac{1}{x^2}$ to the right 2 units. This is shown in graph C.

37. The graph of *f* is obtained by shifting the graph of $y = \dfrac{1}{x}$ to the right 2 units and reflecting it across the x-axis. This is shown in graph B.

39. The vertical asymptote shifts 2 units left to $x = -1$. The horizontal asymptote shifts 1 unit down to $y = 1$. The domain shifts 2 units left to $(-\infty,-1)\cup(-1,\infty)$. The range shifts 1 unit down to $(-\infty,1)\cup(1,\infty)$.

41. Perform long division: $x-2\overline{)\,x-1}$ with quotient 1, subtract $x-2$, remainder 1 $\Rightarrow \dfrac{x-1}{x-2} = 1 + \dfrac{1}{x-2}$. The graph of $y = 1 + \dfrac{1}{x-2}$ is obtained by shifting the graph of $y = \dfrac{1}{x}$ to the right 2 units and 1 unit upward. A sketch is shown in Figure 41a. A calculator graph of $y_1 = (x-1)/(x-2)$ is shown in Figure 41b.

43. Perform long division: $\begin{array}{r} -2 \\ x+3 \overline{)-2x-5} \\ \underline{-2x-6} \\ 1 \end{array}$ $\Rightarrow \dfrac{-2x-5}{x+3} = -2 + \dfrac{1}{x+3}$. The graph of $y = -2 + \dfrac{1}{x+3}$ is obtained by shifting the graph of $y = \dfrac{1}{x}$ to the left 3 units and 2 units downward. A sketch is shown in Figure 43a. A calculator graph of $y_1 = (-2x-5)/(x+3)$ is shown in Figure 43b.

45. Perform long division: $\begin{array}{r} 2 \\ x-3 \overline{)2x-5} \\ \underline{2x-6} \\ 1 \end{array}$ $\Rightarrow \dfrac{2x-5}{x-3} = 2 + \dfrac{1}{x-3}$. The graph of $y = 2 + \dfrac{1}{x-3}$ is obtained by shifting the graph of $y = \dfrac{1}{x}$ to the right 3 units and 2 units upward. A sketch is shown in Figure 45a. A calculator graph of $y_1 = (2x-5)/(x-3)$ is shown in Figure 45b.

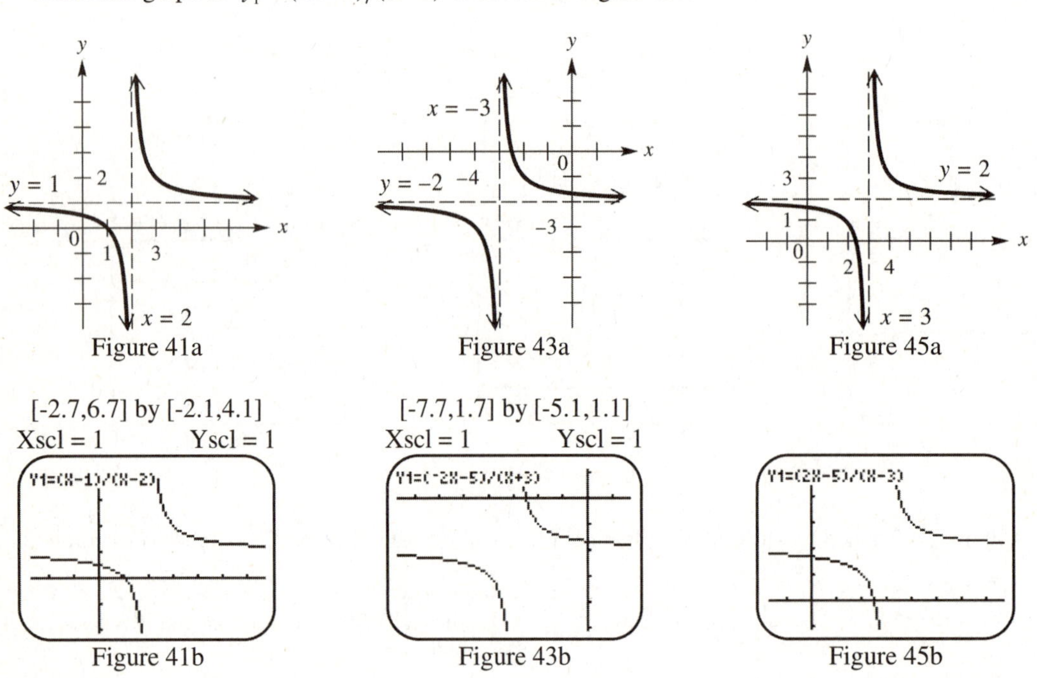

Figure 41a Figure 43a Figure 45a

Figure 41b Figure 43b Figure 45b

4.2: More on Graphs of Rational Functions

1. Because -1 is a zero of the denominator but not of the numerator, the line $x = -1$ is a vertical asymptote: D.
3. The degree of the numerator is less than the degree of the denominator, so the x-axis is a horizontal asymptote: G.
5. Because $f(x) = \dfrac{x^2-16}{x+4} = \dfrac{(x-4)(x+4)}{x+4} = x-4$ for all $x \neq -4$, there is a hole at $x = -4$: E.
7. Because $f(x) = \dfrac{x^2+3x+4}{x-5} = x+8+\dfrac{44}{x-5}$, the line $y = x+8$ is an oblique asymptote: F.
9. Since 5 is a zero of the denominator but not of the numerator, the line $x = 5$ is a vertical asymptote.

Since $f(x) \to 0$ as $|x| \to \infty$, the line $y = 0$ is a horizontal asymptote.

Since the degree of the numerator is less than the degree of the denominator, there is no oblique asymptote.

11. Since $-\frac{1}{2}$ is a zero of the denominator but not of the numerator, the line $x = -\frac{1}{2}$ is a vertical asymptote.

    Since the degrees of the numerator and denominator are equal, the horizontal asymptote is $y = \frac{-3}{2} = -\frac{3}{2}$.

    Since the degree of the numerator is equal to the degree of the denominator, there is no oblique asymptote.

13. Since -3 is a zero of the denominator but not of the numerator, the line $x = -3$ is a vertical asymptote. Since the degree of the numerator is greater than the degree of the denominator, there is no horizontal asymptote.

    Since $\frac{x^2-1}{x+3} = x - 3 + \frac{8}{x+3}$, the oblique asymptote is $y = x - 3$.

15. Since -2 and $\frac{5}{2}$ are zeros of the denominator but not of the numerator, the lines $x = -2$ and $x = \frac{5}{2}$ are vertical asymptotes.

    Since the degrees of the numerator and denominator are equal, the horizontal asymptote is $y = \frac{1}{2}$.

    Since the degree of the numerator is equal to the degree of the denominator, there is no oblique asymptote.

17. Function A, because the denominator can never be equal to 0.

19. From the graph, the vertical asymptote is $x = 2$, the horizontal asymptote is $y = 4$, and there is no oblique asymptote. The function is defined for all $x \neq 2$, therefore the domain is $(-\infty, 2) \cup (2, \infty)$.

21. From the graph, the vertical asymptotes are $x = -2$ and $x = 2$, the horizontal asymptote is $y = -4$, and there is no oblique asymptote. The function is defined for all $x \neq \pm 2$, therefore the domain is $(-\infty, -2) \cup (-2, 2) \cup (2, \infty)$.

23. From the graph, the vertical asymptote is $x = 1$, the horizontal asymptote is $y = 0$ and there is no oblique asymptote. The function is defined for all $x \neq 1$, therefore the domain is $(-\infty, 1) \cup (1, \infty)$.

25. From the graph, the vertical asymptote is $x = -1$, there is no horizontal asymptote, and the oblique asymptote passes through the points $(0, -1)$ and $(1, 0)$. Thus, the equation of the oblique asymptote is $y = x - 1$. The function is defined for all $x \neq -1$, therefore the domain is $(-\infty, -1) \cup (-1, \infty)$.

27. From the graph, there is no vertical asymptote, the horizontal asymptote is $y = 0$, and there is no oblique asymptote. The function is defined for all x, therefore the domain is $(-\infty, \infty)$.

29. See Figure 29.

31. See Figure 31.

33. See Figure 33.

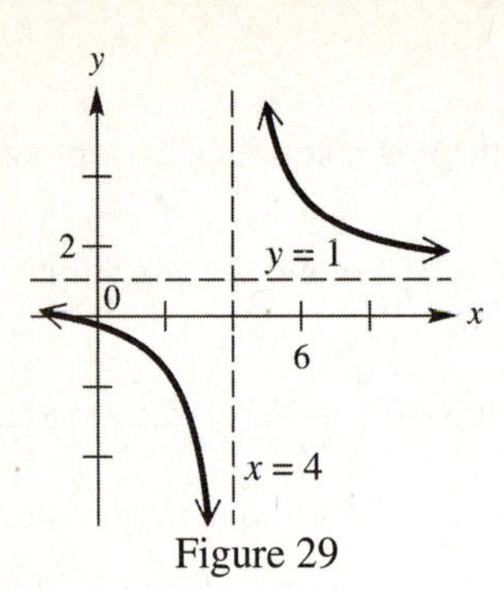

Figure 29

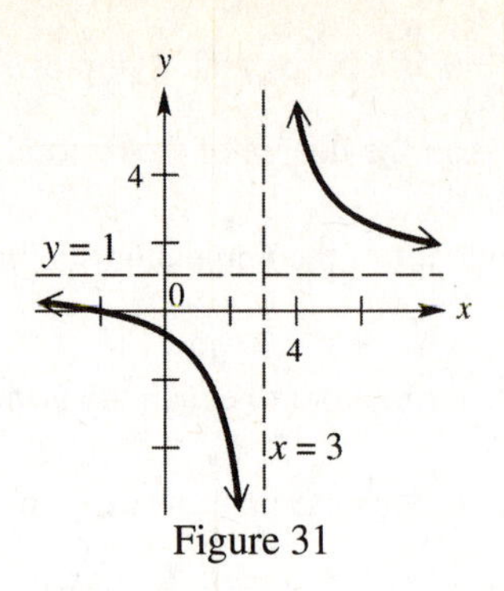

Figure 31

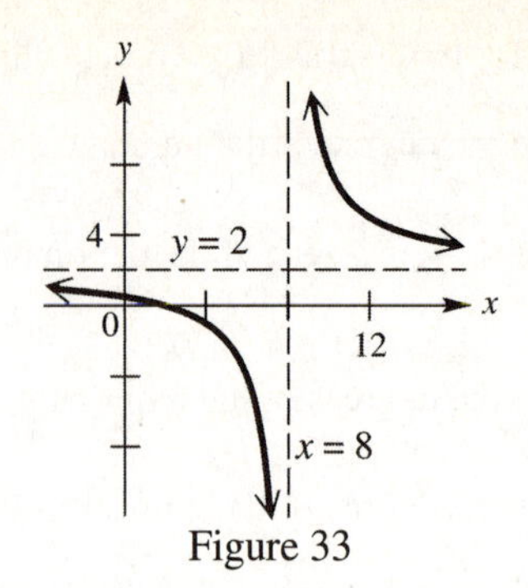

Figure 33

35. See Figure 35.
37. See Figure 37.
39. See Figure 39.

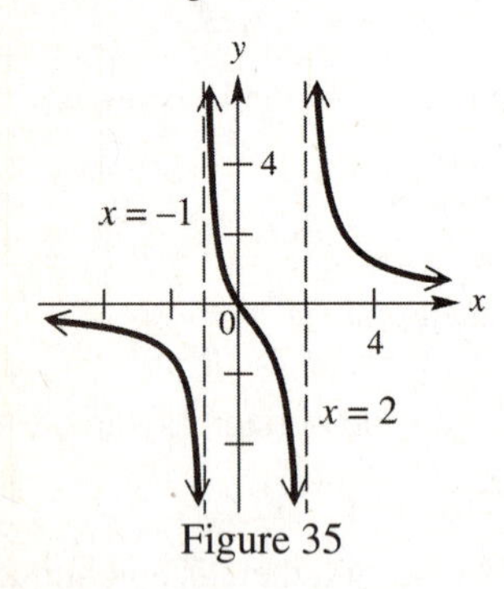

Figure 35

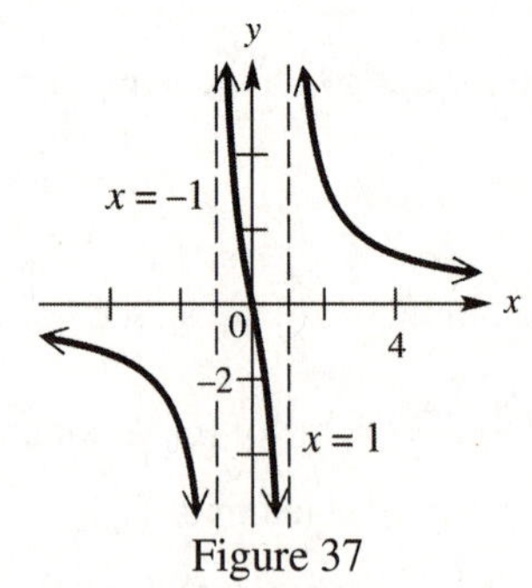

Figure 37

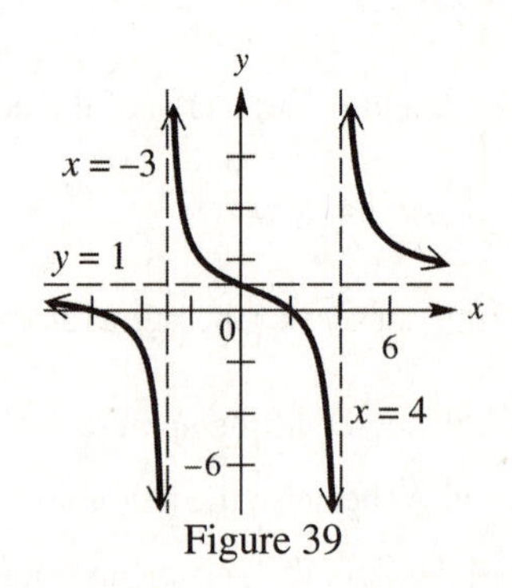

Figure 39

41. See Figure 41.
43. See Figure 43.
45. See Figure 45

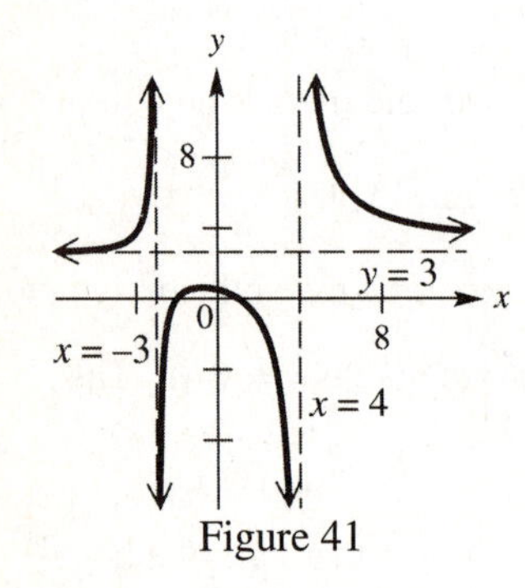

Figure 41

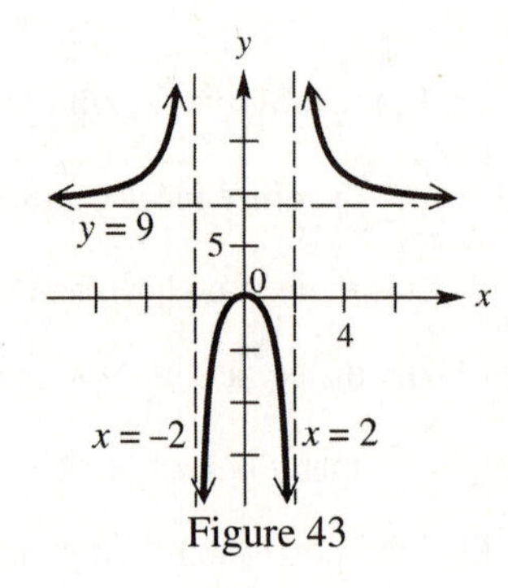

Figure 43

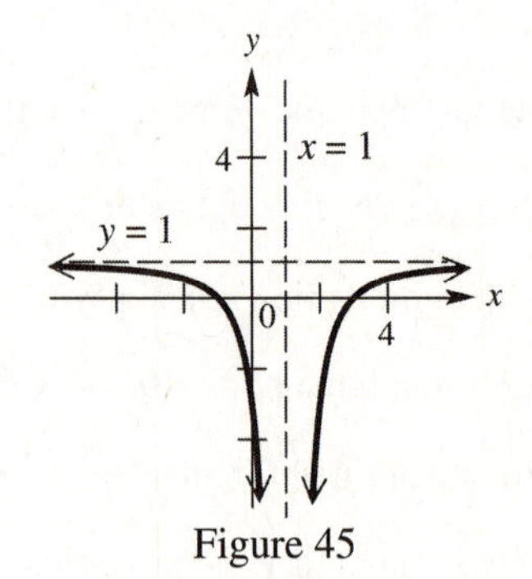

Figure 45

47. See Figure 47.
49. See Figure 49.
51. See Figure 51.

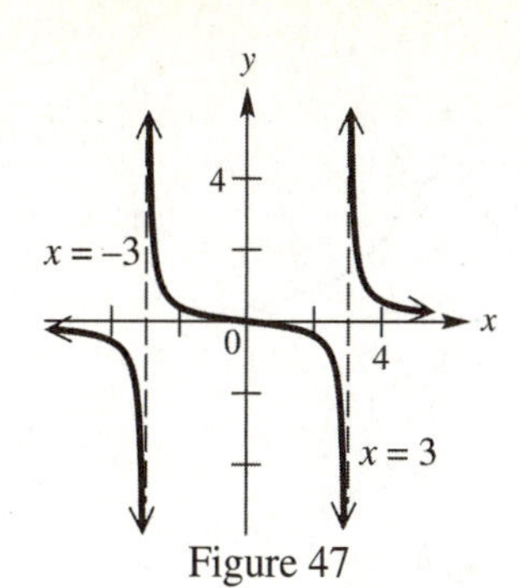

Figure 47

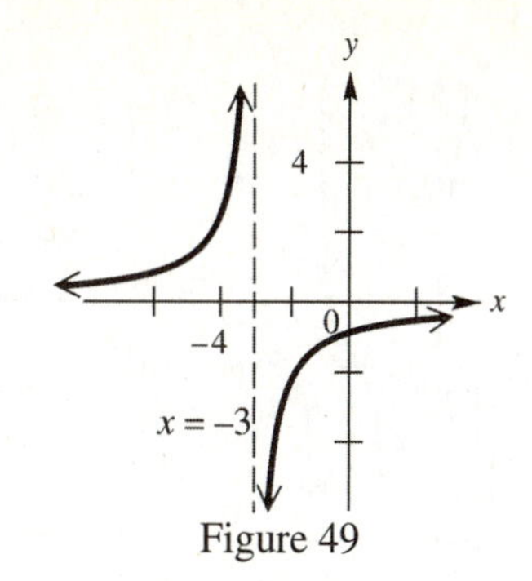

Figure 49

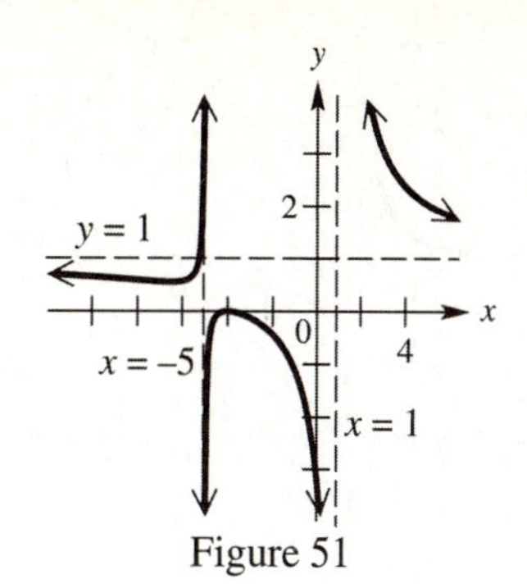

Figure 51

53. See Figure 53.

55. See Figure 55.

57. See Figure 57.

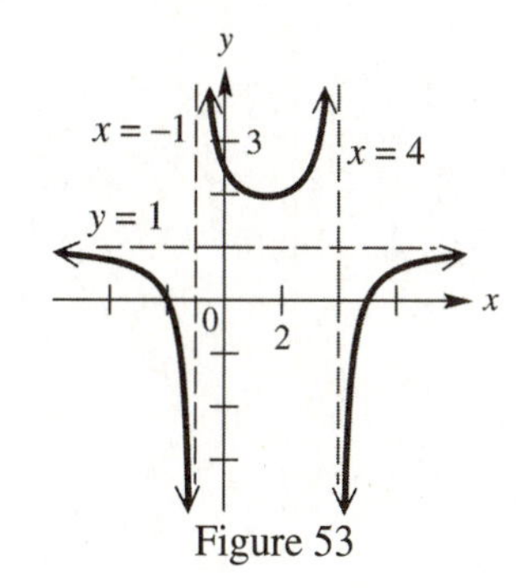

Figure 53

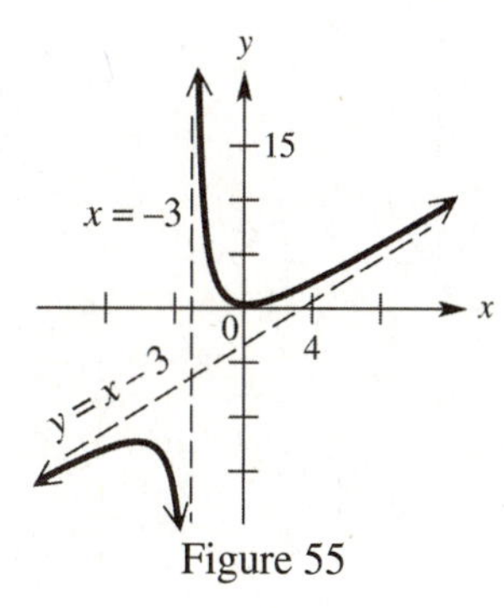

Figure 55

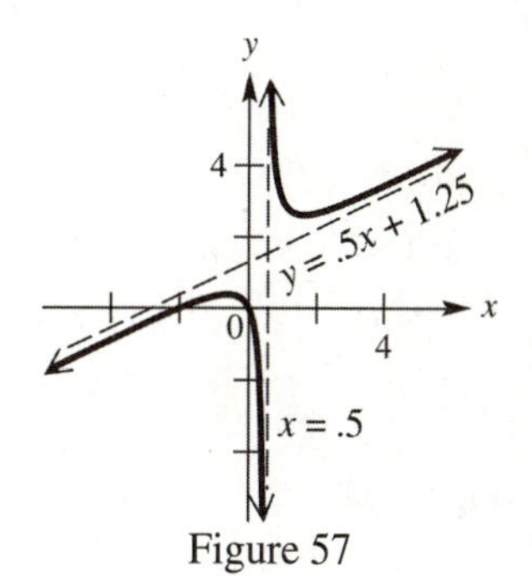

Figure 57

59. See Figure 59.

61. See Figure 61.

63. See Figure 63.

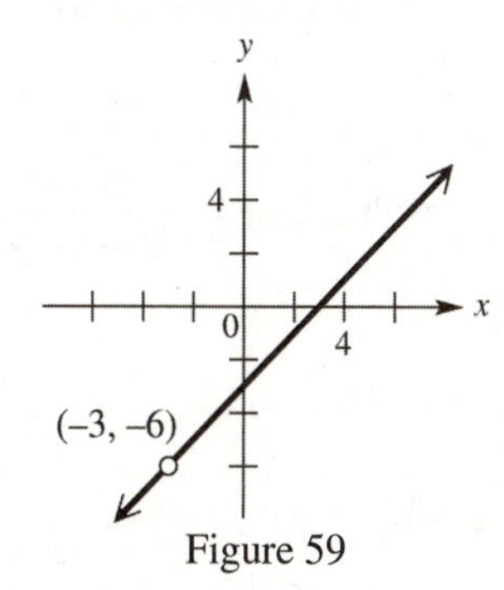

Figure 59

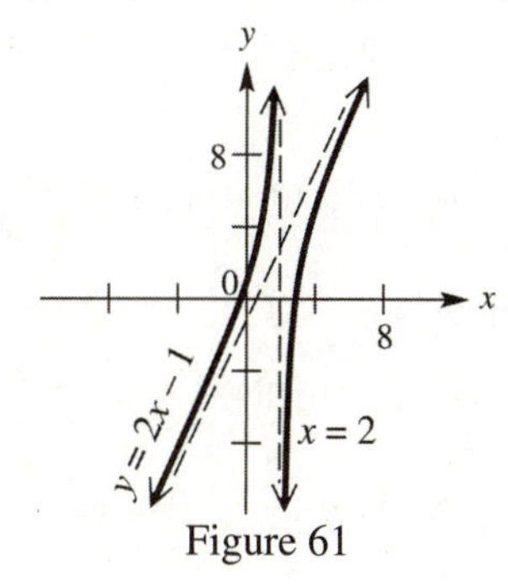

Figure 61

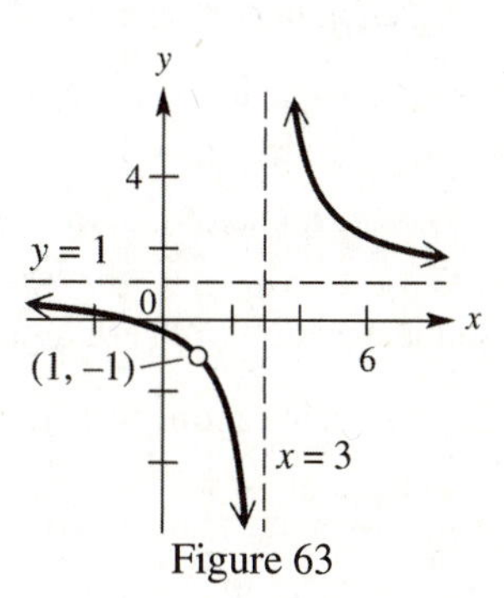

Figure 63

65. See Figure 65.

67. See Figure 67.

69. The domain is $(-\infty, \infty)$ because the denominator as no real zeros, the range is the interval $\left(0, \frac{1}{2}\right]$, and the graph is symmetric with respect to the y-axis. There are no vertical asymptotes and the x-axis is the horizontal asymptote. See Figure 69.

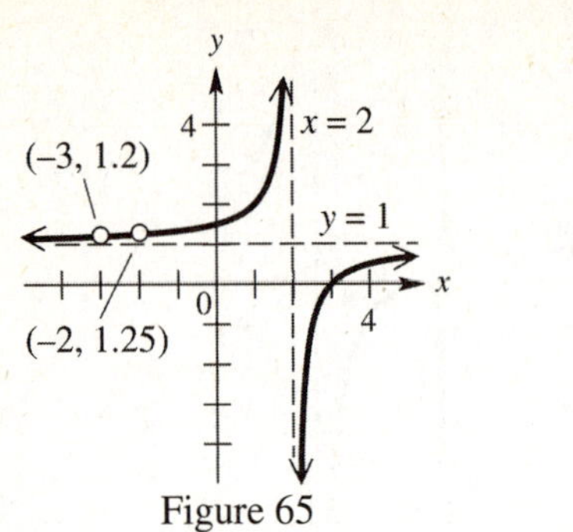

Figure 65

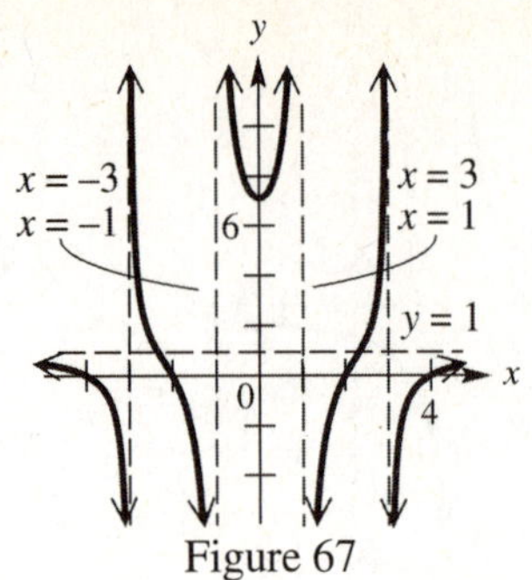

Figure 67

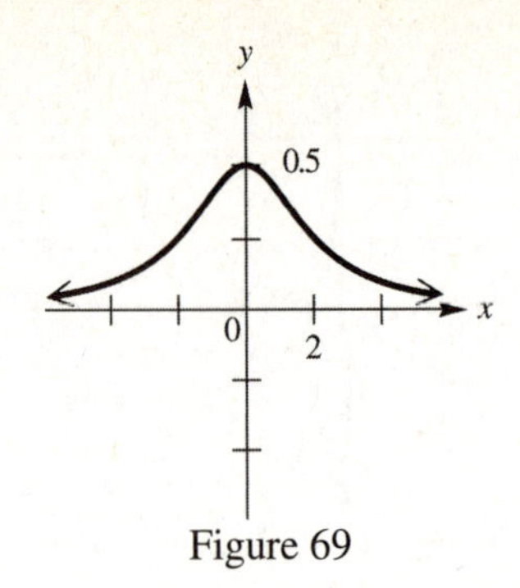

Figure 69

71. $D:(-\infty,\infty)$, $R:(-1,0]$ The function is symmetric to the y-axis and the HA is y = - 1. See Figure 71..

73. $D:(-\infty,\infty)$, $R:[0,1]$ The function is symmetric to the y-axis and the HA is y = 0. See Figure 73.

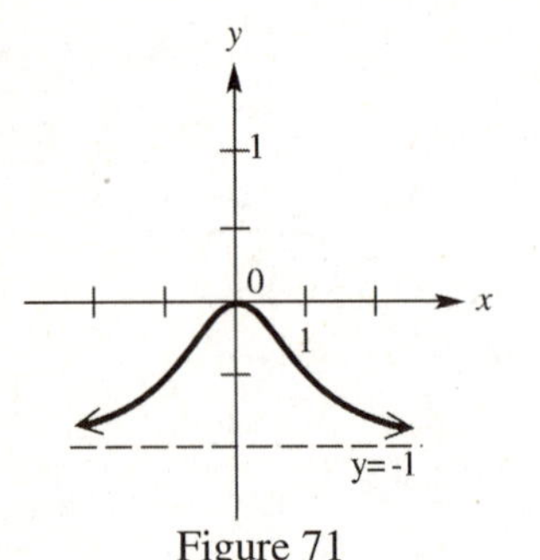

Figure 71

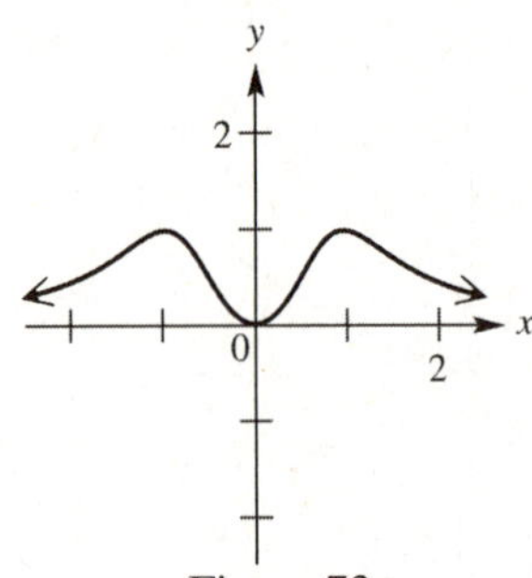

Figure 73

75. (a) The x-intercepts are the solutions of $3x^3+2x^2-12x-8=0 \Rightarrow (3x+2)(x^2-4)=0 \Rightarrow x=-\frac{2}{3},\pm 2$. The y-intercept is $f(0)=-2$.

(b) The denominator has no real zeros because the discriminant is equal to -15. Therefore, there are no vertical asymptotes for $f(x)$.

(c) When the numerator is divided by the denominator using long division the result is $3x-1+\frac{-23x-4}{x^2+x+4}$, so the equation of the oblique asymptote is y=3x-1.

(d) The graph in Figure 75 correctly suggests that both the domain and range are both $(-\infty,\infty)$.

77. (a) The x-intercepts are the solutions of $x^3+4x^2-x-4=0 \Rightarrow (x-4)(x^2-1)=0 \Rightarrow x=4,\pm 1$. The y-intercept is $f(0)=1$.

(b) The denominator has no real zeros because the discriminant is equal to -28. Therefore, there are no vertical asymptotes for $f(x)$.

(c) When the numerator is divided by the denominator using long division the result is $-\frac{1}{2}x-\frac{3}{2}+\frac{-2x-10}{-2x^2-2x-4}$, so the equation of the oblique asymptote is $-\frac{1}{2}x-\frac{3}{2}$.

(d) The graph in Figure 77 correctly suggests that both the domain and range are both $(-\infty,\infty)$.

[-4.7,4.7] by [-3.1,3.1]
Xscl = 1 Yscl = 1

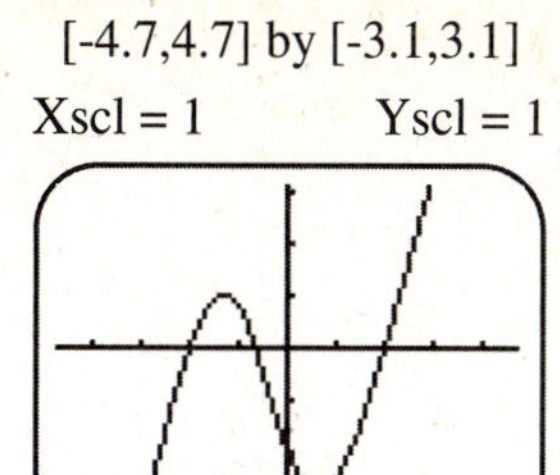

Figure 75

[-4.7,4.7] by [--3.1,3.1]
Xscl = 1 Yscl = 1

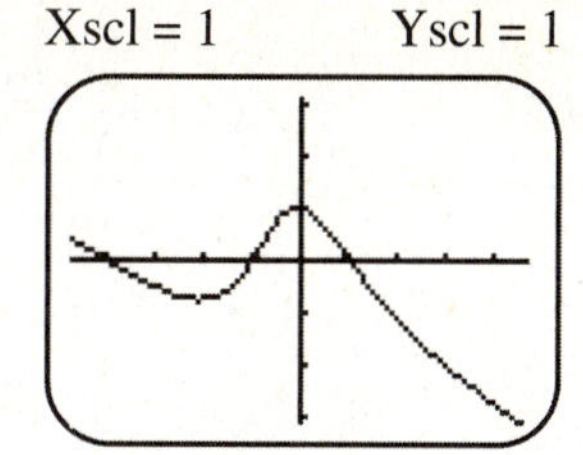

Figure 77

79. The expression is equal to 0 when $x = 4$, thus $p = 4$. The expression is undefined when $x = 2$, thus $q = 2$.

81. The expression is equal to 0 when $x = -2$, thus $p = -2$. The expression is undefined when $x = -1$, thus $q = -1$.

83. A vertical asymptote of $x = 2$ indicates that the denominator should contain the factor $(x-2)$. A hole at $x = -2$ indicates that both the numerator and denominator should contain the factor $(x+2)$. An x-intercept of 3 indicates that the numerator should contain the factor $(x-3)$. One possible equation for the function is $f(x) = \dfrac{(x-3)(x+2)}{(x-2)(x+2)} = \dfrac{x^2 - x - 6}{x^2 - 4}$. Other functions are possible.

85. Vertical asymptotes of $x = 0$ and $x = 4$ indicate that the denominator should contain the factors x and $(x-4)$. An x-intercept of 2 indicates that the numerator should contain the factor $(x-2)$. Since there is a horizontal asymptote at $y = 0$, the degree of the numerator should be less than the degree of the denominator. One possible equation for the function is $f(x) = \dfrac{x-2}{x(x-4)} = \dfrac{x-2}{x^2 - 4x}$. Other functions are possible.

87. (a) Reflect the graph across the x-axis. See Figure 87a.
    (b) Reflect the graph across the y-axis. See Figure 87b.

88. (a) Reflect the graph across the x-axis. See Figure 88a.
    (b) Reflect the graph across the y-axis. See Figure 88b.

89. (a) Reflect the graph across the x-axis. See Figure 89a.
    (b) Reflect the graph across the y-axis. See Figure 89b.

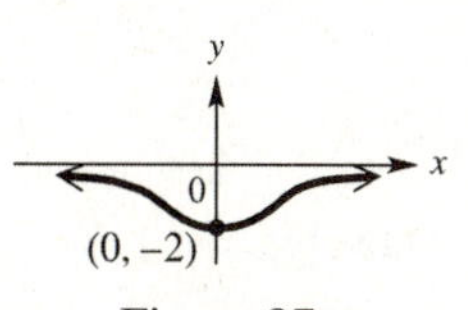

Figure 87a

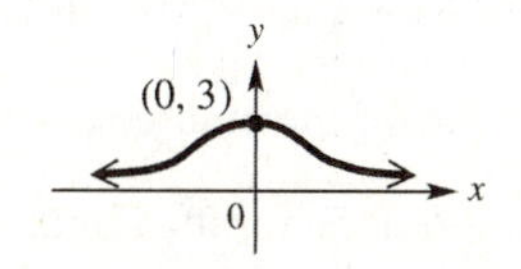

Figure 88a

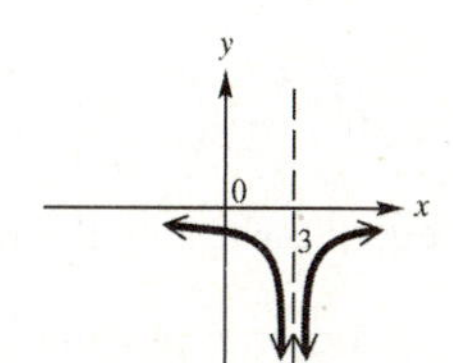

Figure 89a

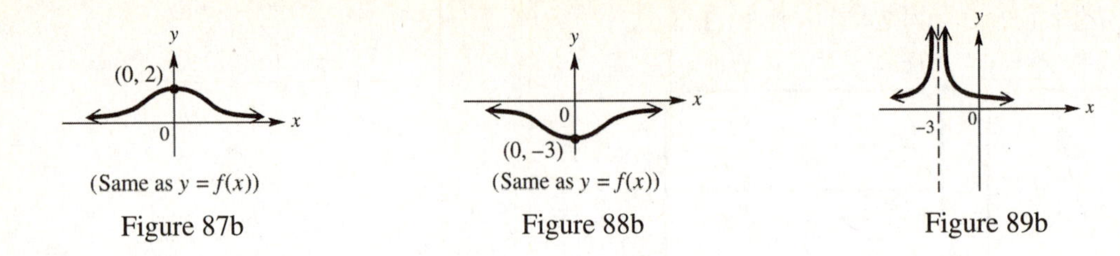

Figure 87b Figure 88b Figure 89b

90. (a) Reflect the graph across the x-axis. See Figure 90a.

(b) Reflect the graph across the y-axis. See Figure 90b.

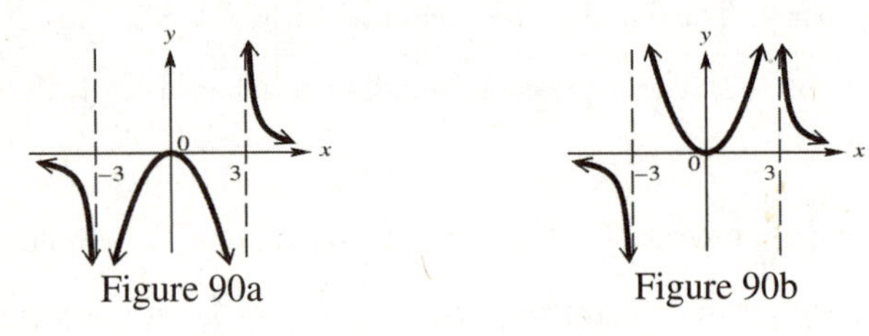

Figure 90a Figure 90b

91. Perform long division:

$$\begin{array}{r} -2x-8 \\ -x+4\overline{)2x^2+0x+3} \\ \underline{2x^2-8x} \\ 8x+\ 3 \\ \underline{8x-32} \\ 35 \end{array}$$

$$\Rightarrow \frac{2x^2+3}{4-x}=-2x-8+\frac{35}{4-x}$$

The oblique asymptote is $y=-2x-8$.

The graphs of $y_1=(2x^2+3)/(4-x)$ and $y_2=-2x-8$ are shown in Figure 91.

93. Perform long division:

$$\begin{array}{r} -x+3 \\ x+2\overline{)-x^2+x+0} \\ \underline{-x^2-2x} \\ 3x+0 \\ \underline{3x+6} \\ -6 \end{array}$$

$$\Rightarrow \frac{x-x^2}{x+2}=-x+3+\frac{-6}{x+2}$$

The oblique asymptote is $y=-x+3$.

The graphs of $y_1=(x-x^2)/(x+2)$ and $y_2=-x+3$ are shown in Figure 93.

95. (a) Disregard the remainder to get the equation of the oblique asymptote, $y=x+1$.

(b) Set the original function equal to the expression for the oblique asymptote and solve:

$$\frac{x^5+x^4+x^2+1}{x^4+1}=x+1 \Rightarrow x^5+x^4+x^2+1=(x^4+1)(x+1) \Rightarrow$$

$$x^5+x^4+x^2+1=x^5+x^4+x+1 \Rightarrow x^2+1=x+1 \Rightarrow x^2-x=0 \Rightarrow x(x-1)=0 \Rightarrow x=0,1.$$

The graph crosses the oblique asymptote at $x=0$ and $x=1$

(c) For large values of $x, x+1<\dfrac{x^5+x^4+x^2+1}{x^4+1}$. Thus, the function approaches its asymptote from above.

97.

$$\begin{array}{r}
x^2-x+8 \\
x^2+x-12\overline{\big)\,x^4+0x^3-5x^2+0x+4} \\
\underline{x^4+x^3-12x^2} \\
-x^3+7x^2+0x \\
\underline{-x^3-x^2+12x} \\
8x^2-12x+4 \\
\underline{8x^2+8x-96} \\
-20x+100
\end{array}
\Rightarrow \frac{x^4-5x^2+4}{x^2+x-12}=x^2-x+8+\frac{-20x+100}{x^2+x-12}$$

The graphs of $y_1=(x^4-5x+4)/(x^2+x-12)$ and $y_2=x^2-x+8$ are shown in figure 97. In this window, the two graphs seem to overlap (coincide), suggesting that as $|x|\to\infty$, the graph of f approaches the graph of g, giving an asymptotic effect.

[-18.8,18.8] by [-50,25]
Xscl = 1 Yscl = 1

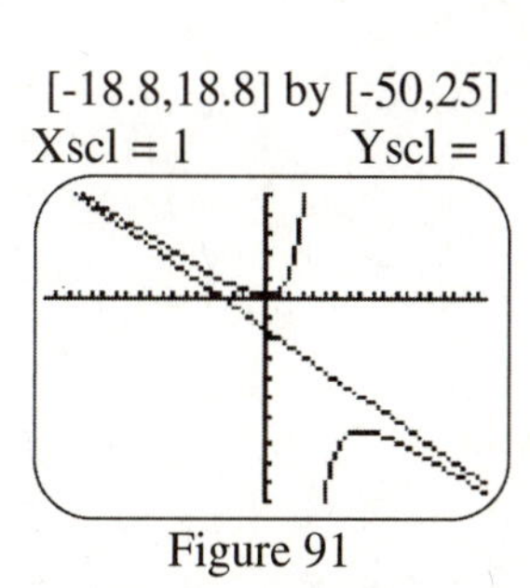

Figure 91

[-9.4,49.4] by [-15,25]
Xscl = 1 Yscl = 1

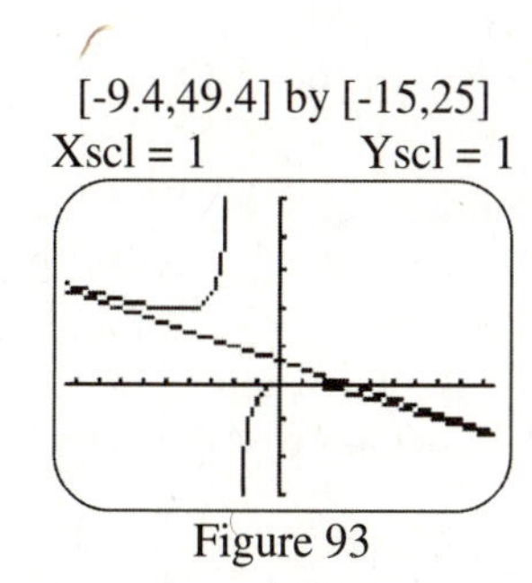

Figure 93

[-50,50] by [0,1000]
Xscl = 10 Yscl = 100

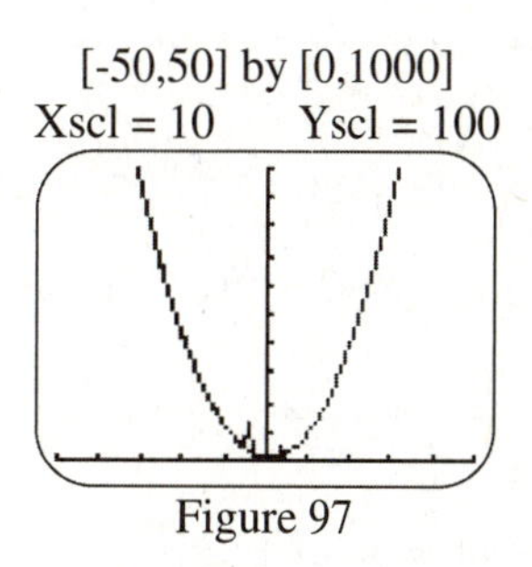

Figure 97

4.3: Rational Equations, Inequalities, Applications, and Models

1. (a) The graph never intersects the x-axis. The solution set of $f(x)=0$ is $\varnothing$.

   (b) The graph is below the x-axis on the interval $(-\infty,-2)$. Thus $f(x)<0$ on $(-\infty,-2)$.

   (c) The graph is above the x-axis on the interval $(-2,\infty)$. Thus $f(x)>0$ on $(-2,\infty)$.

3. (a) The graph intersects the x-axis at $x=-1$. The solution set of $f(x)=0$ is $\{-1\}$.

   (b) The graph is below the x-axis on the interval $(-1,0)$. Thus $f(x)<0$ on $(-1,0)$.

   (c) The graph is above the x-axis on the interval $(-\infty,-1)\cup(0,\infty)$. Thus $f(x)>0$ on $(-\infty,-1)\cup(0,\infty)$.

5. (a) The graph intersects the x-axis at $x=0$. The solution set of $f(x)=0$ is $\{0\}$.

   (b) The graph is below the x-axis on the interval $(-2,0)\cup(2,\infty)$. Thus $f(x)<0$ on $(-2,0)\cup(2,\infty)$.

   (c) The graph is above the x-axis on the interval $(-\infty,-2)\cup(0,2)$. Thus $f(x)>0$ on $(-\infty,-2)\cup(0,2)$.

7. (a) The graph intersects the x-axis at $x=0$. The solution set of $f(x)=0$ is $\{0\}$.

   (b) The graph is below the x-axis on the interval $(-1,0)\cup(0,1)$. Thus $f(x)<0$ on $(-1,0)\cup(0,1)$.

(c) The graph is above the x-axis on the interval $(-\infty,-1)\cup(1,\infty)$. Thus $f(x)>0$ on $(-\infty,-1)\cup(1,\infty)$.

9. (a) The graph intersects the x-axis at $x=0$. The solution set of $f(x)=0$ is $\{0\}$.

(b) The graph is below the x-axis on the interval $(1,2)\cup(2,3)$.. Thus $f(x)<0$ on $(1,2)\cup(2,3)$.

(c) The graph is above the x-axis on the interval $(-\infty,-2)\cup(-2,0)\cup(0,1)\cup(3,\infty)$. Thus $f(x)>0$ on the interval $(-\infty,-2)\cup(-2,0)\cup(0,1)\cup(3,\infty)$.

11. (a) The graph intersects the x-axis at $x=2.5$. The solution set of $f(x)=0$ is $\{2.5\}$.

(b) The graph is below the x-axis on the interval $(2.5,3)$. Thus $f(x)<0$ on $(2.5,3)$.

(c) The graph is above the x-axis on the interval $(-\infty,2.5)\cup(3,\infty)$. Thus $f(x)>0$ on $(-\infty,2.5)\cup(3,\infty)$.

13. Multiply through by the LCD, $(x-1)(x+1)$. $\frac{2x}{x^2-1}\cdot(x-1)(x+1)=\frac{2}{x+1}\cdot(x-1)(x+1)-\frac{1}{x-1}\cdot(x-1)(x+1)\Rightarrow$

$2x=2(x-1)-(x+1)\Rightarrow 2x=2x-2-x-1\Rightarrow 2x=x-3\Rightarrow x=-3$. The solution set is $\{-3\}$.

15. Multiply through by the LCD, $x(x-3)(x+3)$.

$$\frac{4}{x^2-3x}\cdot x(x-3)(x+3)-\frac{1}{x^2-9}\cdot x(x-3)(x+3)=0\cdot x(x-3)(x+3)\Rightarrow 4(x+3)-x=0\Rightarrow$$

$4x+12-x=0\Rightarrow 3x=-12\Rightarrow x=-4$. The solution set is $\{-4\}$.

17. Multiply through by the LCD, x^2.

$$1\cdot x^2-\frac{13}{x}\cdot x^2+\frac{36}{x^2}\cdot x^2=0\cdot x^2\Rightarrow x^2-13x+36=0\Rightarrow(x-4)(x-9)=0\Rightarrow x=4 \text{ or } 9.$$

The solution set is $\{4,9\}$.

19. Multiply through by the LCD, x^2.

$$1\cdot x^2+\frac{3}{x}\cdot x^2=\frac{5}{x^2}\cdot x^2\Rightarrow x^2+3x=5\Rightarrow x^2+3x-5=0\Rightarrow$$

$x=\frac{-3\pm\sqrt{3^2-4(1)(-5)}}{2(1)}=\frac{-3\pm\sqrt{29}}{2}$. The solution set is $\left\{\frac{-3\pm\sqrt{29}}{2}\right\}$.

21. Multiply through by the LCD, $x(2-x)$.

$$\frac{x}{2-x}\cdot x(2-x)+\frac{2}{x}\cdot x(2-x)-5\cdot x(2-x)=0\cdot x(2-x)\Rightarrow x^2+2(2-x)-5x(2-x)=0\Rightarrow$$

$6x^2-12x+4=0\Rightarrow 3x^2-6x+2=0$.

$x=\frac{-(-6)\pm\sqrt{(-6)^2-4(3)(2)}}{2(3)}=\frac{6\pm\sqrt{12}}{6}=\frac{2(3\pm\sqrt{3})}{2(3)}=\frac{3\pm\sqrt{3}}{3}$. The solution set is $\left\{\frac{3\pm\sqrt{3}}{3}\right\}$.

23. Rewrite the equation as $\frac{1}{x^4}-\frac{3}{x^2}-4=0$ and multiply through by the LCD, x^4.

$$\frac{1}{x^4}\cdot x^4-\frac{3}{x^2}\cdot x^4-4\cdot x^4=0\cdot x^4\Rightarrow 1-3x^2-4x^4=0\Rightarrow 4x^4+3x^2-1=0\Rightarrow$$

$$(4x^2-1)(x^2+1)=0\Rightarrow(2x+1)(2x-1)(x^2+1)=0\Rightarrow 2x+1=0 \text{ or } 2x-1=0 \text{ or } x^2+1=0\Rightarrow$$

$$2x=-1 \text{ or } 2x=1 \text{ or } x^2=-1\Rightarrow x=-\frac{1}{2} \text{ or } x=\frac{1}{2} \text{ or } x=\pm i.$$ The solution set is $\left\{\pm\frac{1}{2},\pm i\right\}$.

25. Multiply through by the LCD, $(x+2)(x+7)$.

$$\frac{1}{x+2}\cdot(x+2)(x+7)+\frac{3}{x+7}\cdot(x+2)(x+7)=\frac{5}{x^2+9x+14}\cdot(x+2)(x+7)\Rightarrow$$

$$1(x+7)+3(x+2)=5\Rightarrow x+7+3x+6=5 \Rightarrow 4x=-8 \Rightarrow x=-2.$$

This solution is extraneous. The solution set is $\varnothing$.

27. Multiply through by the LCD, $(x-3)(x+3)$.

$$\frac{x}{x-3}\cdot(x-3)(x+3)+\frac{4}{x+3}\cdot(x-3)(x+3)=\frac{18}{x^2-9}\cdot(x-3)(x+3)\Rightarrow$$

$$x(x+3)+4(x-3)=18\Rightarrow x^2+3x+4x-12=18\Rightarrow x^2+7x-30=0\Rightarrow(x+10)(x-3)=0\Rightarrow x=-10 \text{ or } x=3.$$

The solution $x=3$ is extraneous. The solution set is $\{-10\}$.

29. Rewrite the equation as $\frac{9}{x}+\frac{4}{6x-3}=\frac{2}{6x-3}$ and multiply through by the LCD, $x(6x-3)$.

$$\frac{9}{x}\cdot x(6x-3)+\frac{4}{6x-3}\cdot x(6x-3)=\frac{2}{6x-3}\cdot x(6x-3)\Rightarrow 9(6x-3)+4x=2x\Rightarrow$$

$$54x-27+4x=2x \Rightarrow 56x=27\Rightarrow x=\frac{27}{56}.$$ The solution set is $\left\{\frac{27}{56}\right\}$.

31. (a) $\frac{x-3}{x+5}=0\Rightarrow x-3=0(x+5)\Rightarrow x-3=0\Rightarrow x=3$. The solution set is $\{3\}$.

(b) Graph $y_1=(x-3)/(x+5)$ as shown in Figure 31. With vertical asymptote $x=-5$, the graph is below or intersects the x-axis on the interval $(-5,3]$. The solution set is $(-5,3]$.

(c) Graph $y_1=(x-3)/(x+5)$ as shown in Figure 31. With vertical asymptote $x=-5$, the graph is above or intersects the x-axis on the interval $(-\infty,-5)\cup[3,\infty)$. The solution set is $(-\infty,-5)\cup[3,\infty)$.

33. (a) $\frac{x-1}{x+2}=1\Rightarrow x-1=1(x+2)\Rightarrow x-1=x+2\Rightarrow -1=2$ (false). The solution set is $\varnothing$.

(b) Graph $y_1=(x-1)/(x+2)-1$ as shown in Figure 33. With vertical asymptote $x=-2$, the graph is above the x-axis on the interval $(-\infty,-2)$. The solution set is $(-\infty,-2)$.

(c) Graph $y_1=(x-1)/(x+2)-1$ as shown in Figure 33. With vertical asymptote $x=-2$, the graph is below the x-axis on the interval $(-2,\infty)$. The solution set is $(-2,\infty)$.

35. (a) $\frac{1}{x-1}=\frac{5}{4}\Rightarrow 5(x-1)=4(1)\Rightarrow 5x-5=4\Rightarrow 5x=9\Rightarrow x=\frac{9}{5}$. The solution set is $\left\{\frac{9}{5}\right\}$.

(b) Graph $y_1=1/(x-1)-5/4$ as shown in Figure 35. With vertical asymptote $x=1$, the graph is below the x-axis on the interval $(-\infty,1)\cup\left(\frac{9}{5},\infty\right)$. The solution set is $(-\infty,1)\cup\left(\frac{9}{5},\infty\right)$.

(c) Graph $y_1=1/(x-1)-5/4$ as shown in Figure 35. With vertical asymptote $x=1$, the graph is above the x-axis on the interval $\left(1,\frac{9}{5}\right)$. The solution set is $\left(1,\frac{9}{5}\right)$.

37. (a) $\frac{4}{x-2}=\frac{3}{x-1}\Rightarrow 4(x-1)=3(x-2)\Rightarrow 4x-4=3x-6\Rightarrow x=-2$. The solution set is $\{-2\}$.

(b) Graph $y_1=4/(x-2)-3/(x-1)$ as shown in Figure 37. With vertical asymptotes $x=1$ and $x=2$, the graph is below or intersects the x-axis on $(-\infty,-2]\cup(1,2)$.. The solution set is $(-\infty,-2]\cup(1,2)$.

(c) Graph $y_1=4/(x-2)-3/(x-1)$ as shown in Figure 37. With vertical asymptotes $x=1$ and $x=2$, the graph is above or intersects the x-axis on $[-2,1)\cup(2,\infty)$. The solution set is $[-2,1)\cup(2,\infty)$.

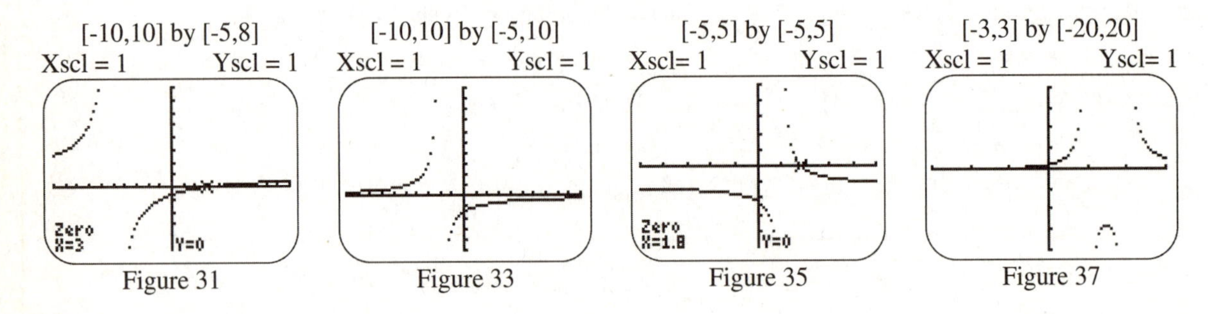

Figure 31 Figure 33 Figure 35 Figure 37

39. (a) $\frac{1}{(x-2)^2}=0\Rightarrow 1=0(x-2)^2\Rightarrow 1=0$ (false). The solution set is $\varnothing$.

(b) Graph $y_1=1/(x-2)^2$ as shown in Figure 39. The graph is never below the x-axis. The solution set is $\varnothing$.

(c) Graph $y_1=1/(x-2)^2$ as shown in Figure 39. With vertical asymptote $x=2$, the graph is above the x-axis on the interval $(-\infty,2)\cup(2,-\infty)$. The solution set is $(-\infty,2)\cup(2,\infty)$.

41. (a) $\frac{5}{x+1}=\frac{12}{x+1}\Rightarrow 5(x+1)=12(x+1)\Rightarrow 5x+5=12x+12\Rightarrow -7x=7\Rightarrow x=-1$. This solution is extraneous. The solution set is $\varnothing$.

(b) Graph $y_1=5/(x+1)-12/(x+1)$ as shown in Figure 41. With vertical asymptote $x=-1$, the graph is above the x-axis on the interval $(-\infty,-1)$. The solution set is $(-\infty,-1)$.

(c) Graph $y_1 = 5/(x+1) - 12/(x+1)$ as shown in Figure 41. With vertical asymptote $x = -1$, the graph is below the x-axis on the interval $(-1, \infty)$. The solution set is $(-1, \infty)$.

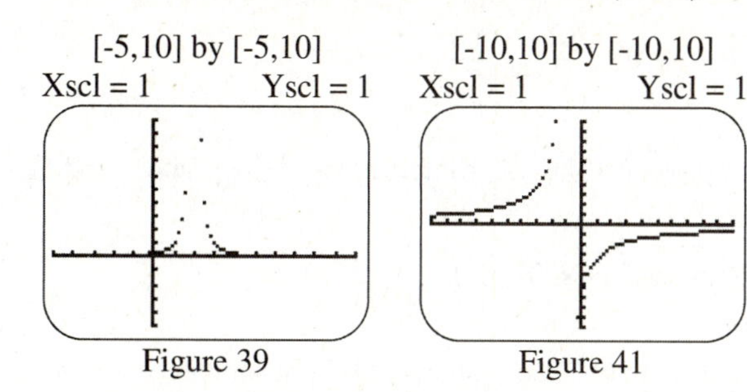

Figure 39 Figure 41

43. The numerator is negative and the denominator is always positive; therefore, the quotient is always negative. Because the inequality requires that the rational expression be less than 0, the solution set is $(-\infty, \infty)$.

45. The numerator is negative and the denominator is always positive; therefore, the quotient is always negative. Because the inequality requires that the rational expression be greater than 0, the solution set is $\varnothing$.

47. The numerator is always positive and the denominator is negative; therefore, the quotient is always negative. Because the inequality requires that the rational expression be greater than or equal to 0, the solution set is $\varnothing$.

49. The numerator is always positive and the denominator is always positive; therefore, the quotient is always positive. Because the inequality requires that the rational expression be greater than 0, the solution set is $(-\infty, \infty)$.

51. The numerator is always positive or zero and the denominator is always positive; therefore, the quotient is always positive or zero. Because the inequality requires that the rational expression be less than or equal to 0, the solution set is $\{1\}$. Note that when $x = 1$, the rational expression equals 0 and the inequality is satisfied.

53. The graph of $y_1 = (3-2x)/(1+x)$ has a vertical asymptote at $x = -1$ and an x-intercept at $x = \frac{3}{2}$. The graph of y_1 is below the x-axis on the interval $(-\infty,\ -1) \cup \left(\frac{3}{2},\ \infty\right)$.

55. The graph of $y_1 = \dfrac{(x+1)(x-2)}{x+3}$ has a vertical asymptote at $x = -3$ and $x-$intercepts at $x = -1$ and $x = 2$. The graph of $P(3) = 0$ is below the $x-$axis on the interval $(-\infty, -3) \cup (-1, 2)$.

57. The graph of $y_1 = (x+1)^2/(x-2)$ has a vertical asymptote at $x = 2$ and an x-intercept at $x = -1$. The graph of y_1 intersects or is below the x-axis on the interval $(-\infty,\ 2)$. Note that 2 can not be included.

59. The graph of $y_1 = \dfrac{2x-5}{(x+1)(x-1)}$ has vertical asymptotes at $x = -1$ and $x = 1$ and an x-intercept at $x = \frac{5}{2}$.

The graph of y_1 is on or above the x-axis on the interval $(-1, 1) \cup \left[\frac{5}{2}, \infty\right)$.

61. First, rewrite the inequality: $\dfrac{1}{x-3} \le \dfrac{5}{x-3} \Rightarrow \dfrac{1}{x-3} - \dfrac{5}{x-3} \le 0 \Rightarrow \dfrac{-4}{x-3} \le 0$

The graph of $y_1 = -4/(x-3)$ has a vertical asymptote at $x = 3$ and no x-intercept.

The graph of y_1 intersects or is below the x-axis on the interval $(3, \infty)$. Note that 3 can not be included.

63. First, rewrite the inequality: $2 - \frac{5}{x} + \frac{2}{x^2} \geq 0 \Rightarrow \frac{2x^2 - 5x + 2}{x^2} \geq 0 \Rightarrow \frac{(2x-1)(x-2)}{x^2} \geq 0$

The graph of $y_1 = (2x-1)(x-2)/x^2$ has a vertical asymptote at $x = 0$ and x-intercepts of $x = \frac{1}{2}$ and $x = 2$.

The graph of y_1 intersects or is above the x-axis on the interval $(-\infty, 0) \cup \left(0, \frac{1}{2}\right] \cup [2, \infty)$.

65. (a) The graph of $y_1 = (\sqrt{2}x + 5)/(x^3 - \sqrt{3})$ is shown in Figure 65. The x-intercept is approximately -3.54. The solution set of the equation is $\{-3.54\}$.

(b) There is a vertical asymptote when $x^3 - \sqrt{3} = 0 \Rightarrow x^3 = \sqrt{3} \Rightarrow x = \sqrt[3]{\sqrt{3}} \approx 1.20$. Thus, the graph is above the x-axis on the interval $(-\infty, -3.54) \cup (1.20, \infty)$. See Figure 65.

(c) The graph is below the x-axis on the interval $(-3.54, 1.20)$. See Figure 65.

67. (a) The graph is shown in Figure 67. The equation of the horizontal asymptote is $y = \frac{10}{1} \Rightarrow y = 10$.

(b) The initial insect population occurs when $x = 0$. Here $f(0) = \frac{10(0)+1}{(0)+1} = 1$ million insects.

(c) After several months, the insect population levels off at 10 million insects.

(d) The horizontal asymptote $y = 10$ represents the limiting population after a long time.

[-4.7,4.7] by [-3.1,3.1]
Xscl = 1 Yscl = 1

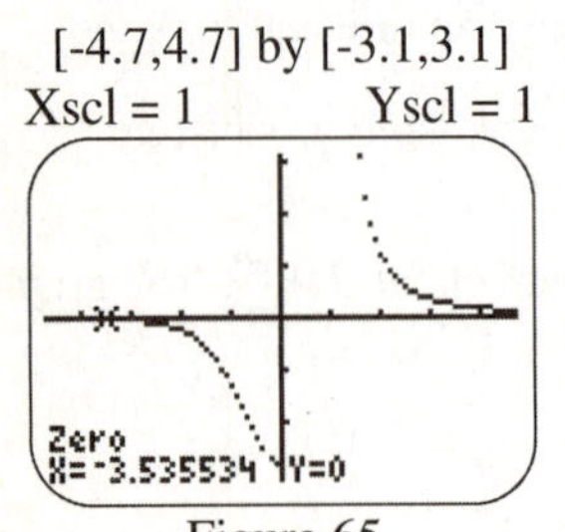

Figure 65

[0,14] by [0,14]
Xscl = 1 Yscl = 1

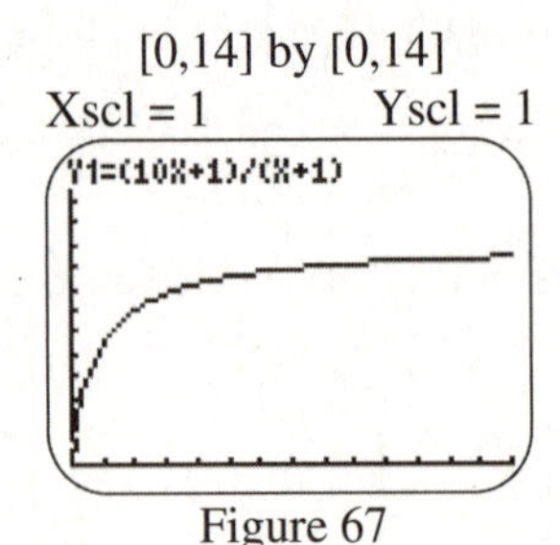

Figure 67

69. (a) $0.25 = \frac{x-5}{x^2 - 10x} \Rightarrow 0.25(x^2 - 10x) = x - 5 \Rightarrow 0.25x^2 - 2.5x = x - 5 \Rightarrow 0.25x^2 - 3.5x + 5 = 0 \Rightarrow$

$x = \frac{-(-3.5) \pm \sqrt{(-3.5)^2 - 4(0.25)(5)}}{2(0.25)} = \frac{3.5 \pm \sqrt{7.25}}{0.5} \approx 1.6$ or 12.4; Since $x > 10$, $x \approx 12.4$ cars/min.

(b) Since 2 attendants can serve only 10 cars/min, 3 attendants are needed.

71. A surface area of 280 square inches indicates that $2lw + 2lh + 2hw = 280$ or $lw + lh + hw = 140$. Solving this equation for h gives $h = \frac{140 - lw}{l + w}$. Since the length is twice the width, $l = 2w$. Then, by substitution,

$h=\dfrac{140-(2w)w}{2w+w}\Rightarrow h=\dfrac{140-2w^2}{3w}$. Since the volume is 196 cubic inches, $lwh=196$. Substituting for l and h yields $(2w)w\left(\dfrac{140-2w^2}{3w}\right)=196\Rightarrow\dfrac{280w-4w^3}{3}=196\Rightarrow 280w-4w^3=588\Rightarrow$

$4w^3-280w+588=0\Rightarrow w^3-70w+147=0$. Graphing this equation as $y_1=x$^$3-70x+147$ in $[0,10]$ by $[-150,450]$ shows the possible values for the width as the x-intercepts (figure not shown). The possible values for the width are 7 inches and approximately 2.266 inches.

If $w\approx 2.266$, then $l\approx 2(2.266)\approx 4.532$ and $h=\dfrac{140-2(2.266)^2}{3(2.266)}\approx 19.084$. In this case, the dimensions are approximately 2.266 by 4.532 by 19.084 inches. If $w=7$, then $l=14$ and $h=\dfrac{140-2(7)^2}{3(7)}=2$ inches. In this case the dimensions are 7 by 14 by 2 inches.

73. (a) $f(400)=\dfrac{2540}{400}=6.35$ in. A curve designed for 60 miles per hour with a radius of 400 feet should have the outer rail elevated 6.35 inches.

(b) See Figure 73. As the radius x of the curve increases, the elevation of the outer rail decreases.

(c) The horizontal asymptote is $y=0$. As the radius of the curve increases without bound $(x\to\infty)$, the tracks become straight and no elevation or banking $(y\to 0)$ is necessary.

(d) $12.7=\dfrac{2540}{x}\Rightarrow 12.7x=2540\Rightarrow x=\dfrac{2540}{12.7}=200$ ft.

[0,600] by [0,50]
Xscl = 100 Yscl = 5

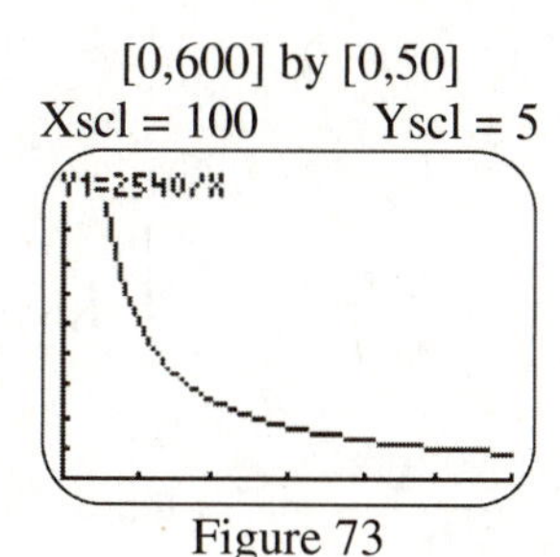

Figure 73

75. (a) $D(0.05)=\dfrac{2500}{30(0.3+0.05)}\approx 238$. The braking distance for a car going 50 mph on a 5% uphill grade is about 238 ft.

(b) As the uphill grade x increases, the braking distance decreases. This agrees with driving experience.

(c) $220=\dfrac{2500}{30(0.3+x)}\Rightarrow 6600(0.3+x)=2500\Rightarrow 0.3+x=\dfrac{2500}{6000}\Rightarrow x=\dfrac{2500}{6600}-0.3\Rightarrow x\approx 0.079$. The grade associated with a braking distance of 220 feet is about 7.9% uphill.

77. Here $r = \frac{km^2}{s}$. By substitution, $12 = \frac{k(6)^2}{4} \Rightarrow 48 = 36k \Rightarrow k = \frac{48}{36} \Rightarrow k = \frac{4}{3}$. That gives $r = \frac{\frac{4}{3}m^2}{s}$. When $m = 4$ and $s = 10$, $r = \frac{\frac{4}{3}(4)^2}{10} = \frac{\frac{64}{3}}{10} = \frac{64}{30} = \frac{32}{15}$.

79. Here $a = \frac{kmn^2}{y^3}$. By substitution, $9 = \frac{k(4)(9)^2}{(3)^3} \Rightarrow 243 = 324k \Rightarrow k = \frac{243}{324} \Rightarrow k = \frac{3}{4}$. That gives $a = \frac{\frac{3}{4}mn^2}{y^3}$. When $m = 6$, $n = 2$ and $y = 5$, $a = \frac{\frac{3}{4}(6)(2)^2}{(5)^3} = \frac{18}{125}$.

81. For $k > 0$, if y varies directly as x, when x increases, y increases, and when x decreases, y decreases.

83. If y is inversely proportional to x, $y = \frac{k}{x}$. If x doubles, $y = \frac{k}{2x} \Rightarrow y = \frac{1}{2} \cdot \frac{k}{x}$. That is, y becomes half as much.

85. If y is directly proportional to the third power of x, $y = kx^3$. If x triples, $y = k(3x)^2 \Rightarrow y = 27 \cdot kx^3$. That is, y becomes 27 times as much.

87. Here $BMI = \frac{kw}{h^2}$. By substitution, $24 = \frac{k(177)}{(72)^2} \Rightarrow 177k \Rightarrow k = \frac{124,416}{177} \Rightarrow k \approx 703$. That gives $BMI \approx \frac{703w}{h^2}$. When $w = 130$ and $h = 66$, $BMI \approx \frac{703(130)}{(66)^2} \approx 21$.

89. Here $R = \frac{k}{d^2}$. By substitution, $0.5 = \frac{k}{(2)^2} \Rightarrow 2 = k$. That gives $R = \frac{2}{d^2}$. When $d = 3$, $R = \frac{2}{(3)^2} \approx \frac{2}{9}$ ohm.

91. Here $W = \frac{k}{d^2}$. By substitution, $160 = \frac{k}{(4000)^2} \Rightarrow 2,560,000,000 = k$. That gives $W = \frac{2,560,000,000}{d^2}$. Note that 8000 miles above Earth's surface, $d = 12,000$. When $d = 12,000$, $W = \frac{2,560,000,000}{(12,000)^2} = 17.8$ lb.

93. Here $V = kr^2h$. By substitution, $300 = k(3)^2(10.62) \Rightarrow \frac{300}{95.58} = k \Rightarrow k \approx 3.1387$. That gives $V = 3.1387r^2h$. When $r = 4$ and $h = 15.92$, $V \approx 3.1387(4)^2(15.92) \approx 799.5$. (Note that the actual value of k is π.)

95. Tooney can clean the entire mess in 15 minutes and can complete $\frac{t}{15}$ of the mess in t minutes. Mudcat can clean the entire mess in12 minutes and can complete $\frac{t}{12}$ of the mess in t minutes. Together they can

complete $\frac{t}{15}+\frac{t}{12}$ of the mess in t minutes. The job is complete when the fraction of the mess reaches 1.

$\frac{t}{15}+\frac{t}{12}=1 \Rightarrow 12t+15t=180 \Rightarrow 27t=180 \Rightarrow t=6$ minutes and 40 seconds.

97. Mrs. Schmulen can grade tests in 5 hours and can complete $\frac{t}{5}$ of the tests in t hours. Mr. Elwyn and Mrs. Schmulen can grade the tests together in 3 hours and can complete $\frac{t}{3}$ of the tests in t hours. The job is complete when the fraction of the completed tests reaches 1.

$\frac{t}{3}-\frac{t}{5}=1 \Rightarrow 5t-3t=15 \Rightarrow 2t=15 \Rightarrow t=7$ hours and 30 minutes.

99. The inlet pipe can fill the vat in 5 hours and can fill $\frac{t}{5}$ of the vat in t hours. The outlet pipe can empty the vat in 10 hours and can empty $\frac{t}{10}$ of the vat in t hours. The job is complete when the fraction of the full tank reaches 1. $\frac{t}{5}-\frac{t}{10}=1 \Rightarrow 10t-5t=50 \Rightarrow 5t=50 \Rightarrow t=10$ hours.

101. Let x = the amount of time required to empty the pool then (x-20) will be the amount of time required to fill the pool. Then the rate of emptying the pool is given by $\frac{1}{x}$ and the rate of filling of the pool is $\frac{1}{x-20}$. When both pipes are open it takes 4 hours (or 240 minutes) to fill the pool. The rate of filling the pool when both pipes are open is given by $\frac{1}{240}$.

$\frac{1}{x-20}-\frac{1}{x}=\frac{1}{240} \Rightarrow 240x-240(x-20)=x(x-20) \Rightarrow x^2-20x-4800=0 \Rightarrow (x-80)(x+60)=0 \Rightarrow x=80.$

Therefore, it will take 80 minutes to empty with the outlet pipe and 60 minutes to fill with the inlet pipe.

Reviewing Basic Concepts (Sections 4.1—4.3)

1. A sketch of the graph is shown in Figure 1a. A graph of $y_1=1/(x+2)-3$ is shown in Figure 1b.
2. For the function to be defined, the denominator cannot equal 0. $x^2-1\neq 0 \Rightarrow x^2\neq 1 \Rightarrow x\neq \pm 1$. The domain is $(-\infty,-1)\cup(-1,1)\cup(1,\infty)$.
3. A vertical asymptote occurs when the denominator equals 0. $x-6=0 \Rightarrow x=6$.
4. Since the degrees of the numerator and the denominator are the same, the ratio of the leading coefficients of the numerator and denominator is used to find the equation of the horizontal asymptote. $y=\frac{1}{1} \Rightarrow y=1$.
5. Perform long division:

$$\begin{array}{r} x-2 \\ x+3\overline{\big)x^2+\ x+5} \\ \underline{x^2+3x} \\ -2x+5 \\ \underline{-2x-6} \\ 11 \end{array} \Rightarrow \frac{x^2+x+5}{x+3}=x-2+\frac{11}{x+3}$$

The oblique asymptote is $y=x-2$.

6. A sketch of the graph is shown in Figure 6a. A graph of $y_1 = (3x+6)/(x-4)$ $y_1 = 1/(x+2)-3$ is shown in Figure 6b.

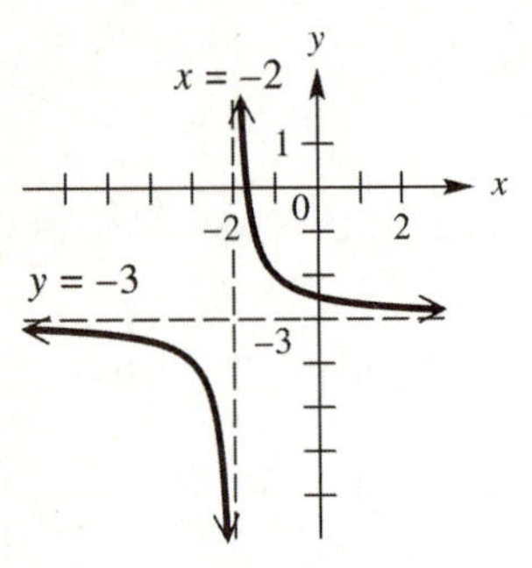

Figure 1a

[-6.7,2.7] by [-8.2,1]

Xscl = 1 Yscl = 1

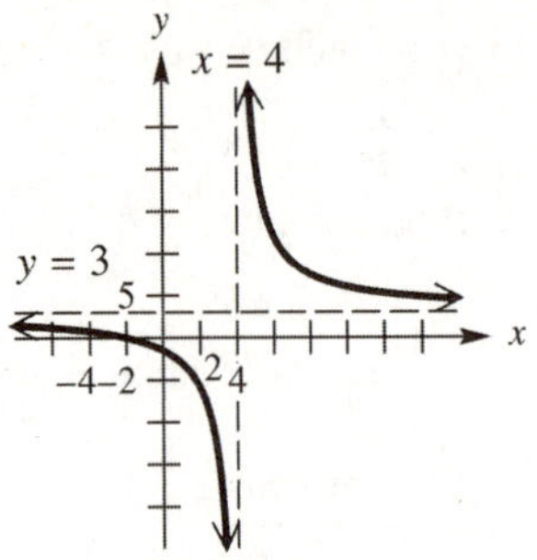

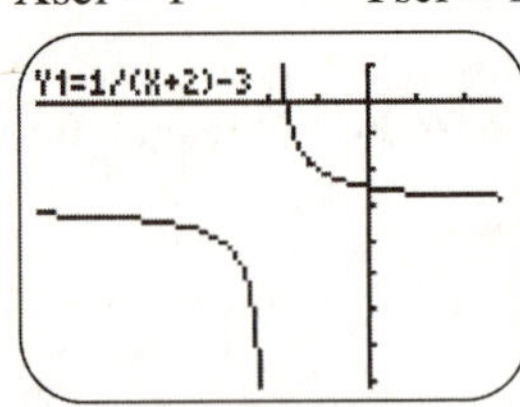

Figure 1b

Figure 6a

[-6.2,12.4] by [-18.6,18.6]

Xscl = 1 Yscl=1

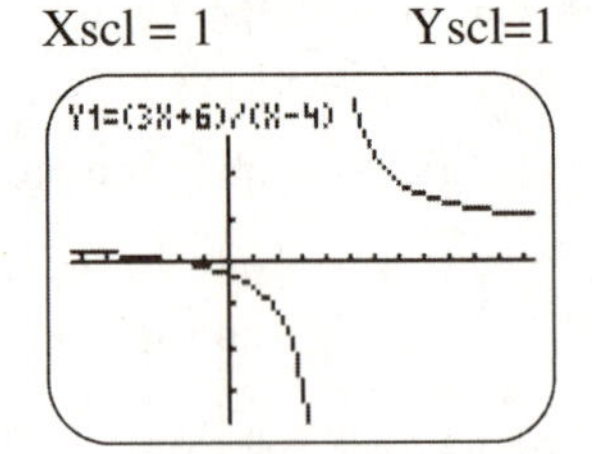

Figure 6b

7. (a) The graph intersects the x-axis at $x = -4$. The solution set is $\{-4\}$.

   (b) The graph is above the x-axis on the interval $(-\infty,-4)\cup(2,\infty)$. The solution set is $(-\infty,-4)\cup(2,\infty)$.

   (c) The graph is below the x-axis on the interval $(-4,2)$. The solution set is $(-4,2)$.

8. First, rewrite the inequality. $\dfrac{x+4}{3x+1} > 1 \Rightarrow \dfrac{x+4}{3x+1} - 1 > 0 \Rightarrow \dfrac{x+4-3x-1}{3x+1} > 0 \Rightarrow \dfrac{-2x+3}{3x+1} > 0$. The graph of $y_1 = (-2x+3)/(3x+1)$ has a vertical asymptote at $x = -\dfrac{1}{3}$ and an x-intercept at $x = \dfrac{3}{2}$. The graph of y_1 is above the x-axis on the interval $\left(-\dfrac{1}{3}, \dfrac{3}{2}\right)$.

9. The base of the parallelogram varies inversely as its height. The constant of variation is 24.

10. Here $v = \dfrac{k\sqrt{t}}{l}$. By substitution, $5 = \dfrac{k\sqrt{225}}{0.60} \Rightarrow 3 = 15k \Rightarrow k = \dfrac{3}{15} \Rightarrow k \approx 0.2$. That is $v = \dfrac{0.2\sqrt{t}}{l}$. When $t = 196$ and $l = 0.65, v = \dfrac{0.2\sqrt{196}}{0.65} \approx 4.3$ vibrations per second.

4.4: Functions Defined by Powers and Roots

1. Since $13^2 = 169, \sqrt{169} = 13$.

3. Since $(-2)^5 = -32, \sqrt[5]{-32} = -2$.

5. $81^{3/2} = (\sqrt{81})^3 = 9^3 = 729$.

7. $125^{-2/3} = \dfrac{1}{125^{2/3}} = \dfrac{1}{\left(\sqrt[3]{125}\right)^2} = \dfrac{1}{5^2} = \dfrac{1}{25}.$

9. $(-1000)^{2/3} = (\sqrt[3]{-1000})^2 = (-10)^2 = 100.$

11. $8^{2/3} = (8^{1/3})^2 = 2^2 = 4.$

13. $16^{3/4} = (16^{1/4})^{-3} = (2)^{-3} = \dfrac{1}{2^3} = \dfrac{1}{8}.$

15. $-81^{0.5} = -81^{1/2} = -\sqrt{81} = -9.$

17. $64^{1/6} = \sqrt[6]{64} = 2.$

19. $(-9^{3/4})^2 = (9^{3/4})^2 = 9^{3/2} = (\sqrt{9})^3 = 3^3 = 27.$

21. $\sqrt[3]{2x} = (2x)^{1/3}.$

23. $\sqrt[3]{z^5} = z^{5/3}.$

25. $(\sqrt[4]{y})^{-3} = (y^{1/4})^{-3} = y^{-3/4} = \dfrac{1}{y^{3/4}}.$

27. $\sqrt{x} \cdot \sqrt[3]{x} = x^{1/2} \cdot x^{1/3} = x^{1/2+1/3} = x^{5/6}.$

29. $\sqrt{y \cdot \sqrt{y}} = (y \cdot y^{1/2})^{1/2} = (y^{3/2})^{1/2} = y^{3/4}.$

31. $\sqrt[3]{(-4)} \approx -1.587401052.$

33. $\sqrt[3]{(-125)} = -5.$

35. $\sqrt[3]{(-17)} \approx -2.571281591.$

37. $\sqrt[6]{(\pi^2)} \approx 1.464591888.$

39. $13^{-1/3} \approx 0.4252903703.$

41. $32^{0.2} = 2.$

43. $(5/6)^{-1.3} \approx 1.267463962.$

45. $\pi^{-3} \approx 0.0322515344.$

47. $f(x) = x^{1.62} \Rightarrow f(1.2) = 1.2^{1.62} \approx 1.34.$

49. $f(x) = x^{3/2} - x^{1/2} \Rightarrow f(50) = 50^{3/2} - 50^{1/2} \approx 346.48.$

51. See Figure 51.

53. See Figure 53.

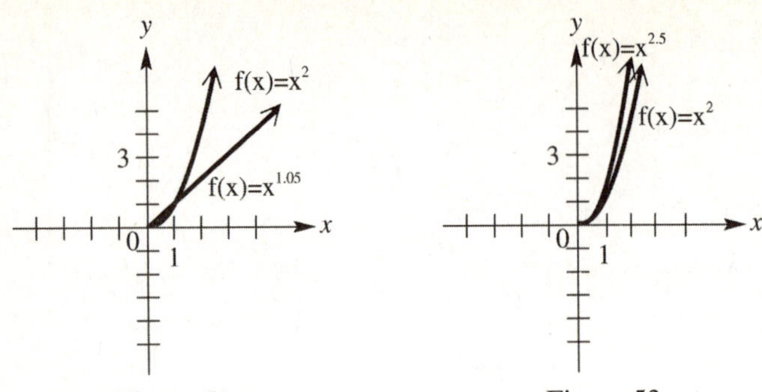

Figure 51 Figure 53

55. (a) $16^{-3/4} = \frac{1}{16^{3/4}} = \frac{1}{(\sqrt[4]{16})^3} = \frac{1}{(2)^3} = \frac{1}{8} = 0.125.$

(b) $16^{-3/4} = (\sqrt[4]{16})^{-3} = 0.125$ and $16^{-3/4} = \sqrt[4]{16^{-3}} = 0.125$. Other expressions are possible.

(c) A calculator will show that $0.125 = \frac{1}{8}$.

57. See Figure 57.
58. See Figure 58.
59. See Figure 59.
60. See Figure 60.

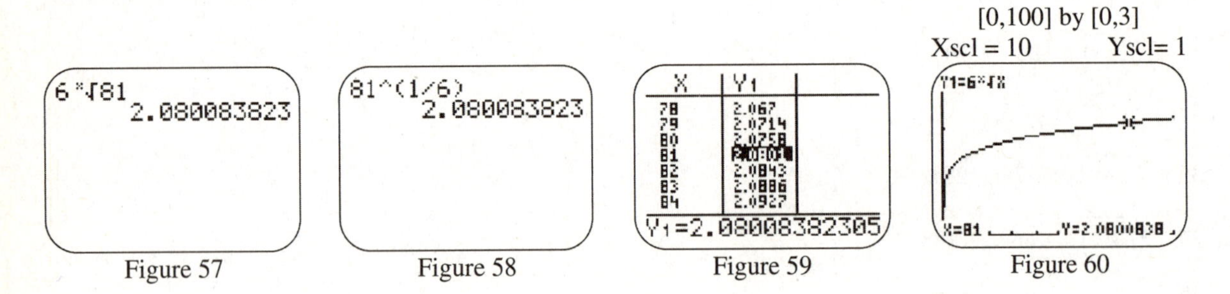

Figure 57 Figure 58 Figure 59 Figure 60

61. $S(4) = 1.27(4)^{2/3} \approx 3.2\ ft^2$.

63. $f(15) = 15^{1.5} \approx 58.1\ yr$.

65. (a) From the hint, $a(1)^b = 1960 \Rightarrow a = 1960$.

(b) Since $f(3) = 525, 1960(3)^b = 525 \Rightarrow 3^b = \frac{525}{1960} \Rightarrow 3^b \approx 0.278$. By trial and error, $b \approx -1.2$.

(c) $f(4) = 1960(4)^{-1.2} \approx 371$. If the zinc ion concentration reaches 371 mg per L, a rainbow trout will survive, on average, 4 min.

67. $f(2) = 0.445(2)^{1.25} \approx 1.06$ g.

69. Using the Power Regression feature on a graphing calculator yields $a \approx 874.54$ and $b \approx -0.49789$..

71. $S(0.5) = 0.2(0.5)^{\frac{2}{3}} \approx 0.126$. The surface area of the wings of a 0.5 kg bird is approximately 0.126 sq. meters.

73. $5+4x \geq 0 \Rightarrow 4x \geq -5 \Rightarrow x \geq -\frac{5}{4}$. The domain is $\left[-\frac{5}{4}, \infty\right)$.

75. $6-x \geq 0 \Rightarrow 6 \geq x \Rightarrow x \leq 6$. The domain is $(-\infty, 6)$.

77. The cube root function is defined for all values of x. The domain is $(-\infty, \infty)$.

79. The graph of $y = 49 - x^2 = (7+x)(7-x)$ is a parabola that opens downward with x-intercepts -7 and

$8 = \frac{k(1^4)}{9^2} \Rightarrow 648 = k \Rightarrow k = 648.$ This graph intersects or is above the x-axis for values of x in the interval

$[-7, 7]$. That is, $49 - x^2 \geq 0$ on the interval $[-7, 7]$. The domain is $[-7, 7]$.

81. The graph of $y = x^3 - x = x(x+1)(x-1)$ is a cubic graph with x-intercepts $-1, 0$ and 1. This graph intersects or is above the x-axis for values of x in the interval $[-1, 0] \cup [1, \infty)$. That is, $x^3 - x \geq 0$ on the interval $[-1, 0] \cup [1, \infty)$. The domain is $[-1, 0] \cup [1, \infty)$.

83. The graph is shown in Figure 83.
   (a) The range is $[0, \infty)$.
   (b) The function is increasing on the interval $\left[-\frac{5}{4}, \infty\right)$.
   (c) The function is never decreasing.
   (d) The graph intersects the x-axis when $x = -1.25$. The solution set of $f(x) = 0$ is $\{-1.25\}$.

85. The graph is shown in Figure 85.
   (a) The range is $(-\infty, 0]$.
   (b) The function is increasing on the interval $(-\infty, 6]$.
   (c) The function is never decreasing.
   (d) The graph intersects the x-axis when $x = 6$. The solution set of $f(x) = 0$ is $\{6\}$.

87. The graph is shown in Figure 87.
   (a) The range is $(-\infty, \infty)$.
   (b) The function is increasing on the interval $(-\infty, \infty)$.
   (c) The function is never decreasing.
   (d) The graph intersects the x-axis when $x = 3$. The solution set of $f(x) = 0$ is $\{3\}$.

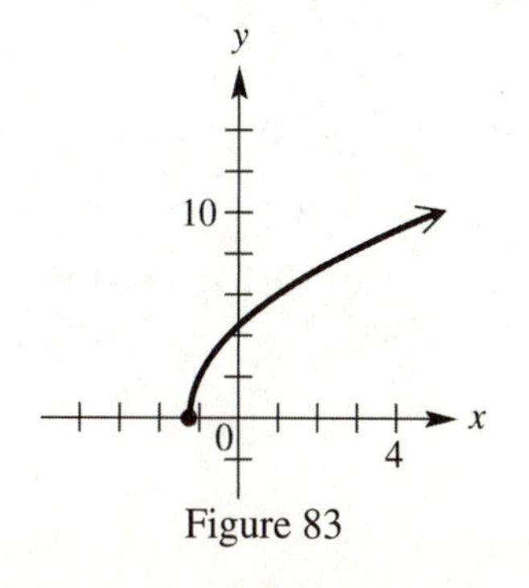

Figure 83

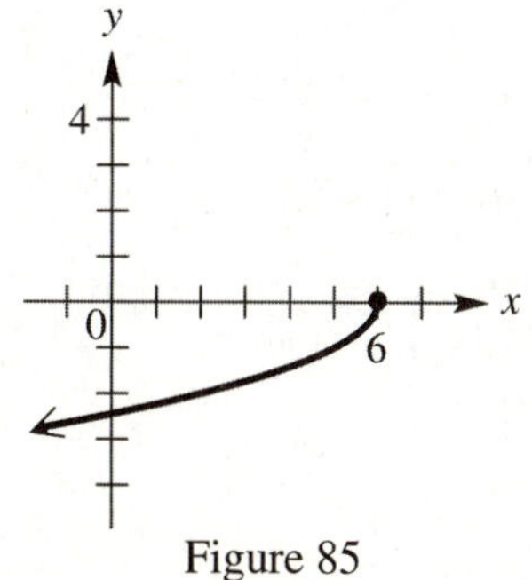

Figure 85

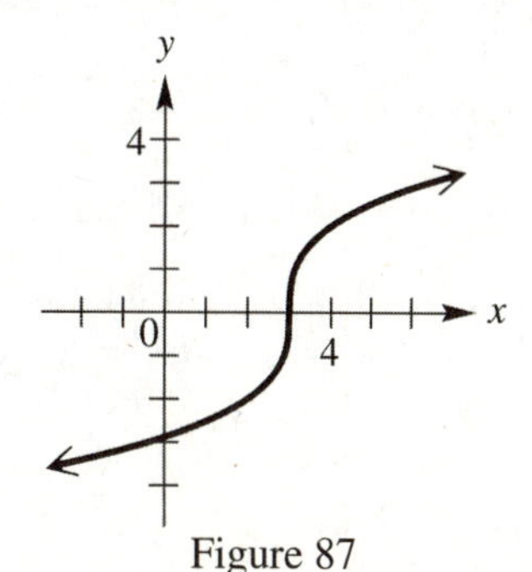

Figure 87

89. The graph is shown in Figure 89.
    (a) The range is $[0,7]$.
    (b) The function is increasing on the interval $[-7,0]$.
    (c) The function is decreasing on the interval $[0,7]$.
    (d) The graph intersects the x-axis when $x=-7$ or $x=7$. The solution set of $f(x)=0$ is $\{-7,7\}$.

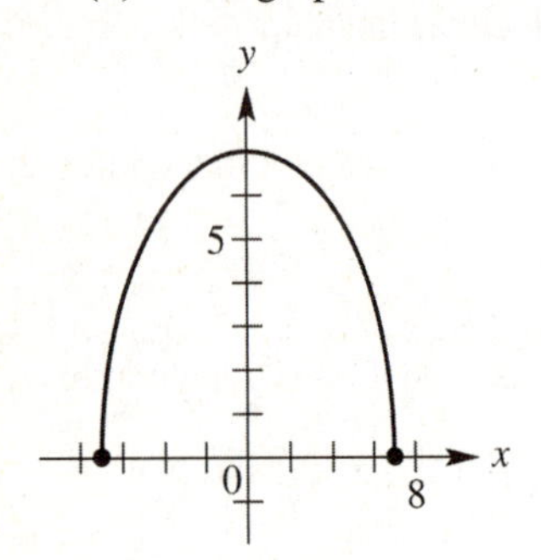

Figure 89

91. Since $y=\sqrt{9x+27}=\sqrt{9(x+3)}=3\sqrt{x+3}$, the graph can be obtained by shifting the graph of $y=\sqrt{x}$ to the left 3 units and vertically stretching by a factor of 3.

93. Since $y=\sqrt{4x+16}+4=\sqrt{4(x+4)}+4=2\sqrt{x+4}+4$, the graph can be obtained by shifting the graph of $y=\sqrt{x}$ to the left 4 units, vertically stretching by a factor of 2, and shifting upward 4 units.

95. Since $y=\sqrt[3]{27x+54}-5=\sqrt[3]{27(x+2)}-5=3\sqrt[3]{x+2}-5$, the graph can be obtained by shifting the graph of $y=\sqrt[3]{x}$ to the left 2 units, vertically stretching by a factor of 3, and shifting downward 5 units.

97. The graph is a circle. See Figure 97.

    $x^2+y^2=100 \Rightarrow y^2=100-x^2 \Rightarrow y=\pm\sqrt{100-x^2}$; Thus $y_1=\sqrt{100-x^2}$ and $y_2=-\sqrt{100-x^2}$.

99. The graph is a (shifted) circle. See Figure 99.

    $(x-2)^2+y^2=9 \Rightarrow y^2=9-(x-2)^2 \Rightarrow y=\pm\sqrt{9-(x-2)^2}$; Thus $y_1=\sqrt{9-(x-2)^2}$ and $y_1=-\sqrt{9-(x-2)^2}$.

101. The graph is a horizontal parabola. See Figure 101.

    $x=y^2+6y+9 \Rightarrow x=(y+3)^2 \Rightarrow \pm\sqrt{x}=y+3 \Rightarrow y=-3\pm\sqrt{x}$; Thus $y_1=-3+\sqrt{x}$ and $y_2=-3-\sqrt{x}$.

103. The graph is a horizontal parabola. See Figure 103.

    $x=2y^2+8y+1 \Rightarrow 0.5x=y^2+4y+0.5 \Rightarrow 0.5x=(y^2+4y+4)+0.5-4 \Rightarrow 0.5x=(y+2)^2-3.5 \Rightarrow$

    $0.5x+3.5=(y+2)^2 \Rightarrow \pm\sqrt{0.5x+3.5}=y+2 \Rightarrow y=-2\pm\sqrt{0.5x+3.5}$

    Thus $y_1=-2+\sqrt{0.5x+3.5}$ and $y_2=-2-\sqrt{0.5x+3.5}$.

[-15,15] by [-10,10] Xscl = 1 Yscl = 1

[-9.4,9.4] by [-6.2,6.2] Xscl = 1 Yscl = 1

[-10,10] by [-10,10] Xscl= 1 Yscl = 1

[-10,10] by [-10,10] Xscl = 1 Yscl= 1

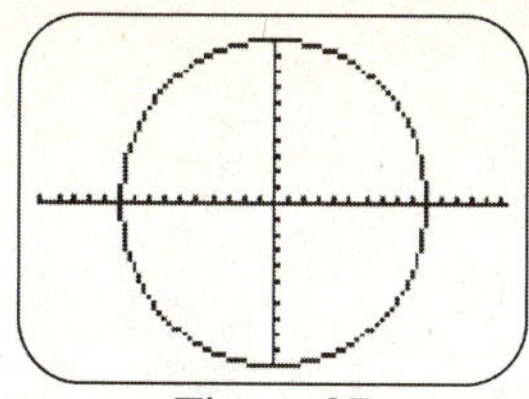

Figure 97

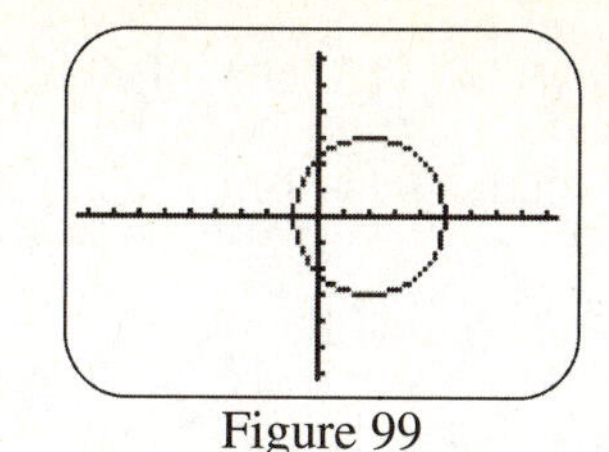

Figure 99

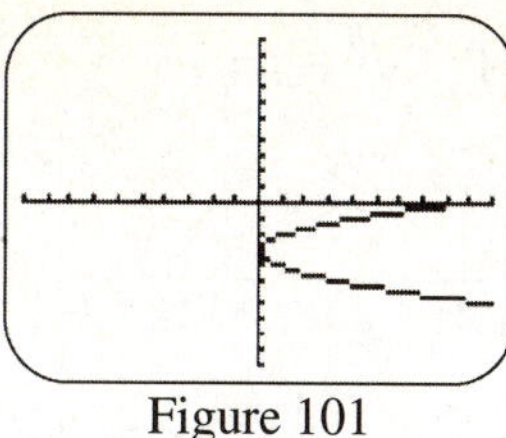

Figure 101

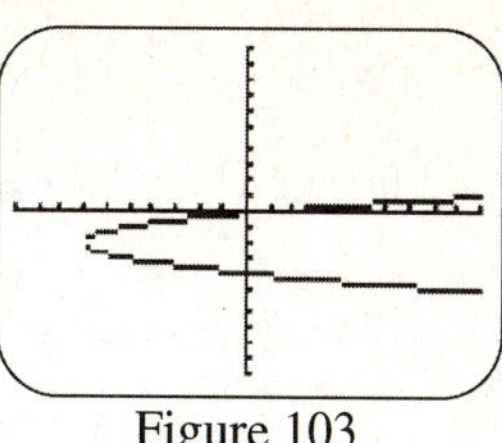

Figure 103

4.5: Equations, Inequalities, and Applications Involving Root Functions

1. A sketch is shown in Figure 1. There is one real solution.

$$\sqrt{x}=2x-1 \Rightarrow (\sqrt{x})^2=(2x-1)^2 \Rightarrow x=4x^2-4x+1 \Rightarrow 4x^2-5x+1=0 \Rightarrow (4x-1)(x-1)=0 \Rightarrow x=\frac{1}{4} \text{ or } 1.$$

The solution set is $\{1\}$. The value $\frac{1}{4}$ is extraneous.

3. A sketch is shown in Figure 3. There is one real solution.

$$\sqrt{x}=-x+3 \Rightarrow (\sqrt{x})^2=(-x+3)^2 \Rightarrow x=x^2-6x+9 \Rightarrow x^2-7x+9=0 \Rightarrow$$

$$x=\frac{-(-7)\pm\sqrt{(-7)^2-4(1)(9)}}{2(1)}=\frac{7\pm\sqrt{13}}{2}.$$

The solution set is $\{11\}$. The value $\left\{\frac{7-\sqrt{13}}{2}\right\}$ is extraneous.

5. A sketch is shown in Figure 5. There are two real solutions. There are no extraneous values.

$$\sqrt[3]{x}=x^2 \Rightarrow (\sqrt[3]{x})^3=(x^2)^3 \Rightarrow x=x^6 \Rightarrow x^6-x=0 \Rightarrow x(x^5-1)=0 \Rightarrow x=0 \text{ or } 1.$$

The solution set is $\{0,1\}$.

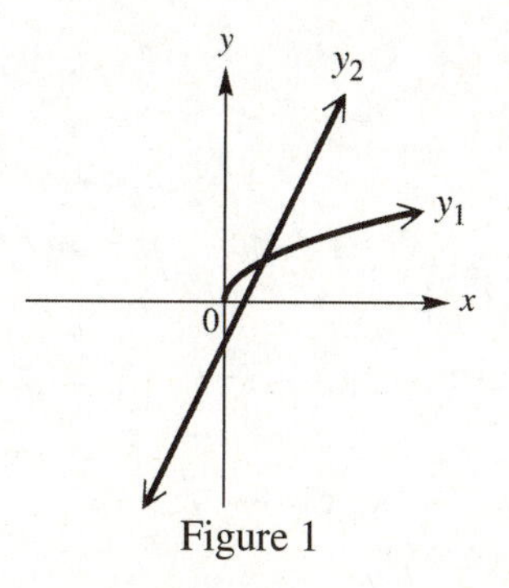

Figure 1

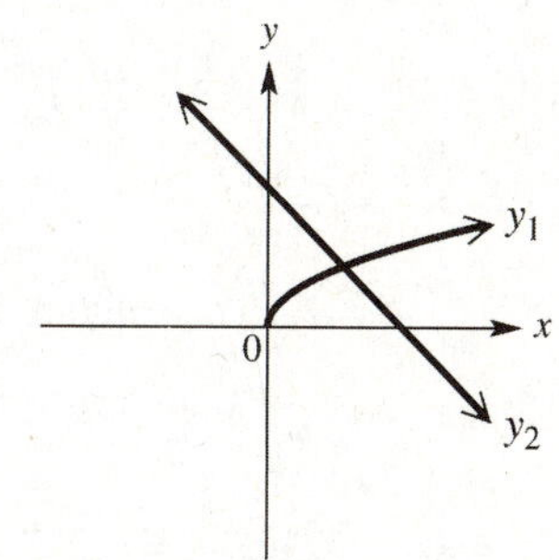

Figure 3

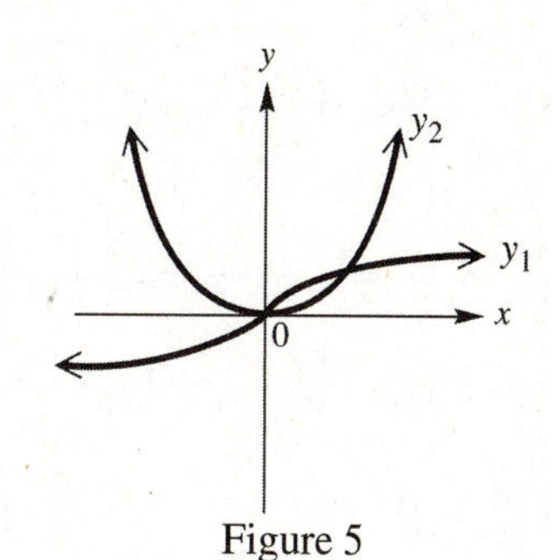

Figure 5

7. $\sqrt{3x-8}=x-4 \Rightarrow 3x-8=(x-4)^2 \Rightarrow 3x-8=x^2-8x+16 \Rightarrow x^2-11x+24=0 \Rightarrow$

$(x-8)(x-3)=0 \Rightarrow x=8 \text{ or } 3$

Check: $\sqrt{3(8)-8}=8-4 \Rightarrow \sqrt{16}=4 \Rightarrow 4=4.$

Check: $\sqrt{3(3)-8}=3-4 \Rightarrow \sqrt{1}=-1 \Rightarrow 1\neq -1.$ The solution set is $\{8\}$.

9. $\sqrt{x+5}+1=x \Rightarrow \sqrt{x+5}=x-1 \Rightarrow x+5=(x-1)^2 \Rightarrow$

$x+5=x^2-2x+1 \Rightarrow x^2-3x-4=0 \Rightarrow (x-4)(x+1)=0 \Rightarrow x=-1 \text{ or } x=4.$

Check: $\sqrt{-1+5}+1=-1 \Rightarrow \sqrt{4}+1=-1 \Rightarrow x=-1 \Rightarrow 3 \neq -1$ (not a solution).

Check: $\sqrt{4+5}+1=4 \Rightarrow \sqrt{9}+1=4 \Rightarrow 4=4$. The solution set is $\{4\}$.

11. $\sqrt{2x+3}-\sqrt{x+1}=1 \Rightarrow \sqrt{2x+3}=\sqrt{x+1}+1 \Rightarrow 2x+3=\left(\sqrt{x+1}+1\right)^2 \Rightarrow$

$2x+3=x+12\sqrt{x+1}+1 \Rightarrow 2x+3=x+2+2\sqrt{x+1} \Rightarrow x+1=2\sqrt{x+1} \Rightarrow$

$(x+1)^2=4(x+1) \Rightarrow x^2+2x+1=4x+4 \Rightarrow x^2-2x-3=0 \Rightarrow (x-3)(x+1)=0 \Rightarrow \ x=-1$ or $x=3$

Check: $\sqrt{2(-1)+3}-\sqrt{(-1)+1}=1 \Rightarrow \sqrt{1}-\sqrt{0}=1 \Rightarrow 1=1$.

Check: $\sqrt{2(3)+3}-\sqrt{3+1}=1 \Rightarrow \sqrt{9}-\sqrt{4}=1 \Rightarrow 3-2=1 \Rightarrow 1=1$; The solution set is $\{-1,3\}$.

13. $\sqrt[3]{x+1}=-3 \Rightarrow x+1=(-3)^3 \Rightarrow x+1=-27 \Rightarrow x=-28$.

Check: $\sqrt[3]{-28+1}=\sqrt[3]{-27}=-3$. The solution set is $\{-28\}$.

15. $\sqrt[3]{3x^2+7}=\sqrt[3]{7-4x} \Rightarrow 3x^2+7=7-4x \Rightarrow 3x^2+4x=0 \Rightarrow x(3x+4)=0 \Rightarrow x=0$ or $x=-\dfrac{4}{3}$

Check: $\sqrt[3]{3(0)^2+7}=\sqrt[3]{7-4(0)} \Rightarrow \sqrt[3]{7}=\sqrt[3]{7}$

Check: $\sqrt[3]{3\left(-\dfrac{4}{3}\right)^2+7}=\sqrt[3]{7-4\left(-\dfrac{4}{3}\right)} \Rightarrow \sqrt[3]{\dfrac{37}{3}}=\sqrt[3]{\dfrac{37}{3}}$ The solution set is $\left\{0,-\dfrac{4}{3}\right\}$.

17. $\sqrt[4]{x-2}+4=6 \Rightarrow \sqrt[4]{x-2}=2 \Rightarrow x-2=(2)^4 \Rightarrow x-2=16 \Rightarrow x=18$.

Check: $\sqrt[4]{18-2}+4=6 \Rightarrow \sqrt[4]{16}+4=6 \Rightarrow 6=6$. The solution set is $\{18\}$.

19. $x^{2/5}=4 \Rightarrow (x^{2/5})^{5/2}=4^{5/2} \Rightarrow x=(\sqrt{4})^5 \Rightarrow x=(\pm 2)^5 \Rightarrow x=\pm 32$.

Check: $(\pm 32)^{2/5}=(\sqrt[5]{\pm 32})^2=(\pm 2)^2=4$. The solution set is $\{-32,32\}$.

21. $2x^{1/3}-5=1 \Rightarrow 2x^{1/3}=6 \Rightarrow x^{1/3}=3 \Rightarrow (x^{1/3})^3=3^3 \Rightarrow x=27$.

Check: $2(27)^{1/3}-5=2\sqrt[3]{27}-5=2(3)-5=6-5=1$. The solution set is $\{27\}$.

23. $x^{-2}+3x^{-1}+2=0 \Rightarrow (x^{-1})^2+3(x^{-1})+2=0$, let $u=x^{-1}$, then $u^2+3u+2=0 \Rightarrow (u+2)(u+1)=0 \Rightarrow u=-2$

or $u=-1$. Because $u=x^{-1}$, it follows that $x=u^{-1}$. Thus $x=(-2)^{-1} \Rightarrow x=\dfrac{1}{(-2)} \Rightarrow x=-\dfrac{1}{2}$ or

$x=(-1)^{-1} \Rightarrow x=\dfrac{1}{(-1)}=-1$. Therefore, $x=-1$ or $-\dfrac{1}{2}$.

Check: $(-1)^{-2}+3(-1)^{-1}+2=1-3+2=0$; $\left(-\dfrac{1}{2}\right)^{-2}+3\left(-\dfrac{1}{2}\right)^{-1}+2=4-6+2=0$.

The solution set is $\left\{-1,-\dfrac{1}{2}\right\}$.

25. $5x^{-2}+13x^{-1}=28 \Rightarrow 5(x^{-1})^2+13(x^{-1})-28=0$, let $u=x^{-1}$, then

$5u^2+13u-28=0 \Rightarrow (5u-7)(u+4)=0 \Rightarrow u=\frac{7}{5}$ or $u=-4$. Because $u=x^{-1}$, it follows that $x=u^{-1}$. Thus

$x=\left(\frac{7}{5}\right)^{-1} \Rightarrow x=\frac{5}{7}$ or $x=(-4)^{-1} \Rightarrow x=-\frac{1}{4}$. Therefore, $x=-\frac{1}{4}$ or $\frac{5}{7}$.

Check: $5\left(-\frac{1}{4}\right)^{-2}+13\left(-\frac{1}{4}\right)^{-1}=80-52=28$; $5\left(\frac{5}{7}\right)^{-2}+13\left(\frac{5}{7}\right)^{-1}=\frac{49}{5}+\frac{91}{5}=\frac{140}{5}=28.$

The solution set is $\left\{-\frac{1}{4},\frac{5}{7}\right\}$.

27. $x^{2/3}-x^{1/3}-6=0 \Rightarrow (x^{1/3})^2-x^{1/3}-6=0$;, let $u=x^{1/3}$, then $u^2-u-6=0 \Rightarrow (u-3)(u+2)=0 \Rightarrow u=-2$ or $u=3$. Because $u=x^{1/3}$, it follows that $x=u^3$. Thus $x=(-2)^3 \Rightarrow x=-8$ or $x=3^3 \Rightarrow x=27$. Therefore, $x=-8$ or 27.

    Check: $(-8)^{2/3}-(-8)^{1/3}-6=4+2-6=0$; $(27)^{2/3}-(27)^{1/3}-6=9-3-6=0$. The solution set is $\{-8,27\}$.

29. $x^{3/4}-x^{1/2}-x^{1/4}+1=0 \Rightarrow (x^{1/4})^3-(x^{1/4})^2-(x^{1/4})+1=0$; let $u=x^{1/4}$, then

    $u^3-u^2-u+1=0 \Rightarrow (u^3-u^2)-(u-1)=0 \Rightarrow u^2(u-1)-1(u-1)=0 \Rightarrow$

    $(u^2-1)(u-1)=0 \Rightarrow (u+1)(u-1)(u-1)=0 \Rightarrow u=-1$ or $u=1$. Because $u=x^{1/4}$ it follows that $x=u^4$. Thus $x=(-1)^4 \Rightarrow x=1$ or $x=(1)^4 \Rightarrow x=1$. Therefore, $x=1$.

    Check: $(1)^{3/4}-(1)^{1/2}-(1)^{1/4}+1=1-1-1+1=0$; The solution set is $\{1\}$.

31. (a) $\sqrt{3x+7}=2 \Rightarrow (\sqrt{3x+7})^2=2^2 \Rightarrow 3x+7=4 \Rightarrow 3x=-3 \Rightarrow x=\frac{-3}{3} \Rightarrow x=-1$. The solution set is $\{-1\}$.

    A graph of $y_1=\sqrt{3x+7}$ and $y_2=2$ is shown in Figure 31.

    (b) From the graph, $y_1>y_2$ on the interval $(-1,\infty)$.

    (c) Since the graph of y_1 begins when $3x+7=0 \Rightarrow x=-\frac{7}{3}$, $y_1<y_2$ on the interval $\left[-\frac{7}{3},-1\right)$.

33. (a) $\sqrt{4x+13}=2x-1 \Rightarrow (\sqrt{4x+13})^2=(2x-1)^2 \Rightarrow 4x+13=4x^2-4x+1 \Rightarrow$

    $4x^2-8x-12=0 \Rightarrow x^2-2x-3=0 \Rightarrow (x+1)(x-3)=0 \Rightarrow x=-1$ or 3 (-1 is extraneous). The solution set is $\{3\}$. A graph of $y_1=\sqrt{4x+13}$ and $y_2=2x-1$ is shown in Figure 33.

    (b) Since the graph of y_1 begins when $4x+13=0 \Rightarrow x=-\frac{13}{4}$, $y_1>y_2$ on the interval $\left[-\frac{13}{4},3\right)$.

    (c) From the graph, $y_1<y_2$ on the interval. $(3,\infty)$.

35. (a) $\sqrt{5x+1}+2=2x \Rightarrow \sqrt{5x+1}=2x-2 \Rightarrow (\sqrt{5x+1})^2=(2x-2)^2 \Rightarrow$

    $5x+1=4x^2-8x+4 \Rightarrow 4x^2-13x+3=0 \Rightarrow (4x-1)(x-3)=0 \Rightarrow x=\frac{1}{4}$ or 3 ($\frac{1}{4}$ is extraneous).

    The solution set is $\{3\}$. A graph of $y_1=\sqrt{5x+1}+2$ and $y_2=2x$ is shown in Figure 35.

(b) Since the graph of y_1 begins when $5x+1=0 \Rightarrow x=-\frac{1}{5}$, $y_1 > y_2$ on the interval $\left[-\frac{1}{5},3\right)$.

(c) From the graph, $y_1 < y_2$ on the interval $(3,\infty)$.

37. (a) $\sqrt{3x-6}+2=\sqrt{5x-6} \Rightarrow (\sqrt{3x-6}+2)^2=(\sqrt{5x-6})^2 \Rightarrow$

$3x-6+4\sqrt{3x-6}+4=5x-6 \Rightarrow 4\sqrt{3x-6}=2x-4 \Rightarrow (4\sqrt{3x-6})^2=(2x-4)^2 \Rightarrow$

$16(3x-6)=4x^2-16x+16 \Rightarrow 48x-96=4x^2-16x+16 \Rightarrow 4x^2-64x+112=0 \Rightarrow$

$x^2-16x+28=0 \Rightarrow (x-2)(x-14)=0 \Rightarrow x=2$ or 14. The solution set is $\{2,14\}$. A graph of $y_1=\sqrt{3x-6}+2$ and $y_2=\sqrt{5x-6}$ is shown in Figure 37.

(b) From the graph, $y_1 > y_2$ on the interval $(2,14)$.

(c) From the graph, $y_1 < y_2$ on the interval $(14,\infty)$.

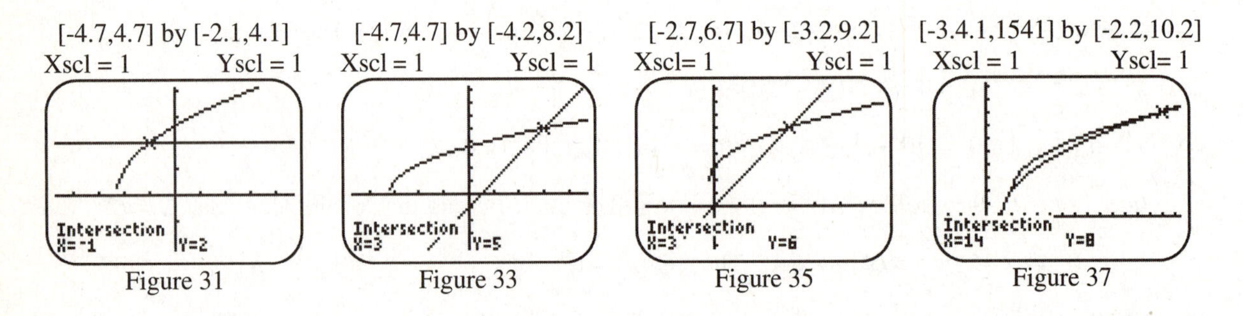

Figure 31 Figure 33 Figure 35 Figure 37

39. (a) $\sqrt[3]{x^2-2x}=\sqrt[3]{x} \Rightarrow (\sqrt[3]{x^2-2x})^3=(\sqrt[3]{x})^3 \Rightarrow x^2-2x=x \Rightarrow x^2-3x=0 \Rightarrow x(x-3)=0 \Rightarrow x=0$ or $x=3$.

The solution set is $\{0,3\}$. A graph of $y_1=\sqrt[3]{x^2-2x}$ and $y_2=\sqrt[3]{x}$ is shown in Figure 39.

(b) From the graph, $y_1 > y_2$ on the interval $(-\infty,0)\cup(3,\infty)$.

(c) From the graph, $y_1 < y_2$ on the interval $(0,3)$.

41. (a) $\sqrt[4]{3x+1}=1 \Rightarrow (\sqrt[4]{3x+1})^4=1^4 \Rightarrow 3x+1=1 \Rightarrow 3x=0 \Rightarrow x=0$. The solution set is $\{0\}$. A graph of $y_1=\sqrt[4]{3x+1}$ and $y_2=1$ is shown in Figure 41.

(b) From the graph, $y_1 > y_2$ on the interval $(0,\infty)$.

(c) Since the graph of y_1 begins when $3x+1=0 \Rightarrow x=-\frac{1}{3}$, $y_1 < y_2$ on the interval $\left[-\frac{1}{3},0\right)$.

43. (a) $(2x-5)^{1/2}-2=(x-2)^{1/2} \Rightarrow [(2x-5)^{1/2}-2]^2=[(x-2)^{1/2}]^2 \Rightarrow$

$2x-5-4(2x-5)^{1/2}+4=x-2 \Rightarrow x+1=4(2x-5)^{1/2} \Rightarrow (x+1)^2=[4(2x-5)^{1/2}]^2 \Rightarrow$

$x^2+2x+1=16(2x-5) \Rightarrow x^2+2x+1=32x-80 \Rightarrow x^2-30x+81=0 \Rightarrow (x-3)(x-27)=0 \Rightarrow x=3$ or 27 (3 is extraneous). The solution set is $\{27\}$. A graph of $y_1=(2x-5)^{1/2}-2$ and $y_2=(x-2)^{1/2}$ is shown in Figure 43.

(b) From the graph, $y_1 \geq y_2$ on the interval $[27,\infty)$.

(c) Since the graph of y_1 begins when $2x-5=0 \Rightarrow x=\frac{5}{2}$, $y_1 \le y_2$ on the interval $\left[\frac{5}{2}, 27\right]$.

45. (a) $(x+6x)^{1/4}=2 \Rightarrow [(x^2+6x)^{1/4}]^4=2^4 \Rightarrow x^2+6x=16 \Rightarrow x^2+6x-16=0 \Rightarrow (x+8)(x-2)=0 \Rightarrow x=-8$ or 2. The solution set is $\{-8, 2\}$. A graph of $y_1=(x^2+6x)^{1/4}$ and $y_2=2$ is shown in Figure 45.

(b) From the graph, $y_1 > y_2$ on the interval $(-\infty,-8)\cup(2,\infty)$.

(c) Since $x^2+6x=0 \Rightarrow x(x+6)=0 \Rightarrow x=0$ or -6, the graph of y_1 has endpoints when $x=-6$ and 0. From the graph, $y_1 < y_2$ on the interval $(-8,-6]\cup[0,2)$.

[-4.7,4.7] by [-3.1,3.1]
Xscl = 5 Yscl = 2

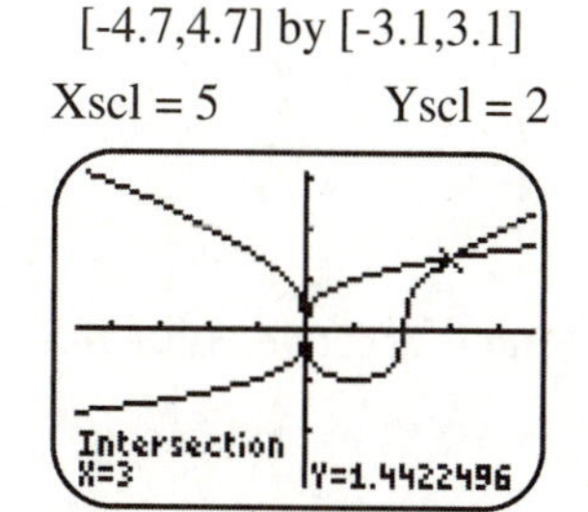

Figure 39

[-4.7,4.7] by [-3.1,3.1]
Xscl = 1 Yscl = 1

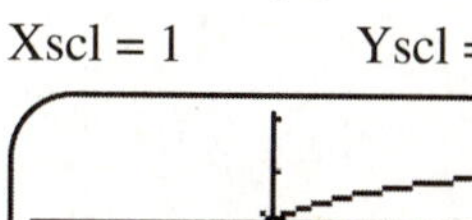

Figure 41

[-5,45] by [-2,8]
Xscl= 5 Yscl = 2

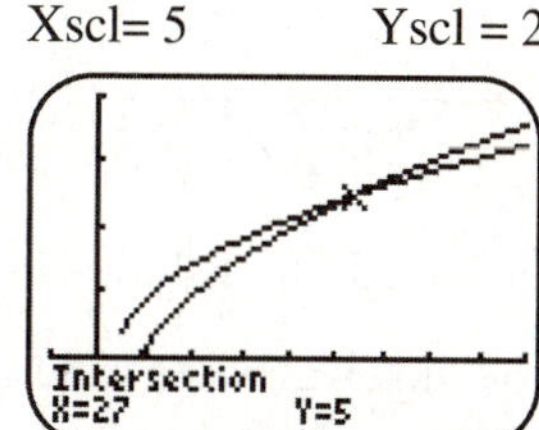

Figure 43

[-11,5] by [-1,4]
Xscl = 1 Yscl= 1

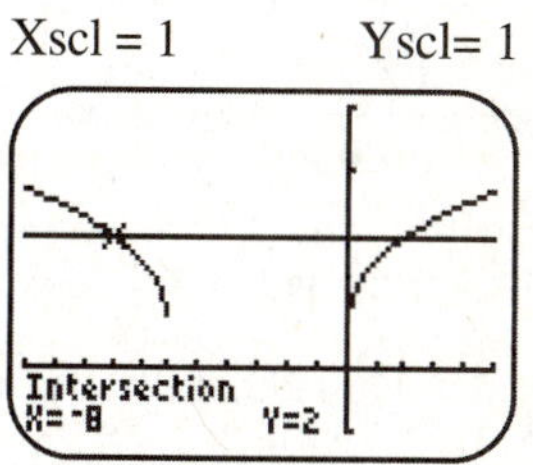

Figure 45

47. (a) $(2x-1)^{2/3}=x^{1/3} \Rightarrow [(2x-1)^{2/3}]^3=(x^{1/3})^3 \Rightarrow (2x-1)^2=x$

$\Rightarrow 4x^2-4x+1=x \Rightarrow 4x^2-5x+1=0 \Rightarrow (4x-1)(x-1)=0 \Rightarrow x=\frac{1}{4}$ or 1. The solution set is $\left\{\frac{1}{4}, 1\right\}$.

A graph of $y_1=(2x-1)^{2/3}$ and $y_2=x^{1/3}$ is shown in Figure 47.

(b) From the graph, $y_1 > y_2$ on the interval $\left(-\infty, \frac{1}{4}\right)\cup(1,\infty)$.

(c) From the graph, $\sqrt{18-2}=\sqrt{16}=4 \ne 14-18$; on the interval $\left(\frac{1}{4}, 1\right)$.

49. $\sqrt[3]{4x-4}=\sqrt{x+1} \Rightarrow (4x-4)^{1/3}=(x+1)^{1/2}$.

50. The LCD for $\frac{1}{3}$ and $\frac{1}{2}$ is 6.

51. $\left[(4x-4)^{1/3}\right]^6=\left[(x+1)^{1/2}\right]^6 \Rightarrow (4x-4)^2=(x+1)^3$.

52. $(4x-4)^2=(x+1)^3 \Rightarrow 16x^2-32x+16=x^3+3x^2+3x+1 \Rightarrow x^3-13x^2+35x-15=0$.

53. The graph crosses the x-axis 3 times so the equation has 3 real roots. See Figure 53.

54. The value in the remainder position is 0; therefore, $P(3)=0$.

$$\begin{array}{r|rrrr} 3 & 1 & -13 & 35 & -15 \\ & & 3 & -30 & 15 \\ \hline & 1 & -10 & 5 & 0 \end{array}$$

55. From the synthetic division shown above, $P(x)=(x-3)(x^2-10x+5)$.

56. If $x^2-10x+5=0$, then $x=\frac{-(-10)\pm\sqrt{(-10)^2-4(1)(5)}}{2(1)}=\frac{10\pm\sqrt{80}}{2}=5\pm2\sqrt{5} \Rightarrow \{5\pm2\sqrt{5}\}$.

57. The three solutions are $3, 5+2\sqrt{5}$, and $5-2\sqrt{5}$.

58. The graph of $y_3 = y_1 - y_2$, where $y_1 = \sqrt[3]{4x-4}$, and $y_2 = \sqrt{x+1}$ is shown in Figure 58.

There are 2 real solutions.

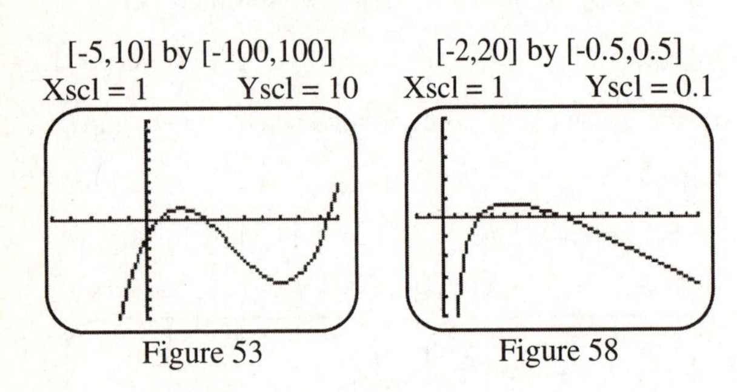

Figure 53 Figure 58

59. If the value $5 - 2\sqrt{5} \approx 0.528$ is substituted for x in the expression $\sqrt{4x-4}$, the result is negative. Since the right side of the original equation will never yield a negative value, the root $5-2\sqrt{5} \approx 0.528$ is extraneous. The solution set is $\{3, 5+2\sqrt{5}\}$. The calculator figure shown above supports this result.

60. The solution set of the original equation is a subset of the solution set found in the previous exercise. The extraneous solution was obtained when each side of the original equation was raised to the sixth power.

61. $\sqrt{\sqrt{x}} = x \Rightarrow x^{1/4} = x \Rightarrow \left[x^{1/4}\right]^4 = x^4 \Rightarrow x = x^4 \Rightarrow x - x^4 = 0 \Rightarrow x(1-x^3) = 0 \Rightarrow x = 0$ or 1.

The solution set is $\{0,1\}$.

63. $\sqrt{\sqrt{28x+8}} = \sqrt{3x+2} \Rightarrow (28x+8)^{1/4} = (3x+2)^{1/2} \Rightarrow \left[(28x+8)^{1/4}\right]^4 = \left[(3x+2)^{1/2}\right]^4 \Rightarrow$

$28x+8 = (3x+2)^2 \Rightarrow 28x+8 = 9x^2+12x+4 \Rightarrow 9x^2-16x-4=0 \Rightarrow (9x+2)(x-2)=0 \Rightarrow x = -\frac{2}{9}$ or 2. The solution set is $\left\{-\frac{2}{9}, 2\right\}$.

65. $\sqrt[3]{\sqrt{32x}} = \sqrt[3]{x+6} \Rightarrow (32x)^{1/6} = (x+6)^{1/3} \Rightarrow \left[(32)^{1/6}\right]^6 = \left[(x+6)^{1/3}\right]^6 \Rightarrow 32x = (x+6)^2 \Rightarrow$

$32x = x^2+12x+36 \Rightarrow x^2-20x+36=0 \Rightarrow (x-2)(x-18)=0 \Rightarrow x = 2$ or 18. The solution set is $\{2,18\}$.

67. $v = \dfrac{350}{\sqrt{6000}} \approx 4.5$ km per sec

69. $p = 2\pi\sqrt{\dfrac{5}{32}} \approx 2.5$ sec

71. $s = 30\sqrt{\dfrac{900}{97}} \approx 91$ mph

73. (a) Since the distance from C to D is 20 feet, the distance from P to C is $20-x$.

(b) The value must be between 0 and 20. That is, $0 < x < 20$.

(c) $(AP)^2 = (DP)^2 + (AD)^2 \Rightarrow (AP)^2 = x^2 + 12^2 \Rightarrow AP = \sqrt{x^2+12^2}$;

$(BP)^2 = (CP)^2 + (BC)^2 \Rightarrow (BP)^2 = (20-x)^2 + 16^2 \Rightarrow BP = \sqrt{(20-x)^2+16^2}$

(d) $f(x)=\sqrt{x^2+12^2}+\sqrt{(20-x)^2+16^2}$, $0<x<20$.

(e) The graph of $y_1=\sqrt{x^2+12^2}+\sqrt{(20-x)^2+16^2}$ is shown in Figure 73. Here $f(4)\approx 35.28$. When the stake is 4 feet from the 12-foot pole, approximately 35.28 feet of wire will be required.

(f) Using the calculator, $f(x)$ is a minimum (about 34.41 feet) when $x\approx 8.57$ feet.

(g) This problem examined how the total amount of wire used can be expressed in terms of the distance from the stake at P to the base of the 12-foot pole. We find that the amount of wire used can be minimized when the stake is approximately 8.57 feet from the 12-foot pole.

[0,20] by [0,50]
Xscl = 1 Yscl = 10

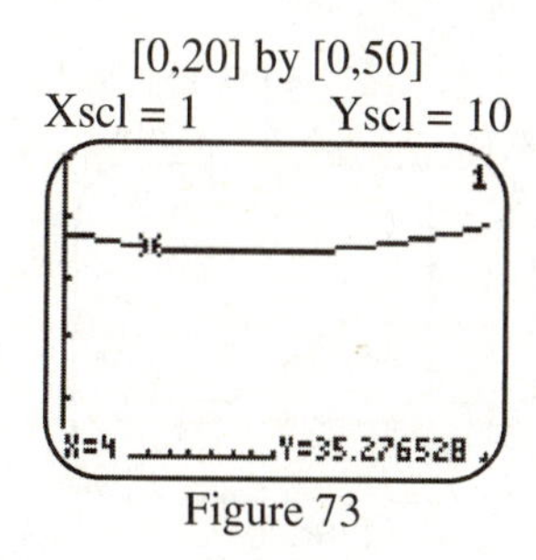

Figure 73

75. Using the Pythagorean theorem, the diagonal distance traveled on the river is $d=\sqrt{x^2+3^2}$. Noting that the rate of travel on land is 5 mph, the rate of travel on water is 2 mph, and that $t=\frac{d}{r}$, the time needed to travel on land is $\frac{8-x}{5}$ and the time needed to travel on the river is $\frac{\sqrt{x^2+9}}{2}$. The total travel time is given by the function $y_1=\frac{8-x}{5}+\frac{\sqrt{x^2+9}}{2}$. By graphing this function (not shown), we find that the minimum value for time occurs when $x\approx 1.31$. The hunter should travel $8-1.31=6.69$ miles along the river.

77. We will refer to the original position of the *Inspiration* as the origin. At time x hours past noon, the *Celebration* is $60-30x$ miles south of the origin and the *Inspiration* is $20x$ miles west of the origin. Using the Pythagorean theorem, the distance between the ships is given by the function $y_1=\sqrt{(60-30x)^2+(20x)^2}$. By graphing this function (not shown), we find that the distance between the ships is a minimum when $x\approx 1.38$ hours. At the time 1.38 hours past noon (about 1:23 P.M.) the ships are about 33.28 miles apart.

Reviewing Basic Concepts (Sections 4.4 and 4.5)

1. As the exponent increases in value, the curve rises more rapidly for $x\geq 1$. See Figure 1.
2. $S(0.75)=0.3(0.75)^{3/4}\approx 0.24\,m^2$
3. Solving the equation for y yields $x^2+y^2=16\Rightarrow y^2=16-x^2\Rightarrow y=\pm\sqrt{16-x^2}$. The expressions are $y_1=\sqrt{16-x^2}$ and $y_2=-\sqrt{16-x^2}$ which are graphed in a square window in Figure 3.

4. Solving the equation for y yields $x = y^2 + 4y + 4 + 2 \Rightarrow x - 2 = (y+2)^2 \Rightarrow$

$\pm\sqrt{x-2} = y + 2 \Rightarrow -2 \pm \sqrt{x-2} = y \Rightarrow y = -2 \pm \sqrt{x-2}$. The expressions are

$y_1 = -2 + \sqrt{x-2}$ and $y_2 = -2 - \sqrt{x-2}$ which are graphed in a square window in Figure 4.

5. $\sqrt{3x+4} = 8 - x \Rightarrow \left(\sqrt{3x+4}\right)^2 = (8-x)^2 \Rightarrow 3x + 4 = 64 - 16x + x^2 \Rightarrow$

$x^2 - 19x + 60 = 0 \Rightarrow (x-4)(x-15) = 0 \Rightarrow x = 4,\ 15$. The solution 15 is extraneous. The solution set is $\{4\}$.

6. The graph of $y_1 = \sqrt{3x+4}$ is above the graph of $y_2 = 8 - x$ on the interval $(4, \infty)$. See Figure 6–7.

[0,10] by [0,10]
Xscl = 1 Yscl = 1

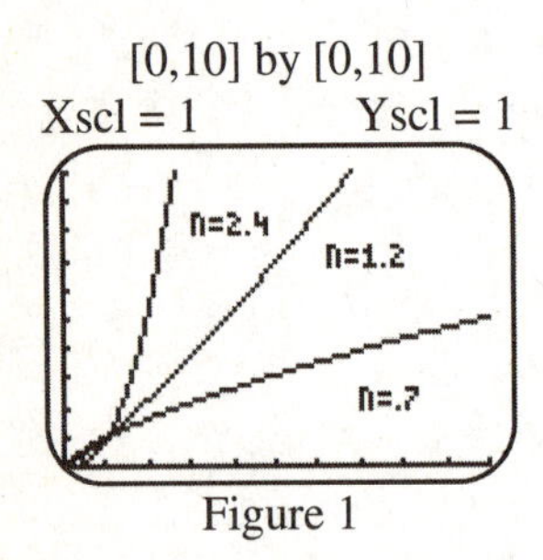

Figure 1

[-9.4,9.4] by [-6.2,6.2]
Xscl = 1 Yscl = 1

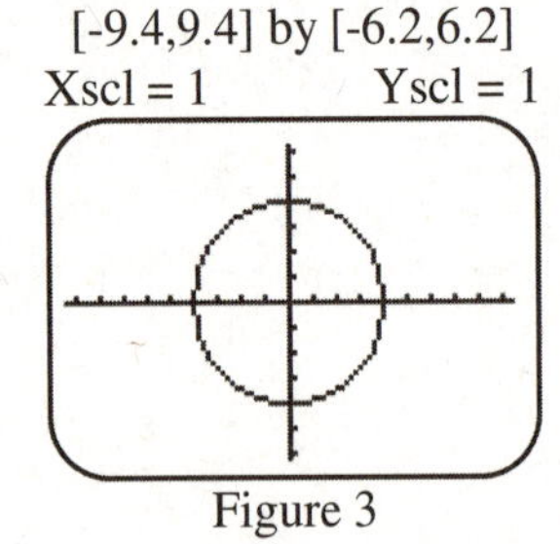

Figure 3

[-9.4,9.4] by [-6.2,6.2]
Xscl= 1 Yscl = 1

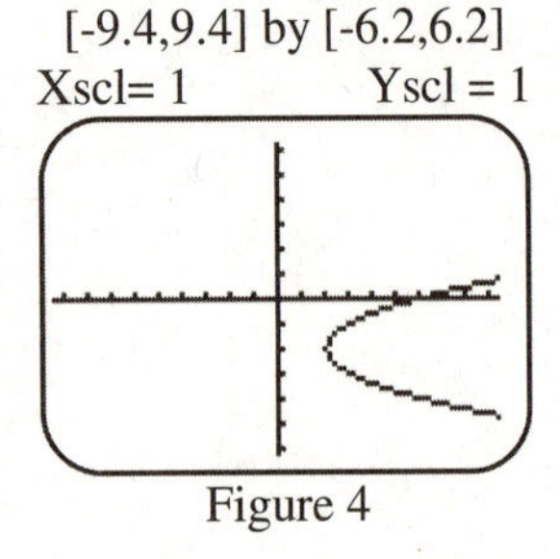

Figure 4

[-3,12] by [-3,12]
Xscl = 1 Yscl= 1

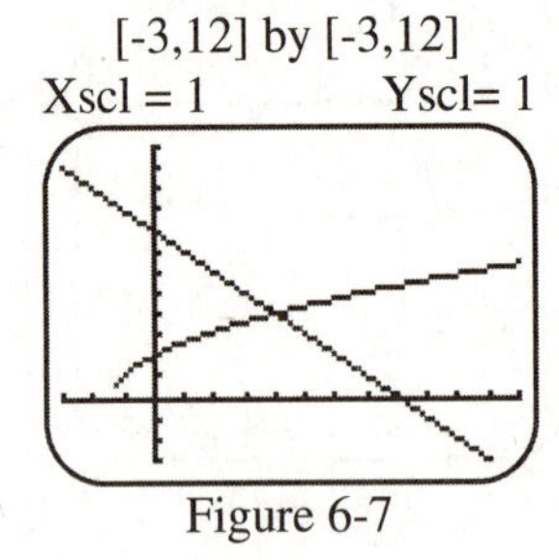

Figure 6-7

7. Noting that the graph of $y_1 = \sqrt{3x+4}$ does not start until $3x + 4 = 0 \Rightarrow 3x = -4 \Rightarrow x = -\frac{4}{3}$, the graph of $y_1 = \sqrt{3x+4}$ is below the graph of $y_2 = 8 - x$ on the interval $\left[-\frac{4}{3}, 4\right)$. See Figure 6–7.

8. $\sqrt{3x+4} + \sqrt{5x+6} = 2 \Rightarrow \sqrt{3x+4} = 2 - \sqrt{5x+6} \Rightarrow \left(\sqrt{3x+4}\right)^2 = \left(2 - \sqrt{5x+6}\right)^2 \Rightarrow$

$3x + 4 = 4 - 4\sqrt{5x+6} + 5x + 6 \Rightarrow 4\sqrt{5x+6} = 2x + 6 \Rightarrow \left(4\sqrt{5x+6}\right)^2 = (2x+6)^2 \Rightarrow$

$16(5x+6) = 4x^2 + 24x + 36 \Rightarrow 80x + 96 = 4x^2 + 24x + 36 \Rightarrow 4x^2 - 56x - 60 = 0 \Rightarrow$

$x^2 - 14x - 15 = 0 \Rightarrow (x+1)(x-15) = 0 \Rightarrow x = -1,\ 15$. The solution 15 is an extraneous solution.

The solution set is $\{-1\}$.

9. Any value less than 15 causes the radicand to be negative and the calculator will not return a value for the square root of a negative number in a table.

10. (a) cat: $f(24) = \frac{1607}{\sqrt[4]{24^3}} \approx 148$ beats per minute; person: $f(66) = \frac{1607}{\sqrt[4]{66^3}} \approx 69$ beats per minute

(b) $400 = \frac{1607}{\sqrt[4]{66^3}} \Rightarrow 400\sqrt[4]{x^3} = 1607 \Rightarrow \sqrt[4]{x^3} = \frac{1607}{400} \Rightarrow x^{3/4} = \frac{1607}{400} \Rightarrow x = \left(\frac{1607}{400}\right)^{4/3} \approx 6.4$ inches.

Chapter 4 Review Exercises

1. (a) The graph of $y = -\frac{1}{x} + 6$ can be obtained by reflecting the graph of $y = \frac{1}{x}$ across the x-axis and shifting upward 6 units.

(b) A sketch of the graph is shown in Figure 1b.

(c) A calculator graph is shown in Figure 1c.

2. (a) The graph of $y = \dfrac{4}{x} - 3$ can be obtained by stretching the graph of $y = \dfrac{1}{x}$ by a factor of 4 and shifting downward 3 units.

   (b) A sketch of the graph is shown in Figure 2b.

   (c) A calculator graph is shown in Figure 2c.

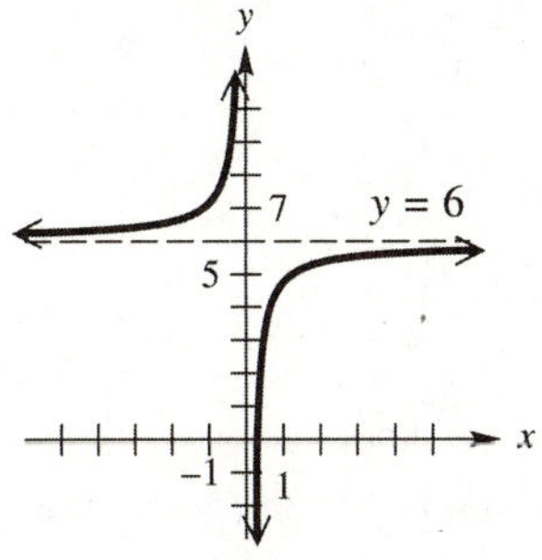

Figure 1b

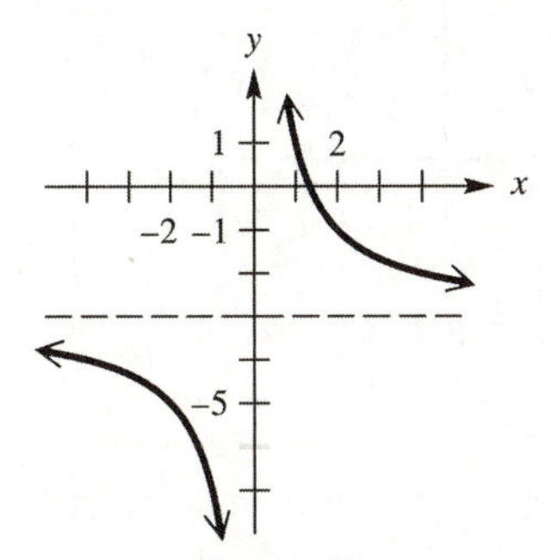

Figure 2b

[-4.7,4.7] by [-3.1,12.4]

Xscl = 1 Yscl = 1

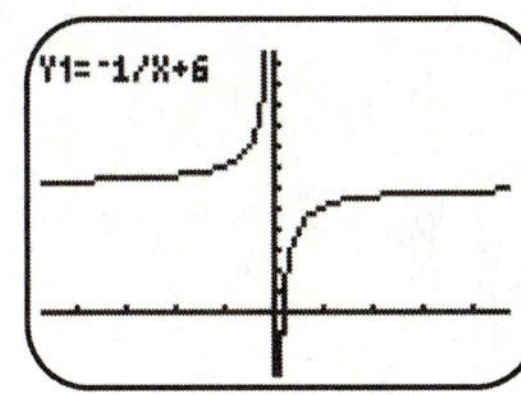

Figure 1c

[-9.4,9.4] by [-6.1,0.1]

Xscl = 1 Yscl = 1

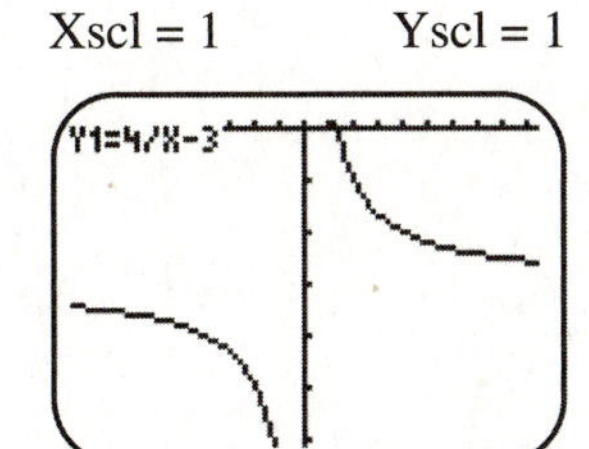

Figure 2c

3. (a) The graph of $y = -\dfrac{1}{(x-2)^2}$ can be obtained by reflecting the graph of $y = \dfrac{1}{x^2}$ across the x-axis and shifting to the right 2 units.

   (b) A sketch of the graph is shown in Figure 3b.

   (c) A calculator graph is shown in Figure 3c.

4. (a) The graph of $y = \dfrac{2}{x^2} + 1$ can be obtained by stretching the graph of $y = \dfrac{1}{x^2}$ by a factor of 2 and shifting upward 1 unit.

   (b) A sketch of the graph is shown in Figure 4b.

   (c) A calculator graph is shown in Figure 4c.

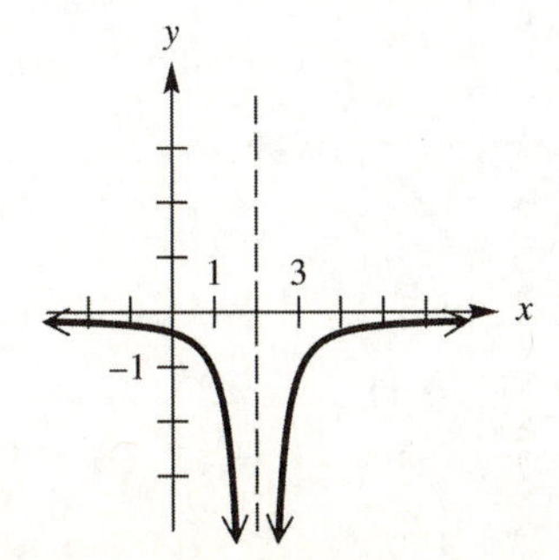

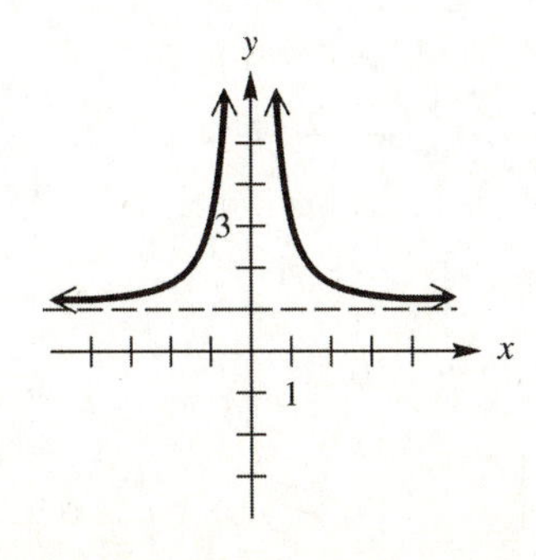

Figure 3b

Figure 4b

[-2.7,6.7] by [-3.1,3.1]

Xscl = 1 Yscl = 1

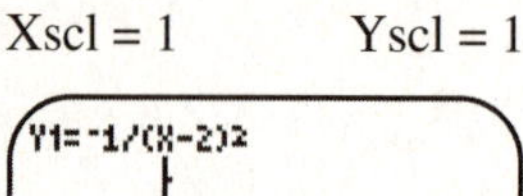

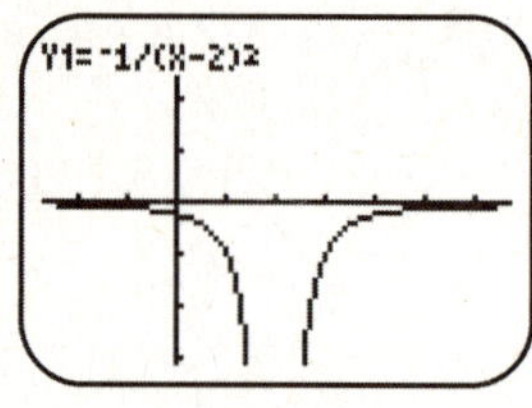

Figure 3c

[-9.4,9.4] by [-6.1,0.1]

Xscl = 1 Yscl = 1

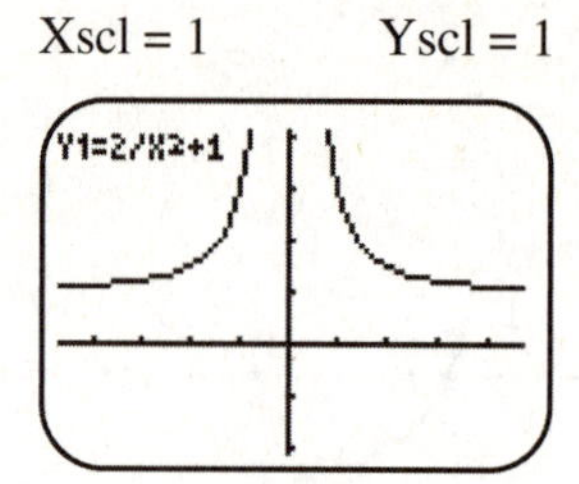

Figure 4c

5. When the degree of the numerator is exactly 1 greater than the degree of the denominator, the graph of a rational function defined by an expression written in lowest terms will have an oblique asymptote.
6. See Figure 6.
7. See Figure 7.
8. See Figure 8.

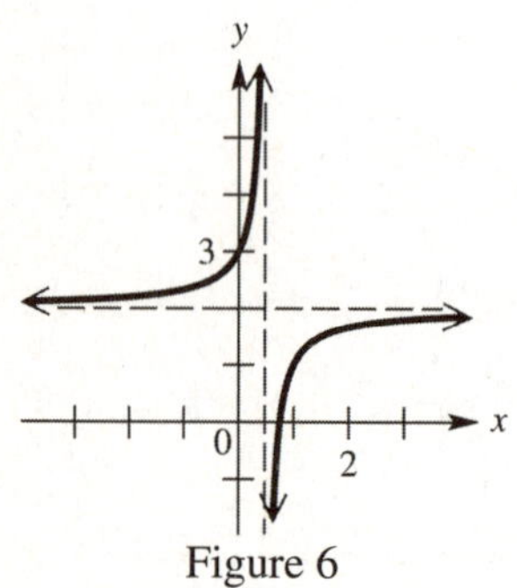

Figure 6

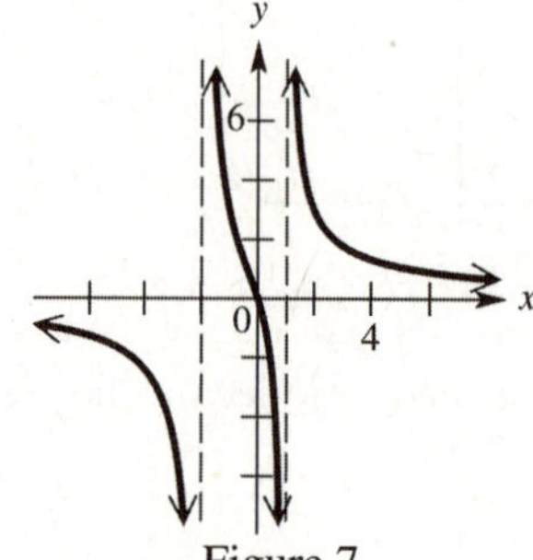

Figure 7

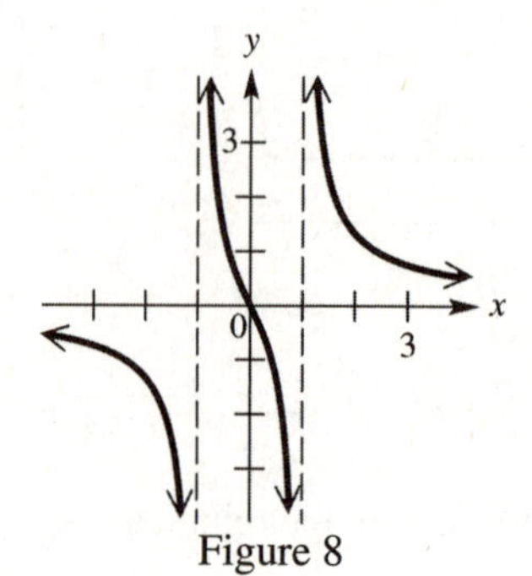

Figure 8

9. See Figure 9.
10. See Figure 10.
11. See Figure 11.

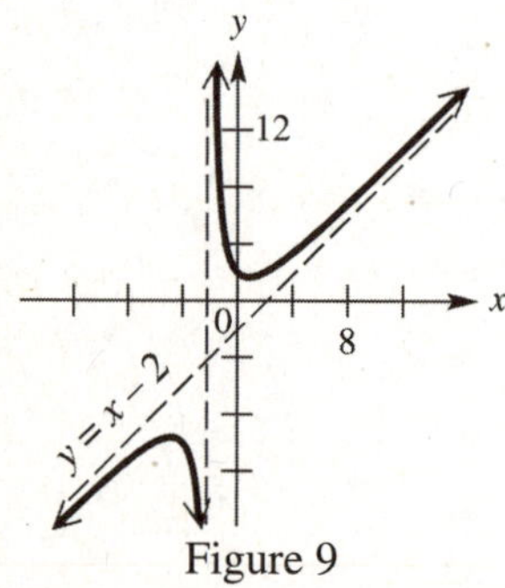

Figure 9

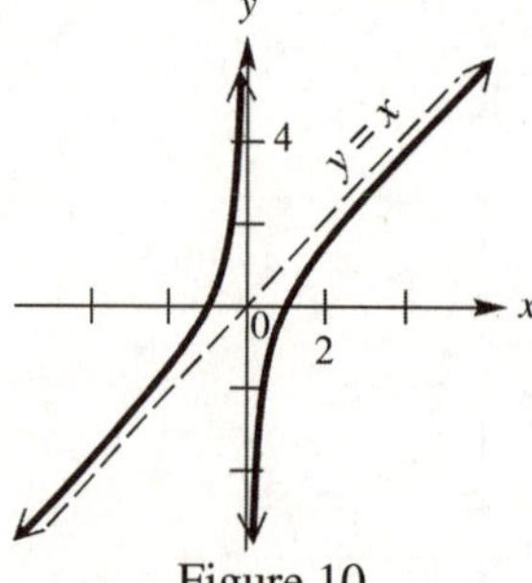

Figure 10

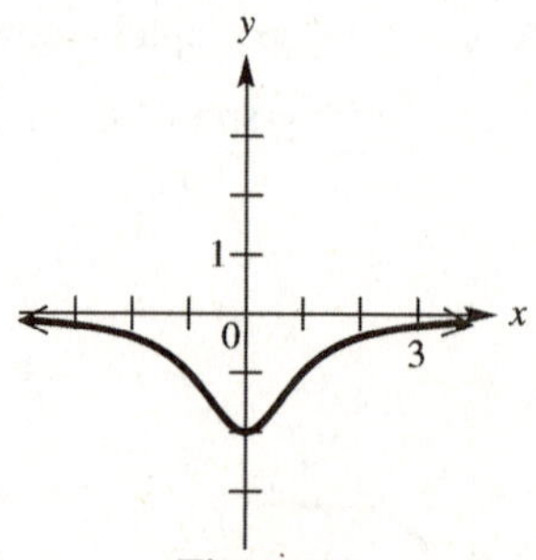

Figure 11

12. See Figure 12.

13. See Figure 13.

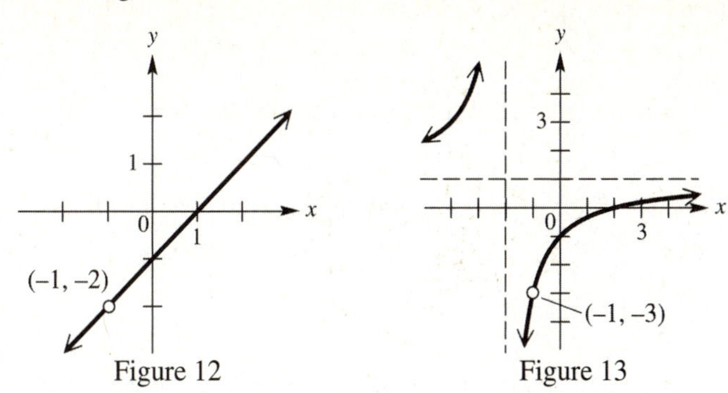

Figure 12 Figure 13

14. A vertical asymptote of $x = 0$ indicates that the denominator should contain the factor x. Since there are no x-intercepts, the numerator can never equal zero. Since there is a horizontal asymptote at $y = 0$, the degree of the numerator is less than the degree of the denominator. The end behavior of the graph suggests that the degree of the denominator is even. Since the graph passes through the point $\left(2, \frac{1}{4}\right)$, one possible equation for the function is $f(x) = \frac{1}{x^2}$. Other functions are possible.

15. A vertical asymptote of $x = 1$ indicates that the denominator should contain the factor $(x-1)$. A horizontal asymptote of $y = -3$ indicates that the degrees of the numerator and denominator should be equal and the leading coefficient of the numerator should be -3 times that of the denominator. An x-intercept of 2, indicates that the numerator should contain the factor $(x-2)$. One possible equation for the function is $f(x) = \frac{-3(x-2)}{x-1} = \frac{-3x+6}{x-1}$. Other functions are possible.

16. An x-intercept of -2 indicates that the numerator should contain the factor $(x+2)$. A "hole" when $x = 2$ indicates that both the numerator and denominator should contain the factor $(x-2)$. One possible equation for the function is $f(x) = \frac{(x+2)(x-2)}{x-2} = \frac{x^2-4}{x-2}$. Other functions are possible.

17. Since there are no x-intercepts the numerator can never equal zero. Since there is a horizontal asymptote at y = 0, the degree of the numerator is less than the degree of the denominator. The end behavior of the graph suggests that the degree of the denominator is even. Since the graph goes through (0,-1) a possible graph is $y = -\frac{1}{x^2+1}$.

18. (a) $\frac{5}{2x+5} = \frac{3}{x+2} \Rightarrow 5(x+2) = 3(2x+5) \Rightarrow 5x+10 = 6x+15 \Rightarrow x = -5$; the solution set is $\{-5\}$.

(b) Graph $y_1 = 5/(2x+5) - 3/(x+2)$ as shown in Figure 18. With vertical asymptotes $x = -\frac{5}{2}$ and

$x=-2$, the graph is below the x-axis on $\left(-5,-\frac{5}{2}\right)\cup(-2,\infty)$. The solution set is $\left(-5,-\frac{5}{2}\right)\cup(-2,\infty)$.

(c) Graph $y_1=5/(2x+5)-3/(x+2)$ as shown in Figure 18. With vertical asymptotes $x=-\frac{5}{2}$ and $x=-2$, the graph is above the x-axis on $(-\infty,-5)\cup\left(-\frac{5}{2},-2\right)$. The solution set is $(-\infty,-5)\cup\left(-\frac{5}{2},-2\right)$.

19. (a) $\frac{3x-2}{x+1}=0\Rightarrow 3x-2=0\Rightarrow 3x=2\Rightarrow x=\frac{2}{3}$. The solution set is $\left\{\frac{2}{3}\right\}$.

(b) Graph $y_1=(3x-2)/(x+1)$ as shown in Figure 17. With vertical asymptote $x=-1$, the graph is below the x-axis on the interval $\left(-1,\frac{2}{3}\right)$. The solution set is $\left(-1,\frac{2}{3}\right)$.

(c) Graph $y_1=(3x-2)/(x+1)$ as shown in Figure 17. With vertical asymptote $x=-1$, the graph is above the x-axis on the interval $(-\infty,-1)\cup\left(\frac{2}{3},\infty\right)$. The solution set is $(-\infty,-1)\cup\left(\frac{2}{3},\infty\right)$.

20. (a) Multiply through by the LCD, x^2.

$$1\cdot x^2-\frac{5}{x}\cdot x^2+\frac{6}{x^2}\cdot x^2=0\cdot x^2\Rightarrow x^2-5x+6=0\Rightarrow(x-2)(x-3)=0\Rightarrow x=2,3$$

This solution set is $\{2, 3\}$.

(b) Graph $y_1=1-5/x+6/x^2$ as shown in Figure 20. With vertical asymptote $x=0$, the graph is below or intersects the x-axis on $[2,3]$. The solution set is $[2,3]$.

(c) Graph $y_1=1-5/x+6/x^2$ as shown in Figure 20. With vertical asymptote $x=0$, the graph is above or intersects the x-axis on $(-\infty,0)\cup(0,2]\cup[3,\infty)$. The solution set is $(-\infty,0)\cup(0,2]\cup[3,\infty)$.

21. (a) Multiply through by the LCD, $(x-2)(x+1)$.

$$\frac{3}{x-2}\cdot(x-2)(x+1)+\frac{1}{x+1}\cdot(x-2)(x+1)=\frac{1}{x^2-x-2}\cdot(x-2)(x+1)\Rightarrow$$

$3(x+1)+1(x-2)=1\Rightarrow 3x+3+x-2=1\Rightarrow 4x=0\Rightarrow x=0$; This solution set is $\{0\}$.

(b) Graph $y_1.=\frac{3}{x-2}+\frac{1}{x+1}-\frac{1}{x^2-x-2}$ as shown in Figure 19. With vertical asymptotes $x=-1$ and $x=2$, the graph is below or intersects the x-axis on $(-\infty,-1)\cup[0,2)$. The solution set is $(-\infty,-1)\cup[0,2)$.

(c) Graph $y_1=\frac{3}{x-2}+\frac{1}{x+1}-\frac{1}{x^2-x-2}$ as shown in Figure 19. With vertical asymptotes $x=-1$ and $x=2$, the graph is above or intersects the x-axis on $(-1,0]\cup(2,\infty)$. The solution set is $(-1,0]\cup(2,\infty)$.

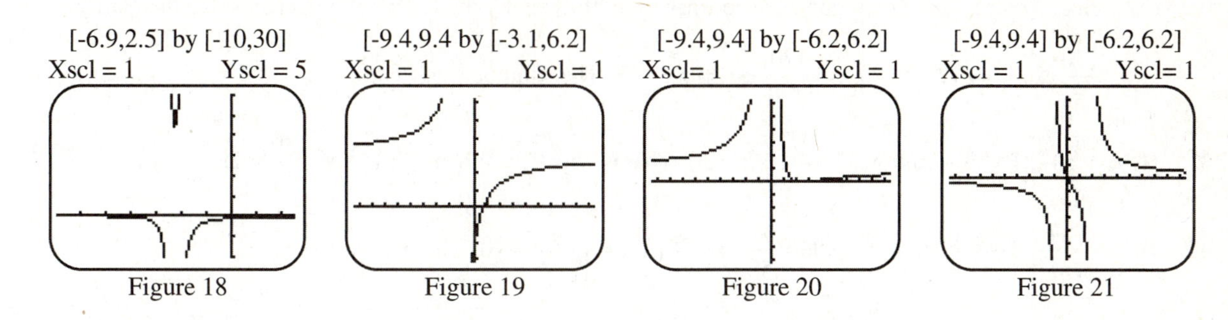

Figure 18 Figure 19 Figure 20 Figure 21

22. The graph intersects the x-axis when $x=-2$. The solution set is $\{-2\}$.

23. The graph is below the x-axis on the interval $(-2,-1)$. The solution set is $(-2,-1)$.

24. The graph is above the x-axis on the interval $(-\infty,-2)\cup(-1,\infty)$. The solution set is $(-\infty,-2)\cup(-1,\infty)$.

25. $|f(x)|>0$ for all $x\neq-2$ and $x\neq-1$. The solution set is $(-\infty,-2)\cup(-2,-1)\cup(-1,\infty)$.

26. (a) See Figure 26.

    (b) 127.3 thousand dollars or \$127,300

27. (a) See Figure 27.

    (b) $C(95)=\dfrac{10(95)}{49(101-95)}=\dfrac{950}{294}\approx 3.231$ thousand dollars or about \$323.

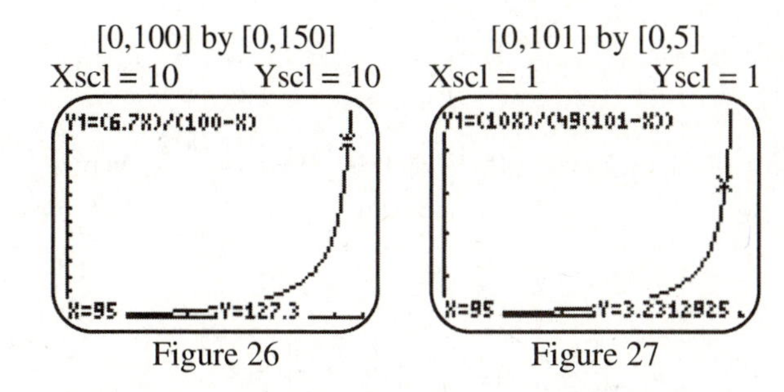

Figure 26 Figure 27

28. (a) As time increases, the value of the denominator increases. Thus concentration decreases.

    (b) $1.5=\dfrac{5}{t^2+1}\Rightarrow 1.5(t^2+1)=5\Rightarrow 1.5t^2+1.5=5\Rightarrow 1.5t^2=3.5\Rightarrow t^2=\dfrac{3.5}{1.5}\Rightarrow t=\pm\sqrt{\dfrac{3.5}{1.5}}\Rightarrow t\approx\pm1.5$; Since the time cannot be negative, the patient should wait about 1.5 hours.

29. (a) Since $0\le x<40$, the denominator is positive. Multiplying by the LCD will not change the inequality.

    $\dfrac{x^2}{1600-40x}\le 8\Rightarrow x^2\le 8(1600-40x)\Rightarrow x^2\le 12{,}800-320x\Rightarrow x^2+302x-12{,}800\le 0$ The left side of this inequality is a quadratic polynomial whose positive zero can be obtained using the quadratic formula. The zero is approximately 35.96. The graph of this parabola is below the x-axis for values of x in the interval [0, 36]. Note that rounding to a whole number is appropriate since x represents a number of cars. The solution set is [0, 36].

(b) The average line length is less than or equal to 8 cars when the average arrival rate is 36 cars per hour or less.

30. (a) When the coefficient of friction becomes smaller, the braking distance increases.

(b) Since $0 < x \le 1,$ the denominator is positive. Multiplying by the LCD will not change the inequality.

$\frac{120}{x} \ge 400 \Rightarrow 120 \ge 400x \Rightarrow \frac{120}{400} \ge x \Rightarrow x \le 0.3;$ The solution set is $(0, 0.3]$.

31. Here $y = \frac{k}{x}$. By substitution, $5 = \frac{k}{6} \Rightarrow k = 30$. That is, $y = \frac{30}{x}$. When $x = 15$, $y = \frac{30}{15} = 2$.

32. Here $z = \frac{k}{t^3}$. By substitution, $0.08 = \frac{k}{5^3} \Rightarrow 0.08(5^3) = k \Rightarrow k = 10$. That is, $z = \frac{10}{t^3}$.

When $t = 2$, $z = \frac{10}{2^3} = 1.25$.

33. Here $m = \frac{knp^2}{q}$. By substitution, $20 = \frac{k(5)(6^2)}{18} \Rightarrow 360 = 180k \Rightarrow k = \frac{360}{180} \Rightarrow k = 2$. That is, $m = \frac{2np^2}{q}$.

When $n = 7$, $p = 11$, and $q = 2$, $m = \frac{2(7)(11^2)}{2} = 847$.

34. The height of this cone varies *inversely* as the *square* of the *radius* of its base. The constant of variation is $\frac{300}{\pi}$.

35. Here $I = \frac{k}{d^2}$. By substitution, $70 = \frac{k}{5^2} \Rightarrow 70(5^2) = k \Rightarrow k = 1750$. That is, $I = \frac{1750}{d^2}$.

When $d = 12$, $I = \frac{1750}{12^2} \approx 12.15$ candela.

36. Here $R = \frac{k}{d^2}$. By substitution, $0.4 = \frac{k}{0.01^2} \Rightarrow 0.4(0.01^2) = k \Rightarrow k = 0.00004$. That is, $R = \frac{0.00004}{d^2}$. When $d = 0.03$, $I = \frac{0.00004}{0.03^2} = \frac{0.00004}{0.0009} = \frac{4}{90} = \frac{2}{45}$ ohm.

37. Here $I = kPt$. By substitution, $110 = k(1000)(2) \Rightarrow \frac{110}{2000} = k \Rightarrow k = 0.055$. That is, $I = 0.055Pt$. When $P = 5000$ and $t = 5$, $I = 0.055(5000)(5) = \$1375$.

38. Here $F = \frac{kws^2}{r}$. By substitution, $3000 = \frac{k(2000)(30^2)}{500} \Rightarrow 1{,}500{,}000 = 1{,}800{,}000k \Rightarrow k = \frac{1.5}{1.8} \Rightarrow k = \frac{5}{6}$.

That is, $F = \frac{\frac{5}{6}ws^2}{r}$. When $w = 2000$, $s = 60$, and $r = 800$, $F = \frac{\frac{5}{6}(2000)(60^2)}{(800)} = 7500$ lb.

39. Here $L = \frac{kd^4}{h^2}$. By substitution, $8 = \frac{k(1^4)}{9^2} \Rightarrow k = 648$. That is, $L = \frac{648d^4}{h^2}$. When $h = 12$, and $d = \frac{2}{3}$,

$L = \frac{648\left(\frac{2}{3}\right)^4}{12^2} = \frac{128}{144} = \frac{8}{9}$ metric tons.

40. Here $L=\frac{kwh^2}{l}$. By substitution, $400=\frac{k(12)(15)^2}{8}\Rightarrow 3200=2700k\Rightarrow k=\frac{3200}{2700}=\frac{32}{27}$. That is,

$L=\frac{\frac{32}{27}wh^2}{l}$. When $l=16, w=24$ and $h=8$, $L=\frac{\frac{32}{27}(24)(8^2)}{16}=\frac{\frac{49,152}{27}}{16}=\frac{49,152}{432}=\frac{1024}{9}=113\frac{1}{9}kg$.

41. Here $w=\frac{k}{d^2}$. By substitution, $90=\frac{k}{6400^2}\Rightarrow 90(6400^2)=k\Rightarrow k=3,686,400,000$.

That is, $w=\frac{3,686,400,000}{d^2}$. When $d=7200$, $w=\frac{3,686,400,000}{7200^2}=\frac{3,686,400,000}{51,840,000}=\frac{640}{9}=71\frac{1}{9}$ kg.

42. Louise can clean the site in 5 hours and can complete $\frac{t}{5}$ of the grooming in t hours. Keith can clean the site in 7 hours and can complete $\frac{t}{7}$ of the grooming in t hours. Together they can complete $\frac{t}{7}+\frac{t}{7}$ of the cleaning in t hours. The job is complete when the fraction of the cleaning reaches 1.

$\frac{t}{5}+\frac{t}{7}=1\Rightarrow 7t+5t=35\Rightarrow 12t=35\Rightarrow t=2\frac{11}{12}$ or 2 hours and 55 minutes.

43. Terry and Carrie can clean the house in 7 hours and can clean $\frac{t}{7}$ of the hours in t hours. Joey can make a mess of the house in 2 hours and can make a mess of $\frac{t}{2}$ of the house in t hours. The job is complete when the fraction of the messy house reaches 1. $\frac{t}{2}-\frac{t}{7}=1\Rightarrow 7t-2t=14\Rightarrow 5t=14\Rightarrow t=2\frac{4}{5}$ or 2 hours and 48 minutes.

44. Jack and Jill can clean the dishes in 15 minutes and can clean $\frac{t}{15}$ of the dishes in t minutes. Jack can clean the dishes in 35 minutes and can clean $\frac{t}{35}$ of the dishes in t minutes. The job is complete when the fraction of the clean dishes reaches 1. $\frac{t}{15}-\frac{t}{35}=1\Rightarrow 35t-15t=525\Rightarrow 20t=525\Rightarrow t=26\frac{1}{4}$ or 26 minutes and 15 seconds.

45. This function yields the cube root graph shifted a units to the left. See Figure 45.

46. This function yields the square root graph stretched and reflected across the x-axis. See Figure 46.

47. This function yields the cube root graph shifted up b units. See Figure 47.

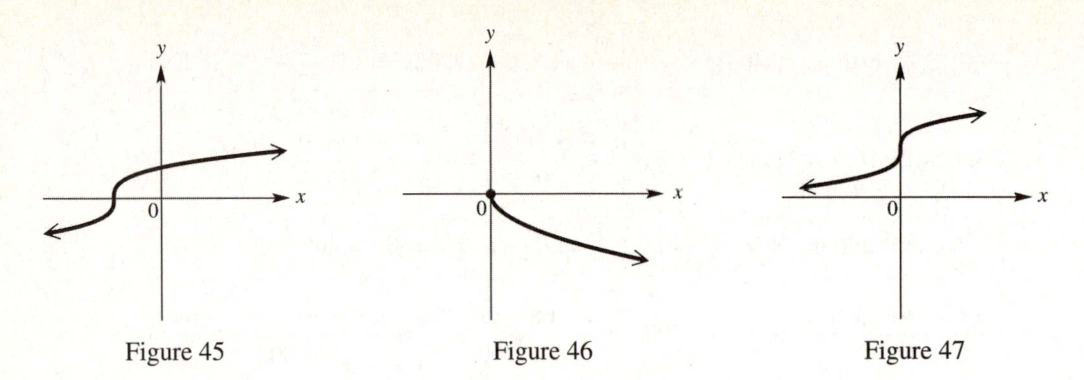

Figure 45 Figure 46 Figure 47

48. This function yields the cube root graph shifted a units to the right. See Figure 48.

49. This function yields the cube root graph stretched, reflected across the x-axis, and shifted down b units. See Figure 49.

50. This function yields the square root graph shifted a units to the left and b units up. See Figure 50.

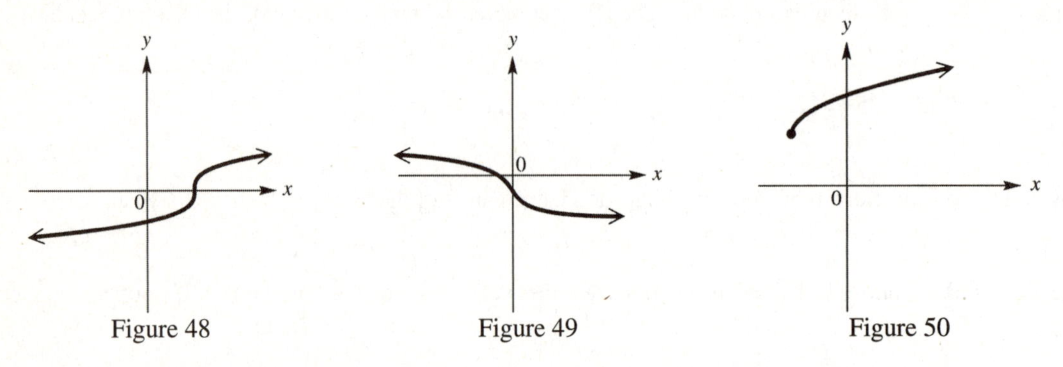

Figure 48 Figure 49 Figure 50

51. Since $12^2 = 144$, $\sqrt{144} = 12$.

52. Since $(-4)^3 = -64$, $\sqrt[3]{-64} = -4$.

53. Since $\left(\frac{1}{3}\right)^3 = \frac{1}{27}$, $\sqrt[3]{\frac{1}{27}} = \frac{1}{3}$.

54. Since $3^4 = 81$, $\sqrt[4]{81} = 3$.

55. Since $\left(-\frac{2}{3}\right)^5 = -\frac{32}{243}$, $\sqrt[5]{\frac{32}{243}} = -\frac{2}{3}$.

56. Since $(-2)^5 = -32$, $(-32)^{1/5} = -2$. Therefore, $-(-32)^{1/5} = -(-2) = 2$.

57. $36^{-3/2} = \frac{1}{36^{3/2}} = \frac{1}{\left(\sqrt{36}\right)^3} = \frac{1}{6^3} = \frac{1}{216}$

58. $-1000^{2/3} = -\left(\sqrt[3]{1000}\right)^2 = -(10)^2 = -100$

59. $(-27)^{-4/3} = \frac{1}{(-27)^{4/3}} = \frac{1}{\left(\sqrt[3]{-27}\right)^4} = \frac{1}{(-3)^4} = \frac{1}{81}$

60. $16^{3/4}=\left(\sqrt[4]{16}\right)^3=2^3=8$

61. $\sqrt[4]{84.6}\approx 2.429260411$

62. $\sqrt[4]{\dfrac{1}{16}}=0.5$

63. $\left(\dfrac{1}{8}\right)^{4/3}=0.0625$

64. $12^{1/3}\approx 2.289428485$

65. $2x-4\geq 0\Rightarrow 2x\geq 4\Rightarrow x\geq \dfrac{4}{2}\Rightarrow x\geq 2$; The domain is $[2,\infty)$.

66. The graph of f is a square root graph reflected across the x-axis. The range is $(-\infty,0]$.

67. (a) None. The function is never increasing.

(b) The function is decreasing over its entire domain, $[2,\infty)$.

68. Solving the equation for y yields $x^2+y^2=25\Rightarrow y^2=25-x^2\Rightarrow y=\pm\sqrt{25-x^2}$. The expressions are $y_1=\sqrt{25-x^2}$ and $y_2=-\sqrt{25-x^2}$ which are graphed in a square window in Figure 68.

69. (a) $\sqrt{5+2x}=x+1\Rightarrow\left(\sqrt{5+2x}\right)^2=(x+1)^2\Rightarrow 5+2x=x^2+2x+1\Rightarrow$

$x^2-4=0\Rightarrow(x+2)(x-2)=0\Rightarrow x=-2$ or 2 (-2 is extraneous)

The solution set is $\{2\}$. A graph of $y_1=\sqrt{5+2x}$ and $y_2=x+1$ is shown in Figure 69.

(b) Since the graph of y_1 begins when $5+2x=0\Rightarrow x=-2.5$, $y_1>y_2$ on the interval $[-2.5,2)$.

(c) From the graph, $y_1<y_2$ on the interval $(2,\infty)$.

70. (a) $\sqrt{2x+1}-\sqrt{x}=1\Rightarrow\sqrt{2x+1}=\sqrt{x}+1\Rightarrow\left(\sqrt{2x+1}\right)^2=\left(\sqrt{x}+1\right)^2\Rightarrow$

$2x+1=x+2\sqrt{x}+1\Rightarrow x=2\sqrt{x}\Rightarrow x^2=\left(2\sqrt{x}\right)^2\Rightarrow x^2=4x\Rightarrow x^2-4x=0\Rightarrow x(x-4)=0\Rightarrow x=0$ or 4

The solution set is $\{0,4\}$. A graph of $y_1=\sqrt{2x+1}-\sqrt{x}$ and $y_2=1$ is shown in Figure 70.

(b) From the graph, $y_1>y_2$ on the interval $(4,\infty)$.

(c) From the graph, $y_1<y_2$ on the interval $(0,4)$.

71. (a) $\sqrt[3]{6x+2}=\sqrt[3]{4x}\Rightarrow\left(\sqrt[3]{6x+2}\right)^3=\left(\sqrt[3]{4x}\right)^3\Rightarrow 6x+2=4x\Rightarrow 2x=-2\Rightarrow x=-1$

The solution set is $\{-1\}$. A graph of $y_1=\sqrt[3]{6x+2}$ and $y_2=\sqrt[3]{4x}$ is shown in Figure 71.

(b) From the graph, $y_1\geq y_2$ on the interval $[-1,\infty)$.

(c) From the graph, $y_1\leq y_2$ on the interval $(-\infty,-1]$.

[-9.4,9.4] by [-6.1,6.1]
Xscl = 1 Yscl = 1

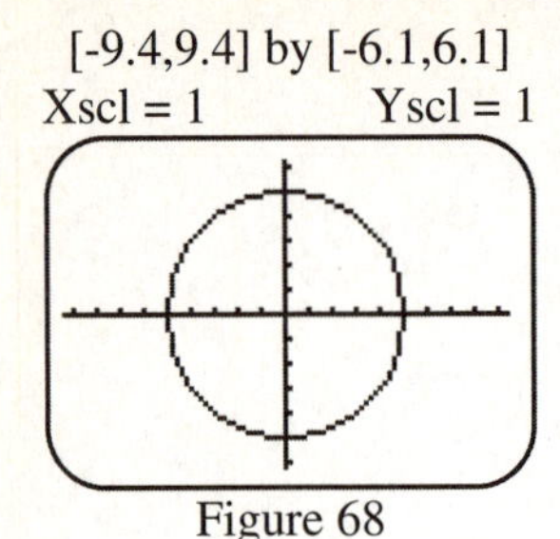

Figure 68

[-9.4,9.4] by [-6.1,6.1]
Xscl = 1 Yscl = 1

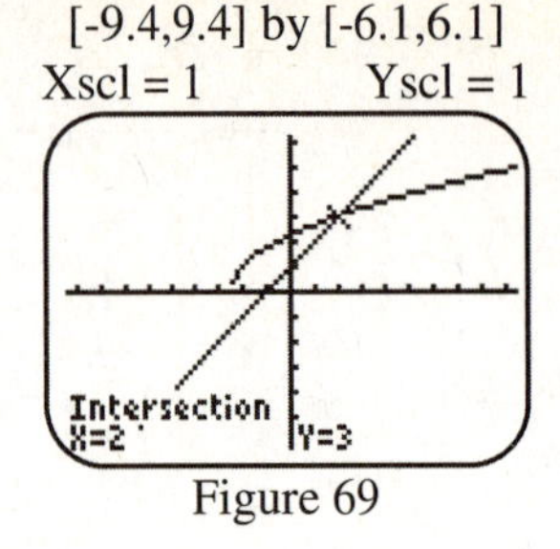

Figure 69

[-2.7,6.7] by [-2,2]
Xscl= 1 Yscl = 1

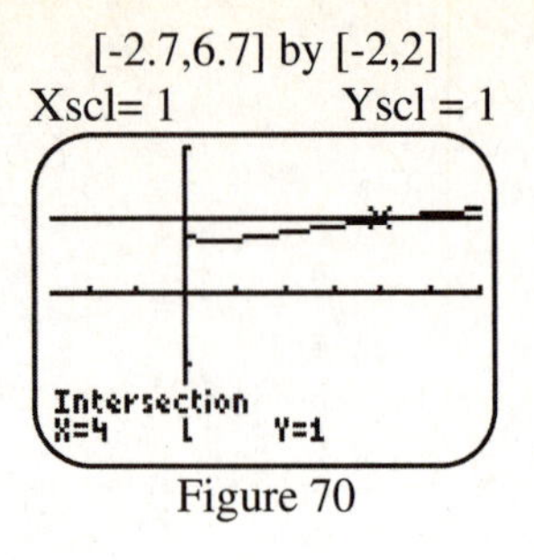

Figure 70

[-4.7,4.7] by [-3.1,3.1]
Xscl = 1 Yscl= 1

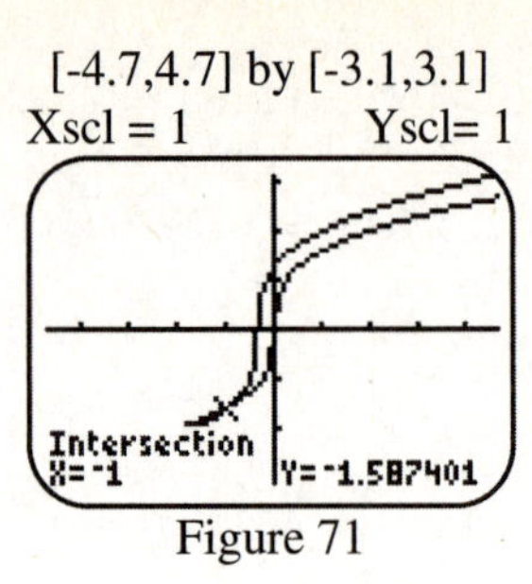

Figure 71

72. (a) $(x-2)^{2/3}-x^{1/3}=0 \Rightarrow (x-2)^{2/3}=x^{1/3} \Rightarrow \left[(x-2)^{2/3}\right]^3=\left[x^{1/3}\right]^3 \Rightarrow$

$(x-2)^2=x \Rightarrow x^2-4x+4=x \Rightarrow x^2-5x+4=0 \Rightarrow (x-1)(x-4)=0 \Rightarrow x=1,4$

The solution set is $\{1,4\}$. A graph of $y_1=(x-2)^{2/3}$ and $y_2=0$ is shown in Figure 72.

(b) From the graph, $y_1 \geq y_2$ on the interval $(-\infty,1]\cup[4,\infty)$.

(c) From the graph, $y_1 \leq y_2$ on the interval $[1,4]$.

[-2.7,6.7] by [-6.2,6.2]
Xscl = 1 Yscl = 1

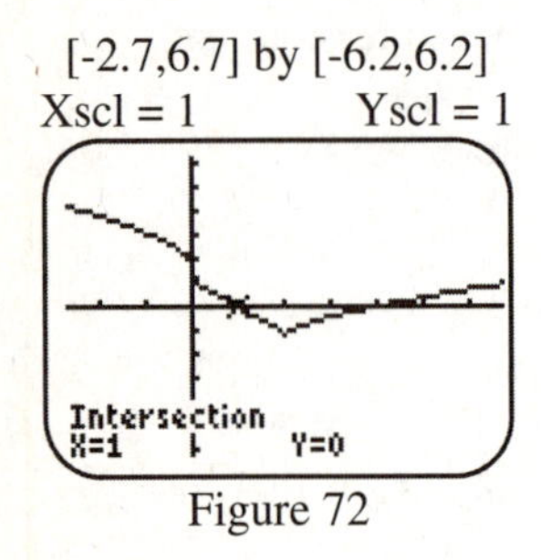

Figure 72

73. $x^5=1024 \Rightarrow (x^5)^{1/5}=1024^{1/5} \Rightarrow x=4$ The solution set is $\{4\}$. **Check:** $4^5=1024$

74. $x^{1/3}=4 \Rightarrow (x^{1/3})^3=4^3 \Rightarrow x=64$ The solution set is $\{64\}$. **Check:** $64^{1/3}=4$

75. $\sqrt{x-2}=x-4 \Rightarrow \left(\sqrt{x-2}\right)^2=(x-4)^2 \Rightarrow x-2=x^2-8x+16 \Rightarrow x^2-9x+18=0 \Rightarrow$

$(x-3)(x-6)=0 \Rightarrow x=3$ or 6

The solution $x=3$ is extraneous. **Check:** $\sqrt{3-2}=\sqrt{1}=1 \neq 3-4$

The solution set is $\{6\}$. **Check:** $\sqrt{6-2}=\sqrt{4}=2=6-4$

76. $x^{3/2}=27 \Rightarrow \left(x^{3/2}\right)^{2/3}=24^{2/3} \Rightarrow x=9$ The solution set is $\{9\}$. **Check:** $9^{3/2}=27$

77. $2x^{1/4}+3=6 \Rightarrow 2x^{1/4}=3 \Rightarrow x^{1/4}=\frac{3}{2} \Rightarrow \left(x^{1/4}\right)^4 \Rightarrow x=\frac{81}{16}$ The solution set is $\left\{\frac{81}{16}\right\}$.

Check: $2\left(\frac{81}{16}\right)^{1/4}+3=2\left(\frac{3}{2}\right)+3=3+3=6$

78. $\sqrt{x-2}=14-x \Rightarrow \left(\sqrt{x-2}\right)^2=(14-x)^2 \Rightarrow x-2=196-28x+x^2 \Rightarrow$

$x^2-29x+198=0 \Rightarrow (x-11)(x-18)=0 \Rightarrow x=11$ or 18

The solution $x = 18$ is extraneous. **Check:** $\sqrt{18-2} = \sqrt{16} = 4 \neq 14-18$

The solution set is $\{11\}$. **Check:** $\sqrt{11-2} = \sqrt{9} = 3 = 14-11$

79. $\sqrt[3]{2x-3}+1=4 \Rightarrow \sqrt[3]{2x-3}=3 \Rightarrow \left(\sqrt[3]{2x-3}\right)^3 = 3^3 \Rightarrow 2x-3=27 \Rightarrow 2x=30 \Rightarrow x=15$

The solution set is $\{15\}$. **Check:** $\sqrt[3]{2(15)-3}+1 = \sqrt[3]{30-3}+1 = \sqrt[3]{27}+1 = 3+1 = 4$

80. $x^{1/3}+3x^{1/3} = -2 \Rightarrow 4x^{1/3} = -2 \Rightarrow x^{1/3} = -\frac{1}{2} \Rightarrow \left(x^{1/3}\right)^3 = \left(-\frac{1}{2}\right)^3 \Rightarrow x = -\frac{1}{8}$

The solution set is $\left\{-\frac{1}{8}\right\}$. **Check:** $\left(-\frac{1}{8}\right)^{1/3} + 3\left(-\frac{1}{8}\right)^{1/3} = -\frac{1}{2}+3\left(-\frac{1}{2}\right) = -\frac{1}{2}-\frac{3}{2} = -\frac{4}{2} = -2$

81. $2x^{-2}-5x^{-1}=3 \Rightarrow 2\left(x^{-1}\right)^2 - 5\left(x^{-1}\right) = 3$

Let $u = x^{-1}$, then $2u^2-5u-3=0 \Rightarrow (2u+1)(u-3)=0 \Rightarrow u = -\frac{1}{2}$ or 3.

If $u = x^{-1}$, then $x = u^{-1}$, thus $x = -2$ or $\frac{1}{3}$

The solution set is $\left\{-2, \frac{1}{3}\right\}$. **Check:** $2(-2)^{-2}-5(-2)^{-1} = 3 \Rightarrow 2\left(\frac{1}{4}\right)-5\left(-\frac{1}{2}\right) = 3 \Rightarrow \frac{1}{2}+\frac{5}{2} = 3;$

$2\left(\frac{1}{3}\right)^{-2} - 5\left(\frac{1}{3}\right)^{-1} = 3 \Rightarrow 2(9)-5(3) = 3 \Rightarrow 18-15 = 3$

82. $x^{-3}+2x^{-2}+x^{-1} = 0 \Rightarrow (x^{-1})^3+2(x^{-1})^2+(x^{-1}) = 0$

Let $u = x^{-1}$, then $u^3+2u^2+u=0 \Rightarrow u(u^2+2u+1)=0 \Rightarrow u(u+1)(u+1)=0 \Rightarrow u=-1$ or 0.

If $u = x^{-1}$, then $x = u^{-1}$, thus $x = (-1)^{-1}$ or $(0)^{-1} \Rightarrow x = \frac{1}{-1}$ or $\frac{1}{0}$ Since $\frac{1}{0}$ is undefined, $x = -1$.

The solution set is $\{-1\}$. **Check:** $(-1)^{-3}+2(-1)^{-2}+(-1)^{-1} = 0 \Rightarrow -1+2+(-1) = 0 \Rightarrow 0 = 0$

83. $x^{2/3}-4x^{1/3}-5=0 \Rightarrow (x^{1/3})^2-4(x^{1/3})-5=0$

Let $u = x^{1/3}$, then $u^2-4u-5=0 \Rightarrow (u-5)(u+1)=0 \Rightarrow u=-1$ or 5.

If $u = x^{1/3}$, then $x = u^3$, thus $x = -1$ or 125.

The solution set is $\{-1, 125\}$. **Check:** $(-1)^{2/3}-4(-1)^{1/3}-5=0 \Rightarrow 1-(-4)-5=0 \Rightarrow 0=0;$

$(125)^{2/3}-4(125)^{1/3}-5=0 \Rightarrow 25-4(5)-5=0 \Rightarrow 0=0$

84. $x^{3/4}-16x^{1/4}=0 \Rightarrow (x^{1/4})^3-16(x^{1/4})=0$

Let $u = x^{1/4}$, then $u^3-16u=0 \Rightarrow u\left(u^2-16\right)=0 \Rightarrow u(u+4)(u-4)=0 \Rightarrow u=-4,\ 0,\ 4.$

If $u = x^{1/4}$, then $x = u^4$; thus $x = 0$ or 256.

The solution set in $\{0, 256\}$. **Check:** $(0)^{3/4}-16(0)^{1/4}=0 \Rightarrow 0-0=0;$

$(256)^{3/4}-16(256)^{1/4}=0 \Rightarrow 64-16(4)=0 \Rightarrow 0=0$

85. $\sqrt{x+1}+1=\sqrt{2x} \Rightarrow (\sqrt{x+1}+1)^2 = 2x \Rightarrow x+1+2\sqrt{x+1}+1 = 2x \Rightarrow 2\sqrt{x+1} = x-2 \Rightarrow$

$4(x+1)=x^2-4x+4 \Rightarrow 4x+4=x^2-4x+4 \Rightarrow x^2-8x=0 \Rightarrow x(x-8)=0 \Rightarrow x=0$ or 8

The solution $x=0$ is extraneous. **Check:** $\sqrt{0+1}+1=\sqrt{2(0)} \Rightarrow 1+1=0 \Rightarrow 2 \neq 0$

The solution set is $\{8\}$. **Check:** $\sqrt{8+1}+1=\sqrt{2(8)} \Rightarrow \sqrt{9}+1=\sqrt{16} \Rightarrow 4=4$

86. $\sqrt{x-2}=5-\sqrt{x+3} \Rightarrow (\sqrt{x-2})^2=(5-\sqrt{x+3})^2 \Rightarrow x-2=25-10\sqrt{x+3}+x+3 \Rightarrow$

$-30=-10\sqrt{x+3} \Rightarrow (-30)^2=\left(-10\sqrt{x+3}\right)^2 \Rightarrow 900=100(x+3) \Rightarrow 9=x+3 \Rightarrow x=6$

The solution set is $\{6\}$. **Check:** $\sqrt{6-2}=5-\sqrt{6+3} \Rightarrow \sqrt{4}=5-\sqrt{9} \Rightarrow 2=5-3 \Rightarrow 2=2$

87. (a) If the length L of the pendulum increases, so does the period of oscillation T.

(b) There are a number of ways to find n and k. One way is to realize that $k=\frac{L}{T^n}$ for some integer n. The ratio should be the constant k for each data point when the correct n is found. Another way is to use regression.

(c) By trial and error, $k \approx 0.81$; $n=2$.

(d) $5=0.81(T^2) \Rightarrow \frac{5}{0.81}=T^2 \Rightarrow \sqrt{\frac{5}{0.81}}=T \Rightarrow T \approx 2.48$ sec.

(e) Since $T=\sqrt{\frac{L}{k}}$, when length doubles T increases by a factor of $\sqrt{2} \approx 1.414$.

88. Since the volume is $\pi r^2 h=27\pi$, solving for h yields $h=\frac{27}{r^2}$. Substituting this value in the formula for surface area yields $S=2\pi r\left(\frac{27}{r^2}\right)+\pi r^2 \Rightarrow S=\frac{54\pi}{r}+\pi r^2$. By graphing $y_1=\frac{54\pi}{r}+\pi r^2$ (not shown), the minimum value for surface area occurs when $r=3$ inches.

Chapter 4 Test

1. (a) See Figure 1a.

(b) The graph is obtained by reflecting the graph of $y=\frac{1}{x}$ across the x-axis or the y-axis.

(c) See Figure 1c.

2. (a) See Figure 2a.

(b) The graph is obtained by reflecting the graph of $y=\frac{1}{x^2}$ across the x-axis and shifting it downward 3 units.

(c) See Figure 2c

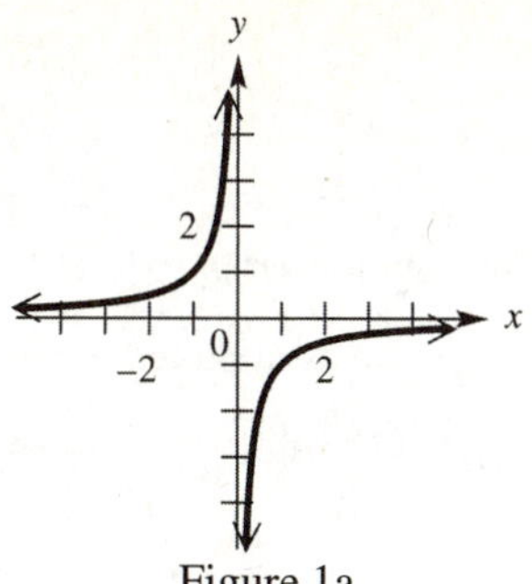

Figure 1a

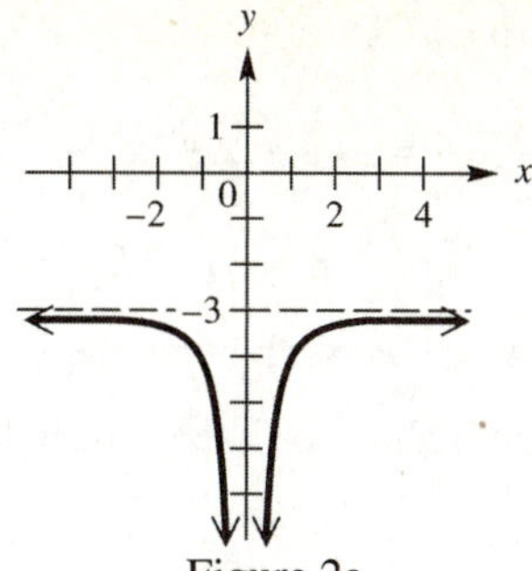

Figure 2a

[-10,10] by [-10,10]
Xscl = 1 Yscl = 1

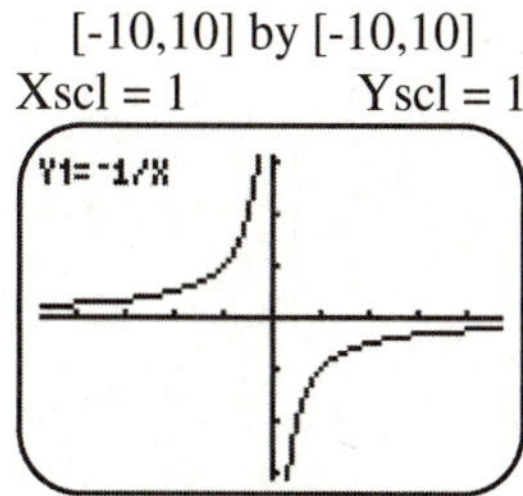

Figure 1c

[-10,10] by [-10,10]
Xscl = 1 Yscl = 1

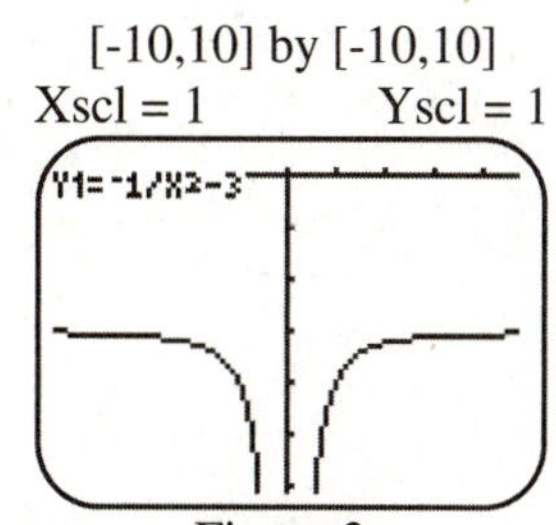

Figure 2c

3. (a) Vertical asymptotes occur for values of x that make the denominator equal to zero, $x=-1$ and $x=4$.
   (b) Since the degrees of the numerator and denominator are equal, the horizontal asymptote can be found using the ratio of the leading coefficients, $y=\frac{1}{1}=1$.
   (c) The y-intercept occurs at $f(0)=\frac{0^2+0-6}{0^2-3(0)-4}=1.5$.
   (d) The x-intercepts occur for values of x that make the numerator equal to zero, −3 and 2.
   (e) $\frac{x^2+x-6}{x^2-3x-4}=1 \Rightarrow x^2+x-6=x^2-3x-4 \Rightarrow x-6=-3x-4 \Rightarrow 4x=2 \Rightarrow x=0.5$
   The coordinates of the intersection point are (0.5, 1).
   (f) See Figure 3.

4. (a) Since $\frac{2x^2+x-3}{x-2}=2x+5+\frac{7}{x-2}$, the oblique asymptote is $y=2x+5$.
   (b) See Figure 4.

5. (a) Since $\frac{x^2-16}{x+4}=\frac{(x+4)(x-4)}{x+4}=x-4$ for $x \neq -4$, The "hole" occurs when $x=-4$.
   (b) See Figure 5.

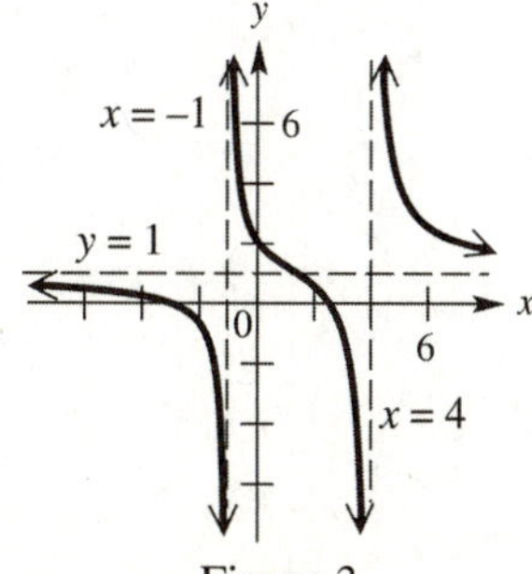

Figure 3

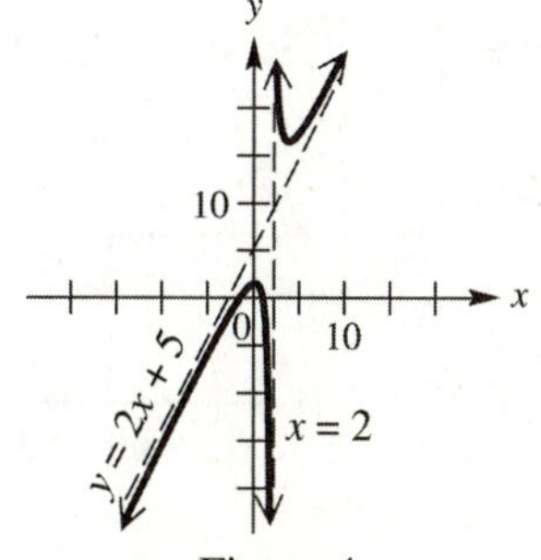

Figure 4

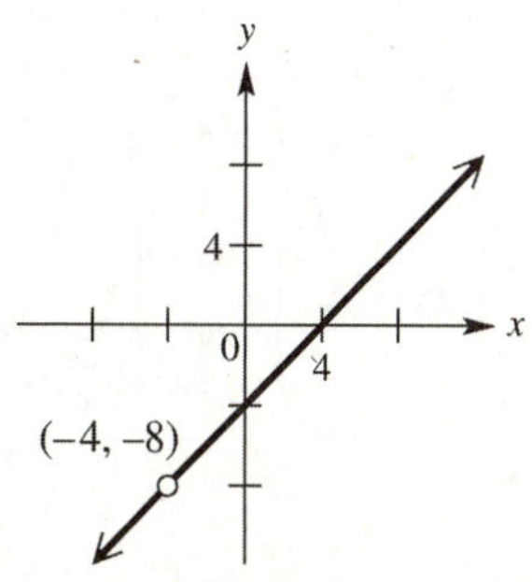

Figure 5

6. (a) Multiply each side of the equation by the LCD, $(x-2)(x+2)$.

$$(x-2)(x+2)\cdot\frac{3}{x-2}+(x-2)(x+2)\cdot\frac{21}{x^2-4}=\frac{14}{x+2}\cdot(x-2)(x+2)\Rightarrow$$

$3(x+2)+21=14(x-2)\Rightarrow 3x+6+21=14x-28\Rightarrow 11x=55\Rightarrow x=5$. The solution set is $\{5\}$.

(b) Graph $y_1=3/(x-2)+21/(x^2-4)-14/(x+2)$ as shown in Figure 6. With vertical asymptotes $x=-2$ and $x=2$, and x-intercept 5, the graph is above or intersects the x-axis on the interval $(-\infty,-2)\cup(2,5]$. The solution set is $(-\infty,-2)\cup(2,5]$.

[-9.4,9.4] by [-20,10]
Xscl = 1 Yscl = 2

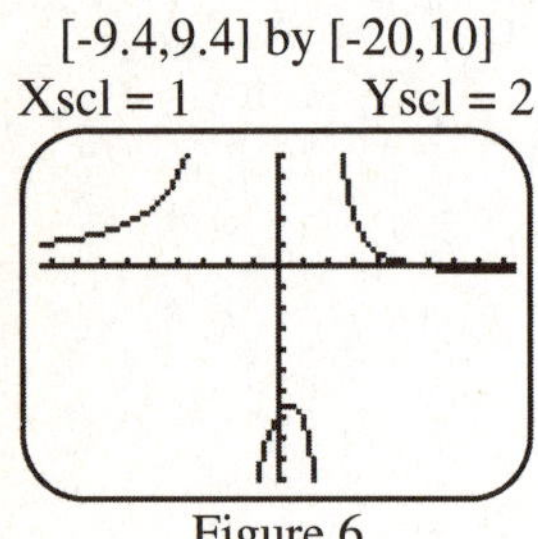

Figure 6

7. (a) $W(30)=\frac{1}{40-30}=\frac{1}{10}$; $W(39)=\frac{1}{40-39}=\frac{1}{1}=1$; $W(39.9)=\frac{1}{40-39.9}=\frac{1}{0.1}=10$.

When the rate is 30 vehicles per minute, the average wait time is $\frac{1}{10}$ minute (6 seconds). The other results are interpreted similarly.

(b) The vertical asymptote occurs when the denominator equals zero, at $x=40$. As x approaches 40, W gets larger and larger without bound. See Figure 7.

(c) $5=\frac{1}{40-x}\Rightarrow 5(40-x)=1\Rightarrow 200-5x=1\Rightarrow 5x=199\Rightarrow x=\frac{199}{5}=39.8$

8. Here $p=\frac{k\sqrt[3]{w}}{h}$. By substitution, $100=\frac{k\sqrt[3]{48,820}}{78.7}\Rightarrow 7870=\sqrt[3]{48,820}k\Rightarrow k=\frac{7870}{\sqrt[3]{48,820}}\Rightarrow k\approx 215.3$. That is, $p=\frac{215.3\sqrt[3]{w}}{h}$. When $w=5,430$ and $h=88.9$, $p=\frac{215.3\sqrt[3]{54,430}}{88.9}\approx 92$. The individual is undernourished.

[0,40] by [-0.5,1]
Xscl = 5 Yscl = 1

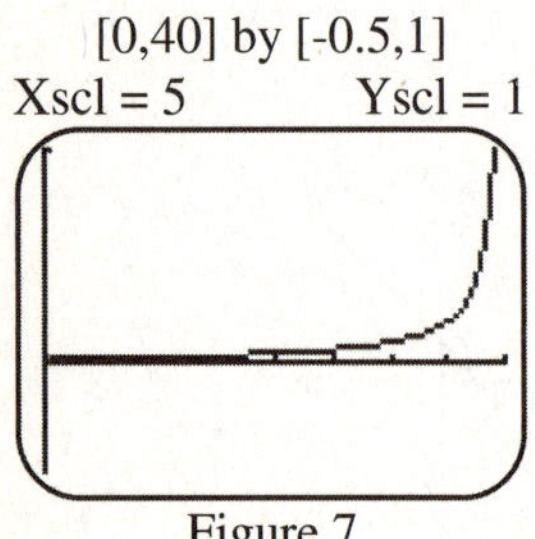

Figure 7

[-5,42] by [-1000,10,000]
Xscl = 5 Yscl = 1000

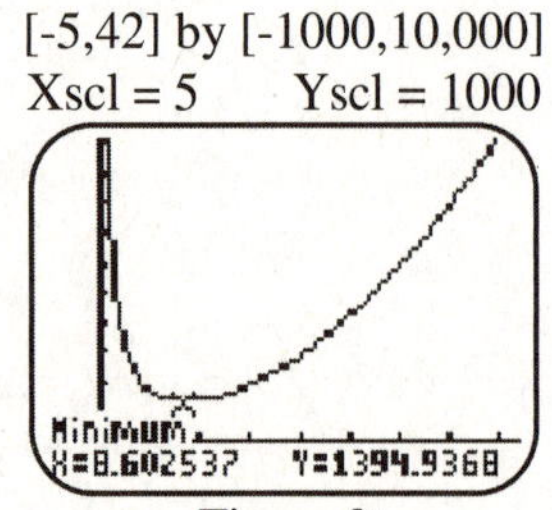

Figure 9

[-10,10] by [-10,10]
Xscl= 1 Yscl = 1

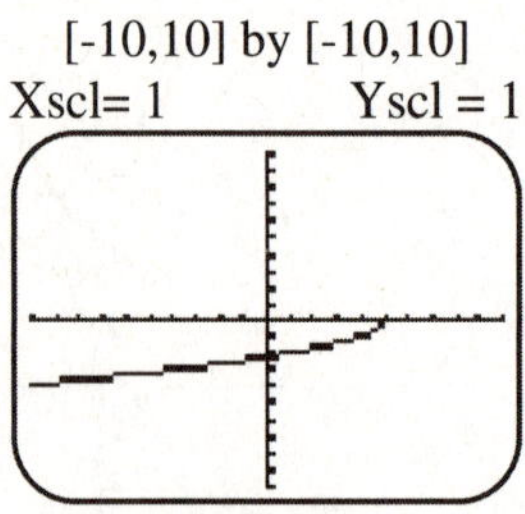

Figure 10

9. Graph $y_1 = (8000 + 2\pi x^3)/x$ as shown in Figure 9. When the radius is approximately 8.6 cm, the amount of aluminum needed will be a minimum of approximately 1394.9 cm^2.

10. The graph of $y_1 = -\sqrt{5-x}$ is shown in Figure 10.

    (a) $5 - x \ge 0 \Rightarrow 5 \ge x \Rightarrow x \le 5$; The domain is $(-\infty, 5]$.

    (b) From the graph, the range is $(-\infty, 0]$.

    (c) This function *increases* over its entire domain.

    (d) The graph intersects the x-axis when $x = 5$. The solution set is $\{5\}$.

    (e) The graph below the x-axis for all values in the domain except 5. The solution set is $(-\infty, 5)$.

11. (a) $\sqrt{4-x} = x + 2 \Rightarrow \left(\sqrt{4-x}\right)^2 = (x+2)^2 \Rightarrow 4 - x = x^2 + 4x + 4 \Rightarrow x^2 + 5x = 0 \Rightarrow x(x+5) = 0 \Rightarrow x = 0, \ -5.$

    The solution $x = -5$ is extraneous. The solution set is $\{0\}$. The graph of $y_1 = \sqrt{4-x}$ and $y_2 = x + 2$ shown in Figure 11 supports this result.

    (b) The graph of y_1 is above the graph of y_2 on the interval $(-\infty, 0)$. The solution set is $(-\infty, 0)$.

    (c) The graph of y_1 is below or intersects the graph of y_2 on the interval $[0, 4]$. The solution set is $[0, 4]$.

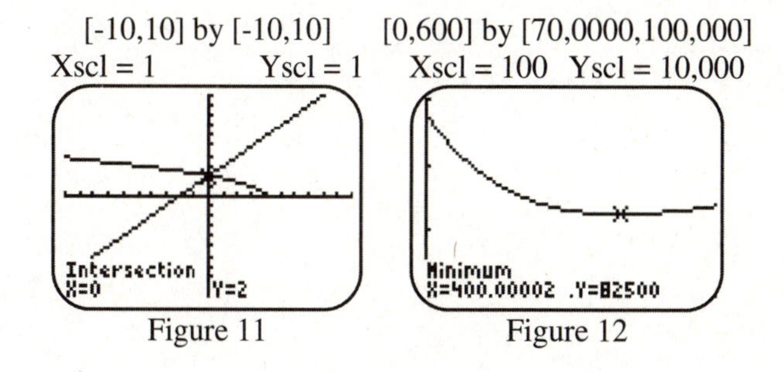

Figure 11 Figure 12

12. Let S be a point on land between Q and R such that the distance from Q to S is x yards. Then the distance from S to R is given by $600 - x$. By the Pythagorean theorem, the distance from P to S is $\sqrt{x^2 + 300^2}$. The total cost of the cable is given by $C = 125\sqrt{x^2 + 300^2} + 100(600 - x)$. The graph of this function is shown in Figure 12. The minimum cost is achieved when the cable is laid underwater from P to S, which is located 400 yards away from Q in the direction of R.

Chapter 5: Inverse, Exponential, and Logarithmic Functions

5.1: Inverse Functions

1. Different x-values always produce different y-values; therefore, yes, it is one-to-one.
3. Choosing 2 and -2 as values for x yields $f(2)=2^2=4$ and $f(-2)=(-2)^2=4$. Since different values of x produce the same value for $f(x)$, the function is not one-to-one.
5. Choosing 6 and -6 as values for x yields $f(6)=\sqrt{36-6^2}=0$ and $f(-6)=\sqrt{36-(6)^2}=0$ .Since different values of x produce the same value for $f(x)$, the function is not one-to-one.
7. Every horizontal line will intersect the graph at exactly one point; therefore, yes, it is one-to-one.
9. There are horizontal lines that will intersect the graph at more than one point; therefore, it is not one-to-one.
11. Every horizontal line will intersect the graph at exactly one point; therefore, yes, it is one-to-one.
13. A certain horizontal line intersects the whole horizontal graph (more than one point); therefore, it is not one-to-one.
15. Every horizontal line will intersect the graph at exactly one point; therefore, yes, it is one-to-one.
17. Choosing 4 and 0 as values for x yields $f(4)=(4-2)^2=4$ and $f(0)=(0-2)^2=4$. Since different values of x produce the same value for $f(x)$, the function is not one-to-one.
19. Different x-values always produce different y-values; therefore, yes, it is one-to-one.
21. Different x-values always produce different y-values; therefore, yes, it is one-to-one.
23. Different x-values always produce different y-values; therefore, yes, it is one-to-one.
25. The graph fails the horizontal line test because the end behavior is either ⋃ or ⋂.
27. For a function to have an inverse, it must be one-to-one.
29. If f and g are inverses, then $(f\circ g)(x)=x$, and $(g\circ f)(x)=x$.
31. If the point (a,b) lies on the graph of f, and f has an inverse, then the point (b,a) lies on the graph f^{-1}.
33. If the function f has an inverse, then the graph of f^{-1} may be obtained by reflecting the graph of f across the line with equation $y=x$.
35. If $f(-4)=16$ and $f(4)=16$, then f does not have an inverse because it is not one-to-one.
37. $(f\circ g)(x)=3\left(\frac{x+7}{3}\right)-7=x+7-7=x$ and $(g\circ f)(x)=\frac{(3x-7)+7}{3}=\frac{3x}{3}=x$. Since $(f\circ g)(x)=x$ and $(g\circ f)(x)=x$, the function f and g are inverses.
39. $(f\circ g)(x)=\left(\sqrt[3]{x-4}\right)^3+4=x-4+4=x$ and $(g\circ f)(x)=\sqrt[3]{(x^3+4)-4}=\sqrt[3]{x^3}=x$. Since $(f\circ g)(x)=x$ and $(g\circ f)(x)=x$, the function f and g are inverses.
41. $(f\circ g)(x)=-(-\sqrt[5]{x})^5=-(-x)=x$ and $(g\circ f)(x)=-\sqrt[5]{(-x^5)}=-(-x)=x$. Since $(f\circ g)(x)=x$ and $(g\circ f)(x)=x$, the function f and g are inverses..

43. Every x-value in f corresponds to only one y-value, and every y-value corresponds to only one x-value, so f is a one-to-one function. The inverse function is found by interchanging the x- and y-values in each ordered pair; therefore, $f^{-1} = \{(4,10),(5,20),(6,30),(7,40)\}$.

45. Every x-value in f corresponds to only one y-value. However, the y-value 5 corresponds to two x-values, 1 and 3. Because some y-values corresponds to more than one x-value, f is not one-to-one.

47. Every x-value in f corresponds to only one y-value, and every y-value corresponds to only one x-value, so f is a one-to-one function. The inverse function is found by interchanging the x- and y-values in each ordered pair. Therefore, $f^{-1} = \{(0^2,0),(1^2,1),(2^2,2),(3^2,3),(4^2,4)\}$.

49. Untying your shoelaces

51. Leaving a room

53. Unwrapping a package

55. For the function $y = 3x-4$, the inverse is $x = 3y-4 \Rightarrow x+4 = 3y \Rightarrow y = \frac{x+4}{3} \Rightarrow f^{-1}(x) = \frac{x+4}{3}$. For the graphs, see Figure 55. Since all real numbers can be input for x in f and f^{-1}, and since all real numbers can be solutions for $f(x)$ and $f^{-1}(x)$, the domain and range of both f and f^{-1} are $(-\infty,\infty)$.

57. For the function $y = x^3+1$, the inverse is $x = y^3+1 \Rightarrow x-1 = y^3 \Rightarrow y = \sqrt[3]{x-1} \Rightarrow f^{-1}(x) = \sqrt[3]{x-1}$. For the graphs, see Figure 57. Since all real numbers can be input for x in f and f^{-1}, and all real numbers can be solutions for $f(x)$ and $f^{-1}(x)$, the domain and range of both f and f^{-1} are $(-\infty,\infty)$.

59. Since when $y = 4$, $x = -2$ or 2, the function is not one-to-one.

61. For the function $y = \frac{1}{x}$, the inverse is $x = \frac{1}{y} \Rightarrow y = \frac{1}{x} \Rightarrow f^{-1}(x) = \frac{1}{x}$ For the graphs, see Figure 61. Since all real numbers except 0 can be input for x in f and f^{-1}, and all real numbers except 0 can be solutions for $f(x)$, the domain and range of both f and f^{-1} are $(-\infty,0)\cup(0,\infty)$.

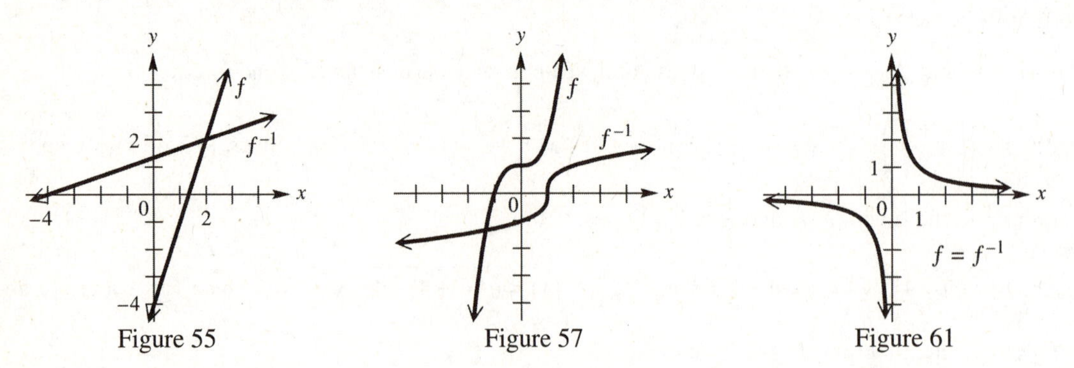

Figure 55 Figure 57 Figure 61

63. For the function $y=\frac{2}{x+3}$, the inverse is $x=\frac{2}{y+3}\Rightarrow(y+3)x=2\Rightarrow y+3=\frac{2}{x}\Rightarrow y=\frac{2}{x}-3\Rightarrow$

$y=\frac{2-3x}{x}\Rightarrow f^{-1}(x)=\frac{2-3x}{x}$. For the graphs, see Figure 63. All real numbers except -3 can be input for x in f and all real numbers except 0 can be input for x in f^{-1}. Therefore, the domain of f is equal to the range of $f^{-1}=(-\infty,-3)\cup(-3,\infty)$, and the domain of f^{-1} is equal to the range of $f=(-\infty,0)\cup(0,\infty)$.

65. For the function $f(x)=\sqrt{6+x},\ x\geq-6$, the inverse is $x=\sqrt{6+y},\ x\geq0$ (A square root can not equal a negative number.) $\Rightarrow x^2=6+y\Rightarrow y=x^2-6\Rightarrow f^{-1}(x)=x^2-6,\ x\geq0$. For the graph, see Figure 65. All real numbers $x\geq-6$ can be input for x in x and all real numbers $x\geq0$ can be input for x in f^{-1}. Also, all real numbers $x\geq0$ can be a solution for f and all real numbers $x\geq-6$ can be a solution for f^{-1}. Therefore, the domain of f is equal to the range of $f^{-1}=[-6,\infty)$, and the domain of f^{-1} is equal to the range of $f=[0,\infty)$.

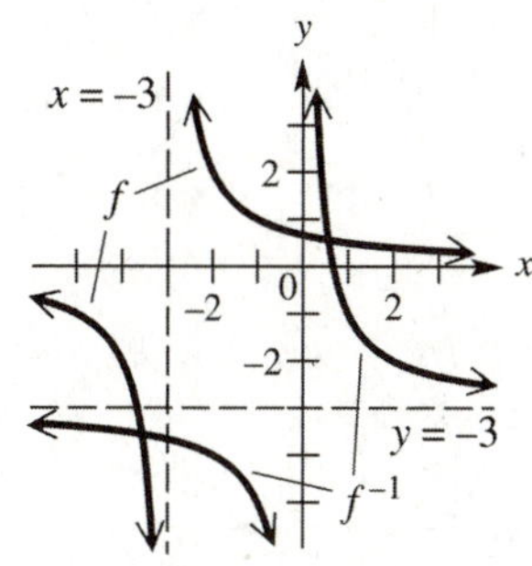

Figure 63

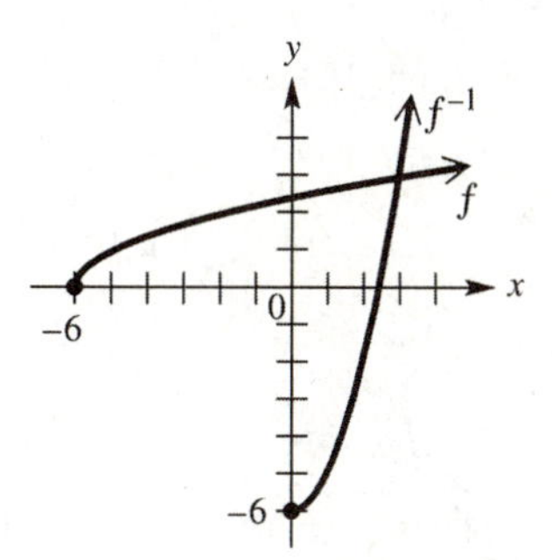

Figure 65

67. $y=\frac{4x}{x+1}\Rightarrow y(x+1)=4x\Rightarrow xy+y=4x\Rightarrow xy-4x=-y\Rightarrow x(y-4)=-y\Rightarrow x=\frac{-y}{y-4}\Rightarrow$

$f^{-1}(x)=\frac{-x}{x-4}\Rightarrow f^{-1}(x)=\frac{x}{4-x}$

69. $y=\frac{1-2x}{3x}\Rightarrow 3xy=1-2x\Rightarrow 3xy+2x=1\Rightarrow x(3y+2)=1\Rightarrow x=\frac{1}{3y+2}\Rightarrow f^{-1}(x)=\frac{1}{3x+2}$

71. $y=\sqrt{x^2-4},x\geq2\Rightarrow y^2=x^2-4\Rightarrow x^2=y^2+4\Rightarrow x=\sqrt{y^2+4}\Rightarrow f^{-1}(x)=\sqrt{x^2+4}$

73. $y=5x^3-7\Rightarrow 5x^3=y+7\Rightarrow x^3=\frac{y+7}{5}\Rightarrow x=\sqrt[3]{\frac{y+7}{5}}\Rightarrow f^{-1}(x)=\sqrt[3]{\frac{x+7}{5}}$

75. $y=\frac{3-4x}{4}\Rightarrow 4y=3-4x\Rightarrow 4x=3-4y\Rightarrow x=\frac{3-4y}{4}\Rightarrow f^{-1}(x)=\frac{3-4x}{4}$

77. $y=\frac{3-x}{2x+1}\Rightarrow y(2x+1)=3-x\Rightarrow 2xy+y=3-x\Rightarrow 2xy+x=3-y\Rightarrow x(2y+1)=3-y\Rightarrow x=\frac{3-y}{2y+1}\Rightarrow$

$f^{-1}(x)=\frac{3-x}{2x+1}$

79. $f(2)=(2)^3=8$.

81. $f(-2)=(-2)^3=-8.$

83. If $f(x)=x^3$, then the inverse is $x=y^3 \Rightarrow y=\sqrt[3]{x} \Rightarrow f^{-1}(x)=\sqrt[3]{x}$, and $f^{-1}(0)=\sqrt[3]{0}=0.$

85. For inverses, when $f^{-1}(4)$, then $f(x)=4$ Therefore, from the graph, $x=4.$

87. For inverses, when $f^{-1}(0)$, then $f(x)=0.$ Therefore, from the graph $x=2.$

89. For inverses, when $f^{-1}(-3)$, then $f(x)=-3.$ Therefore, from the graph $x=-2.$

91. The graphs are reflections of each other through $x=y$; therefore, the functions are inverses.

93. The graphs are not reflections of each other through $x=y$; therefore, the functions are not inverses.

95. Yes, for $f(x)=2x+4$ the inverse is $x=2y+4 \Rightarrow x-4=2y \Rightarrow \frac{x-4}{2}=y \Rightarrow f^{-1}(x)=\frac{1}{2}x-2$, which is equal to $g(x)$.

97. Since $(x_1, y_1)=(y_2, x_2)$, the x's and y's are switched, and the screen suggest that they are linear functions.

99. Reflect the graph across the line $y=x$. See Figure 99.

101. Reflect the graph across the line $y=x$. See Figure 101.

103. Reflect the graph across the line $y=x$. See Figure 103.

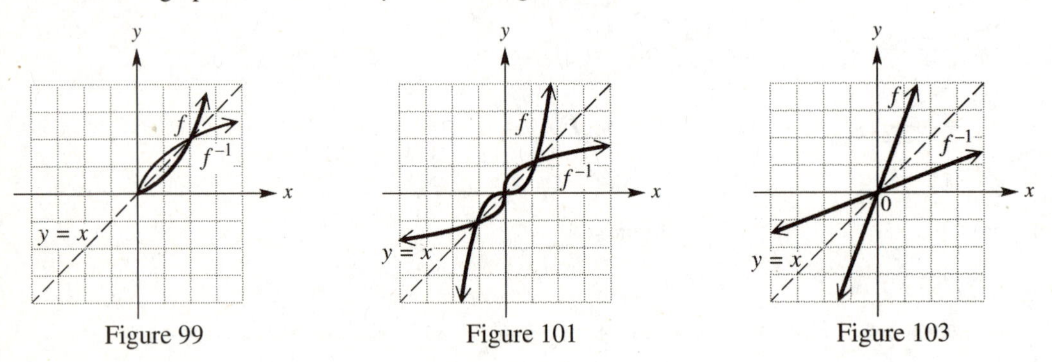

Figure 99 Figure 101 Figure 103

105. It represents the cost in dollars of building 1000 cars.

107. With $m=a=\frac{a}{1}$, the inverse function will switch the x and y terms and the graph of the inverse is a reflection across the line $y=x$ which will have the slope $m=\frac{1}{a}$.

109. From the graph of the function in the given window, the function fails the horizontal line test and is not one-to-one. See Figure 109.

111. From the graph of the function in the given window, the function passes the horizontal line test and is one-to-one. The inverse is $x=\frac{y-5}{y+3} \Rightarrow xy+3x=y-5 \Rightarrow xy-y=-3x-5 \Rightarrow y(x-1)=-3x-5 \Rightarrow y=\frac{-3x-5}{x-1} \Rightarrow$ $f^{-1}(x)=\frac{-3x-5}{x-1}$. For both graphs, see Figure 111.

[-3,2] by [-10,10]
Xscl = 1 Yscl = 1

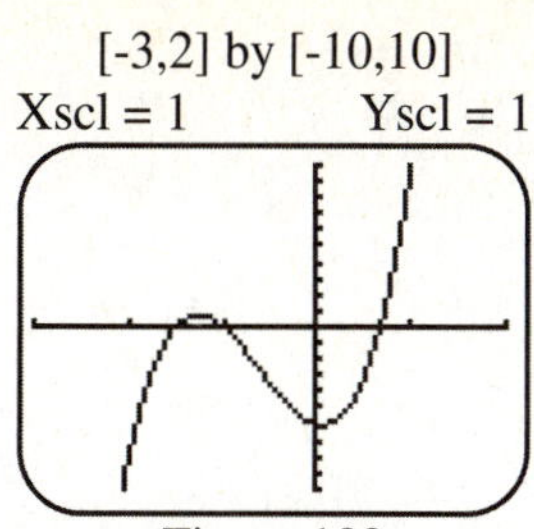

Figure 109

[-9.4,9.4] by [-6.2,6.2]
Xscl = 1 Yscl = 1

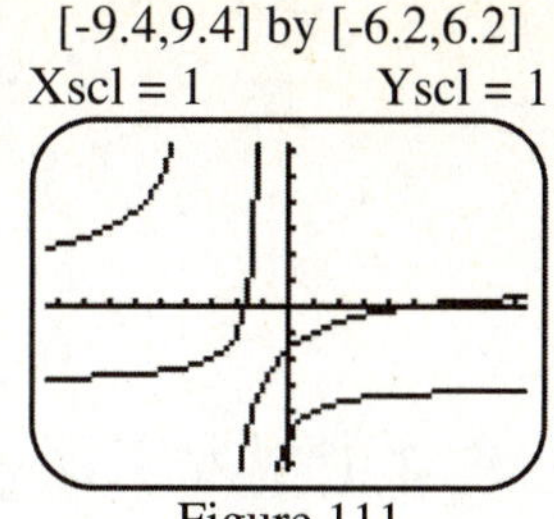

Figure 111

113. From the graph on a calculator, one possible answer that passes the horizontal line test is the domain $[0,\infty)$.

115. From the graph on a calculator, one possible answer that passes the horizontal line test is the domain $[6,\infty)$.

117. From the graph on a calculator, one possible answer that passes the horizontal line test is the domain $[0,\infty)$.

119. If $f(x)=-x^2+4,\ x\geq 0,$ then the inverse is $x=-y^2+4,\ y\geq 0 \Rightarrow x-4=-y^2 \Rightarrow$ $4-x=y^2 \Rightarrow y=\sqrt{4-x} \Rightarrow f^{-1}(x)=\sqrt{4-x}.$

121. If $f(x)=|x-6|,\ x\geq 6,$ then the inverse is $x=|y-6|,\ y\geq 6 \Rightarrow x=y-6 \Rightarrow x+6=y \Rightarrow$ $f^{-1}(x)=x+6,\ x\geq 0.$

123. If $f(x)=4x-1,$ then the inverse is $x=4y-1 \Rightarrow x+1=4y \Rightarrow y=\dfrac{x+1}{4} \Rightarrow f^{-1}(x)=\dfrac{x+1}{4}$. Inputting the numbers given yields numbers that correspond to the letters: TREASURE HUNT IS ON.

125. Using the function $f(x)=x^3+1$ on the letters NO PROBLEM or the corresponding numbers 14, 15, 16, 18, 15, 2, 12, 5, and 13, yields the numbers 2745, 3376, 4097, 5833, 3376, 9, 1729, 126, and 2198. The inverse is $x=y^3+1 \Rightarrow x-1=y^3 \Rightarrow \sqrt[3]{x-1}=y \Rightarrow f^{-1}(x)=\sqrt[3]{x-1}.$

5.2: Exponential Functions

1. See Figure 1
3. See Figure 3

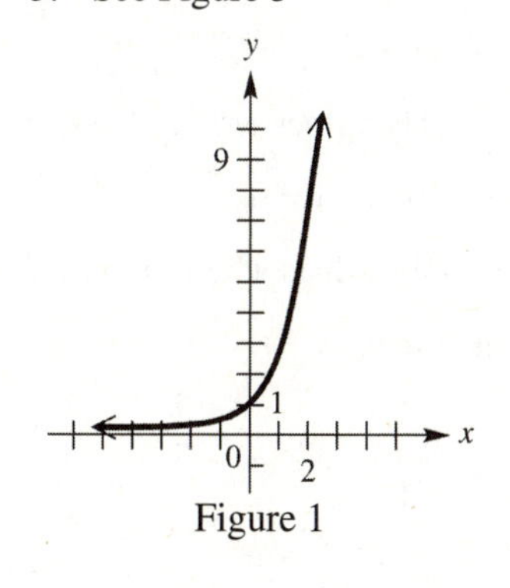

Figure 1

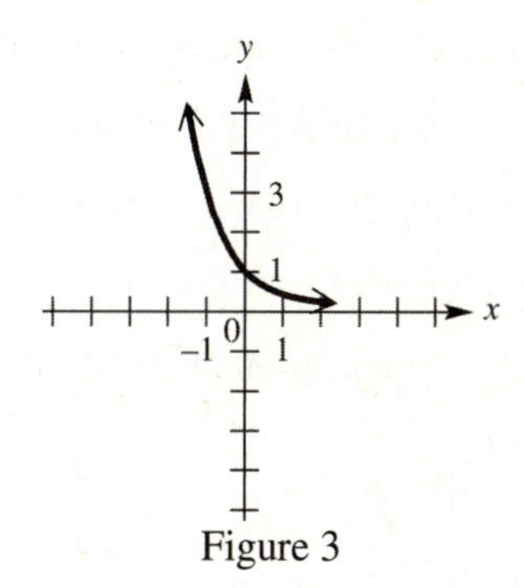

Figure 3

5. $4^x=2 \Rightarrow (2^2)^x=2^1 \Rightarrow 2^{2x}=2^1 \Rightarrow 2x=1 \Rightarrow x=\dfrac{1}{2}.$

7. $\left(\dfrac{1}{2}\right)^x=4 \Rightarrow (2^{-1})^x=2^2 \Rightarrow 2^{-x}=2^2 \Rightarrow -x=2 \Rightarrow x=-2.$

9. From the calculator, $2^{\sqrt{10}}\approx 8.952419619.$

11. From the calculator, $\left(\frac{1}{2}\right)^{\sqrt{2}} \approx 0.3752142272$.

13. From the calculator, $4.1^{-\sqrt{3}} \approx 0.0868214883$.

15. From the calculator, $\sqrt{7}^{\sqrt{7}} \approx 13.1207791$

17. From the calculator, the point $\left(\sqrt{10}, 8.9524196\right)$ lies on the graph of $y = 2^x$.

19. From the calculator, the point $(\sqrt{2}, 0.37521423)$ lies on the graph of $y = \left(\frac{1}{2}\right)^x$.

21. With $a > 0$, all inputs for x yield different solutions. The function is one-to-one and an inverse function exists.

22. See Figure 22.

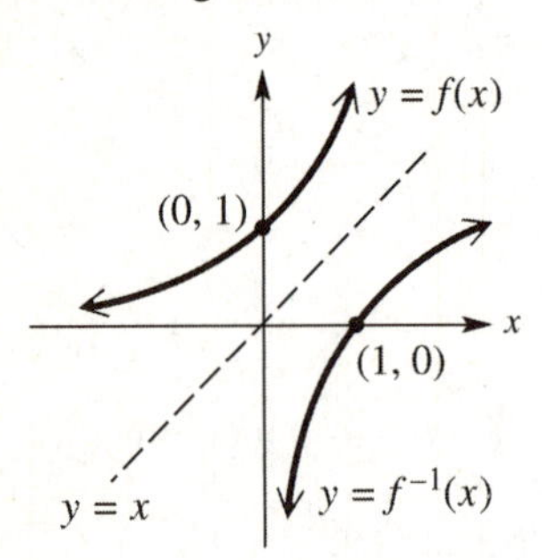

Figure 22

23. To find the inverse, switch the x and y, $x = a^y$.

24. If $a = 10$ then $x = a^y$ will be $x = 10^y$.

25. If $a = e$ then $x = a^y$ will be $x = e^y$.

26. If (p, q) is on f then switching the x and y means (q, p) will be on f^{-1}.

27. See Figure 27. All values can be input for x; therefore, the domain is $(-\infty, \infty)$. Only values where $f(x) > 0$ are possible solutions; therefore, the range is $(-\infty, 0)$. From the graph, the asymptote is $y = 0$.

29. See Figure 29. All values can be input for x; therefore, the domain is $(-\infty, \infty)$. Only values where $f(x) > 0$ are possible solutions; therefore, the range is $(0, \infty)$. From the graph, the asymptote is $y = 0$.

31. See Figure 31. All values can be input for x; therefore, the domain is $(-\infty, \infty)$. Only values where $f(x) > 0$ are possible solutions; therefore, the range is $(0, \infty)$. From the graph, the asymptote is $y = 0$.

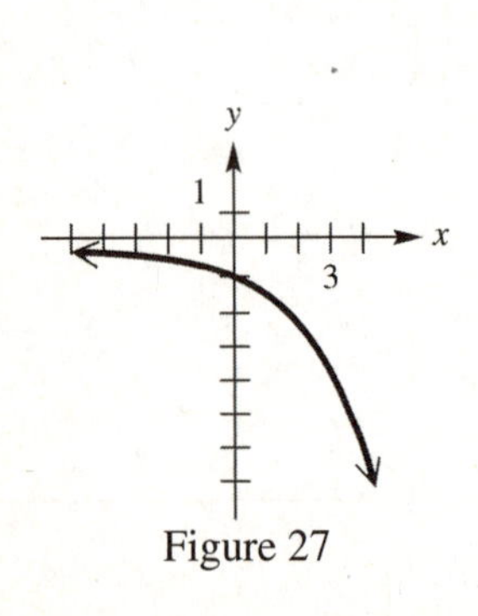

Figure 27

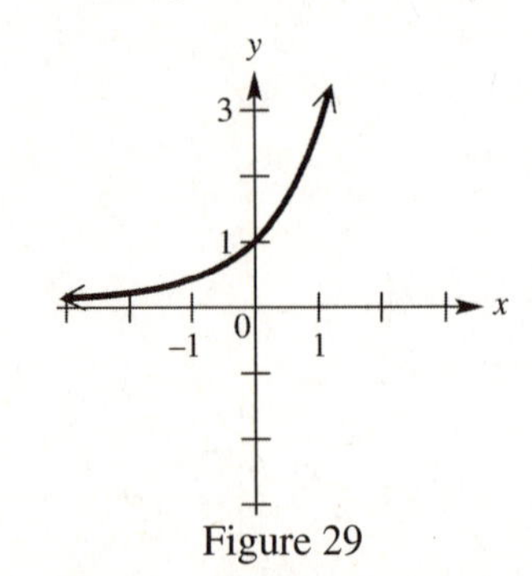

Figure 29

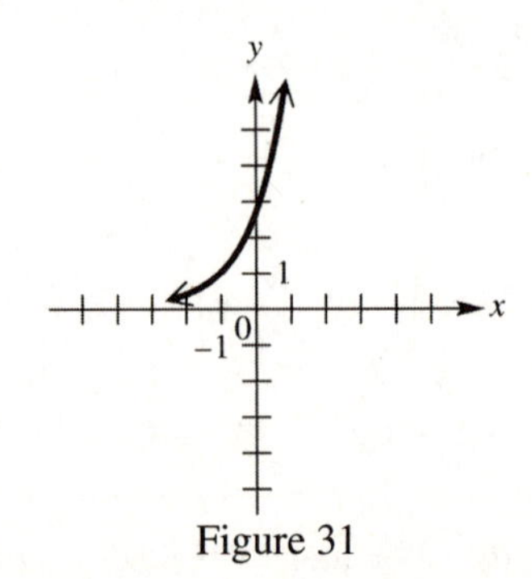

Figure 31

33. See Figure 33. All values can be input for x; therefore, the domain is $(-\infty,\infty)$. Only values where $f(x)>0$ are possible solutions; therefore, the range is $(0,\infty)$. From the graph, the asymptote is $y=0$.

35. See Figure 35. All values can be input for x; therefore, the domain is $(-\infty,\infty)$. Only values where $f(x)>0$ are possible solutions; therefore, the range is $(0,\infty)$. From the graph, the asymptote is $y=0$.

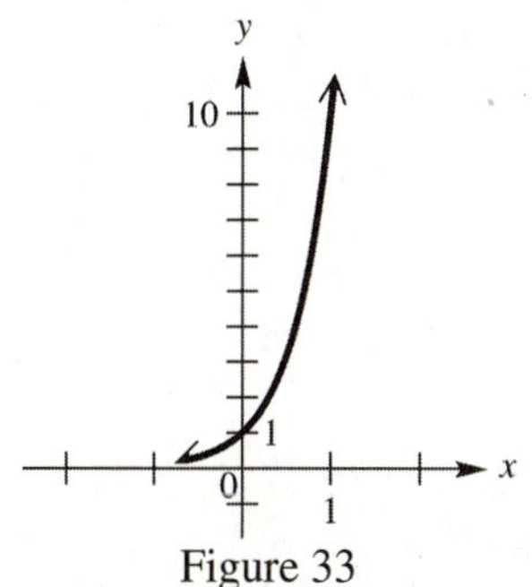

Figure 33

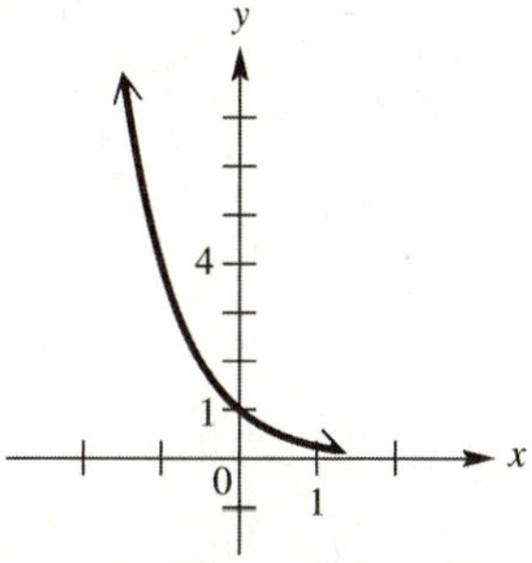

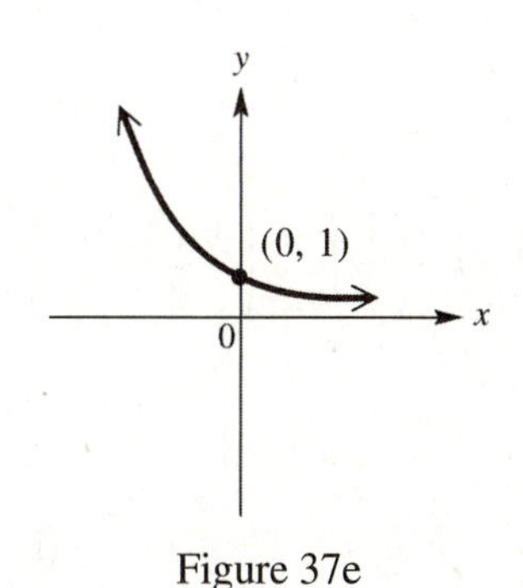

Figure 35

37. (a) Since the graph goes up to the right, $a>1$.

(b) From the graph, the domain is $(-\infty,\infty)$, the range is $(0,\infty)$, and the asymptote is $y=0$.

(c) Reflect f(x) across the x-axis. See Figure 37c.

(d) From the graph, the domain is $(-\infty,\infty)$, the range is $(-\infty,0)$, and the asymptote is $y=0$.

(e) Reflect f(x) across the y-axis. See Figure 37e.

(f) From the graph, the domain is $(-\infty,\infty)$, the range is $(0,\infty)$, and the asymptote is $y=0$.

39. Shift the original graph (not shown) 1 unit upward. See Figure 39.

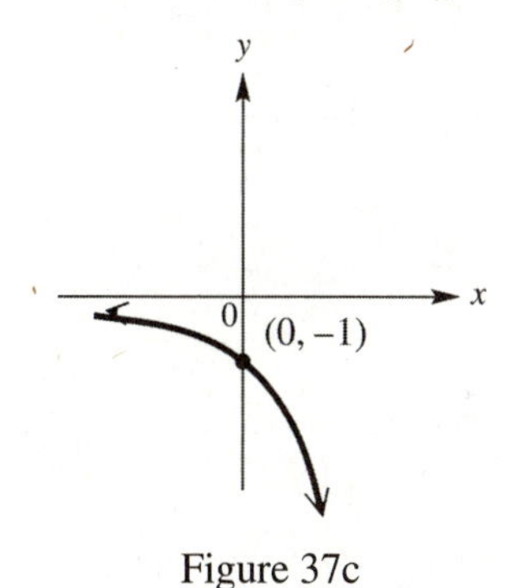

Figure 37c

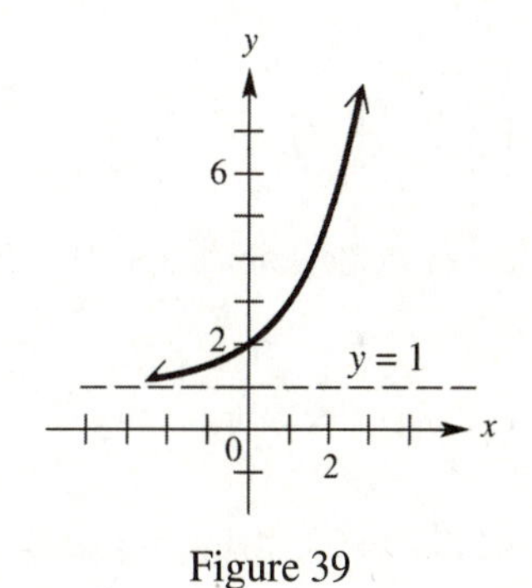

Figure 37e

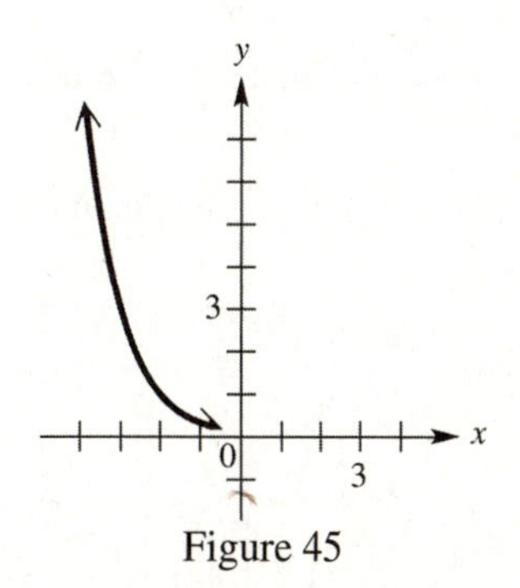

Figure 39

41. Shift the original graph (not shown) 1 unit left. See Figure 41.

43. Shift the original graph (not shown) 2 units downward. See Figure 43.

45. Shift the original graph (not shown) 2 units left. See Figure 45.

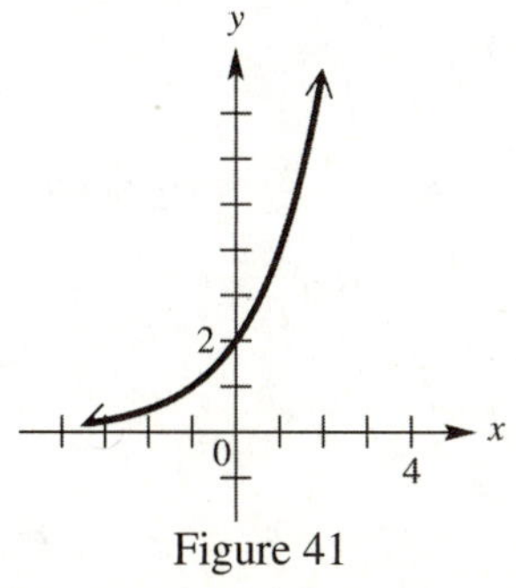

Figure 41

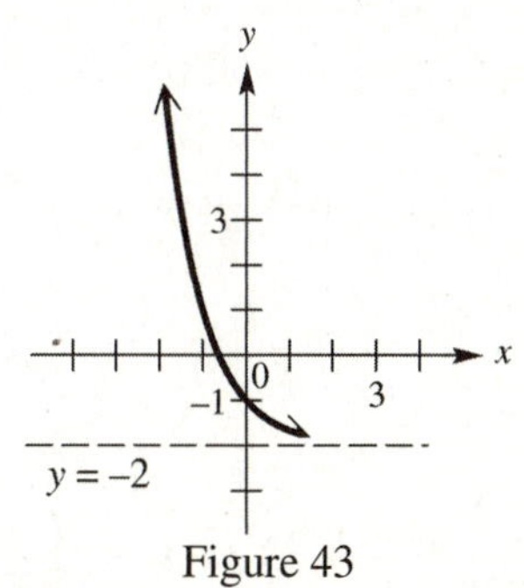

Figure 43

Figure 45

47. $2^{3-x}=8\Rightarrow 2^{3-x}=2^3\Rightarrow 3-x=3\Rightarrow -x=0\Rightarrow x=0.$

49. $12^{x-3}=1\Rightarrow 12^{x-3}=12^0\Rightarrow x-3=0\Rightarrow x=3.$

51. $e^{4x-1}=(e^2)^e\Rightarrow e^{4x-1}=e^{2x}\Rightarrow 4x-1=2x\Rightarrow 2x=1\Rightarrow x=\frac{1}{2}.$

53. $27^{4x}=9^{x+1}\Rightarrow (3^3)^{4x}\Rightarrow (3^2)^{x+1}\Rightarrow 3^{12x}=3^{2x+2}\Rightarrow 12x=2x+2\Rightarrow\ 10x=2\Rightarrow x=\frac{1}{5}.$

55. $4^{x-2}=2^{3x+3}\Rightarrow (2^2)^{x-2}=2^{3x+3}\Rightarrow 2^{2x-4}=2^{3x+3}\Rightarrow 2x-4=3x+3\Rightarrow -7=x\Rightarrow x=-7.$

57. $(\sqrt{2})^{x+4}=4^x\Rightarrow (2^{1/2})^{x+4}=(2^2)^x\Rightarrow 2^{1/2x+2}=2^{2x}\Rightarrow \frac{1}{2}x+2=2x\Rightarrow\ 2=\frac{3}{2}x\Rightarrow 4=3x\Rightarrow x=\frac{4}{3}.$

59. (a) $2^{x+1}=8\Rightarrow 2^{x+1}=2^3\Rightarrow x+1=3\Rightarrow x=2.$

(b) From a calculator graph of $y_1=2^{x+1}$ and $y_2=8,\ 2^{x+1}<8$ for the interval $(2,\infty)$.

(c) From a calculator graph of $y_1=2^{x+1}$ and $y_2=8,\ 2^{x+1}<8$ for the interval $(-\infty,2)$.

61. (a) $27^{4x}=9^{x+1}\Rightarrow (3^3)^{4x}=(3^2)^{x+1}\Rightarrow 3^{12x}=3^{2x+2}\Rightarrow 12x=2x+2\Rightarrow 10x=2\Rightarrow x=\frac{1}{5}.$

(b) From a calculator graph of $y_1=27^{4x}$ and $y_2=9^{x+1},\ 27^{4x}>9^{x+1}$ for the interval $\left(\frac{1}{5},\infty\right)$.

(c) From a calculator graph of $y_1=27^{4x}$ and $y_2=9^{x+1},\ 27^{4x}<9$ for the interval $\left(-\infty,\frac{1}{5}\right)$.

63. (a) $\left(\frac{1}{2}\right)^{-x}=\left(\frac{1}{4}\right)^{x+1}\Rightarrow (2^{-1})^{-x}=(2^{-2})^{x+1}\Rightarrow 2^x=2^{-2x-2}\Rightarrow x=-2x-2\Rightarrow 3x=-2\Rightarrow x=-\frac{2}{3}.$

(b) From a calculator graph of $=\varnothing$ and $y_2=\left(\frac{1}{4}\right)^{x+1},\ \left(\frac{1}{2}\right)^{-x}\geq\left(\frac{1}{4}\right)^{x+1}$ for the interval $\left[-\frac{2}{3},\infty\right)$.

(c) From a calculator graph of $y_1=\left(\frac{1}{2}\right)^{-x}$ and $y_2=\left(\frac{1}{4}\right)^{x+1},\ \left(\frac{1}{2}\right)^{-x}\leq\left(\frac{1}{4}\right)^{x+1}$ for the interval $\left(-\infty,-\frac{2}{3}\right]$.

65. (a) If $f(x)=27,$ then $27=a^3\Rightarrow \sqrt[3]{27}=a\Rightarrow a=3.$ Therefore, $f(1)=3^1=3.$

(b) If $f(x)=27,$ then $27=a^3\Rightarrow \sqrt[3]{27}=a\Rightarrow a=3.$ Therefore, $f(-1)=3^{-1}=\frac{1}{3}.$

(c) If $f(x)=27,$ then $27=a^3\Rightarrow \sqrt[3]{27}=a\Rightarrow a=3.$ Therefore, $f(2)=3^2=9.$

(d) If $f(x)=27,$ then $27=a^3\Rightarrow \sqrt[3]{27}=a\Rightarrow a=3.$ Therefore, $f(0)=3^0=1.$

67. If point (–3, 64) in on the graph of $f(x)=a^x,$ then $64=\frac{1}{a^3}\Rightarrow \frac{1}{64}=a^3\Rightarrow \sqrt[3]{\frac{1}{64}}\Rightarrow a=\frac{1}{4}.$ The equation is $f(x)=\left(\frac{1}{4}\right)^x.$

69. $f(t)=\left(\frac{1}{3}\right)^{1-2t}=\left(\frac{1}{3}\right)^1\cdot\left(\frac{1}{3}\right)^{-2t}=\left(\frac{1}{3}\right)\left(\left(\frac{1}{3}\right)^{-2}\right)^t=\left(\frac{1}{3}\right)(3^2)^t\Rightarrow f(t)=\left(\frac{1}{3}\right)9^t.$

71. (a) Use the formula $A = p\left(1+\frac{r}{n}\right)^{nt}$ to find the amount in the account.

$$A = 20000\left(1+\frac{0.03}{1}\right)^{(1)(4)} = 20000(1.03)^4 \Rightarrow A = \$22,510.18.$$

(b) Use the formula $A = p\left(1+\frac{r}{n}\right)^{nt}$ to find the amount in the account.

$$A = 20000\left(1+\frac{0.03}{2}\right)^{(2)(4)} = 20000(1.015)^8 \Rightarrow A = \$22,529.85.$$

73. (a) Use the formula to find the $A = p\left(1+\frac{r}{n}\right)^{nt}$ amount in the account.

$$A = 27500\left(1+\frac{0.0395}{365}\right)^{(365)(5)} = 27500(1.000108219)^{1825} \Rightarrow A = \$33,504.34.$$

(b) Use the formula $A = Pe^{rt}$ to find the amount in the account.

$$A = 27500\left(e^{(0.0395)(5)}\right) = 27500e^{0.1975} \Rightarrow A = \$33,504.71.$$

75. Plan A: $A = 40000\left(1+\frac{0.045}{4}\right)^{(4)(3)} = 40000(1.01125)^{12} \Rightarrow A = \$45,746.33.$

Plan B: $A = 40000e^{(0.044)(3)} = 40000e^{(0.132)} \Rightarrow A = \$45,644.33.$ Plan A is better by \$102.65.

77. Set $y_1 = 1000\left(1+\frac{0.05}{1}\right)^x \Rightarrow y_1 = 1000(1.05)^x$, and $y_2 = 1000\left(1+\frac{0.05}{12}\right)^{12x}$.

Graphing each on the same calculator screen shows the line y_2 slightly above line y_1. Also, using the table function for y_1, y_2 and $y_2 - y_1$ yields the following differences: (1 year) – \$1.16, (2 years) – \$2.44, (5 years) – \$7.08, (10 years) – \$18.12, (20 years) –\$59.34, (30 years) – \$145.80, and (40 years) – \$318.43.

79. (a) See Figure 79a.

(b) Because the data is graphed in a slight curve, an exponential function would fit the data better.

(c) See Figure 79c.

(d) $p(1500) = 1013e^{-0.0001341(1500)} = 828.4210207 \Rightarrow p(1500) \approx 828$ mb.

$p(11,000) = 1013e^{-0.0001341(11,000)} = 231.7296764 \Rightarrow p(11,000) \approx 232$ mb. $p(1500)$ is slightly lower than the actual, and $p(11,000)$ is slightly higher than the actual.

81. (a) $f(2) = 1 - e^{-5(2)} = 1 - 0.367879441 = 0.632120559 \Rightarrow f(2) \approx 0.63.$ There is a 63% chance that at least one car will enter the intersection during a 2-minute period.

(b) See Figure 81. As time progresses, the probability increases and begins to approach 1. That is, it is almost certain that at least one car will enter the intersection during a 60-minute period.

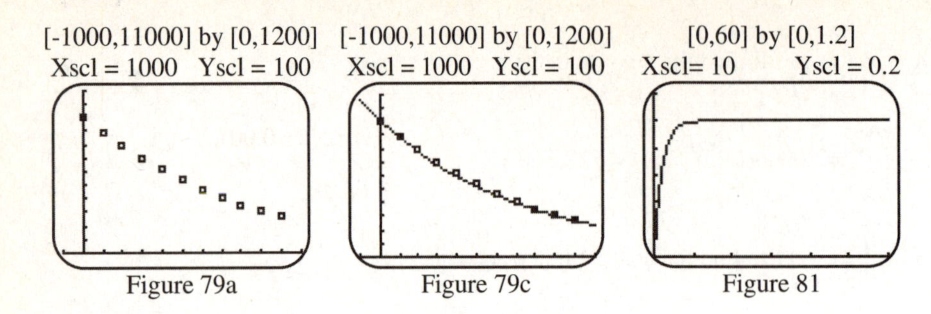

Figure 79a Figure 79c Figure 81

5.3: Logarithms and Their Properties

1. (a) By the definitions of logarithms, $2^4 = 16$ is equivalent to $\log_2 16 = 4$; it matches with C.

   (b) By the definitions of logarithms, $3^0 = 1$ is equivalent to $\log_3 1 = 0$; it matches with A.

   (c) By the definitions of logarithms, $10^{-1} = \frac{1}{10} = 0.1$ is equivalent to $\log_{10} 0.1 = -1$; it matches with E.

   (d) By the definitions of logarithms, $2^{1/2} = \sqrt{2}$ is equivalent to $\log_2 \sqrt{2} = \frac{1}{2}$; it matches with B.

   (e) By the definitions of logarithms, $e^{-2} = \frac{1}{e^2}$ is equivalent to $\log_e \frac{1}{e^2} = -2$; it matches with F.

   (f) By the definitions of logarithms, $\left(\frac{1}{2}\right)^{-3} = 8$ is equivalent to $\log_{1/2} 8 = -3$; it matches with D.

3. By the definitions of logarithms, $3^4 = 81$ is equivalent to $\log_3 81 = 4$.

5. By the definitions of logarithms, $\left(\frac{1}{2}\right)^{-4} = 16$ is equivalent to $\log_{1/2} 16 = -4$.

7. By the definitions of logarithms, $10^{-4} = 0.0001$ is equivalent to $\log 0.0001 = -4$.

9. By the definitions of logarithms and natural logarithms, $e^0 = 1$ is equivalent to $\ln 1 = 0$.

11. By the definitions of logarithms, $\log_6 36 = 2$ is equivalent to $6^2 = 36$.

13. By the definitions of logarithms, $\log_{\sqrt{3}} 81 = 8$ is equivalent to $\left(\sqrt{3}\right)^8 = 81$.

15. By the definitions of logarithms, $\log_{10} 0.001 = -3$ is equivalent to $10^{-3} = 0.001$.

17. By the definitions of logarithms, $\log \sqrt{10} = 0.5$ is equivalent to $10^{0.5} = \sqrt{10}$.

19. By the definitions of logarithms, $\log_5 125 = x \Rightarrow 5^x = 125$. Since we know $5^3 = 125$, it follows that $x = 3$.

21. By the definitions of logarithms, $\log_x 3^{12} = 24 \Rightarrow x^{24} = 3^{12} \Rightarrow \left(x^2\right)^{12} = 3^{12}$. Since $x^2 = 3$, it follows that $x = \sqrt{3}$.

23. By the definitions of logarithms, $\log_6 x = -3 \Rightarrow x = 6^{-3} \Rightarrow x = \frac{1}{6^3} \Rightarrow x = \frac{1}{216}$.

25. By the definitions of logarithms, $\log_x 16 = \frac{4}{3} \Rightarrow x^{4/3} = 16 \Rightarrow \left(x^{4/3}\right)^{3/4} = 16^{3/4} \Rightarrow x = \sqrt[4]{16^3} \Rightarrow x = 8.$

27. By the definitions of logarithms, $\log_x 0.001 = -3 \Rightarrow x^{-3} = 0.001$. Since we know $10^{-3} = 0.001$, it follows that $x = 10.$

29. By the definitions of logarithms, $\log_9 \frac{\sqrt[4]{27}}{3} = x \Rightarrow 9^x = \frac{\sqrt[4]{27}}{3} \Rightarrow \left(3^2\right)^x = \frac{3^{3/4}}{3} \Rightarrow 3^{2x} = 3^{-1/4}$. Since $3^{2x} = 3^{-1/4}$, $2x = -\frac{1}{4} \Rightarrow x = -\frac{1}{8}.$

31. (a) Using the properties of logarithms, $3^{\log_3 7} = 7.$

(b) Using the properties of logarithms, $4^{\log_4 9} = 9.$

(c) Using the properties of logarithms, $12^{\log_{12} 4} = 4.$

(d) Using the properties of logarithms, $a^{\log_a k} (k > 0, a > 0, a \neq 1) = k.$

33. (a) Using the properties of logarithms, $\log_3 1 = 0.$

(b) Using the properties of logarithms, $\log_4 1 = 0.$

(c) Using the properties of logarithms, $\log_{12} 1 = 0.$

(d) Using the properties of logarithms, $\log_a 1 \, (a > 0, a \neq 1) = 0.$

35. By properties of logarithms, $\log 10^{1.5} = \log_{10} 10^{1.5} = 1.5.$

37. By properties of logarithms, $\log 10^{2/5} = \log_{10} 10^{2/5} = \frac{2}{5}.$

39. By properties of logarithms, $\ln e^{2/3} = \frac{2}{3}.$

41. By properties of logarithms, $\ln e^{\pi} = \pi.$

43. By properties of logarithms, $3 \ln e^{1.8} = 3(1.8) = 5.4.$

45. From the calculator, $\log 43 \approx 1.633468456.$

47. From the calculator, $\log 0.783 \approx -1.062382379.$

49. From the calculator, $\log 28^3 = 3 \log 28 = 3(1.447158031) \approx 4.341474094.$

51. From the calculator, $\ln 43 \approx 3.761200116.$

53. From the calculator, $\ln 0.783 \approx -0.244622583.$

55. From the calculator, $\ln 28^3 = 3 \ln 28 = 3(3.33220451) \approx 9.996613531.$

57. Since $\text{pH} = -\log[H_3O^+]$ and grapefruit has $H_3O^+ = 6.3 \times 10^{-4}$, $\text{pH} = -\log(6.3 \times 10^{-4}) \approx 3.20066 \approx 3.2.$

59. Since $\text{pH} = -\log[H_3O^+]$ and crackers have $H_3O^+ = 3.9 \times 10^{-9}$, $\text{pH} = -\log(3.9 \times 10^{-9}) \approx 8.408935 \approx 8.4.$

61. Since $\text{pH} = -\log[H_3O^+]$ and soda pop has a $\text{pH} = 2.7$,

$$2.7 = -\log[H_3O^+] \Rightarrow -2.7 = \log[H_3O^+] \Rightarrow H_3O^+ = 10^{-2.7} \Rightarrow H_3O^+ \approx 0.001995 \approx 2\times10^{-3}.$$

63. Since $\text{pH} = -\log[H_3O^+]$ and beer has a $\text{pH} = 4.8$,

$$4.8 = -\log[H_3O^+] \Rightarrow -4.8 = \log[H_3O^+] \Rightarrow H_3O^+ = 10^{-4.8} \Rightarrow H_3O^+ \approx 0.00001585 \approx 1.6\times10^{-5}.$$

65. Since $A = Pe^{rt}$, we can find the time by solving:

$$3000 = 2500e^{0.0375t} \Rightarrow \frac{3000}{2500} = e^{0.0375t} \Rightarrow \frac{6}{5} = e^{0.0375t} \Rightarrow \ln\frac{6}{5} = 0.0375t \Rightarrow t = \frac{\ln\frac{6}{5}}{0.0375} \approx 4.9 \text{ years.}$$

67. Since $A = Pe^{et}$, we can find time by solving:

$$5000 = 2500e^{.05t} \Rightarrow \frac{5000}{2500} = e^{0.05t} \Rightarrow \frac{2}{1} = e^{0.05t} \Rightarrow \ln 2 = 0.05t \Rightarrow t = \frac{\ln 2}{0.05} \approx 13.9 \text{ years.}$$

69. $\log_3 \frac{2}{5} = \log_3 2 - \log_3 5.$

71. $\log_2 \frac{6x}{y} = \log_2 6 + \log_2 x - \log_2 y.$

73. $\log_5 \frac{5\sqrt{7}}{3m} = \log_5 5 + \log_5 7^{1/2} - (\log_5 3 + \log_5 m) = 1 + \frac{1}{2}\log_5 7 - \log_5 3 - \log_5 m.$

75. $\log_4(2x+5y)$ cannot be rewritten.

77. $\log_k \frac{pq^2}{m} = \log_k p + \log_k q^2 - \log_k m = \log_k p + 2\log_k q - \log_k m.$

79. $\log_m \sqrt{\frac{r^3}{5z^5}} = \log_m \left(\frac{r^3}{5z^5}\right)^{1/2} = \frac{1}{2}[\log_m r^3 - (\log_m 5 + \log_m z^5)] = \frac{1}{2}(3\log_m r - \log_m 5 - 5\log_m z)$ or

$\frac{3}{2}\log_m r - \frac{1}{2}\log_m 5 - \frac{5}{2}\log_m z$

81. $\log_a x + \log_a y - \log_a m = \log_a \frac{xy}{m}$

83. $2\log_m a - 3\log_m b^2 = \log_m a^2 - \log_m \left(b^2\right)^3 = \log_m \frac{a^2}{b^6}$

85. $2\log_a(z-1) + \log_a(3z+2),\ z > 0 \Rightarrow \log_a(z-1)^2 + \log_a(3z+2) = \log_a\left((z-1)^2(3z+2)\right)$

87. $-\frac{2}{3}\log_5 5m^2 + \frac{1}{2}\log_5 25m^2 \Rightarrow \log_5\left(5m^2\right)^{-2/3} + \log_5\left(25m^2\right)^{1/2} \Rightarrow \log_5\left(5^{-2/3}m^{-4/3}\right) + \log_5 5m \Rightarrow$

$\log_5\left(5^{-2/3}m^{-4/3}\right)(5m) \Rightarrow \log_5\left(5^{1/3}m^{-1/3}\right)$ or $\log_5 \sqrt[3]{\frac{5}{m}}$

89. $3\log x - 4\log y = \log x^3 - \log y^4 = \log \frac{x^3}{y^4}$

91. $\ln(a+b)+\ln a-\frac{1}{2}\ln 4=\ln(a+b)+\ln a-\ln\sqrt{4}=\ln\frac{(a+b)a}{2}$

93. $\log_5 10=\frac{\log 10}{\log 5}\approx 1.430676558$

95. $\log_{15} 5=\frac{\log 5}{\log 15}\approx 0.5943161289$

97. $\log_{100} 83=\frac{\log 83}{\log 100}\approx 0.9595390462$

99. $\log_{2.9} 7.5=\frac{\log 7.5}{\log 2.9}\approx 1.892441722$

101. To get the graph of $y=-3^x+7,$ reflect the graph of $y=3^x$ across the x-axis and shift 7 units upward.

102. See Figure 102.

[-5,5] by [-10,10]

Xscl = 1 Yscl = 1

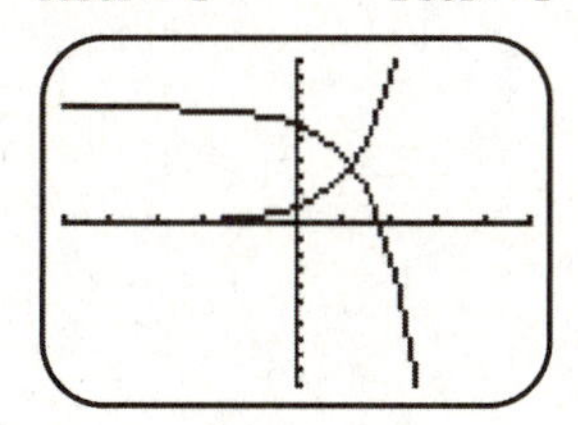

Figure 102

103. From the calculator's capabilities, the x-intercept of $y_2=-3^x+7$ is $x\approx 1.7712437492.$

104. If $0=-3^x+7$ then $3^x=7.$ Now using base 3 logarithm yields $\log_3 3^x=\log_3 7\Rightarrow x=\log_3 7.$

105. $\log_3 7=\frac{\log 7}{\log 3}=1.77124374916$

106. The approximations are close enough to support the conclusion that the x-intercept is equal to $\log_3 7.$

107. With a total of 100 individuals and 50 individuals of each species, $P_1=\frac{50}{100}=\frac{1}{2}$ and $P_2=\frac{50}{100}=\frac{1}{2}.$ Now using the index of diversity, $H=-\left(\frac{1}{2}\log_2\left(\frac{1}{2}\right)+\frac{1}{2}\log_2\left(\frac{1}{2}\right)\right)=-\log_2\left(\frac{1}{2}\right)=-(-1)\Rightarrow H=1.$

109. (a) $S(100)=0.36\ln\left(1+\frac{100}{0.36}\right)\approx 2.0269\approx 2$

(b) $S(200)=0.36\ln\left(1+\frac{200}{0.36}\right)\approx 2.2758\approx 2$

(c) $S(150)=0.36\ln\left(1+\frac{150}{0.36}\right)\approx 2.1725\approx 2$

(d) $S(10)=0.36\ln\left(1+\frac{10}{0.36}\right)\approx 1.2095\approx 1$

Reviewing Basic Concepts (Sections 5.1—5.3)

1. No, because the x-values -2 and 2 both correspond to the y-value 4. In a one-to-one function, each y-value must correspond to exactly one x-value (and each x-value to exactly one y-value).

2. (a) Interchange the x- and y-values:

x	12	21	32	45
y	7	8	9	10

(b) For $f(x)=\dfrac{x+5}{4}$ the inverse is $x=\dfrac{y+5}{4}\Rightarrow 4x=y+5\Rightarrow y=4x-5\Rightarrow f^{-1}(x)=4x-5.$

3. Graph $f(x)=2x+3$ and reflect it across the line $y=x$. See Figure 3.

4. See Figure 4.

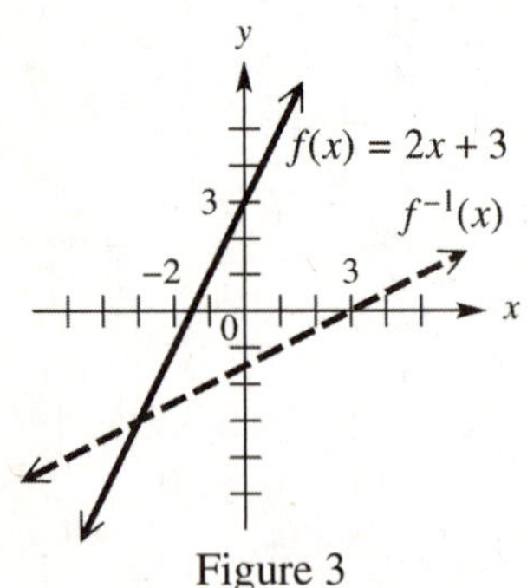

Figure 3

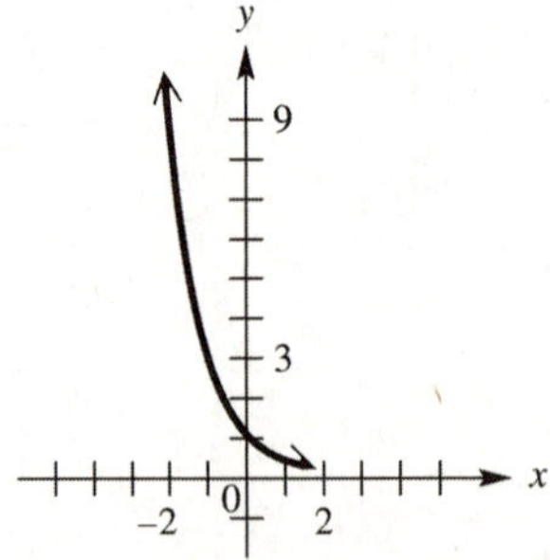

Figure 4

5. $4^{2x}=8\Rightarrow\left(2^2\right)^{2x}=2^3\Rightarrow 2^{4x}=2^3\Rightarrow 4x=3\Rightarrow x=\dfrac{3}{4}.$

6. Using $A=P\left(1+\dfrac{r}{n}\right)^{nt}$, we get $A=600\left(1+\dfrac{0.04}{4}\right)^{4(3)}=600(1.01)^{12}=676.10.$

The interest earned is $\$676.10-\$600=\$76.10.$

7. (a) $\log\left(\dfrac{1}{\sqrt{10}}\right)=\log_{10}10^{-1/2}=-\dfrac{1}{2}$

(b) $2\ln e^{1.5}=\ln e^{2(1.5)}=\log_e e^3=3$

(c) $\log_2 4=2$ because we know that $2^2=4.$

8. From the properties of logarithms, $\log\dfrac{3x^2}{5y}=\log 3+\log x^2-(\log 5+\log y)=\log 3+2\log x-\log 5-\log y.$

9. From the properties of logarithms, $\ln 4+\ln x-3\ln 2=\ln 4+\ln x-\ln 2^3=\ln\dfrac{4x}{8}=\ln\dfrac{x}{2}.$

10. Using $A=Pe^{rt}$, $1650=1500e^{0.045t}\Rightarrow\dfrac{1650}{1500}=e^{0.045t}\Rightarrow 1.1=e^{0.045t}\Rightarrow\ln 1.1=\ln e^{0.045t}\Rightarrow$

$\ln 1.1 = 0.045t \Rightarrow t = \dfrac{\ln 1.1}{0.045} \approx 2.1.$

5.4: Logarithmic Functions

1. Reflect f across the line $y = x$ and interchange the x- and y-coordinates for each point. See Figure 1. From the graph of f^{-1}, Domain: $(0,\infty)$; Range: $(-\infty,\infty)$; the graph increases on its domain $(0,\infty)$; and the vertical asymptote is $x = 0$.
3. Reflect f across the line $y = x$ and interchange the x- and y-coordinates for each point. See Figure 3. From the graph of f^{-1}, Domain: $(0,\infty)$; Range: $(-\infty,\infty)$; the graph decreases on its domain $(0,\infty)$; and the vertical asymptote is $x = 0$.
5. Reflect f across the line $y = x$ and interchange the x- and y-coordinates for each point. See Figure 5. From the graph of f^{-1}, Domain: $(1,\infty)$; Range: $(-\infty,\infty)$; the graph increases on its domain $(1,\infty)$; and the vertical asymptote is $x = 1$.

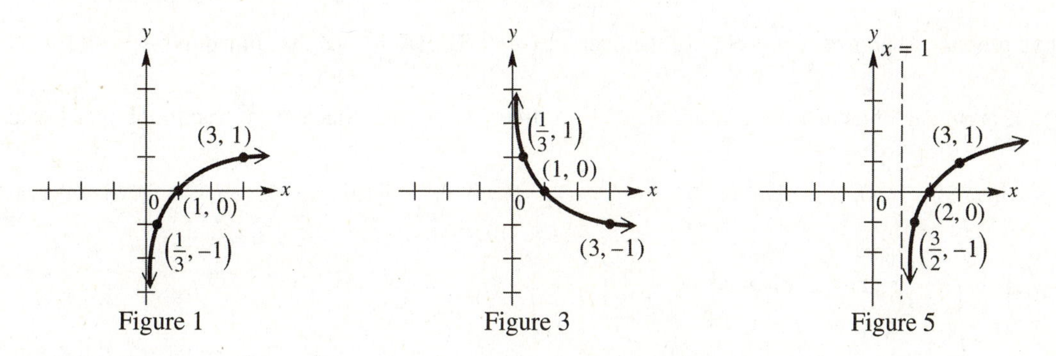

Figure 1 Figure 3 Figure 5

7. Logarithmic
9. Since the argument of a logarithm must be positive, $2x > 0 \Rightarrow x > 0$. The domain is $(0,\infty)$.
11. Since the argument of a logarithm must be positive, $(-x) > 0 \Rightarrow x < 0$. The domain is $(-\infty,0)$.
13. Since the argument of a natural logarithm must be positive, $(x^2 + 7) > 0$, this is true for all values of x, therefore the domain is $(-\infty,\infty)$.
15. Since the argument of a natural logarithm must be positive, $(-x^2 + 4) > 0$. Now set $(-x^2 + 4) = 0 \Rightarrow$ $x^2 - 4 = 0 \Rightarrow x^2 = 4 \Rightarrow x = -2,\ 2$. The endpoints –2 and 2 divide the number line into 3 intervals. The interval $(-\infty,-2)$ has a negative solution; the interval $(-2,2)$ has a positive solution; and the interval $(2,\infty)$ has a negative solution. Therefore, $(-x^2 + 4) > 0$ for the interval $(-2,2)$ and the domain is $y = x$
17. Since the argument of a natural logarithm must be positive, $\left(x^2 - 4x - 21\right) > 0$. Now set $x^2 - 4x - 21 = 0 \Rightarrow (x-7)(x+3) = 0 \Rightarrow x = -3,\ 7$. The endpoints -3 and 7 divide the number line into 3

intervals. The interval $(-\infty,-3)$ has a positive product; the interval $(-3,7)$ has a negative product; and the interval $(7,\infty)$ has a positive product. Therefore, $x^2-4x-21>0$ for the interval $(-\infty,-3)\cup(7,\infty)$ and the domain is $(-\infty,-3)\cup(7,\infty)$.

19. Since the argument of a natural logarithm must be positive, $(x^3-x)>0$. Now set $x^3-x=0\Rightarrow x(x^2-1)=0\Rightarrow x(x+1)(x-1)=0\Rightarrow x=-1,\ 0,\ 1$. The endpoints $-1, 0,$ and 1 divide the number line into 4 intervals. The interval $(-\infty,-1)$ has a negative product; the interval $(-1,0)$ has a positive product; the interval $(0,1)$ has a negative product; and the interval $(1,\infty)$ has a positive product. Therefore, $x^3-x>0$ for the interval $(-1,0)\cup(1,\infty)$ and the domain is $(-1,0)\cup(1,\infty)$.

21. Since the argument of a natural logarithm must be positive, $\frac{x+3}{x-4}>0$. Now set both $x+3=0$ and $x-4=0\Rightarrow x=-3,\ 4$. The endpoints -3 and 4 divide the number line into 3 intervals. The interval $(-\infty,-3)$ has a positive product; the interval $(-3,4)$ has a negative product; and the interval $(4,\infty)$ has a positive product. Therefore, $\frac{x+3}{x-4}>0$ for the interval $(-\infty,-3)\cup(4,\infty)$ and the domain is $(-\infty,-3)\cup(4,\infty)$.

23. Since the argument of a natural logarithm must be positive, $|3x-7|>0$. Since this is true for all real x-values except when $3x-7=0$, we solve for $3x-7=0\Rightarrow 3x=7\Rightarrow x=\frac{7}{3}$. Therefore, $|3x-7|>0$ for the interval $\left(-\infty,\frac{7}{3}\right)\cup\left(\frac{7}{3},\infty\right)$, and the domain is $\left(-\infty,\frac{7}{3}\right)\cup\left(\frac{7}{3},\infty\right)$.

25. Shift the graph of $f(x)=\log_2 x$ upward 3 units to sketch the graph of $f(x)=(\log_2 x)+3$. See Figure 25.

27. Shift the graph of $f(x)=\log_2 x$ left 3 units and reflect all negative y-values across the x-axis to sketch the graph of $f(x)=|\log_2(x+3)|$. See Figure 27.

29. Shift the graph of $f(x)=\log_{1/2} x$ right 2 units to sketch the graph of $f(x)=\log_{1/2}(x-2)$. See Figure 29.

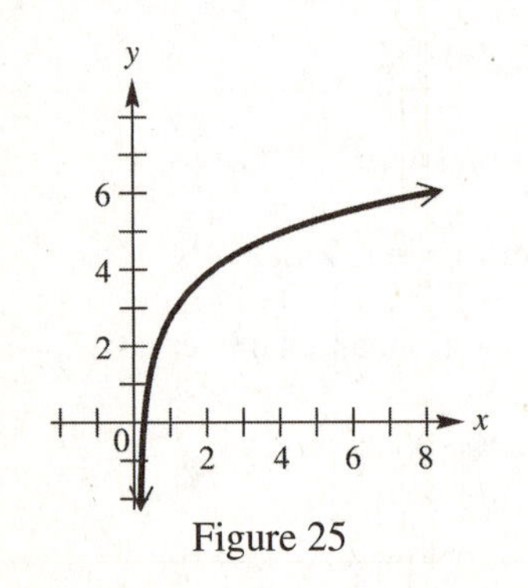

Figure 25

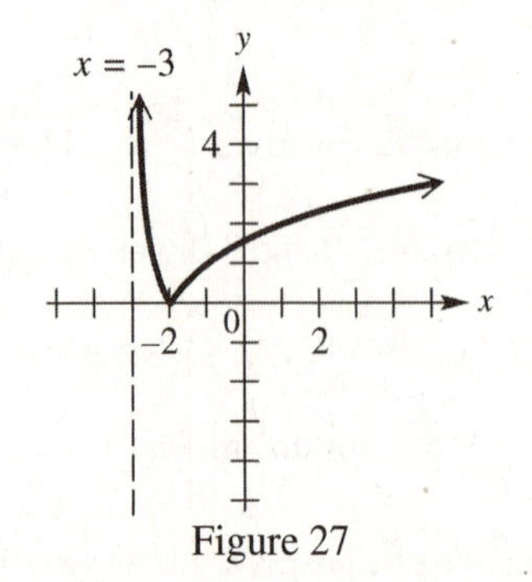

Figure 27

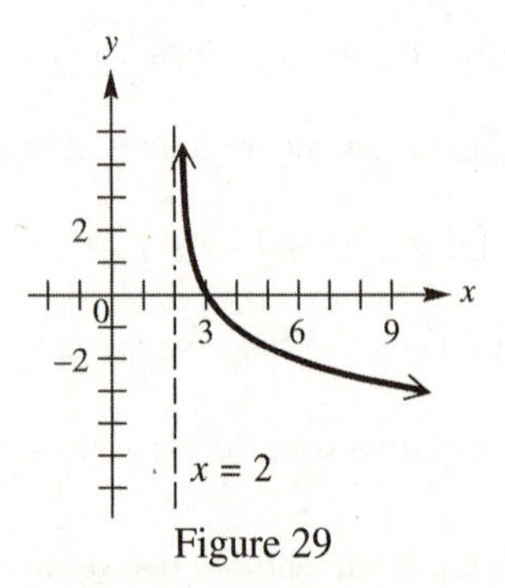

Figure 29

31. The graph of $y=e^x+3$ is the graph of $y=e^x$ shifted 3 units upward. The correct graph is B.

33. The graph of $y = e^{x+3}$ is similar to the graph of $y = e^x$ with a y-intercept of $(0, e^3)$. The correct graph is D..

35. The graph of $y = \ln x + 3$ is the graph of $y = \ln x$ passing through $(1, -3)$. The correct graph is A.

37. The graph of $y = \ln(x-3)$ is the graph of $y = \ln x$ shifted 3 units right. The correct graph is C.

39. Graph a logarithmic function with base greater than 1 that has an asymptote at $x = 0$ and passes through $(1,0)$ and $(5,1)$. See Figure 39.

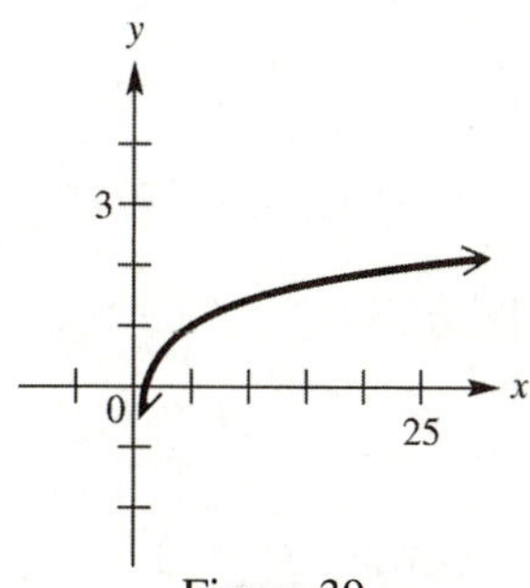

Figure 39

41. Graph a logarithmic function with base between 0 and 1 that is shifted 1 unit left then reflected across the y-axis, has an asymptote at $x = 1$, and passes through $(0,0)$ and $(-1,-1)$. See Figure 41.

43. Graph a logarithmic function with base greater than 1 that is shifted 1 unit right, has an asymptote at $x = 1$, and passes through $(2,0)$ and $(4,1)$. See Figure 43.

45. (a) The graph is shifted 4 units to the left.

    (b) See Figure 45.

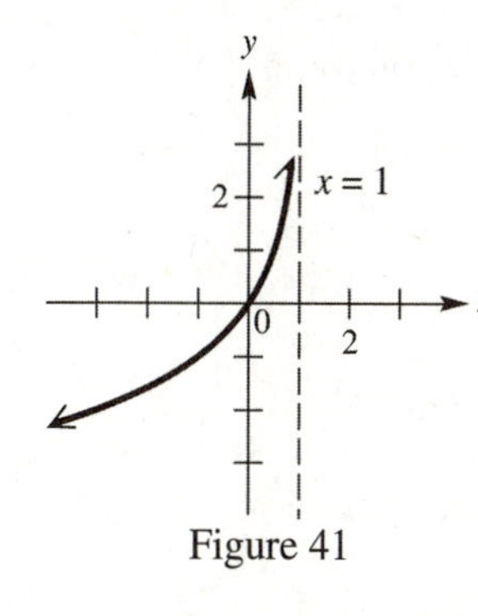

Figure 41

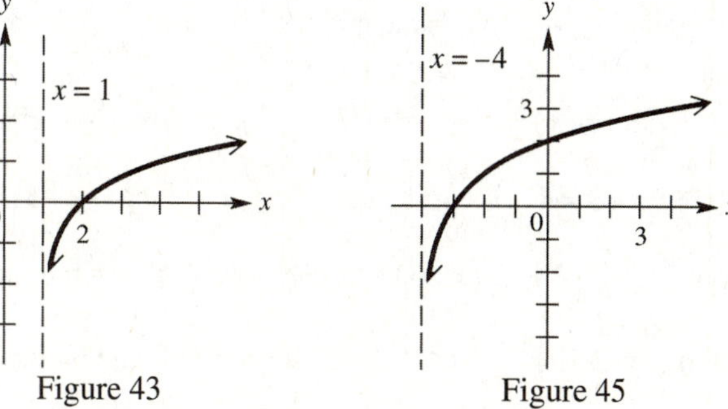

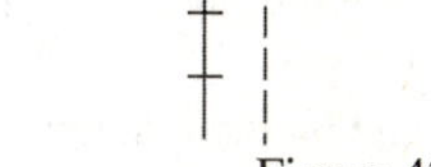

Figure 43

Figure 45

47. (a) The graph is stretched vertically by a factor of 3 and shifted 1 unit upward.

    (b) See Figure 47.

49. (a) The graph is reflected across the y-axis and shifted 1 unit upward.

    (b) See Figure 49.

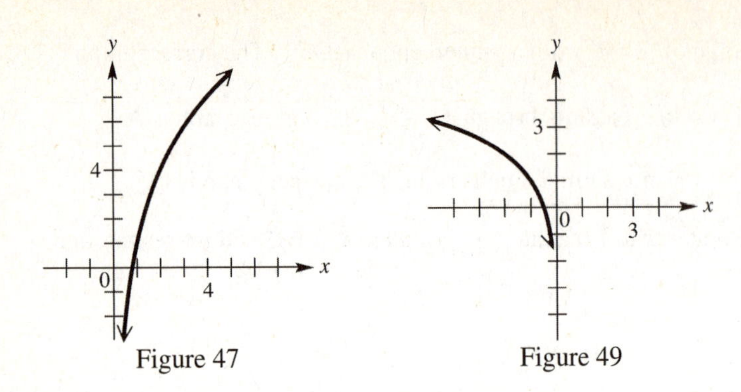

Figure 47 Figure 49

51. The graphs are not the same because the domain of $y=\log x^2$ is $(-\infty,0)\cup(0,\infty)$ while the domain of $y=2\log x$ is $(0,\infty)$. The power rule does not apply if the argument is nonpositive.

53. (a) If $\log_9 27 = x$ then $9^x = 27 \Rightarrow (3^2)^x = 27 \Rightarrow 3^{2x} = 3^3 \Rightarrow 2x = 3 \Rightarrow x = \frac{3}{2}$.

(b) By the change of base rule, $\log_9 27 = \frac{\log 27}{\log 9}$, then by calculator, $\frac{\log 27}{\log 9} \approx \frac{1.43136}{0.95424} = 1.5 = \frac{3}{2}$.

(c) The point $\left(27, \frac{3}{2}\right)$ is on the graph of $y = \log_9 27$, which supports the answer in part a.

55. (a) If $\log_{16}\frac{1}{8} = x$ then $16^x = \frac{1}{8} \Rightarrow (2^4)^x = \frac{1}{2^3} \Rightarrow 2^{4x} = 2^{-3} \Rightarrow 4x = -3 \Rightarrow x = -\frac{3}{4}$.

(b) By the change of base rule, $\log_{16}\frac{1}{8} = \frac{\log\frac{1}{8}}{\log 16}$, then by calculator, $\frac{\log\frac{1}{8}}{\log 16} \approx \frac{-0.90309}{1.20412} = -0.75 = -\frac{3}{4}$.

(c) The point $\left(\frac{1}{8}, -\frac{3}{4}\right)$ is on the graph of $y = \log_{16}\frac{1}{8}$, which supports the answer in part a.

57. For the function $f(x) = 4^x - 3$, the inverse is found by interchanging the x- and y-values.

$x = 4^y - 3 \Rightarrow x + 3 = 4^y \Rightarrow \log_4(x+3) = \log_4 4^y \Rightarrow \log_4(x+3) = y$

Therefore, $f^{-1}(x) = \log_4(x+3)$. Graph f and f^{-1} in the same window. See Figure 57.

59. For the function $f(x) = -10^x + 4$, the inverse is found by interchanging the x- and y-values.

$x = -10^y + 4 \Rightarrow x - 4 = -10^y \Rightarrow -x + 4 = 10^y \Rightarrow \log(-x+4) = \log 10^y \Rightarrow \log(-x+4) = y$

Therefore, $f^{-1}(x) = \log(4-x)$. Graph f and f^{-1} in the same window. See Figure 59.

[-4.7,4.7] by [-3.1,3.1]
Xscl = 1 Yscl = 1

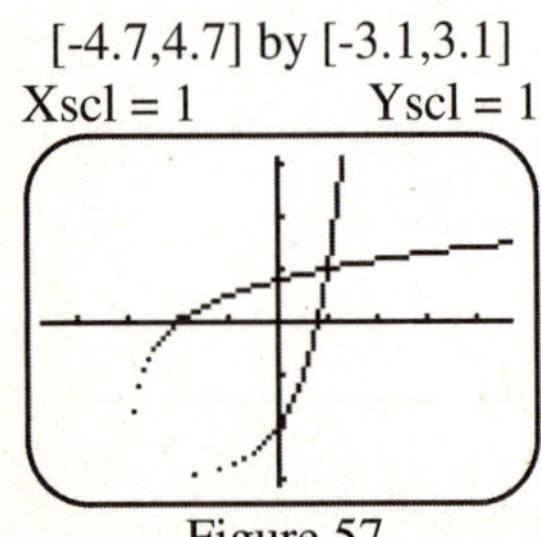

Figure 57

[-6,6] by [-4,4]
Xscl= 1 Yscl = 1

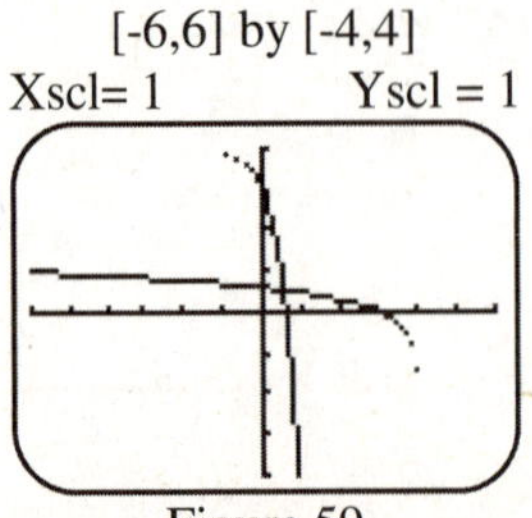

Figure 59

61. (a) First, find a. Using the given information, $2 = \log_a 3 \Rightarrow a^2 = 3 \Rightarrow a = \sqrt{3}.$

Then, $f\left(\frac{1}{9}\right) = \log_{\sqrt{3}}\left(\frac{1}{9}\right) = x \Rightarrow \left(\sqrt{3}\right)^x = \frac{1}{9} \Rightarrow 3^{1/2x} = 3^{-2} \Rightarrow \frac{1}{2}x = -2 \Rightarrow x = -4.$

(b) Since $a = \sqrt{3}$ from part a, $f(27) = \log_{\sqrt{3}} 27 = x \Rightarrow \left(\sqrt{3}\right)^x = 27 \Rightarrow 3^{\frac{1}{2}x} = 3^3 \Rightarrow \frac{1}{2}x = 3 \Rightarrow x = 6.$

(c) Since $a = \sqrt{3}$ from part a, find f^{-1} for the function $f(x) = \log_{\sqrt{3}} x$. The inverse is found by interchanging the x- and y-values: $x = \log_{\sqrt{3}} y \Rightarrow \left(\sqrt{3}\right)^x = y \Rightarrow f^{-1}(x) = \left(\sqrt{3}\right)^x.$

Now $f^{-1}(-2) = \left(\sqrt{3}\right)^{-2} = \frac{1}{3}.$

(d) Since $a = \sqrt{3}$ from part a, find f^{-1} for the function $f(x) = \log_{\sqrt{3}} x$. The inverse is found by interchanging the x- and y-values: $x = \log_{\sqrt{3}} y \Rightarrow \left(\sqrt{3}\right)^x = y \Rightarrow f^{-1}(x) = \left(\sqrt{3}\right)^x.$

Now, $f^{-1}(0) = \left(\sqrt{3}\right)^0 = 1.$

63. Graphing $y_1 = 2^{-x}$ and $y_2 = \log_{10} x$ in the same window yields the intersection point $(1.87, 0.27)$. Therefore, the solution is $x \approx 1.87$.

65. (a) The left side is a reflection of the right side with respect to the axis of the tower. The graph of $f(-x)$ is the reflection of $f(x)$ with respect to the y-axis.

(b) Since the horizontal line on the top has one-half on each side of the y-axis, the x-coordinate on the right side will be $x = \frac{1}{2}(15.7488) \Rightarrow x = 7.8744$. Using $f(7.8744) = -301\ln\left(\frac{7.8744}{207}\right) \approx 984$ feet.

(c) Graphing $y_1 = -301\ln\left(\frac{x}{207}\right)$ and $y_2 = 500$ in the same window yields the approximate intersection point $(39.31, 500)$. Therefore, the height is approximately 39 feet.

67. (a) Using $x = 9$, yields $f(9) = 27 + 1.105\log(9+1) = 27 + 1.105\log 10 = 28.105$ in.

(b) It tells us that at 99 miles from the eye of a typical hurricane, the barometric pressure is 29.21 inches.

5.5: Exponential and Logarithmic Equations and Inequalities

1. $2\left(10^x\right) = 14 \Rightarrow 10^x = 7 \Rightarrow x = \log 7$

3. $3e^{2x} + 1 = 5 \Rightarrow 3e^{2x} = 4 \Rightarrow e^{2x} = \frac{4}{3} \Rightarrow 2x = \ln\frac{4}{3} \Rightarrow x = \frac{1}{2}\ln\frac{4}{3}$

5. $\frac{1}{2}\log_2 x = \frac{3}{4} \Rightarrow \log_2 x = \frac{3}{2} \Rightarrow x = 2^{3/2}$

7. $4\ln 3x = 8 \Rightarrow \ln 3x = 2 \Rightarrow 3x = e^2 \Rightarrow x = \frac{1}{3}e^2$

9. (a) $3^x = 7 \Rightarrow \log 3^x = \log 7 \Rightarrow x\log 3 = \log 7 \Rightarrow x = \frac{\log 7}{\log 3}$ (Other forms of the answer are possible.)

(b) From the calculator, the solution is $x \approx 1.771$.

11. (a) $\left(\frac{1}{2}\right)^x = 5 \Rightarrow \log\left(\frac{1}{2}\right)^x = \log 5 \Rightarrow x\log\left(\frac{1}{2}\right) = \log 5 \Rightarrow x = \frac{\log 5}{\log(1/2)}$

(Other forms of the answer are possible.)

(b) From the calculator, the solution is $x \approx -2.322$.

13. (a) $0.8^x = 4 \Rightarrow \log 0.8^x = \log 4 \Rightarrow x\log 0.8 = \log 4 \Rightarrow x = \frac{\log 4}{\log 0.8}$ (Other forms of the answer are possible.)

(b) From the calculator, the solution is $x \approx -6.213$.

15. (a) $4^{x-1} = 3^{2x} \Rightarrow \log 4^{x-1} = \log 3^{2x} \Rightarrow (x-1)\log 4 = (2x)\log 3 \Rightarrow x\log 4 - \log 4 - 2x\log 3 \Rightarrow$

$x\log 4 - 2x\log 3 = \log 4 \Rightarrow x(\log 4 - 2\log 3) = \log 4 \Rightarrow x = \frac{\log 4}{\log 4 - 2\log 3}$

(Other forms of the answer are possible.)

(b) From the calculator, the solution is $x \approx -1.710$.

17. (a) $6^{x+1} = 4^{2x-1} \Rightarrow \log 6^{x+1} = \log 4^{2x-1} \Rightarrow (x+1)\log 6 = (2x-1)\log 4 \Rightarrow x\log 6 + \log 6 = 2x\log 4 - \log 4 \Rightarrow$

$x\log 6 - 2x\log 4 = -\log 6 - \log 4 \Rightarrow x(\log 6 - 2\log 4) = -\log 6 - \log 4 \Rightarrow x = \frac{-\log 6 - \log 4}{\log 6 - 2\log 4}$

(Other forms of the answer are possible.)

(b) From the calculator, the solution is $x \approx 3.240$.

19. (a) No real number value for x can produce a negative solution. Therefore, the solution set is $\varnothing$.

21. (a) $e^{x-3} = 2^{3x} \Rightarrow \ln e^{x-3} = \ln 2^{3x} \Rightarrow (x-3)\ln e = 3x\ln 2 \Rightarrow x - 3 = 3x\ln 2 \Rightarrow$

$x - 3x\ln 2 = 3 \Rightarrow x(1 - 3\ln 2) = 3 \Rightarrow x = \frac{3}{1 - 3\ln 2}$ (Other forms of the answer are possible.)

(b) From the calculator, the solution is $x \approx -2.779$.

23. (a) No real number value for x can produce a negative solution. Therefore, the solution set is $\varnothing$.

25. (a) $0.05(1.15)^x = 5 \Rightarrow \log 0.05 + \log(1.15)^x = \log 5 \Rightarrow x\log 1.15 = \log 5 - \log 0.05 \Rightarrow$

$x\log 1.15 = \log\frac{5}{0.05} \Rightarrow x\log 1.15 = \log 100 \Rightarrow x\log 1.15 = 2 \Rightarrow x = \frac{2}{\log 1.15}$

(Other forms of the answer are possible.)

(b) From the calculator, the solution is $x \approx 32.950$.

27. (a) $3(2)^{x-2} + 1 = 100 \Rightarrow 3(2)^{x-2} = 99 \Rightarrow \log 3 + \log 2^{x-2} = \log 99 \Rightarrow (x-2)\log 2 = \log 99 - \log 3 \Rightarrow$

$x\log 2 - 2\log 2 = \log 99 - \log 3 \Rightarrow x\log 2 = \log 99 - \log 3 + 2\log 2 \Rightarrow$

$x\log 2 = \log\left(\frac{99}{3}\right) + 2\log 2 \Rightarrow x = \frac{\log 33}{\log 2} + \frac{2\log 2}{\log 2} \Rightarrow x = 2 + \frac{\log 33}{\log 2}$

(Other forms of the answer are possible.)

(b) From the calculator, the solution is $x \approx 7.044$.

29. (a) $2(1.05)^x + 3 = 10 \Rightarrow 2(1.05)^x = 7 \Rightarrow \log 2 + \log 1.05^x = \log 7 \Rightarrow x\log 1.05 = \log 7 - \log 2 \Rightarrow$

$x\log 1.05 = \log\left(\frac{7}{2}\right) \Rightarrow x = \frac{\log 3.5}{\log 1.05}$. (Other forms of the answer are possible.)

(b) From the calculator, the solution is $x \approx 25.677$.

31. (a) $5(1.015)^{x-1980} = 8 \Rightarrow 1.015^{x-1980} = \frac{8}{5} \Rightarrow (x-1980)\log 1.015 = \log 1.6 \Rightarrow$

$x - 1980 = \frac{\log 1.6}{\log 1.015} \Rightarrow x = 1980 + \frac{\log 1.6}{\log 1.015}$ (Other forms of the answer are possible.)

(b) From the calculator, the solution is $x \approx 2011.568$.

33. $5\ln x = 10 \Rightarrow \ln x = \frac{10}{5} \Rightarrow \ln x = 2 \Rightarrow e^{\ln x} = e^2 \Rightarrow x = e^2$

35. $\ln(4x) = 1.5 \Rightarrow e^{\ln 4x} = e^{1.5} \Rightarrow 4x = e^{1.5} \Rightarrow x = \frac{e^{1.5}}{4}$

37. $\log(2-x) = 0.5 \Rightarrow 10^{\log(2-x)} = 10^{0.5} \Rightarrow 2 - x = \sqrt{10} \Rightarrow -x = \sqrt{10} - 2 \Rightarrow x = 2 - \sqrt{10}$

39. $\log_6(2x+4) = 2 \Rightarrow 6^{\log_6(2x+4)} = 6^2 \Rightarrow 2x + 4 = 36 \Rightarrow 2x = 32 \Rightarrow x = 16$

41. $\log_4(x^3+37) = 3 \Rightarrow 4^{\log 4(x^{33}-37)} = 4^3 \Rightarrow x^3 + 37 = 64 \Rightarrow x^3 = 27 \Rightarrow \sqrt[3]{x^3} = \sqrt[3]{27} \Rightarrow x = 3$

43. $\ln x + \ln x^2 = 3 \Rightarrow \ln x(x^2) = 3 \Rightarrow \ln x^3 = 3 \Rightarrow e^{\ln x^3} = e^3 \Rightarrow x^3 = e^3 \Rightarrow \sqrt[3]{x^3} = \sqrt[3]{e^3} \Rightarrow x = e$

45. $\log x + \log(x-21) = 2 \Rightarrow \log x(x-21) = 2 \Rightarrow \log(x^2 - 21x) = 2 \Rightarrow 10^{\log(x^2-21x)} = 10^2 \Rightarrow$

$x^2 - 21x = 100 \Rightarrow x^2 - 21x - 100 = 0 \Rightarrow (x-25)(x+4) = 0 \Rightarrow x = -4,\ 25$

Since the argument of a logarithm can not be negative, the solution is $x = 25$

47 $\ln(4x-2) - \ln 4 = -\ln(x-2) \Rightarrow \ln(4x-2) + \ln(x-2) = \ln 4 \Rightarrow \ln(4x-2)(x-2) = \ln 4 \Rightarrow$

$\ln(4x^2 - 10x + 4) = \ln 4 \Rightarrow e^{\ln(4x^2-10x+4)} = e^{\ln 4} \Rightarrow 4x^2 - 10x + 4 = 4 \Rightarrow 4x^2 - 10x = 0 \Rightarrow$

$(2x)(2x-5) = 0 \Rightarrow x = 0,\ \frac{5}{2}$

Since the argument of a logarithm cannot be zero, the solution is $x = 2.5$.

49. $\log_5(x+2) + \log_5(x-2) = 1 \Rightarrow \log_5(x+2)(x-2) = 1 \Rightarrow \log_5(x^2-4) = 1 \Rightarrow 5^{\log_5(x^2-4)} = 5^1 \Rightarrow$

$x^2 - 4 = 5 \Rightarrow x^2 = 9 \Rightarrow x = -3,\ 3$

Since the argument of a logarithm cannot be negative, the solution is $x = 3$.

51. $\log_7(4x) - \log_7(x+3) = \log_7 x \Rightarrow \log_7 4x = \log_7 x + \log_7(x+3) \Rightarrow \log_7 4x = \log_7 x(x+3) \Rightarrow$

$\log_7 4x = \log_7(x^2+3x) \Rightarrow 7^{\log 74x} = 7^{\log 7(x^2+3x)} \Rightarrow 4x = x^2 + 3x \Rightarrow x^2 - x = 0 \Rightarrow x(x-1) = 0 \Rightarrow x = 0,\ 1$

Since the argument of a logarithm can not be zero, the solution is $x = 1$.

53. $\ln e^x - 2\ln e = \ln e^4 \Rightarrow x - 2(1) = 4 \Rightarrow x = 6$

55. $\log x = \sqrt{\log x} \Rightarrow (\log x)^2 = (\sqrt{\log x})^2 \Rightarrow (\log x)^2 = \log x \Rightarrow (\log x)^2 - \log x = 0 \Rightarrow \log x(\log x - 1) = 0,$

then $\log x = 0 \Rightarrow x = 1$ or $\log x - 1 = 0 \Rightarrow \log x = 1 \Rightarrow x = 10$. The solution are $x = 1,\ 10$.

57. (a) Graph on a calculator $y_1 = e^{-2\ln x}$ and $y_2 = \frac{1}{16}$. The graphs intersect at $x = 4$.

(b) From the graph, $e^{-2\ln x} < \frac{1}{16}$ for the interval $(4, \infty)$.

(c) From the graph, $e^{-2\ln x} > \frac{1}{16}$ for the interval $(0, 4)$.

59. The statement is incorrect. We must reject any solution that is not in the domain of any logarithmic function in the equation.

61. If $1.5^{\log x} = e^5$ then $1.5^{\log x} - e^5 = 0$. From a calculator graph of $y_1 = 1.5^{\log x} - e^5$, the x-intercept or solution is $x \approx 19.106$.

63. $r = p - k\ln t \Rightarrow r - p = -k\ln t \Rightarrow \frac{r-p}{-k} = \ln t \Rightarrow \ln t = \frac{p-r}{k} \Rightarrow e^{\ln t} = e^{(p-r)/k} \Rightarrow t = e^{(p-r)/k}$

65. $T = T_0 + (T_1 - T_0)10^{-kt} \Rightarrow T - T_0 = (T_1 - T_0)10^{-kt} \Rightarrow \frac{T - T_0}{T_1 - T_0} = 10^{-kt} \Rightarrow$

$$\log\left(\frac{T - T_1}{T_1 - T_0}\right) = \log 10^{-kt} \Rightarrow \log\left(\frac{T - T_1}{T_1 - T_0}\right) = -kt \Rightarrow t = -\frac{1}{k}\log\left(\frac{T - T_1}{T_1 - T_0}\right)$$

67. $A = T_0 + Ce^{-kt} \Rightarrow A - T_0 = Ce^{-kt} \Rightarrow \frac{A - T_0}{C} = e^{-kt} \Rightarrow \ln\left(\frac{A - T_0}{C}\right) = \ln e^{-kt} \Rightarrow$

$$-kt = \ln\left(\frac{A - T_0}{C}\right) \Rightarrow k = \frac{\ln\left(\frac{A - T_0}{C}\right)}{-t}$$

69. $y = A + B(1 - e^{-Cx}) \Rightarrow y = A + B - Be^{-Cx} \Rightarrow y - A - B = -Be^{-Cx} \Rightarrow \frac{-y + A + B}{B} = e^{-Cx} \Rightarrow$

$$\ln\left(\frac{A + B - y}{B}\right) = \ln e^{-Cx} \Rightarrow \ln\left(\frac{A + B - y}{B}\right) = -Cx \Rightarrow x = -\frac{\ln\left(\frac{A + B - y}{B}\right)}{C}$$

71. $\log A = \log B - C\log x \Rightarrow \log A = \log B - \log x^C \Rightarrow \log A = \log\frac{B}{x^C} \Rightarrow 10^{\log A} = 10^{\log(B/x^C)} \Rightarrow A = \frac{B}{x^C}$

73. $A = P\left(1 + \frac{r}{n}\right)^{nt} \Rightarrow \frac{A}{P} = \left(1 + \frac{r}{n}\right)^{nt} \log\left(\frac{A}{P}\right) = \log\left(1 + \frac{r}{n}\right)^{nt} \Rightarrow \log\left(\frac{A}{P}\right) = nt\log\left(1 + \frac{r}{n}\right) \Rightarrow t = \frac{\log\left(\frac{A}{P}\right)}{n\log\left(1 + \frac{r}{n}\right)}$

75. $e^{2x} - 6e^x + 8 = 0 \Rightarrow (e^x - 4)(e^x - 2) = 0 \Rightarrow (e^x - 4) = 0 \text{ or } (e^x - 2) = 0 \Rightarrow x = \{\ln 4, \ln 2\}$.

77. $2e^{2x} + e^x = 6 \Rightarrow 2e^{2x} + e^x - 6 = 0 \Rightarrow (2e^x - 3)(e^x + 2) = 0 \Rightarrow (2e^x - 3) = 0 \text{ or } (e^x + 2) = 0 \Rightarrow x = \{\ln\frac{3}{2}\}$.

79. $\frac{1}{2}e^{2x}+e^x=1\Rightarrow\frac{1}{2}e^{2x}+e^x-1=0.$ Let $a=e^x\Rightarrow\frac{1}{2}a^2+a-1=0$ and use the quadratic formula to solve:

$$a=\frac{-1\pm\sqrt{1^2-(4)(\frac{1}{2})(-1)}}{(2)(\frac{1}{2})}=\frac{-1\pm\sqrt{1+2}}{1}=-1\pm\sqrt{3}\Rightarrow a=\sqrt{3}-1\Rightarrow\sqrt{3}-1=e^x\Rightarrow x=\ln\left(\sqrt{3}-1\right)$$

81. $3^{2x}+35=12\left(3^x\right)\Rightarrow 3^{2x}-12\left(3^x\right)+35=0\Rightarrow\left(3^x-7\right)\left(3^x-5\right)=0\Rightarrow\left(3^x-7\right)\ or\ \left(3^x-5\right)=0\Rightarrow$

$x=\{\log_3 7,\log_3 5\}$

83. $(\log_2 x)^2+\log_2 x=2\Rightarrow(\log_2 x)^2+\log_2 x-2=0.$ Let $a=\log_2 x\Rightarrow a^2+a-2=0\Rightarrow$

$(a+2)(a-1)=0\Rightarrow a=1,-2.\ \ 1=\log_2 x\Rightarrow x=2\ or\ -2=\log_2 x\Rightarrow x=\frac{1}{4}\Rightarrow x=\{2,\frac{1}{4}\}.$

85. $(\ln x)^2+16=10\ln x\Rightarrow(\ln x)^2-10\ln x+16=0.$ Let $a=\ln x\Rightarrow a^2-10a+16=0\Rightarrow$

$(a-8)(a-2)=0\Rightarrow a=8,2.\ \ 8=\ln x\Rightarrow x=e^8\ or\ 2=\ln x\Rightarrow x=e^2\Rightarrow x=\{e^8,e^2\}$

87. $-2e^x+5=0\Rightarrow-2e^x=-5\Rightarrow e^x=\frac{5}{2}\Rightarrow x=\ln\frac{5}{2}$. From the graph $f(x)<0$ when x is in the interval

$\left(\ln\frac{5}{2},\infty\right)$ and $f(x)\ge 0$ when x is in the interval $\left(-\infty,\ln\frac{5}{2}\right]$.

89. $2\left(3^x\right)-18=0\Rightarrow 2\left(3^x\right)=18\Rightarrow 3^x=9\Rightarrow x=2$. From the graph $f(x)<0$ when x is in the interval

$(-\infty,2)$ and $f(x)\ge 0$ when x is in the interval $[2,\infty)$.

91. $3^{2x}-9^{x+1}=0\Rightarrow 3^{2x}=9^{x+1}\Rightarrow 3^{2x}=3^{2x+2}\Rightarrow 2x=2x+2\Rightarrow$ No solution. From the graph $f(x)<0$ when x is in the interval $(-\infty,\infty)$ and $f(x)\ge 0$ for no values of x.

93. $8-4\log_5(x)=0\Rightarrow 8=4\log_5(x)\Rightarrow 2=\log_5(x)\Rightarrow x=25.$ From the graph $f(x)<0$ when x is in the interval $(25,\infty)$ and $f(x)\ge 0$ for values of x in the interval $(0,25]$.

95. $\ln(x+2)=0\Rightarrow x+2=1\Rightarrow x=-1$. From the graph $f(x)<0$ when x is in the interval $(-2,-1)$ and $f(x)\ge 0$ for values of x in the interval $[-1,\infty)$.

97. $7-5\log x=0\Rightarrow 5\log x=7\Rightarrow\log x=\frac{7}{5}\Rightarrow x=10^{7/5}$. From the graph $f(x)<0$ when x is in the interval

$(10^{7/5},\infty)$ and $f(x)\ge 0$ for values of x in the interval $(0,10^{7/5}]$.

99. For $x^2=2^x$, graph on a calculator $y_1=x^2-2^x$. From the calculator, the x-intercepts or solutions are -0.767, 2, and 4.

101. For $\log x=x^2-8x+14$, graph on a calculator $y_1=\log x-x^2+8x-14$, From the calculator the x-intercepts or solutions are 2.454 and 5.659.

103. For $e^x=\frac{1}{x+2}$, graph on a calculator $y_1=e^x-\frac{1}{x+2}$. From the calculator, the x-intercept or solution

is -0.443.

105. $\log_2 \sqrt{2x^2} - 1 = 0.5 \Rightarrow \log_2 \sqrt{2x^2} = 1.5 \Rightarrow 2^{\log_2 \sqrt{2x^2}} = 2^{3/2} \Rightarrow \sqrt{2x^2} = \sqrt{2^3} \Rightarrow$

$2x^2 = 2^3 \Rightarrow 2x^2 = 8 \Rightarrow x^2 = 4 \Rightarrow x = -2, 2.$

107. $\ln(\ln e^{-x}) = \ln 3 \Rightarrow \ln(-x) = \ln 3 \Rightarrow -x = 3 \Rightarrow x = -3$

109. If $y = 2$, then $2 = \dfrac{2 - \log(100 - x)}{0.42} \Rightarrow 2(0.42) = 2 - \log(100 - x) \Rightarrow$

$0.84 = 2 - \log(100 - x) \Rightarrow -1.16 = -\log(100 - x) \Rightarrow 1.16 = \log(100 - x) \Rightarrow$

$10^{1.16} = 10^{\log(100-x)} \Rightarrow 14.454398 \approx 100 - x \Rightarrow -85.5 \approx -x \Rightarrow x \approx 85.5\%$

111. If $f(x) = 33$, then $33 = 31.5 + 1.1\log(x+1) \Rightarrow 1.5 = 1.1\log(x+1) \Rightarrow \dfrac{1.5}{1.1} = \log(x+1) \Rightarrow$

$10^{1.5/1.1} = 10^{\log(x+1)} \Rightarrow 23.1013 \approx x + 1 \Rightarrow x \approx 22.1013 \Rightarrow x \approx 22$.

Reviewing Basic Concepts (Sections 5.4 and 5.5)

1. If $f(x) = 3^x$ and $f(x) \approx 10.98(1.14)^x$ then *f* and *g* are inverse functions, and their graphs are symmetric with respect to the line with equation $y = x$. The domain of *f* is the range of *g* and vice versa.

2. See Figure 2.

[-1,10] by [-5,5]
Xscl = 1 Yscl = 1

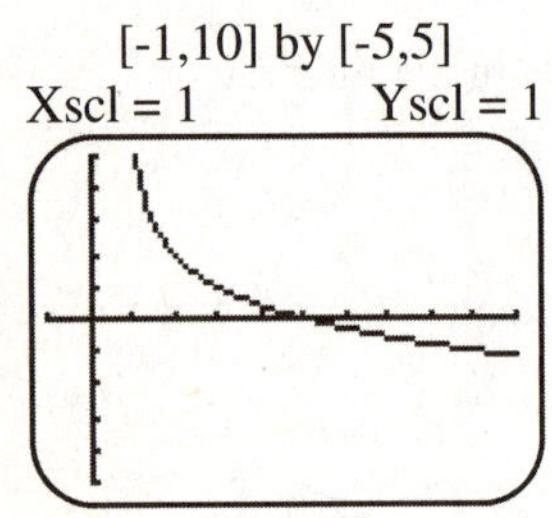

Figure 2

3. From the graph, the asymptote is $x = 5$ the *x*-intercept is 5; and there is no *y*-intercept.

4. The graph of $f(x)$ is the same as the graph of $g(x)$, reflected across the *x*-axis, shifted 1 unit to the right, and shifted two units upward.

5. For $f(x) = 2 - \log_2(x-1)$, the inverse is determined by interchanging x and y:

$x = 2 - \log_2(y-1) \Rightarrow x - 2 = -\log_2(y-1) \Rightarrow 2 - x = \log_2(y-1) \Rightarrow 2^{(2-x)} = 2^{\log(y-1)} \Rightarrow$

$2^{(2-x)} = y - 1 \Rightarrow y = 1 + 2^{(2-x)} \Rightarrow f^{-1}(x) = 1 + 2^{2-x}$

6. $3^{2x-1} = 4^x \Rightarrow \log 3^{2x-1} = \log 4^x \Rightarrow (2x-1)\log 3 = x\log 4 \Rightarrow 2x\log 3 - \log 3 = x\log 4 \Rightarrow$

$2x\log 3 - x\log 4 = \log 3 \Rightarrow x(2\log 3 - \log 4) = \log 3 \Rightarrow x = \dfrac{\log 3}{2\log 3 - \log 4} = \dfrac{\log 3}{\log \frac{9}{4}}$

7. $\ln 5x - \ln(x+2) = \ln 3 \Rightarrow \ln\left(\dfrac{5x}{x+2}\right) = \ln 3 \Rightarrow \dfrac{5x}{x+2} = 3 \Rightarrow 3(x+2) = 5x \Rightarrow 3x + 6 = 5x \Rightarrow 6 = 2x \Rightarrow x = 3$

8. $10^{5\log x} = 32 \Rightarrow \log 10^{5\log x} = \log 32 \Rightarrow 5\log x = \log 32 \Rightarrow \log x^5 = \log 32 \Rightarrow x^5 = 32 \Rightarrow x = 2$

9. $H = 1000(1 - e^{-kn}) \Rightarrow \dfrac{H}{1000} = 1 - e^{-kN} \Rightarrow 1 - \dfrac{H}{1000} = e^{-kN} \Rightarrow \ln\left(1 - \dfrac{H}{100}\right) = \ln e^{-kN} \Rightarrow$

$\ln\left(1 - \dfrac{H}{1000}\right) = -kN \Rightarrow N = -\dfrac{1}{k}\ln\left(1 - \dfrac{H}{1000}\right)$

10. If $f(x) = 2300$, then $2300 = 280\ln(x+1) + 1925 \Rightarrow 375 = 280\ln(x+1) \Rightarrow$

$e^{375/280} = e^{\ln(x+1)} \Rightarrow e^{375/280} = x + 1 \Rightarrow x = e^{375/280} - 1 \Rightarrow x \approx 2.8$ acres.

5.6: Further Applications and Modeling with Exponential and Logarithmic Functions

1. Using the given function, $\dfrac{1}{3}A_0 = A_0 e^{-0.0001216t} \Rightarrow \dfrac{1}{3} = e^{-0.0001216t} \Rightarrow \ln\dfrac{1}{3} = \ln e^{-0.0001216t} \Rightarrow$

$\ln\dfrac{1}{3} = -0.0001216t \Rightarrow t = \dfrac{\ln\left(\dfrac{1}{3}\right)}{-0.0001216} \Rightarrow t \approx 9034.6 \approx 9{,}000$ years ago.

3. Using the given function, $0.15A_0 = A_0 e^{-0.0001216t} \Rightarrow 0.15 = e^{-0.0001216t} \Rightarrow \ln 0.15 = \ln e^{-0.0001216t} \Rightarrow$

$\ln 0.15 = -0.0001216t \Rightarrow t = \dfrac{\ln(0.15)}{-0.0001216} \Rightarrow t \approx 15601.3 \approx 16{,}000$ years old.

5. (a) With a half-life of 21.7 years and using the exponential decay function, our model is

$\dfrac{1}{2}A_0 = A_0 e^{-21.7k} \Rightarrow 0.5 = e^{-21.7k} \Rightarrow \ln 0.5 = \ln e^{-21.7k} \Rightarrow \ln 0.5 = -21.7k \Rightarrow k = \dfrac{\ln 0.5}{-21.7}$

$k \approx 0.032$. The exponential decay model is $A(t) = A_0 e^{-0.032t}$.

(b) $400 = 500e^{-0.032t} \Rightarrow \dfrac{400}{500} = e^{-0.032t} \Rightarrow \ln 0.8 = \ln e^{-0.032t} \Rightarrow \ln 0.8 = -0.032t \Rightarrow \dfrac{\ln 0.8}{-0.032} = t \Rightarrow t \approx 6.97$ yr.

(c) $A(10) = 500e^{-0.032(10)} \Rightarrow A(10) = 500e^{-0.32} \Rightarrow A(10) \approx 363$ grams.

7. (a) Since the amount remaining is less than half of the original 2 milligrams, the half-life is less than 200 days.

(b) Using the exponential decay function, $A_0 = 2$, and the ordered pair (100,1.22) from the table, yields the formula: $1.22 = 2e^{-100k} \Rightarrow 0.61 = e^{-100k} \Rightarrow \ln 0.61 = \ln e^{-100k} \Rightarrow$

$\ln 0.61 = -100k \Rightarrow k = \dfrac{\ln 0.61}{100} \Rightarrow k \approx .005$. The formula that models the data is $A = 2e^{-0.005t}$.

(c) Graph $y_1 = 2e^{-0.005t}$ and $y_2 = 1$ on the calculator. The intersection of the graphs is the approximate half-life: 140 days.

9. (a) $R(x) = \log\dfrac{I}{I_o} \Rightarrow 6.7 = \log\dfrac{I}{I_o} \Rightarrow \dfrac{I}{I_0} = 10^{6.7} \Rightarrow I = 10^{6.7}I_o \approx 5{,}012{,}000I_0$

(b) $R(x) = \log\dfrac{I}{I_o} \Rightarrow 8.0 = \log\dfrac{I}{I_o} \Rightarrow \dfrac{I}{I_0} = 10^{8.0} \Rightarrow I = 10^{8.0}I_o \approx 100{,}000{,}000I_0$

(c) $\dfrac{10^8}{10^{6.7}} = 10^{1.3} \approx 20$ times more intense.

11. Use the equation to find the intensity of a star magnitude 1:

$1=6-\frac{5}{2}\log\frac{I}{I_0} \Rightarrow -5=-\frac{5}{2}\log\frac{I}{I_0} \Rightarrow \log\frac{I}{I_0} \Rightarrow 10^2=10^{\log I/I_0} \Rightarrow 100=\frac{I}{I_0} \Rightarrow I=100I_0$. Now use the equation to find the intensity of a star magnitude 3:

$3=6-\frac{5}{2}\log\frac{I}{I_0} \Rightarrow -3=-\frac{5}{2}\log\frac{I}{I_0} \Rightarrow \frac{6}{5}=\log\frac{I}{I_0} \Rightarrow 10^{6/5}=10^{\log I/I_0} \Rightarrow 10^{6/5}=\frac{I}{I_0} \Rightarrow I\approx 15.85I_0$. Comparing intensities of the magnitude 1 star to the magnitude 3 star yields $\frac{100I_0}{15.85I_0}\approx 6.31$. The magnitude 1 star is approximately 6.3 times more intense then the magnitude 3 star.

13. (a) First solve for k, using $f(x)=60, T_0=20, t=1 C=100-20=80; 60=20+80e^{-k(1)} \Rightarrow$

$40=80e^{-k} \Rightarrow 0.5=e^{-k} \Rightarrow \ln 0.5=\ln e^{-k} \Rightarrow \ln 0.5=-k \Rightarrow k=-\ln 0.5 \Rightarrow k\approx 0.693$. The equation is $f(t)=20+80x^{-0.693t}$.

(b) $f(0.5)=20+80e^{-0.693(0.5)} \Rightarrow f(0.5)\approx 76.6^\circ C$.

(c) Solve for t, with $f(t)=50, 50=20+80e^{-0.693t} \Rightarrow 30=80e^{-0.69et} \Rightarrow \frac{30}{80}=e^{-0.693t} \Rightarrow$

$\ln 0.375=\ln e^{-0.693t} \Rightarrow \ln 0.375=-0.693t \Rightarrow t=\frac{\ln 0.375}{-0.693} \Rightarrow t\approx 1.415$ or about 1 hour 25 minutes.

15. (a) Set $L_1 \to$ Year and $L_2 \to CFC\ 12(ppb)$ and use the exponential regression capabilities of the calculator to find the equation: $y\approx 0.72(1.041)^x$. Therefore the values are $C\approx 0.72$ and $a\approx 1.041$.

(b) Use the equation from part a, to find $f(13): f(13)=0.72(1.041)^{13} \Rightarrow f(13)\approx 1.21$.

17. (a) Use the Compound Interest Formula

$A=P\left(1+\frac{r}{n}\right)^{nt}$, $5000=1000\left(1+\frac{0.035}{4}\right)^{4t} \Rightarrow 5=1.00875^{4t} \Rightarrow \log 5=\log 1.00875^{4t} \Rightarrow$

$t=\frac{\log 5}{4\log 1.00785} \Rightarrow\approx 46.2$ years.

(b) Use the Continuous Compound Interest Formula

$A=Pe^{rt}$, $5000=1000e^{0.035t} \Rightarrow 5=e^{0.035t} \Rightarrow \ln 5=\ln e^{0.035t} \Rightarrow \ln 5=0.035t \Rightarrow$

$t=\frac{\ln 5}{0.035} \Rightarrow t\approx 46.0$ years.

19. Use the Compound Interest Formula

$A=P\left(1+\frac{r}{n}\right)^{nt}$, $30000=27000\left(1+\frac{0.06}{4}\right)^{4t} \Rightarrow \frac{30000}{27000}=1.015^{4t} \Rightarrow \frac{10}{9}=1.015^{4t} \Rightarrow$

$\log\frac{10}{9}=\log 1.015^{4t} \Rightarrow \log\frac{10}{9}=4t\cdot\log 1.015 \Rightarrow t=\frac{\log\frac{10}{9}}{4\log 1.015} \Rightarrow t\approx 1.8$ years.

21. Use the Compound Interest Formula

$$A = P\left(1+\frac{r}{n}\right)^{nt},\ A = 60000\left(1+\frac{0.07}{4}\right)^{4(5)} \Rightarrow A = 60000(1.0175)^{20} \Rightarrow A = \$84,086.69.$$

Now use the Continuous Compound Interest Formula $A = Pe^{rt}$, $A = 60000e^{0.0675(5)} \Rightarrow A \approx \$84,086.38.$

The better investment is the 7% rate compounded quarterly; it will earn \$800.31 more.

23. $R = \left(1+\frac{0.03}{4}\right)^{4} - 1 \Rightarrow R \approx 1.075^{4} - 1 \Rightarrow R \approx 0.030339$ or $R \approx 3.03\%$.

25. $P = 10,000\left(1+\frac{0.03}{2}\right)^{-2(5)} \Rightarrow P = 10,000(1.015)^{-10} \Rightarrow P = \$8616.67.$

27. $25000 = 30416\left(1+\frac{r}{1}\right)^{-1(5)} \Rightarrow \frac{25000}{30416} = (1+r)^{-5} \Rightarrow \left(\frac{25000}{30416}\right)^{-1/5} = 1+r \Rightarrow r \approx 0.03993$ or about 4%. .

29. (a) Using $R = \dfrac{P}{\dfrac{1-(1+i)^{-n}}{i}}$, and $n = 12(4) = 48$, $i = \frac{0.075}{12} = 0.00625$ yields

$$R = \frac{8500}{\dfrac{1-(1+0.00625)^{-48}}{0.00625}} \Rightarrow R \approx \$205.52.$$

(b) The total interest paid will be $I = nR - P \Rightarrow I = 48(205.52) - 8500 \Rightarrow I \approx \$1364.96.$

(b) The total interest paid will be $I = nR - P \Rightarrow I = 60(182.04) - 9600 \Rightarrow I \approx \$1322.69.$

31. (a) Using $R = \dfrac{P}{\dfrac{1-(1+i)^{-n}}{i}}$, and $n = 12(30) = 360$, $i = \frac{0.0725}{12}$ yields

$$R = \frac{125000}{\dfrac{1-(1+0.0725/12)^{-360}}{0.0725/12}} \Rightarrow R \approx \$852.72.$$

(b) The total interest paid will be $I = nR - P \Rightarrow I = 360(852.72) - 125000 \Rightarrow I \approx \$181,979.20.$

33. (a) First enter $y_1 = 1500\left(1+\frac{0.0575}{365}\right)^{365t}$ into the calculator, now using the table feature, the investment will triple when $y_1 = 4500$. From the table, $t \approx 19.1078 \Rightarrow t \approx 19$ years $+ 0.1078(365)$ days $\Rightarrow t \approx 19$ years, 39 days. Analytically, for $y = 4500$, the solution is

$$4500 = 1500\left(1+\frac{0.0575}{365}\right)^{365t} \Rightarrow 3 = (1.000157534)^{365t} \Rightarrow \ln 3 = 365t \ln 1.000157534$$

$$\Rightarrow t = \frac{\ln 3}{365 \ln 1.000157534} \Rightarrow t \approx 19.1078 \text{ or 19 years and 39 days.}$$

(b) First enter $y_1 = 2000\left(1+\frac{0.08}{365}\right)^{365t}$ into the calculator, now using the table feature, the investment will triple when $y_1 = 2000$. From the table, $t \approx 11.455 \Rightarrow t \approx 11$ years $+ 0.455(365)$ days

$\Rightarrow t \approx 11$ years 166 days.

35. (a) $A(5) = 2,400,000e^{0.023(5)} \Rightarrow A(5) \approx 2,692,496 \Rightarrow A(5) \approx 2,700,000.$

(b) $A(10) = 2,400,000e^{0.023(5)} \Rightarrow A(10) \approx 3,020,640 \Rightarrow A(10) \approx 3,000,000.$

(c) $A(60) = 2,400,000e^{0.023(5)} \Rightarrow A(60) \approx 9,539,764 \Rightarrow A(60) \approx 9,500,000.$

37. (a) $5.33e^{0.0133x} = 5.9 \Rightarrow e^{0.0133x} = \dfrac{5.9}{5.33} \Rightarrow 0.0133x = \ln\left(\dfrac{5.9}{5.33}\right) \Rightarrow x = \dfrac{\ln\left(\dfrac{5.9}{5.33}\right)}{0.0133} \Rightarrow x \approx 7.64$. The population will reach 5.9 million in the approximate year of 2003.

(b) See Figure 37.

(c) $5.33e^{0.0133x} = 6.25 \Rightarrow e^{0.0133x} = \dfrac{6.25}{5.33} \Rightarrow 0.0133x = \ln\left(\dfrac{6.25}{5.33}\right) \Rightarrow x = \dfrac{\ln\left(\dfrac{6.25}{5.33}\right)}{0.0133} \Rightarrow x \approx 11.97$. The population will reach 6.25 million in the approximate year of 2007.

39. (a) $P(x) = Ca^{x-2007} \Rightarrow P(2007) = Ca^{2007-2007} \Rightarrow 164 = Ca^0 \Rightarrow C = 164$, so $P(x) = 164a^{x-2007}$

$$P(2025) = 164a^{2025-2007} \Rightarrow 250 = 164a^{18} \Rightarrow \left(\frac{250}{164}\right) = a^{18} \Rightarrow a = \left(\frac{250}{164}\right)^{1/18} \approx 1.0237$$

(b) $P(2015) = 164(1.0237)^{2015-2007} \approx 197.8$ million. The predicted value is low by about 6 million.

(c) $212 = 164(1.0237)^{x-2007} \Rightarrow \dfrac{212}{164} = (1.0237)^{x-2007} \Rightarrow \ln\left(\dfrac{212}{164}\right) = \ln(1.0237)^{x-2007} \Rightarrow$

$$\ln\left(\frac{212}{164}\right) = (x-2007)\ln(1.0237) \Rightarrow \frac{\ln\left(\frac{212}{164}\right)}{\ln(1.0237)} = x - 2007 \Rightarrow$$

$$x = 2007 + \frac{\ln\left(\frac{212}{164}\right)}{\ln(1.0237)} \approx 2017.96 \Rightarrow 2018.$$

41. $N(t) = 100,000e^{rt} \Rightarrow 200,000 = 100,000e^{r(2)} \Rightarrow 2 = e^{r(2)} \Rightarrow \ln 2 = 2r \Rightarrow r = \dfrac{\ln 2}{2} \Rightarrow r \approx 0.3466.$

$350,000 = 100,000e^{0.3466t} \Rightarrow 3.5 = e^{0.3466t} \Rightarrow \ln 3.5 = 0.3466t \Rightarrow t = \dfrac{\ln 3.5}{0.3466} \Rightarrow t \approx 3.6$ hours.

43. Drug effectiveness can be modeled by $E(t) = D_0P^{t-1}$ where D_0 = original dose, P = percent of effectiveness for each hour, and t = hours since taken. Using this yields:

$E(6) = 250(0.75)^{6-1} \Rightarrow E(6) = 250(0.75)^5 \Rightarrow E(6) \approx 59.32617$ or $E(6) \approx 59$ mg remain.

45. (a)

x	0	15	30	45	60	75	90	125
g(x)	7	21	57	111	136	158	164	178

(b) Since there are 261 people in the village the equation is $g(x) = 261 - f(x)$.

(c) Graph the data on the same calculator graph as $y_1 = \dfrac{171}{1+18.6e^{-0.0747x}}$ and $y_2 = 18.3(1.024)^x$.

From the graph, y_1 is the better fit.

(d) Using $g(x) = 261 - f(x)$ and $g(x) = \dfrac{171}{1+18.6e^{-0.0747x}}$ yields $f(x) = 261 - \dfrac{171}{1+18.6e^{-0.0747x}}$.

47. (a) Set $L_{1\to}$ Year and $L_{2\to}$ Number of Midair Collisions and use the exponential regression capabilities of the calculator to find C and a. From the regression equation, C=236.16 and a = 0.878.

(b) Graph the data and the equation $f(x) = 236.16(0.878)^{x-2000}$ in the same window. See Figure 47.

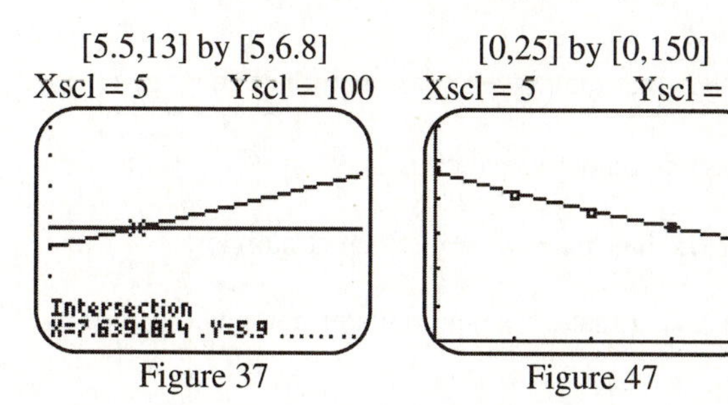

Figure 37 Figure 47

49. (a) $f(25) = \dfrac{9}{1+271e^{-0.122(25)}} \Rightarrow f(25) \approx 0.065$; and $f(65) = \dfrac{9}{1+271e^{-0.122(25)}} \Rightarrow f(65) \approx 0.82$. Among people age 25, 6.5% have some CHD, while among people age 65, 82% have some CHD.

(b) $0.50 = \dfrac{0.9}{1+271e^{-0.122x}} \Rightarrow 0.5(1+271e^{-0.122x}) = 0.9 \Rightarrow 1+271e^{-0.122x} = \dfrac{0.9}{0.5} \Rightarrow$

$$271e^{-0.122x} = \frac{0.9}{0.5} - 1 \Rightarrow e^{-0.122x} = \frac{\frac{0.9}{0.5}-1}{271} \Rightarrow \ln e^{-0.122x} = \ln\left(\frac{\frac{0.9}{0.5}-1}{271}\right) \Rightarrow$$

$$-0.122x = \ln\left(\frac{\frac{0.9}{0.5}-1}{271}\right) \Rightarrow x = \frac{\ln\left(\frac{\frac{0.9}{0.5}-1}{271}\right)}{-0.122} \Rightarrow x \approx 47.75.$$ The likelihood is 50% at about age 48.

Summary Exercises on Functions: Domains and Defining Equations

1. Domain: $(-\infty, \infty)$

3. Domain: $(-\infty, \infty)$

5. Domain: $(-\infty, \infty)$ The domain is the set of all real numbers that make the denominator not equal to zero.

7. Domain: $(-\infty, -3) \cup (-3, 3) \cup (3, \infty)$ The domain is the set of all real numbers that make the denominator not equal to zero.

9. Domain: $(-4,4)$ The domain is the set of all real numbers that make $16-x^2>0$.

11. Domain: $(-\infty,-1]\cup[8,\infty)$ The domain is the set of all real numbers that make $x^2-7x-8\geq 0$.

Interval	Test Point	Value of x^2-7x-8	Sign of x^2-7x-8
$(-\infty,-1)$	-2	10	Positive
$(-1, 8)$	0	-8	Negative
$(8, \infty)$	10	22	Positive

13. Domain: $(-\infty,\infty)$ The domain is the set of all real numbers that make the denominator not equal to zero.

15. Domain: $[1,\infty)$ The domain is the set of all real numbers that make $x^3-1\geq 0$.

17. Domain: $(-\infty,\infty)$ The domain is the set of all real numbers that make x^2+x+4 a real number.

19. Domain: $(-\infty,1)$ The domain is the set of all real numbers that make the denominator not equal to zero or

$\frac{-1}{x^3-1}\geq 0$.

Interval	Test Point	Value of $\frac{-1}{x^3-1}$	Sign of $\frac{-1}{x^3-1}$
$(-\infty, 1)$	0	1	Positive
$(1, \infty)$	2	$-\frac{1}{7}$	Negative

21. Domain: $(-\infty,\infty)$ The domain is the set of all real numbers that make $x^2+1>0$

23. Domain: $(-\infty,-2)\cup(-2,3)\cup(3,\infty)$ Since $\left(\frac{x+2}{x-3}\right)^2\geq 0$ for all real numbers, the domain of $f(x)$ is the set of all real numbers such that $\frac{x+2}{x-3}\neq 0$.

25. Domain: $(-\infty,0)\cup(0,\infty)$ The domain is the set of all real numbers such that $\frac{1}{x}$ is defined.

27. Domain: $(-\infty,\infty)$

29. Domain: $[-2,2]$ The domain is the set of all real numbers such that $16-x^4 \geq 0$

Interval	Test Point	Value of $16-x^4$	Sign of $16-x^4$
$(-\infty, -2)$	-3	-65	Negative
$(-2, 2)$	0	16	Positive
$(2, \infty)$	3	-65	Negative

31. Domain: $(-\infty,-7]\cup(-4,3)\cup[9,\infty)$ The domain is the set of real numbers such that $\frac{x^2-2x-63}{x^2+x-12} \geq 0$.

$\frac{x^2-2x-63}{x^2+x-12}$ is not defined for $x^2+x-12=0 \Rightarrow (x+4)(x-3)=0 \Rightarrow x \neq -4$ or $x \neq 3$. Solve $\frac{x^2-2x-63}{x^2+x-12}=0$

to find the test intervals: $\frac{x^2-2x-63}{x^2+x-12}=0 \Rightarrow x^2-2x-63=0 \Rightarrow (x-9)(x+7)=0 \Rightarrow x=9$ or $x=-7$.

Interval	Test Point	Value of $\frac{x^2-2x-63}{x^2+x-12}$	Sign
$(-\infty, -7)$	-10	$\frac{19}{26}$	Positive
$(-7, -4)$	-5	$-\frac{7}{2}$	Negative
$(-4, 3)$	0	$\frac{21}{4}$	Positive
$(3, 9)$	5	$-\frac{8}{3}$	Negative
$(9, \infty)$	10	$\frac{17}{98}$	Positive

33. Domain: $(-\infty,5]$ The domain is the set of real numbers such that $5-x \geq 0 \Rightarrow 5 \geq x$

35. Domain: $(-\infty,4)\cup(4,\infty)$ The domain is the set of real numbers such that

$\left|\frac{1}{4-x}\right| > 0 \Rightarrow \frac{1}{4-x} > 0 \Rightarrow 4-x > 0 \Rightarrow 4 > x$ or $-\frac{1}{4-x} < 0 \Rightarrow -4+x < 0 \Rightarrow x < 4$

37. Domain: $(-\infty,-5]\cup[5,\infty)$ The domain is the set of real numbers such that $\sqrt{x^2-25}$ is a real number. $x^2-25 \geq 0 \Rightarrow x^2 \geq 25 \Rightarrow x \geq 5$ or $x \leq -5$

39. Domain: $(-2,6)$ The domain is the set of real numbers such that $\frac{-3}{(x+2)(x-6)}>0$ and $(x+2)(x-6)\neq 0 \Rightarrow$ $x\neq -2$ or $x\neq 6$.

Interval	Test Point	Value of $\frac{-3}{(x+2)(x-6)}$	Sign of $\frac{-3}{(x+2)(x-6)}$
$(-\infty, -2)$	-3	$-\frac{1}{3}$	Negative
$(-2, 6)$	0	$\frac{1}{4}$	Positive
$(6, \infty)$	7	$-\frac{1}{3}$	Negative

41. Choice A can be written as a function of x. $3x+2y=6 \Rightarrow y=f(x)=-\frac{3}{2}x+3$

43. Choice C can be written as a function of x. $x^3+y^3=5 \Rightarrow y=f(x)=\sqrt[3]{5-x^3}$

45. Choice A can be written as a function of x. $x=\frac{2-y}{y+3} \Rightarrow y=f(x)=\frac{2-3x}{x+1}$

47. Choice D can be written as a function of x. $2x=\frac{1}{y^3} \Rightarrow y=f(x)=\sqrt[3]{\frac{1}{2x}}$

49. Choice C can be written as a function of x. $\frac{x}{4}-\frac{y}{9}=0 \Rightarrow y=f(x)=\frac{9x}{4}$

51. (a) $(f\circ g)(x)=f(g(x))=f(5x+7)=-6(5x+7)+9=-30x-42+9=-30x-33$

The domain and range of both f and g are $(-\infty,\infty)$, so the domain of $f\circ g$ is $(-\infty,\infty)$.

(b) $(g\circ f)(x)=g(f(x))=g(-6x+9)=5(-6x+9)+7=-30x+45+7=-30x+52$

The domain of $g\circ f$ is $(-\infty,\infty)$.

53. (a) $(f\circ g)(x)=f(g(x))=f(x+3)=\sqrt{x+3}$ The domain and range of g are $(-\infty,\infty)$, however, the domain and range of f are $[0,\infty)$. So, $x+3\geq 0 \Rightarrow x\geq -3$. Therefore, the domain of $f\circ g$ is $[-3,\infty)$.

(b) $(g\circ f)(x)=g(f(x))=g\left(\sqrt{x}\right)=\sqrt{x}+3$ The domain and range of g are $(-\infty,\infty)$, however, the domain and range of f are $[0,\infty)$.Therefore, the domain of $g\circ f$ is $[0,\infty)$.

55. (a) $(f\circ g)(x)=f(g(x))=f(x^2+3x-1)=(x^2+3x-1)^3$

The domain and range of f and g are $(-\infty,\infty)$, so the domain of $f\circ g$ is $(-\infty,\infty)$.

(b) $(g\circ f)(x)=g(f(x))=g(x^3)=\left(x^3\right)^2+3\left(x^3\right)-1=x^6+3x^3-1$

The domain and range of f and g are $(-\infty,\infty)$, so the domain of $g \circ f$ is $(-\infty,\infty)$.

57. (a) $(f \circ g)(x) = f(g(x)) = f(3x) = \sqrt{3x-1}$ The domain and range of g are $(-\infty,\infty)$, however, the domain and range of f are $[1,\infty)$. So, $3x-1 \geq 0 \Rightarrow x \geq \frac{1}{3}$. Therefore, the domain of $f \circ g$ is $\left[\frac{1}{3},\infty\right)$.

(b) $(g \circ f)(x) = g(f(x)) = g\left(\sqrt{x-1}\right) = 3\sqrt{x-1}$ The domain and range of g are $(-\infty,\infty)$, however, the range of f is $[0,\infty)$. So $x-1 \geq 0 \Rightarrow x \geq 1$. Therefore, the domain of $g \circ f$ is $[1,\infty)$.

59. (a) $(f \circ g)(x) = f(g(x)) = f(x+1) = \frac{2}{x+1}$ The domain and range of g are $(-\infty,\infty)$, however, the domain of f is $(-\infty,0) \cup (0,\infty)$. So, $x+1 \neq 0 \Rightarrow x \neq -1$. Therefore, the domain of $f \circ g$ is $(-\infty,-1) \cup (-1,\infty)$.

(b) $(g \circ f)(x) = g(f(x)) = g\left(\frac{2}{x}\right) = \frac{2}{x}+1$ The domain and range of f is $(-\infty,0) \cup (0,\infty)$, however, the domain and range of g are $(-\infty,\infty)$. So $x \neq 0$. Therefore, the domain of $g \circ f$ is $(-\infty,0) \cup (0,\infty)$.

61. (a) $(f \circ g)(x) = f(g(x)) = f\left(-\frac{1}{x}\right) = \sqrt{-\frac{1}{x}+2}$ The domain and range of g are $(-\infty,0) \cup (0,\infty)$, however, the domain of f is $[-2,\infty)$. So, $-\frac{1}{x}+2 \geq 0 \Rightarrow$ $x < 0$ or $x \geq \frac{1}{2}$ (using test intervals). Therefore, the domain of $f \circ g$ is $(-\infty,0) \cup \left[\frac{1}{2},\infty\right)$.

(b) $(g \circ f)(x) = g(f(x)) = g\left(\sqrt{x+2}\right) = -\frac{1}{\sqrt{x+2}}$ The domain of f is $[-2,\infty)$ and its range is $(-\infty,\infty)$. The domain and range of g are $(-\infty,0) \cup (0,\infty)$. So $x+2 > 0 \Rightarrow x > -2$. Therefore, the domain of $g \circ f$ is $(-2,\infty)$.

63. (a) $(f \circ g)(x) = f(g(x)) = f\left(\frac{1}{x+5}\right) = \sqrt{\frac{1}{x+5}}$ The domain of g is $(-\infty,-5) \cup (-5,\infty)$, and the range of g is $(-\infty,0) \cup (0,\infty)$. The domain of f is $[0,\infty)$. Therefore, the domain of $f \circ g$ is $(-5,\infty)$.

(b) $(g \circ f)(x) = g(f(x)) = g\left(\sqrt{x}\right) = \frac{1}{\sqrt{x}+5}$ The domain and range of f is $[0,\infty)$. The domain of g is $(-\infty,-5) \cup (-5,\infty)$. Therefore, the domain of $g \circ f$ is $[0,\infty)$.

65. (a) $(f \circ g)(x) = f(g(x)) = f\left(\frac{1}{x}\right) = \frac{1}{1/x-2} = \frac{x}{1-2x}$ The domain and range of g are $(-\infty,0) \cup (0,\infty)$. The domain of f is $(-\infty,-2) \cup (-2,\infty)$, and the range of f is $(-\infty,0) \cup (0,\infty)$. So, $\frac{x}{1-2x} < 0 \Rightarrow x < 0$ or $0 < x < \frac{1}{2}$ or $x > \frac{1}{2}$ (using test intervals). Thus, $x \neq 0$ and $x \neq \frac{1}{2}$. Therefore, the domain of $f \circ g$ is $(-\infty,0) \cup \left(0,\frac{1}{2}\right) \cup \left(\frac{1}{2},\infty\right)$.

(b) $(g \circ f)(x) = g(f(x)) = g\left(\frac{1}{x-2}\right) = \frac{1}{1/(x-2)} = x - 2$ The domain and range of g are $(-\infty, 0) \cup (0, \infty)$.

The domain of f is $(-\infty, -2) \cup (-2, \infty)$, and the range of f is $(-\infty, 0) \cup (0, \infty)$. Therefore, the domain of $g \circ f$ is $(-\infty, -2) \cup (-2, \infty)$.

67. (a) $(f \circ g)(x) = f(g(x)) = f\left(\sqrt{x}\right) = \log\left(\sqrt{x}\right)$ The domain of g is $(-\infty, \infty)$ and range of g is $[0, \infty)$. The domain of f is $(0, \infty)$, and the range of f is $(-\infty, \infty)$. So the domain of $f \circ g$ is $(0, \infty)$.

(b) $(g \circ f)(x) = g(f(x)) = g(\log x) = \left(\sqrt{\log x}\right)$ The domain of g is $(-\infty, \infty)$ and range of g is $[0, \infty)$. The domain of f is $(0, \infty)$, and the range of f is $(-\infty, \infty)$. So the domain of $g \circ f$ is $[1, \infty)$.

69. (a) $(f \circ g)(x) = f(g(x)) = f\left(\sqrt{x}\right) = e^{\sqrt{x}}$ The domain of g is $(-\infty, \infty)$ and range of g is $[0, \infty)$. The domain of f is $(-\infty, \infty)$, and the range of f is $(0, \infty)$. So the domain of $f \circ g$ is $[0, \infty)$.

(b) $(g \circ f)(x) = g(f(x)) = g\left(e^x\right) = \left(\sqrt{e^x}\right)$ The domain of g is $(-\infty, \infty)$ and range of g is $[0, \infty)$. The domain of f is $(-\infty, \infty)$, and the range of f is $(0, \infty)$. So the domain of $g \circ f$ is $(-\infty, \infty)$.

Chapter 5 Review Exercises

1. The function is not one-to-one, the graph fails the horizontal line test.
2. The function is not one-to-one, the graph fails the horizontal line test.
3. The function is not one-to-one, the graph fails the horizontal line test.
4. The function is one-to-one, different x-values always produce different y-values.
5. The function is not one-to-one, the graph fails the horizontal line test.
6. The function is not one-to-one, the graph fails the horizontal line test.
7. All real numbers can be input for x, therefore the domain is $(-\infty, \infty)$.
8. All real numbers can be solutions to $f(x)$, therefore the range is $(-\infty, \infty)$.
9. Since f is one-to-one, it has an inverse.
10. For $f(x) = \sqrt[3]{2x-7}$ the inverse is

$$x = \sqrt[3]{2y-7} \Rightarrow x^3 = 2y - 7 \Rightarrow x^3 + 7 = 2y \Rightarrow y = \frac{x^3+7}{2} \Rightarrow f^{-1}(x) = \frac{x^3+7}{2}.$$

11. See Figure 11. The graphs are reflections across the line $y = x$.

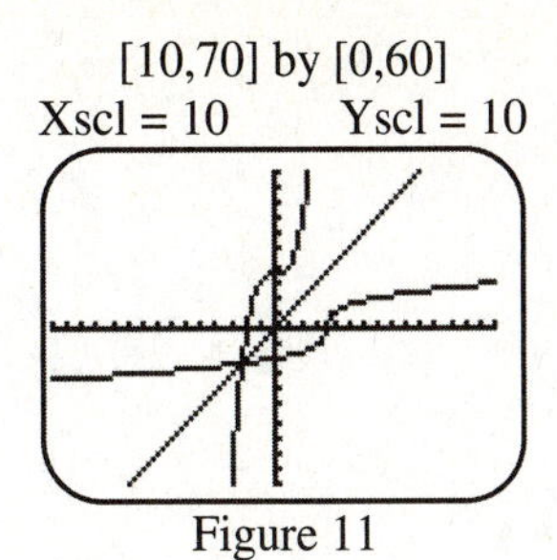

Figure 11

12. $(f \circ f^{-1})(x) = \sqrt[3]{2\left(\frac{x^3+7}{2}\right) - 7} = \sqrt[3]{x^3} = x$ and $(f^{-1} \circ f)(x) = \frac{\left(\sqrt[3]{2x-7}\right)^3 + 7}{2} = \frac{2x-7+7}{2} = \frac{2x}{2} = x$.

13. $y = a^{x+2}$ is the graph of $y = a^x$ shifted 2 units left with a y-intercept of a^2. Therefore, it matches C.

14. $y = a^x + 2$ is the graph of $y = a^x$ shifted 2 units upward with a y-intercept of 3. Therefore, it matches A.

15. $y = -a^x + 2$ is the graph of $y = a^x$ reflected across the x-axis, then shifted 2 units upward, with a y-intercept of 1. Therefore, it matches D.

16. $y = a^{-x} + 2$ is the graph of $y = a^x$ reflected across the y-axis, then shifted 2 units upward, with a y-intercept of 3. Therefore, it matches B.

17. Because the graph goes up to the left, the value of a is $0 < x < 1$.

18. All real numbers can be input for x, therefore the domain is $(-\infty, \infty)$.

19. From the graph, only values where $f(x) > 0$, are possible solutions. Therefore, the range is $(0, \infty)$.

20. Since every real number a raised to the zero power equals 1, $f(0) = a^0 = 1$.

21. Reflect $f(x)$ across the line $y = x$. See Figure 21.

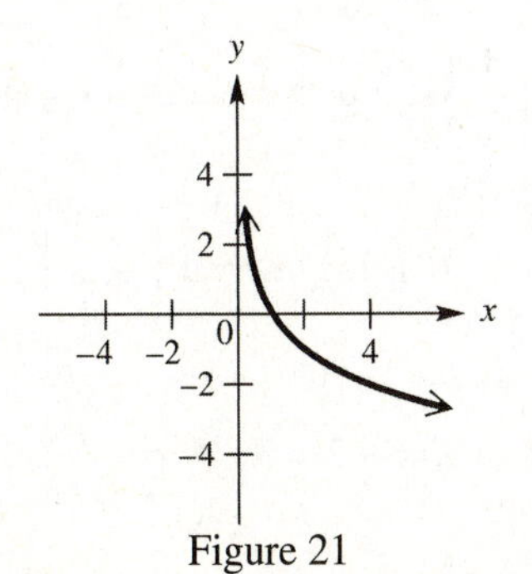

Figure 21

22. For $f(x)=a^x$, the inverse is, $x=a^y \Rightarrow y=\log_a x \Rightarrow f^{-1}(x)=\log_a x$.

23. See Figure 23. From the graph, the domain is $(-\infty,\infty)$; and the range is $(0,\infty)$.

24. See Figure 24. From the graph, the domain is $(-\infty,\infty)$; and the range is $(2,\infty)$.

25. See Figure 25. From the graph, the domain is $(-\infty,\infty)$; and the range is $(-\infty,0)$.

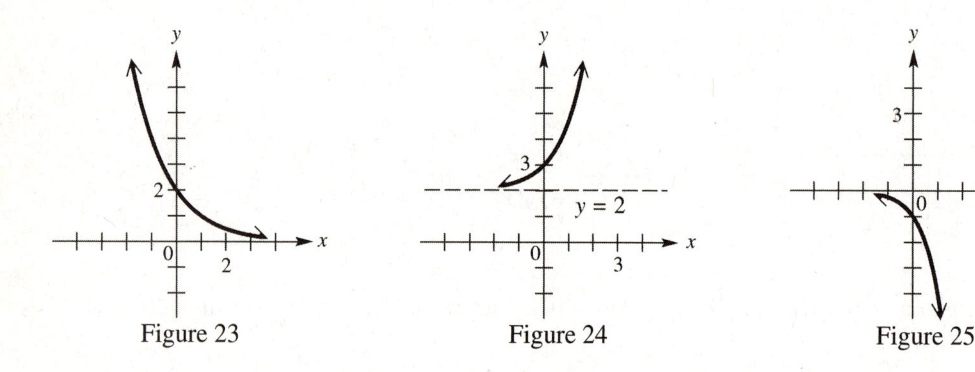

Figure 23 Figure 24 Figure 25

26. The graph of $f(x)=a^x$ with $0<x<1$ would have a range that decreases as the domain increases. Since the negative is on the exponent for $f(x)=a^{-x}$, the graph is reflected across the y-axis and is therefore now increasing as the domain increases. It is increasing on its domain.

27. (a) $\left(\frac{1}{8}\right)^{-2x}=2^{x+3} \Rightarrow (2^{-3})^{-2x}=2^{x+3} \Rightarrow 2^{6x}=2^{x+3} \Rightarrow 6x=x+3 \Rightarrow 5x=3 \Rightarrow x=\frac{3}{5}$

(b) If $\left(\frac{1}{8}\right)^{-2x} \geq 2^{x+3}$, then $\left(\frac{1}{8}\right)^{-2x}-2^{x+3} \geq 0$. Now graph $y_1=\left(\frac{1}{8}\right)^{-2x}-2^{x+3}$ on a calculator, use the answer from part (a), and from the graph the solution to $\left(\frac{1}{8}\right)^{-2x}-2^{x+3} \geq 0$ is the interval $\left[\frac{3}{5},\infty\right)$.

28. (a) $3^{-x}=\left(\frac{1}{27}\right)^{1-2x} \Rightarrow 3^{-x}=(3^{-3})^{1-2x} \Rightarrow 3^{-x}=3^{-3+6x} \Rightarrow -x=-3+6x \Rightarrow 3=7x \Rightarrow x=\frac{3}{7}$

(b) If $3^{-x}<\left(\frac{1}{27}\right)^{1-2x}$, then $3^{-x}-\left(\frac{1}{27}\right)^{-2x}<0$. Now graph $y_1=3^{-x}-\left(\frac{1}{27}\right)^{1-2x}$ on a calculator, use the answer from part (a), and from the graph the solution to $3^{-x}-\left(\frac{1}{27}\right)^{1-2x}<0$ is the interval $\left(\frac{3}{7},\infty\right)$.

29. (a) $0.5^{-x}=0.25^{x+1} \Rightarrow 0.5^{-x}=(0.5^2)^{x+1} \Rightarrow 0.5^{-x}=0.5^{2x+2} \Rightarrow -x=2x+2 \Rightarrow -2=3x \Rightarrow x=-\frac{2}{3}$

(b) If $0.5^{-x} > 0.25^{x+1}$, then $0.5^{-x} - 0.25^{x+1} > 0$. Now graph $y_1 = 0.5^{-x} - 0.25^{x+1}$ on a calculator, use the answer from part (a), and from the graph the solution to $0.5^{-x} - 0.25^{x+1} > 0$ is the interval $\left(-\frac{2}{3}, \infty\right)$.

30. (a) $0.4x = 2.5^{1-x} \Rightarrow \log 0.4^x = \log 2.5^{1-x} \Rightarrow x \log 0.4 = (1-x)\log 2.5 \Rightarrow x \log 0.4 = \log 2.5 - x \log 2.5 \Rightarrow$ $x \log 0.4 + x \log 2.5 = \log 2.5 \Rightarrow x(\log 0.4 + \log 2.5) = \log 2.5 \Rightarrow x = \dfrac{\log 2.5}{\log 0.4 + \log 2.5}$, using a calculator, this is equal to $\dfrac{\log 2.5}{0}$, which is undefined. The solution is $\varnothing$.

(b) If $0.4^x < 2.5^{1-x}$ then $0.4x - 2.5^{1-x} < 0$. Now graph $y_1 = 0.4x - 2.5^{1-x}$ on a calculator, from the graph the solution to $0.4x - 2.5^{1-x} < 0$ is the interval $(-\infty, \infty)$.

31. We can find the intersection points of $y = x^2$ and $y = 2^x$ by using the intersection of graphs method and graph the equation $y_1 = x^2 - 2^x$ to finding the x-intercepts. Using the calculator the intercepts are $2, 4$, and -0.766664696. The missing point in common is $(-0.766664696, 0.58777475603)$.

32. Using the intersection of graphs method, graph $y = 3^x - \pi$ and find the x-intercept. From the calculator, the x-intercepts is $x = \{1.041978046\}$.

33. From the calculator, $\log 58.3 \approx 1.7656685547633 \approx 1.7657$.

34. From the calculator, $\log 0.00233 \approx -2.63264407897 \approx -2.6326$.

35. From the calculator, In $58.3 \approx 4.06560209336 \approx 4.0656$.

36. From the calculator, $\log_2 0.00233 = \dfrac{\log 0.00233}{\log 2} = -8.7455$.

37. By the definitions of logarithms, $\log_{13} 1 = x \Rightarrow 13^x = 1$ since we know $13^0 = 1$ it follows that $x = 0$.

38. By the properties of logarithms, $\ln e^{\sqrt{6}} = \log_e e^{\sqrt{6}} = \sqrt{6}$.

39. By the properties of logarithms, $\log_5 5^{12} = 12$.

40. By the properties of logarithms, $7^{\log_7 13} = 13$.

41. By the properties of logarithms, $3^{\log_3 5} = 5$.

42. By the change of base rule, $\log_4 9 = \dfrac{\log 9}{\log 4} \approx 1.58496250072 \approx 1.5850.$

43. The x-intercept is 1, and the graph is increasing. The correct graph is E.

44. Since $f(x) = \log_2(2x) = \log_2 + \log_2 x = 1 + \log_2 x$, the graph will be similar to that of $f(x) = \log_2 x$, but will have a vertical shift up 1. The x-intercept will be $\dfrac{1}{2}$, since $\log_2 2\left(\dfrac{1}{2}\right) = \log_2 1 = 0.$

The correct graph is D.

45. Since $f(x) = \log_2\left(\dfrac{1}{x}\right) = \log_2 x^{-1} = -\log_2 x$, the graph will be similar to that of $f(x) = \log_2 x$, but will be reflected across the x-axis. The x-intercept is 1. The correct graph is B.

46. Since $f(x) = \log_2\left(\dfrac{x}{2}\right) = \log_2\left(\dfrac{1}{2}x\right) = \log_2\left(\dfrac{1}{2}\right) + \log_2 x = -1 + \log_2 x$, the graph will be similar to that of $f(x) = \log_2 x$, but will have a vertical shift down 1 unit. The x-intercept will be 2, since $\log_2\left(\dfrac{2}{2}\right) = \log_2 1 = 0$. The correct graph is C.

47. With $f(x) = \log_2(x-1)$, the graph will be similar to that of $f(x) = \log_2 x$, but will be shifted 1 unit to the right; the x-intercept will be 2. The correct graph is F.

48. With $f(x) = \log_2(-x)$, the graph will be similar to that of $f(x) = \log_2 x$, but will be reflected across the y-axis. The x-intercept is -1, since $\log_2(-(-1)) = \log_2 1 = 0$. The correct graph is A.

49. See Figure 49. From the graph, the domain is $(0,\infty)$ and the range is $(-\infty,\infty)$.

50. See Figure 50. From the graph, the domain is $(2,\infty)$ and the range is $(-\infty,\infty)$.

51. See Figure 51. From the graph, the domain is $(-\infty,0)$ and the range is $(-\infty,\infty)$.

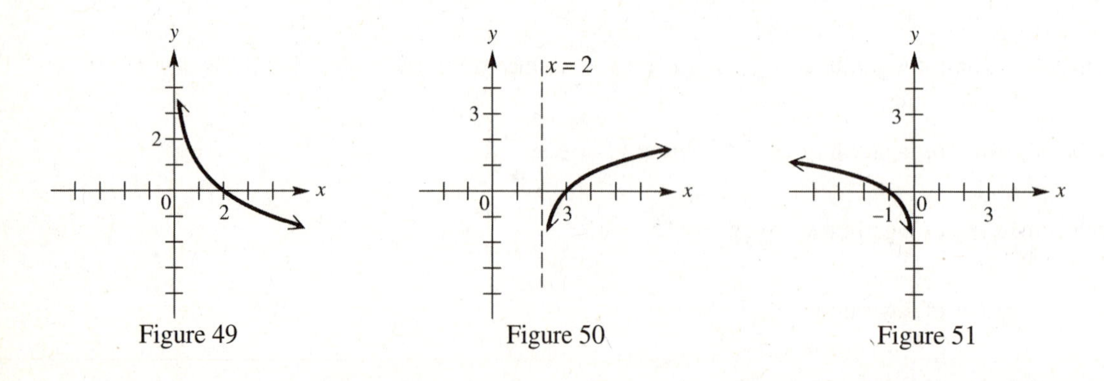

Figure 49 Figure 50 Figure 51

52. The functions in Exercises 49, 50, and 51 are the inverses, respectively, of those in Exercises 23, 24, and 25.

53. If the function has a point $(81,4)$, then $f(81)=4$. Therefore, $\log_a 81=4 \Rightarrow a^4=81 \Rightarrow a^4=3^4 \Rightarrow a=3$. The base is 3.

54. If the function has a point $\left(-4,\frac{1}{16}\right)$, then $f(-4)=\frac{1}{16}$. Therefore, $a^{-4}=\frac{1}{16} \Rightarrow a^{-4}=2^{-4} \Rightarrow a=2$. The base is 2.

55. $\log_3 \frac{mn}{5r}=\log_3 mn-\log_3 5r=\log_3 m+\log_3 n-(\log_3 5+\log_3 r)=\log_3 m+\log_3 n-\log_3 5-\log_3 r$

56. $\log_2 \frac{\sqrt{7}}{15}=\log_2 \sqrt{7}-\log_2 15=\log_2 7^{1/2}-\log_2 15=\frac{1}{2}\log_2 7-\log_2 15.$

57. $\log_5(x^2y^4\sqrt[5]{m^3p})=\log_5 x^2+\log_5 y^4+\log_5(m^3p)^{1/5}=2\log_5 x+4\log_5 y+\frac{1}{5}\log_5(m^3p)=$

$2\log_5 x+4\log_5 y+\frac{1}{5}(\log_5 m^3+\log_5 p)=2\log_5 x+4\log_5 y+\frac{3}{5}\log_5 m+\frac{1}{5}\log_5 p$

58. The properties of logarithms do not apply to polynomials.

59. (a) $\log(x+3)+\log x=1 \Rightarrow \log(x+3)(x)=1 \Rightarrow 10^{\log(x+3)(x)}=10^1 \Rightarrow (x+3)(x)=10 \Rightarrow$
$x^2+3x-10=0 \Rightarrow (x+5)(x-2)=0 \Rightarrow x=-5,2.$

Since the argument of a logarithm cannot be negative, the solution is $x=2$.

(b) Graph $y_1=\log(x+3)+\log x$ and $y_2=1$ in the same window. From the graph, $y_1>y_2$ for the interval $(2,\infty)$.

60. (a) $\ln e^{\ln e}-\ln(x-4)=\ln 3 \Rightarrow \ln x-\ln(x-4)=\ln 3 \Rightarrow \ln\left(\frac{x}{x-4}\right)=\ln 3 \Rightarrow e^{\ln(x/x-4)}=e^{\ln x} \Rightarrow$

$\frac{x}{x-4}=3 \Rightarrow x=3x-12 \Rightarrow -2x=-12 \Rightarrow x=6$

(b) Graph $y_1=\ln e^{\ln x}-\ln(x-4)$ and $y_2=\ln 3$ in the same window. From the graph, $y_1 \le y_2$ for the interval $[6,\infty)$.

61. (a) $\ln e^{\ln 2}-\ln(x-1)=\ln 5 \Rightarrow \ln 2-\ln(x-1)=\ln 5 \Rightarrow \ln\left(\frac{2}{x-1}\right)=\ln 5 \Rightarrow e^{\ln(2/x-1)}=e^{\ln 5} \Rightarrow$

$\frac{2}{x-1}=5 \Rightarrow 2=5x-5 \Rightarrow 5x=7 \Rightarrow x=\frac{7}{5}$ or 1.4

(b) Graph $y_1 = \ln e^{\ln 2} - \ln(x-1)$ and $y_2 = \ln 5$ in the same window. From the graph, $y_1 \geq y_2$ for the interval $(1, 1.4]$.

62. (a) $8^x = 32 \Rightarrow (2^3)^x = 2^5 \Rightarrow 2^{3x} = 2^5 \Rightarrow 3x = 5 \Rightarrow x = \frac{5}{3}$

63. (a) $\frac{8}{27} = x^{-3} \Rightarrow \left(\frac{2}{3}\right)^3 = x^{-3} \Rightarrow \left(\frac{2}{3}\right)^3 = \left(\frac{1}{x}\right)^3 \Rightarrow \frac{2}{3} = \frac{1}{x} \Rightarrow 2x = 3 \Rightarrow x = \frac{3}{2}$

64. (a) $10^{2x-3} = 17 \Rightarrow \log 10^{2x-3} = \log 17 \Rightarrow (2x-3)\log 10 = \log 17$

$\Rightarrow 2x - 3 = \log 17 \Rightarrow 2x = 3 + \log 17 \Rightarrow x = \frac{3 + \log 17}{2}$

(b) From the calculator the solutions is $x \approx 2.115$.

65. (a) $e^{x+1} = 10 \Rightarrow \ln e^{x+1} = \ln 10 \Rightarrow (x+1)\ln e = \ln 10 \Rightarrow x + 1 = \ln 10 \Rightarrow x = -1 + \ln 10$

(Other forms of the answer are possible.)

(b) From the calculator, the solutions is $x \approx 1.303$.

66. (a) $\log_{64} x = \frac{1}{3} \Rightarrow 64^{\log_{64} x} = 64^{1/3} \Rightarrow x = 64^{1/3} \Rightarrow x = 4$

67. (a) $\ln(6x) - \ln(x+1) = \ln 4 \Rightarrow \ln \frac{6x}{x+1} = \ln 4 \Rightarrow e^{\ln 6x/x+1} = e^{\ln 4} \Rightarrow \frac{6x}{x+1} = 4 \Rightarrow 4x + 4 = 6x \Rightarrow$

$4 = 2x \Rightarrow x = 2$

68. (a) $\log_{12}(2x) + \log_{12}(x-1) = 1 \Rightarrow \log_{12}(2x)(x-1) = 1 \Rightarrow \log_{12} 2x^2 - 2x = 1 \Rightarrow$

$12^{\log_{12} 2x^2 - 2x} = 12^1 \Rightarrow 2x^2 - 2x = 12 \Rightarrow 2x^2 - 2x - 12 = 0 \Rightarrow 2(x^2 - x - 6) = 0 \Rightarrow$

$2(x-3)(x+2) = 0 \Rightarrow x = -2, 3$. Since the argument of a logarithm cannot be negative, the solution is $x = 3$.

69. (a) $\log_{16} \sqrt{x+1} = \frac{1}{4} \Rightarrow 16^{\log_{16} \sqrt{x+1}} = 16^{1/4} \Rightarrow \sqrt{x+1} = 16^{1/4} \Rightarrow x + 1 = (16^{1/4})^2 \Rightarrow$

$x + 1 = 16^{1/2} \Rightarrow x + 1 = 4 \Rightarrow x = 3$

70. (a) $\ln x + 3\ln 2 = \ln \frac{2}{x} \Rightarrow \ln x + \ln 2^3 = \ln \frac{2}{x} \Rightarrow \ln[(x)(2^3)] = \ln \frac{2}{3} \Rightarrow$

$\ln 8x = \ln \frac{2}{x} \Rightarrow e^{\ln 8x} = e^{\ln 2/x} \Rightarrow 8x = \frac{2}{x} \Rightarrow 8x^2 = 2 \Rightarrow x^2 = \frac{2}{8} \Rightarrow x^2 = \frac{1}{4} \Rightarrow x = -\frac{1}{2}, \frac{1}{2}$

Since the argument of a natural logarithm can not be negative, the solution is $x = \frac{1}{2}$.

71. (a) $\ln[\ln(e^{-x})] = \ln 3 \Rightarrow \ln(-x) = \ln 3 \Rightarrow e^{\ln(-x)} = e^{\ln 3} \Rightarrow -x = 3 \Rightarrow x = -3$

72. (a) $\ln e^x - \ln e^3 = \ln e^5 \Rightarrow x - 3 = 5 \Rightarrow x = 8$

73. No real number value for x can produce a negative solution; therefore, $x = \varnothing$.

74. $N = a + b\ln\left(\frac{c}{d}\right) \Rightarrow N - a = b\ln\left(\frac{c}{d}\right) \Rightarrow \frac{N-a}{b} = \ln\left(\frac{c}{d}\right) \Rightarrow$

$e^{(N-a)/b} = e^{\ln(c/d)} \Rightarrow e^{(N-a)/b} = \frac{c}{d} \Rightarrow c = de^{(N-a)/b}$

75. $y = y_0 e^{-kt} \Rightarrow \frac{y}{y_0} = e^{-kt} \Rightarrow \ln\left(\frac{y}{y_0}\right) = \ln e^{-kt} \Rightarrow \ln\left(\frac{y}{y_0}\right) = -kt \Rightarrow t = \frac{\ln(y/y_0)}{-k}$

76. Graph $y_1 = \log_{10} x$ and $y_2 = x - 2$ in the same window. By the capabilities of the calculator, the x-coordinates of the intersections are $x \approx 0.010$ and $x \approx 2.376$.

77. Graph $y_1 = 2^{-x}$ and $y_2 = \log_{10} x$ in the same window. By the capabilities of the calculator, the x-coordinate of the intersection is $x \approx 1.874$.

78. Graph $y_1 = x^2 - 3$ and $y_2 = \log x$ in the same window. By the capabilities of the calculator, the x-coordinates of the intersections are $x \approx 0.001$ and $x \approx 1.805$.

79. Use the Compound Interest Formula: $A = P\left(\frac{1+r}{n}\right)^{nt}$ with $A = \$4613$, $P = \$3500$, $t = 10$, and $n = 1$.

$4613 = 3500\left(1 + \frac{r}{1}\right)^{1(10)} \Rightarrow \frac{4613}{3500} = (1+r)^{10} \Rightarrow \left(\frac{4613}{3500}\right)^{0.1} = 1 + r \Rightarrow r = \left(\frac{4613}{3500}\right)^{0.1} - 1 \Rightarrow$

$r \approx 0.027996$. The annual interest rate, rounded to the nearest tenth, is 2.8%.

80. Use the Compound Interest Formula: $A = P\left(1 + \frac{r}{n}\right)^{nt}$ with $A = \$58344$, $P = \$48000$, $r = 0.05$, and $n = 2$.

$58,344 = 48,000\left(1 + \frac{0.05}{2}\right)^{2t} \Rightarrow 58,344 = 48,000(1.025)^{2t} \Rightarrow \left(\frac{58,344}{48,000}\right) = 1.025^{2t} \Rightarrow 1.2155 = 1.025^{2t}$

$\Rightarrow \ln 1.2155 = \ln 1.025^{2t} \Rightarrow \ln 1.2155 = 2t \ln 1.025 \Rightarrow t = \frac{\ln 1.2155}{2\ln 1.025} \Rightarrow t \approx 3.951698$ or about 4 years.

81. For the first 8 years, use the Compound Interest Formula:

$A=P\left(1+\frac{r}{n}\right)^{nt}$ with $P=\$12000,\ t=8,\ r=0.05,$ and $n=1.$

$A=12{,}000\left(1+\frac{0.05}{1}\right)^{8(1)} \Rightarrow A=12{,}000(1.05)^8 \Rightarrow A\approx \$17{,}729.47$

Now for the second 6 years, use the Compound Interest Formula:

$A=P\left(1+\frac{r}{n}\right)^{nt}$ with $P=\$17{,}729.47,\ t=6,\ r=0.06,$ and $n=1.$

$A=17{,}729.47\left(1+\frac{0.06}{1}\right)^{1(6)} \Rightarrow A=17{,}729.47(1.06)^6 \Rightarrow A\approx \$25{,}149.59$

At the end of the 14 year period, $25,149.59 would be in the account.

82. (a) Use the Compound Interest Formula: $A=P\left(1+\frac{r}{n}\right)^{nt}$ with $P=\$2000,\ r=.03,\ t=5,$ and $n=4.$

$A=2000\left(1+\frac{0.03}{4}\right)^{4(5)} \Rightarrow A=2000(1.0075)^{20} \Rightarrow A\approx \$2{,}322.37$

(b) Use the Continuous Compound Interest Formula: $A=Pe^{rt}$, with $P=\$2000,\ r=0.03,$ and $t=5.$

$A=2000e^{(0.03)(.5)} \Rightarrow A\approx \$2323.67$.

(c) Use the Continuous Compound Interest Formula:

$A=Pe^{rt}$ with $A=\$6000,\ P=\$2000,$ and $r=0.03.$

$6000=2000e^{0.03t} \Rightarrow 3=e^{0.03t} \Rightarrow \ln 3=\ln e^{0.03t} \Rightarrow \ln 3=0.03t \Rightarrow t=\frac{\ln 3}{0.03} \Rightarrow t\approx 36.6204$

It would take about 36.6 years to triple.

83. (a) See Figure 83. Since L is increasing, heavier planes require longer runways.

(b) We can find the answer by solving $L(10)$ and $L(100)$. $L(10)=3\log 10=3(1)=3$ or 3000 feet; $L(100)=3\log 100=3(2)=6$ or 6000 feet. No, it does not increase by a factor of 10, but rather it increases by a factor of 2 to 6000 feet.

84. $2200=280\ln(x+1)+1925 \Rightarrow 275=280(x+1) \Rightarrow \frac{275}{280}=\ln(x+1) \Rightarrow e^{275/280}=e^{\ln(x+1)} \Rightarrow$

$e^{275/280}=x+1 \Rightarrow x=e^{275/280}-1 \Rightarrow x\approx 1.67$ acres.

85. (a) $P(0.5) = 0.04e^{-4(0.5)} \Rightarrow P(0.5) \approx 0.0054$ grams/liter

(b) $P(1) = 0.4e^{-4(1)} \Rightarrow P(1) \approx 0.00073$ grams/liter

(c) $P(2) = 0.04e^{-4(2)} \Rightarrow P(2) \approx 0.000013$ grams/liter

(d) $0.002 = 0.04e^{-4x} \Rightarrow 0.05 = e^{-4x} \Rightarrow \ln 0.05 = \ln e^{-4x} \Rightarrow \ln 0.05 = -4x \Rightarrow$

$x = \frac{\ln 0.05}{-4} \Rightarrow x \approx 0.75$ miles.

86. (a) $p(2) = 250 - 120(2.8)^{-0.5(2)} \Rightarrow p(2) \approx 207$

(b) $p(10) = 250 - 120(28)^{-0.5(10)} \Rightarrow p(10) \approx 249$

(c) See Figure 86. When $x = 2$, the graph supports the answer of 207 from part (a).

[0,50] by [0,6]
Xscl = 10 Yscl = 1

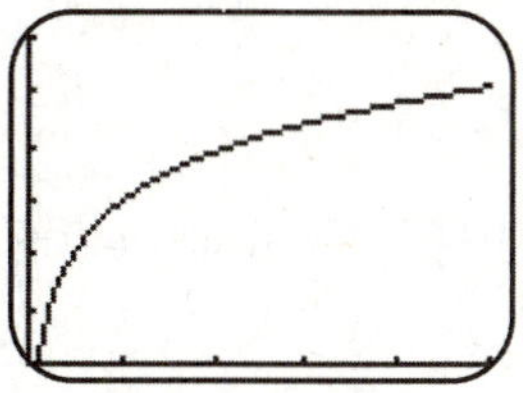

Figure 83

[0,10] by [0,300]
Xscl = 1 Yscl = 50

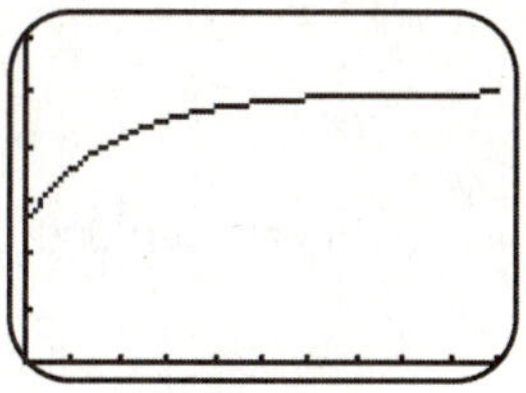

Figure 86

87. Find t when $v(t) = 147$.

$$147 = 176\left(1 - e^{-0.18t}\right) \Rightarrow \frac{147}{176} = 1 - e^{-0.18t} \Rightarrow e^{-0.18t} = 1 - \frac{147}{176} \Rightarrow e^{-0.18t} = \frac{29}{176} \Rightarrow$$

$$\ln e^{-0.18t} = \ln\left(\frac{29}{176}\right) \Rightarrow -0.18t = \ln\left(\frac{29}{176}\right) \Rightarrow t = \frac{\ln\left(\frac{29}{176}\right)}{-0.18} = t \approx 10.017712$$

It will take the skydiver about 10 seconds to attain the speed of 147 feet/second (100 mph).

Chapter 5 Test

1. (a) This function has a decreasing graph with a y-axis asymptote. This matches with B.

(b) This function has an increasing graph with a y-intercept of 1. This matches with A.

(c) This function has an increasing graph with a y-axis asymptote. This matches with C.

(d) This function has a decreasing graph with a y-intercept of 1. This matches with D.

2. (a) See Figure 2.

(b) From the graph, the domain is $(-\infty, \infty)$, and the range is $(-\infty, 8)$.

(c) From the graph, it has a horizontal asymptote with equation $y = 8$.

(d) Set $y = 0$ to find the x-intercept. $0 = -2^{x-1} + 8 \Rightarrow 2^{x-1} = 8 \Rightarrow 2^{x-1} = 2^3 \Rightarrow x - 1 = 3 \Rightarrow x = 4.$

The x-intercept is (4, 0). Set $x = 0$ to find the y-intercept. $y = -2^{0-1} + 8 \Rightarrow y = -\dfrac{1}{2} + 8 \Rightarrow y = 7.5.$

The y-intercept is (0, 7.5).

(e) To find the inverse, interchange the x and y variables and solve for y. $x = -2^{y-1} + 8 \Rightarrow$

$x - 8 = -2^{y-1} \Rightarrow 8 - x = 2^{y-1} \Rightarrow \log(8 - x) = \log 2^{y-1} \Rightarrow \log(8 - x) = (y - 1)\log 2 \Rightarrow$

$\log(8 - x) = y\log 2 - 1\log 2 \Rightarrow \log(8 - x) + \log 2 = y\log 2 \Rightarrow y = \dfrac{\log(8 - x) + \log 2}{\log 2}$ or

$y = \dfrac{\log(8 - \mathrm{x})}{\log 2} + 1.$

3. $(\frac{1}{8})^{2x+3} = 16^{x+1} \Rightarrow (2^{-3})^{2x-3} = (2^4)^{x+1} \Rightarrow 2^{-6x+9} = 2^{4x+4} \Rightarrow -6x + 9 = 4x + 4 \Rightarrow 5 = 10x \Rightarrow \quad x = \dfrac{1}{2}$ or $x = 0.5$.

4. (a) Use the Compound Interest Formula $A = P\left(1 + \dfrac{r}{n}\right)^{nt}$, with $P = \$10000$, $r = 0.035$, $t = 4$, and $n = 4$.

$A = 10000\left(1 + \dfrac{0.035}{4}\right)^{4(4)} \Rightarrow A \approx \$11,495.74.$

(b) Use the Continuous Compound Interest Formula $A = Pe^{rt}$, with $P = \$10000$, $r = 0.035$, and $t = 4$.

$A = 10,000e^{0.035(4)} \Rightarrow A \approx \$11,502.74.$

5. The expression, $\log_5 27$, is the exponent to which 5 must be raised in order to obtain 27. To find an approximation with a calculator, use the change-of-base rule: $\log_5 27 = \dfrac{\log 27}{\log 5}$.

6. (a) From the calculator, $\log 45.6 \approx 1.659$.

(b) From the calculator, $\ln 470 \approx 6.153$.

(c) From the calculator, $\log_3 769 = \dfrac{\log 769}{\log 3} \approx 6.049$.

7. $\log \dfrac{m^3 n}{\sqrt{y}} = \log m^3 + \log n - \log\sqrt{y} = \log m^3 + \log n - \log y^{1/2} \Rightarrow 3\log m + \log n - \dfrac{1}{2}\log y.$

8. $A = Pe^{rt} \Rightarrow \dfrac{A}{P} = e^{rt} \Rightarrow \ln\dfrac{A}{P} = \ln e^{rt} \Rightarrow \ln A - \ln P = rt \Rightarrow t = \dfrac{\ln A - \ln P}{r}.$

9. (a) $\log_2 x + \log_2(x+2) = 3;\ x > 0 \Rightarrow \log_2[x(x+2)] = 3 \Rightarrow 2^{\log_2[x(x+2)]} = 2^3 \Rightarrow x(x+2) = 2^3 \Rightarrow$

$x^2 + 2x - 8 = 0 \Rightarrow (x+4)(x-2) = 0 \Rightarrow x = -4, 2.$ Since the argument of a logarithm can not be negative, the solution cannot be the extraneous solution -4, so the solution is $x = 2$.

(b) Using the change-of-base rule on $y_1 = \log_2 x + \log_2(x+2) - 3$ yields $y_1 = \dfrac{\log x}{\log 2} + \dfrac{\log(x+2)}{\log 2} - 3.$

Graph the equation. See Figure 9. The x-intercept is $x = 2$.

(c) From the graph, $\log_2 x + \log_2(x+2) > 3$ for the interval $(2, \infty)$.

[-10,10] by [-10,10]

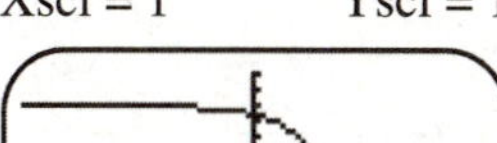

Xscl = 1 Yscl = 1

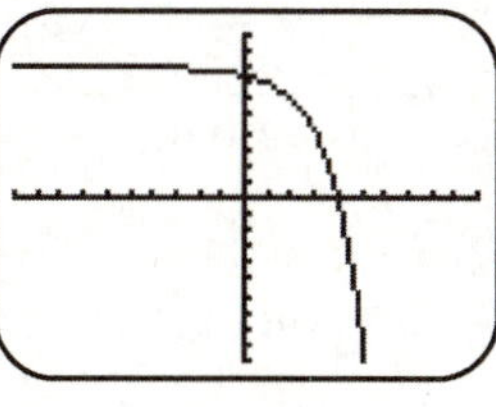

Figure 2

[-10,10] by [-10,10]

Xscl = 1 Yscl = 1

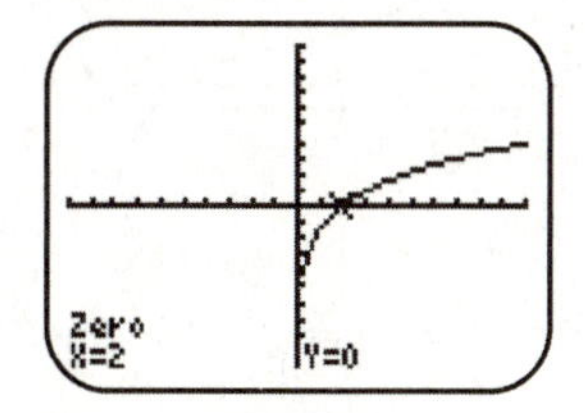

Figure 9

10. (a) $2e^{5x+2} = 8 \Rightarrow e^{5x+2} = 4 \Rightarrow \ln e^{5x+2} = \ln 4 \Rightarrow 5x + 2 = \ln 4 \Rightarrow\ 5x = \ln 4 - 2 \Rightarrow \mathrm{x} = \dfrac{\ln 4\ - 2}{5}.$

(Other forms of the answer are possible.)

(b) From the calculator, the solution is $x = -0.123$.

11. (a) $6^{2-x} = 2^{3x+1} \Rightarrow \log 6^{2-x} = \log 2^{3x+1} \Rightarrow (2-x)\log 6 = (3x+1)\log 2 \Rightarrow$

$2\log 6 - x\log 6 = 3x\log 2 + \log 2 \Rightarrow\ 2\log 6 - \log 2 = 3x\log 2 + x\log 6 \Rightarrow$

$\log\dfrac{6^2}{2} = x(\ \log(2^3)(6)) \Rightarrow \log 18 = x(\log 48) \Rightarrow x = \dfrac{\log 18}{\log 48}.$

(b) From the calculator, the solution is $x = 0.747$.

12. (a) $\log(\ln x) = 1 \Rightarrow 10^{\log(\ln x)} = 10^1 \Rightarrow \ln x = 10 \Rightarrow e^{\ln x} = e^{10} \Rightarrow x = e^{10}.$

(b) From the calculator, the solution is $x = 22{,}026.466$.

13. (a) The 2 in the equation is the original population; tripling this would yield $y = 3\cdot 2$ The match is B.

(b) In this function, $y = 3$; therefore, the match is D.

(c) In this function, $t = 3$; therefore, the match is C.

(d) Since 4 months equals $\dfrac{1}{3}$ year, in this function $t = \dfrac{1}{3}$; therefore, the match is A.

14. (a) See Figure 14a.

(b) See Figure 14b.

(c) See Figure 14c.

(d) See Figure 14d.

Function (c) is the best at describing $A(t)$ because it starts at 350 and gradually decreases as time increases.

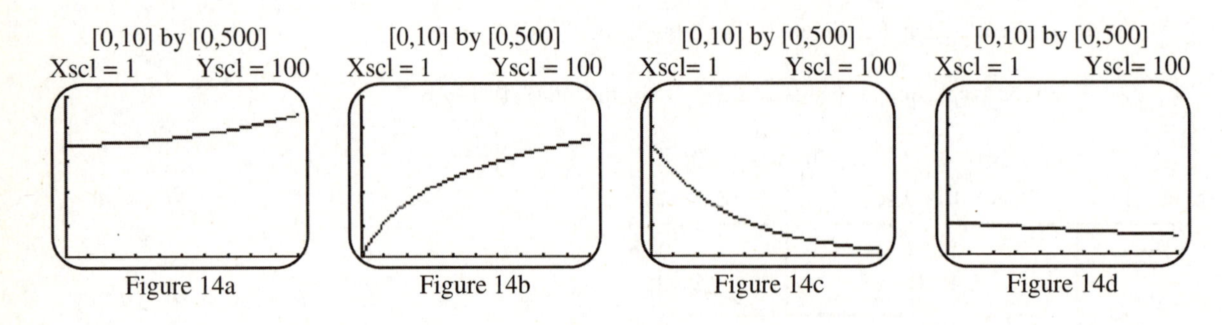

Figure 14a Figure 14b Figure 14c Figure 14d

15. (a) Using the exponential decay model $A(t) = A_0 e^{-kt}$, we can solve for k with $A_0 = 2$ and $t = 1600$.

$$\frac{1}{2}(2) = 2e^{-k(1600)} \Rightarrow 1 = 2e^{-k(1600)} \Rightarrow \frac{1}{2} = e^{-k(1600)} \Rightarrow \ln\frac{1}{2} = \ln e^{-k(1600)} \Rightarrow \ln\frac{1}{2} = -k(1600) \Rightarrow$$

$$k = \frac{\ln(5)}{-1600} \Rightarrow k \approx 0.000433.$$ The model is $A(t) = 2e^{-0.000433t}$.

(b) $A(9600) = 2e^{-0.000433(9600)} \Rightarrow A(9600) \approx 0.03$ grams.

(c) $0.5 = 2e^{-0.000433t} \Rightarrow 0.25 = e^{-0.000433t} \Rightarrow \ln 0.25 = \ln e^{-0.000433t} \Rightarrow$

$$\ln 0.25 = -0.000433\text{t} \Rightarrow \text{t} = \frac{\ln 0.25}{-0.000433} \Rightarrow t \approx 3200 \text{ years.}$$

Chapter 6: Analytic Geometry

6.1: Circles and Parabolas

1. E. Since $x = 2y^2$. is equivalent to $y^2 = 4\left(\frac{1}{8}\right)x$, this is a parabola that opens to the right $(c > 0)$.

3. H. Since $x^2 = -3y$ is equivalent to $x^2 = 4\left(-\frac{3}{4}\right)y$, this is a parabola that opens downward $(c < 0)$.

5. F. This is the equation of a circle centered at the origin with radius $\sqrt{5}$.

7. D. This is the equation of a circle centered at the point $(-3,4)$ with radius $\sqrt{25} = 5$.

9. Here $h = 1$, $k = 4$ and $r^2 = 3^2 = 9$. The equation is $(x-1)^2 + (y-4)^2 = 9$.

11. A circle that is centered at the origin with $r^2 = 1^2 = 1$ has equation $x^2 + y^2 = 1$.

13. Here $h = \frac{2}{3}$, $k = -\frac{4}{5}$ and $r^2 = \left(\frac{3}{7}\right)^2 = \frac{9}{49}$. The equation is $\left(x - \frac{2}{3}\right)^2 + \left(y + \frac{4}{5}\right)^2 = \frac{9}{49}$.

15. The radius is the distance between $(-1,2)$ and $(2,6)$: $r = \sqrt{(2-(-1))^2 + (6-2)^2} = \sqrt{9+16} = 5$. Here $h = -1$, $k = 2$ and $r^2 = 5^2 = 25$. The equation is $(x+1)^2 + (y-2)^2 = 25$.

17. If the center is $(-3,-2)$, the circle must touch the x-axis at the point $(-3,0)$. The radius is 2. Here $h = -3$, $k = -2$ and $r^2 = 2^2 = 4$. The equation is $(x+3)^2 + (y+2)^2 = 4$.

19. The equation is that of a circle with center (3, 3) and radius 0. That is, the graph is the point (3, 3).

21. Midpoint: $\left(\frac{5+(-1)}{2}, \frac{-9+3}{2}\right) = (2,-3) \Rightarrow$ The center of the circle is (2,-3).

    Distance: $d = \sqrt{(2-5)^2 + (-3-(-9))^2} = \sqrt{9+36} = \sqrt{45} \Rightarrow$ The radius of the circle is $\sqrt{45}$ units. The equation of the circle is $(x-2)^2 + (y+3)^2 = 45$.

23. Midpoint: $\left(\frac{-5+(1)}{2}, \frac{-7+1}{2}\right) = (-2,-3) \Rightarrow$ The center of the circle is (-2,-3).

    Distance: $d = \sqrt{(-2-1)^2 + (-3-(1))^2} = \sqrt{9+16} = \sqrt{25} = 5 \Rightarrow$ The radius of the circle is 5 units. The equation of the circle is $(x+2)^2 + (y+3)^2 = 25$.

25. Midpoint: $\left(\frac{-5+(5)}{2}, \frac{0+0}{2}\right) = (0,0) \Rightarrow$ The center of the circle is (0,0).

    Distance: $d = \sqrt{(0-(-5))^2 + (0-0)^2} = \sqrt{25} = 5 \Rightarrow$ The radius of the circle is 5 units.

    The equation of the circle is $(x)^2 + (y)^2 = 25$.

27. In a circle, the radius is the distance from the center to any point on the circle.

29. This is the equation of a circle centered at the origin with radius $\sqrt{4}=2$. See Figure 29. From the figure, the domain is $[-2,2]$, and the range is $[-2,2]$.

31. This is the equation of a circle centered at the origin with radius $\sqrt{0}=0$. The graph is only the point (0, 0). See Figure 31. From the figure, the domain is $\{0\}$, and the range is $\{0\}$.

33. This is the equation of a circle centered at $(2,0)$ with radius $\sqrt{36}=6$. See Figure 33. From the figure, the domain is $[-4,8]$, and the range is $[-6,6]$.

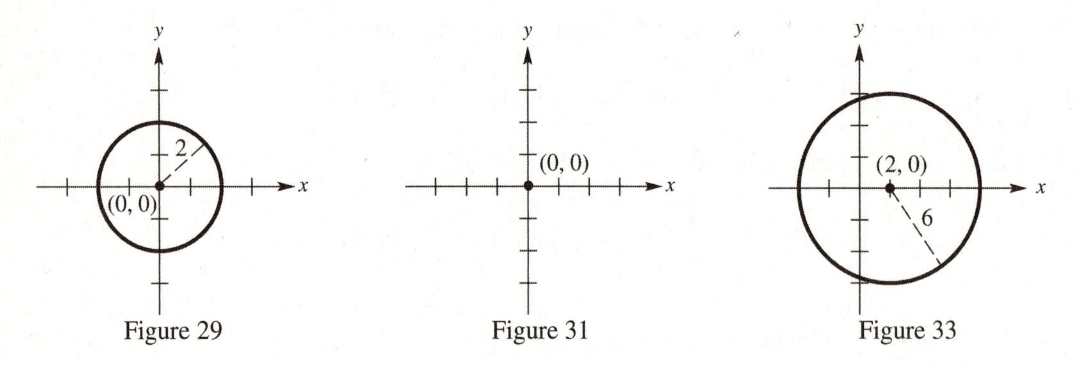

Figure 29 Figure 31 Figure 33

35. This is the equation of a circle centered at $(5,-4)$ with radius $\sqrt{49}=7$. See Figure 35. From the figure, the domain is $[-2,12]$, and the range is $[-11,3]$.

37. This is the equation of a circle centered at $(-3,-2)$ with radius $\sqrt{36}=6$. See Figure 37. From the figure, the domain is $[-9,3]$, and the range is $[-8,4]$.

39. This is the equation of a circle centered at $(1,-2)$ with radius $\sqrt{16}=4$. See Figure 39. From the figure, the domain is $[-3,5]$, and the range is $[-6,2]$.

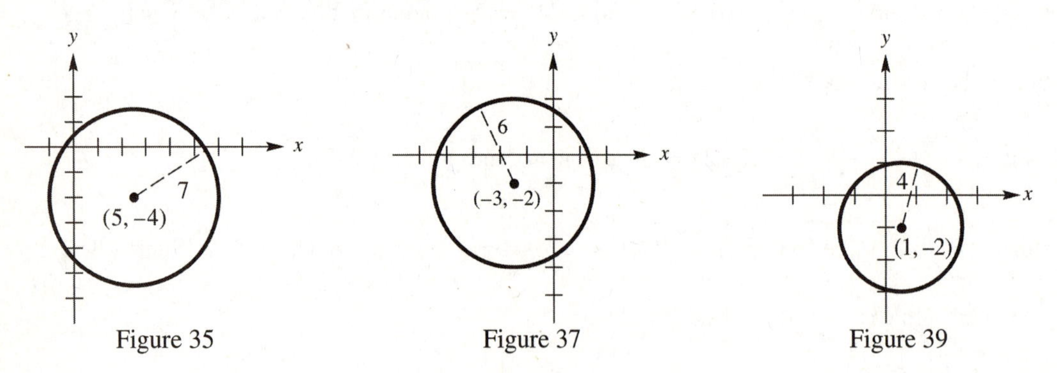

Figure 35 Figure 37 Figure 39

41. $x^2+y^2=81 \Rightarrow y^2=81-x^2 \Rightarrow y=\pm\sqrt{81-x^2}$ Graph $y_1=\sqrt{81-x^2}$ and $y_2=\sqrt{81-x^2}$ as shown in Figure 41. From the figure, the domain is $[-9,9]$, and the range is $[-9,9]$.

43. $(x-3)^2+(y-2)^2=25 \Rightarrow (y-2)^2=25-(x-3)^2 \Rightarrow y-2=\pm\sqrt{25-(x-3)^2} \Rightarrow y=2\pm\sqrt{25-(x-3)^2}$.

Graph $y_1=2-\sqrt{25-(x-3)^2}$ and $y_2=2+\sqrt{25-(x-3)^2}$ as shown in Figure 43. From the figure, the domain is $[-2,8]$, and the range is $[-3,7]$.

[-14.1,14.1] by [-9.3,9.3]
Xscl = 1 Yscl = 1

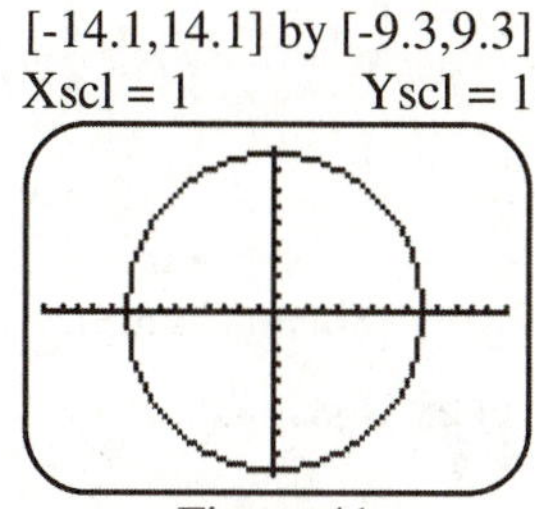

Figure 41

[-9.4,9.4] by [-4.2,8.2]
Xscl = 1 Yscl = 1

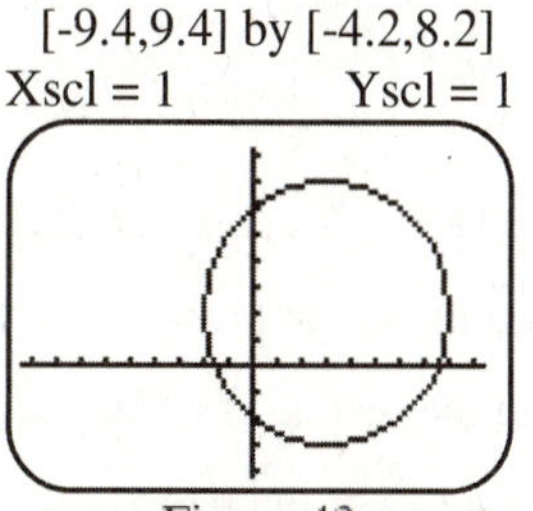

Figure 43

45. $x^2+6x+y^2+8y+9=0 \Rightarrow (x^2+6x+9)+(y^2+8y+16)=-9+9+16 \Rightarrow (x+3)^2+(y+4)^2=16$. The graph is a circle with center $(-3,-4)$, and radius $r=4$.

47. $x^2-4x+y^2+12y=-4 \Rightarrow (x^2-4x+4)+(y^2+12y+36)=-4+4+36 \Rightarrow (x-2)^2+(y+6)^2=36$.

The graph is a circle with center $(2,-6)$, and radius $r=6$.

49. $4x^2+4x+4y^2-16y-19=0 \Rightarrow 4\left(x^2+x+\frac{1}{4}\right)+4(y^2-4y+4)=19+1+16 \Rightarrow$

$4\left(x+\frac{1}{2}\right)^2+4(y-2)^2=36 \Rightarrow \left(x+\frac{1}{2}\right)^2+(y-2)^2=9$. The graph is a circle with center $\left(-\frac{1}{2},2\right)$, and radius $r=3$.

51. $x^2+2x+y^2-6y+14=0 \Rightarrow (x^2+2x+1)+(y^2-6y+9)=-14+1+9 \Rightarrow (x+1)^2+(y-3)^2=-4$. The graph does not exist since the value for the radius is not a real number.

53. $x^2-2x+y^2+4y+9=0 \Rightarrow (x^2-2x+1)+(y^2+4y+4)=-9+1+4 \Rightarrow (x-1)^2+(y+2)^2=-4$. The graph does not exist since the value for the radius is not a real number.

55. $b^2=\frac{1}{2}$. $(x+2)^2+y^2=\frac{4}{9}$. The graph is a circle with center $(-2,0)$, and radius $r=\frac{2}{3}$.

57. D. Since $(x-4)^2=y+2$ is equivalent to $(x-4)^2=4\left(\frac{1}{4}\right)(y+2)$, the parabola has vertex $(4,-2)$, and it opens upward $(c>0)$.

59. C. Since $y+2=-(x-4)^2$ is equivalent to $(x-4)^2=4\left(-\frac{1}{4}\right)(y+2)$, the parabola has vertex $(4,-2)$, and it opens downward $(c<0)$.

61. F. Since $(y-4)^2 = x+2$ is equivalent to $(y-4)^2 = 4\left(\frac{1}{4}\right)(x+2)$, the parabola has vertex $(-2,4)$, and it opens to the right $(c>0)$.

63. E. Since $x+2=-(y-4)^2$ is equivalent to $(y-4)^2 = 4\left(-\frac{1}{4}\right)(x+2)$, the parabola has vertex $(-2,4)$, and it opens to the left $(c<0)$.

65. (a) If both coordinates of the vertex are negative, the vertex is in quadrant III.
(b) If the first coordinate of the vertex is negative and the second is positive, the vertex is in quadrant II.
(c) If the first coordinate of the vertex is positive and the second is negative, the vertex is in quadrant IV.
(d) If both coordinates of the vertex are positive, the vertex is in quadrant I.

67. Since $x^2 = 16y$ is equivalent to $x^2 = 4(4)y$, the equation is in the form $x^2 = 4cy$ with $c = 4$. The focus is $(0,4)$, and the equation of the directrix is $y = -4$. The axis is $x = 0$, or the y-axis.

69. Since $x^2 = -\frac{1}{2}y$ is equivalent to $x^2 = 4\left(-\frac{1}{8}\right)y$, the equation is in the form $x^2 = 4cy$ with $c = -\frac{1}{8}$. The focus is $\left(0,-\frac{1}{8}\right)$, and the equation of the directrix is $y = \frac{1}{8}$. The axis is $x = 0$, or the y-axis.

71. Since $y^2 = \frac{1}{16}x$ is equivalent to $y^2 = 4\left(\frac{1}{64}\right)x$, the equation is in the form $y^2 = 4cx$ with $c = \frac{1}{64}$. The focus is $\left(\frac{1}{64},0\right)$, and the equation of the directrix is $x = -\frac{1}{64}$. The axis is $y = 0$, or the x-axis.

73. Since $y^2 = -16x$ is equivalent to $y^2 = 4(-4)x$, the equation is in the form $y^2 = 4cx$ with $c = -4$. The focus is $(-4,0)$. and the equation of the directrix is $x = 4$. The axis is $y = 0$, or the x-axis.

75. If the vertex is $(0,0)$ and the focus is $(0,-2)$, then the parabola opens downward and $c = -2$. The equation is $x^2 = 4cy \Rightarrow x^2 = -8y$.

77. If the vertex is $(0,0)$ and the focus is $\left(-\frac{1}{2},0\right)$, then the parabola opens to the left and $c = -\frac{1}{2}$. The equation is $y^2 = 4cx \Rightarrow y^2 = -2x$.

79. If the vertex is $(0,0)$ and the parabola opens to the right, the equation is in the form $y^2 = 4cx$. Find the value or c by using the fact that the parabola passes through $(2,-2,\sqrt{2})$. Thus, $(-2,\sqrt{2})^2 = 4c(2) \Rightarrow c = 1$. The equation is $y^2 = 4cx \Rightarrow y^2 = 4x$.

81. If the vertex is (0, 0) and the parabola opens downward, the equation is in the form $x^2 = 4cy$. Find the value of c by using the fact that the parabola passes through $\left(\sqrt{10},-5\right)$. Thus, $\left(\sqrt{10}\right)^2 = 4c(-5) \Rightarrow c = -\frac{1}{2}$. The equation is $x^2 = 4cy \Rightarrow x^2 = -2y$.

83. If the vertex is (0, 0) and the parabola has y-axis symmetry, the equation is in the form $x^2 = 4cy$. Find the value of c by using the fact that the parabola passes through $(2,-4)$. Thus, $(2)^2 = 4c(-4) \Rightarrow c = -\frac{1}{4}$. The equation is $x^2 = 4cy \Rightarrow x^2 = -y$.

85. If the focus is (0,2) and the vertex is (0,1), the parabola opens upward and $c = 1$. Substituting in $(x-h)^2 = 4c(y-k)$, we get $(x-0)^2 = 4(1)(y-1)$ or $x^2 = 4(y-1)$.

87. If the focus is (0,0) and the directrix has equation $x = -2$, the vertex is (-1,0) and $c = 1$. The parabola opens to the right. Substituting in $(y-k)^2 = 4c(x-h)$, we get $(y-0)^2 = 4(1)(x-(-1))$ or $y^2 = 4(x+1)$.

89. If the focus is (-1,3) and the directrix has equation $y = 7$, the vertex is (-1,5) and $c = -2$. The parabola opens downward. Substituting in $(y-k)^2 = 4c(x-h)$, we get $(x+1)^2 = 4(-2)(y-5)$ or $(x+1)^2 = -8(y-5)$.

91. Since the parabola has a horizontal axis, the equation is in the form $(y-k)^2 = 4c(x-h)$. Find the value of c by using the fact that the parabola passes through (-4,0) and the vertex is (-2,3). Substituting $x=-4$, $y=0$, $h=-2$, and $k=3$ yields $(0-3)^2 = 4c(-4(-2)) \Rightarrow c = -\frac{9}{8}$. The equation is $(y-3)^2 = -\frac{9}{2}(x+2)$.

93. The equation $y = (x+3)^2 - 4$ can be written as $(x+3)^2 = 4\left(\frac{1}{4}\right)(y+4)$. The vertex is $(-3,-4)$. The vertical axis has equation $x = -3$, and the parabola opens upward. See Figure 93. From the figure, the domain is $(-\infty,\infty)$, and the range is $[-4,\infty)$.

95. The equation $y = -2(x+3)^2 + 2$ can be written as $(x+3)^2 = 4\left(-\frac{1}{8}\right)(y-2)$. The vertex is (-3,2). The vertical axis has equation $x = -3$, and the parabola opens downward. See Figure 95. From the figure, the domain is $(-\infty,\infty)$, and the range is $(-\infty,2]$.

97. Rewrite the equation: $y = x^2 - 2x + 3 \Rightarrow y - 3 + 1 = x^2 - 2x + 1 \Rightarrow y - 2 = (x-1)^2$. The equation $y - 2 = (x-1)^2$ can be written as $(x-1)^2 = 4\left(\frac{1}{4}\right)(y-2)$. The vertex is (1,2).The vertical axis has equation

$x = 1$, and the parabola opens upward. See Figure 97. From the figure, the domain is $(-\infty, \infty)$, and the range is $[2, \infty)$.

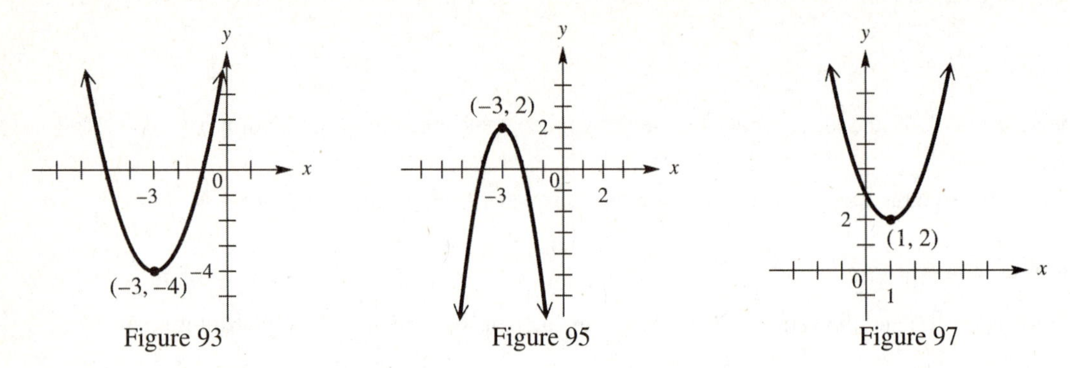

Figure 93 Figure 95 Figure 97

99. Rewrite the equation: $y = 2x^2 - 4x + 5 \Rightarrow y - 5 + 2 = 2(x^2 - 2x + 1) \Rightarrow y - 3 = 2(x-1)^2$. The equation $y - 3 = 2(x-1)^2$ can be written as $(x-1)^2 = 4\left(\frac{1}{8}\right)(y-3)$. The vertex is (1,3). The vertical axis has equation $x = 1$, and the parabola opens upward. See Figure 99. From the figure, the domain is $(-\infty, \infty)$, and the range is $[3, \infty)$.

101. The equation $x = y^2 + 2$ can be written as $(y-0)^2 = 4\left(\frac{1}{4}\right)(x-2)$. The vertex is (2,0). The horizontal axis has equation $y = 0$, and the parabola opens to the right. See Figure 101. From the figure, the domain is $[2, \infty)$ and the range is $(-\infty, \infty)$.

103. The equation $x = (y-3)^2$ can be written as $(y-3)^2 = 4\left(\frac{1}{4}\right)(x-0)$. The vertex is (0,3). The horizontal axis has equation $y = 3$, and the parabola opens to the right. See Figure 103. From the figure, the domain is $[0, \infty)$, and the range is $(-\infty, \infty)$.

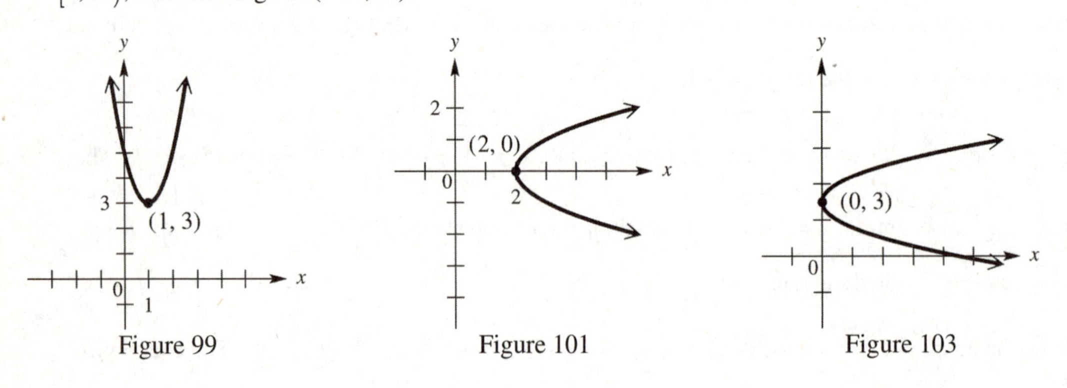

Figure 99 Figure 101 Figure 103

105. The equation $x=(y-4)^2+2$ can be written as $(y-4)^2=4\left(\frac{1}{4}\right)(x-2)$. The vertex is (2,4). The horizontal axis has equation $y=4,$ and the parabola opens to the right. See Figure 105. From the figure, the domain is $[2,\infty)$, and the range is $(-\infty,\infty)$.

107. Rewrite the equation: $x=\frac{2}{3}y^2-4y+8 \Rightarrow \frac{3}{2}x=y^2-6y+12 \Rightarrow \frac{3}{2}x-12+9=y^2-6y+9 \Rightarrow$ $\frac{3}{2}x-3=(y-3)^2 \Rightarrow \frac{3}{2}(x-2)=(y-3)^2$. The equation $\frac{3}{2}(x-2)=(y-3)^2$ can be written $(y-3)^2=4\left(\frac{3}{8}\right)(x-2)$. The vertex is $(2,3)$. The horizontal axis has equation $y=3$ and the parabola opens to the right. See Figure 107. From the figure, the domain is $[2,\infty)$ and the range is $(-\infty,\infty)$.

109. Rewrite the equation: $x=-4y^2-4y-3 \Rightarrow x+3-1=-4\left(y^2+y+\frac{1}{4}\right) \Rightarrow x+2=-4\left(y+\frac{1}{2}\right)^2$. The equation $x+2=-4\left(y+\frac{1}{2}\right)^2$ can be written $\left(y+\frac{1}{2}\right)^2=4\left(-\frac{1}{16}\right)(x+2)$. The vertex is $\left(-2,-\frac{1}{2}\right)$. The horizontal axis has equation $y=-\frac{1}{2}$ and the parabola opens to the left. See Figure 109. From the figure, the domain is $(-\infty,-2]$ and the range is $(-\infty,\infty)$.

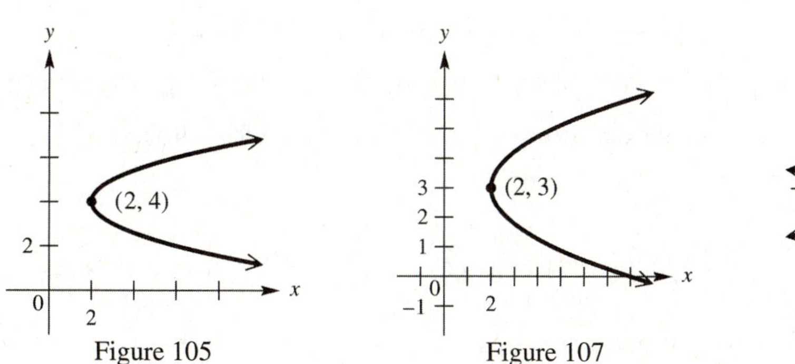

Figure 105

Figure 107

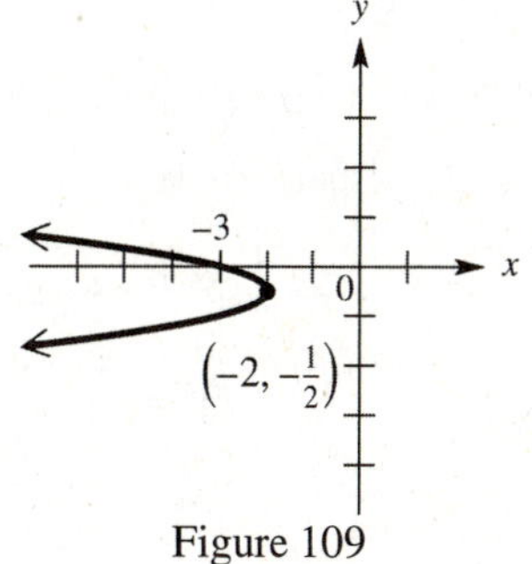

Figure 109

111. Rewrite the equation: $x=-2y^2+2y-3 \Rightarrow x+3-\frac{1}{2}=-2\left(y^2-y+\frac{1}{4}\right) \Rightarrow x+\frac{5}{2}=-2\left(y-\frac{1}{2}\right)^2$. The equation $x+\frac{5}{2}=-2\left(y-\frac{1}{2}\right)^2$ can be written $\left(y-\frac{1}{2}\right)^2=4\left(-\frac{1}{8}\right)\left(x+\frac{5}{2}\right)$. The vertex is $\left(-\frac{5}{2},\frac{1}{2}\right)$. The horizontal axis has equation $y=\frac{1}{2}$ and the parabola opens to the left. See Figure 111. From the figure, the domain is $\left(-\infty,-\frac{5}{2}\right]$ and the range is $(-\infty,\infty)$.

113. Rewrite the equation: $2x = y^2 - 2y + 9 \Rightarrow 2x - 9 + 1 = y^2 - 2y + 1 \Rightarrow 2(x-4) = (y-1)^2$. The equation $2(x-4) = (y-1)^2$ can be written $(y-1)^2 = 4\left(\frac{1}{2}\right)(x-4)$. The vertex is $(4, 1)$. The horizontal axis has equation $y = 1$ and the parabola opens to the right. See Figure 113. From the figure, the domain is $[4, \infty)$ and the range is $(-\infty, \infty)$.

115. Rewrite the equation: $y^2 - 4y + 4 = 4x + 4 \Rightarrow (y-2)^2 = 4(x+1)$. The equation $(y-2)^2 = 4(x+1)$ can be written $(y-2)^2 = 4(1)(x+1)$. The vertex is $(-1, 2)$. The horizontal axis has equation $y = 2$ and the parabola opens to the right. See Figure 115. From the figure, the domain is $[-1, \infty)$ and the range is $(-\infty, \infty)$.

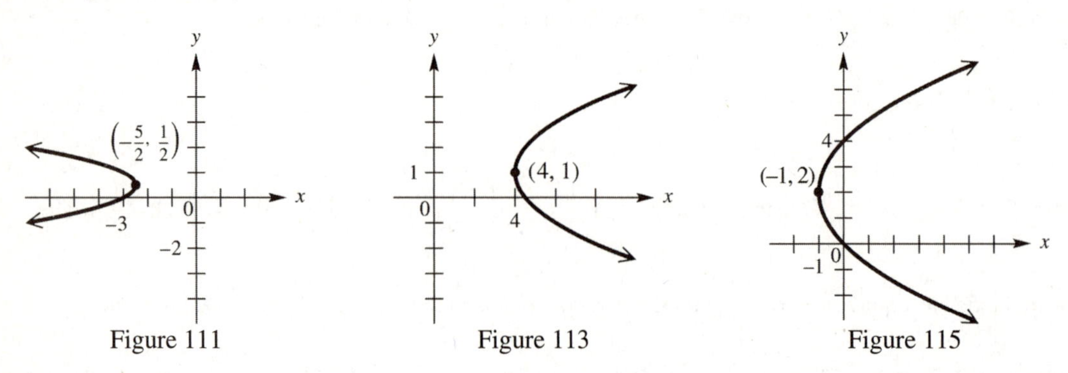

Figure 111 Figure 113 Figure 115

117. Since the directrix has equation $x = -c$, a point on the directrix has the form $(-c, y)$. Let (x, y) be a point on the parabola. By definition, the distance from the focus $(c, 0)$ to point (x, y) on the parabola, must be equal to the distance from point $(-c, y)$ on the directrix to point (x, y) on the parabola. That is

$$\sqrt{(x-c)^2 + (y-0)^2} = \sqrt{(x+c)^2 + (y-y)^2} \Rightarrow$$

$$(x-c)^2 + y^2 = (x+c)^2 \Rightarrow x^2 - 2xc + c^2 + y^2 = x^2 + 2xc + c^2 \Rightarrow -2xc + y^2 = 2xc \Rightarrow y^2 = 4xc.$$

119. (a) For Mars, $y = \frac{19}{11}x - \frac{12.6}{3872}x^2$. For the moon, $y = \frac{19}{11}x - \frac{5.2}{3872}x^2$. Graph

$y_1 = (19/11)x - (12.6/3872)x^2$ **and** $y_2 = (19/11)x - (5.2/3872)x^2$ as shown in Figure 119.

(b) From the graph, the ball thrown on Mars reaches a maximum height of $y \approx 229$ and the ball thrown on the moon reaches a maximum height of $y \approx 555$.

[0,1500] by [0,1000]
Xscl = 500 Yscl = 500

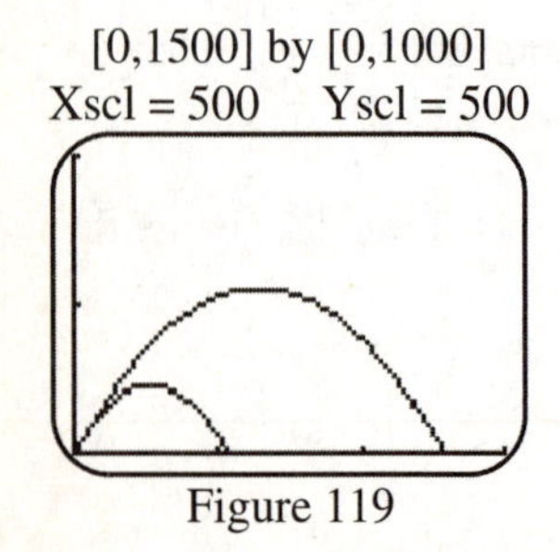

Figure 119

121. $y=-\dfrac{5\times10^{-9}}{2(10^{7})}(0.4)^2=-4\times10^{-17}$; the alpha particle is deflected 4×10^{-17} meter downward.

123. Let the vertex of the parabola be (0, 12). The equation of the parabola is of the form $(x-h)^2=4c(y-k)$.

By substitution, the equation is $(x-0)^2=4c(y-12)\Rightarrow\ x^2=4c(y-12)$. Since the parabola passes through the point (6, 0), the value of c can be found by substitution:

$6^2=4c(0-12)\Rightarrow 36=4c(-12)\Rightarrow -3=4c\Rightarrow c=-\dfrac{3}{4}$. The equation is $x^2=-3(y-12)$. Noting that the y-coordinate 9 feet up is 9, half the width can be found by substitution: $x^2=-3(9-12)\Rightarrow x^2=9\Rightarrow x=3$; The width is 6 feet.

6.2: Ellipses and Hyperbolas

1. G. This is an ellipse with $a^2=16,\ b^2=4,$ and $c=\sqrt{16-4}=\sqrt{12}=2\sqrt{3}$. The Foci are $\left(0,\pm2\sqrt{3}\right)$.

3. F. This is a hyperbola centered at (0, 0) with a horizontal transverse axis.

5. E. Since $h=-2$ and $k=4$, this is an ellipse centered at $(-2, 4)$.

7. D. Since $h=-2$ and $k=4$, this is a hyperbola centered at $(-2,4)$.

9. A circle can be interpreted as an ellipse whose foci have the same coordinates. The "coinciding foci" give the center of the circle.

11. $\dfrac{x^2}{9}+\dfrac{y^2}{4}=1\Rightarrow a=3$ and $b=2$. $a^2-b^2=3^2-2^2=5=c^2\Rightarrow c=\sqrt{5}$. The foci are $(\pm\sqrt{5},0)$. The endpoints of the major axis (vertices) are $(\pm3,0)$ so the domain is $[-3,3]$. The endpoints of the minor axis are $(0,\pm2)$, so the range is $[-2,2]$. The ellipse is graphed in Figure 11.

13. $9x^2+6y^2=54\Rightarrow\dfrac{x^2}{6}+\dfrac{y^2}{9}=1\Rightarrow a=3$ and $b=\sqrt{6}$. $a^2-b^2=9-6=3=c^2\Rightarrow\ c=\sqrt{3}$. The foci are $\left(0,\pm\sqrt{3}\right)$. The endpoints of the major axis (vertices) are $(0,\pm3)$ so the range is $[-3,3]$. The endpoints of the minor axis are $\left(\pm\sqrt{6},0\right)$ so the domain is $\left[-\sqrt{6},\sqrt{6}\right]$. The ellipse is graphed in Figure 13.

15. $\dfrac{25y^2}{36}+\dfrac{64x^2}{9}=1\Rightarrow\dfrac{y^2}{\frac{36}{25}}+\dfrac{x^2}{\frac{9}{64}}=1\Rightarrow a=\sqrt{\dfrac{36}{25}}=\dfrac{6}{5}$ and $b=\sqrt{\dfrac{9}{64}}=\dfrac{3}{8}$

The endpoints of the major axis (vertices) are $\left(0,\pm\dfrac{6}{5}\right)$ so the range is $\left[-\dfrac{6}{5},\dfrac{6}{5}\right]$. The endpoints of the minor axis are $\left(\pm\dfrac{3}{8},0\right)$ so the domain is $\left[-\dfrac{3}{8},\dfrac{3}{8}\right]$. See Figure 15.

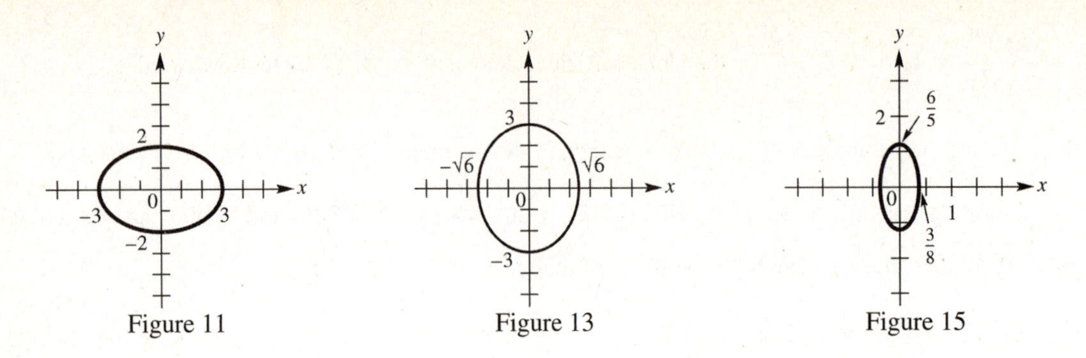

Figure 11 Figure 13 Figure 15

17. The ellipse is centered at $(1,-3)$. The major axis is vertical and has length $2a=10$. The length of the minor axis is $2b=6$ The graph is shown in Figure 17. The domain is $[-2,4]$ and the range is $[-8, 2]$.

19. The ellipse is centered at $(2,1)$. The major axis is horizontal and has length $2a=8$. The length of the minor axis is $2b=6$. The graph is shown in Figure 19. The domain is $[-2,6]$ and the range is $[-2,4]$.

21. The ellipse is centered at $(-1,2)$. The major axis is horizontal and has length $2a=16$. The length of the minor axis is $2b=14$. The graph is shown in Figure 21. The domain is $[-9,7]$ and the range is $[-5,9]$.

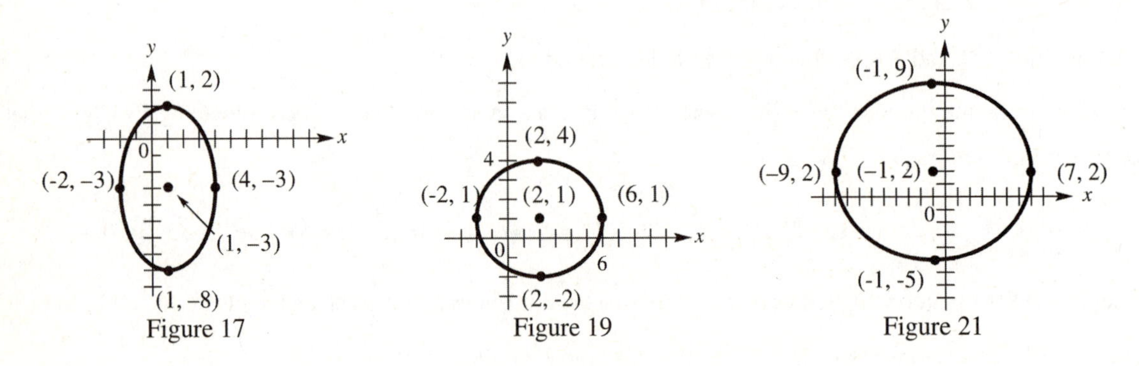

Figure 17 Figure 19 Figure 21

23. The ellipse is centered between the foci at $(0,0)$. The major axis is horizontal with $a=4$. Since the foci are $(\pm 2, 0)$, we know that $c=2$. Since $c^2=a^2-b^2$, the value of b can be found by substitution:

$b^2=a^2-c^2=4^2-2^2=16-4=12 \Rightarrow b=\sqrt{12}$. The equation is $\frac{x^2}{16}+\frac{y^2}{12}=1$.

25. The ellipse is centered between the foci at $(0,0)$. The major axis is vertical with $a=2\sqrt{2}$. Since the foci are $(0,\pm 2)$, we know that $c=2$. Since $c^2=a^2-b^2$, the value of b can be found by substitution:

$b^2=a^2-c^2=\left(2\sqrt{2}\right)^2-2^2=8-4=4 \Rightarrow b=2$. The equation is $\frac{x^2}{4}+\frac{y^2}{8}=1$.

27. The ellipse is centered between the endpoint of the major axis of $(0,0)$. The major axis is horizontal with $a=4$ and the minor axis is vertical with $b=2$. The equation is $\frac{x^2}{16}+\frac{y^2}{4}=1$.

29. The ellipse is centered between the endpoints of the major axis at $(0,0)$. The major axis is horizontal with $a=6$.

Since $c^2=a^2-b^2$, the value of b can be found by substitution:

$b^2=a^2-c^2=6^2-4^2=36-16=20\Rightarrow b=\sqrt{20}$. The equation is $\dfrac{x^2}{36}+\dfrac{y^2}{20}=1$.

31. Since the center is $(3,-2)$, we know that $h=3$ and $k=-2$. Since $c^2=a^2-b^2$, the value of b can be found by substitution: $b^2=a^2-c^2=5^2-3^2=25-9=16\Rightarrow b=4$. The major axis is vertical so the equation is $\dfrac{(x-3)^2}{16}+\dfrac{(y+2)^2}{25}=1$.

33. The ellipse is centered between the foci at $(0,0)$. The major axis is vertical with $a=3$. Since the foci are $(0,\pm 2)$, we know that $c=2$. Since $c^2=a^2-b^2$, the value of b can be found by substitution:

$b^2=a^2-c^2=3^2-2^2=9-4=5\Rightarrow b=\sqrt{5}$. The equation is $\dfrac{y^2}{9}+\dfrac{x^2}{5}=1$.

35. Since the center is $(5,2)$, we know that $h=5$ and $k=2$. Since the minor axis is horizontal and has length 8, $b=4$. Since $c^2=a^2-b^2$ and $c=3$, the value of a can be found by substitution:

$3^2+4^2=a^2\Rightarrow a^2=25\Rightarrow a=5$. The equation is $\dfrac{(x-5)^2}{25}+\dfrac{(y-2)^2}{16}=1$.

37. The ellipse is centered between the vertices at $(4,5)$ and $a=\dfrac{9-1}{2}=4$. The major axis is vertical. Since the minor axis has length 6, $b=3$. The equation is $\dfrac{(x-4)^2}{9}+\dfrac{(y-5)^2}{16}=1$.

39. $9x^2+18x+4y^2-8y-23=0\Rightarrow 9(x^2+2x)+4(y^2-2y)=23\Rightarrow 9(x^2+2x+1)+4(y^2-2y+1)=$

$23+9+4\Rightarrow 9(x+1)^2+4(y-1)^2=36\Rightarrow \dfrac{(x+1)^2}{4}+\dfrac{(y-1)^2}{9}=1$

The center is $(-1,1)$. The vertices are $(-1,1-3),(-1,1+3)$ or $(-1,-2),(-1,4)$.

41. $4x^2+8x+y^2+2y+1=0\Rightarrow 4(x^2+2x)+(y^2+2y)=-1\Rightarrow 4(x^2+2x+1)+(y^2+2y+1)=$

$-1+4+1\Rightarrow 4(x+1)^2+(y+1)^2=4\Rightarrow \dfrac{(x+1)^2}{1}+\dfrac{(y+1)^2}{4}=1$

The center is $(-1,-1)$. The vertices are $(-1,-1-2)$, $(-1,-1+2)$ or $(-1,-3),(-1,1)$.

43. $4x^2+16x+5y^2-10y+1=0\Rightarrow 4(x^2+4x)+5(y^2-2y)=-1\Rightarrow 4(x^2+4x+4)+5(y^2-2y+1)=$

$-1+16+5\Rightarrow 4(x+2)^2+5(y-1)^2=20\Rightarrow \dfrac{(x+2)^2}{5}+\dfrac{(y-1)^2}{4}=1$

The center is $(-2,1)$. The vertices are $(-2-\sqrt{5},1),(-2+\sqrt{5},1)$.

45. $16x^2-16x+4y^2+12y=51 \Rightarrow 16(x^2-x)+4(y^2+3y)=51 \Rightarrow 16\left(x^2-x+\frac{1}{4}\right)+4\left(y^2+3y+\frac{9}{4}\right)=$

$51+4+9 \Rightarrow 16\left(x-\frac{1}{2}\right)^2+4\left(y+\frac{3}{2}\right)^2=64 \Rightarrow \frac{\left(x-\frac{1}{2}\right)^2}{4}+\frac{\left(y+\frac{3}{2}\right)^2}{16}=1$

The center is $\left(\frac{1}{2},-\frac{3}{2}\right)$. The vertices are $\left(\frac{1}{2},-\frac{3}{2}-4\right),\left(\frac{1}{2},-\frac{3}{2}+4\right)$ or $\left(\frac{1}{2},-\frac{11}{2}\right),\left(\frac{1}{2},\frac{5}{2}\right)$.

47. The transverse axis is horizontal with $a=4$ and $b=3$. The asymptotes are $y=\pm\frac{3}{4}x$. See Figure 47. The domain is $(-\infty,-4)\cup[4,\infty)$ and the range is $(-\infty,\infty)$.

49. $49y^2-36x^2=1764 \Rightarrow \frac{y^2}{36}-\frac{x^2}{49}=1$ The transverse axis is vertical with $a=6$ and $b=7$. The asymptotes are $y=\pm\frac{6}{7}x$. See Figure 49. The domain is $(-\infty,\infty)$ and the range is $(-\infty,-6]\cup[6,\infty)$.

51. $\frac{4x^2}{9}-\frac{25y^2}{16}=1 \Rightarrow \frac{x^2}{\frac{9}{4}}-\frac{y^2}{\frac{16}{25}}=1$. The transverse axis is horizontal with $a=\frac{3}{2}$ and $b=\frac{4}{5}$. The asymptotes are $y=\pm\frac{18}{15}x$. See Figure 51. The domain is $\left(-\infty,-\frac{3}{2}\right]\cup\left[\frac{3}{2},\infty\right)$ and the range is $(-\infty,\infty)$.

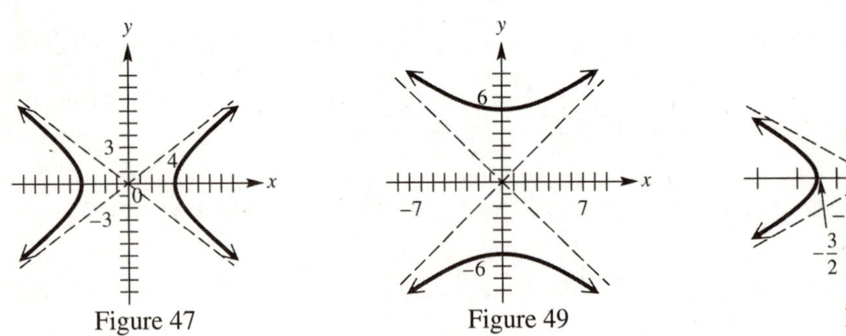

Figure 47

Figure 49

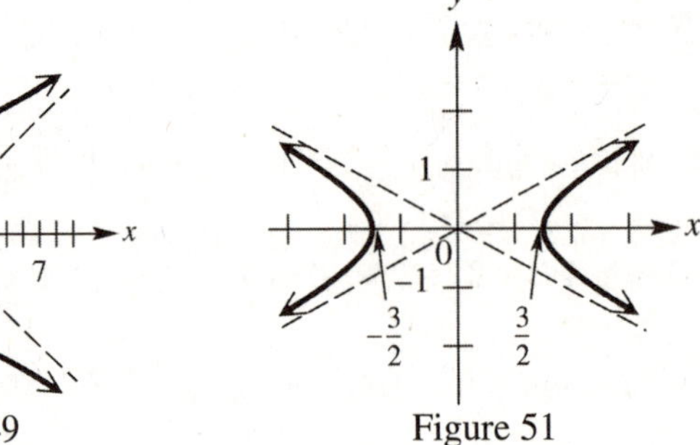

Figure 51

53. $9x^2-4y^2=1 \Rightarrow \frac{x^2}{\frac{1}{9}}-\frac{y^2}{\frac{1}{4}}=1$. The transverse axis is horizontal with $a=\frac{1}{3}$ and $b=\frac{1}{2}$. The asymptotes are $y=\pm\frac{3}{2}x$. See Figure 53. The domain is $\left(-\infty,-\frac{1}{3}\right]\cup\left[\frac{1}{3},\infty\right)$ and the range is $(-\infty,\infty)$.

55. The center is $(1,-3)$ and the transverse axis is horizontal with $a=3$ and $b=5$. See Figure 55. The domain is $(-\infty,-2]\cup[4,\infty)$ and the range is $(-\infty,\infty)$.

57. The center is $(-1,5)$ and the transverse axis is vertical with $a=2$ and $b=3$. See Figure 57. The domain is $(-\infty,\infty)$ and the range is $(-\infty,3]\cup[7,\infty)$.

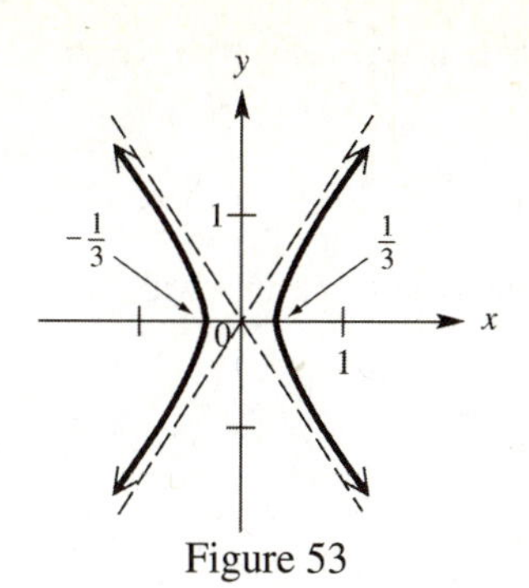

Figure 53

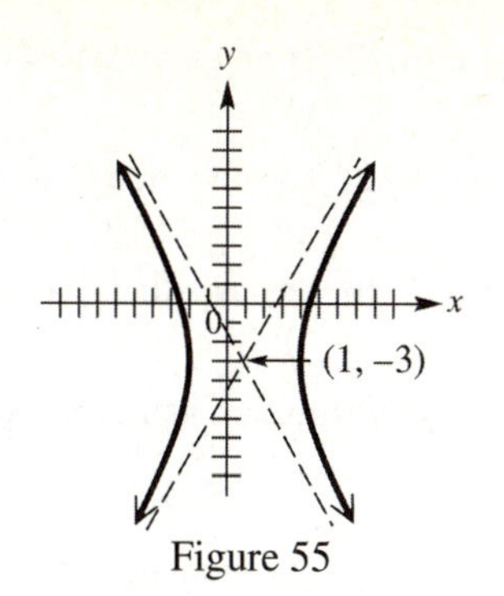

Figure 55

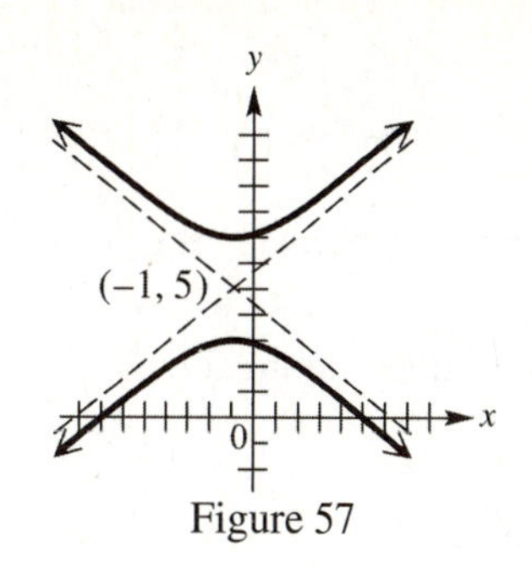

Figure 57

59. $16(x+5)^2-(y-3)^2=1 \Rightarrow \dfrac{(x+5)^2}{\frac{1}{16}}-\dfrac{(y-3)^2}{1}=1$ The center is (–5, 3) and the transverse axis is horizontal with $a=\dfrac{1}{4}$ and $b=1$. See Figure 59. The domain is $\left(-\infty,-5-\dfrac{1}{4}\right]\cup\left[-5+\dfrac{1}{4},\infty\right)$ or $\left(-\infty,-\dfrac{21}{4}\right]\cup\left[-\dfrac{19}{4},\infty\right)$ and the range is $(-\infty,\infty)$.

61. $9(x-2)^2-4(y+1)^2=36 \Rightarrow \dfrac{(x-2)^2}{4}-\dfrac{(y+1)^2}{9}=1$. The center is (2, –1) and the transverse axis is vertical with $a=2$ and $b=3$. See Figure 61. The domain is $(-\infty,0]\cup[4,\infty)$ and the range is $(-\infty,\infty)$.

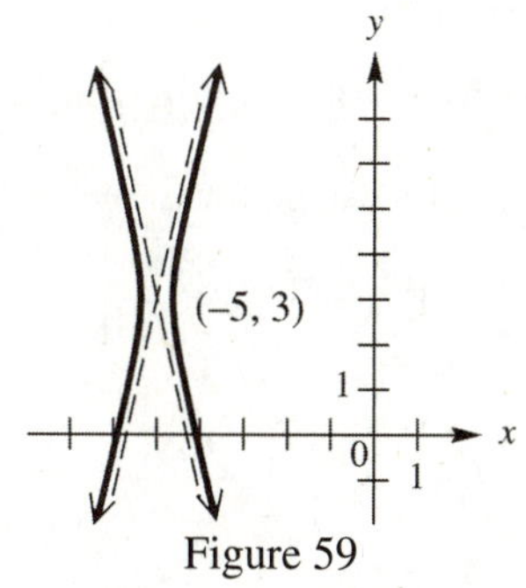

Figure 59

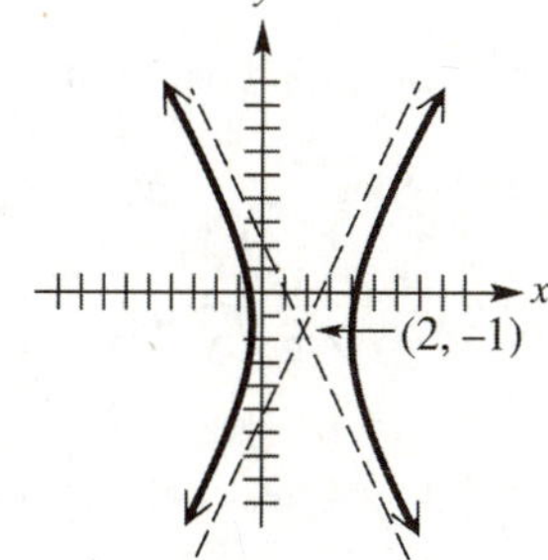

Figure 61

63. The hyperbola has a horizontal transverse axis with $c=4$. The x-intercepts coincide with the vertices, so $a=3$. The center is located between the foci at (0, 0). Since $c^2=a^2+b^2$, the value of b can be found by substitution: $b^2=c^2-a^2=4^2-3^2=16-9=7 \Rightarrow b=\sqrt{7}$. The equation is $\dfrac{x^2}{9}-\dfrac{y^2}{7}=1$.

65. The asymptotes intersect at the origin so the center is (0, 0). The hyperbola has a vertical transverse axis and the y-intercepts coincide with the vertices so $a=3$. From the asymptotes, $\dfrac{a}{b}=\dfrac{3}{5}$ with $a=3 \Rightarrow b=5$. The equation is $\dfrac{y^2}{9}-\dfrac{x^2}{25}=1$.

67. The center is located between the vertices at (0, 0). The hyperbola has a vertical transverse axis with $a = 6$.

From the asymptotes, $\frac{a}{b} = \frac{1}{2}$ with $a = 6 \Rightarrow b = 12$. The equation is $\frac{y^2}{36} - \frac{x^2}{144} = 1$.

69. The center is located between the vertices at (0, 0). The hyperbola has a horizontal transverse axis with $a = 3$.

The equation is of the form $\frac{x^2}{a^2} - \frac{y^2}{b^2} = 1$. By substitution using the point (6, 1),

$\frac{6^2}{3^2} - \frac{1^2}{b^2} = 1 \Rightarrow \frac{36}{9} - 1 = \frac{1}{b^2} \Rightarrow 3 = \frac{1}{b^2} \Rightarrow b^2 = \frac{1}{3}$. The equation is $\frac{x^2}{3^2} - \frac{y^2}{\frac{1}{3}} = 1$ or $\frac{x^2}{9} - 3y^2 = 1$.

71. The center is located between the foci at (0, 0). The hyperbola has a vertical transverse axis with $c = \sqrt{13}$. From the asymptotes, $\frac{a}{b} = 5$. Also, from $c^2 = a^2 + b^2 \Rightarrow a^2 + b^2 = 13$. Solving these equations simultaneously results in $a^2 = \frac{25}{2}$ and $b^2 = \frac{1}{2}$. The equation is $\frac{y^2}{\frac{25}{2}} - \frac{x^2}{\frac{1}{2}} = 1$ or $\frac{2y^2}{25} - 2x^2 = 1$.

73. The center is located between the vertices at (4, 3). The hyperbola has a vertical transverse axis with $a = 2$.

From the asymptotes, $\frac{a}{b} = 7$ with $a = 2 \Rightarrow b = \frac{2}{7}$. The equation is $\frac{(y-3)^2}{4} - \frac{(x-4)^2}{\frac{4}{49}} = 1$ or

$\frac{(y-3)^2}{4} - \frac{49(x-4)^2}{4} = 1$.

75. With center (1, –2) and vertex (3, –2), we know the hyperbola has a horizontal transverse axis with $a = 2$. With center (1, –2) and focus (4, –2), we know $c = 3$. Since $c^2 = a^2 + b^2$, the value of b can be found by substitution: $b^2 = c^2 - a^2 = 3^2 - 2^2 = 9 - 4 = 5 \Rightarrow b = \sqrt{5}$. The equation is $\frac{(x-1)^2}{4} - \frac{(y+2)^2}{5} = 1$.

77. $x^2 - 2x - y^2 + 2y = 4 \Rightarrow (x^2 - 2x + 1) - (y^2 - 2y + 1) = 4 + 1 - 1 \Rightarrow (x-1)^2 - (y-1)^2 = 4 \Rightarrow$

$\frac{(x-1)^2}{4} - \frac{(y-1)^2}{4} = 1$. The center is (1, 1). The vertices are (1–2, 1), (1+2, 1) or (–1, 1), (3, 1).

79. $3y^2 + 24y - 2x^2 + 12x + 24 = 0 \Rightarrow 3(y^2 + 8y) - 2(x^2 - 6x) = -24 \Rightarrow$

$3(y^2 + 8y + 16) - 2(x^2 - 6x + 9) = -24 + 48 - 18 \Rightarrow 3(y+4)^2 - 2(x-3)^2 = 6 \Rightarrow \frac{(y+4)^2}{2} - \frac{(x-3)^2}{3} = 1$.

The center is $(3, 4)$. The vertices are $(3, -4-\sqrt{2})$, $(3, -4+\sqrt{2})$.

81. $x^2 - 6x - 2y^2 + 7 = 0 \Rightarrow (x^2 - 6x + 9) - 2y^2 = -7 + 9 \Rightarrow (x-3)^2 - 2(y-0)^2 = 2 \Rightarrow$

$\frac{(x-3)^2}{2} - \frac{(y-0)^2}{1} = 1$. The center is $(3, 0)$. The vertices are $(3-\sqrt{2}, 0)$, $(3+\sqrt{2}, 0)$.

83. $4y^2 + 32y - 5x^2 - 10x + 39 = 0 \Rightarrow 4(y^2 + 8y) - 5(x^2 + 2x) = -39 \Rightarrow$

$$4(y^2+8y+16)-5(x^2+2x+1)=-39+64-5 \Rightarrow 4(y+4)^2-5(x+1)^2=20 \Rightarrow \frac{(y+4)^2}{5}-\frac{(x+1)^2}{4}=1.$$

The center is $(-1,-4)$. The vertices are $(-1,-4-\sqrt{5}),\ (-1,-4+\sqrt{5})$.

85. $c^2=a^2-b^2 \Rightarrow c^2=16-12 \Rightarrow c^2=4 \Rightarrow c=2$. The foci are $F_1(-2,0)$ and $F_2(2,0)$.

86. See Figure 86. The point (3, 2.2912878) is shown. Other points include (0, 3.4641016) and (–3, –2.291288).

87. $\sqrt{(3-(-2))^2+(2.2912878-0)^2}+\sqrt{(3-2)^2+(2.2912878-0)^2} \approx 7.999999937 \approx 8.$

$\sqrt{(0-(-2))^2+(3.4641016-0)^2}+\sqrt{(0-2)^2+(3.4641016-0)^2} \approx 7.999999974 \approx 8.$

$\sqrt{(-3-(-2))^2+(-2.291288-0)^2}+\sqrt{(-3-2)^2+(-2.291288-0)^2} \approx 8.000000203 \approx 8.$ Note that the sums calculated here do not equal exactly 8 because of rounding.

88. $c^2=a^2+b^2 \Rightarrow c^2=4+12 \Rightarrow c^2=16 \Rightarrow c=4$. The foci are $a=2$ and $F_2(4,0)$. See Figure 88. The point(–3, 3.8729833) is shown. Other points include (–2, 0), (2, 0) and (4, 6).

[-9.4,9.4] by [-6.2,6.2]
Xscl = 1 Yscl = 1

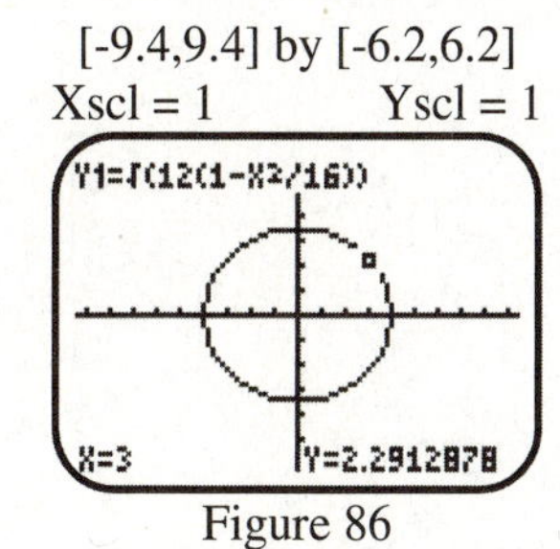

Figure 86

[-9.4,9.4] by [-6.2,6.2]
Xscl = 1 Yscl = 1

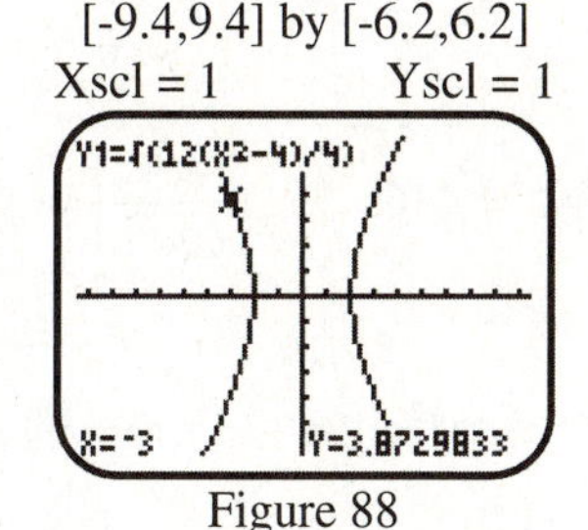

Figure 88

89. $\left|\sqrt{(-3-(-4))^2+(3.8729833-0)^2}-\sqrt{(-3-4)^2+(3.8729833-0)^2}\right| \approx 4.000000022 \approx 4.$

$\left|\sqrt{(-2-(-4))^2+(0-0)^2}-\sqrt{(-2-4)^2+(0-0)^2}\right|=4.$

$\left|\sqrt{(2-(-4))^2+(0-0)^2}-\sqrt{(2-4)^2+(0-0)^2}\right|=4.$

$\left|\sqrt{(4-(-4))^2+(6-0)^2}-\sqrt{(4-4)^2+(6-0)^2}\right|=4.$ Note that the first difference calculated here does not equal exactly 4 because of rounding.

90. Exercise 87 demonstrates that the points on the graph satisfy the definition of an ellipse for that particular ellipse. Exercise 89 demonstrates that the points on the graph satisfy the definition of a hyperbola for that particular hyperbola.

91. The patient and the emitter are 12 units apart. These positions represent the foci of the ellipse so $c=6$. With the minor axis measuring 16 units, $b=8$. Since $c^2=a^2-b^2$, the value of a can be found by substitution: $a^2=b^2+c^2=6^2+8^2=36+64=100 \Rightarrow a=10$. The equation is $\frac{x^2}{100}+\frac{y^2}{64}=1$.

93. A major axis measuring 620 feet indicates that $a = 310$. A minor axis measuring 513 feet indicates that $b = 256.5$. Then $5c^2 = a^2 - b^2 = 310^2 - 256.5^2 \Rightarrow c = \sqrt{310^2 - 256.5^2} \approx 174.1$. The distance between the foci is $2c \approx 2(174.1) \approx 348.2$ feet.

95. Using a vertical major axis, $a = 15$. The minor axis has length 20, so $b = 10$. The equation is $\frac{y^2}{225} + \frac{x^2}{100} = 1$. Assuming the truck drives exactly in the middle of the road, we want to find y when $x = 6$.
$\frac{y^2}{225} + \frac{6^2}{100} = 1 \Rightarrow \frac{y^2}{225} = 1 - \frac{36}{100} \Rightarrow y^2 = 225\left(1 - \frac{36}{100}\right) \Rightarrow y = \sqrt{225\left(1 - \frac{36}{100}\right)} = 12$. The truck must be just under 12 feet high to pass through.

97. (a) Since $c = \sqrt{a^2 - b^2} = \sqrt{4465^2 - 4462^2} \approx 163.6$, one focus is located at the point $(163.6, 0)$. The graph representing Earth is a circle with radius 3960 with center (163.6, 0). The equation for Earth is $(x - 163.6)^2 + y^2 = 3690^2$. To graph this equation, solve for y and graph two parts.
$(x - 163.6)^2 + y^2 = 3690^2 \Rightarrow y^2 = 3690^2 - (x - 163.6)^2 \Rightarrow y = \pm\sqrt{3690^2 - (x - 163.6)^2}$ To graph the ellipse, solve for y and graph two parts.
$$\frac{x^2}{4465^2} + \frac{y^2}{4462^2} = 1 \Rightarrow \frac{y^2}{4462^2} = 1 - \frac{x^2}{4465^2} \Rightarrow y^2 = 4462^2\left(1 - \frac{x^2}{4465^2}\right) \Rightarrow y = \pm\sqrt{4462^2\left(1 - \frac{x^2}{4465^2}\right)}$$
The graphs are shown in Figure 97.

(b) The minimum distance is $4465 - (3960 + 163.6) \approx 341$ miles. The maximum distance is $4465 - (3960 + 163.6) \approx 669$ miles.

[-6750,6750] by [-4500,4500]
Xscl = 1000 Yscl = 1000

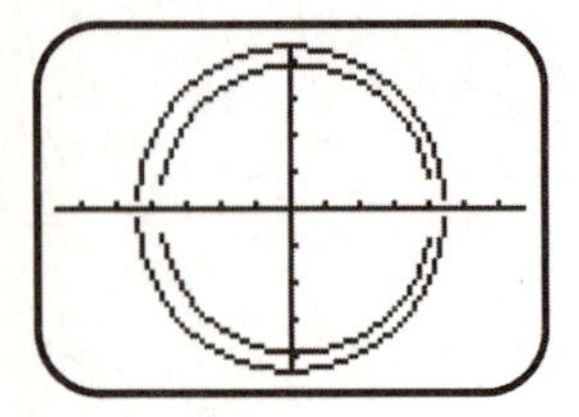

Figure 97

99. (a) Find a and b in the equation $\frac{x^2}{a^2} - \frac{y^2}{b^2} = 1$. Because the equations of the asymptotes of a hyperbola with horizontal transverse axes are $y = \pm\frac{b}{a}x$, and the given asymptotes are $y = \pm x$, it follows that $\frac{a}{b} = 1$ or $a = b$. Since the line $y = x$ intersects the x-axis at a 45° angle, the triangle shown in the third quadrant is a 45°- 45°- 90° right triangle and both legs must have length d. Then, by the Pythagorean theorem, $c^2 = d^2 + d^2 = 2d^2$. That gives $c = d\sqrt{2}$. Also, for a hyperbola $c^2 = a^2 + b^2$, and since $a = b$, $c^2 = a^2 + a^2 = 2a^2$. That gives $c = a\sqrt{2}$. From these two equations, $a\sqrt{2} = d\sqrt{2}$

and so $a = d$. That is, $a = b = d = 5\times10^{-14}$. Thus the equation of the trajectory of A, where $x > 0$, is given by $\dfrac{x^2}{\left(5\times10^{-14}\right)^2} - \dfrac{y^2}{\left(5\times10^{-14}\right)^2} = 1$. Solving for x yields $x^2 - y^2 = \left(5\times10^{-14}\right)^2 \Rightarrow$ $x^2 = y^2 + 2.5\times10^{-27} \Rightarrow x = \sqrt{y^2 + 2.5\times10^{-27}}$. This equation represents the right half of the hyperbola, as shown in the textbook.

(b) Since $a = 5\times10^{-14}$, the distance from the origin to the vertex is 5×10^{-14}. The distance from N to the origin can be found using the Pythagorean theorem. Let h represent this distance, then $h^2 = d^2 + d^2$. That is, $h^2 = \left(5\times10^{-14}\right)^2 + \left(5\times10^{-14}\right)^2 \Rightarrow h^2 = 5\times10^{-27} \Rightarrow h \approx 7\times10^{-14}$. The minimum distance between the centers of the alpha particle and the gold nucleus is $5\times10^{-14} + 7\times10^{-14} \approx 1.2\times10^{-13}$.

101. Let (x, y) be any point on the ellipse and start with the distance formula.

$\sqrt{(x+3)^2 + y^2} + \sqrt{(x-3)^2 + y^2} = 10$	Given Equation.
$\sqrt{(x+3)^2 + y^2} = 10 - \sqrt{(x-3)^2 + y^2}$	Subtract $\sqrt{(x-3)^2 + y^2}$.
$(x+3)^2 + y^2 = 100 - 20\sqrt{(x-3)^2 + y^2} + (x-3)^2 + y^2$	Square each side.
$(x+3)^2 - (x-3)^2 - 100 = -20\sqrt{(x-3)^2 + y^2}$	Subtract $(x-3)^2$ and 100.
$x^2 + 6x + 9 - x^2 + 6x - 9 - 100 = -20\sqrt{(x-3)^2 + y^2}$	Expand the binomials.
$12x - 100 = -20\sqrt{(x-3)^2 + y^2}$	Simplify.
$25 - 3x = 5\sqrt{(x-3)^2 + y^2}$	Divide each side by -4.
$625 - 150x + 9x^2 = 25\left(x^2 - 6x + 9 + y^2\right)$	Square each side.
$625 - 150x + 9x^2 = 25x^2 - 150x + 225 + 25y^2$	Multiply the right side.
$-16x^2 - 25y^2 = -400$	Simplify.
$\dfrac{-16x^2}{-400} + \dfrac{-25y^2}{-400} = \dfrac{-400}{-400}$	Divide each side by $x = \ln(t-1) \Rightarrow t = e^x + 1$.
$\dfrac{x^2}{25} + \dfrac{y^2}{16} = 1$	Simplify.

Reviewing Basic Concepts (Sections 6.1 and 6.2)

1. (a) The circle is defined in B.

   (b) The parabola is defined in D.

   (c) The ellipse is defined in A.

   (d) The hyperbola is defined in C.

2. $12x^2-4y^2=48 \Rightarrow \frac{x^2}{4}-\frac{y^2}{12}=1$. The transverse axis is horizontal with $a=2$ and $b=\sqrt{12}$. The asymptotes are $y=\pm\frac{\sqrt{12}}{2}x$. See Figure 2.

3. Rewrite the equation. $y=2x^2+3x-1 \Rightarrow y+1+\frac{9}{8}=2\left(x^2+\frac{3}{2}x+\frac{9}{16}\right) \Rightarrow y+\frac{17}{8}=2\left(x+\frac{3}{4}\right)^2$. The equation $y+\frac{17}{8}=2\left(x+\frac{3}{4}\right)^2$ can be written $y+\frac{17}{8}=4\left(\frac{1}{2}\right)\left(x+\frac{3}{4}\right)^2$. The vertex is $\left(-\frac{3}{4},\frac{17}{8}\right)$.

   See Figure 3.

4. $x^2+y^2-2x+2y-2=0 \Rightarrow \left(x^2-2x+1\right)+\left(y^2+2y+1\right)=2+1+1 \Rightarrow (x-1)^2+(y+1)^2=4$. The graph is a circle with center $(1,-1)$ and radius $\left(-4+\frac{1}{16},1\right) \Rightarrow \left(-\frac{63}{16},-1\right)$ See Figure 4.

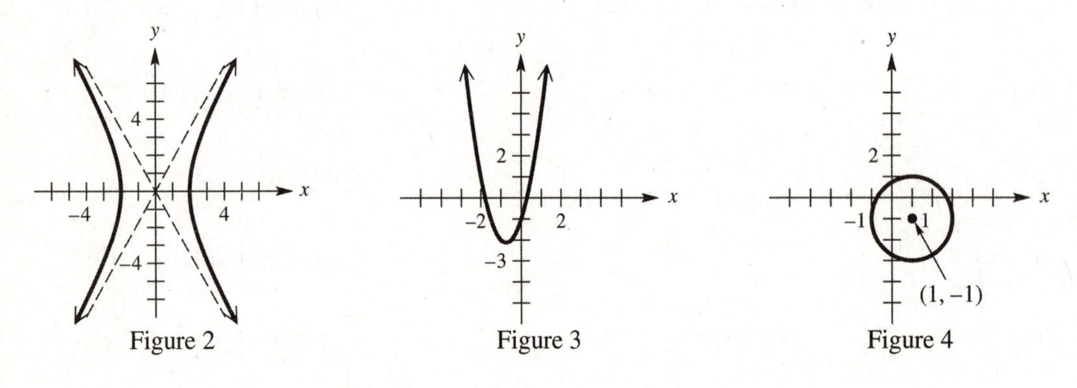

Figure 2 Figure 3 Figure 4

5. $4x^2+9y^2=36 \Rightarrow \frac{x^2}{9}+\frac{y^2}{4}=1 \Rightarrow a=3$ and $b=2$. The ellipse is graphed in Figure 5.

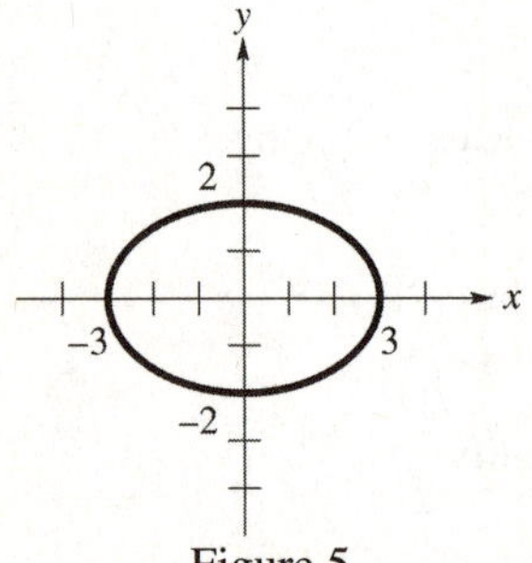

Figure 5

6. If $c<a$, it is an ellipse. If $c>a$, it is a hyperbola.

7. A circle with center $(2,-1)$ and radius 3 has equation $(x-2)^2+(y+1)^2=9$.

8. The ellipse is centered between the foci at $(0,0)$. The major axis is horizontal with $a=6$. Since the foci are $(\pm 4,0)$, we know that $c=4$. Since $c^2=a^2-b^2$, the value of b can be found by substitution: $b^2=a^2-c^2=6^2-4^2=36-16=20 \Rightarrow b=\sqrt{20}$. The equation is $\frac{x^2}{36}+\frac{y^2}{20}=1$.

9. The hyperbola has a vertical transverse axis with $c=4$. The vertices are $(0,\pm 2)$ so $a=2$. The center is located between the foci at $(0,0)$. Since $c^2=a^2+b^2$, the value of b can be found by substitution: $b^2=c^2-a^2=4^2-2^2=16-4=12 \Rightarrow b=\sqrt{12}$. The equation is $\frac{y^2}{4}-\frac{x^2}{12}=1$.

10. If the vertex is $(0,0)$ and the focus is $\left(0,\frac{1}{2}\right)$, then the parabola opens upward and $c=\frac{1}{2}$. The equation is $x^2=4cy \Rightarrow x^2=2y$.

Section 6.3 Summary of Conic Sections

1. $x^2+y^2=144 \Rightarrow (x-0)^2+(y-0)^2=12^2$ The graph of this equation is a circle with center $(0,0)$ and radius 12. Also, note in our original equation, the x^2- and y^2-terms have the same positive coefficient.

3. $y=2x^2+3x-4 \Rightarrow y=2\left(x^2+\frac{3}{2}x\right)-4 \Rightarrow y=2\left(x^2+\frac{3}{2}x+\frac{9}{16}-\frac{9}{16}\right)-4 \Rightarrow y=2\left(x+\frac{3}{4}\right)^2+2\left(-\frac{9}{16}\right)-4$

$\Rightarrow y=2\left(x+\frac{3}{4}\right)^2-\frac{9}{8}-4 \Rightarrow y=2\left(x+\frac{3}{4}\right)^2-\frac{41}{8} \Rightarrow y-\left(-\frac{41}{8}\right)=2\left[x-\left(-\frac{3}{4}\right)\right]^2$ The graph of this equation is a parabola opening upwards with a vertex of $\left(-\frac{3}{4},-\frac{41}{8}\right)$. Also, note our original equation has an x^2- term, but no y^2-term.

5. $x-1=-3(y-4)^2$ The graph of this equation is a parabola opening to the left with a vertex of $(1,4)$. Also, note when expanded, our original equation has a y^2-term, but no x^2-term.

7. $\frac{x^2}{49}+\frac{y^2}{100}=1 \Rightarrow \frac{x^2}{7^2}+\frac{y^2}{10^2}=1$ The graph of this equation is an ellipse centered at the origin and x-intercepts of 7 and –7, and y-intercepts of 10 and –10. Also, note in our original equation, the x^2- and y^2-terms both have different positive coefficients.

9. $\frac{x^2}{4}-\frac{y^2}{16}=1 \Rightarrow \frac{x^2}{2^2}-\frac{y^2}{4^2}=1$ The graph of this equation is a hyperbola centered at the origin with x-intercepts of 2 and –2, and asymptotes of $y=\pm\frac{4}{2}x=\pm 2x$. Also, note in our original equation, the x^2- and y^2-terms have coefficient that are opposite in sign.

11. $\frac{x^2}{25}-\frac{y^2}{25}=1 \Rightarrow \frac{x^2}{5^2}-\frac{y^2}{5^2}=1$ The graph of this equation is a hyperbola centered at the origin with x-intercepts of 5 and –5, and asymptotes of $y=\pm\frac{5}{5}x=\pm x$. Also, note in our original equation, the x^2- and y^2-terms have coefficients that are opposite in sign.

13. $\frac{x^2}{4}=1-\frac{y^2}{9} \Rightarrow \frac{x^2}{4}+\frac{y^2}{9}=1 \Rightarrow \frac{(x-0)^2}{2^2}+\frac{(y-0)^2}{3^2}=1$ The equation is of the form $\frac{(x-h)^2}{b^2}+\frac{(y-k)^2}{a^2}=1$ with $a=3$, $b=2$, $h=0$, and $k=0$, so the graph of the given equation is an ellipse.

15. $\frac{(x+3)^2}{16}+\frac{(y-2)^2}{16}=1 \Rightarrow (x+3)^2+(y-2)^2=16 \Rightarrow [x-(-3)]^2+(y-2)^2=4^2$ The equation is of the form $(x-h)^2+(y-k)^2=r^2$ with $r=4$, $h=-3$, and $k=2$, so the graph of the given equation is a circle.

17. $x^2-6x+y=0 \Rightarrow y=-x^2+6x \Rightarrow y=-(x^2-6x+9-9) \Rightarrow y=-(x-3)^2+9 \Rightarrow y-9=-(x-3)^2$

The equation is of the form $y-k=a(x-h)^2$ with $a=-1$, $h=3$, and $k=9$, so the graph of the given equation is a parabola.

19. $4(x-3)^2+3(y+4)^2=0 \Rightarrow \frac{4(x-3)^2}{12}+\frac{3(y+4)^2}{12}=0 \Rightarrow \frac{(x-3)^2}{3}+\frac{[y-(-4)]^2}{4}=0$

The graph is the point $(3,-4)$.

21. $x-4y^2-8y=0 \Rightarrow x=4y^2+8y \Rightarrow x=4(y^2+2y+1-1) \Rightarrow x=4(y+1)^2-4 \Rightarrow x-(-4)=4[y-(-1)]^2$

The equation is of the form $x-h=a(y-k)^2$ with $a=4$, $h=-4$, and $k=-1$, so the graph of the given equation is a parabola.

23. $6x^2-12x+6y^2-18y+25=0 \Rightarrow 6(x^2-2x+1-1)+6(y^2-3y+\frac{9}{4}-\frac{9}{4})=-25 \Rightarrow$

$6(x^2-2x+1)-6+6(y^2-3y+\frac{9}{4})-\frac{27}{2}=-25 \Rightarrow 6(x-1)^2+6(y-\frac{3}{2})^2=-25+6+\frac{27}{2}$

$6(x-1)^2+6(y-\frac{3}{2})^2=-\frac{50}{2}+\frac{12}{2}+\frac{27}{2} \Rightarrow 6(x-1)^2+6(y-\frac{3}{2})^2=-\frac{11}{2} \Rightarrow (x-1)^2+(y-\frac{3}{2})^2=-\frac{11}{12}$

A sum of squares can never be negative. This equation has no graph.

25. $x^2 = 4y - 8 \Rightarrow x^2 = 4(y-2) \Rightarrow y - 2 = \frac{1}{4}(x-0)^2$ The equation is of the form $y - k = a(x-h)^2$ with $a = \frac{1}{4}$, $h = 0$, and $k = 2$, so the graph of the given equation is a parabola with vertex (0, 2) and vertical axis $x = 0$ (the y-axis). Use the vertex and axis and plot a few additional points.

27. $x^2 = 25 + y^2 \Rightarrow x^2 - y^2 = 25 \Rightarrow \frac{x^2}{25} - \frac{y^2}{25} = 1 \Rightarrow \frac{(x-0)^2}{5^2} - \frac{(y-0)^2}{5^2} = 1$ The equation is of the form $\frac{(x-h)^2}{a^2} - \frac{(y-k)^2}{b^2} = 1$ with $a = 5$, $b = 5$, $h = 0$, and $k = 0$, so the graph of the given equation is a hyperbola with center (0, 0), vertices (–5, 0) and (5, 0), and asymptotes $y = \pm x$.

29. $\frac{x^2}{4} + \frac{y^2}{4} = -1 \Rightarrow x^2 + y^2 = -4$ A sum of squares can never be negative. This equation has no graph.

31. $y^2 - 4y = x + 4 \Rightarrow y^2 - 4y + 4 - 4 = x + 4 \Rightarrow (y-2)^2 - 4 = x + 4 \Rightarrow x + 8 = (y-2)^2 \Rightarrow$ $x - (-8) = (y-2)^2$ The equation is of the form $x - h = a(y-k)^2$ with $a = 1$, $h = -8$, and $k = 2$, so the graph of the given equation is a parabola with vertex (–8, 2) and horizontal axis $y = 2$.

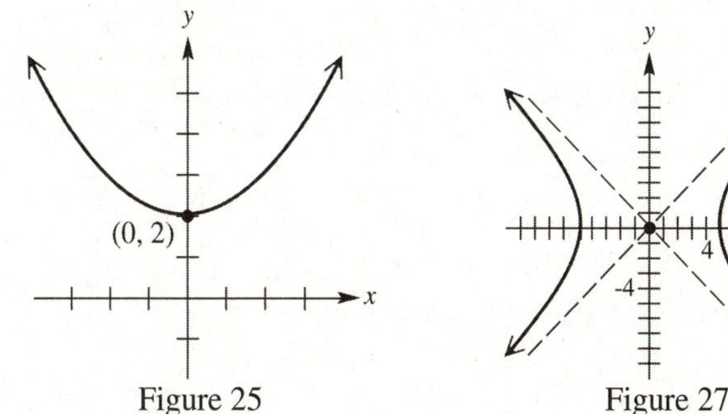

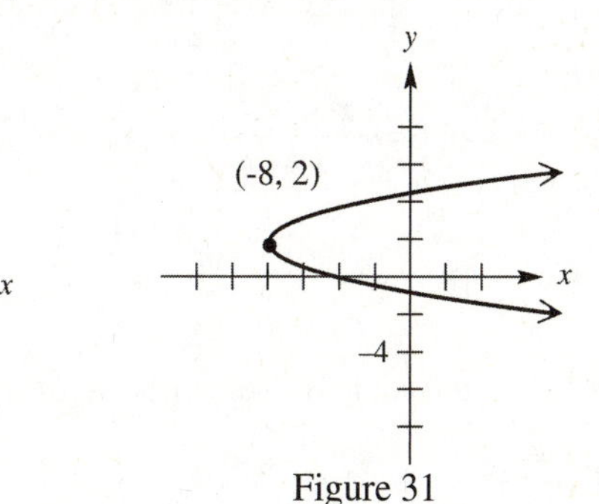

Figure 25 Figure 27 Figure 31

33. $3x^2 + 6x + 3y^2 - 12y = 12 \Rightarrow x^2 + 2x + y^2 - 4y = 4 \Rightarrow (x^2 + 2x + 1 - 1) + (y^2 - 4y + 4 - 4) = 4$

$(x+1)^2 - 1 + (y-2)^2 - 4 = 4 \Rightarrow (x-1)^2 + (y-2)^2 = 4 + 1 + 4 \Rightarrow (x+1)^2 + (y-2)^2 = 9$

$\Rightarrow [x-(-1)]^2 + (y-2)^2 = 3^2$ The equation is of the form $(x-h)^2 + (y-k)^2 = r^2$ with $r = 3$, $h = -1$, and $k = 2$, so the graph of the given equation is a circle with center (–1, 2) and radius 3.

35. $4x^2 - 8x + 9y^2 - 36y = -4 \Rightarrow 4(x^2 - 2x + 1 - 1) + 9(y^2 - 4y + 4 - 4) = -4 \Rightarrow$

$4(x-1)^2 - 4 + 9(y-2)^2 - 36 = -4 \Rightarrow 4(x-1)^2 + 9(y-2)^2 = 36 \Rightarrow$

$\frac{4(x-1)^2}{36} + \frac{9(y-2)^2}{36} = 1 \Rightarrow \frac{(x-1)^2}{9} + \frac{(y-2)^2}{4} = 1 \Rightarrow \frac{(x-1)^2}{3^2} + \frac{(y-2)^2}{2^2} = 1$

The equation is of the form $\frac{(x-h)^2}{a^2} + \frac{(y-k)^2}{b^2} = 1$ with $a = 3$, $b = 2$, $h = 1$, and $k = 2$, so the graph of the given equation is an ellipse with center (1, 2) and vertices (–2, 2), (4, 2), (1, 0) and (1, 4).

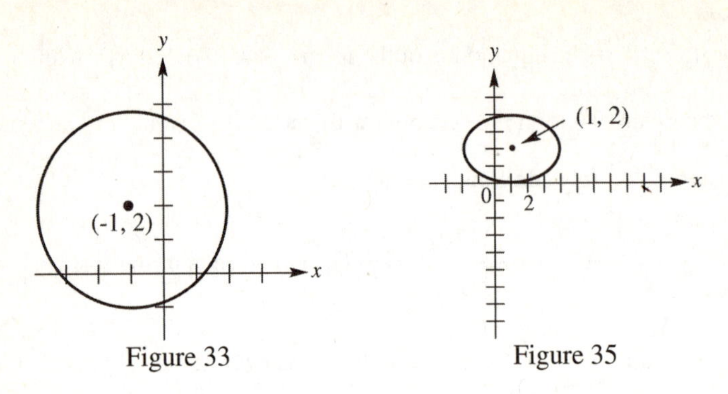

Figure 33 Figure 35

37. Since the sum of the distances from two points (foci) is a constant, the conic section is an ellipse.

39. Since the ratio of the distance from a point to (3, 0) and the distance from a point to the line $x=\frac{4}{3}$ is 1.5, the eccentricity is greater than 1. The conic section is a hyperbola.

41. $12x^2+9y^2=36 \Rightarrow \frac{x^2}{3}+\frac{y^2}{4}=1 \Rightarrow a=2,\ b=\sqrt{3},$ and $c=\sqrt{4-3}=1;\ d=\frac{c}{a}=\frac{1}{2}.$

43. $x^2-y^2=4 \Rightarrow \frac{x^2}{4}-\frac{y^2}{4}=1 \Rightarrow a=2,\ b=2,$ and $c=\sqrt{4+4}=\sqrt{8};\ e=\frac{c}{a}=\frac{\sqrt{8}}{2}=\sqrt{2}.$

45. $4x^2+7y^2=28 \Rightarrow \frac{x^2}{7}+\frac{y^2}{4}=1 \Rightarrow a=\sqrt{7},\ b=2,$ and $c=\sqrt{7-4}=\sqrt{3};\ e=\frac{c}{a}=\frac{\sqrt{3}}{\sqrt{7}}=\frac{\sqrt{21}}{7}.$

47. $x^2-9y^2=18 \Rightarrow \frac{x^2}{18}-\frac{y^2}{2}=1 \Rightarrow a=\sqrt{18},\ b=\sqrt{2},$ and $c=\sqrt{18+2}=\sqrt{20};\ e=\frac{c}{a}=\frac{\sqrt{20}}{\sqrt{18}}=\frac{\sqrt{10}}{3}.$

49. Since $e=1$, the conic is a parabola. With center $(0,0)$ and focus $(0,8)$, the equation is $x^2=4cy \Rightarrow x^2=32y.$

51. Since $0<e<1$, the conic is an ellipse with $c=3$. Now $\frac{c}{a}=e \Rightarrow \frac{3}{a}=\frac{1}{2} \Rightarrow a=6$. For an ellipse, $b^2=a^2-c^2=36-9=27$. The equation is $\frac{x^2}{36}+\frac{y^2}{27}=1.$

53. Since $e>1$, the conic is a hyperbola with $a=6$. Now $\frac{c}{a}=e \Rightarrow \frac{c}{6}=2 \Rightarrow c=12$. For a hyperbola, $b^2=c^2-a^2=144-36=108$. The equation is $\frac{x^2}{36}-\frac{y^2}{108}=1.$

55. Since $e=1$, the conic is a parabola. With center $(0,0)$ and focus $(0,-1)$, the equation is $x^2=4cy \Rightarrow x^2=-4y.$

57. Since $0<e<1$, the conic is an ellipse with $a=3$. Now $\frac{c}{a}=e \Rightarrow \frac{c}{3}=\frac{4}{5} \Rightarrow c=\frac{12}{5}$. For an ellipse, $b^2=a^2-c^2=9-\frac{144}{25}=\frac{81}{25}$. The equation is $\frac{x^2}{\frac{81}{25}}+\frac{y^2}{9}=1$ or $\frac{25x^2}{81}+\frac{y^2}{9}=1.$

59. From the graph, the coordinates of P (a point on the graph) are (–3, 8), the coordinates of F (a focus) are (3, 0), the equation of L (the directrix) is $x = 27$. By the distance formula, the distance from P to F is $\sqrt{(x_2-x_1)^2+(y_2-y_1)^2}=\sqrt{[3-(-3)]^2+(0-8)^2}=\sqrt{6^2+(-8)^2}=\sqrt{36+64}=\sqrt{100}=10$

The distance from a point to a line is defined as the perpendicular distance, so the distance from P to L is $|27-(-3)|=30$. Thus, $e=\frac{\text{Distance of } P \text{ from } F}{\text{Distance of } P \text{ from } L}=\frac{10}{30}=\frac{1}{3}$.

61. From the graph, we see that $F=(\sqrt{2},0)$ and L is the vertical line $x=-\sqrt{2}$. Choose $(0,0)$, the vertex of the parabola, as P. Distance of P from $F=\sqrt{2}$, and distance of P from $L=\sqrt{2}$. Thus, we have $e=\frac{\text{Distance of } P \text{ from } F}{\text{Distance of } P \text{ from } L}=\frac{\sqrt{2}}{\sqrt{2}}=1$.

63. From the graph, we see that $P = (9, -7.5)$, $F = (9, 0)$ and L is the vertical line $x = 4$. Distance of P from $F=7.5$, and distance of P from $L = 5$. Thus, $e=\frac{\text{Distance of } P \text{ from } F}{\text{Distance of } P \text{ from } L}=\frac{7.5}{5}=1.5$.

65. For an ellipse, $c=\sqrt{a^2-b^2}=\sqrt{5013-4970}=\sqrt{43}$. The eccentricity is $e=\frac{c}{a}=\frac{\sqrt{43}}{\sqrt{5013}}\approx .093$.

67. (a) Earth orbits every $365\cdot 24\cdot 60\cdot 60=31{,}536{,}000$ seconds. Thus $\frac{y^2}{16}-\frac{x^2}{9}=1$ The maximum velocity of Earth is $v_{\max}=\frac{2\pi(1.496\times 10^8)}{31{,}536{,}000}\sqrt{\frac{1+.0167}{1-.0167}}\approx 30.3$ km per sec. The maximum velocity of Earth is

$$v_{\min}=\frac{2\pi(1.496\times 10^8)}{31{,}536{,}000}\sqrt{\frac{1+.0167}{1-.0167}}\approx 29.3 \text{ km per sec.}$$

(b) The minimum and maximum velocities are equal. Therefore, the planet's velocity is constant.

(c) A planet is at its maximum and minimum distances from a focus when it is located at the vertices of the ellipse. Thus the minimum and maximum velocities of a planet will occur at the vertices of the elliptical orbit, which are $a+c$ for the minimum and $a-c$ for the maximum.

69. Here $a+c=94.6$ and $a-c=91.4$. Solving these equations simultaneously results in $a=93$ and $c=1.6$.

The eccentricity is $e=\frac{c}{a}=\frac{1.6}{93}\approx 0.0172$.

6.4: Parametric Equations

1.

t	−2	−1	0	1	2
x	−3	−1	1	3	5
y	−4	−3	−2	−1	0

3.

t	−2	−1	0	1	2
x	−1	0	1	2	3
y	3	0	−1	0	3

5.

t	-2	-1	0	1	2
x	6	3	2	3	6
y	3	2	1	0	1

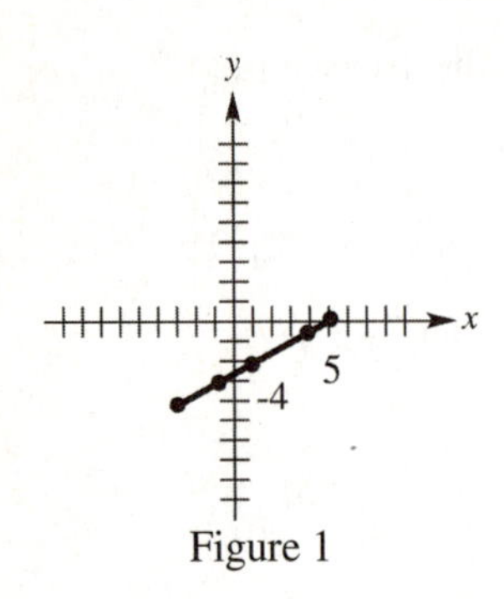

Figure 1

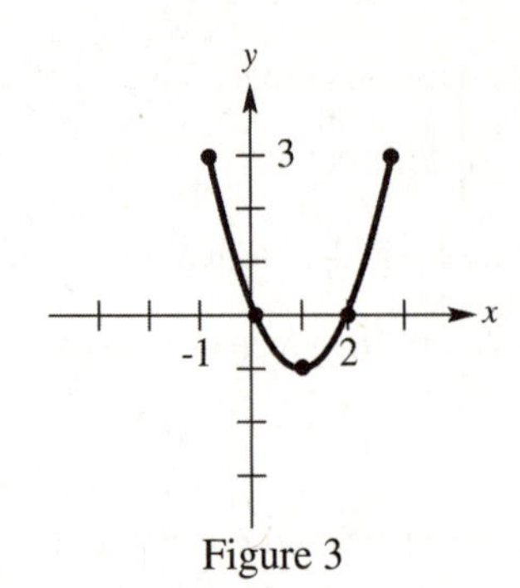

Figure 3

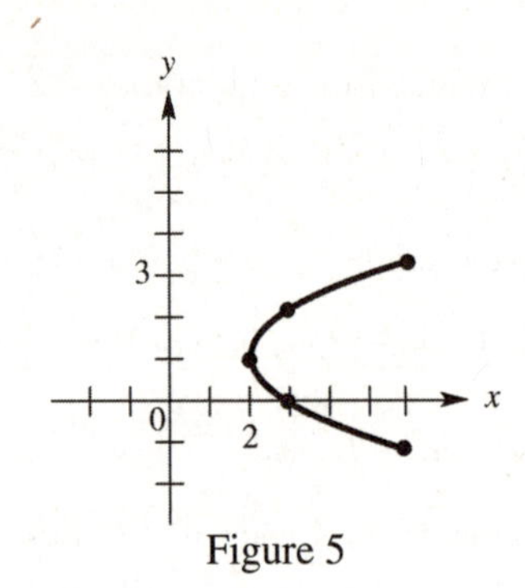

Figure 5

7. See Figure 7. From the first equation, $x=2t \Rightarrow t=\frac{1}{2}x$. By substitution in the second equation, $y=\frac{1}{2}x+1$.

   When t is in $[-2,3]$, the range of $x=2t$ is x in $[-4,6]$.

9. See Figure 9. From the first equation, $x=\sqrt{t} \Rightarrow t=x^2$. By substitution in the second equation,

   $y=3x^2-4$. When t is in $[0,4]$, the range of $x=\sqrt{t}$ is x in $[0,2]$.

11. See Figure 11. From the first equation, $x=t^3+1 \Rightarrow t^3=x-1$. By substitution in the second equation,

    $y=x-2$. When t is in $[-3,3]$, the range of $x=t^3+1$ is x in $[-26,28]$.

13. See Figure 13. From the second equation, $y=\sqrt{3t-1} \Rightarrow t=\frac{y^2+1}{3}$. By substitution in the second equation,

    $x=2^{(y^2+1)/3}$. When t is in $\left[\frac{1}{3},4\right]$, the range of $y=\sqrt{3t=-1}$ is y in $[0,\sqrt{11}]$.

[-8,8] by [-8,8]
Xscl = 1 Yscl = 1

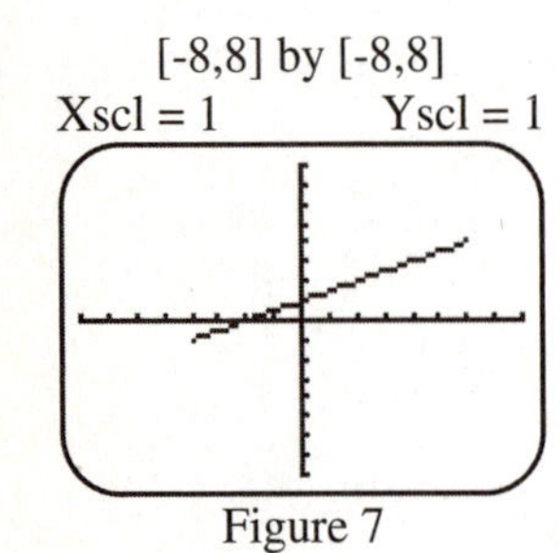
Figure 7

[-6,6] by [-6,10]
Xscl = 5 Yscl = 10

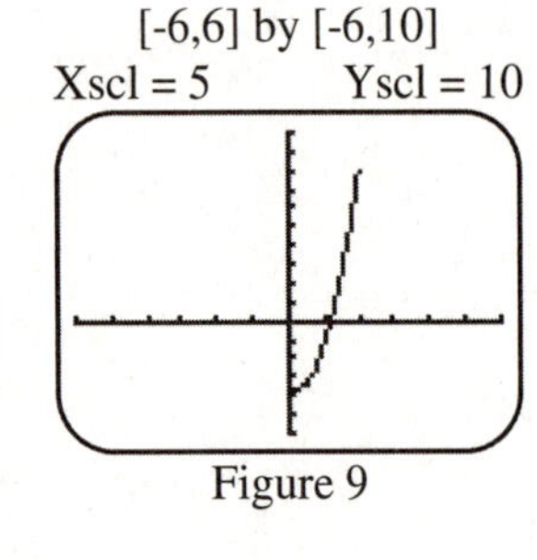
Figure 9

[-30,30] by [-30,30]
Xscl= 2 Yscl = 2

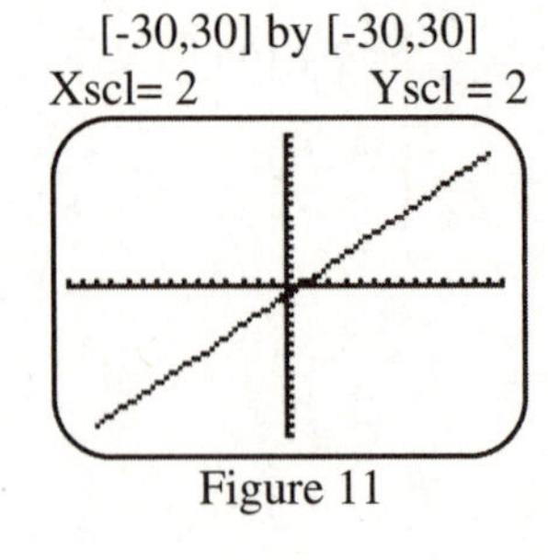
Figure 11

[-2,30] by [-2,10]
Xscl = 2 Yscl= 1

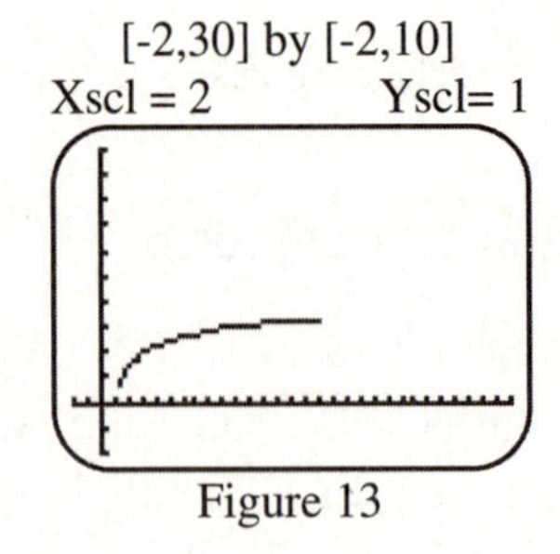
Figure 13

15. See Figure 15. From the first equation, $x=t+2 \Rightarrow t=x-2$. By substitution in the second equation,

    $y=-\frac{1}{2}\sqrt{9-(x-2)^2}$. When t is in $[-3,3]$, the range of $x=t+2$ is x in $[-1,5]$.

17. See Figure 17. From the first equation, $x=t \Rightarrow t=x$. By substitution in the second equation, $y=\frac{1}{x}$. When

    t is in $(-\infty,0)\cup(0,\infty)$, the range of $x=t$ is x in $(-\infty,0)\cup(0,\infty)$.

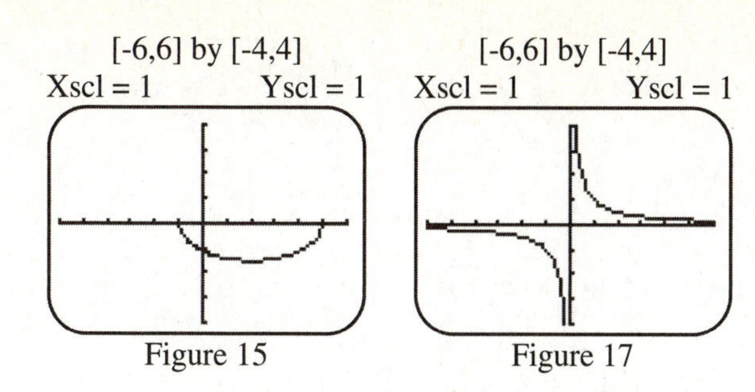

Figure 15 Figure 17

19. From the first equation, $x = 3t \Rightarrow t = \frac{1}{3}x.$ By substitution in the second equation, $y = \frac{1}{3}x - 1.$ When t is in $(-\infty, \infty),$ the range of $x = 3t$ is x in $(-\infty, \infty).$

21. From the second equation, $y = t + 1 \Rightarrow t = y - 1.$ By substitution in the first equation, $x = 3(y-1)^2.$ When t is in $(-\infty, \infty),$ the range of $y = t + 1$ is y in $(-\infty, \infty).$

23. From the second equation, $y = 4t^3 \Rightarrow t = \sqrt[3]{\frac{y}{4}}.$ By substitution in the first equation, $x = 3\left(\frac{y}{4}\right)^{2/3}.$ When t is in $(-\infty, \infty),$ the range of $y = 4t^3$ is y in $(-\infty, \infty).$

25. From the first equation, $x = t \Rightarrow t = x.$ By substitution in the second equation, $y = \sqrt{x^2 + 2}.$ When t is in $(-\infty, \infty),$ the range of $x = t$ is x in $(-\infty, \infty).$

27. From the first equation, $x = e^t \Rightarrow t = \ln x.$ By substitution in the second equation, $y = \frac{1}{x}.$ When t is in $(-\infty, \infty),$ the range of $x = e^t$ is x in $(0, \infty).$

29. From the first equation, $x = \frac{1}{\sqrt{t+2}} \Rightarrow t = \frac{1}{x^2} - 2.$ By substitution in the second equation, $y = 1 - 2x^2.$

When t is in $(-2, \infty),$ the range of $x = \frac{1}{\sqrt{t+2}}$ is x in $(0, \infty).$

31. From the first equation, $x = t + 2 \Rightarrow t = x - 2.$ By substitution in the second equation, $y = \frac{1}{x}.$ When t has the restriction $t \neq 2,$ the range of $x = t + 2$ has the restriction $x \neq 0.$

33. From the first equation, $x = t^2 \Rightarrow t = \sqrt{x}.$ By substitution in the second equation, $y = \ln x.$ When t is in $(0, \infty),$ the range of $x = \ln t$ is x in $(0, \infty).$

35. For $x = \frac{1}{2}t,$ $y = 2\left(\frac{1}{2}t\right) + 3 \Rightarrow y = t + 3.$ For $x = \frac{t+3}{2},$ $y = 2\left(\frac{t+3}{2}\right) + 3 \Rightarrow y = t + 6.$

37. For $x=\frac{1}{3}t$, $y\sqrt{3\left(\frac{1}{3}t\right)+2} \Rightarrow y=\sqrt{t+2}$ for t in $[-2,\infty)$. For $x=\frac{t-2}{3}$, $y=\sqrt{3\left(\frac{t-2}{3}\right)+2} \Rightarrow y=\sqrt{t}$ for t in $[0,\infty)$.

39. For $x=t^3+1$, $t^3+1=y^3+1 \Rightarrow y=t$. For $x=t$, $t=y^3+1 \Rightarrow y=\sqrt[3]{t-1}$.

41. For $x=\sqrt{t+1}$, $\sqrt{t+1}=\sqrt{y+1} \Rightarrow y=t$ for t in $[-1,\infty)$. For $x=\sqrt{t}$,
$\sqrt{t}=\sqrt{y+1} \Rightarrow y+1=t \Rightarrow y=t-1$ for t in $[0,\infty)$.

43. (a) Find t when $y=0$. $400 \bullet \frac{\sqrt{2}}{2}t-16t^2=0 \Rightarrow 16t^2-200\sqrt{2}t=0 \Rightarrow t(16t-200\sqrt{2})=0 \Rightarrow$

$$16t-200\sqrt{2}=0 \Rightarrow 16t=200\sqrt{2} \Rightarrow t=\frac{200\sqrt{2}}{16} \approx 17.7 \text{ seconds.}$$

(b) Find x when $t \approx 17.7$. $x=400 \cdot \frac{\sqrt{2}}{2}(17.7) \approx 5000$ feet.

(c) Find y when $x \approx 8.85$ (half the total time). $y=400 \cdot \frac{\sqrt{2}}{2}(8.85)-16(8.85)^2 \approx 1250$ feet.

45. See Figure 45. From the first equation, $x=60t \Rightarrow t=\frac{x}{60}$. By substitution in the second equation,

$$y=80\left(\frac{x}{60}\right)-16\left(\frac{x}{60}\right)^2 \Rightarrow y=\frac{4}{3}x-\frac{x^2}{225}.$$

[0,300] by [0,200]
Xscl = 50 Yscl = 50

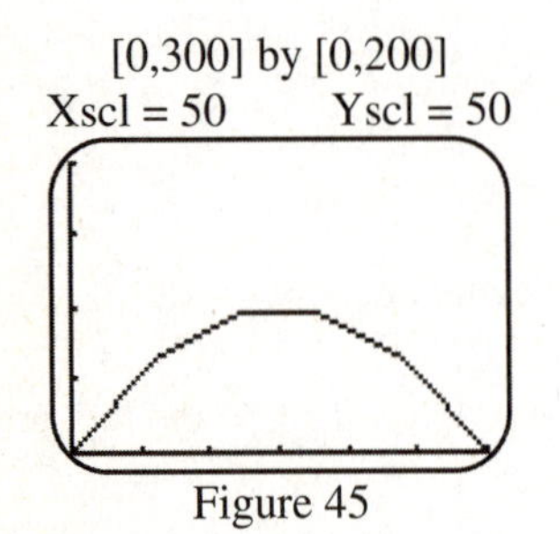

Figure 45

47. From the first equation, $x=v_0\frac{\sqrt{2}}{2}t \Rightarrow t=\frac{2x}{v_0\sqrt{2}}$. By substitution in the second equation,

$$y=v_0\frac{\sqrt{2}}{2}\left(\frac{2x}{v_0\sqrt{2}}\right)-16\left(\frac{2x}{v_0\sqrt{2}}\right)^2 \Rightarrow y=x-\frac{32}{(v_0)^2}x^2.$$

49. A line through (x_1, y_1) with slope m is given by $y-y_1=m(x-x_1)$. For $x=t$, $y-y_1=m(t-x_1)$.

For $t=x-x_1$, $y-y_1=mt \Rightarrow y=mt+y_1$. Many answers are possible.

50. For $x=t,\ y=a(t-h)^2+k$. For $x=t+h,\ y=at^2+k$. Many answers are possible.

Reviewing Basic Concepts (Sections 6.3 and 6.4)

1. $3x^2+y^2-6x+6y=0 \Rightarrow 3(x^2-2x+1)+(y^2+6y+9)=0+3+9 \Rightarrow$

   $3(x-1)^2+(y+3)^2=12 \Rightarrow \frac{(x-1)^2}{4}+\frac{(y+3)^2}{12}=1$; ellipse centered at $(1,-3)$.

2. $y^2-2x^2+8y-8x-4=0 \Rightarrow (y^2+8y+16)-2(x^2+4x+4)=4+16-8 \Rightarrow$

   $(y+4)^2-2(x+2)^2=12 \Rightarrow \frac{(y+4)^2}{12}-\frac{(x+2)^2}{6}=1$; hyperbola centered at $(-4,-2)$.

3. $3y^2+12y+5x=3 \Rightarrow 3(y^2+4y+4)=-5x+3+12 \Rightarrow 3(y+2)^2=-5(x-3) \Rightarrow (y+2)^2=\frac{5}{3}(x-3)$;

   parabola with center $(3,-2)$ that opens to the left.

4. $x^2+25y^2=25 \Rightarrow \frac{x^2}{25}+\frac{y^2}{1}=1 \Rightarrow a=5,\ b=1,$ and $c=\sqrt{25-1}=\sqrt{24}$. $e=\frac{c}{a}=\frac{\sqrt{24}}{5}=\frac{2\sqrt{6}}{5}$.

5. $8y^2-4x^2=8 \Rightarrow \frac{y^2}{1}-\frac{x^2}{2}=1 \Rightarrow a=1,\ b=\sqrt{2},$ and $c=\sqrt{1+2}=\sqrt{3}$. $e=\frac{c}{a}=\frac{\sqrt{3}}{1}=\sqrt{3}$.

6. $3x^2+4y^2=108 \Rightarrow \frac{x^2}{36}+\frac{y^2}{27}=1 \Rightarrow a=6,\ b=\sqrt{27},$ and $c=\sqrt{36-27}=\sqrt{9}$. $e=\frac{c}{a}=\frac{3}{6}=\frac{1}{2}$.

7. Since $e=1$, the conic is a parabola. With center $(0,0)$ and focus $(-2,0)$, the equation is

   $y^2=4cx \Rightarrow y^2=-8x$.

8. The ellipse is centered between the foci at $(0,0)$. The major axis is horizontal with $a=5$. Since the foci are

   $(\pm 3,0)$, we know that $c=3$. Since $c^2=a^2-b^2$, the value of b can be found by substitution.

   $b^2=a^2-c^2=5^2-3^2=25-9=16 \Rightarrow b=4$. The equation is $\frac{x^2}{25}+\frac{y^2}{16}=1$.

9. The hyperbola has a vertical transverse axis with $c=5$. The vertices are $(0,\pm 4)$, so $a=4$. The center is

   located between the foci at $(0,0)$. Since $c^2=a^2+b^2$, the value of b can be found by substitution.

   $b^2=c^2-a^2=5^2-4^2=25-16=9 \Rightarrow b=3$. The equation is $\frac{y^2}{16}-\frac{x^2}{9}=1$.

10. The center is located between the vertices at (0, 0). The hyperbola has a horizontal transverse axis with $a=3$.

From the asymptotes, $\frac{b}{a}=\frac{2}{3}$ with. $a=3\Rightarrow b=2$. The equation is $\frac{x^2}{9}-\frac{y^2}{4}=1$.

11. Using a vertical minor axis, $b=9$. The major axis has length 30, so $a=15$. The equation is

$\frac{x^2}{225}+\frac{y^2}{81}=1$. When $x=6$, $\frac{6^2}{225}+\frac{y^2}{81}=1\Rightarrow\frac{y^2}{81}=1-\frac{36}{225}\Rightarrow y^2=81\left(1-\frac{36}{225}\right)\Rightarrow$

$y=\sqrt{81\left(1-\frac{36}{225}\right)}\approx 8.25$ feet.

12. (a) See Figure 12.

(b) From the first equation, $x=2t\Rightarrow t=\frac{x}{2}$. By substitution in the second equation, $y=\sqrt{\frac{x^2}{4}+1}$.

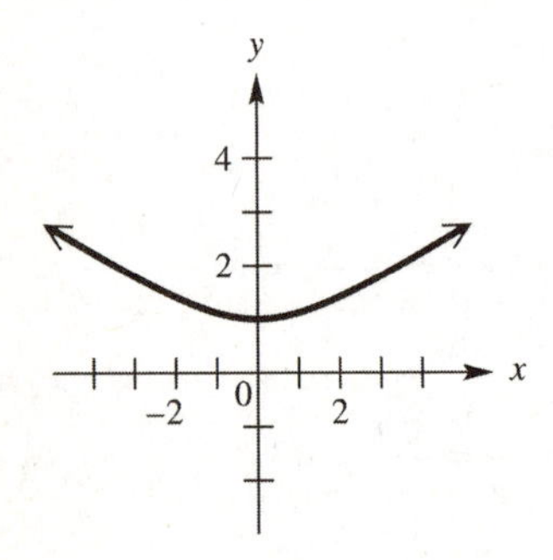

Figure 12

Chapter 6 Review Exercises

1. Here $h=-2$, $k=3$, and $r^2=5^2=25$. The equation is $(x+2)^2+(y-3)^2=25$. See Figure 1. From the figure, the domain is $[-7,3]$, and the range is $[-2,8]$.

2. Here $h=\sqrt{5}$, $k=-\sqrt{7}$, and $r^2=\left(\sqrt{3}\right)^2=3$. The equation is $\left(x-\sqrt{5}\right)^2+\left(y+\sqrt{7}\right)^2=3$. See Figure 2. From the figure, the domain is $\left[\sqrt{5}-\sqrt{3},\sqrt{5}+\sqrt{3}\right]$, and the range is $\left[-\sqrt{7}-\sqrt{3},-\sqrt{7}+\sqrt{3}\right]$.

3. The radius is the distance between $(-8,1)$ and $(0,16)$. $r=\sqrt{(0-(-8))^2+(16-1)^2}=\sqrt{64+225}=17$.

Here $h=-8$, $k=1$, and $r^2=17^2=289$. The equation is $(x+8)^2+(y-1)^2=289$. See Figure 3. From the figure, the domain is $[-25,9]$, and the range is $[-16,18]$.

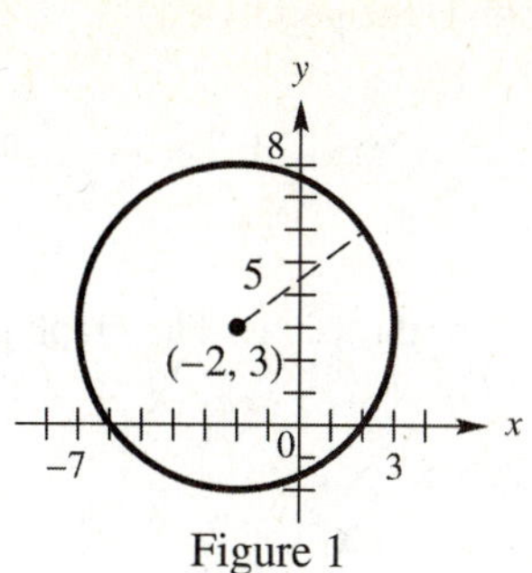

Figure 1

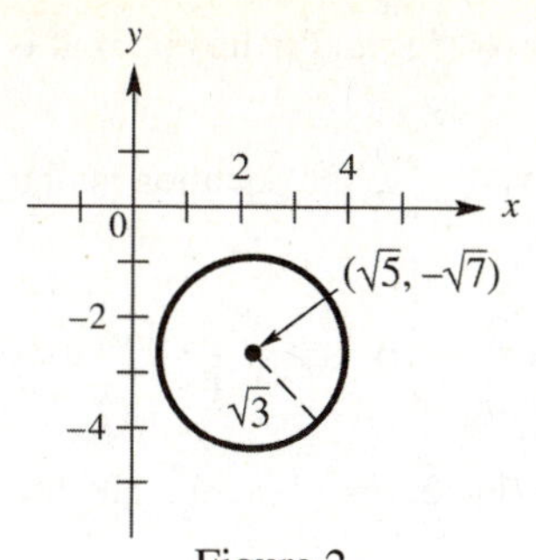

Figure 2

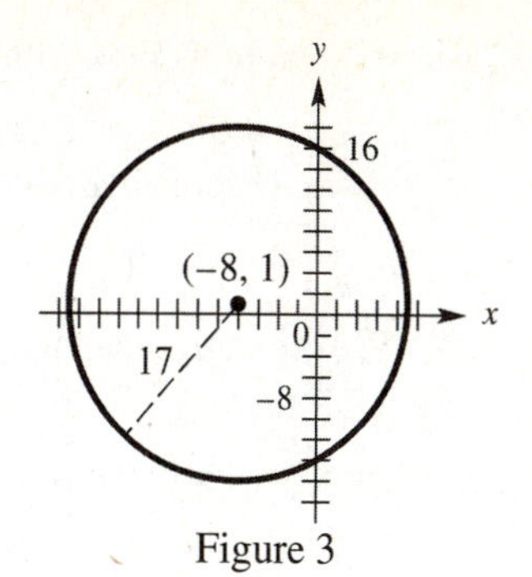

Figure 3

4. If the center is $(3,-6)$, the circle must touch the x-axis at the point $(3,0)$. The radius is 6. Here $h=3$, $k=-6$, and $r^2=6^2=36$. The equation is $(x-3)^2+(y+6)^2=36$. See Figure 4. From the figure, the domain is $[-3,9]$, and the range is $[-12,0]$.

5. $x^2-4x+y^2+6y+12=0 \Rightarrow (x^2-4x+4)+\left(y^2+6y+9\right)=-12+4+9 \Rightarrow (x-2)^2+(y+3)^2=1.$

The circle has center $(2,-3)$, and radius $r=1$.

6. $x^2-6x+y^2-10y+30=0 \Rightarrow (x^2-6x+9)+\left(y^2-10y+25\right)=-30+9+25 \Rightarrow (x-3)^2+(y-5)^2=4.$

The circle has center $(3,5)$, and radius $r=2$.

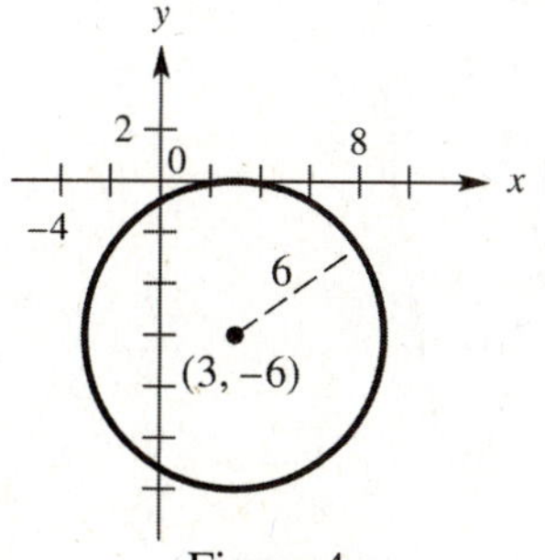

Figure 4

7. $2x^2+14x+2y^2+6y=-2 \Rightarrow x^2+7x+y^2+3y=-1 \Rightarrow$

$$\left(x^2+7x+\frac{49}{4}\right)+\left(y^2+3y+\frac{9}{4}\right)=-1+\frac{49}{4}+\frac{9}{4} \Rightarrow \left(x+\frac{7}{2}\right)^2+\left(y+\frac{3}{2}\right)^2=\frac{54}{4}.$$

The circle has center $\left(-\frac{7}{2},-\frac{3}{2}\right)$, and radius $r=\frac{\sqrt{54}}{2}=\frac{3\sqrt{6}}{2}$.

8. $3x^2+3y^2+33x-15y=0 \Rightarrow x^2+11x+y^2-5=0 \Rightarrow$

$$\left(x^2+11x+\frac{121}{4}\right)+\left(y^2-5y+\frac{25}{4}\right)=0+\frac{121}{4}+\frac{25}{4} \Rightarrow \left(x+\frac{11}{2}\right)^2+\left(y-\frac{5}{2}\right)^2=\frac{146}{4}.$$

The circle has center $\left(-\frac{11}{2},\frac{5}{2}\right)$, and radius $r=\frac{\sqrt{146}}{2}$.

9. The equation is that of a circle with center $(4,5)$, and radius 0. That is, the graph is the point $(4,5)$.

10. Since $y^2 = -\frac{2}{3}x$ is equivalent to $y^2 = 4\left(-\frac{1}{6}\right)x$, the equation is in the form $y^2 = 4cx$ with $c = -\frac{1}{6}$. The focus is $\left(-\frac{1}{6}, 0\right)$, and the equation of the directrix is $x = \frac{1}{6}$. The axis is $y = 0$, or the x-axis. The graph is shown in Figure 10. From the figure, the domain is $(-\infty, 0]$, and the range is $(-\infty, \infty)$.

11. Since $y^2 = 2x$ is equivalent to $y^2 = 4\left(\frac{1}{2}\right)x$, the equation is in the form $y^2 = 4cx$ with $c = \frac{1}{2}$. The focus is $\left(\frac{1}{2}, 0\right)$, and the equation of the directrix is $x = -\frac{1}{2}$. The axis is $y = 0$, or the x-axis. The graph is shown in Figure 11. From the figure, the domain is $[0, \infty)$, and the range is $(-\infty, \infty)$.

12. Since $3x^2 - y = 0$ is equivalent to $x^2 = 4\left(\frac{1}{12}\right)y$, the equation is in the form $x^2 = 4cy$ with $c = \frac{1}{12}$. The focus is $\left(0, \frac{1}{12}\right)$, and the equation of the directrix is $y = -\frac{1}{12}$. The axis is $x = 0$, or the y-axis. The graph is shown in Figure 12. From the figure, the domain is $(-\infty, \infty)$, and the range is $[0, \infty)$.

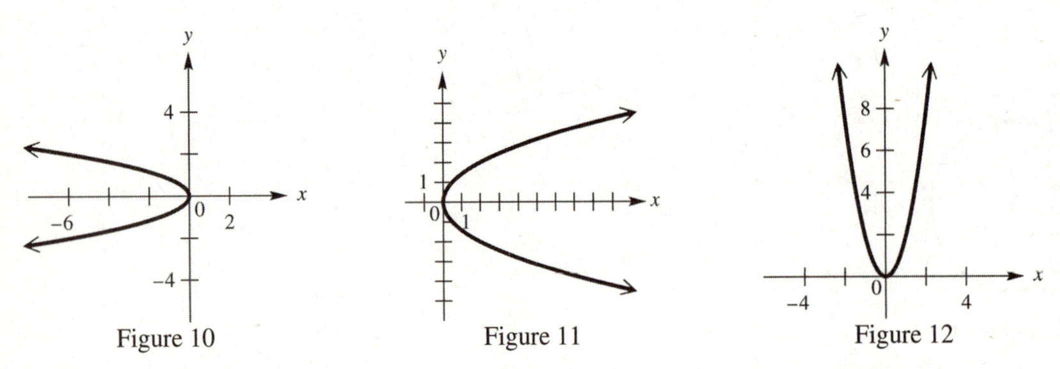

Figure 10 Figure 11 Figure 12

13. Since $x^2 + 2y = 0$ is equivalent to $x^2 = 4\left(-\frac{1}{2}\right)y$, the equation is in the form $x^2 = 4cy$ with $c = -\frac{1}{2}$. The focus is $\left(0, -\frac{1}{2}\right)$, and the equation of the directrix is $y = \frac{1}{2}$. The axis is $x = 0$, or the y-axis. The graph is shown in Figure 13. From the figure, the domain is $(-\infty, \infty)$, and the range is $(-\infty, 0]$.

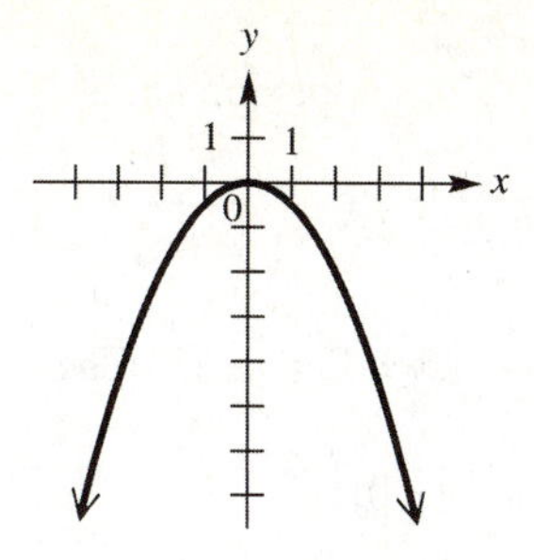

Figure 13

14. If the vertex is $(0,0)$, and the focus is $(4,0)$, then the parabola opens to the right and $c=4$. The equation is $y^2=4cx \Rightarrow y^2=16x$.

15. If the vertex is $(0,0)$, and the parabola opens to the right, the equation is in the form $y^2=4cx$. Find the value of c by using the fact that the parabola passes through $(2,5)$. Thus, $(5)^2=4c(2) \Rightarrow c=\frac{25}{8}$. The equation is $y^2=4cx \Rightarrow y^2=\frac{25}{2}x$.

16. If the vertex is $(0,0)$, and the parabola opens downward, the equation is in the form $x^2=4cy$. Find the value of c by using the fact that the parabola passes through $(3,-4)$. Thus, $(3)^2=4c(-4) \Rightarrow c=-\frac{9}{16}$. The equation is $x^2=4cy \Rightarrow x^2=-\frac{9}{4}y$.

17. If the vertex is $(0,0)$, and the focus is $(0,-3)$, then the parabola opens downward and $c=-3$. The equation is $x^2=4cy \Rightarrow x^2=-12y$.

18. If the equation has the x-term squared, it has a vertical axis, and opens up if the coefficient of x^2 is positive or down if the coefficient is negative. If the y-term is squared, it has a horizontal axis, and opens to the right if the coefficient of y^2 is positive or to the left if the coefficient is negative.

19. If the focus is $(2,6)$, and the vertex is t $(-5,6)$, the parabola opens to the right and $c=7$. Substituting in $(y-k)^2=4c(x-h)$, we get $(y-6)^2=-28(x+5)$.

20. If the focus is $(4,5)$, and the vertex is $(4,3)$, the parabola opens upward and $c=2$. Substituting in $(x-h)^2=4c(y-k)$, we get $(x-4)^2=8(y-3)$.

21. $\frac{x^2}{5}+\frac{y^2}{9}=1 \Rightarrow a=3$ and $b=\sqrt{5}$. The endpoints of the major axis (vertices) are $(0,\pm 3)$, so the range is $[-3,3]$, The endpoints of the minor axis are $(\pm\sqrt{5},0)$, so the domain is $[-\sqrt{5},\sqrt{5}]$. See Figure 21.

22. $\frac{x^2}{16}+\frac{y^2}{4}=1 \Rightarrow a=4$ and $b=2$. The endpoints of the major axis (vertices) are $(\pm 4,0)$, so the domain is $[-4,4]$. The endpoints of the minor axis are $(0,\pm 2)$, so the range is $[-2,2]$. See Figure 22.

23. The transverse axis is horizontal with $a=8$ and $b=6$. The asymptotes are $y=\pm\frac{3}{4}x$. See Figure 23. The domain is $(-\infty,-8]\cup[8,\infty)$, and the range is $(-\infty,\infty)$. The vertices are $(-8,0)$ and $(8,0)$.

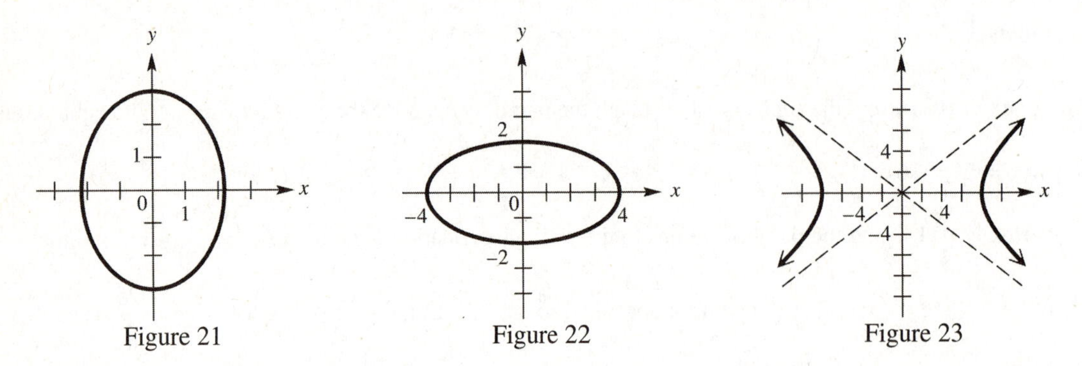

24. The transverse axis is vertical with $a=5$ and $b=3$. The asymptotes are $y=\pm\frac{5}{3}x$. See Figure 24. The domain is $(-\infty,\infty)$, and the range is $(-\infty,-5]\cup[5,\infty)$. The vertices are $(0,-5)$ and $(0,5)$.

25. The ellipse is centered at $(3,-1)$. The major axis is horizontal and has length $2a=4$, so the vertices are $(1,-1)$ and $(5,-1)$. The length of the minor axis is $2b=2$. The graph is shown in Figure 25. The domain is $[1,5]$, and the range is $[-2,0]$.

26. The ellipse is centered at $(2,-3)$. The major axis is horizontal and has length $2a=6$, so the vertices are $(-1,-3)$ and $(5,-3)$. The length of the minor axis is $2b=4$. The graph is shown in Figure 26. The domain is $[-1,5]$ and the range is $[-5,-1]$.

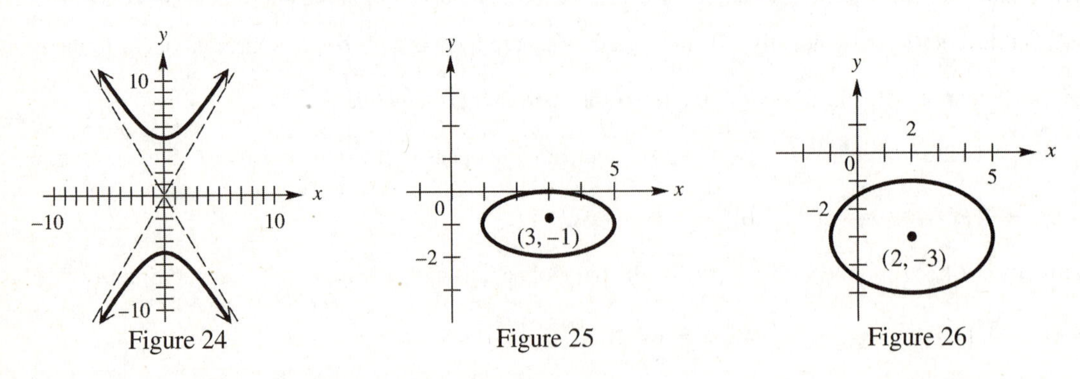

27. The center is $(-3,-2)$ and the transverse axis is vertical with $a=2$ and $b=3$. See Figure 27. The domain is $(-\infty,\infty)$ and the range is $(-\infty,-4]\cup[0,\infty)$. The vertices are $(-3,-4)$ and $(-3,0)$.

28. The center is $(-1,2)$ and the transverse axis is horizontal with $a=4$ and $b=2$. See Figure 28. The domain is $(-\infty,-5]\cup[3,\infty)$ and the range is $(-\infty,\infty)$. The vertices are $(-5,2)$ and $(3,2)$.

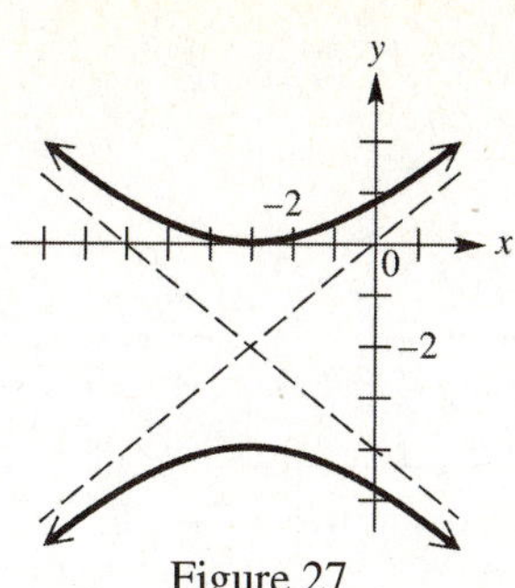

Figure 27

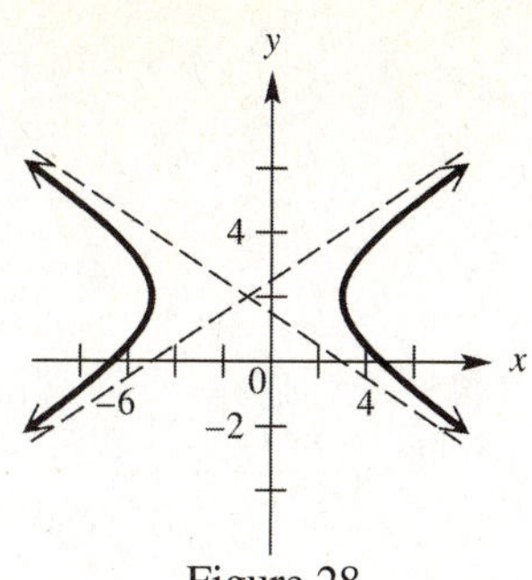

Figure 28

29. The major axis is vertical with $a = 4$. Since one focus is $(0,2)$, we know that $c = 2$. Since $c^2 = a^2 - b^2$, the value of b can be found by substitution: $b^2 = a^2 - c^2 = 4^2 - 2^2 = 16 - 4 = 12 \Rightarrow b = \sqrt{12}$. The equation is $\frac{x^2}{12} + \frac{y^2}{16} = 1$.

30. The major axis is horizontal with $a = 6$. Since one focus is $(-2,0)$, we know that $c = 2$. Since $c^2 = a^2 - b^2$, the value of b can be found by substitution:

$b^2 = a^2 - c^2 = 6^2 - 2^2 = 36 - 4 = 32 \Rightarrow b = \sqrt{32}$. The equation is $\frac{x^2}{36} + \frac{y^2}{32} = 1$.

31. The hyperbola has a vertical transverse axis with $c = 5$. The y-intercepts coincide with the vertices so $a = 4$. Since $c^2 = a^2 + b^2$, the value of b can be found by substitution:

$b^2 = c^2 - a^2 = 5^2 - 4^2 = 25 - 16 = 9 \Rightarrow b = 3$. The equation is $\frac{y^2}{16} - \frac{x^2}{9} = 1$.

32. The hyperbola has a vertical transverse axis. The y-intercept coincides with a vertex so $a = 2$. The equation is of the form $\frac{y^2}{a^2} - \frac{x^2}{b^2} = 1$. By substitution using the point (2, 3):

$\frac{3^2}{2^2} - \frac{2^2}{b^2} = 1 \Rightarrow \frac{9}{4} - 1 = \frac{4}{b^2} \Rightarrow \frac{5}{4} = \frac{4}{b^2} \Rightarrow 5b^2 = 16 \Rightarrow b^2 = \frac{16}{5}$. The equation is $\frac{y^2}{4} - \frac{x^2}{\frac{16}{5}} = 1$ or

$\frac{y^2}{4} - \frac{5x^2}{16} = 1$.

33. Since $0 < e < 1$, the conic is an ellipse with $c = 3$. Now $\frac{c}{a} = e \Rightarrow \frac{3}{a} = \frac{2}{3} \Rightarrow a = \frac{9}{2}$. For an ellipse,

$b^2 = a^2 - c^2 = \frac{81}{4} - 9 = \frac{45}{4}$. The equation is $\frac{x^2}{\frac{45}{4}} + \frac{y^2}{\frac{81}{4}} = 1$ or $\frac{4x^2}{45} + \frac{4y^2}{81} = 1$.

34. Since $e > 1$, the conic is a hyperbola with $c = 5$. Now $\frac{c}{a} = e \Rightarrow \frac{5}{a} = \frac{5}{2} \Rightarrow a = 2$. For a hyperbola,

$b^2 = c^2 - a^2 = 25 - 4 = 21$. The equation is $\frac{x^2}{4} - \frac{y^2}{21} = 1$.

35. (a) $x^2+y^2+2x+6y-15=0 \Rightarrow x^2+2x+1+y^2+6y+9=15+9+1 \Rightarrow (x+1)^2+(y+3)^2=25;$

The center is $(-1,-3)$.

(b) The radius is $r=\sqrt{25}=5$.

(c) $(x+1)^2+(y+3)^2=25 \Rightarrow (y+3)^2=25-(x+1)^2 \Rightarrow y+3=\pm\sqrt{25-(x+1)^2} \Rightarrow$

$y=-3\pm\sqrt{25-(x+1)^2}$; Graph $y=-3-\sqrt{25-(x+1)^2}$ and $y=-3+\sqrt{25-(x+1)^2}$.

36. D. $4x^2+y^2=36 \Rightarrow \frac{x^2}{9}+\frac{y^2}{36}=1$; This is an ellipse with major axis on the y-axis.

37. E. $x=2y^2+3 \Rightarrow 4\left(\frac{1}{8}\right)(x-3)=(y-0)^2$; This is a parabola that opens to the right.

38. A. $(x-1)^2+(y+2)^2=36$; This is a circle with center $(1,-2)$ and radius 6.

39. C. $\frac{x^2}{36}+\frac{y^2}{9}=1$; This is an ellipse with major axis on the x-axis.

40. B. $(y-1)^2-(x-2)^2=36 \Rightarrow \frac{(y-1)^2}{36}-\frac{(x-2)^2}{36}=1$; This is a hyperbola with center $(2,1)$.

41. F. $y^2=36+4x^2 \Rightarrow y^2-4x^2=36 \Rightarrow \frac{y^2}{36}-\frac{x^2}{9}=1$; This is a hyperbola with transverse axis on the y-axis.

42. $4x^2+8x+25y^2-250y=-529 \Rightarrow 4(x^2+2x)+25(y^2-10y)=-529 \Rightarrow$

$4(x^2+2x+1)+25(y^2-10y+25)=-529+4+625 \Rightarrow 4(x+1)^2+25(y-5)^2=100 \Rightarrow$

$\frac{(x+1)^2}{25}+\frac{(y-5)^2}{4}=1$; The center is $(-1,5)$. The vertices are $(-1-5,5), (-1+5,5)$ or $(-6,5),(4,5)$.

43. $5x^2+20x+2y^2-8y=-18 \Rightarrow 5(x^2+4x)+2(y^2-4y)=-18 \Rightarrow$

$5(x^2+4x+4)+2(y^2-4y+4)=-18+20+8 \Rightarrow 5(x+2)^2+2(y-2)^2=10 \Rightarrow$

$\frac{(x+2)^2}{2}+\frac{(y-2)^2}{5}=1$; The center is $(-2,2)$. The vertices are $\left(-2,2-\sqrt{5}\right)$, $\left(-2,2+\sqrt{5}\right)$.

44. $x^2+4x-4y^2+24y=36 \Rightarrow (x^2+4x+4)-4(y^2-6y+9)=36+4-36 \Rightarrow$

$(x+2)^2-4(y-3)^2=4 \Rightarrow \frac{(x+2)^2}{4}-\frac{(y-3)^2}{1}=1$. The center is $(-2,3)$. The vertices are

$(-2-2,3)$, $(-2+2,0)$ or $(-4,3)$, $(0,3)$.

45. $4y^2+8y-3x^2+6x=11 \Rightarrow 4(y^2+2y+1)-3(x^2-2x+1)=11+4-3 \Rightarrow$

$4(y+1)^2-3(x-1)^2=12 \Rightarrow \frac{(y+1)^2}{3}-\frac{(x-1)^2}{4}=1$. The center is $(1,-1)$.

The vertices are $\left(1, -1-\sqrt{3}\right), \left(1, -1+\sqrt{3}\right)$.

46. $9x^2+25y^2=225 \Rightarrow \frac{x^2}{25}+\frac{y^2}{9}=1 \Rightarrow a=5, b=3,$ and $c=\sqrt{25-9}=4; e=\frac{c}{a}=\frac{4}{5}$

47. $4x^2+9y^2=36 \Rightarrow \frac{x^2}{9}+\frac{y^2}{4}=1 \Rightarrow a=3, b=2,$ and $c=\sqrt{9-4}=\sqrt{5};\ e=\frac{c}{a}=\frac{\sqrt{5}}{3}$

48. $9x^2-y^2=9 \Rightarrow \frac{x^2}{1}-\frac{y^2}{9}=1 \Rightarrow a=1, b=3,$ and $c=\sqrt{1+9}=\sqrt{10};\ e=\frac{c}{a}=\frac{\sqrt{10}}{1}=\sqrt{10}$

49. The parabola opens to the right so the equation has the form $(y-k)^2=4c(x-h)$. With vertex $(-3,2)$ and y-intercept (0, 5), $(5,-2)^2=4c(0+3) \Rightarrow 9=12c \Rightarrow c=\frac{3}{4}$. The equation is $(y-2)^2=3(x+3)$. This can also be written $x=\frac{1}{3}(y-2)^2-3$.

50. The center is located between the foci at (0, 0). The hyperbola has a vertical transverse axis with $c=12$. From the asymptotes, $\frac{a}{b}=1$. Also $c^2=a^2+b^2 \Rightarrow a^2+b^2=144$. Solving these equations simultaneously results in $a^2=72$ and $b^2=72$. The equation is $\frac{y^2}{72}-\frac{x^2}{72}=1$.

51. The foci are (0, 0) and (4, 0), so the center is (2, 0) and $c=2$. The sum of the distances is 8 so $2a=8 \Rightarrow a=4$. For an ellipse, $b^2=a^2-c^2=16-4=12$. The equation is $\frac{(x-2)^2}{16}+\frac{y^2}{12}=1$.

52. The foci are (0, 0) and (0, 4), so the center is (0, 2) and $c=2$. The difference of the distances is 2 so $2a=2 \Rightarrow a=1$. For a hyperbola, $b^2=c^2-a^2=4-1=3$. The equation is $\frac{(y-2)^2}{1}-\frac{x^2}{3}=1$.

53. See Figure 53.

54. See Figure 54.

[-10,10] by [-10,10]
Xscl = 1 Yscl = 1

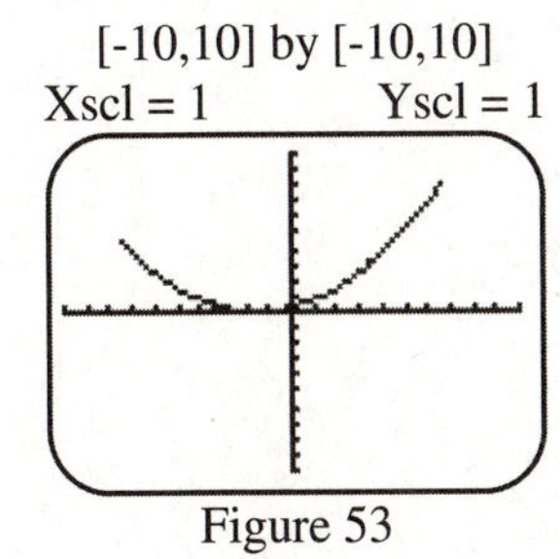

Figure 53

[-10,10] by [-10,10]
Xscl = 1 Yscl = 1

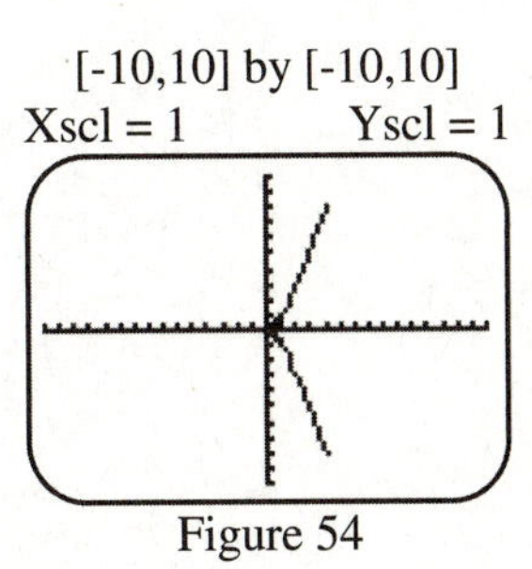

Figure 54

55. From the 1st equation, $x=\sqrt{t-1} \Rightarrow t=x^2+1$. By substitution in the 2nd equation $y=\sqrt{x^2+1}$. This is equivalent to $y^2-x^2=1$. When t is in $[1,\infty)$, the range of $x=\sqrt{t-1}$ is x in $[0,\infty)$.

56. From the 1st equation, $x = 3t + 2 \Rightarrow t = \frac{x-2}{3}$. By substitution in the 2nd equation $y = \frac{x-2}{3} - 1$. This is equivalent to $x - 3y = 5$. When t is in $[-5, 5]$, the range of $x = 3t + 2$ is x in $[-13, 17]$.

57. Since the major axis has length 134.5 million miles, $2a = 134.5 \Rightarrow a = 67.25$. From the given eccentricity, $\frac{c}{a} = 0.006775 \Rightarrow \frac{c}{67.25} = 0.006775 \Rightarrow c = 67.25(0.006775) = 0.4456$ million miles. The smallest distance is $67.25 - 0.4456 \approx 66.8$ million miles and the largest distance is $67.25 + 0.4456 \approx 67.7$ million miles.

58. Since the smallest distance between the comet and the sun is 89 million miles, $a - c = 89$. The given eccentricity provides the equation $\frac{c}{a} = 0.964$. Solving these two equations simultaneously gives $a \approx 2472.222$ and $c \approx 2383.222$. Then $b^2 = a^2 - c^2 \Rightarrow b^2 = 2472.222 - 2383.222 \Rightarrow b^2 \approx 432{,}135$. The equation is $\frac{x^2}{a^2} + \frac{y^2}{b^2} = 1 \Rightarrow \frac{x^2}{6{,}111{,}882} + \frac{y^2}{432{,}135} = 1$.

59. The value of $\frac{k}{\sqrt{D}}$ is $\frac{2.82 \times 10^7}{\sqrt{42.5 \times 10^6}} \approx 4326$. Since $2090 < 4326$, $V < \frac{k}{\sqrt{D}}$. The trajectory is elliptic.

60. The velocity must be more than 4326 m per sec. The minimum increase is $4326 - 2090 \approx 2236$ m per sec.

61. The required increase in velocity is less when D is larger.

62. $Ax^2 + Cy^2 + Dx + Ey + F = 0 \Rightarrow A\left(x^2 + \frac{D}{A}x\right) + C\left(y^2 + \frac{E}{C}y\right) = -F \Rightarrow$

$$A\left(x^2 + \frac{D}{A}x + \frac{D^2}{4A^2}\right) + C\left(y^2 + \frac{E}{C}y + \frac{E^2}{4C^2}\right) = -F + \frac{D^2}{4A} + \frac{E^2}{4C} \Rightarrow A\left(x + \frac{D}{2A}\right)^2 + C\left(y + \frac{E}{2C}\right)^2$$

$$= \frac{CD^2 + AE^2 - 4ACF}{4AC} \Rightarrow \frac{\left(x + \frac{D}{2A}\right)^2}{\frac{CD^2 + AE^2 - 4ACF}{4A^2C}} + \frac{\left(y + \frac{E}{2C}\right)^2}{\frac{CD^2 + AE^2 - 4ACF}{4AC^2}} = 1.$$

The center is $\left(-\frac{D}{2A}, -\frac{E}{2C}\right)$.

Chapter 6 Test

1. (a) B. This is a hyperbola with center $(-3, -2)$.

   (b) A. This is a circle with center (3, 2) and radius 4.

   (c) D. This is a circle with center $(-3, 2)$ and radius 4.

   (d) E. This is a parabola that opens downward.

   (e) F. This is a parabola that opens to the right.

   (f) C. This is an ellipse with center $(-3, -2)$.

2. $y^2 = \frac{1}{8}x \Rightarrow (y - 0)^2 = 4\left(\frac{1}{32}\right)(x - 0)$; this is a parabola with vertex (0, 0) that opens to the right. Since

$c = \frac{1}{32}$, the focus is located at $\left(\frac{1}{32}, 0\right)$ and the equation of the directrix is $x = -\frac{1}{32}$.

3. See Figure 3. This is the graph of a function with domain $[-6, 6]$ and range $[-1, 0]$.

[-9,9] by [-3,3]
Xscl = 1 Yscl = 1

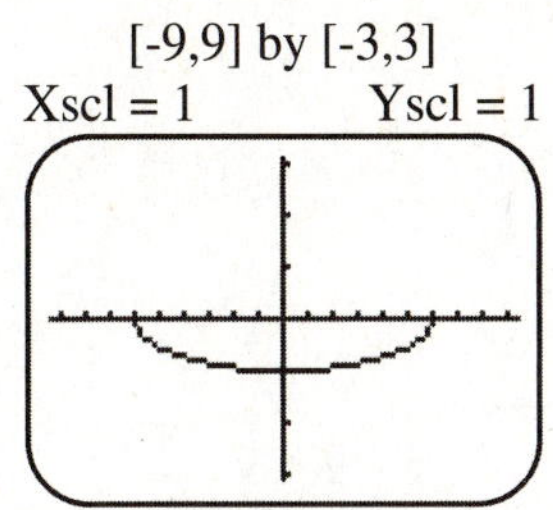

Figure 3

4. $\frac{x^2}{25} - \frac{y^2}{49} = 1 \Rightarrow \frac{y^2}{49} = \frac{x^2}{25} - 1 \Rightarrow y^2 = 49\left(\frac{x^2}{25} - 1\right) \Rightarrow y = \pm\sqrt{49\left(\frac{x^2}{25} - 1\right)} \Rightarrow \; y = \pm 7\sqrt{\frac{x^2}{25} - 1}$ The equations are $y_1 = 7\sqrt{\frac{x^2}{25} - 1}$ and $y_2 = -7\sqrt{\frac{x^2}{25} - 1}$.

5. This is a hyperbola with vertical transverse axis. Here $a = 2$ and $b = 3$. The asymptotes are $y = \pm\frac{2}{3}x$. See Figure 5. Since $c = \sqrt{a^2 + b^2} = \sqrt{2^2 + 3^2} = \sqrt{13}$, the foci are $(0, -\sqrt{13})$ and $(0, \sqrt{13})$. The center is (0, 0). The vertices are $(0, -2)$ and (0, 2).

6. $x^2 + 4y^2 + 2x - 16y + 17 = 0 \Rightarrow x^2 + 2x + 1 + 4(y^2 - 4y + 4) = -17 + 1 + 16 \Rightarrow \; (x+1)^2 + 4(y-2)^2 = 0$; The only point that satisfies this equation is $(-1, 2)$. See Figure 6.

7. $y^2 - 8y - 2x + 22 = 0 \Rightarrow y^2 - 8y + 16 = 2x - 22 + 16 \Rightarrow (y-4)^2 = 2(x-3) \Rightarrow (y-4)^2 = 4\left(\frac{1}{2}\right)(x-3)$; This is a parabola with vertex (3, 4) and focus (3.5, 4). See Figure 7.

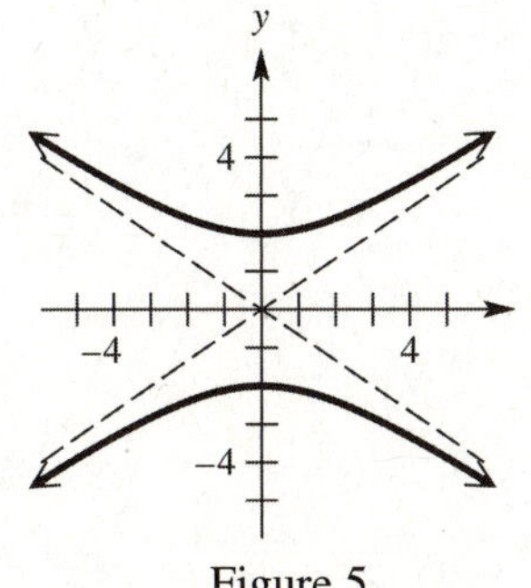

Figure 5

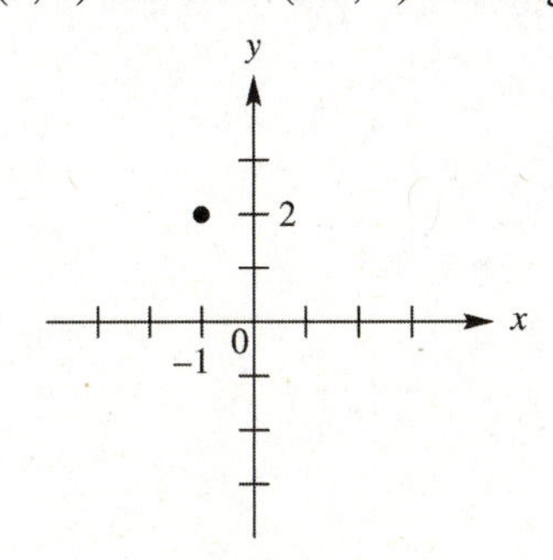

Figure 6

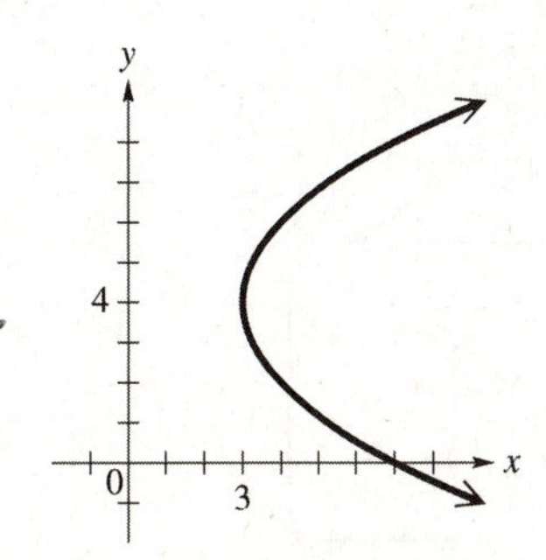

Figure 7

8. $x^2 + (y-4)^2 = 9 \Rightarrow (x-0)^2 + (y-4)^2 = 9$; This is a circle with center (0, 4) and radius 3. See Figure 8.

9. This is an ellipse with horizontal major axis. Here $a = 7$ and $b = 4$ so $c = \sqrt{7^2 - 4^2} = \sqrt{33}$. The center is $(3, -1)$. The vertices are $(3-7, -1)$ and $(3+7, -1)$ or $(-4, -1)$ and $(10, -1)$. The foci are $\left(3 + \sqrt{33}, -1\right)$ and $\left(3 - \sqrt{33}, -1\right)$. See Figure 9.

10. From the 2nd equation $y = t - 1 \Rightarrow t = y + 1$. Substituting in the 1st equation yields $x = 4(y+1)^2 - 4$. This equation can be written $(y+1)^2 = 4\left(\frac{1}{16}\right)(x+4)$. This is a parabola that opens to the right. The vertex is $(-4, -1)$ and the focus is $\left(-4 + \frac{1}{16}, -1\right) \Rightarrow \left(-\frac{63}{16}, -1\right)$. See Figure 10.

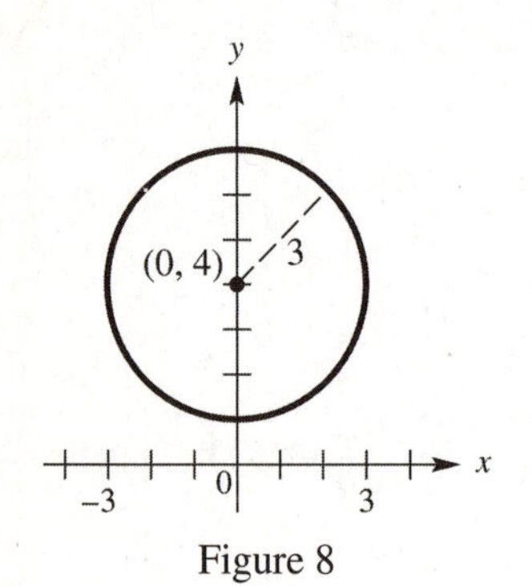

Figure 8

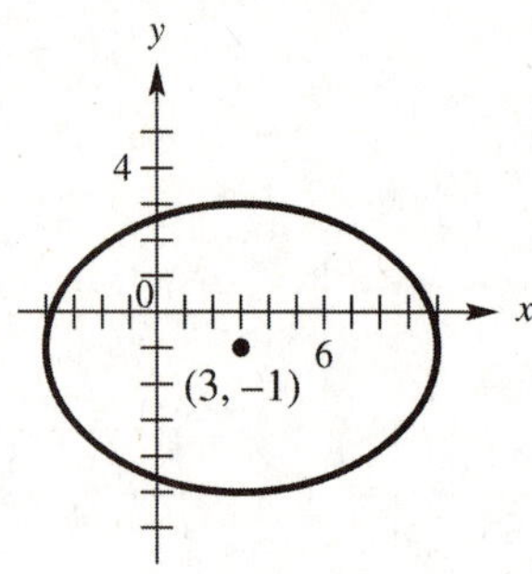

Figure 9

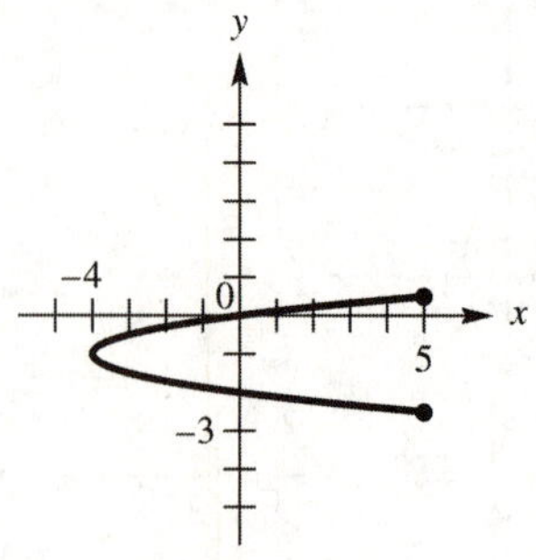

Figure 10

11. (a) Since $e = 1$, the conic is a parabola. With center (0, 0) and focus $(0, -2)$, we know that $c = -2$. The equation is $x^2 = 4cy \Rightarrow x^2 = -8y \Rightarrow y = -\frac{1}{8}x^2$.

(b) Since $0 < e < 1$, the conic is an ellipse with $a = 3$. Now $\frac{c}{a} = e \Rightarrow \frac{c}{3} = \frac{5}{6} \Rightarrow c = \frac{15}{6}$. For an ellipse, $b^2 = a^2 - c^2 = 9 - \frac{225}{36} = \frac{11}{4}$. The equation is $\frac{x^2}{\frac{11}{4}} + \frac{y^2}{9} = 1$ or $\frac{4x^2}{11} + \frac{y^2}{9} = 1$.

12. Using a vertical minor axis, $b = 12$. The major axis has length 40 so $a = 20$. The equation is $\frac{x^2}{400} + \frac{y^2}{144} = 1$. When $x = 10$, $\frac{10^2}{400} + \frac{y^2}{144} = 1 \Rightarrow \frac{y^2}{144} = 1 - \frac{1}{4} \Rightarrow y^2 = 144\left(\frac{3}{4}\right) \Rightarrow y = 12\sqrt{\left(\frac{3}{4}\right)} \approx 10.39\ ft$.

13. See Figure 13.

[-5,5] by [0,10]
Xscl = 1 Yscl = 1

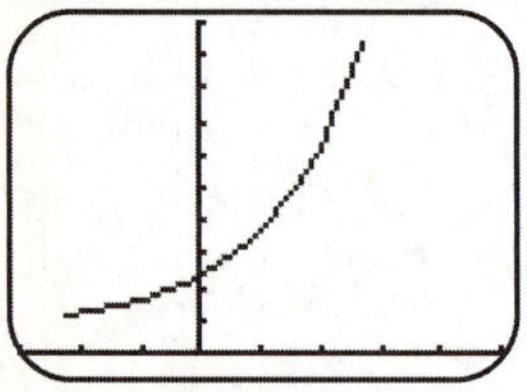

Figure 13

14. $x = \frac{1}{t+3} \Rightarrow t = \frac{1}{x} - 3$. Substituting this in $y = t + 3 \Rightarrow y = \frac{1}{x}, x \neq 0$.

Chapter 7: Systems of Equations and Inequalities; Matrices

7.1: Systems of Equations

1. From the graph, the trend appears to be higher in the year 2005.

3. From the graph, the intersection points are approximately (1998,0.060), (2000, 0.059), (2001, 0.058), (2003, 0.059), (2005, 0.056).

5. The variable t would represent the year and the variable $f(t)$ would represent the ozone level in ppm.

7. From the graph, the solution is $(2,2)$. Using substitution, first solve [equation 1] for $x, x-y=0 \Rightarrow x=y$ [equation 3]. Substitute y for x in [equation 2]: $y+y=4 \Rightarrow 2y=4 \Rightarrow y=2$. Now substitute 2 for y in [equation 3]: $x=y \Rightarrow x=2$. The solution is $(2,2)$.

9. From the graph, the solution is $\left(\frac{1}{2},-2\right)$. Using substitution, first solve [equation 2] for x: $2x-3y=7 \Rightarrow 2x=3y+7 \Rightarrow x=\frac{3}{2}y+\frac{7}{2}$ [equation 3]. Substitute $\frac{3}{2}y+\frac{7}{2}$ for x in [equation 1]: $6\left(\frac{3}{2}y+\frac{7}{2}\right)+4y=-5 \Rightarrow 9y+21+4y=-5 \Rightarrow 13y=-26 \Rightarrow y=-2$. Now substitute -2 in for y in [equation 3]: $x=\frac{3}{2}(-2)+\frac{7}{2} \Rightarrow x=-3+\frac{7}{2} \Rightarrow x=\frac{1}{2}$. The solution is $\left(\frac{1}{2},-2\right)$.

11. Since $y=x$ in [equation 2], we can use substitution. Substitute x for y in [equation 1]: $6x-(x)=5 \Rightarrow 5x=5 \Rightarrow x=1$. Now substitute 1 for x in [equation 2]: $y=x \Rightarrow y=1$. The solution is (1,1).

13. Using substitution, first solve [equation 1] for $x, x+2y=-1 \Rightarrow x=-2y-1$ [equation 3]. Substitute $-2y-1$ for x in [equation 2]: $2(-2y-1)+y=4 \Rightarrow -3y-2=4 \Rightarrow -3y=6 \Rightarrow y=-2$. Now substitute -2 for y in [equation 3]: $x=-2(-2)-1 \Rightarrow x=3$. The solution is $(3,-2)$.

15. Since $y=2x+3$ in [equation 1], we can use substitution. Substitute $2x+3$ for y in [equation 2]: $3x+4(2x+3)=78 \Rightarrow 11x+12=78 \Rightarrow 11x=66 \Rightarrow x=6$. Now substitute 6 for x in [equation 1]: $y=2(6)+3 \Rightarrow y=15$. The solution is $(6,15)$.

17. Using substitution, first solve [equation 1] for x: $3x-2y=12 \Rightarrow 3x=2y+12 \Rightarrow x=\frac{2}{3}y+4$ [equation 3]. Substitute $\frac{2}{3}y+4$ for x in [equation 2]: $5\left(\frac{2}{3}y+4\right)=4-2y \Rightarrow \frac{10}{3}y+2y=4-20 \Rightarrow \frac{16}{3}y=-16$ $\Rightarrow y=-3$. Now substitute -3 for y in [equation 3]: $x=\frac{2}{3}(-3)+4 \Rightarrow x=2$. The solution is $(2,-3)$.

19. Using substitution, first solve [equation 2] for y: $2x+y=5 \Rightarrow y=-2x+5$ [equation 3]. Substitute $-2x+5$ for y in [equation 1]: $4x-5(-2x+5)=-11 \Rightarrow 4x+10x-25=-11 \Rightarrow 14x=14 \Rightarrow x=1$. Now substitute 1 for y in [equation 3]: $y=-2(1)+5 \Rightarrow y=3$. The solution is (1,3).

21. Using substitution, first solve [equation 2] for y: $9y = 31 + 2x \Rightarrow y = \dfrac{31+2x}{9}$ [equation 3]. Substitute

$\dfrac{31+2x}{9}$ for y in [equation 1]: $4x + 5\left(\dfrac{31+2x}{9}\right) = 7 \Rightarrow 9\left[4x + 5\left(\dfrac{31+2x}{9}\right) = 7\right] \Rightarrow$

$36x + 5(31+2x) = 63 \Rightarrow 36x + 155 + 10x = 63 \Rightarrow 46x = -92 \Rightarrow x = -2.$

Now substitute -2 for x in [equation 3]: $y = \dfrac{31+2(-2)}{9} \Rightarrow y = 3.$ The solution is $(-2,3)$.

23. Using substitution, first solve [equation 1] for x: $3x - 7y = 15 \Rightarrow 3x = 7y + 15 \Rightarrow x = \dfrac{7y+15}{3}$ [equation 3].

Substitute $\dfrac{7y+15}{3}$ for x in [equation 2]: $3\left(\dfrac{7y+15}{3}\right) + 7y = 15 \Rightarrow 7y + 15 + 7y = 15 \Rightarrow 14y = 0 \Rightarrow y = 0.$

Now substitute 0 for y in [equation 3]: $x = \dfrac{7(0)+15}{3} \Rightarrow x = 5.$ The solution is $(5,0)$.

25. Using substitution, first solve [equation 1] for x: $2x - 7y = 8 \Rightarrow 2x = 7y + 8 \Rightarrow x = \dfrac{7y+8}{2}$ [equation 3].

Substitute $\dfrac{7y+8}{2}$ for x in [equation 2]: $-3\left(\dfrac{7y+8}{2}\right) + \dfrac{21}{2}y = 5 \Rightarrow 2\left[-3\left(\dfrac{7y+8}{2}\right) + \dfrac{21}{2}y = 5\right] \Rightarrow$

$-3(7y+8) + 21y = 10 \Rightarrow -21y - 24 + 21y = 10 \Rightarrow -24 = 10.$

Since this is a false statement, the solution set is $\varnothing$.

27. Using substitution, first solve [equation 1] for x: $x - 2y = 4 \Rightarrow x = 4 + 2y$ [equation 3]. Substitute $x = 4 + 2y$ for x in [equation 2]: $-2(4+2y) + 4y = -8 \Rightarrow -8 - 4y + 4y = -8 \Rightarrow -8 = -8.$ Since this is a true statement, the system is dependent and the solutions are $(4+2y, y)$ or $\left(x, \dfrac{x-4}{2}\right)$.

29. Multiply the first equation by 3 and add to eliminate the y-variable:

$$\begin{array}{r} 9x - 3y = -12 \\ x + 3y = \ \ 12 \\ \hline 10x \quad = \ \ \ 0 \end{array}$$

$\Rightarrow x = 0.$ Substitute 0 for x in the second equation: $(0) + 3y = 12 \Rightarrow y = 4.$ The solution is $(0,4)$.

31. Multiply the second equation by 2, and subtract to eliminate the x-variable:

$$\begin{array}{r} 4x + 3y = -1 \\ 4x + 10y = \ \ 6 \\ \hline -7y = -7 \end{array}$$

$\Rightarrow y = 1.$ Substitute 1 for y in the first equation: $4x + 3(1) = -1 \Rightarrow 4x = -4 \Rightarrow x = -1.$

The solution is $(-1,1)$.

33. Multiply the second equation by 4, and subtract to eliminate the x-variable:

$$\begin{array}{rl} 12x - 5y = & 9 \\ 12x - 32y = & -72 \\ \hline 27y = & 81 \end{array}$$

$\Rightarrow y = 3$. Substitute 3 for y in the first equation: $12x - 5(3) = 9 \Rightarrow 12x = 24 \Rightarrow x = 2$.

The solution is $(2,3)$.

35. Multiply the first equation by 2 and add to eliminate both variables.

$$\begin{array}{rl} 8x - 2y = & 18 \\ -8x + 2y = & -18 \\ \hline 0 = & 0 \end{array}$$

$\Rightarrow$ infinite number of solutions. Therefore, the solutions have the following relationship: $4x - y = 9 \Rightarrow -y = -4x + 9 \Rightarrow y = 4x - 9$. The solution set is $\{(x, 4x-9)\}$ or $\left\{\left(\frac{y+9}{4}, y\right)\right\}$.

37. Multiply the first equation by 2 and add to eliminate both variables:

$$\begin{array}{rl} 18x - 10y = & 2 \\ -18y + 10y = & 1 \\ \hline 0 = & 3 \end{array}$$

$\Rightarrow \varnothing$

39. Multiply the first equation by 2 and subtract to eliminate both variables:\

$$\begin{array}{rl} 6x + 2y = & 12 \\ 6y + 2y = & 1 \\ \hline 0 = & 13 \end{array}$$

$\Rightarrow \varnothing$

41. First, multiplying both equations by 6 yields $3x + 2y = 48$ and $4x + 9y = 102$. Now multiply the first equation by 4, the second equation by 3, and subtract to eliminate the x-variable:

$$\begin{array}{rl} 12x + 8y = & 192 \\ 12x + 27y = & 306 \\ \hline -19y = & -114 \end{array}$$

$\Rightarrow y = 6$. Substitute 6 for y in the first equation (multiplied by 6):

$3x + 2(6) = 48 \Rightarrow 3x = 36 \Rightarrow x = 12$. The solution is $(12, 6)$.

43. Multiplying the first equation by 12 and the second by 6 yields $8x + 3y = 46$ and $x + 2y = 9$. Now multiply the second equation by 8, and subtract to eliminate the x-variable:

$$\begin{array}{rl} 8x + 3y = & 46 \\ 8x + 16y = & 72 \\ \hline -13y = & -26 \end{array}$$

$\Rightarrow y = 2$. Substitute 2 for y in the second equation (multiplied by 6):

$x + 2(2) = 9 \Rightarrow x = 5$. The solution is $(5, 2)$.

45. Graph each function in the same window. The intersection or solution is $(0.138, -4.762)$.

47. Graph each function in the same window. The intersection or solution is $(-8.708, -15.668)$.

49. An inconsistent system will conclude with no variables and a false statement, such as $0 = 1$. A system with dependent equations will conclude with no variables and a true statement, such as $0 = 0$. $k = -6$.

51. See Figure 51.

53. See Figure 53

55. See Figure 55.

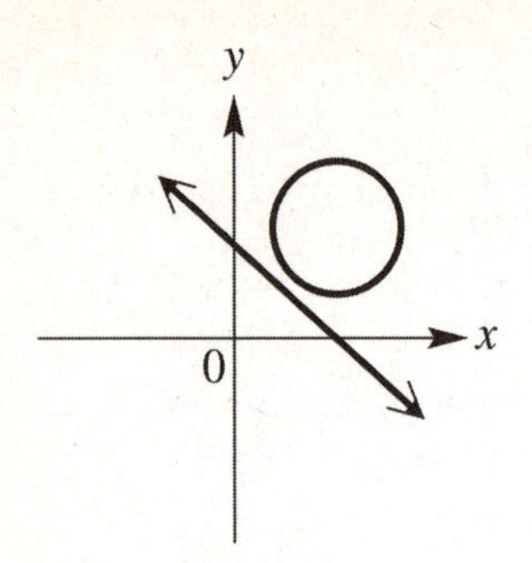

Figure 51

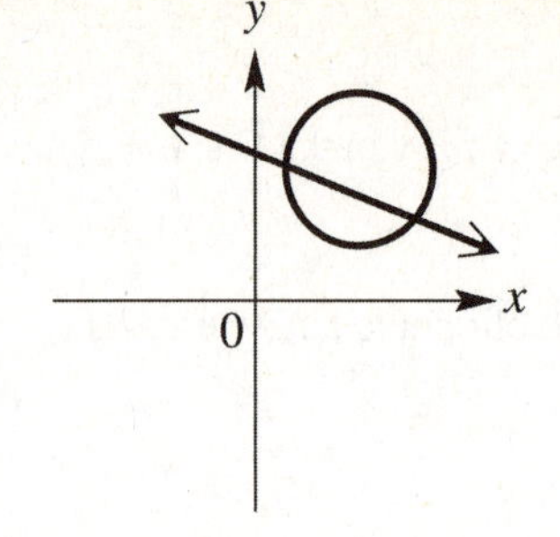

Figure 53

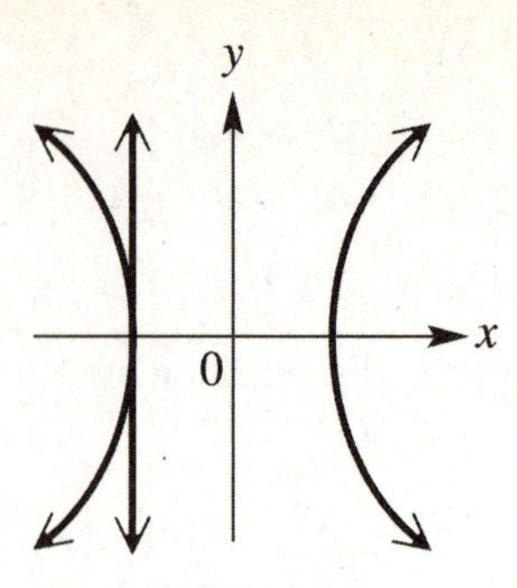

Figure 55

57. See Figure 57.

59. See Figure 59.

61. Check: $1+2=3 \Rightarrow 3=3,\quad 2(1)-2=0 \Rightarrow 0=0.$ See Figure 61

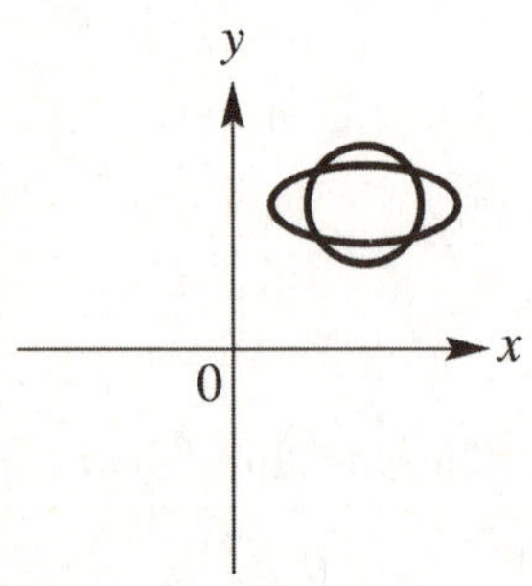

Figure 57

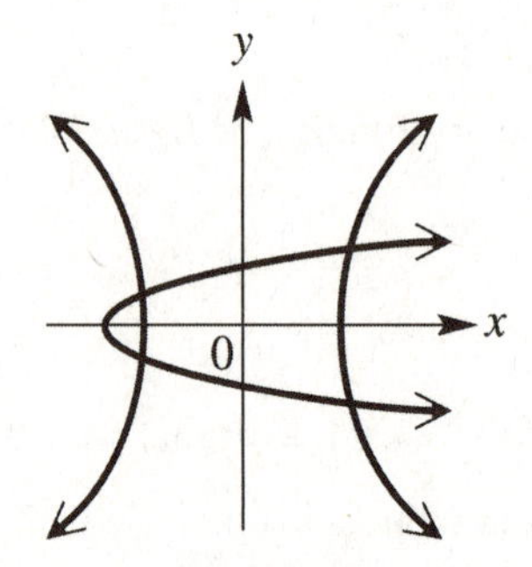

Figure 59

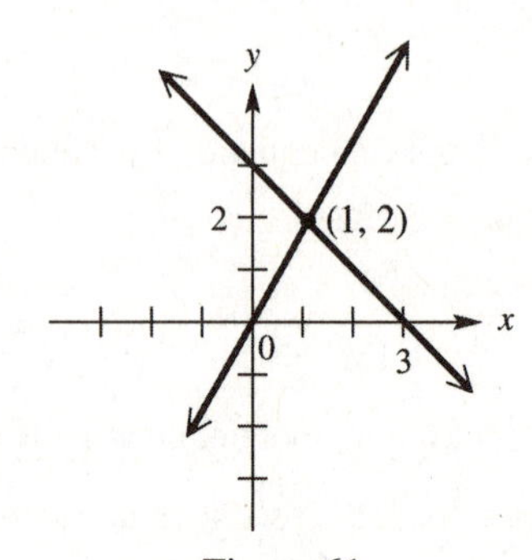

Figure 61

63. Check: $1^2+2^2=5 \Rightarrow 5=5,\ 2=3-1 \Rightarrow 2=2,\ 2^2+1^2=5 \Rightarrow 5=5,\ 1=3-2 \Rightarrow 1=1.$

65. Check: $1=(-1)^2 \Rightarrow 1=1,\ (-1)^2+1^2=2,\ 1=1^2 \Rightarrow 1=1,\ 1^2+1^2=2 \Rightarrow 2=2.$

67. Check: $2^2-0^2=4 \Rightarrow 4-0=4 \Rightarrow 4=4,\ 2+0=2 \Rightarrow 2=2.$

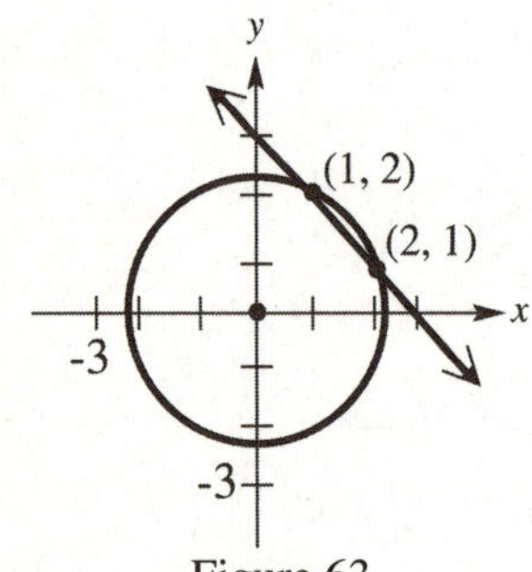

Figure 63

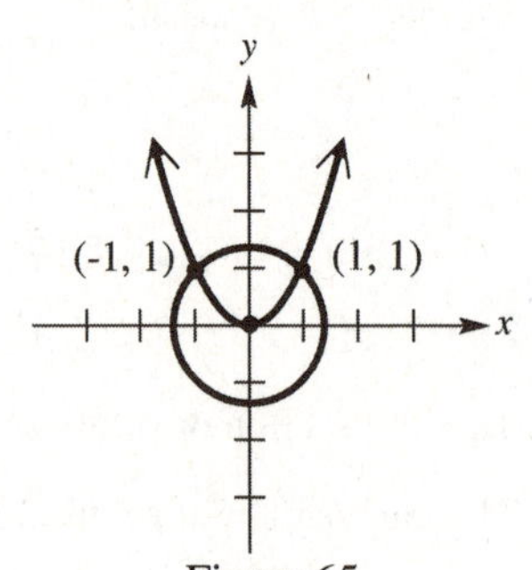

Figure 65

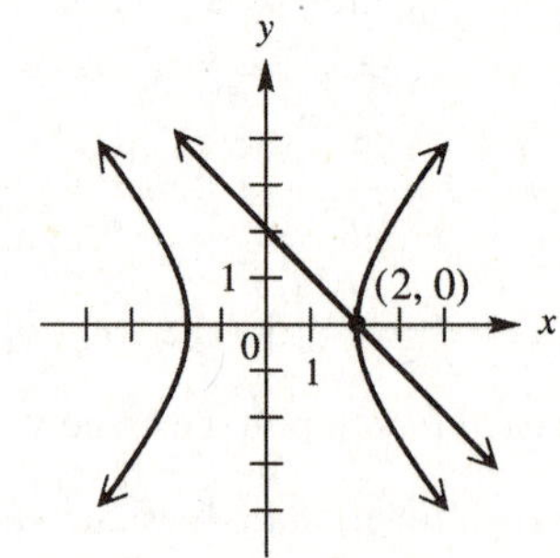

Figure 67

69. Since [equation 1] is solved for y, we substitute $-x^2+2$ for y in [equation 2]:

$x-(-x^2+2)=0 \Rightarrow x^2+x-2 \Rightarrow (x+2)(x-1)=0 \Rightarrow x=-2$ or 1. Substituting these values into [equation 1] yields $y=-(-2)^2+2 \Rightarrow y=-4+2 \Rightarrow y=-2$ and $y=-(1)^2+2 \Rightarrow y=-1+2 \Rightarrow y=1$.

The solution set is $\{(-2,-2),(1,1)\}$.

71. First solve [equation 2] for x: $x-y=-2 \Rightarrow x=y-2$ [equation 3]. Now substitute $y-2$ for x in

[equation 1]: $3(y-2)^2+2y^2=5 \Rightarrow 3(y^2-4y+4)+2y^2=5 \Rightarrow 3y^2-12y+12+2y^2-5=0 \Rightarrow$

$5y^2-12y+7=0 \Rightarrow (5y-7)(y-1)=0 \Rightarrow y=1$ or $\frac{7}{5}$. Substituting these values into [equation 3]

yields $x=(1)-2 \Rightarrow x=-1$ and $x=\left(\frac{7}{5}\right)-2 \Rightarrow x=-\frac{3}{5}$. The solution set is $\left\{\left(-\frac{3}{5},\frac{7}{5}\right),(-1,1)\right\}$.

73. Add the equations to eliminate the y^2:

$$\begin{aligned} x^2+y^2&=10\\ 2x^2-y^2&=17\\ \hline 3x^2\quad&=27 \Rightarrow x^2=9 \Rightarrow x=\pm 3. \end{aligned}$$

Substituting these values into [equation 1] yields $(-3)^2+y^2=10 \Rightarrow y^2=1 \Rightarrow y=\pm 1$ and

$(3)^2+y^2=10 \Rightarrow y^2=1 \Rightarrow y=\pm 1$. The solution set is $\{(-3,1),(-3,-1),(3,1),(3,-1)\}$.

75. Multiply [equation 1] by 3 and subtract to eliminate the x^2:

$$\begin{aligned} 3x^2+6y^2&=27\\ 3x^2-4y^2&=27\\ \hline 10y^2&=0 \Rightarrow 10y^2=0 \Rightarrow y=0. \end{aligned}$$

Substituting 0 into [equation 1] for y, yields $x^2+2(0)^2=9 \Rightarrow x^2=9 \Rightarrow x=\pm 3$. The solution set is

$\{(3,0),(-3,0)\}$.

77. Multiply [equation 1] by 3 and [equation 2] by 2 and subtract to eliminate both x^2 and y^2:

$$\begin{aligned} 6x^2+6y^2&=60\\ 6x^2+6y^2&=60\\ \hline 0\quad&=0 \Rightarrow \end{aligned}$$

infinite number of solutions. The solutions have the following relationship:

$2x^2+2y^2=20 \Rightarrow 2y^2=20-2x^2 \Rightarrow y^2=10-x^2 \Rightarrow y=\pm\sqrt{10-x^2}$. The solution set is

$\left\{\left(x,-\sqrt{10-x^2}\right),\left(x,\sqrt{10-x^2}\right)\right\}$.

79. If $y=|x-1|$ and $y=x^2-4$ then $|x-1|=x^2-4$. Solving this yields $x-1=x^2-4 \Rightarrow x^2-x-3=0$ or

$x-1=-(x^2-4) \Rightarrow x-1=-x^2+4 \Rightarrow -x^2+x-5=0$. Solve both of these quadratic equations using the

quadratic formula: Solve for x, $x^2-x-3=0 \Rightarrow x=\dfrac{-(-1)\pm\sqrt{(-1)^2-4(1)(-3)}}{2(1)} \Rightarrow x=\dfrac{1\pm\sqrt{13}}{2}$.

Solve for x, $x^2+x-5=0 \Rightarrow x=\dfrac{-1\pm\sqrt{1^2-4(1)(-5)}}{2(1)} \Rightarrow x=\dfrac{-1\pm\sqrt{21}}{2}$. Now substitute each of these values of x into [equation 2] and solve for y: For $x=\dfrac{1+\sqrt{13}}{2}$,

$$y=\left(\frac{1+\sqrt{13}}{2}\right)^2-4=\frac{1+2\sqrt{13}+13}{4}-\frac{16}{4}=\frac{2\sqrt{13}-2}{4} \Rightarrow y=\frac{\sqrt{13}-1}{2}.$$

For $x=\dfrac{1-\sqrt{13}}{2}$, $y=\left(\dfrac{1-\sqrt{13}}{2}\right)^2-4=\dfrac{1-2\sqrt{13}+13}{4}-\dfrac{16}{4}=\dfrac{-2\sqrt{13}-2}{4} \Rightarrow y=\dfrac{-\sqrt{13}-1}{2}$. From [equation 1], we know $y>0$. Since $y=\dfrac{\sqrt{13}-1}{2}<0$, there is no solution with this value of x. For $x=\dfrac{-1+\sqrt{21}}{2}$, $y=\left(\dfrac{-1+\sqrt{21}}{2}\right)^2-4=\dfrac{1-2\sqrt{21}+21}{4}-\dfrac{16}{4}=\dfrac{-2\sqrt{21}+6}{4} \Rightarrow y=\dfrac{\sqrt{21}+3}{2}$. From [equation 1] we know $y>0$. Since $y=\dfrac{-\sqrt{21}+3}{2}<0$, there is no solution with this value of x. For $x=\dfrac{-1-\sqrt{21}}{2}$, $y=\left(\dfrac{-1-\sqrt{21}}{2}\right)^2-4=\dfrac{1+2\sqrt{21}+21}{4}-\dfrac{16}{4}=\dfrac{2\sqrt{21}+3}{4} \Rightarrow \dfrac{\sqrt{21}+3}{2}$.

The solution set is $\left\{\left(\dfrac{1+\sqrt{13}}{2},\dfrac{-1+\sqrt{13}}{2}\right),\left(\dfrac{-1-\sqrt{21}}{2},\dfrac{3+\sqrt{21}}{2}\right)\right\}$.

81. Graph each function in the same window. The intersections or solutions are $(-0.79,0.62)$ and $(0.88,0.77)$.

83. Graph each function in the same window. The intersection or solution is $(0.06,2.88)$.

85. (a) Let x represent the number of robberies in 2000, and let y represent the number of robberies in 2001. The required system of equations is $x+y=831{,}000$ and $x-y=15{,}000$.

(b) Adding the two equations results in $2x=846{,}000 \Rightarrow x=423{,}000$. From the first equation, $423{,}000+y=831{,}000 \Rightarrow y=408{,}000$. The solution set is $\{(423{,}000,\ 408{,}000)\}$.

(c) There were 423,000 robberies in 2000 and 408,000 in 2001.

87. Let x = hours spent on the internet in 2001. Then, $1.1x=11 \Rightarrow x=10$. Therefore, 10 hours were spent on the internet in 2001 and $10+11$ or 21 hours in 2005.

89. Let x = the width of the base; therefore, the length $2x$, and let y = the height of the box. Since the volume is 588 in^2, $2x^2y=588$ and so $y=\dfrac{588}{2x^2}=\dfrac{294}{x^2}$. Since the surface area is 448 in^2, $2(2x)(x)+2(x)(y)+2(2x)(y)=448 \Rightarrow 4x^2+6xy=448$. Substituting for y in this equation yields $4x^2+6x\cdot\dfrac{294}{x^2}=448$. Simplifying we get $4x^3-448x+1764=0$. By graphing the left side of the equation, the x-intercepts are approximately 5.17 and 7. See Figure 89. When $x\approx 5.17$, the dimensions are 5.17 by

$2(5.17) \approx 10.34$ by $\dfrac{294}{(5.17)^2} \approx 11.00$ inches. When $x = 7$, the dimensions are 7 by $2(7) = 14$

by $\dfrac{294}{(7)^2} = 6$ inches.

91. First let $H = 180$ in each equation and solve for y:

$180 = 0.491x + 0.468y + 11.2 \Rightarrow 0.468y = 180 - 11.2 - 0.491x \Rightarrow y_1 = \dfrac{168.8 - 0.491x}{0.468}$

$180 = -0.981x + 1.872y + 26.4 \Rightarrow 1.872y = 180 - 26.4 - 0.981x \Rightarrow y_2 = \dfrac{153.6 + 0.981x}{1.872}$. Now graph each equation in the same window and the intersection is approximately $(177.1, 174.9)$. This means that if an athlete's maximum heart rate is 180 beats per minute (bpm), then it will be about 177 bpm after 5 seconds and 175 bpm after 10 seconds.

93. (a) Let x = the amount of natural gas consumed and y = the amount of petroleum consumed. Then $x + y = 3.74$ and $x - y = 3.02$.

(b)
$$\begin{array}{l} x + y = 3.74 \\ \underline{x - y = 3.02} \\ 2x \quad = 6.76 \Rightarrow x = 3.38. \end{array}$$

Then, $x - y = 3.02 \Rightarrow 3.38 - y = 3.02 \Rightarrow y = 0.36$. The solution is $(3.38, 0.36)$.

95. If $V = \pi r^2 h$ and $V = 50$ then $50 = \pi r^2 h$; and $S.A = 2\pi rh$ and $S.A. = 65$ then $65 = 2\pi rh$. Now solve each for $h: 50 = \pi r^2 h \Rightarrow h_1 = \dfrac{50}{\pi r^2}$; and $65 = 2\pi rh \Rightarrow h_2 = \dfrac{65}{2\pi r}$. Graph each equation in the same window. See Figure 95. The intersection is approximately: (1.538, 6.724); therefore, $r \approx 1.538$ and $h \approx 6.724$.

[0,10] by [-500,1500]
Xscl = 1 Yscl = 500

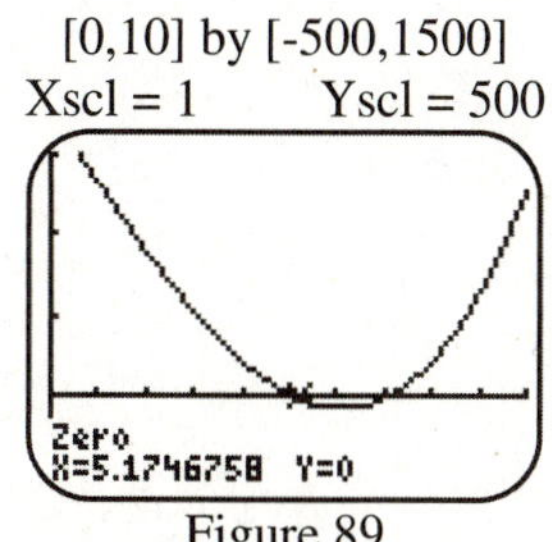

Figure 89

[0,10] by [0,10]
Xscl = 1 Yscl = 1

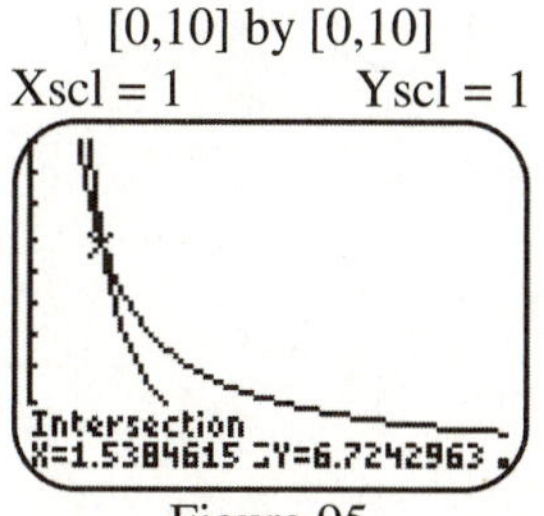

Figure 95

97. Add the equations to eliminate the w_2-variable:

$$\begin{array}{r} w_1 + \sqrt{2}w_2 = 300 \\ \underline{\sqrt{3}w_1 - \sqrt{2}w_2 = \quad 0} \\ (1+\sqrt{3})w_1 = 300 \end{array} \quad \Rightarrow w_1 = \frac{300}{(1+\sqrt{3})} \approx 109.8.$$

Now substitute 109.8 for w_1 in the first equation:

$\left(\dfrac{300}{(1+\sqrt{3})}\right)+\sqrt{2}\,w_2=300 \Rightarrow \sqrt{2}w_2=300-\dfrac{300}{(1+\sqrt{3})} \Rightarrow \sqrt{2}w_2=\dfrac{300(1+\sqrt{3})-300}{1+\sqrt{3}} \Rightarrow$

$\sqrt{2}\,w_2=\dfrac{300\sqrt{3}}{1+\sqrt{3}} \Rightarrow w_2=\dfrac{300\sqrt{3}}{1+\sqrt{3}}\times\dfrac{1}{\sqrt{2}}=\dfrac{300\sqrt{3}}{\sqrt{2}+\sqrt{6}} \Rightarrow w_2 \approx 134.5$. The solution is (109.8, 134.5).

99. The total number of vehicles entering intersection A is $500+150=650$ vehicles per hour. The expression $x+y$ represents the number of vehicles leaving intersection A each hour. Therefore, we have $x+y=650$. The total number of vehicles leaving intersection B is $50+400=450$ vehicles per hour. There are 100 vehicles entering intersection B from the south and y vehicles entering intersection B from the west. Thus $y+100=450 \Rightarrow y=350$. Using substitution, substitute 350 in for y into the equation $x+y=650$: $x+350=650 \Rightarrow x=300$. At intersection A, a stoplight should allow for 300 vehicles per hour to travel south and 350 vehicles per hour to continue traveling east.

101. Find where $C=R$, $4x+125=9x-200 \Rightarrow 325=5x \Rightarrow x=65$. Now substitute 65 in for x into the first equation: $C=4(65)+125=260+125 \Rightarrow C=385$. When $x=65$; $R=C=\$385$.

102. If $t=\dfrac{1}{x}$ and $u=\dfrac{1}{y}$, then $5\left(\dfrac{1}{x}\right)+15\left(\dfrac{1}{y}\right)=16 \Rightarrow 5t+15u=16$ and $5\left(\dfrac{1}{x}\right)+4\left(\dfrac{1}{y}\right)=5 \Rightarrow 5t+4u=5$. The equations are $5t+4u=16$ and $5t+4u=5$.

103. Subtract the two equations to eliminate the x-variable:

$$\begin{array}{l} 5t+15u=16 \\ \underline{5t+\ \ 4u=\ \ 5} \\ \quad\ 11u=11 \quad \Rightarrow u=1. \end{array}$$

Now substitute 1 for u in the first equation: $5t+15(1)=16 \Rightarrow 5t=1 \Rightarrow t=\dfrac{1}{5}$. Therefore, $u=1$ and $t=\dfrac{1}{5}$.

104. If $t=\dfrac{1}{x}$ and $t=\dfrac{1}{5}$, then $\dfrac{1}{x}=\dfrac{1}{5}$ and $x=5$. If $u=\dfrac{1}{y}$ and $u=1$, then $\dfrac{1}{y}=1 \Rightarrow y=1$.

105. $\dfrac{5}{x}+\dfrac{15}{y}=16 \Rightarrow xy\left(\dfrac{5}{x}+\dfrac{15}{y}=16\right) \Rightarrow 5y+15x=16xy \Rightarrow 5y-16xy=-15x \Rightarrow$

$y(5-16x)=-15x \Rightarrow y=\dfrac{-15x}{5-16x}$.

106. $\dfrac{5}{x}+\dfrac{4}{y}=5 \Rightarrow xy\left(\dfrac{5}{x}+\dfrac{4}{y}=5\right) \Rightarrow 5y+4x=5xy \Rightarrow 5y-5xy=-4x \Rightarrow y(5-5x)=-4x \Rightarrow\ y=\dfrac{-4x}{5-5x}$.

107. See Figure 107. The graphs intersect at (5, 1), which supports the indicated exercise.

[0,10] by [0,2]
Xscl = 1 Yscl = 0.5

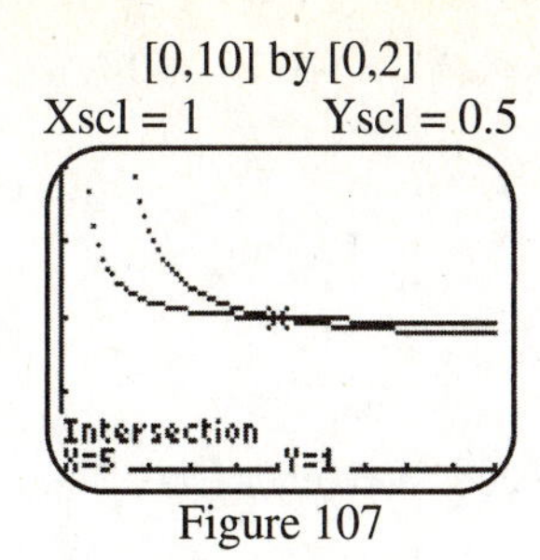

Figure 107

109. If we let $t=\frac{1}{x}$ and $u=\frac{1}{y}$, then our new equations are $2t+u=11$ and $3t-5u=10$.

Multiply the first equation by 5 and add to eliminate the u-variable:

$$\begin{aligned} 10t+5u&=55 \\ 3t-5u&=10 \\ \hline 13t\quad&=65 \end{aligned}$$

$\Rightarrow t=5$. If $t=\frac{1}{x}$ and $t=5$, then $\frac{1}{x}=5\Rightarrow 5x=1\Rightarrow x=\frac{1}{5}$. Now substitute $t=5$ into

$3t-5u=10$, which yields $3(5)-5u=10\Rightarrow -5u=-5\Rightarrow u=1$. If $u=\frac{1}{y}$ and $u=1$, then $\frac{1}{y}=1\Rightarrow y=1$.

The solution is $\left(\frac{1}{5},1\right)$.

111. If we let $t=\frac{1}{x}$ and $u=\frac{1}{y}$, then our new equations are $t+3u=\frac{16}{5}$ and $5t+4u=5$. Multiply the first

equation by 5 and subtract to eliminate the t-variable.

$$\begin{aligned} 5t+15u&=16 \\ 5t+\ 4u&=\ 5 \\ \hline 11u&=11 \end{aligned}$$

$\Rightarrow u=1$. If $x=2$ and $u=1$, then $\frac{1}{y}=1\Rightarrow y=1$. Now substitute $u=1$ into

$5t+4u=5$, which yields $5t+4(1)=5\Rightarrow 5t=1\Rightarrow t=\frac{1}{5}$. If $t=\frac{1}{x}$ and $t=\frac{1}{5}$, then $\frac{1}{x}=\frac{1}{5}\Rightarrow x=5$.

The solution is $(5,1)$.

7.2: Solution of Linear Systems in Three Variables

1. For $2x+y-z=-1$ and $(-3,6,1): 2(-3)+(6)-1=-1\Rightarrow -6+6-1=-1\Rightarrow -1=-1$.

   For $x-y+3z=-6$ and $(-3,6,1): (-3)-6+3(1)=-6\Rightarrow -3-6+3=-6\Rightarrow -6=-6$.

   For $-4x+y+z=19$ and $(-3,6,1): -4(-3)+6+1=19\Rightarrow 12+6+1=19\Rightarrow 19=19$.

3. For $5x-y+2z=-0.4$ and $(-0.2,0.4,0.5)$: $5(-0.2)-(0.4)+2(0.5)=-0.4\Rightarrow$

   $-1-0.4+1=0.4\Rightarrow -0.4=-0.4$.

   For $x+4z=1.8$ and $(-0.2,0.4,0.5)$: $(-0.2)+4(0.5)=1.8\Rightarrow -0.2+4(0.5)=1.8\Rightarrow$

   $-0.2+2.0=1.8\Rightarrow 1.8=1.8$.

For $-3y+z=-0.7$ and $(-0.2,0.4,0.5)$: $-3(0.4)+(0.5)=-0.7 \Rightarrow -1.2+0.5=-0.7 \Rightarrow -0.7=-0.7$.

5. For $x-y+z=-2$ and $(-2,-1,3)$: $(-2)-(-1)+(3)=2 \Rightarrow -2+1+3=2 \Rightarrow 2=2$.

   For $3x-2y+z=-1$ and $(-2,-1,3)$: $3(-2)-2(-1)+(3)=-1 \Rightarrow -6+2+3=-1 \Rightarrow -1=-1$.

   For $x+y=-3$ and $(-2,-1,3)$: $(-2)+(-1)=-3 \Rightarrow -2+(-1)=-3 \Rightarrow -3=-3$.

7. First add [equation 1] and [equation 2] to eliminate z and produce [equation 4]. Second add [equation 2] and [equation 3] to eliminate y and z and produce [equation 5].

$$\begin{array}{rl} x+y+z=2 & [1] \\ 2x+y-z=5 & [2] \\ \hline 3x+2y=7 & [4] \end{array} \qquad \begin{array}{rl} 2x+y-z=5 & [2] \\ x-y+z=-2 & [3] \\ \hline 3x=3 & [5] \end{array}$$

   Solve [equation 5] for $x: 3x=3 \Rightarrow x=1$. Substitute $x=1$ into [equation 4] to solve for y: $3(1)+2y=7 \Rightarrow 2y=4 \Rightarrow y=2$. Finally, substitute $x=1$ and $y=2$ into [equation 1] to solve for z: $(1)+(2)+z=2 \Rightarrow z=-1$. The solution is $(1,2,-1)$.

9. First, add [equation 1] and [equation 2] to eliminate y and produce [equation 4]. Second, multiply [equation 3] by 3 and add [equation 1] to eliminate y and produce[equation 5].

$$\begin{array}{rl} x+3y+4z=14 & [1] \\ 2x-3y+2z=10 & [2] \\ \hline 3x+6z=24 & [4] \end{array} \qquad \begin{array}{rl} 9x-3y+3z=27 & 3[3] \\ x+3y+4z=14 & [1] \\ \hline 10x+7z=41 & [5] \end{array}$$

   Next multiply [equation 4] by 10 and add to [equation 5] multiplied by -3 to eliminate x and produce [equation 6]

$$\begin{array}{rl} 30x+60z=240 & 10[4] \\ -30x-21z=-123 & -3[5] \\ \hline 39z=117 & [6] \end{array}$$

   Solve [equation 6] for $z: 39z=117 \Rightarrow z=3$. Substitute $z=3$ into [equation 4] to solve for x: $3x+6(3)=24 \Rightarrow 3x+6 \Rightarrow x=2$. Finally, substitute $x=2$ and $z=3$ into [equation 1] to solve for y: $(2)+3y+4(3)=14 \Rightarrow 3y=0 \Rightarrow y=0$. The solution is $(2,0,3)$.

11. First, multiply [equation 1] by 2 and add [equation 3] to eliminate all the variables.

$$\begin{array}{rl} 2x+4y+6z=16 & 2[1] \\ -2x-4y-6z=5 & [3] \\ \hline 0=21 & \end{array}$$

   Since this is false, the solution set is $\varnothing$.

13. First, add [equation 1] and [equation 2] to eliminate z and produce [equation 4]. Second, multiply [equation 1] by 3 and add [equation 3] to eliminate z and produce [equation 5].

$$\begin{array}{rl} x+4y-z=6 & [1] \\ 2x-y+z=3 & [2] \\ \hline 3x+3y=9 & [4] \end{array} \qquad \begin{array}{rl} 3x+12y-3z=18 & 3[1] \\ 3x+2y+3z=16 & [2] \\ \hline 6x+14y=34 & [5] \end{array}$$

   Next multiply [equation 4] by -2 and add to [equation 5] to eliminate x and produce [equation 6].

$$\begin{array}{rrl} -6x- \ 6y= & -18 & -2[4] \\ 6x+14y= & 34 & [5] \\ \hline 8y= & 16 & [6] \end{array}$$

Solve [equation 6] for y: $y: 8y=16 \Rightarrow y=2$. Substitute $y=2$ into [equation 4] to solve for x: $3x+3(2)=9 \Rightarrow 3x=3 \Rightarrow x=1$. Finally, substitute $x=1$ and $y=2$ into [equation 1] to solve for z: $(1)+4(2)-z=6 \Rightarrow -z=-3 \Rightarrow z=3$. The solution is $(1,2,3)$.

15. First, multiply [equation 1] by -3 and add [equation 2] to eliminate y and produce [equation 4]. Second, multiply [equation 1] by 2 and add [equation 3] to eliminate y and produce [equation 5]

$$\begin{array}{rrrrl} -15x- & 3y+ & 9z= & 18 & -3[1] \\ 2x+ & 3y+ & z= & 5 & [2] \\ \hline -13x & & +10z= & 23 & [4] \end{array}$$

$$\begin{array}{rrrrl} 10x+ & 2y- & 6z= & -12 & 2[1] \\ -3x- & 2y+ & 4z= & 3 & [3] \\ \hline 7x & & -\ 2z= & -9 & [5] \end{array}$$

Next multiply [equation 5] by 5 and add [equation 4] to eliminate z and produce [equation 6].

$$\begin{array}{rrl} 35x-10z= & -45 & 5[5] \\ -13x+10z= & 23 & [4] \\ \hline 22x \quad = & -22 & [6] \end{array}$$

Solve [equation 6] for x: $22x=-22 \Rightarrow x=-1$. Substitute $x=-1$ into [equation 4] to solve for z: $-13(-1)+10z=23 \Rightarrow 10z=10 \Rightarrow z=1$. Finally, substitute $x=-1$ and $z=1$ into [equation 1] to solve for y: $5(-1)+y-3(1)=-6 \Rightarrow y=2$. The solution is $(-1,2,1)$.

17. First, add [equation 1] and [equation 3] to eliminate x and produce [equation 4]. Second, multiply [equation 3] by 3 and add [equation 2] to eliminate x and produce [equation 5].

$$\begin{array}{rrrrl} x- & 3y- & 2z= & -3 & [1] \\ -x- & y+ & 4z= & 3 & [3] \\ \hline & -4y+ & 2z= & 0 & [4] \end{array}$$

$$\begin{array}{rrrrl} -3x- & 3y+ & 12z= & 9 & 3[3] \\ 3x+ & 2y- & z= & 12 & [2] \\ \hline & -y+ & 11z= & 21 & [5] \end{array}$$

Next multiply [equation 5] by -4 and add [equation 4] to eliminate y and produce [equation 6].

$$\begin{array}{rrrl} 4y- & 44z= & -84 & -4[5] \\ -4y+ & 2z= & 0 & [4] \\ \hline & -42z= & -84 & [6] \end{array}$$

Solve [equation 6] for z: $-42z=-84 \Rightarrow z=2$. Substitute $z=2$ into [equation 4] to solve for y: $-4y+2(2)=0 \Rightarrow -4y=-4 \Rightarrow y=1$. Finally, substitute $y=1$ and $z=2$ into [equation 1] to solve for x: $x-3(1)-2(2)=-3 \Rightarrow x=4$. The solution is $(4,1,2)$.

19. First eliminate x by multiplying [equation 1] by -2 and adding [equation 2].

$$\begin{array}{rrrrl} -2x+ & 4y- & 6z= & -12 & -2[1] \\ 2x- & y+ & 2z= & 5 & [2] \\ \hline & 3y- & 4z= & -7 & \end{array}$$

Now solve this for y. $3y-4z=-7 \Rightarrow 3y=4z-7 \Rightarrow y=\frac{4z-7}{3}$. Now substitute this into [equation 2].

$$2x-\frac{4z-7}{3}+2z=5 \Rightarrow 3\left(2x-\frac{4z-7}{3}+2z=5\right) \Rightarrow 6x-4z+7+6z=15 \Rightarrow$$

$6x+2z=8 \Rightarrow x=\frac{8-2z}{6} \Rightarrow x=\frac{4-z}{3}$. We have infinitely many solutions $\left(\frac{4-z}{3},\frac{4z-7}{3},z\right)$.

21. First eliminate y by adding [equation 1] to [equation 2] multiplied by 4:

$$\begin{array}{rl} 5x-4y+z=9 & [1] \\ 4x+4y=60 & 4[2] \\ \hline 9x+z=69 & \end{array}$$

Now solve this for x. $9x+z=69 \Rightarrow 9x=69-z \Rightarrow x=\frac{69-z}{9}$. Now substitute this into [equation 2].

$$\frac{69-z}{9}+y=15 \Rightarrow 9\left(\frac{69-z}{9}+y=15\right) \Rightarrow 69-z+9y=135 \Rightarrow -z+9y=66 \Rightarrow$$

$9y=z+66 \Rightarrow y=\frac{z+66}{9}$. We have infinitely many solutions $\left(\frac{69-z}{9},\frac{z+66}{9},z\right)$.

23. First, multiply [equation 1] by 3 and add [equation 2] to eliminate y and produce [equation 4], then simplify. Second, multiply [equation 3] by 3 and add [equation 2] to eliminate y and produce [equation 5] then simplify.

$$\begin{array}{rrrrl} 24x- & 9y+ & 18z= & -6 & 3[1] \\ 4x+ & 9y+ & 4z= & 18 & [2] \\ \hline 28x & & +22z= & 12 & [4] \\ =14x & & +11z= & 6 & [4a] \end{array} \qquad \begin{array}{rrrrl} 36x- & 9y+ & 24z= & -6 & 3[2] \\ 4x+ & 9y+ & 4z= & 18 & [2] \\ \hline 40x & & +28z= & 12 & [5] \\ =10x & & +7z= & 3 & [5a] \end{array}$$

Next multiply [equation 4a] by 5 and add [equation 5a] multiplied by -7 to eliminate x and produce [equation 6].

$$\begin{array}{rrl} 70x+55z= & 30 & 5[4a] \\ -70x-49z= & -21 & -7[5a] \\ \hline 6z= & 9 & [6] \end{array}$$

Solve [equation 6] for $z: 6z=9 \Rightarrow z=\frac{3}{2}$. Substitute $z=\frac{3}{2}$ into [equation 5a] to solve for x:

$10x+7\left(\frac{3}{2}\right)=3 \Rightarrow 2\left[10x+7\left(\frac{3}{2}\right)=3\right] \Rightarrow 20x+21=6 \Rightarrow 20x=-15 \Rightarrow x=-\frac{3}{4}$. Finally,

substitute $x=-\frac{3}{4}$ and $z=\frac{3}{2}$ into [equation 2] to solve for y: $4\left(-\frac{3}{4}\right)+9y+4\left(\frac{3}{2}\right)=18 \Rightarrow$

$-3+9y+6=18 \Rightarrow 9y=15 \Rightarrow y=\frac{15}{9}=\frac{5}{3}$. The solution is $\left(-\frac{3}{4},\frac{5}{3},\frac{3}{2}\right)$.

25. First, add [equation 1] and [equation 2] multiplied by -1 to eliminate x and produce [equation 4]. Second, add [equation 3] and [equation 4] to eliminate y and produce [equation 5].

$$\begin{array}{rll} x - z = & 2 & [1] \\ -x - y = & 3 & -1[2] \\ \hline -y - z = & 5 & [4] \end{array} \qquad \begin{array}{rll} y - \ z = & 1 & [3] \\ -y - \ z = & 5 & [4] \\ \hline -2z = & 6 & [5] \end{array}$$

Solve [equation 5] for $z: -2z = 6 \Rightarrow z = -3$. Substitute $z = -3$ into [equation 1] to solve for x: $x - (-3) = 2 \Rightarrow x = -1$. Finally, substitute $z = -3$ into [equation 3] to solve for y: $y - (-3) = 1 \Rightarrow y = -2$. The solution is $(-1, -2, -3)$.

27. First, multiply [equation 1] by 3 and add [equation 2] multiplied by -2 to eliminate y and produce [equation 4]. Second, add [equation 4] and [equation 3] multiplied by -9 to eliminate x.

$$\begin{array}{rrrrl} 9x + & 6y - & 3z = & -3 & 3[1] \\ & -6y - & 2z = & -24 & -2[2] \\ \hline 9x & - & 5z = & -27 & [4] \end{array} \qquad \begin{array}{rrrl} 9x & - & 5z = -27 & [4] \\ -9x & + & 27z = \ 27 & [3] \\ \hline & & 22z = \ \ 0 & \Rightarrow z = 0. \end{array}$$

Substitute $z = 0$ to solve [equation 2] for y: $3y + 0 = 12 \Rightarrow y = 4$. Finally, substitute $z = 0$ into [equation 3] to solve for x: $x - 3(0) = -3 \Rightarrow x = -3$. The solution is $(-3, 4, 0)$.

29. First eliminate y by adding [equation 1] to [equation 2].

$$\begin{array}{rrrl} x - & y + & z = & -6 \\ 4x + & y + & z = & 7 \\ \hline 5x & & +2z = & 1 \end{array}$$

Now solve this for x. $5x + 2z = 1 \Rightarrow 5x = 1 - 2x \Rightarrow x = \dfrac{1-2x}{5}$. Now substitute this into [equation 1].

$\dfrac{1-2z}{5} - y + z = -6 \Rightarrow 5\left(\dfrac{1-2z}{5} - y + z = -6\right) \Rightarrow 1 - 2z - 5y + 5z = -30 \Rightarrow -5y + 3z = -31 \Rightarrow$

$-5y = -3z - 31 \Rightarrow y = \dfrac{3z+31}{5}$. We have infinitely many solutions $\left(\dfrac{1-2z}{5}, \dfrac{3z+31}{5}, z\right)$.

31. First, multiply [equation 1] by -3 and add [equation 2] to eliminate x and produce [equation 4]. Second, add [equation 4] and [equation 3] to eliminate y.

$$\begin{array}{rrrrl} -6x - & 9y - & 12z = & -9 & -3[1] \\ 6x + & 3y + & 8z = & 6 & -2[2] \\ \hline & -6y - & 4z = & -3 & [4] \end{array} \qquad \begin{array}{rrrl} -6y - & 4z = & -3 & [4] \\ 6y - & 4z = & 1 & [3] \\ \hline & -8z = & -2 & \Rightarrow z = \frac{1}{4}. \end{array}$$

Substitute $z = \dfrac{1}{4}$ to solve [equation 3] for y: $6y - 4\left(\dfrac{1}{4}\right) = 1 \Rightarrow 6y = 2 \Rightarrow y = \dfrac{1}{3}$. Finally, substitute $y = \dfrac{1}{3}$ and $z = \dfrac{1}{4}$ into [equation 1] to solve for x: $2x + 3\left(\dfrac{1}{3}\right) + 4\left(\dfrac{1}{4}\right) = 3 \Rightarrow 2x = 1 \Rightarrow x = \dfrac{1}{2}$.

The solution is $\left(\frac{1}{2},\frac{1}{3},\frac{1}{4}\right)$.

33. If we let $t=\frac{1}{x}$, $u=\frac{1}{y}$, and $v=\frac{1}{z}$, then our new equations are $t+u-v=\frac{1}{4}$ [1], $2t-u+3v=\frac{9}{4}$ [2], and $-t-2u+4v=1$ [3]. First, add [equation 1] and [equation 3] to eliminate t and produce [equation 4]. Second, add [equation 2] and [equation 3] multiplied by 2 to eliminate t and produce [equation 5]. $x^2-3x=4\Rightarrow x^2-3x-4=0\Rightarrow x=(x-4)(x+1)=0\Rightarrow x=4,-1.$

$$\begin{array}{rl} t+u-v=\frac{1}{4} & [1] \\ -t-2u+4v=1 & [2] \\ \hline -u+3v=\frac{5}{4} & [4] \end{array} \qquad \begin{array}{rl} 2t-u+3v=\frac{4}{9} & [2] \\ -2t-4u+8v=2 & 2[3] \\ \hline -5u+11v=\frac{17}{4} & [5] \end{array}$$

Next multiply [equation 4] by -5 and add [equation 5] to eliminate u and produce [equation 6].

$$\begin{array}{rl} 5u-15v=-\frac{25}{4} & -5[4] \\ -5u+11v=\frac{17}{4} & [5] \\ \hline -4v=-\frac{8}{4} & [6] \end{array}$$

Solve [equation 6] for v. $-4v=-2\Rightarrow v=-\frac{1}{2}$. If $v=-\frac{1}{2}$ and $v=\frac{1}{z}$, then $-\frac{1}{2}=\frac{1}{z}\Rightarrow z=-2.$

Substitute $v=\frac{1}{2}$ into [equation 4] to solve for u. $-u+3\left(\frac{1}{2}\right)=\frac{5}{4}\Rightarrow -u=-\frac{1}{4}$. If $u=\frac{1}{4}$ and $u=\frac{1}{y}$ then $\frac{1}{4}=\frac{1}{y}\Rightarrow y=4$. Finally, substitute $v=\frac{1}{2}$ and $u=\frac{1}{4}$ into [equation 3] to solve for t.

$$-t-2\left(\frac{1}{4}\right)+4\left(\frac{1}{2}\right)=1\Rightarrow -2t-1+4=2\Rightarrow -2t=-1\Rightarrow t=\frac{1}{2}.$$

If $t=\frac{1}{2}$ and $t=\frac{1}{x}$ then $\frac{1}{2}=\frac{1}{x}\Rightarrow x=2$. The solution is $(2,4,2)$.

35. If we let $t=\frac{1}{x}$, $u=\frac{1}{y}$, and $v=\frac{1}{z}$ then our new equations are $2t-2u+v=-1$ [1], $4t+u-2v=-9$ [2], and $t+u-3v=-9$ [3]. First, add [equation 1] and [equation 2] multiplied by 2 to eliminate u and produce [equation 4]. Second, add [equation 1] and [equation 3] multiplied by 2 to eliminate u and produce [equation 5].

$$\begin{array}{rl} 2t-2u+v=-1 & [1] \\ 8t+2u-4v=-18 & 2[2] \\ \hline 10t-3v=-19 & [4] \end{array} \qquad \begin{array}{rl} 2t-2u+v=-1 & [1] \\ 8t+2u-6v=-18 & 2[3] \\ \hline 4t-5v=-19 & [5] \end{array}$$

Next multiply [equation 4] by 2 and add [equation 5] multiplied by -5 to eliminate t and produce [equation 6].

$$\begin{array}{rcll} 20t - 6v &=& -38 & 2[2] \\ -20t + 25v &=& 95 & -5[5] \\ \hline 19v &=& 57 & [6] \Rightarrow v = 3 \end{array}$$

If $v=3$ and $v=\frac{1}{z}$ then $3=\frac{1}{z}\Rightarrow 3z=1\Rightarrow z=\frac{1}{3}$. Substitute $v=3$ into [equation 5] to solve for t.

$4t+5(3)=-19\Rightarrow 4t=-4\Rightarrow t=-1$. If $t=-1$ and $t=\frac{1}{x}$ then $-1=\frac{1}{x}\Rightarrow -x=1\Rightarrow x=-1$. Finally,

substitute $v=3$ and $t=-1$ into [equation 1] to solve for u. $2(-1)-2u+(3)=-1\Rightarrow -2u=-2\Rightarrow u=1$.

If $u=1$ and $=\frac{1}{y}$ then $1=\frac{1}{y}\Rightarrow y=1$. The solution is $\left(-1,1,\frac{1}{3}\right)$.

37. Add [equation 1] and [equation 2] to eliminate z and produce [equation 4].

$$\begin{array}{rl} x-4y+2z=-2 & [1] \\ x+2y-2z=-3 & [2] \\ \hline 2x-2y \quad =-5 & [4] \end{array}$$

Add [equation 4] to [equation 3] multiplied by -2.

$$\begin{array}{rl} 2x-2y=-5 & [4] \\ -2x+2y=-8 & -2[3] \\ \hline 0=-13 & \end{array}$$

Since this is a false statement, the solution is $\varnothing$.

39. Subtract [equation 3] from [equation 1] to eliminate x and produce equation [4].

$$\begin{array}{rl} x+y+z=0 & [1] \\ x+3y+3z=5 & [3] \\ \hline -2y-2z=-5 & [4] \end{array}$$

Subtract [equation 3] from [equation 2] to eliminate x and produce equation [5].

$$\begin{array}{rl} x-y-z=3 & [2] \\ x+3y+3z=5 & [3] \\ \hline -4y-4z=-2 & [5] \end{array}$$

Multiply [equation 4] by 2 and subtract [equation 5].

$$\begin{array}{rl} -4y-4z=-10 & 2[4] \\ -4y-4z=-2 & [5] \\ \hline 0=-8 & \end{array}$$

Since this is a false statement, the solution is $\varnothing$.

41. Add the first two equations.

$$\begin{array}{r} 2x-y+2z=6 \\ -x+y+z=0 \\ \hline x \quad +3z=6 \end{array}$$

Add this equation to the third equation.

$$\begin{array}{r} x+3z = 6 \\ -x-3z=-6 \\ \hline 0 = 0 \end{array}$$

Therefore, we have infinitely many solutions. $x+3z=6 \Rightarrow x=-3z+6$.

Multiply the second equation by 2 and add to the first equation:

$$\begin{array}{r} 2x- \ y+2z=6 \\ -2x+2y+2z=0 \\ \hline y+4z=6 \Rightarrow y=-4z+6. \end{array}$$

We have infinitely many solutions $(-3z+6, -4z+6, z)$.

43. There are many other examples, one is given.

(a) $x-y-z=0$ when added to $x+y+z=4$ yields $2x=4 \Rightarrow x=2$. $-x+y-z=0$ when added to $x+y+z=4$ yields $2y=4 \Rightarrow y=2$. This one solution is $(2,2,0)$.

(b) $x-y-z=1$ and $-x-y-z=0$ when added to $69x+27z=15 \Rightarrow 69x=12-27z \Rightarrow$ $x=\dfrac{15-27z}{69} \Rightarrow x=\dfrac{5-9z}{23}$ yield $4=0$, which means no solution.

(c) $x-y-z=2$ and $2x+2y+2z=8$ when added to $x+y+z=4$ multiplied by -2, yields $0=0$, which has infinitely many solutions.

45. Let x = number of gallons of \$9 grade, y = number of gallons of \$3 grade, and z = number of gallons of \$4.50 grade. From the information the equations are $x+y+z=300$ [1], $9x+3y+4.5z=6(300)$ [2], and $z=2y$ [3]. Substitute [equation 3] into both [equation 1] and [equation 2] to produce equations [4] and [5]. [3] into [1] produces $x+y+2y=300 \Rightarrow x+3y=300$ [4]. [3] into [2] produces $9x+3y+4.5(2y)=1800 \Rightarrow 9x+12y=1800$ [5]. Now add [equation 5] to [equation 4] multiplied by -9.

$$\begin{array}{rl} 9x+12y = \ \ 1800 & [5] \\ -9x-27y=-2700 & -9\ [4] \\ \hline -15y = \ -900 & \Rightarrow y=60 \end{array}$$

Substitute $y=60$, into [3]. $z=2(60) \Rightarrow z=120$. Finally, substitute $y=60$, and $z=120$ into [1]. $x+60+120=300 \Rightarrow x=120$. She should use 120 gallons of \$9 water, 60 gallons of \$3 water, and 120 gallons of \$4.50 water.

47. Let x = price of Up Close tickets, y = price of Middle tickets, and z = price of Farther Back tickets. From the information the equations are $x=6+y$ [1], $y=3+z$ [2], and $2x=3+3z$ [3]. First substitute [equation 1] into [equation 3]. $2(6+y)=3+3z$. Now substitute [equation 2] into this new equation and solve for z. $2(6+(3+z))=3+3z \Rightarrow 2(9+z)=3+3z \Rightarrow 18+2z=3+3z \Rightarrow z=2$. Substitute z = 15 into [2]. $y=3+(15) \Rightarrow y=18$. Finally, substitute y = 18 into [1]. $x=6+18 \Rightarrow x=24$. The Up Close ticket was \$24, the Middle ticket was \$18, and the Farther Back ticket was \$15.

49. Let x = measure of the largest angle, y = measure of the medium angle, and z = measure of the smallest angle. From the information the equations are $a+b+c=180$ [1], $a=2b-55$ [2], and $c=b-25$ [3]. Substitute both [2] and [3] into [1]. $(2b-55)+b+(b-25)=180 \Rightarrow 4b-80=180 \Rightarrow 4b=260 \Rightarrow b=65$. Substitute $b=65$ into [3]. $c=65-25 \Rightarrow c=40$, and substitute $b=65$ into [2]. $a=2(65)-55 \Rightarrow a=130-55 \Rightarrow a=75$. The angles are 75°, 65°, and 40°.

51. Let x = the amount invested at 4%, y = the amount invested at 4.5%, and z = the amount invested at 2.5%. From the information the equations are $x+y+z=10000$ [1], $0.04x+0.025y+0.045z=415$ [2], and $y=2x$ [3]. Substitute [equation 3] into [equation 1] and [equation 3] into [equation 2] multiplied by 1000, to produce equations [4] and [5]. [3] into [1] $\Rightarrow x+2x+z=10000 \Rightarrow 3x+z=10000$ [4]. [3] into 1000[2] $\Rightarrow 40x+45(2x)+25z=415000 \Rightarrow 130x+25z=415000 \Rightarrow 26x+5z=83000$ [5]. Now multiply [equation 4] by -5 and add to [equation 5].

$$\begin{array}{rcll} -15x-5y &=& -50000 & -5[4] \\ 26x+5y &=& 83000 & [5] \\ \hline 11x &=& 33000 & \Rightarrow x=3000 \end{array}$$

Substitute x =3000, into [3]. $y=2(3000) \Rightarrow y=6000$. Finally, substitute x = 3000 and y = 6000 into [1]. $3000+6000+z=10000 \Rightarrow z=1000$. He invested \$3000 at 4%, \$6000 at 4.5%, and \$1000 at 2.5%.

53. Let x = the number of EZ models, y = the number of Compact models, and z = the number of Commercial models. From the information, the two equations are derived:
(weight) $10x+20y+60z=440$ [1] and (volume) $10x+8y+28z=248$ [2]. First, subtract [2] from [1].

$$\begin{array}{rl} 10x+20y+60z=440 & [1] \\ 10x+8y+28z=248 & [2] \\ \hline 12y+32z=192 & \Rightarrow 12y=-32z+192 \Rightarrow y=-\frac{8}{3}z+16. \end{array}$$

Now substitute into [1]: $10x+20\left(-\frac{8}{3}z+16\right)+60z=440 \Rightarrow 3\left[10x+20\left(-\frac{8}{3}z+16\right)+60z=440\right] \Rightarrow$

$30x-160z+960+180z=1320 \Rightarrow 30x=-20z+360 \Rightarrow x=-\frac{2}{3}z+12$. From the equations for x and y, z must be a multiple of 3 for x and y to be whole numbers. There are three possibilities:

If z = 0, then $y=-\frac{8}{3}(0)+16=16$ and $x=-\frac{2}{3}(0)+12=12$. Therefore, 12 EZ, 16 Comp., and 0 Comm.

If z = 3, then $y=-\frac{8}{3}(3)+16=8$ and $x=-\frac{2}{3}(3)+12=10$. Therefore, 10 EZ, 8 Comp, and 3 Comm.

If z = 6, then $y=-\frac{8}{3}(6)+16=0$ and $x=-\frac{2}{3}(6)+12=8$. Therefore, 8 EZ, 0 Comp., and 6 Comm.

If $z \geq 9$, then y is less than zero; therefore, no solution is possible.

55. (a)

$$\begin{aligned} a+20b+2c&=190\\ a+\ \ 5b+3c&=320\\ a+40b+\ \ c&=\ \ 50 \end{aligned}$$

(b) Using technology, the solution is (30, –2, 100). So the equation is $P=30-2A+100S$.

(c) When A = 10 and S = 2500, P = 30 – 2(10) + 100(2.5) = $260,000.

57. Using the form $y=ax^2+bx+c$ and the given points yields the following equations.

For $(1.5, 6.25)$: $6.25=a(1.5)^2+b(1.5)+c \Rightarrow 2.25a+1.5b+c=6.25$. [equation 1]

For $(0,-2)$: $-2=a(0)^2+b(0)+c \Rightarrow c=-2$. [equation 2]

For $(-1.5, 3.25)$: $3.25=a(-1.5)^2+b(-1.5)+c \Rightarrow 2.25a-1.5b+c=3.25$. [equation 3]

Substitute c = –2 into [1]. $2.25a+1.5b-2=6.25 \Rightarrow 2.25a+1.5b=8.25$. [equation 4]

Substitute c = –2 into [3]. $2.25a-1.5b-2=3.25 \Rightarrow 2.25a-1.5b=5.25$. [equation 5]

Now add [equation 4] and [equation 5]

$$\begin{array}{ll} 2.25a+1.5b=8.25 & [4]\\ \underline{2.25a-1.5b=5.25} & [5]\\ 4.5a \qquad\quad =13.5 & \Rightarrow a=3 \end{array}$$

Finally, substitute a = 3 and c = –2 into [1]. $6.25=2.25(3)+1.5b+(-2) \Rightarrow b=1$.

The equation is $y=3x^2+x-2$.

59. Using the form $y=ax^2+bx+c$, and the given points yields the following equations.

For $(-1,4)$: $4=a(-1)^2+b(-1)+c \Rightarrow a-b+c=4$. [equation 1]

For $(1,2)$: $2=a(1)^2+b(1)+c \Rightarrow a+b+c=2$. [equation 2]

For $(3,8)$: $8=a(3)^2+b(3)+c \Rightarrow 9a+3b+c=8$. [equation 3]

First, add [equation 1] and [equation 2] multiplied to eliminate both *b*.

Second, add [equation 2] and [equation 3] multiplied by –3 to eliminate *b* and produce [equation 5].

$$\begin{array}{llllll} a-b+c=4 & [1] & & -3a-3b-3c=-6 & -3[2]\\ \underline{a+b+c=2} & [2] & & \underline{9a+3b+\ \ c=\ \ 8} & [3]\\ 2a \quad +2c=6 & [4] & & 6a \qquad\quad -2c=2 & [5] \end{array}$$

$$\begin{array}{ll} 2a+2c=6 & [4]\\ \underline{6a-2c=2} & [5]\\ 8a \qquad =8 & \Rightarrow a=1 \end{array}$$

Finally, substitute a = 1 into [5]. $6(1)-2c=2 \Rightarrow c=2$.

Finally, substitute a = 1 and c = 2 into [1]. $(1)-b+(2)=4 \Rightarrow -b=1 \Rightarrow b=-1$.

The equation is $y=x^2-x+2$.

61. Using the form $y=ax^2+bx+c$, and the given points yields the following equations.

For $(0,1)$: $1=a(0)^2+b(0)+c \Rightarrow c=1$. [equation 1]

For $(1,0)$: $0=a(1)^2+b(1)^2+c \Rightarrow a+b+c=0$. [equation 2]

For $(2,-5)$: $-5=a(2)^2+b(2)+c \Rightarrow 4a+2b+c=-5$. [equation 3]

Substitute c = 1 into [1]. $a+b+1=0 \Rightarrow a+b=-1$. [equation 4]

Substitute c = 1 into [3]. $4a+2b+1=-5 \Rightarrow 4a+2b=-6$. [equation 5]

Now add [equation 4] and [equation 5] multiplied by -2.

$$\begin{array}{lll} -2a-2b= & 2 & -2[4] \\ \underline{4a-2b=} & \underline{-6} & [5] \\ 2a \quad = & -4 & \Rightarrow a=-2 \end{array}$$

Finally, substitute a = -2 and c = 1 into [1]. $(-2)+b+(1)=0 \Rightarrow b=1$.

The equation is $y=-2x^2+x+1$.

63. Using the form $x^2+y^2+ax+by+c=0$ and the given points yields the following equations.

For $(-1,3)$: $(-1)^2+(3)^2+a(-1)+b(3)+c=0 \Rightarrow 1+9-a+3b+c=0 \Rightarrow -a+3b+c=-10$. [equation 1]

For $(6,2)$: $(6)^2+(2)^2+a(6)+b(2)+c=0 \Rightarrow 36+4+6a+2b+c=0 \Rightarrow 6a+2b+c=-40$. [equation 2]

For $(-2,-4)$: $(-2)^2+(-4)^2+a(-2)+b(-4)+c=0 \Rightarrow 4+16-2a-4b+c=0 \Rightarrow -2a-4b+c=-20$. [equation 3]

Now, subtract [equation 1] from [equation 2] to eliminate c and produce [equation 4].
Then, subtract [equation 3] from [equation 2] to eliminate c and produce [equation 5].

$$\begin{array}{lll} 6a+2b+c=-40 & [2] \\ \underline{a-3b-c=\ 10} & -1[1] \\ 7a-b \quad =-30 & [4] \end{array} \qquad \begin{array}{lll} 6a+2b+c=-40 & [2] \\ \underline{2a+4b-c=\ 20} & -1[3] \\ 8a+6b \quad =-20 & [5] \end{array}$$

Next multiply [equation 4] by 6 and add [equation 5] to eliminate b.

$$\begin{array}{lll} 42a-6b=-180 & 6[4] \\ \underline{8a+6b=\ -20} & [5] \\ 50a \quad =-200 & \Rightarrow a=-4 \end{array}$$

Substitute $a=-4$ into [4]. $7(-4)-b=-30 \Rightarrow -b=-2 \Rightarrow b=2$. Finally, substitute $a=-4$ and $b=2$ into [1]. $-(-4)+3(2)+c=-10 \Rightarrow c=-20$. The equation is $x^2+y^2-4x+2y-20=0$.

65. Using the form $x^2+y^2+ax+by+c=0$ and the given points yields the following equations.

For $(2,1)$: $(2)^2+(1)^2+a(2)+b(1)+c=0 \Rightarrow 4+1+2a+b+c=0 \Rightarrow 2a+b+c=-5$. [equation 1]

For $(-1,0)$: $(-1)^2+(0)^2+a(-1)+b(0)+c=0 \Rightarrow 1+0-a+0b+c=0 \Rightarrow -a+c=-1$. [equation 2]

For $(3,3)$: $(3)^2+(3)^2+a(3)+b(3)+c=0 \Rightarrow 9+9+3a+3b+c=0 \Rightarrow 3a+3b+c=-18$. [equation 3]

Now, multiply [equation 1] by –3 and add [equation 3] to eliminate b and produce [equation 4].
Next multiply [equation 2] by 2 and add [equation 4] to eliminate c.

$$\begin{array}{rll} -6a-3b-3c= & 15 & -3\,[1] \\ 3a+3b+\ c= & -18 & [3] \\ \hline -3a \qquad -2c= & -3 & [4] \end{array} \qquad \begin{array}{rll} -2a+2c= & -2 & 2\,[2] \\ -3a-2c= & -3 & [4] \\ \hline -5a \qquad = & -5 & \Rightarrow a=1. \end{array}$$

Substitute $a=1$ into [2]. $-(1)+c=-1 \Rightarrow c=0$. Finally, substitute $a=1$ and $c=0$ into [1].

$2(1)+b+0=-5 \Rightarrow b=-7$. The equation is $x^2+y^2+x-7y=0$.

67. Using the form $s(t)=at^2+bt+c$ and the given points yields the following equations.

For $(0,5)$: $5=a(0)^2+b(0)+c \Rightarrow c=5$.

For $(1,23)$: $23=a(1)^2+b(1)+c \Rightarrow a+b+c=23$. [equation 1]

For $(2,37)$: $37=a(2)^2+b(2)+c \Rightarrow 4a+2b+c=37$. [equation 2]

First, multiply [equation 1] by -2 and add [equation 2] to eliminate b and produce [equation 3].

$$\begin{array}{rll} -2a-2b-2c= & -46 & -2\,[1] \\ 4a+2b+\ c= & 37 & [2] \\ \hline 2a \qquad -c= & -9 & [3] \end{array}$$

Substitute $c=5$ into [3]. $2a-(5)=-9 \Rightarrow 2a=-4 \Rightarrow a=-2$.

Finally, substitute $a=-2$ and $c=5$ into [1]. $(-2)+b+(5)=23 \Rightarrow b=20$.

The equation is $s(t)=-2t^2+20t+5$, and $s(8)=-2(8)^2+20(8)+5=-128+160+5=37$.

7.3: Solution of Linear Systems by Row Transformations

1. For $\begin{bmatrix} 2 & 4 \\ 4 & 7 \end{bmatrix}$, $\frac{1}{2}R_1 \rightarrow \begin{bmatrix} 1 & 2 \\ 4 & 7 \end{bmatrix}$.

3. For $\begin{bmatrix} 1 & 5 & 6 \\ -2 & 3 & -1 \\ 4 & 7 & 0 \end{bmatrix}$, $2R_1+R_2 \rightarrow \begin{bmatrix} 1 & 5 & 6 \\ 0 & 13 & 11 \\ 4 & 7 & 0 \end{bmatrix}$.

5. For $\begin{bmatrix} -3 & 1 & -4 \\ 2 & 1 & 3 \\ 10 & 5 & 2 \end{bmatrix}$, $-5R_2+R_3 \rightarrow \begin{bmatrix} -3 & 1 & -4 \\ 2 & 1 & 3 \\ 0 & 0 & -13 \end{bmatrix}$.

7. The augmented matrix is $\left[\begin{array}{cc|c} 2 & 3 & 11 \\ 1 & 2 & 8 \end{array}\right]$.

9. The augmented matrix is $\left[\begin{array}{cc|c} 1 & 5 & 6 \\ 1 & 0 & 3 \end{array}\right]$.

11. The augmented matrix is $\left[\begin{array}{ccc|c} 2 & 1 & 1 & 3 \\ 3 & -4 & 2 & -7 \\ 1 & 1 & 1 & 2 \end{array}\right]$.

13. The augmented matrix is $\left[\begin{array}{ccc|c} 1 & 1 & 0 & 2 \\ 0 & 2 & 1 & -4 \\ 0 & 0 & 1 & 2 \end{array}\right]$.

15. The system of equations is $\begin{aligned} 2x + \ \ y &= 1 \\ 3x - 2y &= -9. \end{aligned}$

17. The system of equations is $\begin{aligned} x &= 2 \\ y &= 3 \\ z &= -2. \end{aligned}$

19. The system of equations is $\begin{aligned} 3x + 2y + z &= 1 \\ 2y + 4z &= 22 \\ -x - 2y + 3z &= 15 \end{aligned}$

21. The system can be written as $x + 2y = 3$ and $y = -1$. Substituting $y = -1$ into the first equation gives $x + 2(-1) = 3 \Rightarrow x = 5$. The solution is (5, –1).

23. The system can be written as $x - y = 2$ and $y = 0$. Substituting $y = 0$ into the first equation gives $x - 0 = 2 \Rightarrow x = 2$. The solution is (2, 0).

25. The system can be written as $x + y - z = 4$, $y - z = 2$, and $z = 1$. Substituting $z = 1$ into the second equation gives $y - (1) = 2 \Rightarrow y = 3$ Substituting $y = 3$ and $z = 1$ into the first equation gives $x + (3) - (1) = 4 \Rightarrow x = 2$. The solution is (2, 3, 1).

27. The system can be written as $x + 2y - z = 5$, $y - 2z = 1$, and $0 = 0$. Since $0 = 0$, there are an infinite number of solutions. The second equation gives $y = 1 + 2z$. Substituting this into the first equation gives $x + 2(1 + 2z) - z = 5 \Rightarrow x = 3 - 3z$. The solution can be written as $\{\,(3 - 3z, 1 + 2z, z) \mid z$ is a real number$\}$.

29. The system can be written as $x + 2y + z = -3$, $y - 3z = \frac{1}{2}$, and $0 = 4$. Since $0 = 4$ is false, there are no solutions.

31. Using the Row Echelon Method, the given system of equations yields the matrix $\left[\begin{array}{cc|c} 1 & 1 & 5 \\ 1 & -1 & -1 \end{array}\right]$, which by

$-R_1 + R_2 \rightarrow \left[\begin{array}{cc|c} 1 & 1 & 5 \\ 0 & -2 & -6 \end{array}\right] \Rightarrow \left(-\frac{1}{2}R_2\right) \rightarrow \left[\begin{array}{cc|c} 1 & 1 & 5 \\ 0 & 1 & 3 \end{array}\right]$. From this matrix, we have the resulting equation $y = 3$.

Now substitute $y = 3$ into the resulting R_1 equation. $x + (3) = 5 \Rightarrow x = 2$. The solution is $(2, 3)$.

33. Using the Row Echelon Method, the given system of equations yields the matrix $\left[\begin{array}{cc|c} 1 & 1 & -3 \\ 2 & -5 & -6 \end{array}\right]$, which by

$-2R_1 + R_2 \rightarrow \left[\begin{array}{cc|c} 1 & 1 & -3 \\ 0 & -7 & 0 \end{array}\right] \Rightarrow \left(-\frac{1}{7}R_2\right) \rightarrow \left[\begin{array}{cc|c} 1 & 1 & -3 \\ 0 & 1 & 0 \end{array}\right]$. From this matrix, we have the resulting equation $y = 0$. Now substitute $y = 0$ into the resulting R_1 equation. $x + (0) = -3 \Rightarrow x = -3$.

The solution is $(-3, 0)$.

35. Using the Row Echelon Method, the given system of equations yields the matrix $\begin{bmatrix} 2 & -3 & | & 10 \\ 2 & 2 & | & 5 \end{bmatrix}$, which by

$-R_1 + R_2 \rightarrow \begin{bmatrix} 2 & -3 & | & 10 \\ 0 & 5 & | & -5 \end{bmatrix} \Rightarrow \left(\frac{1}{2}R_1\right)$ and $\left(\frac{1}{5}R_2\right) \rightarrow \begin{bmatrix} 1 & -\frac{3}{2} & | & 5 \\ 0 & 1 & | & -1 \end{bmatrix}$. From this matrix, we have the resulting

equation $y = -1$. Now substitute $y = -1$ into the resulting R_1 equation.

$x - \frac{3}{2}(-1) = 5 \Rightarrow 2x + 3 = 10 \Rightarrow 2x = 7 \Rightarrow x = \frac{7}{2}$. The solution is $\left(\frac{7}{2}, -1\right)$.

37. Using the Row Echelon Method, the given system of equations yields the matrix $\begin{bmatrix} 2 & -3 & | & 2 \\ 4 & -6 & | & 1 \end{bmatrix}$, which by

$-2R_1 + R_2 \rightarrow \begin{bmatrix} 2 & -3 & | & 2 \\ 0 & 0 & | & -3 \end{bmatrix}$. Since this matrix yields the equation $0 = -3$, which is false; therefore,

the solution is $\varnothing$.

39. Using the Row Echelon Method, the given system of equations yields the matrix $\begin{bmatrix} 6 & -3 & | & 1 \\ -12 & 6 & | & -2 \end{bmatrix}$, which by

$2R_1 + R_2 \rightarrow \begin{bmatrix} 6 & -3 & | & 1 \\ 0 & 0 & | & 0 \end{bmatrix}$. The matrix yields the equation $0 = 0$. Since this is true, there are ∞ solutions.

From the resulting R_1, the solutions have the relationship: $-3y = 1 - 6x \Rightarrow y = \frac{6x-1}{3}$; $\left[\text{or } x = \frac{3y+1}{6}\right]$.

The solution is $\left(x, \frac{6x-1}{3}\right)$ or $\left(\frac{3y+1}{6}, y\right)$.

41. Using the Row Echelon Method, the given system of equations yields the matrix $\begin{bmatrix} 1 & 1 & 0 & | & -1 \\ 0 & 1 & 1 & | & 4 \\ 1 & 0 & 1 & | & 1 \end{bmatrix}$, which by

$-R_1 + R_3 \rightarrow \begin{bmatrix} 1 & 1 & 0 & | & -1 \\ 0 & 1 & 1 & | & 4 \\ 0 & -1 & 1 & | & 2 \end{bmatrix} \Rightarrow R_2 + R_3 \rightarrow \begin{bmatrix} 1 & 1 & 0 & | & -1 \\ 0 & 1 & 1 & | & 4 \\ 0 & 0 & 2 & | & 6 \end{bmatrix} \Rightarrow \left(\frac{1}{2}R_3\right) \rightarrow \begin{bmatrix} 1 & 1 & 0 & | & -1 \\ 0 & 1 & 1 & | & 4 \\ 0 & 0 & 1 & | & 3 \end{bmatrix}$. From this matrix,

we have the resulting equation $z = 3$. Now use back-substitution. Substituting $z = 3$ into the resulting R_2 yields

$y = 3 = 4 \Rightarrow y = 1$. Finally, substituting $y = 1$ and $z = 3$ into the resulting R_1 yields $x + 1 = -1 \Rightarrow x = -2$.

The solution is $(-2, 1, 3)$.

43. Using the Row Echelon Method, the given system of equations yields the matrix $\left[\begin{array}{ccc|c} 1 & 1 & -1 & 6 \\ 2 & -1 & 1 & -9 \\ 1 & -2 & 3 & 1 \end{array}\right]$, which

by $(-2R_1+R_2)$ and $(-R_1+R_3)$ $\rightarrow \left[\begin{array}{ccc|c} 1 & 1 & -1 & 6 \\ 0 & -3 & 3 & -21 \\ 0 & -3 & 4 & -5 \end{array}\right] \Rightarrow -R_2+R_3 \rightarrow \left[\begin{array}{ccc|c} 1 & 1 & -1 & 6 \\ 0 & -3 & 3 & -21 \\ 0 & 0 & 1 & 16 \end{array}\right] \Rightarrow$

$\left(-\frac{1}{3}R_2\right) \rightarrow \left[\begin{array}{ccc|c} 1 & 1 & -1 & 6 \\ 0 & 1 & -1 & 7 \\ 0 & 0 & 1 & 16 \end{array}\right]$. From this matrix, we have the resulting equation $z=16$. Now use back substitution, substituting $z=16$ into the resulting R_2. This yields $y-16=7 \Rightarrow y=23$. Finally, substituting $y=23$ and $z=16$ into the resulting R_1 yields $x+23-16=6 \Rightarrow x=-1$. The solution is $(-1,23,16)$.

45. Using the Row Echelon Method, the given system of equations yields the matrix $\left[\begin{array}{ccc|c} -1 & 1 & 0 & -1 \\ 0 & 1 & -1 & 6 \\ 1 & 0 & 1 & -1 \end{array}\right]$, which

by $R_1+R_3 \rightarrow \left[\begin{array}{ccc|c} -1 & 1 & 0 & -1 \\ 0 & 1 & -1 & 6 \\ 0 & 1 & 1 & -2 \end{array}\right] \Rightarrow -R_2+R_3 \rightarrow \left[\begin{array}{ccc|c} -1 & 1 & 0 & -1 \\ 0 & 1 & -1 & 6 \\ 0 & 0 & 2 & -8 \end{array}\right] \Rightarrow$

$(-R_1)$ and $\left(\frac{1}{2}R_2\right) \rightarrow \left[\begin{array}{ccc|c} 1 & -1 & 0 & 1 \\ 0 & 1 & -1 & 6 \\ 0 & 0 & 1 & -4 \end{array}\right]$. From this matrix, we have the resulting equation $z=-4$. Now use back-substitution. Substituting $z=-4$ into the resulting R_2 yields $y+4=6 \Rightarrow y=2$. Finally, substituting $y=2$ and $z=-4$ into the resulting R_1 yields $-x+2=-1 \Rightarrow x=3$. The solution is $(3,2,-4)$.

47. Using the Row Echelon Method, the given system of equations yields the matrix $\left[\begin{array}{ccc|c} 2 & -1 & 3 & 0 \\ 1 & 2 & -1 & 5 \\ 0 & 2 & 1 & 1 \end{array}\right]$, which by

$R_1 \leftrightarrow R_2 \rightarrow \left[\begin{array}{ccc|c} 1 & 2 & -1 & 5 \\ 2 & -1 & 3 & 0 \\ 0 & 2 & 1 & 1 \end{array}\right] \Rightarrow -2R_1+R_2 \rightarrow \left[\begin{array}{ccc|c} 1 & 2 & -1 & 5 \\ 0 & -5 & 5 & -10 \\ 0 & 2 & 1 & -1 \end{array}\right] \Rightarrow 2R_2+5R_3 \rightarrow \left[\begin{array}{ccc|c} 1 & 2 & -1 & 5 \\ 0 & -5 & 5 & -10 \\ 0 & 0 & 15 & -15 \end{array}\right] \Rightarrow$

$\left(-\frac{1}{5}R_2\right)+\left(\frac{1}{15}R_3\right) \rightarrow \left[\begin{array}{ccc|c} 1 & 2 & -1 & 5 \\ 0 & 1 & -1 & 2 \\ 0 & 0 & 1 & -1 \end{array}\right]$. From this matrix, we have the resulting equation $z=-1$. Now use back-substitution. Substituting $(x+1)(x-3) \Rightarrow 5x-3=A(x-3)+B(x+1)$. into the resulting R_2 yields $y+1=2 \Rightarrow y=1$. Finally, substituting $y=1$ and $x=3 \Rightarrow 12=B(4) \Rightarrow B=3$. into the resulting R_1 yields $x+2+1=5 \Rightarrow x=2$. The solution is $(2,1,-1)$.

49. Using the capabilities of the calculator, the Reduced Row Echelon Method of

$$\left[\begin{array}{ccc|c} 2.1 & 0.5 & 1.7 & 4.9 \\ -2 & 1.5 & -1.7 & 3.1 \\ 5.8 & -4.6 & 0.8 & 9.3 \end{array}\right] = \left[\begin{array}{ccc|c} 1 & 0 & 0 & 5.21127 \\ 0 & 1 & 0 & 3.73944 \\ 0 & 0 & 1 & -4.65493 \end{array}\right].$$ The solution is approximately $(5.211, 3.739, -4.655)$.

51. Using the capabilities of the calculator, the Reduced Row Echelon Method of

$$\left[\begin{array}{ccc|c} 53 & 95 & 12 & 108 \\ 81 & -57 & -24 & -92 \\ -9 & 11 & -78 & 21 \end{array}\right] = \left[\begin{array}{ccc|c} 1 & 0 & 0 & -0.24997 \\ 0 & 1 & 0 & 1.2838 \\ 0 & 0 & 1 & -0.05934 \end{array}\right].$$ The solution is approximately $(-0.250, 1.284, -0.059)$.

53. In both cases, we simply write the coefficients and do not write the variables. This is possible because we agree on the order in which the variables appear (descending degree).

55. The given system of equations yields the matrix, $\left[\begin{array}{ccc|c} 1 & -3 & 2 & 10 \\ 2 & -1 & -1 & 8 \end{array}\right]$, which by

$-2R_1 + R_2 \rightarrow \left[\begin{array}{ccc|c} 1 & -3 & 2 & 10 \\ 0 & 5 & -5 & -12 \end{array}\right] \Rightarrow \frac{1}{5}R_2 \left[\begin{array}{ccc|c} 1 & -3 & 2 & 10 \\ 0 & 1 & -1 & -\frac{12}{5} \end{array}\right]$. Since we cannot eliminate more than one coefficient there are an infinite number of solutions and they will be in the following relationship. From the result of R_2, $y - z = -\frac{12}{5} \Rightarrow 5y - 5z = -12 \Rightarrow 5y = 5z - 12 \Rightarrow y = \frac{5z-12}{5}$. Now substitute this into the result of R_1: $x - 3\left(\frac{5z-12}{5}\right) + 2z = 10 \Rightarrow 5x - 5z + 36 + 10z = 50 \Rightarrow 5x - 5z = 14 \Rightarrow$

$5x = 5z + 14 \Rightarrow x = \frac{5z+14}{5}$. The solution is $\left(\frac{5z+14}{5}, \frac{5z-12}{5}, z\right)$.

57. The given system of equations yields the matrix $\left[\begin{array}{ccc|c} 1 & 2 & -1 & 0 \\ 3 & -1 & 1 & 6 \\ -2 & -4 & 2 & 0 \end{array}\right]$, which by

$(-3R_1 + R_2)$ and $(2R_1 + R_3) \rightarrow \left[\begin{array}{ccc|c} 1 & 2 & -1 & 0 \\ 0 & -7 & 4 & 6 \\ 0 & 0 & 0 & 0 \end{array}\right]$ since $-\frac{32}{11}z = -\frac{96}{11} \Rightarrow z = 3.$ is $0 = 0$, there are infinitely many solutions and the solutions are in the following relationship: From the result of R_2,

$-7y + 4z = 6 \Rightarrow -7y = 6 - 4z \Rightarrow y = \frac{4z-6}{7}$. Now substitute this into the result of R_1:

$x + 2\left(\frac{4z-6}{7}\right) - z = 0 \Rightarrow 7x + 8z - 12 - 7z = 0 \Rightarrow 7x + z = 12 \Rightarrow 7x = 12 - z \Rightarrow x = \frac{12-z}{7}$.

The solution is $\left(\frac{12-z}{7}, \frac{4z-6}{7}, z\right)$.

59. The given system of equations yields the matrix $\begin{bmatrix} 1 & -2 & 1 & | & 5 \\ -2 & 4 & -2 & | & 2 \\ 2 & 1 & -1 & | & 2 \end{bmatrix}$. By $(2R_1 + R_2)$ and

$(-2R_1 + R_3) \rightarrow \begin{bmatrix} 1 & -2 & 1 & | & 5 \\ 0 & 0 & 0 & | & 12 \\ 0 & 5 & -3 & | & -8 \end{bmatrix}$. Since R_2 is $0 = 12$, there is no solution or $\varnothing$.

61. The given system of equations yields the matrix $\begin{bmatrix} 1 & 3 & -2 & -1 & | & 9 \\ 4 & 1 & 1 & 2 & | & 2 \\ -3 & -1 & 1 & -1 & | & -5 \\ 1 & -1 & -3 & -2 & | & 2 \end{bmatrix}$, which by $\left.\begin{matrix} -4R_1 + R_2 \\ 3R_1 + R_3 \\ -R1 + R_4 \end{matrix}\right\} \rightarrow$

$$\begin{bmatrix} 1 & 3 & -2 & -1 & | & 9 \\ 0 & -11 & 9 & 6 & | & -34 \\ 0 & 8 & -5 & -4 & | & 22 \\ 0 & -4 & -1 & -1 & | & -7 \end{bmatrix} \Rightarrow \left.\begin{matrix} R_3 + 2R_4 \\ 8R_2 + 11R_3 \end{matrix}\right\} \rightarrow \begin{bmatrix} 1 & 3 & -2 & -1 & | & 9 \\ 0 & -11 & 9 & 6 & | & -34 \\ 0 & 0 & 17 & 4 & | & -30 \\ 0 & 0 & -7 & -6 & | & 8 \end{bmatrix} \Rightarrow$$

$$7R_3 + 17R_4 \rightarrow \begin{bmatrix} 1 & 3 & -2 & -1 & | & 9 \\ 0 & -11 & 9 & 6 & | & -34 \\ 0 & 0 & 17 & 4 & | & -30 \\ 0 & 0 & 0 & -74 & | & -74 \end{bmatrix} \Rightarrow \left.\begin{matrix} -\frac{1}{11}R_2 \\ -\frac{1}{17}R_3 \\ -\frac{1}{74}R_4 \end{matrix}\right\} \rightarrow \begin{bmatrix} 1 & 3 & -2 & -1 & | & 9 \\ 0 & 1 & \frac{-9}{11} & \frac{-6}{11} & | & \frac{34}{11} \\ 0 & 0 & 1 & \frac{4}{17} & | & -\frac{30}{17} \\ 0 & 0 & 0 & 1 & | & 1 \end{bmatrix}$$

. From this matrix, we have the resulting equation $w = 1$. Now use back-substitution. Substituting $w = 1$ into the resulting R_3 yields $z + \frac{4}{17}(1) = -\frac{30}{17} \Rightarrow z = -\frac{34}{17} \Rightarrow z = -2$. Substituting $w = 1$ and $z = -2$ into the resulting R_2 yields $y - \frac{9}{11}(-2) - \frac{6}{11}(1) = \frac{34}{11} \Rightarrow y + \frac{18-6}{11} = \frac{34}{11} \Rightarrow y = \frac{34-12}{11} = \frac{22}{11} \Rightarrow y = 2$. Finally, substitute $y = 2, z = -2,$ and $w = 1$ into R_1: $x + 3(2) - 2(-2) - 1 = 9 \Rightarrow x + 9 = 9 \Rightarrow x = 0$.

The solution is $(0, 2, -2, 1)$.

63. (a) Using the given equation model, the given system of equations and the capabilities of the calculator, the Reduced Row Echelon Method of: $\begin{bmatrix} 1800 & 5000 & 1 & | & 1300 \\ 3200 & 12000 & 1 & | & 5300 \\ 4500 & 13000 & 1 & | & 6500 \end{bmatrix} = \begin{bmatrix} 1 & 0 & 0 & | & 0.5714286 \\ 0 & 1 & 0 & | & 0.4571429 \\ 0 & 0 & 1 & | & -2014.2857 \end{bmatrix}$.

Therefore, the equation is $F = 0.5714N + 0.4571R - 2014$.

(b) Let $N = 3500$ and $R = 12{,}500$. Then $F = 0.5714(3500) + 0.4571(12{,}500) - 2014 \Rightarrow F \approx 5699.65$.

This model predicts monthly food costs of approximately \$5700.

65. Let x represent the number of model A, and y the number of model B. Using the information yields

$$\begin{cases} 2x + 3y = 34 \\ 25x + 30y = 365 \end{cases} \rightarrow \begin{bmatrix} 2 & 3 & | & 34 \\ 25 & 30 & | & 365 \end{bmatrix}, \text{ which by } \left.\begin{matrix} \frac{1}{2}R_1 \\ \frac{1}{5}R_2 \end{matrix}\right\} \rightarrow \begin{bmatrix} 1 & \frac{3}{2} & | & 17 \\ 5 & 6 & | & 73 \end{bmatrix} \Rightarrow -5R_1 + R_2 \rightarrow$$

$\begin{bmatrix} 1 & \frac{3}{2} & 17 \\ 0 & -\frac{3}{2} & -12 \end{bmatrix} \Rightarrow -\frac{2}{3}R_2 \rightarrow \begin{bmatrix} 1 & \frac{3}{2} & 17 \\ 0 & 1 & 8 \end{bmatrix}$. From this matrix, we have the resulting equation $y = 8$. Now substituting $y = 8$ into R_1 yields $x + \frac{3}{2}(8) = 7 \Rightarrow x + 12 = 17 \Rightarrow x = 5$. The maximum number is Model A: 5 bicycles, and Model B: 8 bicycles.

67. Let $x =$ the amount of money at 4%, $y =$ the amount of money at 6%, and $z =$ the amount of money at 5%. Using the information yields $\begin{cases} x + y + z = 25{,}000 \\ y = 2000 + \frac{1}{2}x \Rightarrow -\frac{1}{2}x + y = 2000. \\ 0.04x + 0.06y + 0.05z = 1220 \end{cases}$ This system yields

$\begin{bmatrix} 1 & 1 & 1 & 25{,}000 \\ -0.5 & 1 & 0 & 2{,}000 \\ 0.04 & 0.06 & 0.05 & 1{,}220 \end{bmatrix}$, which by $\frac{1}{2}R_1 + R_2$ and $-4R_1 + R_3$, $\begin{bmatrix} 1 & 1 & 1 & 25{,}000 \\ 0 & 1.5 & .5 & 14{,}500 \\ 0 & 2 & 1 & 22{,}000 \end{bmatrix}$ and $-\frac{2}{3}R_2$

$-2R_2 + R_3 \rightarrow \begin{bmatrix} 1 & 1 & 1 & 25{,}000 \\ 0 & 1 & 0.\overline{33} & 9{,}666.\overline{66} \\ 0 & 0 & 0.33 & 2{,}666.66 \end{bmatrix} \Rightarrow -R_3 + R_2 \rightarrow \begin{bmatrix} 1 & 1 & 1 & 25{,}000 \\ 0 & 1 & 0 & 7{,}000 \\ 0 & 0 & -0.333 & -1333.\overline{33} \end{bmatrix} \Rightarrow$

$-3R_3 \rightarrow \begin{bmatrix} 1 & 1 & 1 & 25{,}000 \\ 0 & 1 & 0 & 7{,}000 \\ 0 & 0 & 1 & 8{,}000 \end{bmatrix}$. From this matrix, we have the resulting equation $z = 8{,}000$ and $y = 7{,}000$

Finally, substitute $y = 7000$ and $z = 8000$ into R_1: $x + 7000 + 8000 = 25{,}000 \Rightarrow x = 10{,}000$. The loans were: \$10,000 at 4%, \$7000 at 6%, and \$8000 at 5%.

69. (a) The three equations can be written as: $3 = 0^2 + 0 + c, 55 = 2^2a + 2b + c$, and

$150 = 4^2a + 4b + c \Rightarrow 16a + 4b + c = 150, 4a + 2b + c = 55, \textit{ and } c = 3.$

Therefore, the matrix is: $\begin{bmatrix} 16 & 4 & 1 & 150 \\ 4 & 2 & 1 & 55 \\ 0 & 0 & 1 & 3 \end{bmatrix}$

(b) Using technology, $a = 5.375, b = 15.25, c = 3$. $f(x) = 5.375x^2 + 15.25x + 3$.

(c) See Figure 69.

(d) For example, in 2012 the predicted sales is $f(8) = 5.375(8)^2 + 15.25(8) + 3 = 469$. Answers may vary.

71. (a) Using $f(x) = ax^2 + bx + c$ and the data from the table yields:

From the point $(1990, 11)$: $11 = a(1990)^2 + b(1990) + c \Rightarrow 1990^2a + |1990b + c = 11$.

From the point $(2010, 10)$: $10 = a(2010)^2 + b(2010) + c \Rightarrow 2010^2a + 2010b + c = 10$.

From the point $(2030, 6)$: $6 = a(2030)^2 + b(2030) + c \Rightarrow 2030^2a + 2030b + c = 6$.

(b) Now using the capabilities of the calculator, the Reduced Row Echelon Method of:

$$\left[\begin{array}{ccc|c} 1990^2 & 1990 & 1 & 11 \\ 2010^2 & 2010 & 1 & 10 \\ 2030^2 & 2030 & 1 & 6 \end{array}\right] = \left[\begin{array}{ccc|c} 1 & 0 & 0 & -0.00375 \\ 0 & 1 & 0 & 14.95 \\ 0 & 0 & 1 & -14,889.125 \end{array}\right].$$

Therefore, the equation is $f(x) = -0.00375x^2 + 14.95x - 14889.125$.

(c) See Figure 71.

(d) Answers will vary, for example in 2015 the predicted ratio is $f(2015) \approx 9.3$.

$f(2010) \approx 394$.

[0.5,2035] by [-10,175]
Xscl = 5 Yscl = 25

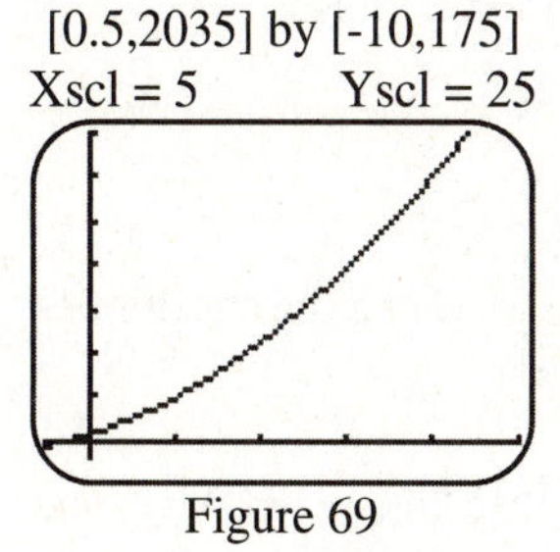

Figure 69

[1985,2035] by [5,12]
Xscl = 5 Yscl = 1

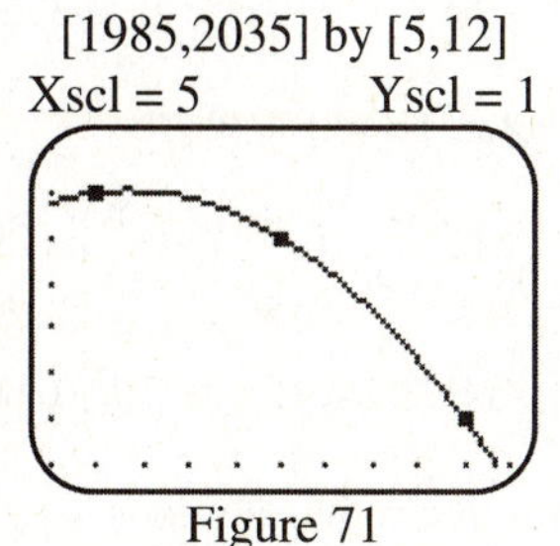

Figure 71

73. (a) At intersection A, incoming traffic is equal to $x+5$. The outgoing traffic is given by $y+7$. Therefore, $x+5 = y+7$. The incoming traffic at intersection B is $z+6$ and the outgoing traffic is $x+3$, so $z+6 = x+3$. Finally, at intersection C, the incoming flow is $y+3$ and the outgoing flow is $z+4$, so $y+3 = z+4$.

(b) The three equations are $\begin{cases} x+5 = y+7 \Rightarrow x-y=2 \\ z+6 = x+3 \Rightarrow x-z=3 \\ y+3 = z+4 \Rightarrow y-z=1 \end{cases}$ which can be represented by $\left[\begin{array}{ccc|c} 1 & -1 & 0 & 2 \\ 1 & 0 & -1 & 3 \\ 0 & 1 & -1 & 1 \end{array}\right]$.

Then by $-R_1 + R_2 \rightarrow \left[\begin{array}{ccc|c} 1 & -1 & 0 & 2 \\ 0 & 1 & -1 & 1 \\ 0 & 1 & -1 & 1 \end{array}\right] \Rightarrow -R_2 + R_3 \rightarrow \left[\begin{array}{ccc|c} 1 & -1 & 0 & 2 \\ 0 & 1 & -1 & 1 \\ 0 & 0 & 0 & 0 \end{array}\right]$. Since in the result of R_3 $0 = 0$, there are infinite solutions. They are $y - z = 1 \Rightarrow y = z+1$ and $x - z = 3 \Rightarrow x = z+3$. The solution is $(z+3, z+1, z)$, where $z \geq 0$.

(c) There are infinitely many solutions since some cars could be driving around the block continually.

75. (a) Using the given model and information from the table the equations are

$$x = \frac{Dx}{D} = \frac{-114}{-114} = 1,\ y = \frac{Dy}{D} = \frac{228}{-114} = -2,$$

(b) The equations can be represented by $\left[\begin{array}{cccc|c} 1 & 871 & 11.5 & 3 & 239 \\ 1 & 847 & 12.2 & 2 & 234 \\ 1 & 685 & 10.6 & 5 & 192 \\ 1 & 969 & 14.2 & 1 & 343 \end{array}\right]$. Using the capabilities of the calculator, the Reduced Row Echelon Method produces the solutions $a \approx -715.457,\ b \approx 0.34756,\ c \approx 48.6585,$ and $d \approx 30.71951$.

(c) The equation is $F = -715.457 + 0.34756A + 48.6585P + 30.71951W$.

(d) Input the given data into the equation.

$F = -715.457 + 0.34756(960) + 48.6585(12.6) + 30.71951(3) \Rightarrow F \approx 323.45623$. This is approximately 323, which is close to the actual value of 320.

77. The given system of equations can be represented by $\left[\begin{array}{ccc|c} 1 & 1 & 1 & 50 \\ -2 & 0 & 1 & 0 \\ 15.99 & 12.99 & 10.19 & 618.5 \end{array}\right]$. Using the capabilities of the calculator, the Reduced Row Echelon Method produces the solutions $a \approx 11.92$ lb, $b \approx 14.23$ lb, , and $c \approx 23.85$ lb.

Reviewing Basic Concepts (Section 7.1-7.3)

1. Multiply the first equation by 5, the second equation by 2, and subtract to eliminate the x-variable.

$10x - 15y = 90$
$10x + 4y = 14$

$-19y = 76 \Rightarrow y = -4$ Substitute (-4) for y in the second equation:

$10x + 4(-4) = 14 \Rightarrow 10x = 30 \Rightarrow x = 3$. The solution is $\{(3,-4)\}$.

2. The equations are: $2x + y = -4 \Rightarrow y_1 = -2x - 4$ and $-x + 2y = 2 \Rightarrow 2y = x + 2 \Rightarrow y = \frac{x+2}{2} \Rightarrow$

$y_2 = \frac{1}{2}x + 1$. Graph y_1 and y_2 in the same window, the intersection is $\{(-2,0)\}$. See Figure 2.

3. Using substitution, first solve [equation 2] for x, $x + 2y = 2 \Rightarrow x = -2y + 2$ [equation 3]. Substitute $(-2y + 2)$ in for x, in [equation 1]: $5(-2y + 2) + 10y = 10 \Rightarrow -10y + 10 + 10y = 10 \Rightarrow 10 = 10$. Since this is true, there is an infinite number of solutions given as $(-2y + 2, y)$ or $\left(x, \frac{2-x}{2}\right)$.

4. Subtract [equation 2] from [equation 1] to eliminate both variables.

$x - y = 6$
$x + y = 4$

$0 = 2 \Rightarrow$ Since this is false the solution is $\varnothing$.

5. First solve [equation 1] for y and substitute into [equation 2]: $6x + 2y = 10 \Rightarrow 2y = 10 - 6x \Rightarrow y = 5 - 3x$.

Now substitute: $2x^2-3(-3x+5)=11 \Rightarrow 2x^2+9x-15=11 \Rightarrow 2x^2+9x-26=0 \Rightarrow$

$(2x+13)(x-2)=0 \Rightarrow x=-6.5$ and $x = 2$. If $x = -6.5$, then by substitution:

$y=-3(-6.5)+5 \Rightarrow y=19.5+5 \Rightarrow y=24.5$. If $x = 2$, then by substitution $y=-3(2)+5 \Rightarrow y=-1$.

The solution is: $\{(2,-1),(-6.5,24.5)\}$. See Figure 5 for graphical support.

[-10,10] by [-10,10]
Xscl = 1 Yscl = 1

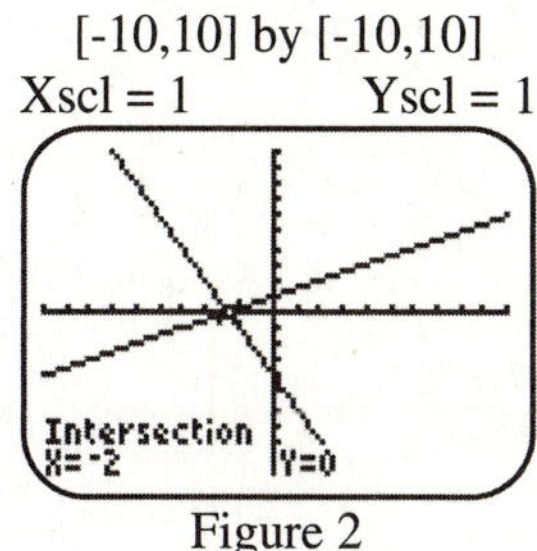

Figure 2

[-10,10] by [-10,10]
Xscl = 1 Yscl = 1

Intersection
X=-6.5
Y=24.5

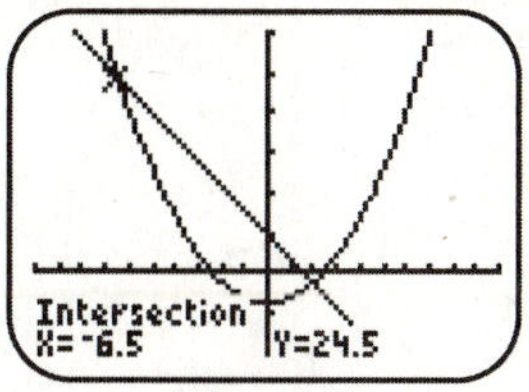

Figure 5

6. The equations can be represented by: $\left[\begin{array}{ccc|c} 1 & 1 & 1 & 1 \\ -1 & 1 & 1 & 5 \\ 0 & 1 & 2 & 5 \end{array}\right]$. $R_1+R_2 \rightarrow \left[\begin{array}{ccc|c} 1 & 1 & 1 & 1 \\ 0 & 2 & 2 & 6 \\ 0 & 1 & 2 & 5 \end{array}\right] \Rightarrow$

$R_2-2R_1 \rightarrow \left[\begin{array}{ccc|c} 1 & 1 & 1 & 1 \\ 0 & 2 & 2 & 6 \\ 0 & 0 & -2 & -4 \end{array}\right]$.

Now solve the resulting equation $R_3: -2z=-4 \Rightarrow z=2$. Substitute $z=2$ into the resulting equation R_2:

$2y+2(2)=6 \Rightarrow 2y=2 \Rightarrow y=1$. Finally, substitute $y=1$ *and* $z=2$ into R_1:

$x+(1)+(2)=1 \Rightarrow x=-2$. The solution is $\{(-2,1,2)\}$.

7. Using the Reduced Row Echelon method, the given system of equations yields the matrix, $\left[\begin{array}{ccc|c} 2 & 4 & 4 & 4 \\ 1 & 3 & 1 & 4 \\ -1 & 3 & 2 & -1 \end{array}\right]$.

Using the capabilities of the calculator, the Reduced Row Echelon Method produces the solutions $x=2$, $y=1$, and $z=-1$.

8. Solve the augmented matrix $\left[\begin{array}{ccc|c} 2 & 1 & 2 & 10 \\ 1 & 0 & 2 & 5 \\ 1 & -2 & 2 & 1 \end{array}\right]$, which by $\left.\begin{array}{c} R_1-2R_2 \\ R_1-2R_3 \end{array}\right\} \rightarrow \left[\begin{array}{ccc|c} 2 & 1 & 2 & 10 \\ 0 & 1 & -2 & 0 \\ 0 & 5 & -2 & 8 \end{array}\right] \Rightarrow$

$-5R_2+R_3 \rightarrow \left[\begin{array}{ccc|c} 2 & 1 & 2 & 10 \\ 0 & 1 & -2 & 0 \\ 0 & 0 & 8 & 8 \end{array}\right]$. Now solve the resulting equation R_3: $8z=8 \Rightarrow z=1$. Substitute $z=1$

into the resulting equation R_2: $y-2(1)=0 \Rightarrow y=2$. Finally, substitute $y=2$ and $z=1$ into R_1:

$2x+(2)+2(1)=10 \Rightarrow 2x=6 \Rightarrow x=3$. The solution is $(3,2,1)$.

9. Let x represent the number of LCD monitors and y represent the number of CRT monitors.

Then from the information the equations are $\begin{array}{ll} x+y=133 & [1] \\ x-y=43 & [2] \end{array}$ Add [equation 1] and [equation 2]:

$2x=176 \Rightarrow x=88$. Substitute x=88 into [1]: $88+y=143 \Rightarrow y=45$.. There were approximately 88 million LCD monitors sold and 45 million CRT monitors sold in 2006.

10. Let x = amount at 8%, y = amount at 11%, and x = amount at 14%. Then from the information:

$x+y+z=5000$
$0.03x+0.04y+0.06z=240$
$z=x+y \Rightarrow x+y-z=0$

Using the Reduced Row Echelon method, the given system of equations yields the matrix

$$\left[\begin{array}{ccc|c} 1 & 1 & 1 & 5000 \\ 0.03 & 0.04 & 0.06 & 595 \\ 1 & 1 & -1 & 0 \end{array}\right].$$

Using the capabilities of the calculator, the Reduced Row Echelon Method produces the solutions $x=1000,\ y=1500,$ and $z=2500$.

There was \$1000 invested at 8%, \$1500 invested at 11%, and \$2500 invested at 14%.

7.4: Matrix Properties and Operations

1. Because the number of rows equals the number of columns, this is a 2×2 square matrix.

3. This is a 3×4 matrix.

5. Because there is only one column, this is a 2×1 column matrix.

7. Because the number of rows equals the number of columns, there is only one row and column, this is a 1×1 square matrix (one row and one column).

9. For $\begin{bmatrix} w & x \\ y & z \end{bmatrix} = \begin{bmatrix} 3 & 2 \\ -1 & 4 \end{bmatrix}$, w = 3, x = 2, y = −1, and z = 4.

11. For $\begin{bmatrix} 0 & 5 & x \\ -1 & 3 & y+2 \\ 4 & 1 & z \end{bmatrix} = \begin{bmatrix} 0 & w+3 & 6 \\ -1 & 3 & 0 \\ 4 & 1 & 8 \end{bmatrix}$, $w+3=5 \Rightarrow w=2,\ x=6,\ y+2=0 \Rightarrow y=-2$, and $z=8$.

13. For $\begin{bmatrix} -7+z & 4r & 8s \\ 6p & 2 & 5 \end{bmatrix} + \begin{bmatrix} -9 & 8r & 3 \\ 2 & 7 & 4 \end{bmatrix} = \begin{bmatrix} 2 & 36 & 27 \\ 20 & 7 & 12a \end{bmatrix}$,

$-7+z-9=2 \Rightarrow z=18;\ 4r+8r=36 \Rightarrow r=3;\ 8s+3=27 \Rightarrow 8s=24 \Rightarrow s=3$

$6p+2=20 \Rightarrow 6p=18 \Rightarrow p=3;\ 5+4=12a \Rightarrow 9=12a \Rightarrow a=\dfrac{3}{4}$

15. The two matrices must have the same dimensions. To find the sum, add the corresponding entries. The sum will be a matrix with the same dimension.

17. $\begin{bmatrix} 6 & -9 & 2 \\ 4 & 1 & 3 \end{bmatrix} + \begin{bmatrix} -8 & 2 & 5 \\ 6 & -3 & 4 \end{bmatrix} = \begin{bmatrix} 6-8 & -9+2 & 2+5 \\ 4+6 & 1-3 & 3+4 \end{bmatrix} = \begin{bmatrix} -2 & -7 & 7 \\ 10 & -2 & 7 \end{bmatrix}$

19. $\begin{bmatrix}-6 & 8\\ 0 & 0\end{bmatrix}-\begin{bmatrix}0 & 0\\ -4 & -2\end{bmatrix}=\begin{bmatrix}-6-0 & 8-0\\ 0-(-4) & 0-(-2)\end{bmatrix}=\begin{bmatrix}-6 & 8\\ 4 & 2\end{bmatrix}$

21. $\begin{bmatrix}6 & -2\\ 5 & 4\end{bmatrix}+\begin{bmatrix}-1 & 7\\ 7 & -4\end{bmatrix}=\begin{bmatrix}6-1 & -2+7\\ 5+7 & 4-4\end{bmatrix}=\begin{bmatrix}5 & 5\\ 12 & 0\end{bmatrix}$

23. $\begin{bmatrix}-8 & 4 & 0\\ 2 & 5 & 0\end{bmatrix}+\begin{bmatrix}6 & 3\\ 8 & 9\end{bmatrix}$ Matrices must be the same size to add or subtract. We cannot add $(2\times 3)+(2\times 2)$.

25. $\begin{bmatrix}9 & 4 & 1 & -2\\ 5 & -6 & 3 & 4\\ 2 & -5 & 1 & 2\end{bmatrix}-\begin{bmatrix}-2 & 5 & 1 & 3\\ 0 & 1 & 0 & 2\\ -8 & 3 & 2 & 1\end{bmatrix}+\begin{bmatrix}2 & 4 & 0 & 3\\ 4 & -5 & 1 & 6\\ 2 & -3 & 0 & 8\end{bmatrix}=$

$\begin{bmatrix}9+2+2 & 4-5+4 & 1-1+0 & -2-3+3\\ 5-0+4 & -6-1-5 & 3-0+1 & 4-2+6\\ 2+8+2 & -5-3-3 & 1-2+0 & 2-1+8\end{bmatrix}=\begin{bmatrix}13 & 3 & 0 & -2\\ 9 & -12 & 4 & 8\\ 12 & -11 & -1 & 9\end{bmatrix}$

27. $2\begin{bmatrix}2 & -1\\ 5 & 1\\ 0 & 3\end{bmatrix}+\begin{bmatrix}5 & 0\\ 7 & -3\\ 1 & 1\end{bmatrix}-\begin{bmatrix}9 & -4\\ 4 & 4\\ 1 & 6\end{bmatrix}=\begin{bmatrix}4+5-9 & -2+0+4\\ 1+7-4 & 2-3-4\\ 0+1-1 & 6+1-6\end{bmatrix}=\begin{bmatrix}0 & 2\\ 13 & -5\\ 0 & 1\end{bmatrix}$

29 $2\begin{bmatrix}2 & -1 & -1\\ -1 & 2 & -1\\ -1 & -1 & 2\end{bmatrix}+3\begin{bmatrix}1 & 2 & 3\\ 2 & 1 & 3\\ 2 & 3 & 1\end{bmatrix}=\begin{bmatrix}4+3 & -2+6 & -2+9\\ -2+6 & 4+3 & -2+9\\ -2+6 & -2+9 & 4+3\end{bmatrix}=\begin{bmatrix}7 & 4 & 7\\ 4 & 7 & 7\\ 4 & 7 & 7\end{bmatrix}.$

31. $3\begin{bmatrix}6 & -1 & 4\\ 2 & 8 & -3\\ -4 & 5 & 6\end{bmatrix}+5\begin{bmatrix}-2 & -8 & -6\\ 4 & 1 & 3\\ 2 & -1 & 5\end{bmatrix}=\begin{bmatrix}18 & -3 & 12\\ 6 & 24 & -9\\ -12 & 15 & 18\end{bmatrix}+\begin{bmatrix}-10 & -40 & -30\\ 20 & 5 & 15\\ 10 & -5 & 25\end{bmatrix}=$

$\begin{bmatrix}18-10 & -3-40 & 12-30\\ 6+20 & 24+5 & -9+15\\ -12+10 & 15-5 & 18+25\end{bmatrix}=\begin{bmatrix}8 & -43 & -18\\ 26 & 29 & 6\\ -2 & 10 & 43\end{bmatrix}$

33. $2[A]=2\begin{bmatrix}-2 & 4\\ 0 & 3\end{bmatrix}=\begin{bmatrix}-4 & 8\\ 0 & 6\end{bmatrix}$

35. $2[A]-[B]=2\begin{bmatrix}-2 & 4\\ 0 & 3\end{bmatrix}-\begin{bmatrix}-6 & 2\\ 4 & 0\end{bmatrix}=\begin{bmatrix}-4+6 & 8-2\\ 0-4 & 6-0\end{bmatrix}=\begin{bmatrix}2 & 6\\ -4 & 6\end{bmatrix}$

37. $5[A]+0.5[B]=5\begin{bmatrix}-2 & 4\\ 0 & 3\end{bmatrix}+0.5\begin{bmatrix}-6 & 2\\ 4 & 0\end{bmatrix}=\begin{bmatrix}-10-3 & 20+1\\ 0+2 & 15+0\end{bmatrix}=\begin{bmatrix}-13 & 21\\ 2 & 15\end{bmatrix}$

39. $[A]=([A]+[B])-[B]=\begin{bmatrix}6 & 12 & 0\\ -10 & -4 & 11\end{bmatrix}-\begin{bmatrix}4 & 6 & -5\\ -6 & 3 & 2\end{bmatrix}=\begin{bmatrix}6-4 & 12-6 & 0+5\\ -10+6 & -4-3 & 11-2\end{bmatrix}=\begin{bmatrix}2 & 6 & 5\\ -4 & -7 & 9\end{bmatrix}$

41. A is 4×2 and B is 2×4. For AB, the number of columns of A is the same as the number of rows of B; therefore, AB has dimensions: $4(\text{rows of } A)\times 4(\text{columns of } B)$ or AB is 4×4. For BA, the number of columns of B is the same as the number of rows of A; therefore, BA has dimensions: $2(\text{rows of } B)\times 2(\text{columns of } A)$ or BA is 2×2.

43. A is 3×5 and B is 5×2. For AB, the number of columns of A is the same as the number of rows of B; therefore, AB has dimensions: $3(\text{rows of } A)\times2(\text{columns of } B)$ or AB is 3×2. For BA, the number of columns of B is not the same as the number of rows of A; therefore, BA is not defined.

45. A is 4×3 and B is 2×5. For AB, the number of columns of A is not the same as the number of rows of B; therefore, AB is not defined. For BA, the number of columns of B is not the same as the number of rows of A; therefore, BA is also not defined.

47. ... the number of columns of M equal the number of rows of N.

49. A and B are both 2×2 so AB and BA are also both 2×2.

$$AB=\begin{bmatrix}1&-1\\2&0\end{bmatrix}\begin{bmatrix}-2&3\\1&2\end{bmatrix}=\begin{bmatrix}-3&1\\-4&6\end{bmatrix}; BA=\begin{bmatrix}-2&3\\1&2\end{bmatrix}\begin{bmatrix}1&-1\\2&0\end{bmatrix}=\begin{bmatrix}4&2\\5&-1\end{bmatrix}$$

51. Since both A and B are 2×3, the number of rows in B is not equal to the number of columns in A, so AB is undefined. Also, the number of rows in A is not equal to the number of columns in B so BA is undefined.

53. $$AB=\begin{bmatrix}3&-1\\1&0\\-2&-4\end{bmatrix}\begin{bmatrix}-2&5&-3\\9&-7&0\end{bmatrix}=$$

$$\begin{bmatrix}3(-2)+(-1)(9)&3(5)+(-1)(-7)&3(-3)+(-1)(0)\\1(-2)+0(9)&1(5)+0(-7)&1(-3)+0(0)\\-2(-2)+(-4)(9)&-2(5)+(-4)(-7)&-2(-3)+(-4)(0)\end{bmatrix}\Rightarrow AB=\begin{bmatrix}-15&22&-9\\-2&5&-3\\-32&18&6\end{bmatrix}$$

$$BA=\begin{bmatrix}-2&5&-3\\9&-7&0\end{bmatrix}\begin{bmatrix}3&-1\\1&0\\-2&-4\end{bmatrix}=\begin{bmatrix}-2(3)+5(1)+(-3)(-2)&-2(-1)+5(0)+(-3)(-4)\\9(3)+(-7)(1)+0(-2)&9(-1)+(-7)(0)+0(-4)\end{bmatrix}\Rightarrow BA=\begin{bmatrix}5&14\\20&-9\end{bmatrix}$$

55. AB is undefined, we cannot multiply a 3×3 by a 2×3.

$$BA=\begin{bmatrix}-1&3&-1\\7&-7&1\end{bmatrix}\begin{bmatrix}1&-1&0\\2&-1&5\\6&1&-4\end{bmatrix}=$$

$$\begin{bmatrix}-1(1)+3(2)+(-1)(6)&-1(-1)+3(-1)+(-1)(1)&-1(0)+3(5)+(-1)(-4)\\7(1)+(-7)(2)+1(6)&7(-1)+-7(-1)+1(1)&7(0)+(-7)(-5)+1(-4)\end{bmatrix}\Rightarrow BA=\begin{bmatrix}-1&-3&19\\-1&1&-39\end{bmatrix}$$

57. $$\begin{bmatrix}3&-4&1\\5&0&2\end{bmatrix}\begin{bmatrix}-1\\4\\2\end{bmatrix}=\begin{bmatrix}3(1)-4(4)+1(2)\\5(-1)+0(4)+2(2)\end{bmatrix}=\begin{bmatrix}-3-16+2\\-5+0+4\end{bmatrix}=\begin{bmatrix}-17\\-1\end{bmatrix}$$

59. $$\begin{bmatrix}5&2\\-1&4\end{bmatrix}\begin{bmatrix}3&-2\\1&0\end{bmatrix}=\begin{bmatrix}5(3)+2(1)&5(-2)+2(0)\\-1(3)+4(1)&-1(-2)+4(0)\end{bmatrix}=\begin{bmatrix}15+2&-10+0\\-3+4&2+0\end{bmatrix}=\begin{bmatrix}17&-10\\1&2\end{bmatrix}$$

61. $$\begin{bmatrix}2&2&-1\\3&0&1\end{bmatrix}\begin{bmatrix}0&2\\-1&4\\0&2\end{bmatrix}=\begin{bmatrix}2(0)+2(-1)+(-1)(0)&2(2)+2(4)+(-1)(2)\\3(0)+0(-1)+1(0)&3(2)+0(4)+1(2)\end{bmatrix}=\begin{bmatrix}0-2+0&4+8-2\\0+0+0&6+0+2\end{bmatrix}=\begin{bmatrix}-2&10\\0&8\end{bmatrix}$$

63. $\begin{bmatrix} -2 & -3 & -4 \\ 2 & -1 & 0 \\ 4 & -2 & 3 \end{bmatrix}\begin{bmatrix} 0 & 1 & 4 \\ 1 & 2 & -1 \\ 3 & 2 & -2 \end{bmatrix} =$

$$\begin{bmatrix} -2(0)+(-3)(1)+(-4)(3) & -2(1)+(-3)(2)+(-4)(2) & -2(4)+(-3)(-1)+(-4)(-2) \\ 2(0)+(-1)(1)+0(3) & 2(1)+(-1)(2)+0(2) & 2(4)+(-1)(-1)+0(-2) \\ 4(0)+(-2)(1)+3(3) & 4(1)+(-2)(2)+3(2) & 4(4)+(-2)(-1)+3(-2) \end{bmatrix} =$$

$$\begin{bmatrix} 0-3-12 & -2-6-8 & -8+3-8 \\ 0-1+0 & 2-2+0 & 8+1+0 \\ 0-2+9 & 4-4+6 & 16+2-6 \end{bmatrix} = \begin{bmatrix} -15 & -16 & 3 \\ -1 & 0 & 9 \\ 7 & 6 & 12 \end{bmatrix}$$

65. $\begin{bmatrix} -2 & 4 & 1 \end{bmatrix}\begin{bmatrix} 3 & -2 & 4 \\ 2 & 1 & 0 \\ 0 & -1 & 4 \end{bmatrix} = \begin{bmatrix} -2(3)+4(2)+1(0) & -2(-2)+4(1)+1(-1) & -2(4)+4(0)+1(4) \end{bmatrix} = \begin{bmatrix} -6+8+0 & 4+4-1 & -8+0+4 \end{bmatrix} = \begin{bmatrix} 2 & 7 & -4 \end{bmatrix}$

67. $\begin{bmatrix} p & q \\ r & s \end{bmatrix}\begin{bmatrix} a & c \\ b & d \end{bmatrix} = \begin{bmatrix} pa+qb & pc+qd \\ ra+sb & rc+sd \end{bmatrix}$

69. $BA = \begin{bmatrix} 5 & 1 \\ 0 & -2 \\ 3 & 7 \end{bmatrix}\begin{bmatrix} 4 & -2 \\ 3 & 1 \end{bmatrix} = \begin{bmatrix} 5(4)+1(3) & 5(-2)+1(1) \\ 0(4)+(-2)(3) & 0(-2)+(-2)(1) \\ 3(4)+7(3) & 3(-2)+7(1) \end{bmatrix} = \begin{bmatrix} 23 & -9 \\ -6 & -2 \\ 33 & 1 \end{bmatrix}$

71. $BC = \begin{bmatrix} 5 & 1 \\ 0 & -2 \\ 3 & 7 \end{bmatrix}\begin{bmatrix} -5 & 4 & 1 \\ 0 & 3 & 6 \end{bmatrix} = \begin{bmatrix} 5(-5)+1(0) & 5(4)+1(3) & 5(1)+1(6) \\ 0(-5)-2(0) & 0(4)-2(3) & 0(1)-2(6) \\ 3(-5)+7(0) & 3(4)+7(3) & 3(1)+7(6) \end{bmatrix} = \begin{bmatrix} -25 & 23 & 11 \\ 0 & -6 & -12 \\ -15 & 33 & 45 \end{bmatrix}$

73. We cannot multiply A and B because we cannot multiply a (2×2) by a (3×2).

75. $A^2 = \begin{bmatrix} 4 & -2 \\ 3 & 1 \end{bmatrix}\begin{bmatrix} 4 & -2 \\ 3 & 1 \end{bmatrix} = \begin{bmatrix} 4(4)-2(3) & 4(-2)-2(1) \\ 3(4)+1(3) & 3(-2)+1(1) \end{bmatrix} = \begin{bmatrix} 10 & -10 \\ 15 & -5 \end{bmatrix}$

77. $AB \neq BA; BC \neq CB; AC \neq CA.$. Matrix multiplication is not commutative.

79. (a) From the information the matrix is $\begin{bmatrix} 50 & 100 & 30 \\ 10 & 90 & 50 \\ 60 & 120 & 40 \end{bmatrix}$.

(b) The income matrix is $\begin{bmatrix} 12 \\ 10 \\ 15 \end{bmatrix}$.

(c) $\begin{bmatrix} 50 & 100 & 30 \\ 10 & 90 & 50 \\ 60 & 120 & 40 \end{bmatrix}\begin{bmatrix} 12 \\ 10 \\ 15 \end{bmatrix} = \begin{bmatrix} 50(12)+100(10)+30(15) \\ 10(12)+90(10)+50(15) \\ 60(12)+120(10)+40(15) \end{bmatrix} = \begin{bmatrix} 2050 \\ 1770 \\ 2520 \end{bmatrix}$

(d) $2050+1770+2520=\$6340$

81. (a) From R_2 the equation is $d_{n+1} = -0.05m_n + 1.05d_n$ (hundredths). The deer population will grow at 5 %/year.

(b) Let $m_n = 2000$ and $d_n = 5000$ and multiply the given matrices:

$$\begin{bmatrix} m_{n+1} \\ d_{n+1} \end{bmatrix} = \begin{bmatrix} 0.51 & 0.4 \\ -0.05 & 1.05 \end{bmatrix}\begin{bmatrix} 2000 \\ 5000 \end{bmatrix} = \begin{bmatrix} 1020+2000 \\ -100+5250 \end{bmatrix} = \begin{bmatrix} 3020 \\ 5150 \end{bmatrix}$$

After 1 year there will be 3020 mountain lions and 515,000 deer.

Now multiply this 1 year matrix by the given matrix to find year 2.

$$\begin{bmatrix} m_{n+2} \\ d_{n+2} \end{bmatrix} = \begin{bmatrix} 0.51 & 0.4 \\ -0.05 & 1.05 \end{bmatrix}\begin{bmatrix} 3020 \\ 5150 \end{bmatrix} = \begin{bmatrix} 1540.2+2060 \\ -151+5407.5 \end{bmatrix} = \begin{bmatrix} 3600.2 \\ 5256.5 \end{bmatrix}$$

After 2 years there will be 3600 mountain lions and approximately 525,700 deer.

(c) Let $m_n = 4000$ and $d_n = 5000$ and multiply the given matrices:

$$\begin{bmatrix} m_{n+1} \\ d_{n+1} \end{bmatrix} = \begin{bmatrix} 0.51 & 0.4 \\ -0.05 & 1.05 \end{bmatrix}\begin{bmatrix} 4000 \\ 5000 \end{bmatrix} = \begin{bmatrix} 2040+2000 \\ -200+5250 \end{bmatrix} = \begin{bmatrix} 4040 \\ 5050 \end{bmatrix}$$

After 1 year there will be 4040 mountain lions and 505,000 deer. Now multiply this 1 year matrix by the given matrix to find year 2. $\begin{bmatrix} m_{n+2} \\ d_{n+2} \end{bmatrix} = \begin{bmatrix} 0.51 & 0.4 \\ -0.05 & 1.05 \end{bmatrix}\begin{bmatrix} 4040 \\ 5050 \end{bmatrix} = \begin{bmatrix} 2060.4+2020 \\ -202+5302.5 \end{bmatrix} = \begin{bmatrix} 4080.4 \\ 5100.5 \end{bmatrix}$

After 2 years there will be 4080 mountain lions and approximately 510,050 deer. Mountain Lions: after

$$1\text{ year} = \frac{4040}{4000} = 1.01; \text{ after 2 years} = \frac{4080}{4040} = 1.01$$

$$\text{Deer: after 1 year} = \frac{505{,}000}{500{,}000} = 1.01; \text{ after 2 years} = \frac{510{,}050}{500{,}000} = 1.01$$

83. $A+B = B+A$: $A+B = \begin{bmatrix} a_{11}+b_{11} & a_{12}+b_{12} \\ a_{21}+b_{21} & a_{22}+b_{22} \end{bmatrix} = B+A.$

85. $(AB)C = A(BC)$:

$$(AB)C = \begin{bmatrix} a_{11}b_{11}+a_{12}b_{21} & a_{11}b_{12}+a_{12}b_{22} \\ a_{21}b_{11}+a_{22}b_{21} & a_{21}b_{12}+a_{22}b_{22} \end{bmatrix}\begin{bmatrix} c_{11} & c_{12} \\ c_{21} & c_{22} \end{bmatrix} =$$

$$\begin{bmatrix} (a_{11}b_{11}c_{11}+a_{12}b_{21}c_{11})+(a_{11}b_{12}c_{21}+a_{12}b_{22}c_{21}) & (a_{11}b_{11}c_{12}+a_{12}b_{21}c_{12})+(a_{11}b_{12}c_{22}+a_{12}b_{22}c_{22}) \\ (a_{21}b_{11}c_{11}+a_{22}b_{21}c_{11})+(a_{21}b_{12}c_{21}+a_{22}b_{22}c_{21}) & (a_{21}b_{11}c_{12}+a_{22}b_{21}c_{12})+(a_{21}b_{12}c_{22}+a_{22}b_{22}c_{22}) \end{bmatrix}$$

$$A(BC) = \begin{bmatrix} a_{11} & a_{12} \\ a_{21} & a_{22} \end{bmatrix}\begin{bmatrix} b_{11}c_{11}+b_{12}c_{21} & b_{11}c_{12}+b_{12}c_{22} \\ b_{21}c_{11}+b_{22}c_{21} & b_{21}c_{12}+b_{22}c_{22} \end{bmatrix} =$$

$$\begin{bmatrix} (a_{11}b_{11}c_{11}+a_{11}b_{12}c_{21})+(a_{12}b_{21}c_{11}+a_{12}b_{22}c_{21}) & (a_{11}b_{11}c_{12}+a_{11}b_{12}c_{22})+(a_{12}b_{21}c_{12}+a_{12}b_{22}c_{22}) \\ (a_{21}b_{11}c_{11}+a_{21}b_{21}c_{21})+(a_{22}b_{21}c_{11}+a_{22}b_{22}c_{21}) & (a_{21}b_{11}c_{12}+a_{21}b_{12}c_{22})+(a_{22}b_{21}c_{12}+a_{22}b_{22}c_{22}) \end{bmatrix}$$

Therefore, $(\text{AB})\text{C} = \text{A}(\text{BC})$.

87. $c(A+B) = cA+cB$:

$$c(A+B) = \begin{bmatrix} c(a_{11}+b_{11}) & c(a_{12}+b_{12}) \\ c(a_{21}+b_{21}) & c(a_{22}+b_{22}) \end{bmatrix} = \begin{bmatrix} ca_{11}+cb_{11} & ca_{12}+cb_{12} \\ ca_{21}+cb_{21} & ca_{22}+cb_{22} \end{bmatrix} = cA+cB$$

89. $(cA)d = (cd)A$:

$$(cA)d = \begin{bmatrix} ca_{11} & ca_{12} \\ ca_{21} & ca_{22} \end{bmatrix} \bullet d = \begin{bmatrix} cda_{11} & cda_{12} \\ cda_{21} & cda_{22} \end{bmatrix} = (cd)A.$$

7.5: Determinants and Cramer's Rule

1. $\det\begin{bmatrix} -5 & 9 \\ 4 & -1 \end{bmatrix} = 5 - 36 = -31$

3. $\det\begin{bmatrix} -1 & -2 \\ 5 & 3 \end{bmatrix} = -3 - (-10) = 7$

5. $\det\begin{bmatrix} 9 & 3 \\ -3 & -1 \end{bmatrix} = -9 - (-9) = 0$

7. $\det\begin{bmatrix} 3 & 4 \\ 5 & -2 \end{bmatrix} = -6 - 20 = -26$

9. Use $A_{ij} = (-1)^{i+j} \cdot M_{ij}$, to find the cofactor of each element a.

For a_{21}, find $A_{21} = (-1)^{2+1} \cdot \left(\det\begin{bmatrix} 0 & 1 \\ 2 & 1 \end{bmatrix}\right) = -1(0-2) = 2$

For $a_{22,}$ find $A_{22} = (-1)^{2+2} \cdot \left(\det\begin{bmatrix} -2 & 1 \\ 4 & 1 \end{bmatrix}\right) = 1(-2-4) = -6$

For $a_{23,}$ find $A_{23} = (-1)^{2+3} \cdot \left(\begin{bmatrix} -2 & 0 \\ 4 & 2 \end{bmatrix}\right) = -1(-4-0) = 4$

11. Use $A_{ij} = (-1)^{i+j} \cdot M_{ij}$, to find the cofactor of each element a.

For a_{21}, find $A_{21} = (-1)^{2+1} \cdot \left(\det\begin{bmatrix} 2 & -1 \\ 4 & 1 \end{bmatrix}\right) = -1(2+4) = -6$

For a_{22}, find $A_{22} = (-1)^{2+2} \cdot \left(\det\begin{bmatrix} 1 & -1 \\ 1 & 1 \end{bmatrix}\right) = 1(1-1) = 0$

For a_{23}, find $A_{23} = (-1)^{2+3} \cdot \left(\begin{bmatrix} 1 & 2 \\ -1 & -4 \end{bmatrix}\right) = -1(4+2) = -6$

13. Evaluate, expand by the second row. Therefore, $\det = (-)(a_{21})(M_{21}) + (a_{22})(M_{22}) - (a_{23})(M_{23})$:

$$-2\left(\det\begin{bmatrix} -7 & 8 \\ 3 & 0 \end{bmatrix}\right) + 1\left(\det\begin{bmatrix} 4 & 8 \\ -6 & 0 \end{bmatrix}\right) - 3\left(\det\begin{bmatrix} 4 & -7 \\ -6 & 3 \end{bmatrix}\right) = -2(-24) + 1(48) - 3(-30) = 48 + 48 + 90 = 186.$$

15. Evaluate, expand by the first column. Therefore, $\det = (a_{11})(M_{11}) - (a_{21})(M_{21}) + (a_{31})(M_{31})$:

$$1\left(\det\begin{bmatrix} 2 & -1 \\ 1 & 4 \end{bmatrix}\right) - (-1)\left(\det\begin{bmatrix} 2 & 0 \\ 1 & 4 \end{bmatrix}\right) + 0\left(\det\begin{bmatrix} 2 & 0 \\ 2 & -1 \end{bmatrix}\right) = 1(8+1) + 1(8-0) + 0 = 9 + 8 = 17.$$

17. Evaluate, expand by the first row. Therefore, $\det = (a_{11})(M_{11}) - (a_{12})(M_{12}) + (a_{13})(M_{13})$:

$$10\left(\det\begin{bmatrix}4 & 3\\8 & 10\end{bmatrix}\right)-2\left(\det\begin{bmatrix}-1 & 3\\-3 & 10\end{bmatrix}\right)+1\left(\det\begin{bmatrix}-1 & 4\\-3 & 8\end{bmatrix}\right)=$$

$10(40-24)-2(-10+9)+1(-8+12)=160+2+4=166.$

19. Evaluate, expand by the second row will give a determinant of 0.

21. Evaluate, expand by the third column. Therefore, $\det=(a_{13})(M_{13})-(a_{23})(M_{23})+(a_{33})(M_{33})$:

$$-1\left(\det\begin{bmatrix}2 & 6\\-6 & -6\end{bmatrix}\right)-0\left(\det\begin{bmatrix}3 & 3\\-6 & -6\end{bmatrix}\right)+2\left(\det\begin{bmatrix}3 & 3\\2 & 6\end{bmatrix}\right)=-1(-12+36)-0+2(18-6)=-24+24=0.$$

23. Evaluate, expand by the second row. Therefore, $\det=(-)(a_{21})(M_{21})+(a_{22})(M_{22})-(a_{23})(M_{23})$:

$$-0.3\left(\det\begin{bmatrix}-0.8 & 0.6\\4.1 & -2.8\end{bmatrix}\right)+0.9\left(\det\begin{bmatrix}0.4 & -0.6\\3.1 & -2.8\end{bmatrix}\right)-0.7\left(\det\begin{bmatrix}0.4 & -0.8\\3.1 & 4.1\end{bmatrix}\right)=$$

$-0.3(2.24-2.46)+0.9(-1.12-1.86)-0.7(1.64+2.48)=.066-2.682-2.884=-5.5.$

25. Evaluate, expand by the second row. Therefore, $\det=(a_{31})(M_{31})-(a_{32})(M_{32})+(a_{33})(M_{33})$:

$$7(\det)\begin{bmatrix}-4 & 3\\5 & -15\end{bmatrix}+9(\det)\begin{bmatrix}17 & 3\\11 & -15\end{bmatrix}+23(\det)\begin{bmatrix}17 & -4\\11 & 5\end{bmatrix}=$$

$7(60-15)+9(-255-33)+23(85+44)=690$

27. If $\det\begin{bmatrix}-0.5 & 2\\x & x\end{bmatrix}=0$, then $-0.5x-2x=0\Rightarrow-2.5x=0\Rightarrow x=0$. The solution set is $\{0\}$.

29. If $\det\begin{bmatrix}2x & 3\\11 & x\end{bmatrix}=6$, then $2x^2-11x=6\Rightarrow 2x^2-11x-6=0\Rightarrow x=(x-6)(2x+1)=0\Rightarrow x=6,-\frac{1}{2}$. The solution set is $\left\{\frac{1}{2},6\right\}$.

31. Evaluate, expand by the second column. Therefore, $\det=(a_{13})(M_{13})-(a_{23})(M_{23})+(a_{33})(M_{33})$:

$$0-1\left(\det\begin{bmatrix}4 & 3\\-3 & x\end{bmatrix}\right)+(-1)\left(\det\begin{bmatrix}4 & 3\\2 & 0\end{bmatrix}\right)=5-1(4x+9)-1(0-6)=5\Rightarrow-4x-9+6=5\Rightarrow$$

$-4x=8\Rightarrow x=-2$. The solution set is $\{-2\}$

33. Evaluate, expand by the second column. Therefore, $\det=(-)(a_{12})(M_{12})+(a_{22})(M_{22})-(a_{32})(M_{32})$:

$$-1\left(\det\begin{bmatrix}0 & x\\3 & 2\end{bmatrix}\right)+4\left(\det\begin{bmatrix}2x & -1\\3 & 2\end{bmatrix}\right)+0=x\Rightarrow-1(0-3x)+4(4x+3)=x\Rightarrow$$

$3x+16x+12=x\Rightarrow 18x=-12\Rightarrow x=-\frac{12}{18}$. The solution set is $\left\{-\frac{2}{3}\right\}$.

35. Using the calculator, the determinant is 298.

37. Using the calculator, the determinant is -88.

39. With the given points find: $A=\frac{1}{2}\det\begin{bmatrix}0 & 0 & 1\\ 0 & 2 & 1\\ 1 & 4 & 1\end{bmatrix}$. Using the calculator the determinant is -2.

Therefore, $A=\left|\frac{1}{2}(-2)\right| \Rightarrow A=1$ square units.

41. With the given points find $A=\frac{1}{2}\det\begin{bmatrix}2 & 5 & 1\\ -1 & 3 & 1\\ 4 & 0 & 1\end{bmatrix}$. Using the calculator the determinant is 19. Therefore,

$A=\left|\frac{1}{2}(19)\right| \Rightarrow A=\frac{19}{2}$ square units.

43. With the given points find $A=\frac{1}{2}\det\begin{bmatrix}1 & 2 & 1\\ 4 & 3 & 1\\ 3 & 5 & 1\end{bmatrix}$. Using the calculator the determinant is 7. Therefore,

$A=\left|\frac{1}{2}(7)\right| \Rightarrow A=\frac{7}{2}$ square units.

45. If the three points form a triangle with no area ($D=0$ using determinants), then the points must be collinear.

$D=\frac{1}{2}\det\begin{bmatrix}1 & -3 & 2\\ 3 & 11 & 1\\ 1 & 1 & 1\end{bmatrix}=0$; The points are collinear.

47. If the three points form a triangle with no area ($D=0$ using determinants), then the points must be collinear.

$D=\frac{1}{2}\det\begin{bmatrix}-2 & 4 & 2\\ -5 & 4 & 3\\ 1 & 1 & 1\end{bmatrix}=6\neq 0$; the points are not collinear.

49. If the three points form a triangle with no area ($D=0$ using determinants), then the points must be collinear.

$D=\frac{1}{2}\det\begin{bmatrix}4 & 6 & 12\\ -1 & 0 & 4\\ 1 & 1 & 1\end{bmatrix}=1\neq 0$; the points are not collinear.

51. Since the second column is all zeros, by Determinant Theorem 1, the determinant is 0.

53. Use Determinant Theorem 6; multiplying column 1 by 2 and adding the result to column 3 yields the equivalent determinant, $\det\begin{bmatrix}6 & 8 & 0\\ -1 & 0 & 0\\ 4 & 0 & 0\end{bmatrix}$. Since column 3 has all zeros, the determinant is 0

(Determinant Theorem 1) and $\det\begin{bmatrix}6 & 8 & -12\\ -1 & 0 & 2\\ 4 & 0 & -8\end{bmatrix}=0$.

55. Use Determinant Theorem 6; multiplying column 2 by 4 and adding the result to column 1, and multiplying column 2 by –4 and adding the result to column 3, yields the equivalent determinant $\det\begin{bmatrix} 0 & 1 & 0 \\ 2 & 0 & 1 \\ 8 & 2 & -4 \end{bmatrix}$. Now expand this about row 1: $\det\begin{bmatrix} 0 & 1 & 0 \\ 2 & 0 & 1 \\ 8 & 2 & -4 \end{bmatrix} = 0(0-2) - 1(-8-8) + 0(4-0) = 16 \Rightarrow \det\begin{bmatrix} -4 & 1 & 4 \\ 2 & 0 & 1 \\ 0 & 2 & 4 \end{bmatrix} = 16.$

57. Find the determinants: $D = \det\begin{bmatrix} 1 & 1 \\ 2 & -1 \end{bmatrix} = -3,\ D_x = \det\begin{bmatrix} 4 & 1 \\ 2 & -1 \end{bmatrix} = -6,$ and $D_y = \det\begin{bmatrix} 1 & 4 \\ 2 & 2 \end{bmatrix} = -6.$

Then $x = \frac{Dx}{D} = \frac{-6}{-3} = 2$ and $y = \frac{D_y}{D} = \frac{-6}{-3} = 2$. The solution set is $\{(2,2)\}$.

59. Find the determinants: $D = \det\begin{bmatrix} 4 & 3 \\ 2 & 3 \end{bmatrix} = 6,\ D_x = \det\begin{bmatrix} -7 & 3 \\ -11 & 3 \end{bmatrix} = 12,$ and $D_y = \det\begin{bmatrix} 4 & -7 \\ 2 & -11 \end{bmatrix} = -30.$

Then $x = \frac{D_x}{D} = \frac{12}{2} = 2$ and $y = \frac{D_y}{D} = \frac{-30}{6} = -5$. The solution set is $\{(2,-5)\}$.

61. Find the determinant D: $D = \det\begin{bmatrix} 3 & 2 \\ 6 & 4 \end{bmatrix} = 0$; therefore, there are infinitely many solutions or no solutions.

Now use Row Echelon Method on the augmented matrix: $\left[\begin{array}{cc|c} 3 & 2 & 4 \\ 6 & 4 & 8 \end{array}\right],\ -2R_1 + R_2 \rightarrow \left[\begin{array}{cc|c} 3 & 2 & 4 \\ 0 & 0 & 0 \end{array}\right]$. Since $0 = 0$, there are infinitely many solutions and using R_1, they are in the following relationship: $3x + 2y = 4 \Rightarrow 3x = 4 - 2y \Rightarrow x = \frac{4-2y}{3}$. The solution is $\left(\frac{4-2y}{3}, y\right)$ or $\left(x, \frac{4-3x}{2}\right)$.

63. Find the determinants: $D = \det\begin{bmatrix} 2 & -3 \\ 1 & 5 \end{bmatrix} = 13,\ D_x = \det\begin{bmatrix} -5 & -3 \\ 17 & 5 \end{bmatrix} = 26,$ and $D = \det\begin{bmatrix} 2 & -5 \\ 1 & 17 \end{bmatrix} = 39.$

Then $x = \frac{D_x}{D} = \frac{26}{13} = 2$ and $y = \frac{D_y}{D} = \frac{39}{13} = 3$. The solution set is $\{(2,3)\}$.

65. Using your calculator, find the following determinants:

$D = \det\begin{bmatrix} 4 & -1 & 3 \\ 3 & 1 & 1 \\ 2 & -1 & 4 \end{bmatrix} = 15,\ D_x = \det\begin{bmatrix} -3 & -1 & 3 \\ 0 & 1 & 1 \\ 0 & -1 & 4 \end{bmatrix} = -15,\ D_y = \det\begin{bmatrix} 4 & -3 & 3 \\ 3 & 0 & 1 \\ 2 & 0 & 4 \end{bmatrix} = 30,$ and

$D_z = \det\begin{bmatrix} 4 & -1 & -3 \\ 3 & 1 & 0 \\ 2 & -1 & 0 \end{bmatrix} = 15.$ Then $x = \frac{D_x}{D} = \frac{-15}{15} = -1,\ y = \frac{D_y}{D} = \frac{30}{15} = 2,$ and

$z = \frac{D_z}{D} = \frac{15}{15} = 1$. The solution set is $\{(-1,2,1)\}$.

67. Using your calculator, find the following determinants:

$$D = \det\begin{bmatrix} 2 & -1 & 4 \\ 3 & 2 & -1 \\ 1 & 4 & 2 \end{bmatrix} = 63,\ D_x = \det\begin{bmatrix} -2 & -1 & 4 \\ -3 & 2 & -1 \\ 17 & 4 & 2 \end{bmatrix} = -189,\ D_y = \det\begin{bmatrix} 2 & -2 & 4 \\ 3 & -3 & -1 \\ 1 & 17 & 2 \end{bmatrix} = 252,$$

$$D_z = \det\begin{bmatrix} 2 & -1 & -2 \\ 3 & 2 & -3 \\ 1 & 4 & 17 \end{bmatrix} = 126. \text{ Then } x = \frac{D_x}{D} = \frac{-189}{63} = -3,\ y = \frac{D_y}{D} = \frac{252}{63} = 4, \text{ and } z = \frac{Dz}{D} = \frac{126}{63} = 2.$$

The solution set is $\{(-3,4,2)\}$.

69. Using your calculator, find the following determinants:

$$D = \det\begin{bmatrix} 5 & -1 & 0 \\ 3 & 0 & 2 \\ 0 & 4 & 3 \end{bmatrix} = -31,\ D_x = \det\begin{bmatrix} -4 & -1 & 0 \\ 4 & 0 & 2 \\ 22 & 4 & 3 \end{bmatrix} = 0,\ D_y = \det\begin{bmatrix} 5 & -4 & 0 \\ 3 & 4 & 2 \\ 0 & 22 & 3 \end{bmatrix} = -124, \text{ and}$$

$$D_z = \det\begin{bmatrix} 5 & -1 & -4 \\ 3 & 0 & 4 \\ 0 & 4 & 22 \end{bmatrix} = -62. \text{ Then } x = \frac{D_x}{D} = \frac{0}{-31} = 0,\ y = \frac{D_y}{D} = \frac{-124}{-31} = 4, \text{ and } z = \frac{Dz}{D} = \frac{-62}{-31} = 2.$$

The solution set is $\{(0,4,2)\}$.

71. Using your calculator, find the following determinant: $D = \det\begin{bmatrix} 2 & -1 & 3 \\ -2 & 1 & -3 \\ 5 & -1 & 1 \end{bmatrix} = 0$; therefore, there are infinitely many solutions or no solutions. If we add [equation 1] + [equation 2] the result is $0 \neq 3$; therefore, no solution or $\varnothing$.

73. Using your calculator, find the following determinants:

$$D = \det\begin{bmatrix} 3 & 2 & 0 & -1 \\ 2 & 0 & 1 & 2 \\ 1 & 2 & -1 & 0 \\ 2 & -1 & 1 & 1 \end{bmatrix} = -9;\ D_x = \det\begin{bmatrix} 0 & 2 & 0 & -1 \\ 5 & 0 & 1 & 2 \\ -2 & 2 & -1 & 0 \\ 2 & -1 & 1 & 1 \end{bmatrix} = 9$$

$$D_y = \det\begin{bmatrix} 3 & 0 & 0 & -1 \\ 2 & 5 & 1 & 2 \\ 1 & -2 & -1 & 0 \\ 2 & 2 & 1 & 1 \end{bmatrix} = -18;\ D_z = \det\begin{bmatrix} 3 & 2 & 0 & -1 \\ 2 & 0 & 5 & 2 \\ 1 & 2 & -2 & 0 \\ 2 & -1 & 2 & 1 \end{bmatrix} = -45;\ D_w = \det\begin{bmatrix} 3 & 2 & 0 & 0 \\ 2 & 0 & 1 & 5 \\ 1 & 2 & -1 & -2 \\ 2 & -1 & 1 & 2 \end{bmatrix} = -9$$

Then $x = \frac{D_x}{D} = \frac{9}{-9} = -1;\ y = \frac{D_y}{D} = \frac{-18}{-9} = 2;\ z = \frac{D_z}{D} = \frac{-45}{-9} = 5;\ w = \frac{D_w}{D} = \frac{-9}{-9} = 1.$

The solution set is $\{(-1,2,5,1\}$.

75. If $D = 0$, Cramer's rule cannot be applied because there is no unique solution. There are either no solutions or infinitely many solutions.

7.6: Solution of Linear Systems by Matrix Inverses

1. Yes, $AB = BA = \begin{bmatrix} 1 & 0 \\ 0 & 1 \end{bmatrix}$.

$$\begin{bmatrix} 5 & 7 \\ 2 & 3 \end{bmatrix}\begin{bmatrix} 3 & -7 \\ -2 & 5 \end{bmatrix} = \begin{bmatrix} 15-14 & -35+35 \\ 6-6 & -14+15 \end{bmatrix} = \begin{bmatrix} 1 & 0 \\ 0 & 1 \end{bmatrix};$$

$$\begin{bmatrix} 3 & -7 \\ -2 & 5 \end{bmatrix}\begin{bmatrix} 5 & 7 \\ 2 & 3 \end{bmatrix} = \begin{bmatrix} 15-14 & 21-21 \\ -10+10 & -14+15 \end{bmatrix} = \begin{bmatrix} 1 & 0 \\ 0 & 1 \end{bmatrix}$$

3. No, $AB \neq \begin{bmatrix} 1 & 0 \\ 0 & 1 \end{bmatrix}$.

$$\begin{bmatrix} -1 & 2 \\ 3 & -5 \end{bmatrix}\begin{bmatrix} -5 & -2 \\ -3 & -1 \end{bmatrix} = \begin{bmatrix} 5-6 & 2-2 \\ -15+15 & -6+5 \end{bmatrix} = \begin{bmatrix} -1 & 0 \\ 0 & -1 \end{bmatrix}$$

5. No, $AB \neq \begin{bmatrix} 1 & 0 & 0 \\ 0 & 1 & 0 \\ 0 & 0 & 1 \end{bmatrix}$. Use your calculator to multiply the matrices. $\begin{bmatrix} 0 & 1 & 0 \\ 0 & 0 & -2 \\ 1 & -1 & 0 \end{bmatrix}\begin{bmatrix} 1 & 0 & 1 \\ 1 & 0 & 0 \\ 0 & -1 & 0 \end{bmatrix} = \begin{bmatrix} 1 & 0 & 0 \\ 0 & 2 & 0 \\ 0 & 0 & 1 \end{bmatrix}$

7. Yes. Using the calculator, multiply the matrices. $AB = \begin{bmatrix} 1 & 0 & 0 \\ 0 & 1 & 0 \\ 0 & 0 & 1 \end{bmatrix} = BA$

9. Using $A^{-1} = \dfrac{1}{\det A}\begin{bmatrix} d & -b \\ -c & a \end{bmatrix}$ to find the inverse yields $A^{-1} = \dfrac{1}{15-14}\begin{bmatrix} 5 & -7 \\ -2 & 3 \end{bmatrix} = \begin{bmatrix} 5 & -7 \\ -2 & 3 \end{bmatrix}$.

11. Using $A^{-1} = \dfrac{1}{\det A}\begin{bmatrix} d & -b \\ -c & a \end{bmatrix}$ to find the inverse yields

$$A^{-1} = \frac{1}{-4+6}\begin{bmatrix} 4 & 2 \\ -3 & -1 \end{bmatrix} = \left(\frac{1}{2}\right)\begin{bmatrix} 4 & 2 \\ -3 & -1 \end{bmatrix} = \begin{bmatrix} 2 & 1 \\ -\frac{3}{2} & -\frac{1}{2} \end{bmatrix}.$$

13. The $\det A = -12-(-12) = 0$; therefore, No inverse.

15. Using $A^{-1} = \dfrac{1}{\det A}\begin{bmatrix} d & -b \\ -c & a \end{bmatrix}$ to find the inverse yields

$$A^{-1} = \frac{1}{0.06-0.1}\begin{bmatrix} 0.1 & -0.2 \\ -0.5 & 0.6 \end{bmatrix} = (-25)\begin{bmatrix} 0.1 & -0.2 \\ -0.5 & 0.6 \end{bmatrix} = \begin{bmatrix} -2.5 & 5 \\ 12.5 & -15 \end{bmatrix}.$$

17. $A|I_3 = \left[\begin{array}{ccc|ccc} 0 & 0 & 1 & 1 & 0 & 0 \\ 1 & 0 & 0 & 0 & 1 & 0 \\ 0 & 1 & 0 & 0 & 0 & 1 \end{array}\right] \begin{array}{c} R_2 \to \\ R_3 \to \\ R_1 \to \end{array} \left[\begin{array}{ccc|ccc} 1 & 0 & 0 & 0 & 1 & 0 \\ 0 & 1 & 0 & 0 & 0 & 1 \\ 0 & 0 & 1 & 1 & 0 & 0 \end{array}\right]; A^{-1} = \begin{bmatrix} 0 & 1 & 0 \\ 0 & 0 & 1 \\ 1 & 0 & 0 \end{bmatrix}$

19. $A|I_3 = \left[\begin{array}{ccc|ccc} 1 & 0 & 1 & 1 & 0 & 0 \\ 2 & 1 & 3 & 0 & 1 & 0 \\ -1 & 1 & 1 & 0 & 0 & 1 \end{array}\right] \begin{array}{l} \\ R_2 - 2R_1 \rightarrow \\ R_3 + R_1 \rightarrow \end{array} \left[\begin{array}{ccc|ccc} 1 & 0 & 1 & 1 & 0 & 0 \\ 0 & 1 & 1 & -2 & 1 & 0 \\ 0 & 1 & 2 & 1 & 0 & 1 \end{array}\right] \begin{array}{l} \\ \\ R_3 - R_2 \rightarrow \end{array} \left[\begin{array}{ccc|ccc} 1 & 0 & 0 & 1 & 0 & 0 \\ 0 & 1 & 1 & -2 & 1 & 0 \\ 0 & 0 & 1 & 3 & -1 & 1 \end{array}\right]$

$\begin{array}{l} R_1 - R_3 \rightarrow \\ R_2 - R_3 \rightarrow \\ \\ \end{array} \left[\begin{array}{ccc|ccc} 1 & 0 & 0 & -2 & 1 & -1 \\ 0 & 1 & 0 & -5 & 2 & -1 \\ 0 & 0 & 1 & 3 & -1 & 1 \end{array}\right]; \ A^{-1} = \begin{bmatrix} -2 & 1 & -1 \\ -5 & 2 & -1 \\ 3 & -1 & 1 \end{bmatrix}$

21. The inverse will not exist if its determinant is equal to 0.

23. Solve using the calculator. Since $\det A = -1$, the inverse exists and $A^{-1} = \begin{bmatrix} 1 & 0 & 0 \\ 0 & -1 & 0 \\ -1 & 0 & 1 \end{bmatrix}$.

25. Solve using the calculator. Since $\det A = -1$, the inverse exists and $A^{-1} = \begin{bmatrix} 15 & 4 & -5 \\ -12 & -3 & 4 \\ -4 & -1 & 1 \end{bmatrix}$.

27. Solve using the calculator. Since $\det A = 0.036$, the inverse exists and $A^{-1} = \begin{bmatrix} -\frac{10}{3} & \frac{5}{9} & -\frac{10}{9} \\ \frac{20}{3} & \frac{5}{9} & \frac{80}{9} \\ -5 & \frac{5}{6} & -\frac{20}{3} \end{bmatrix}$.

29. Solve using the calculator. Since $\det A = 0$, there is No inverse.

31. Solve using the calculator. Since $\det A \approx 9.207$, the inverse exists and

$A^{-1} = \begin{bmatrix} 0.0543058761 & -0.054358761 \\ 1.846399787 & 0.153600213 \end{bmatrix}$.

33. Solve using the calculator. Since $\det A = 1.478$, the inverse exists and

$A^{-1} = \begin{bmatrix} 0.9987635516 & -0.252092087 & -0.3330564627 \\ -0.5037783375 & 1.007556675 & -0.2518891688 \\ -0.2481013617 & -0.255676976 & 1.003768868 \end{bmatrix}$.

35. Using the matrix inverse method, put the system into the proper matrix form: $\begin{bmatrix} 2 & -1 \\ 3 & 1 \end{bmatrix} \begin{bmatrix} x \\ y \end{bmatrix} = \begin{bmatrix} -8 \\ -2 \end{bmatrix} \Rightarrow$

$\begin{bmatrix} x \\ y \end{bmatrix} = \begin{bmatrix} 2 & -1 \\ 3 & 2 \end{bmatrix}^{-1} \begin{bmatrix} -8 \\ -2 \end{bmatrix} \Rightarrow \frac{1}{5} \begin{bmatrix} 1 & 1 \\ -3 & 2 \end{bmatrix} \begin{bmatrix} -8 \\ -2 \end{bmatrix} \Rightarrow \begin{bmatrix} \frac{1}{5} & \frac{1}{5} \\ -\frac{3}{5} & \frac{2}{5} \end{bmatrix} \begin{bmatrix} -8 \\ -2 \end{bmatrix} \Rightarrow \begin{bmatrix} -\frac{8}{5} - \frac{2}{5} \\ -\frac{24}{5} - \frac{4}{5} \end{bmatrix} = \begin{bmatrix} -2 \\ 4 \end{bmatrix}$. The solution set is $\{(-2, 4)\}$.

37. Using the matrix inverse method, put the system into the proper matrix form: $\begin{bmatrix} 2 & 3 \\ 3 & 4 \end{bmatrix} \begin{bmatrix} x \\ y \end{bmatrix} = \begin{bmatrix} -10 \\ -12 \end{bmatrix} \Rightarrow$

$\begin{bmatrix} x \\ y \end{bmatrix} = \begin{bmatrix} 2 & 3 \\ 3 & 3 \end{bmatrix}^{-1} \begin{bmatrix} -10 \\ -12 \end{bmatrix} \Rightarrow -1 \begin{bmatrix} 4 & -3 \\ -3 & 2 \end{bmatrix} \begin{bmatrix} -10 \\ -12 \end{bmatrix} \Rightarrow \begin{bmatrix} -4 & 3 \\ 3 & -2 \end{bmatrix} \begin{bmatrix} -10 \\ -12 \end{bmatrix} \Rightarrow \begin{bmatrix} 40 - 36 \\ -30 + 24 \end{bmatrix} = \begin{bmatrix} 4 \\ -6 \end{bmatrix}$.

The solution set is $\{(4, -6)\}$.

39. Using the matrix inverse method, put the system into the proper matrix form: $\begin{bmatrix} 2 & -5 \\ 4 & -5 \end{bmatrix}\begin{bmatrix} x \\ y \end{bmatrix} = \begin{bmatrix} 10 \\ 15 \end{bmatrix} \Rightarrow$

$$\begin{bmatrix} x \\ y \end{bmatrix} = \begin{bmatrix} 2 & -5 \\ 4 & -5 \end{bmatrix}^{-1}\begin{bmatrix} 10 \\ 15 \end{bmatrix} \Rightarrow \frac{1}{10}\begin{bmatrix} -5 & 5 \\ -4 & 2 \end{bmatrix}\begin{bmatrix} 10 \\ 15 \end{bmatrix} \Rightarrow \begin{bmatrix} -0.5 & 0.5 \\ -0.4 & 0.2 \end{bmatrix}\begin{bmatrix} 10 \\ 15 \end{bmatrix} \Rightarrow \begin{bmatrix} -5+7.5 \\ -4+3 \end{bmatrix} = \begin{bmatrix} 2.5 \\ -1 \end{bmatrix}.$$

The solution set is $\{(2.5, -1)\}$.

41. Solve on the calculator using the matrix inverse method: $\begin{bmatrix} x \\ y \\ z \end{bmatrix} = \begin{bmatrix} 2 & 0 & 4 \\ 3 & 1 & 5 \\ -1 & 1 & -2 \end{bmatrix}^{-1} \bullet \begin{bmatrix} 14 \\ 19 \\ -7 \end{bmatrix} = \begin{bmatrix} 3 \\ 0 \\ 2 \end{bmatrix}$; therefore, the solution set is $\{(3, 0, 2)\}$.

43. Solve on the calculator using the matrix inverse method: $\begin{bmatrix} x \\ y \\ z \end{bmatrix} = \begin{bmatrix} 1 & 3 & 1 \\ 1 & -2 & 3 \\ 2 & -3 & -1 \end{bmatrix}^{-1} \bullet \begin{bmatrix} 2 \\ -3 \\ 34 \end{bmatrix} = \begin{bmatrix} 12 \\ -\frac{15}{11} \\ -\frac{65}{11} \end{bmatrix}$; therefore, the solution set is $\left\{\left(12, -\frac{15}{11}, -\frac{65}{11}\right)\right\}$.

45. Solve on the calculator using the matrix inverse method: $\begin{bmatrix} x \\ y \\ z \\ w \end{bmatrix} = \begin{bmatrix} 1 & 3 & -2 & -1 \\ 4 & 1 & 1 & 2 \\ -3 & -1 & 1 & -1 \\ 1 & -1 & -3 & -2 \end{bmatrix}^{-1} \bullet \begin{bmatrix} 9 \\ 2 \\ -5 \\ 2 \end{bmatrix} = \begin{bmatrix} 0 \\ 2 \\ -2 \\ 1 \end{bmatrix}$; therefore, the solution set is $\{(0, 2, -2, 1)\}$.

47. Solve on the calculator using the matrix inverse method: $\begin{bmatrix} x \\ y \end{bmatrix} = \begin{bmatrix} 1 & -\sqrt{2} \\ 0.75 & 1 \end{bmatrix}^{-1} \bullet \begin{bmatrix} 2.6 \\ -7 \end{bmatrix} = \begin{bmatrix} -3.542308934 \\ -4.343268299 \end{bmatrix}$; therefore the solution set is $\{(-3.542308934, -4.343268299)\}$.

49. Solve on the calculator using the matrix inverse method: $\begin{bmatrix} x \\ y \\ z \end{bmatrix} = \begin{bmatrix} \pi & e & \sqrt{2} \\ e & \pi & \sqrt{2} \\ \sqrt{2} & e & \pi \end{bmatrix}^{-1} \bullet \begin{bmatrix} 1 \\ 2 \\ 3 \end{bmatrix} = \begin{bmatrix} -.9704156969 \\ 1.391914631 \\ .1874077432 \end{bmatrix}$; therefore, the solution set is $\{(-0.9704156959, 1.391914631, 0.1874077432)\}$.

51. If $P(x) = ax^3 + bx^2 + cx + d$, then the ordered pair $(-1, 14)$ yields the equation $-a + b - c + d = 14$, the ordered pair $(1.5, 1.5)$ yields the equation $3.375a + 2.25b + 1.5c + d = 1.5$, the ordered pair $(2, -1)$ yields the equation $8a + 4b + 2c + d = -1$, and the ordered pair $(3, -18)$ yields the equation $27a + 9b + 3c + d = -18$. Now solve on the calculator using the matrix inverse method:

$$\begin{bmatrix} a \\ b \\ c \\ d \end{bmatrix} = \begin{bmatrix} -1 & 1 & -1 & 1 \\ 3.375 & 2.25 & 1.5 & 1 \\ 8 & 4 & 2 & 1 \\ 27 & 9 & 3 & 1 \end{bmatrix}^{-1} \bullet \begin{bmatrix} 14 \\ 1.5 \\ -1 \\ -18 \end{bmatrix} = \begin{bmatrix} -2 \\ 5 \\ -4 \\ 3 \end{bmatrix}$$; therefore, the equation is $P(x) = -2x^3 + 5x^2 - 4x + 3$.

53. If $P(x)=ax^4+bx^3+cx^2+dx+e,$ then the ordered pair (−2, 13) yields the equation $16a-8b+4c-2d+e=13,$ the ordered pair (−1, 2) yields the equation $a-b+c-d+e=2,$ the ordered pair (0, −1) yields the equation $e=-1,$ the ordered pair (1, 4) yields the equation $a+b+c+d+e=4,$ and the ordered pair (2, 41) yields the equation $16a+8b+4c+2d+e=41.$ Now solve on the calculator using the matrix inverse method: $\begin{bmatrix} a \\ b \\ c \\ d \\ e \end{bmatrix}=\begin{bmatrix} 16 & -8 & 4 & -2 & 1 \\ 1 & -1 & 1 & -1 & 1 \\ 0 & 0 & 0 & 0 & 1 \\ 1 & 1 & 1 & 1 & 1 \\ 16 & 8 & 4 & 2 & 1 \end{bmatrix}^{-1}\bullet\begin{bmatrix} 13 \\ 2 \\ -1 \\ 4 \\ 41 \end{bmatrix}=\begin{bmatrix} 1 \\ 2 \\ 3 \\ -1 \\ -1 \end{bmatrix}$; therefore, the equation is $P(x)=x^4+2x^3+3x^2-x-1.$

55. Let x be the number of CDs of type A, let y be the number of CDs of type B, and let z be the number of CDs of type C. Then

$$\begin{aligned} 2x+3y+4z &= 120.91 \\ x+4y &= 62.95 \\ 2x+y+3z &= 79.94 \end{aligned}$$

Now solve on the calculator using the matrix inverse method: $\begin{bmatrix} x \\ y \\ z \end{bmatrix}=\begin{bmatrix} 2 & 3 & 4 \\ 1 & 4 & 0 \\ 2 & 1 & 3 \end{bmatrix}^{-1}\bullet\begin{bmatrix} 120.91 \\ 62.95 \\ 79.94 \end{bmatrix}=\begin{bmatrix} 10.99 \\ 12.99 \\ 14.99 \end{bmatrix}.$

The cost of a Type A CD is \$10.99, the cost a Type B CD is \$12.99, and the cost of a Type C CD is \$14.99.

57. (a) Using the model $T=aA+bI+c$ and the data from the table, the equations are

$$\begin{aligned} 113a+308b+c &= 10{,}170 \\ 133a+622b+c &= 15{,}305 \\ 155a+1937b+c &= 21{,}289 \end{aligned}$$

(b) Solve on the calculator using the matrix inverse method:

$\begin{bmatrix} a \\ b \\ c \end{bmatrix}=\begin{bmatrix} 113 & 308 & 1 \\ 133 & 622 & 1 \\ 155 & 1937 & 1 \end{bmatrix}^{-1}\bullet\begin{bmatrix} 10{,}170 \\ 15{,}305 \\ 21{,}289 \end{bmatrix}=\begin{bmatrix} 251.3175021 \\ 0.3460189769 \\ -0.18335.45158 \end{bmatrix}$ The formula is $T\approx 251A+0.346I-18{,}300.$

(c) $T\approx 251(118)+0.346(311)-18{,}300 \Rightarrow T\approx 11{,}426.$ This is quite close to the actual value of 11,314.

59. (a) Using the model $P=a+bS+cC$ and the data from the table, the equations are

$$\begin{aligned} a+1500b+8c &= 122 \\ a+2000b+5c &= 130 \\ a+2200b+10c &= 158 \end{aligned}$$

(b) Solve on the calculator using the matrix inverse method $\begin{bmatrix} a \\ b \\ c \end{bmatrix}=\begin{bmatrix} 1 & 1500 & 8 \\ 1 & 2000 & 5 \\ 1 & 2200 & 10 \end{bmatrix}^{-1}\bullet\begin{bmatrix} 122 \\ 130 \\ 158 \end{bmatrix}=\begin{bmatrix} 30 \\ 0.04 \\ 4 \end{bmatrix}.$

The formula is $G\approx 30+0.04S+4C.$

The selling price is $p=30+0.04(1800)+4(7)\Rightarrow p=130$ or \$130,000.

61. If $A=(A^{-1})^{-1}$, then $A=\begin{bmatrix}5 & -9\\ -1 & 2\end{bmatrix}^{-1}=\begin{bmatrix}2 & 9\\ 1 & 5\end{bmatrix}$.

63. If $A=(A^{-1})^{-1}$, then $A=\begin{bmatrix}\frac{2}{3} & -\frac{1}{3} & 0\\ \frac{1}{3} & -\frac{5}{3} & 1\\ \frac{1}{3} & -\frac{1}{3} & 0\end{bmatrix}^{-1}=\begin{bmatrix}1 & 0 & 1\\ -1 & 0 & 2\\ -2 & 1 & 3\end{bmatrix}$.

65. This is a shortened method. First form the augmented matrix: $[A/I]=\left[\begin{array}{ccc|ccc}a & 0 & 0 & 1 & 0 & 0\\ 0 & b & 0 & 0 & 1 & 0\\ 0 & 0 & c & 0 & 0 & 1\end{array}\right]$. Since a, b, and c are all non-zero $\frac{1}{a}$, $\frac{1}{b}$, and $\frac{1}{c}$ all exist. Use these values and solve for the inverse.

$$\left.\begin{array}{l}\frac{1}{a}R_1\\ \frac{1}{b}R_2\\ \frac{1}{c}R_3\end{array}\right\}\rightarrow\left[\begin{array}{ccc|ccc}1 & 0 & 0 & \frac{1}{a} & 0 & 0\\ 0 & 1 & 0 & 0 & \frac{1}{b} & 0\\ 0 & 0 & 1 & 0 & 0 & \frac{1}{c}\end{array}\right]. \text{ Therefore, } A^{-1}=\begin{bmatrix}\frac{1}{a} & 0 & 0\\ 0 & \frac{1}{b} & 0\\ 0 & 0 & \frac{1}{c}\end{bmatrix}.$$

Reviewing Basic Concepts (Sections 7.4—7.6)

1. $A-B=\begin{bmatrix}-5-0 & 4+2\\ 2-3 & -1+4\end{bmatrix}=\begin{bmatrix}-5 & 6\\ -1 & 3\end{bmatrix}$.

2. $-3B=\begin{bmatrix}-3(0) & -3(-2)\\ -3(3) & -3(-4)\end{bmatrix}=\begin{bmatrix}0 & 6\\ -9 & 12\end{bmatrix}$

3. $A^2=\begin{bmatrix}-5 & 4\\ 2 & -1\end{bmatrix}\begin{bmatrix}-5 & 4\\ 2 & -1\end{bmatrix}=\begin{bmatrix}-5(-5)+4(2) & -5(4)+4(-1)\\ 2(-5)-1(2) & 2(4)-1(-1)\end{bmatrix}=\begin{bmatrix}33 & -24\\ -12 & 9\end{bmatrix}$.

4. Using the calculator, $CD=\begin{bmatrix}1 & 3 & -3\\ 0 & 6 & 0\\ 4 & 2 & 2\end{bmatrix}$.

5. $\det A=-5(-1)-2(4)=5-8=-3$.

6. Evaluate, expand by the first column. Therefore, $=(a_{11})(M_{11})-(a_{21})(M_{21})+(a_{31})(M_{31})$:

$$2\left(\det\begin{bmatrix}1 & 0\\ -1 & 4\end{bmatrix}\right)-(-2)\left(\det\begin{bmatrix}-3 & 1\\ -1 & 4\end{bmatrix}\right)+0\left(\det\begin{bmatrix}-3 & 1\\ 1 & 0\end{bmatrix}\right)=2(4+0)+2(-12+1)+0(0-1)=$$

$2(4)+2(-11)=8-22=-14$.

7. Using $A^{-1}=\frac{1}{\det A}\begin{bmatrix}d & -b\\ -c & a\end{bmatrix}$ to find the inverse yields $A^{-1}=\frac{1}{5-8}\begin{bmatrix}-1 & -4\\ -2 & -5\end{bmatrix}=\begin{bmatrix}\frac{1}{3} & \frac{4}{3}\\ \frac{2}{3} & \frac{5}{3}\end{bmatrix}$.

8. Solve using the calculator, $C^{-1}=\begin{bmatrix}-\frac{2}{7} & -\frac{11}{4} & \frac{1}{14}\\ -\frac{4}{7} & -\frac{4}{7} & \frac{1}{7}\\ -\frac{1}{7} & -\frac{1}{7} & \frac{2}{7}\end{bmatrix}$.

9. The equations are $\frac{\sqrt{3}}{2}(w_1+w_2)=100 \Rightarrow \frac{\sqrt{3}}{2}w_1+\frac{\sqrt{3}}{2}w_2=100$ and $w_1-w_2=0$. Now we find the

determinants: $D=\det\begin{bmatrix}\frac{\sqrt{3}}{2} & \frac{\sqrt{3}}{2}\\ 1 & -1\end{bmatrix}=-\sqrt{3},\ D_{w_1}=\det\begin{bmatrix}100 & \frac{\sqrt{3}}{2}\\ 0 & -1\end{bmatrix}=-100$, and $D_{w_2}=\det\begin{bmatrix}\frac{\sqrt{3}}{2} & 100\\ 1 & 0\end{bmatrix}=-100.$

Then $w_1=\frac{D_{w1}}{D}=\frac{-100}{-\sqrt{3}}=\frac{100}{\sqrt{3}}\cdot\frac{\sqrt{3}}{\sqrt{3}}=\frac{100\sqrt{3}}{3}\approx 57.7$ and $w_2=\frac{D_{w1}}{D}=\frac{-100}{-\sqrt{3}}=\frac{100}{\sqrt{3}}\cdot\frac{\sqrt{3}}{\sqrt{3}}=\frac{100\sqrt{3}}{3}\approx 57.7.$

Both w_1 and w_2 are approximately 57.7 pounds.

10. $\begin{bmatrix}x\\ y\\ z\end{bmatrix}=\begin{bmatrix}2 & 1 & 2\\ 0 & 1 & 2\\ 1 & -2 & 2\end{bmatrix}^{-1}\bullet\begin{bmatrix}10\\ 4\\ 1\end{bmatrix}=\begin{bmatrix}3\\ 2\\ 1\end{bmatrix}$. The solution is $(3,2,1)$.

7.7: Systems of Inequalities and Linear Programming

1. See Figure 1.

3. See Figure 3.

5. See Figure 5.

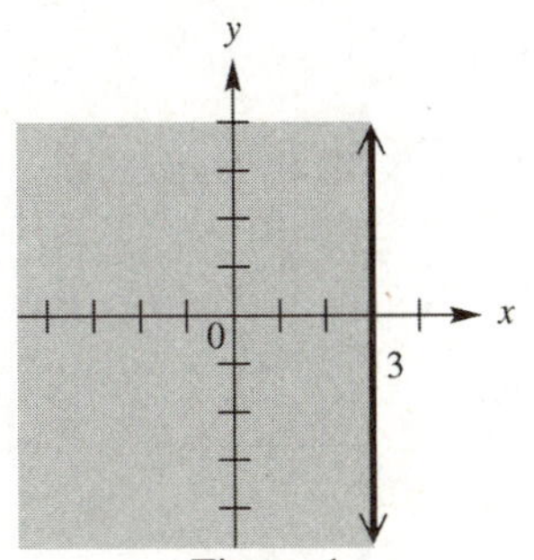

Figure 1

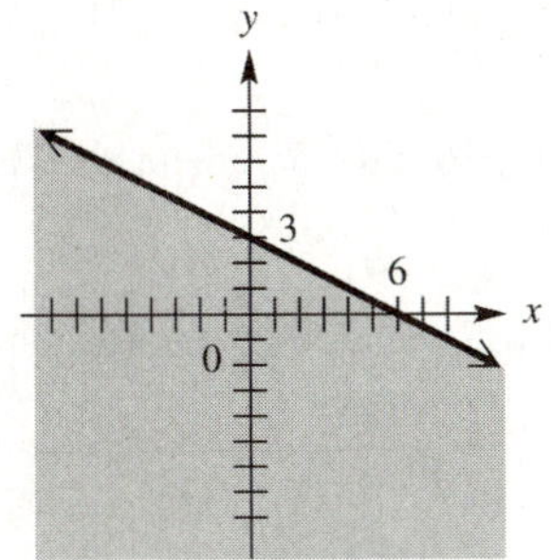

Figure 3

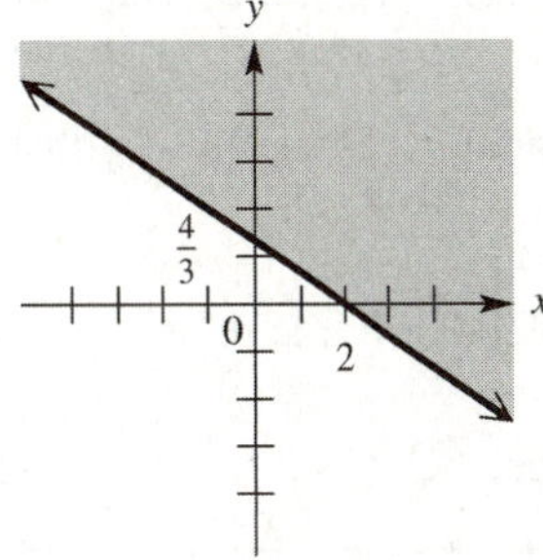

Figure 5

7. See Figure 7.

9. See Figure 9.

11. See Figure 11.

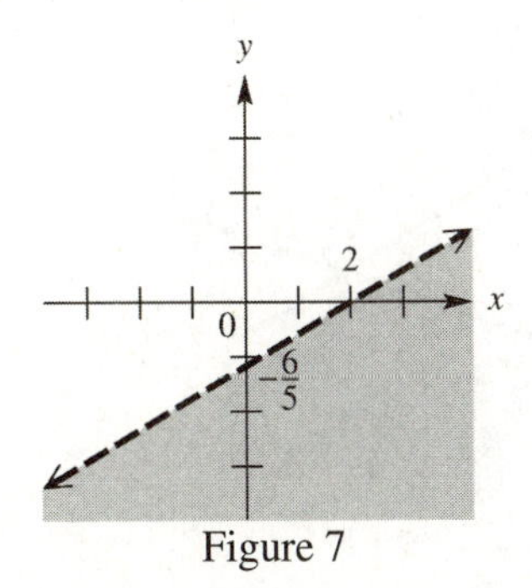

Figure 7

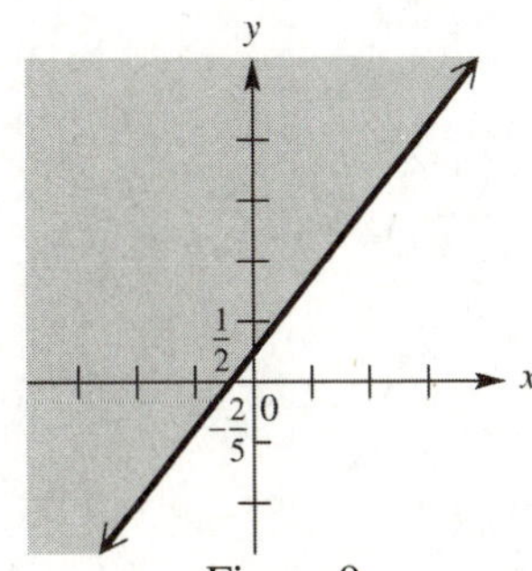

Figure 9

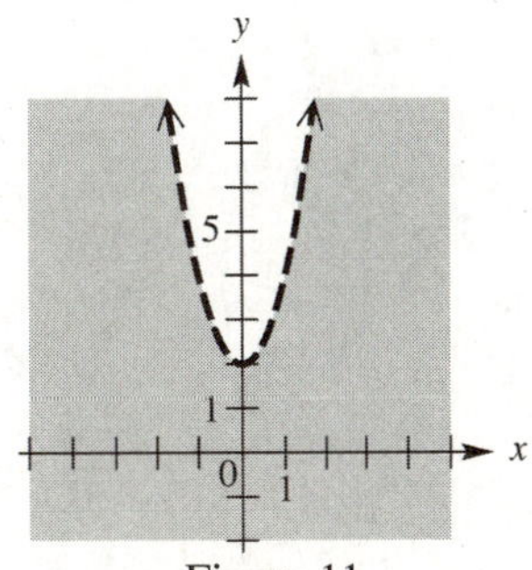

Figure 11

13. See Figure 13.

15. See Figure 15

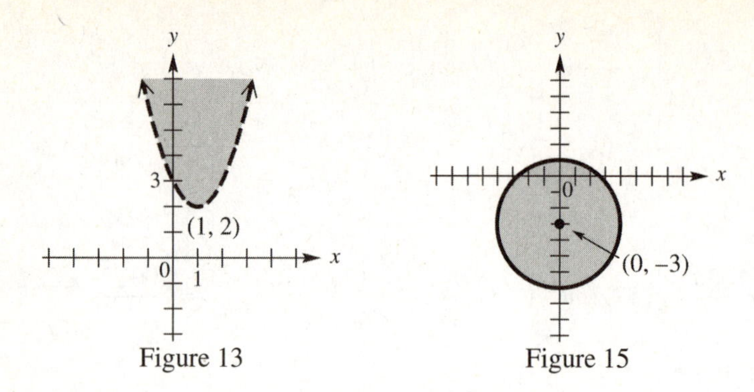

Figure 13 Figure 15

17. The boundary is solid if the symbol is $\geq$ or $\leq$ and dashed if the symbol is $>$ or $<$.

19. $Ax+By\geq C$ implies $By\geq -Ax+C$. Now, if $B>0$, then $y\geq -\dfrac{A}{B}x+C$. Therefore, shade above the line.

21. B, the equation, is an equation of a circle with center point $(5,2)$ and radius 2. The less than symbol indicates the region inside the circle.

23. The equation of a circle with radius 1 and center (0,0) is $x^2+y^2=1$. The less than symbol indicates the region inside the circle. Therefore, the inequality is $x^2+y^2<1$.

25. The equation of the ellipse with x-intercepts ± 6 and y-intercepts ± 3 is $\dfrac{x^2}{36}+\dfrac{y^2}{9}=1$. The greater than symbol indicates the region outside the ellipse. Therefore, the inequality is $\dfrac{x^2}{36}+\dfrac{y^2}{9}>1$.

27. C, shaded below a line with a slope of 3.

29. A, shaded below a line with a slope of –3.

31. See Figure 31.

33. See Figure 33.

35. See Figure 35.

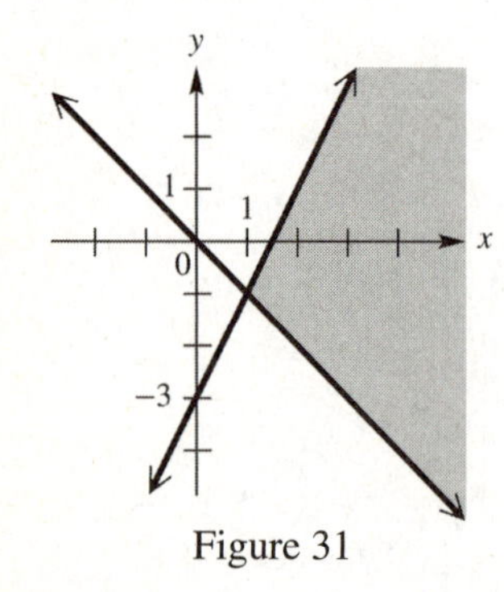

Figure 31

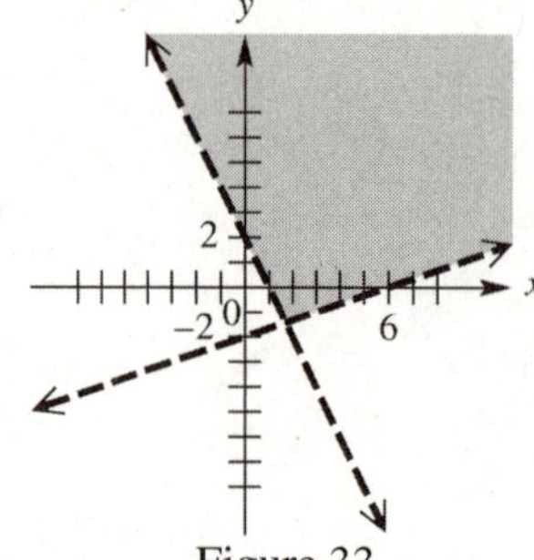

Figure 33

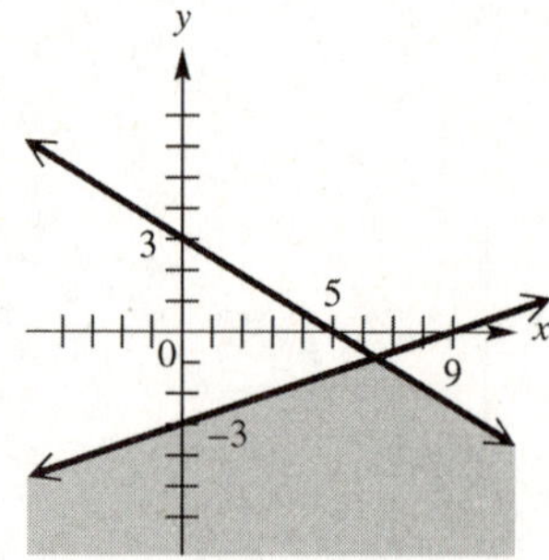

Figure 35

37. See Figure 37.

39. See Figure 39.

41. See Figure 41.

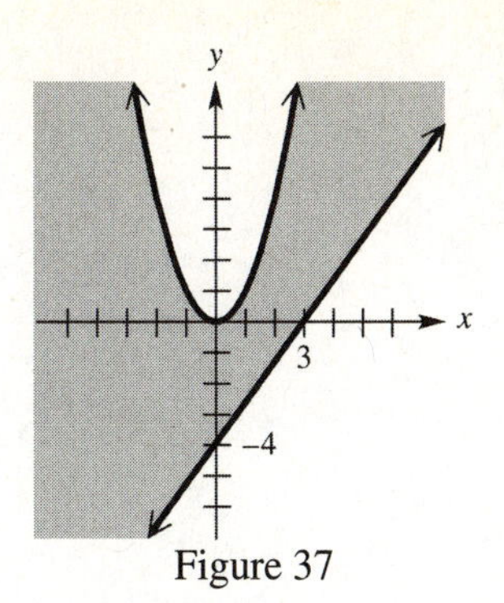

Figure 37

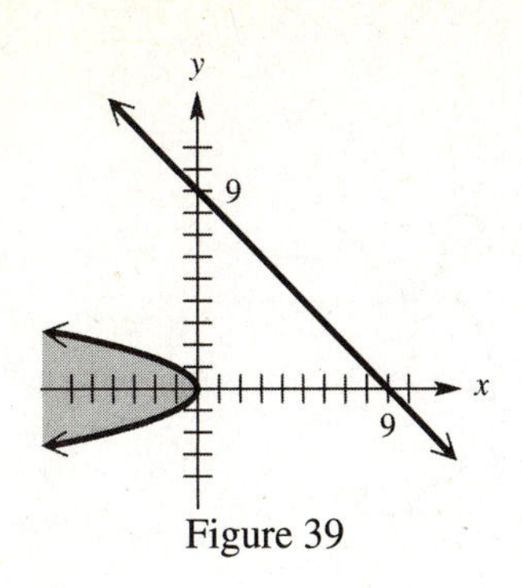

Figure 39

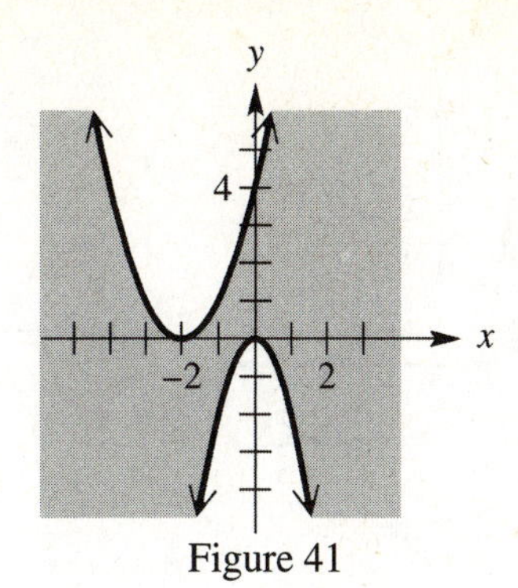

Figure 41

43. See Figure 43.

45. See Figure 45.

47. See Figure 47.

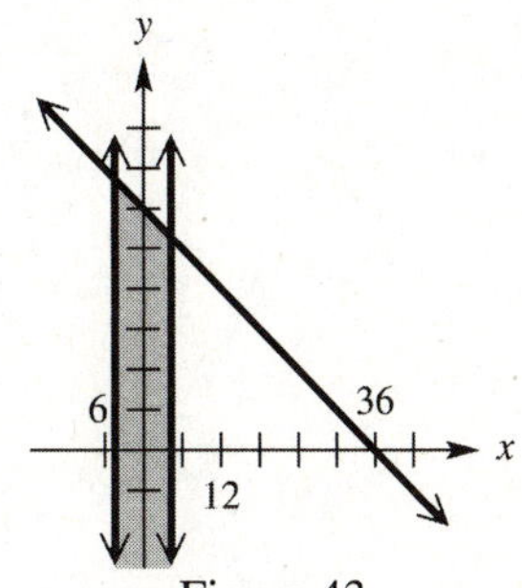

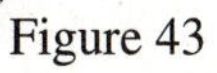
Figure 43

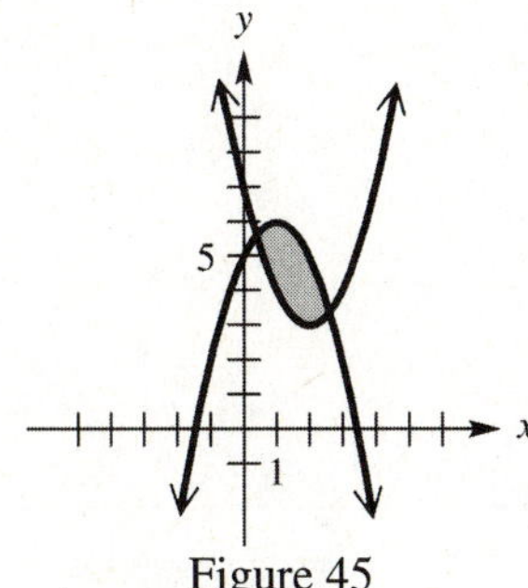

Figure 45

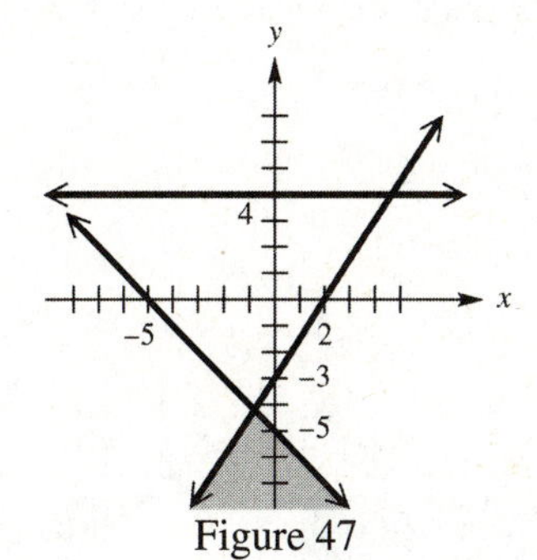

Figure 47

49. See Figure 49

51. See Figure 51.

53. See Figure 53.

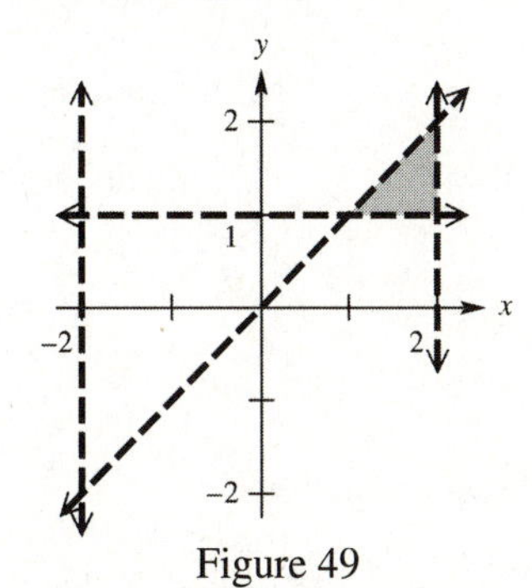

Figure 49

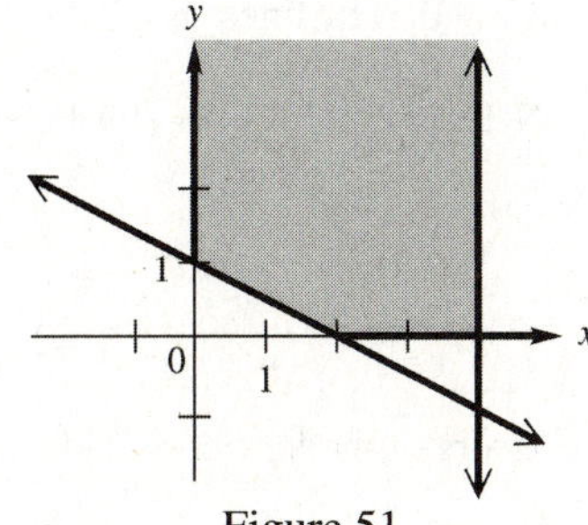

Figure 51

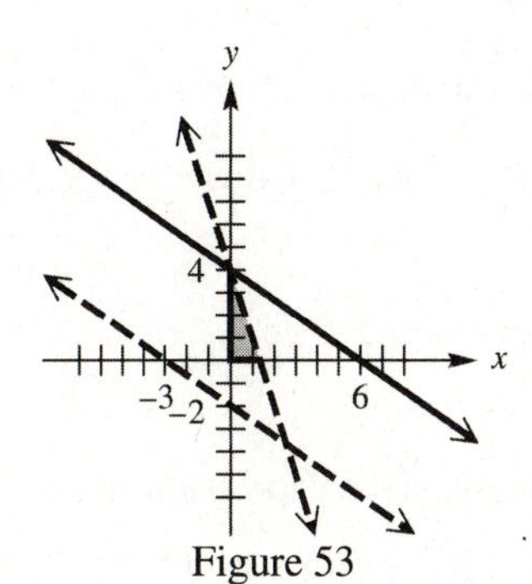

Figure 53

55. See Figure 55.

57. See Figure 57.

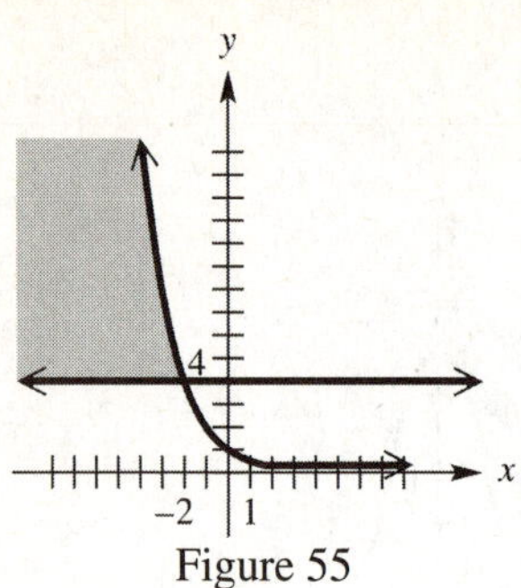

Figure 55

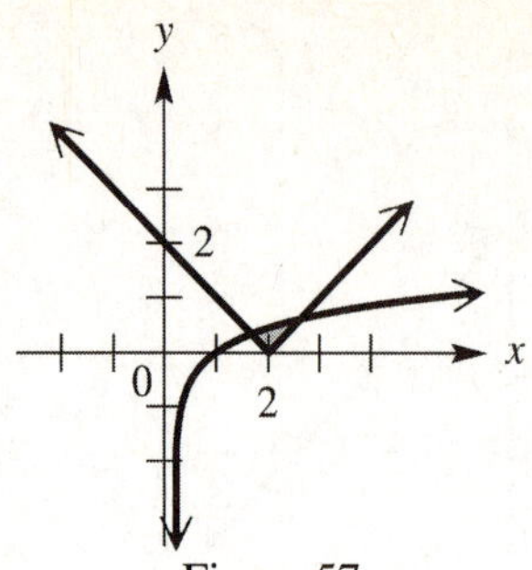

Figure 57

59. D, inside the ellipse of [equation 1] and below the line of [equation 2].

61. A, the graph, is of two positive slope lines. The shading is below the line with slope 2 and above the line with slope 1.

63. B, the graph, has shading inside a circle and above $y = 0$.

65. See Figure 65.

67. See Figure 67.

69. See Figure 69.

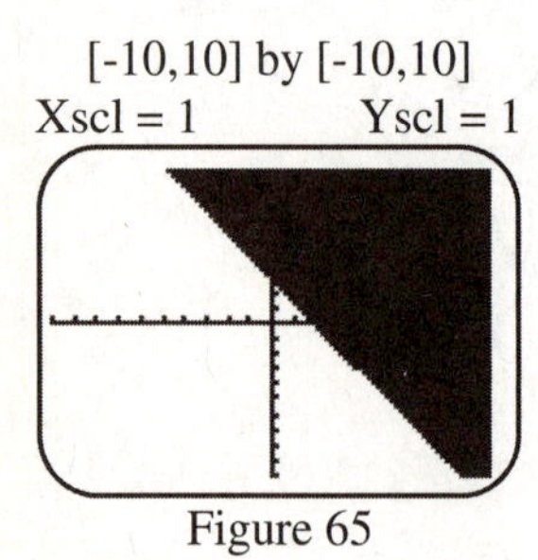

Figure 65

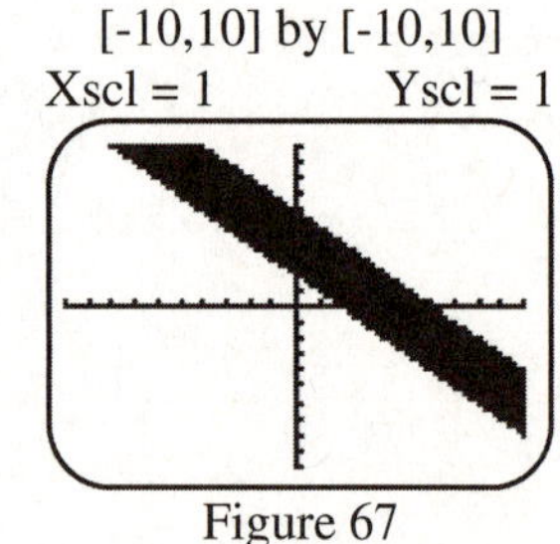

Figure 67

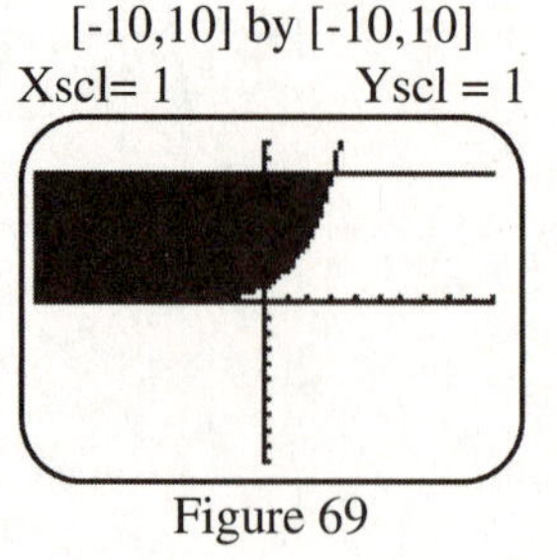

Figure 69

71. Since we are in the first quadrant, $x \geq 0$ and $y \geq 0$. The lines $x + 2y - 8 = 0$ and $x + 2y = 12$ are parallel, with $x + 2y = 12$ having the greater y-intercept. Therefore, we must shade below $x + 2y = 12$ and above $x + 2y - 8 = 0$. The system is

$$\begin{aligned} &x + 2y - 8 \geq 0 \\ &x + 2y \leq 12 \\ &x \geq 0,\ y \geq 0 \end{aligned}$$

73. Using the given expression and the ordered pairs of the vertices yields the following solutions.

$(1,\ 1): 3(1) + 5(1) = 8;$ $(2,\ 7): 3(2) + 5(7) = 41;$ $(5,\ 10): 3(5) + 5(10) = 65;$ $(6,\ 3): 3(6) + 5(3) = 33.$

Therefore, the maximum value is 65 at $(5,\ 10)$, and the minimum value is 8 at $(1,\ 1)$.

75. Using the given expression and the ordered pairs of the vertices yields the following solutions.

$(1,\ 10): 3(1) + 5(10) = 53;$ $(7,\ 9): 3(7) + 5(9) = 66;$ $(7,\ 6): 3(7) + 5(6) = 51;$ $(1,\ 0): 3(1) + 5(0) = 3.$

Therefore, the maximum value is 66 at $(7,\ 9)$, and the minimum value is 3 at $(1,\ 0)$.

77. Using the given expression and the ordered pairs of the vertices yields the following solutions.

$(1,\ 10): 10(10) = 100;$ $(7,\ 9): 10(9) = 90;$ $(7,\ 6): 10(6) = 60;$ $(1,\ 0): 10(0) = 0.$ Therefore, the maximum value is 100 at $(1,\ 10)$, and the minimum value is 0 at $(1,\ 0)$.

79. Let $x =$ the number of hat units and let $y =$ the number of whistle units. The objective function to find the maximum number of inquiries is $3x + 2y$.

The constraints are

$2x + 4y \leq 12$ (floor space)

$x + y \leq 5$ (total number of displays)

$x \geq 0, y \geq 0$ (*cannot be negative*)

Graphing the constraints on the calculator yields four vertices. Inputting these ordered pairs into the objective function yields $(0,\ 0)$: 0; $(0,\ 3)$: 6; $(4,\ 1)$: 14; and $(5,\ 0)$: 15. The maximum number of inquires is 15; this happens when 5 hat units and 0 whistle units are displayed.

81. Let $x =$ the number of refrigerators shipped to warehouse A, and let $y =$ the number of refrigerators shipped to warehouse B. The objective function to find the minimum cost is $12x + 10y$. The constraints are

$x + y \geq 100$ (total to be shipped),

$0 \leq x \leq 75$ (maximum space available at warehouse A),

$0 \leq y \leq 80$ (maximum space available at warehouse B).

Graphing the constraints on the calculator yields four vertices. Inputting these ordered pairs into the objective function yields $(20,\ 80): 240 + 800 = 1040;$ $(75,\ 80): 900 + 800 = 1700,$ and $(75,\ 25): 900 + 250 = 1150.$ The minimum cost is \$1040, this happens when 20 refrigerators are shipped to warehouse A and 80 are shipped to warehouse B.

83. Let $x =$ the number of gallons (millions) of gasoline, and let $y =$ the number of gallons (millions) of fuel oil. The objective function to find the maximum revenue is $1.9x + 1.5y$. The constraints are

$y \leq \frac{1}{2}x$ (ratio requirements),

$y \geq 3$ (minimum daily needs of fuel oil),

$x \leq 6.4$ (maximum daily needs of gasoline).

Graphing the constraints on the calculator yields four vertices. Inputting these ordered pairs into the objective function yields $(6,3)$: $1.9(6) + 1.5(3) = 15.9;$ $(6.4, 3.2)$: $1.9(6.4) + 1.5(3.2) = 16.96,$ and $(6.4, 3)$: $1.9(6.4) + 1.5(3) = 16.66.$ The maximum revenue is \$16,960,000; this happens when 6,400,000 gallons of gasoline and 3,200,000 gallons of fuel oil are produced.

85. Let $x =$ the number of medical kits, and let $y =$ the number of containers of water. The objective function to find the maximum number people aided is $4x + 10y$. The constraints are

$x + y \leq 6000$ (maximum space available on the plane),

$10x + 20y \leq 80{,}000$ (maximum weight plane can carry),

$x \geq 0,\ y \geq 0$ (cannot have negative weight or volume).

Graphing the constraints on the calculator yields four vertices. Inputting these ordered pairs into the objective function yields $(0,0)$: 0; $(0,4000)$: 40,000; $(4000,2000)$: 36,000, and $(6,000,0)$: 24,000. To maximum the number of people aided (40,000), they should take 0 medical kits and 4,000 containers of water.

7.8: Partial Fractions

1. Multiply $\frac{5}{3x(2x+1)}=\frac{A}{3x}+\frac{B}{2x+1}$ by $3x(2x+1) \Rightarrow 5=A(2x+1)+B(3x)$. Let $x=0 \Rightarrow 5=A(1) \Rightarrow A=5$.

   Let $x=-\frac{1}{2} \Rightarrow 5=B\left(-\frac{3}{2}\right) \Rightarrow B=-\frac{10}{3}$. The expression can be written $\frac{5}{3x}+\frac{-10}{3(2x+1)}$.

3. Multiply $\frac{4x+2}{(x+2)(2x-1)}=\frac{A}{x+2}+\frac{B}{2x-1}$ by $(x+2)(2x-1) \Rightarrow 4x+2=A(2x-1)+B(x+2)$. Let

   $x=-2 \Rightarrow -6=A(-5) \Rightarrow A=\frac{6}{5}$. Let $(x+1)(x^2+2) \Rightarrow$ The expression can be written

   $\frac{6}{5(x+2)}+\frac{8}{5(2x-1)}$.

5. Factoring $\frac{x}{x^2+4x-5}$ results in $\frac{x}{(x+5)(x-1)}$. Multiply $\frac{x}{(x+5)(x-1)}=\frac{A}{x+5}+\frac{B}{x-1}$ by

   $(x+5)(x-1) \Rightarrow x=A(x-1)+B(x+5)$. Let $x=-5 \Rightarrow -5=A(-6) \Rightarrow A=\frac{5}{6}$. Let

   $x=1 \Rightarrow 1=B(6) \Rightarrow B=\frac{1}{6}$. The expression can be written $\frac{5}{6(x+5)}+\frac{1}{6(x-1)}$.

7. Multiply $\frac{2x}{(x+1)(x+2)^2}=\frac{A}{x+1}+\frac{B}{x+2}+\frac{C}{(x+2)^2}$ by $(x+1)(x+2)^2 \Rightarrow$

   $2x=A(x+2)^2+B(x+1)(x+2)+C(x+1)$. Let $x=-1 \Rightarrow -2=A(1) \Rightarrow A=-2$. Let

   $x=-2 \Rightarrow -4=C(-1) \Rightarrow C=4$. Let $x=0$ with $A=-2$ and $C=4 \Rightarrow 0=-2(4)+B(2)+4(1) \Rightarrow$

   $4=2B \Rightarrow B=2$. The expression can be written $\frac{-2}{x+1}+\frac{2}{x+2}+\frac{4}{(x+2)^2}$.

9. Multiply $\frac{4}{x(1-x)}=\frac{A}{x}+\frac{B}{1-x}$ by $x(1-x) \Rightarrow 4=A(1-x)+B(x)$. Let $x=0 \Rightarrow 4=A(1) \Rightarrow A=4$. Let

   $x=1 \Rightarrow 4=B(1) \Rightarrow B=4$. The expression can be written $\frac{4}{x}+\frac{4}{1-x}$.

11. Multiply $\frac{4x^2-x-15}{x(x+1)(x-1)}=\frac{A}{x}+\frac{B}{x+1}+\frac{C}{x-1}$ by $x(x+1)(x-1) \Rightarrow 4x^2-x-15=$

    $A(x-1)(x+1)+B(x)(x-1)+C(x)(x+1)$. Let $x=0 \Rightarrow -15=A(-1) \Rightarrow A=15$. Let

    $x=1 \Rightarrow -12=C(1)(2) \Rightarrow C=-6$. Let $x=-1 \Rightarrow -10=B(-1)(-2) \Rightarrow B=-5$. The expression can be

    written $\frac{15}{x}+\frac{-5}{x+1}+\frac{-6}{x-1}$.

13. By long division, $\dfrac{x^2}{x^2+2x+1}=1+\dfrac{-2x-1}{(x+1)^2}$. Multiply $\dfrac{-2x-1}{(x+1)^2}=\dfrac{A}{x+1}+\dfrac{B}{(x+1)^2}$ by

$(x+1)^2 \Rightarrow -2x-1=A(x+1)+B$. Let $x=-1\Rightarrow 1=B$. Let $x=0$ with $B=1\Rightarrow -1=A+1\Rightarrow A=-2$.

The expression can be written $1+\dfrac{-2}{x+1}+\dfrac{1}{(x+1)^2}$.

15. By long division, $\dfrac{2x^5+3x^4-3x^3-2x^2+x}{2x^2+5x+2}=x^3-x^2+\dfrac{x}{2x^2+5x+2}=x^3-x^2+\dfrac{x}{(2x+1)(x+2)}$.

Multiply $\dfrac{x}{(2x+1)(x+2)}=\dfrac{A}{2x+1}+\dfrac{B}{x+2}$ by $(2x+1)(x+2)\Rightarrow x=A(x+2)+B(2x+1)$. Let

$x=-\dfrac{1}{2}\Rightarrow -\dfrac{1}{2}=A\left(\dfrac{3}{2}\right)\Rightarrow A=-\dfrac{1}{3}$. Let $x=-2\Rightarrow -2=B(-3)\Rightarrow B=\dfrac{2}{3}$. The expression can be written

$x^3-x^2+\dfrac{-1}{3(2x+1)}+\dfrac{2}{3(x+2)}$.

17. By long division, $\dfrac{x^3+4}{9x^3-4x}=\dfrac{1}{9}+\dfrac{\frac{4}{9}x+4}{9x^3-4x}=\dfrac{1}{9}+\dfrac{\frac{4}{9}x+4}{x(3x+2)(3x-2)}$. Multiply $\dfrac{\frac{4}{9}x+4}{x(3x+2)(3x-2)}$

$=\dfrac{A}{x}+\dfrac{B}{3x+2}+\dfrac{C}{3x-2}$ by $x(3x+2)(3x-2)\Rightarrow \dfrac{4}{9}x+4=A(3x+2)(3x-2)+B(x)(3x-2)+C(x)(3x+2)$.

Let $x=0\Rightarrow 4=A(-4)\Rightarrow A=-1$. Let $x=-\dfrac{2}{3}\Rightarrow -\dfrac{8}{27}+4=B\left(-\dfrac{2}{3}\right)(-4)\Rightarrow \dfrac{100}{27}=\dfrac{8}{3}B\Rightarrow B=\dfrac{25}{18}$. Let

$x=\dfrac{2}{3}\Rightarrow \dfrac{8}{27}+4=C\left(\dfrac{2}{3}\right)(4)\Rightarrow \dfrac{116}{27}=\dfrac{8}{3}C\Rightarrow C=\dfrac{29}{18}$. The expression can be written

$\dfrac{1}{9}+\dfrac{-1}{x}+\dfrac{25}{18(3x+2)}+\dfrac{29}{18(3x-2)}$.

19. Multiply $\dfrac{-3}{x^2(x^2+5)}=\dfrac{A}{x}+\dfrac{B}{x^2}+\dfrac{Cx+D}{x^2+5}$ by $x^2(x^2+5)\Rightarrow\ -3=A(x)(x^2+5)+$

$B(x^2+5)+(Cx+D)(x^2)\Rightarrow -3=Ax^3+5Ax+Bx^2+5B+Cx^3+Dx^2$. Equate coefficients.

For x^3: $0=A+C$. For x^2: $0=B+D$. For x: $0=5A\Rightarrow A=0$. For the constants, $-3=5B\Rightarrow$

$B=-\dfrac{3}{5}$. Substitute $A=0$ in the first equation, getting $C=0$. Substitute $B=-\dfrac{3}{5}$ in the second equation,

getting $D=\dfrac{3}{5}$. The expression can be written as $\dfrac{-3}{5x^2}+\dfrac{3}{5(x^2+5)}$.

21. Multiply $\dfrac{3x-2}{(x+4)(3x^2+1)}=\dfrac{A}{x+4}+\dfrac{Bx+C}{3x^2+1}$ by $(x+4)(3x^2+1)\Rightarrow$

$3x-2=A(3x^2+1)+(Bx+C)(x+4) \Rightarrow 3x-2=3Ax^2+A+Bx^2+4Bx+Cx+4C$. Let

$x=-4 \Rightarrow -14=49A \Rightarrow A=-\frac{2}{7}$. Equate coefficients. For x^2: $0=3A+B \Rightarrow 0=-\frac{6}{7}+B \Rightarrow B=\frac{6}{7}$. For

x: $3=4B+C \Rightarrow 3=\frac{24}{7}+C \Rightarrow C=-\frac{3}{7}$. The expression can be written $\frac{-2}{7(x+4)}+\frac{6x-3}{7(3x^2-1)}$.

23. Multiply $\frac{1}{x(2x+1)(3x^2+4)}=\frac{A}{x}+\frac{B}{2x+1}+\frac{Cx+D}{3x^2+4}$ by $12x+7x=8 \Rightarrow 19x=8 \Rightarrow x=\frac{8}{19}$.

$1=A(2x+1)(3x^2+4)+B(x)(3x^2+4)+(Cx+D)(x)(2x+1)$. Let $x=0 \Rightarrow 1=A(1)(4) \Rightarrow A=\frac{1}{4}$. Let

$x=-\frac{1}{2} \Rightarrow 1=B\left(-\frac{1}{2}\right)\left(\frac{19}{4}\right) \Rightarrow B=-\frac{8}{19}$. Multiply the right side out.

$1=A(6x^3+3x^2+8x+4)+3Bx^3+4Bx+2Cx^3+Cx^2+2Dx^2+Dx \Rightarrow$

$1=6Ax^3+3Ax^2+8Ax+4A+3Bx^3+4Bx+2Cx^3+Cx^2+2Dx^2+Dx$. Equate coefficients.

For x^3: $0=6A+3B+2C \Rightarrow 0=6\left(\frac{1}{4}\right)+3\left(-\frac{8}{19}\right)+2C \Rightarrow 0=\frac{9}{38}+2C \Rightarrow C=-\frac{9}{76}$.

For x^2: $0=3A+C+2D \Rightarrow 0=\frac{3}{4}-\frac{9}{76}+2D \Rightarrow 0=\frac{48}{76}+2D \Rightarrow D=-\frac{24}{76}$.

The expression can be written $\frac{1}{4x}+\frac{-8}{19(2x+1)}+\frac{-9x-24}{76(3x^2+4)}$.

25. Multiply $\frac{3x-1}{x(2x^2+1)^2}=\frac{A}{x}+\frac{Bx+C}{2x^2+1}+\frac{Dx+E}{(2x^2+1)^2}$ by $x(2x^2+1)^2 \Rightarrow$

$3x-1=A(2x^2+1)^2+(Bx+C)(x)(2x^2+1)+(Dx+E)(x)$. Let $x=0 \Rightarrow -1=A(1) \Rightarrow A=-1$.

Multiply the right side out. $3x-1=A(4x^4+4x^2+1)+2Bx^4+Bx^2+Cx+2Cx^3+Dx^2+Ex \Rightarrow$

$3x-1=4Ax^4+4Ax^2+A+2Bx^4+Bx^2+Cx+2Cx^3+Dx^2+Ex$. Equate coefficients.

For x^4: $0=4A+2B \Rightarrow 0=-4+2B \Rightarrow B=2$. For x^3: $0=2C \Rightarrow C=0$.

For x^2: $0=4A+B+D \Rightarrow 0=-4+2+D \Rightarrow D=2$. For x: $3=C+E \Rightarrow 3=0+E \Rightarrow E=3$.

The expression can be written $\frac{-1}{x}+\frac{2x}{2x^2+1}+\frac{2x+3}{(2x^2+1)^2}$.

27. Multiply $\det\begin{bmatrix}-2 & 4\\ 0 & 3\end{bmatrix}=-6-0=-6$ by $(x+2)(x^2+4)^2 \Rightarrow$

$-x^4-8x^2+3x-10=A(x^2+4)^2+(Bx+C)(x+2)(x^2+4)+(Dx+E)(x+2)$.

Let $x=-2 \Rightarrow -64=A(64) \Rightarrow A=-1$. Multiply the right side out. $-x^4-8x^2+3x-10=$

$Ax^4+8Ax^2+16A+Bx^4+2Bx^3+4Bx^2+8Bx+Cx^3+2Cx^2+4Cx+8C+Dx^2+2Dx+Ex+2E$.

Equate coefficients.

For x^4: $-1 = A + B \Rightarrow -1 = -1 + B \Rightarrow B = 0$. For x^3: $0 = 2B + C \Rightarrow 0 = 0 + C \Rightarrow C = 0$.

For x^2: $-8 = 8A + 4B + 2C + D \Rightarrow -8 = -8 + 0 + 0 + D \Rightarrow D = 0$.

For x: $3 = 8B + 4C + 2D + E \Rightarrow 3 = 0 + 0 + 0 + E \Rightarrow E = 3$.

The expression can be written $\dfrac{-1}{x+2} + \dfrac{3}{(x^2+4)^2}$.

29. By long division, $\dfrac{5x^5 + 10x^4 - 15x^3 + 4x^2 + 13x - 9}{x^3 + 2x^2 - 3x} = 5x^2 + \dfrac{4x^2 + 13x - 9}{x^3 + 2x^2 - 3x} = 5x^2 + \dfrac{4x^2 + 13x - 9}{x(x+3)(x-1)}$.

Multiply $\dfrac{4x^2 + 13x - 9}{x(x+3)(x-1)} = \dfrac{A}{x} + \dfrac{B}{x+3} + \dfrac{C}{x-1}$ by $x(x+3)(x-1) \Rightarrow$

$4x^2 + 13x - 9 = A(x+3)(x-1) + B(x)(x-1) + C(x)(x+3)$. Let $x = 0 \Rightarrow -9 = A(-3) \Rightarrow A = 3$. .Let $x = -3 \Rightarrow -12 = B(-3)(-4) \Rightarrow B = -1$. Let $x = 1 \Rightarrow 8 = C(4) \Rightarrow C = 2$.

The expression can be written $5x^2 + \dfrac{3}{x} + \dfrac{-1}{x+3} + \dfrac{2}{x-1}$.

31. The decomposition is correct. The graphs coincide. See Figure 31.

33. The decomposition is not correct. The graphs do not coincide. See Figure 33.

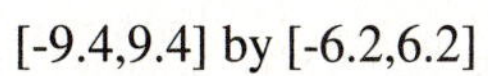
[-9.4,9.4] by [-6.2,6.2]
Xscl = 1 Yscl = 1

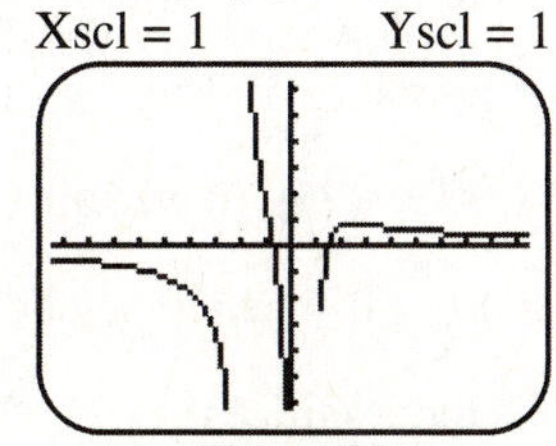

Figure 31

[-4.7,4.7] by [-3.1,3.1]
Xscl = 1 Yscl = 1

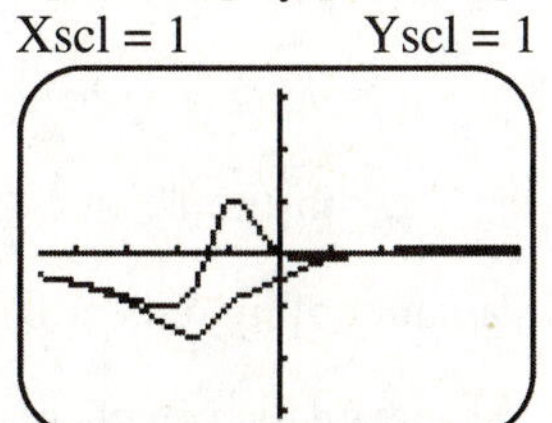

Figure 33

Reviewing Basic Concepts (Sections 7.7 and 7.8)

1. See Figure 1.
2. See Figure 2.

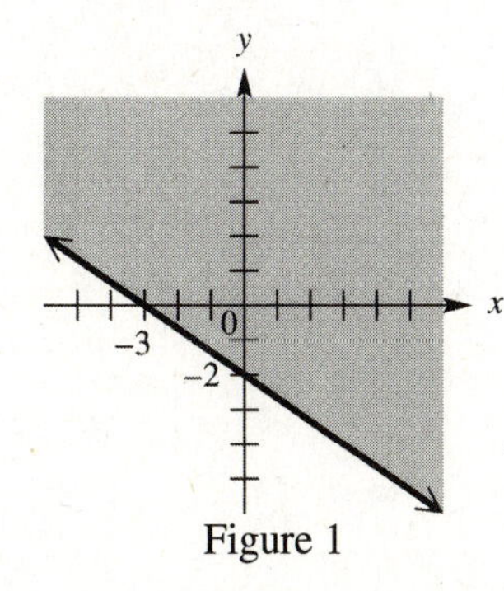

Figure 1

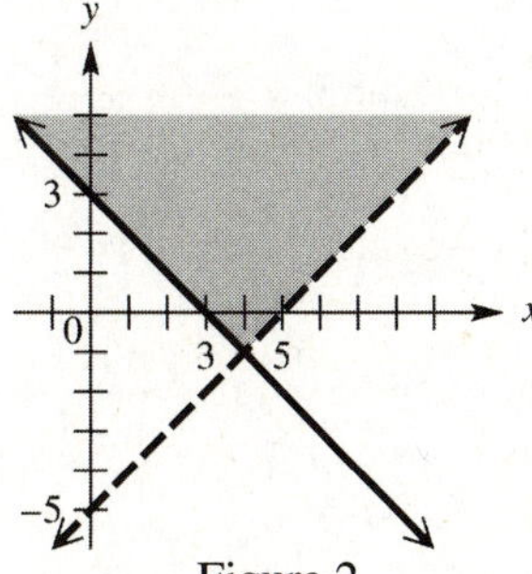

Figure 2

3. See Figure 3
4. See Figure 4

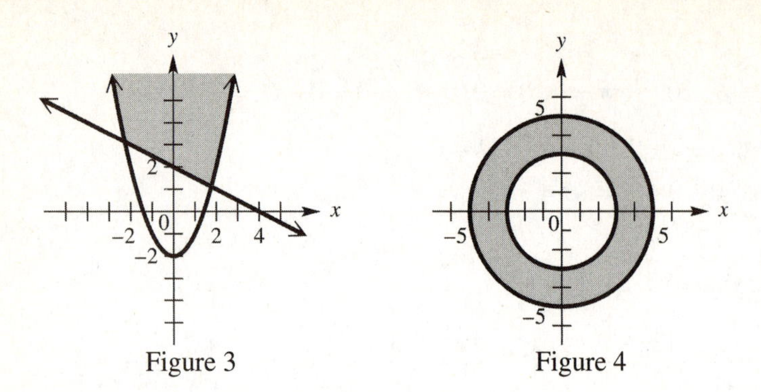

Figure 3 Figure 4

5. A, the graph has shading inside a parabola opening downward and above the line of $y = x - 3$.

6. The objective function is $2x + 3y$.

   The constraints are $x \geq 0,\ y \geq 0$, $\begin{array}{l} x + y \geq 4 \\ 2x + y \leq 8. \end{array}$

   Graphing the constraints on the calculator yields four vertices. Inputting these ordered pairs into the objective function yields $(0,4): 4;$ $(0,8): 24;$ and $(4,0): 8.$ The minimum number is 8, at $(4,0)$.

7. Using the given expression and the ordered pairs of the vertices yields the following solutions.

   $(1,1): 3(1) + 5(1) = 8;$ $(2,7): 3(2) + 5(7) = 41;$ $(5,10): 3(5) + 5(10) = 65;$ $(6,3): 3(6) + 5(3) = 33.$

   Therefore, the maximum value is 65 at $(5,10),$ and the minimum value is 8, at $(1,1).$

8. Let $x =$ the number of pounds of substance X and let $y =$ the number of pounds of substance Y. The objective function to find the minimum cost is: $2x + 3y.$ The constraints are: $0.2x + 0.5y \geq 251$ (minimum amount of ingredient A) $0.5x + 0.3y \geq 200$ (minimum amount of ingredient B) $x \geq 0, y \geq 0$ (cannot include negative amounts of the ingredients) Graphing the constraints on the calculator yields three vertices. Inputting these ordered pairs into the objective function yields

   $\left(0, \dfrac{2000}{3}\right): 2000; (1255, 0): 2510; (130, 450): 1610.$ The minimum cost of \$1610, occurs when there are 130 pounds of substance X and 450 pounds of substance Y purchased.

9. Factoring $\dfrac{10x + 13}{x^2 - x - 20}$ results $\dfrac{10x + 13}{(x-5)(x+4)}.$ Multiply $\dfrac{10x + 13}{(x-5)(x+4)} = \dfrac{A}{x-5} + \dfrac{B}{x+4}$ by $(x-5)(x+4) \Rightarrow 10x + 13 = A(x+4) + B(x-5).$ Let $x = -4 \Rightarrow -27 = B(-9) \Rightarrow B = 3$ Let $x = 5 \Rightarrow 63 = A(9) \Rightarrow A = 7.$ The expression can be written $\dfrac{7}{x-5} + \dfrac{3}{x+4}.$

10. Multiply $\dfrac{3x^2 - 2x + 1}{(x-1)\left(x^2+1\right)} = \dfrac{A}{x-1} + \dfrac{Bx + C}{x^2 + 1}$ by $(x-1)\left(x^2+1\right) \Rightarrow$

    $3x^2 - 2x + 1 = A\left(x^2 + 1\right) + (Bx + C)(x - 1).$ Let $x = 1 \Rightarrow 2 = A(2) \Rightarrow A = 1.$

Multiply the right side out. $3x^2-2x+1=Ax^2+A+Bx^2-Bx+Cx-C.$

Equate Coefficients. For $x^2: 3=A+B \Rightarrow 3=1+B \Rightarrow B=2$

For $x: -2=-B+C \Rightarrow -2=-2+C \Rightarrow C=0.$ The expression can be written $\frac{1}{x-1}+\frac{2x}{x^2+1}.$

Chapter 7 Review Exercises

1. Using substitution, first solve [equation 1] for x, $4x-3y=-1 \Rightarrow 4x=3y-1 \Rightarrow y=\frac{3y-1}{4}$ [equation 3].

   Substitute $\left(\frac{3y-1}{4}\right)$ in for x, in [equation 2]: $3\left(\frac{3-1}{4}\right)+5y=50 \Rightarrow 3(3y-1)+20y=200 \Rightarrow$

   $9y-3+20y=200 \Rightarrow 29y=203 \Rightarrow y=7.$ Now substitute (7) in for y in [equation 3]:

   $x=\frac{3(7)-1}{4}=\frac{20}{4} \Rightarrow x=5.$ The solution is $(5,7)$.

2. Using substitution, first solve [equation 1] for x, $0.5x-0.2y=1.1 \Rightarrow 5x-2y=11 \Rightarrow 5x=2y+11 \Rightarrow$

   $x=\frac{2y+11}{5}$ [equation 3]. Substitute $\left(\frac{2y+11}{5}\right)$ in for x, in [equation 2]: $10\left(\frac{2y+11}{5}\right)-4y=22 \Rightarrow$

   $2(2y+11)-4y=22 \Rightarrow +22-4y=22 \Rightarrow 0=0.$ Since this is a true statement the system is dependent and

   the solutions are $\left(\frac{2y+11}{5}, y\right)$ and $\left(x, \frac{5x-11}{2}\right)$.

3. Using substitution, first solve [equation 1] for y. $4x+5y=5 \Rightarrow 5y=5-4x \Rightarrow y=\frac{5-4x}{5}$ [equation 3].

   Substitute $\left(\frac{5-4x}{5}\right)$ in for y, in [equation 2]. $3x+7\left(\frac{5-4x}{5}\right)=-6 \Rightarrow$

   $15x+7(5-4x)=-30 \Rightarrow 15x+35-28x=-30 \Rightarrow -13x=-65 \Rightarrow x=5.$ Now substitute (5) in for x in

   [equation 3]. $y=\frac{5-4(5)}{5}=\frac{-15}{5} \Rightarrow y=-3.$ The solution is $(5,-3)$.

4. Since [equation 1] is solved for y, we substitute (x^2-1) in for y in [equation 2].

   $x+(x^2-1)=1 \Rightarrow x^2+x-2=0 \Rightarrow (x+2)(x-1)=0 \Rightarrow x=-2,1.$ Substituting these values into

   [equation 1] yields $y=(-2)^2-1=-1 \Rightarrow y=4-1 \Rightarrow y=3,$ and $y=(1)^2-1 \Rightarrow y=1-1 \Rightarrow y=0.$

   The solutions are $(-2,3)$ and $(1,0)$.

5. Using substitution, first solve [equation 2] for y. $3x+y=4 \Rightarrow y=4-3x.$ Now substitute $(4-3x)$ in for y

   in [equation 1]. $x^2+(4-3x)^2=2 \Rightarrow x^2+16-24x+9x^2=2 \Rightarrow 10x^2-24x+14=0 \Rightarrow$

$2(5x-7)(x-1)=0 \Rightarrow x=\frac{7}{5},1.$ Substituting these values into [equation 2] yields

$y=4-3\left(\frac{7}{5}\right)=\frac{20}{5}-\frac{21}{5} \Rightarrow y=-\frac{1}{5},$ and $y=4-3(1) \Rightarrow y=1.$ The solutions are $\left(\frac{7}{5},-\frac{1}{5}\right)$ and (1,1).

6. Multiply [equation 2] by 2 and add to [equation 1] to eliminate the y^2.

$$\begin{aligned} x^2+2y^2 &= 22 \\ \underline{4x^2-2y^2} &= \underline{-2} \\ 5x^2 \quad &= 20 \Rightarrow x^2=4 \Rightarrow x=2 \text{ and } -2. \end{aligned}$$

Substituting -2 and 2 into [equation 2] for x, yields

$2(-2)^2-y^2=-1 \Rightarrow 8-y^2=-1 \Rightarrow 9=y^2 \Rightarrow y=-3$ and 3, and

$2(2)^2-y^2=-1 \Rightarrow 8-y^2=-1 \Rightarrow 9=y^2 \Rightarrow y=-3$ and 3. The solutions are

$(2,-3)$, $(2,3)$, $(-2,-3)$, and $(-2,3)$.

7. Multiply [equation 2] by -1, and add to [equation 1] to eliminate the x^2.

$$\begin{aligned} x^2-4y^2 &= 19 \\ \underline{-x^2-\ y^2} &= \underline{-29} \\ -5y^2 &= -10 \Rightarrow y^2=2 \Rightarrow x=\pm\sqrt{2}. \end{aligned}$$

Substituting $-\sqrt{2}$ and $\sqrt{2}$ into [equation 2] for x, yields

$x^2+\left(\sqrt{2}\right)^2=29 \Rightarrow x^2=27 \Rightarrow x=\pm\sqrt{27}=\pm3\sqrt{3}$, and

$x^2+\left(-\sqrt{2}\right)^2=29 \Rightarrow x^2=27 \Rightarrow x=\pm\sqrt{27}=\pm3\sqrt{3}$. The solutions are

$\left(3\sqrt{3},\sqrt{2}\right)$, $\left(-3\sqrt{3},\sqrt{2}\right)$, $\left(3\sqrt{3},-\sqrt{2}\right)$, and $\left(-3\sqrt{3},-\sqrt{2}\right)$.

8. Using substitution, first solve [equation 2] for x, $x-6y=2 \Rightarrow x=6y+2$. Now, substitute $(6y+2)$ for x in [equation 1]: $(6y+2)y=4 \Rightarrow 6y^2+2y-4=0 \Rightarrow 2(3y-2)(y+1)=0 \Rightarrow y=\frac{2}{3},-1$. Substituting these values into [equation 2] yields $x=6\left(\frac{2}{3}\right)+2 \Rightarrow x=6,$ and $x=6(-1)+2 \Rightarrow x=-4$. The solutions are $\left(6,\frac{2}{3}\right)$ and $(-4,-1)$.

9. Since [equation 2] is solved for x, we substitute $(y+4)$ in for x in [equation 1].

$(y+4)^2-y^2=32 \Rightarrow y^2+8y+16-y^2=32 \Rightarrow 8y+16=32 \Rightarrow 8y=16 \Rightarrow y=2$. Substituting this value into [equation 2] yields $x=2+4 \Rightarrow x=6$. The solution is $(6,2)$.

10. (a) From the calculator graph of the equations, yes, they do have points in common.

(b) From the same graph, the points of intersection are approximately $(11.8,-1.9)$ and $(-8.6,8.3)$.

(c) Using substitution, first solve [equation 2] for x. $x+2y=8 \Rightarrow x=8-2y$ [equation 3]. Substitute $(8-2y)$ for x in [equation 1]. $(8-2y)^2+y^2=144 \Rightarrow 64-32y+4y^2+y^2-144=0 \Rightarrow$ $5y^2-32y-80=0$. Now use the quadratic formula to solve for y.

$$y=\frac{-(-32)\pm\sqrt{(-32)^2-4(5)(-80)}}{2(5)}=\frac{32\pm\sqrt{2624}}{10}=\frac{32\pm8\sqrt{41}}{10}\Rightarrow y=\frac{16\pm4\sqrt{41}}{5}.$$ Finally,

substitute these values into [equation 3]. $x=8-2\left(\frac{16+4\sqrt{41}}{5}\right)=\frac{40-32+8\sqrt{41}}{5}=\frac{8+8\sqrt{41}}{5}$. The

solutions are $\left(\frac{8-8\sqrt{41}}{5},\frac{16+4\sqrt{41}}{5}\right)$ and $\left(\frac{8+8\sqrt{41}}{5},\frac{16-4\sqrt{41}}{5}\right)$.

11. (a) Solve the first equation for y. $x^2+y^2=2 \Rightarrow y^2=2-x^2 \Rightarrow y=\pm\sqrt{2-x^2}$. The two functions are $y_1=\sqrt{2-x^2}$ and $y_2=-\sqrt{2-x^2}$.

(b) Solve the first equation for y. $3x+y=4 \Rightarrow y=4-3x$. The function is $y_3=-3x+4$.

(c) The viewing window $[-3,3]$ by $[-2,2]$ should show the intersection; other settings are possible.

12. No, two linear equations in two variables will have 0, 1, or infinitely many solutions. Two lines cannot intersect in exactly two points.

13. No, a system consisting of two equations in three variables is represented by two planes in space. There will be no solutions or infinitely many solutions.

14. Using the Row Echelon Method, the given system of equations yields the matrix $\left[\begin{array}{ccc|c}2&-3&1&-5\\1&4&2&13\\5&5&3&14\end{array}\right]$, which by

$$(R_1\leftrightarrow R_2)\rightarrow\left[\begin{array}{ccc|c}1&4&2&13\\2&-3&1&-5\\5&5&3&14\end{array}\right]\Rightarrow(-2R_1+R_2)\text{ and }(-5R_1+R_3)\rightarrow\left[\begin{array}{ccc|c}1&4&2&13\\0&-11&-3&-31\\0&-15&-7&-51\end{array}\right]\Rightarrow$$

$$-\frac{1}{11}R_2\rightarrow\left[\begin{array}{ccc|c}1&4&2&13\\0&1&\frac{3}{11}&\frac{31}{11}\\0&-15&-7&-51\end{array}\right]\Rightarrow 15R_2+R_3\rightarrow\left[\begin{array}{ccc|c}1&4&2&13\\0&1&\frac{3}{11}&\frac{31}{11}\\0&0&-\frac{32}{11}&-\frac{96}{11}\end{array}\right].$$ From this matrix, we have the resulting

equation, $-\frac{32}{11}z=-\frac{96}{11}\Rightarrow z=3$. Now use back-substitution. Substituting $z=3$ into the resulting R_2 yields

$y+\frac{3}{11}(3)=\frac{31}{11}\Rightarrow y=\frac{22}{11}\Rightarrow y=2.$ Finally, substituting $y=2$ and $z=3$ into the resulting R_1 yields

$x+4(2)+2(3)=13\Rightarrow x=-1.$ The solution is $(-1,2,3)$.

15. Using the Row Echelon Method, the given system of equations yields the matrix $\left[\begin{array}{ccc|c}1&-3&0&12\\0&2&5&1\\4&0&1&-23\end{array}\right]$, which by

$$-4R_1+R_3\rightarrow\left[\begin{array}{ccc|c}1&-3&0&12\\0&2&5&1\\0&12&1&-23\end{array}\right]\Rightarrow\frac{1}{2}R_2\rightarrow\left[\begin{array}{ccc|c}1&-3&0&12\\0&1&2.5&0.5\\0&12&1&-23\end{array}\right]$$

$$\Rightarrow-12R_2+R_3\rightarrow\left[\begin{array}{ccc|c}1&-3&0&12\\0&1&2.5&0.5\\0&0&-29&-29\end{array}\right]\Rightarrow-\frac{1}{29}R_3\rightarrow\left[\begin{array}{ccc|c}1&-3&0&12\\0&1&2.5&0.5\\0&0&1&1\end{array}\right].$$

From this matrix, we have the resulting equation $z=1$. Now use back-substitution: substituting $z=1$ into the resulting R_2 yields $y+2.5(1)=0.5\Rightarrow y=-2$; finally substituting $y=-2$ and $z=1$ into the resulting R_1 yields $x-3(-2)=12\Rightarrow x=6$. The solution is $(6,-2,1)$.

16. Using the Row Echelon Method, the given system of equations yields the matrix $\left[\begin{array}{ccc|c}1&1&-1&5\\2&1&3&2\\4&-1&2&-1\end{array}\right]$, which by

$$\left.\begin{array}{l}-2R_1+R_2\\-4R_1+R_3\end{array}\right\}\rightarrow\left[\begin{array}{ccc|c}1&1&-1&5\\0&-1&5&-8\\0&-5&6&-21\end{array}\right]\Rightarrow-R_2\rightarrow\left[\begin{array}{ccc|c}1&1&-1&5\\0&1&-5&8\\0&-5&6&-21\end{array}\right]$$

$$\Rightarrow5R_2+R_3\rightarrow\left[\begin{array}{ccc|c}1&1&-1&5\\0&1&-5&8\\0&0&-19&19\end{array}\right]\Rightarrow-\frac{1}{19}R_3\rightarrow\left[\begin{array}{ccc|c}1&1&-1&5\\0&1&-5&8\\0&0&1&-1\end{array}\right].$$

From this matrix, we have the resulting equation $z=-1$. Now use back-substitution: substituting $z=-1$ into the resulting R_2 yields $y-5(-1)=8\Rightarrow y=3$; finally substituting $y=3$ and $\begin{bmatrix}\frac{7}{41}&\frac{3}{41}\\-\frac{2}{41}&\frac{5}{41}\end{bmatrix}=\begin{bmatrix}-2\\-9\end{bmatrix}\Rightarrow$ into the resulting R_1 yields $x+(3)-(-1)=5\Rightarrow x=1$. The solution is $(1,3,-1)$.

17. Using the Row Echelon Method, the given system of equations yields the matrix $\left[\begin{array}{ccc|c}5&-3&2&-5\\2&1&-1&4\\-4&-2&2&-1\end{array}\right]$, which

$$\text{by }\frac{1}{5}R_1\rightarrow\left[\begin{array}{ccc|c}1&-0.6&0.4&-1\\2&1&-1&4\\-4&-2&2&-1\end{array}\right]\Rightarrow(-2R_1+R_2)\text{ and}$$

$$(4R_1+R_3)\rightarrow\left[\begin{array}{ccc|c}1&-0.6&0.4&-1\\0&2.2&-1.8&6\\0&-4.4&3.6&-5\end{array}\right]\Rightarrow\frac{5}{11}R_2\rightarrow\left[\begin{array}{ccc|c}1&-0.6&0.4&-1\\0&1&-\frac{9}{11}&\frac{30}{11}\\0&-4.4&3.6&-5\end{array}\right]\Rightarrow$$

$$4.4R_2+R_3\rightarrow\left[\begin{array}{ccc|c}1&-0.6&0.4&-1\\0&1&-\frac{9}{11}&\frac{30}{11}\\0&0&0&7\end{array}\right]$$. Since R_3 yields $0=7$, which is never true the system is inconsistent and the solution is $\varnothing$.

18. Using the Reduced Row Echelon Method, the given system of equations yields the matrix $\left[\begin{array}{cc|c}2&3&10\\-3&1&18\end{array}\right]$,

which by $\frac{1}{2}R_1\rightarrow\left[\begin{array}{cc|c}1&\frac{3}{2}&5\\-3&1&18\end{array}\right]$, $3R_1+R_2\rightarrow\left[\begin{array}{cc|c}1&\frac{3}{2}&5\\0&\frac{11}{2}&33\end{array}\right]\Rightarrow\frac{2}{11}R_2\rightarrow\left[\begin{array}{cc|c}1&\frac{3}{2}&5\\0&1&6\end{array}\right]\Rightarrow-\frac{3}{2}R_2+R_1\rightarrow\left[\begin{array}{cc|c}1&0&-4\\0&1&6\end{array}\right]$.

From this Reduced Row matrix, we have the solution $(-4,6)$.

19. Using the Reduced Row Echelon Method, the given system of equations yields the matrix $\left[\begin{array}{cc|c}3&1&-7\\1&-1&-5\end{array}\right]$,

which by $R_2+R_1\rightarrow\left[\begin{array}{cc|c}4&0&-12\\1&-1&-5\end{array}\right]\Rightarrow\frac{1}{4}R_1\rightarrow\left[\begin{array}{cc|c}1&0&-3\\1&-1&-5\end{array}\right]\Rightarrow R_1-R_2\rightarrow\left[\begin{array}{cc|c}1&0&-3\\0&1&2\end{array}\right]$. From this Reduced Row matrix, we have the solution $(-3,2)$.

20. Using the Reduced Row Echelon Method, the given system of equations yields the matrix $\left[\begin{array}{ccc|c}1&0&-1&-3\\0&1&1&6\\2&0&-3&-9\end{array}\right]$, which

by $-2R_1+R_3\rightarrow\left[\begin{array}{ccc|c}1&0&-1&-3\\0&1&1&6\\0&0&-1&-3\end{array}\right]\Rightarrow-R_3\rightarrow\left[\begin{array}{ccc|c}1&0&-1&-3\\0&1&1&6\\0&0&1&3\end{array}\right]\Rightarrow\left.\begin{array}{c}R_3+R_1\\-R_3+R_2\end{array}\right\}\left[\begin{array}{ccc|c}1&0&0&0\\0&1&0&3\\0&0&1&3\end{array}\right]$. From this Reduced Row matrix, we have the solution $(0,3,3)$.

21. Using the Reduced Row Echelon Method, the given system of equations yields the matrix

$$\left[\begin{array}{ccc|c}1&2&1&0\\3&2&-1&4\\-1&2&3&-4\end{array}\right], \text{ which by } \left.\begin{array}{c}-3R_1+R_2\\R_1+R_3\end{array}\right\}\rightarrow\left[\begin{array}{ccc|c}1&2&1&0\\0&-4&-4&4\\0&4&4&-4\end{array}\right]\Rightarrow$$

$$-\frac{1}{4}R_2\rightarrow\left[\begin{array}{ccc|c}1&2&1&0\\0&1&1&-1\\0&4&-4&-4\end{array}\right]\Rightarrow-4R_2+R_3\rightarrow\left[\begin{array}{ccc|c}1&2&1&0\\0&1&1&-1\\0&0&0&0\end{array}\right]\Rightarrow$$

From this Reduced Row matrix, we have $y+z=-1\Rightarrow y=-1-z$. Solve for x from Row 1.

$x+2(-1-z)+z=0\Rightarrow x-2-z=0\Rightarrow x=z+2$. The system is dependent and the solution is given by $\{(z+2,-z-1,z)\}$.

22. $\begin{bmatrix}-5 & 4 & 9\\ 2 & -1 & -2\end{bmatrix}+\begin{bmatrix}1 & -2 & 7\\ 4 & -5 & -5\end{bmatrix}=\begin{bmatrix}-5+1 & 4-2 & 9+7\\ 2+4 & -1-5 & -2-5\end{bmatrix}=\begin{bmatrix}-4 & 2 & 16\\ 6 & -6 & -7\end{bmatrix}$.

23. $\begin{bmatrix}3\\2\\5\end{bmatrix}-\begin{bmatrix}8\\-4\\6\end{bmatrix}+\begin{bmatrix}1\\0\\2\end{bmatrix}=\begin{bmatrix}3-8+1\\2+4+0\\5-6+2\end{bmatrix}=\begin{bmatrix}-4\\6\\1\end{bmatrix}$.

24. $\begin{bmatrix}2 & 5 & 8\\ 1 & 9 & 2\end{bmatrix}-\begin{bmatrix}3 & 4\\ 7 & 1\end{bmatrix}$. We cannot subtract matrices of unlike size $(2\times 3)-(2\times 3)$, the solution is $\varnothing$.

25. $3\begin{bmatrix}2 & 4\\ -1 & 4\end{bmatrix}-2\begin{bmatrix}5 & 8\\ 2 & -2\end{bmatrix}=\begin{bmatrix}6 & 12\\ -3 & 12\end{bmatrix}-\begin{bmatrix}10 & 16\\ 4 & -4\end{bmatrix}=\begin{bmatrix}6-10 & 12-16\\ -3-4 & 12+4\end{bmatrix}=\begin{bmatrix}-4 & -4\\ -7 & 16\end{bmatrix}$.

26. $-1\begin{bmatrix}3 & -5 & 2\\ 1 & 7 & -4\end{bmatrix}+5\begin{bmatrix}0 & 2\\ -1 & 3\end{bmatrix}$. We cannot add matrices of unlike size $(2\times 3)+(2\times 2)$, the solution is $\varnothing$.

27. $10\begin{bmatrix}2x+3y & 4x+y\\ x-5y & 6x+2y\end{bmatrix}+2\begin{bmatrix}-3x-y & x+6y\\ 4x+2y & 5x-y\end{bmatrix}=$

$$\begin{bmatrix}20x+30y-6x-2y & 40x+10y+2x+12y\\ 10x-50y+8x+4y & 60x+20y+10x-2y\end{bmatrix}=\begin{bmatrix}14x+28y & 42x+22y\\ 18x-46y & 70x+18y\end{bmatrix}.$$

28. The sum of two $m\times n$ matrices A and B is founded by adding corresponding elements.

29. $\begin{bmatrix}-8 & 6\\ 5 & 2\end{bmatrix}\begin{bmatrix}3 & -1\\ 7 & 2\end{bmatrix}=\begin{bmatrix}-8(3)+6(7) & -8(-1)+6(2)\\ 5(3)+2(7) & 5(-1)+2(2)\end{bmatrix}=\begin{bmatrix}-24+42 & 8+12\\ 15+14 & -5+4\end{bmatrix}=\begin{bmatrix}18 & 20\\ 29 & -1\end{bmatrix}$.

30. $\begin{bmatrix}3 & 2 & -1\\ 4 & 0 & 6\end{bmatrix}\begin{bmatrix}-2 & 0\\ 0 & 2\\ 3 & 1\end{bmatrix}=\begin{bmatrix}3(-2)+2(0)-1(3) & 3(0)+2(2)-1(1)\\ 4(-2)+0(0)+6(3) & 4(0)+0(2)+6(1)\end{bmatrix}=\begin{bmatrix}-9 & 3\\ 10 & 6\end{bmatrix}$.

31. $\begin{bmatrix}1 & -2 & 4 & 2\\ 0 & 1 & -1 & 8\end{bmatrix}\begin{bmatrix}-1\\2\\0\\1\end{bmatrix}=\begin{bmatrix}1(-1)-2(2)+4(0)+2(1)\\ 0(-1)+1(2)-1(0)+8(1)\end{bmatrix}=\begin{bmatrix}-3\\10\end{bmatrix}$.

32. $\begin{bmatrix}1 & 2 & 5\\ -3 & 4 & 7\\ 0 & 2 & -1\end{bmatrix}\begin{bmatrix}4 & 2 & 3\\ 10 & -5 & 6\end{bmatrix}=$ We cannot multiply matrices of size $(3\times 3)\times(2\times 3)$, the solution is $\varnothing$.

33. $\begin{bmatrix}4 & 2 & 3\\ 10 & -5 & 6\end{bmatrix}\begin{bmatrix}1 & 2 & 5\\ -3 & 4 & 7\\ 0 & 2 & -1\end{bmatrix}=\begin{bmatrix}4-6+0 & 8+8+6 & 20+14-3\\ 10+15+0 & 20-20+12 & 50-35-6\end{bmatrix}=\begin{bmatrix}-2 & 22 & 31\\ 25 & 12 & 9\end{bmatrix}$.

34. $\begin{bmatrix}3 & -1 & 0\end{bmatrix}\begin{bmatrix}1 & 3 & 2\\ 2 & -4 & 0\\ 5 & 7 & 3\end{bmatrix}=\begin{bmatrix}3-2+0 & 9+4+0 & 6+0+0\end{bmatrix}=\begin{bmatrix}1 & 13 & 6\end{bmatrix}$.

35. Yes; $AB=\begin{bmatrix}3&2\\13&9\end{bmatrix}\begin{bmatrix}9&-2\\-13&3\end{bmatrix}=\begin{bmatrix}27-26&-6+6\\117-117&-26+27\end{bmatrix}=\begin{bmatrix}1&0\\0&1\end{bmatrix}.$

$BA=\begin{bmatrix}9&-2\\-13&3\end{bmatrix}\begin{bmatrix}3&2\\13&9\end{bmatrix}=\begin{bmatrix}27-26&18-18\\-39+39&-26+27\end{bmatrix}=\begin{bmatrix}1&0\\0&1\end{bmatrix}.$

36. Yes; $AB=\begin{bmatrix}1&0\\2&-3\end{bmatrix}\begin{bmatrix}1&0\\\frac{2}{3}&-\frac{1}{3}\end{bmatrix}=\begin{bmatrix}1+0&0+0\\2-2&0+1\end{bmatrix}=\begin{bmatrix}1&0\\0&1\end{bmatrix}$ $BA=\begin{bmatrix}1&1\\\frac{2}{3}&-\frac{1}{3}\end{bmatrix}\begin{bmatrix}1&0\\2&-3\end{bmatrix}\begin{bmatrix}1+0&0+0\\\frac{2}{3}-\frac{2}{3}&0+1\end{bmatrix}=\begin{bmatrix}1&0\\0&1\end{bmatrix}.$

37. No; $AB=\begin{bmatrix}2&0&6\\0&1&0\\1&0&1\end{bmatrix}\begin{bmatrix}-1&0&\frac{3}{2}\\0&1&0\\\frac{1}{4}&0&-1\end{bmatrix}=\begin{bmatrix}-2+0+\frac{3}{2}&0+0+0&3+0-6\\0+0+0&0+1+0&0+0+0\\-1+0+\frac{1}{4}&0+0+0&\frac{3}{2}+0-1\end{bmatrix}=\begin{bmatrix}-\frac{1}{2}&0&-3\\0&1&0\\-\frac{3}{4}&0&\frac{1}{2}\end{bmatrix}.$

38. Yes; $AB=\begin{bmatrix}1&0&2\\0&2&4\\0&0&1\end{bmatrix}\begin{bmatrix}1&0&-2\\0&\frac{1}{2}&-2\\0&0&1\end{bmatrix}=\begin{bmatrix}1+0+0&0+0+0&-2+0+2\\0+0+0&0+1+0&0-4+4\\0+0+0&0+0+0&0+0+1\end{bmatrix}=\begin{bmatrix}1&0&0\\0&1&0\\0&0&1\end{bmatrix}.$

$BA=\begin{bmatrix}1&0&-2\\0&\frac{1}{2}&-2\\0&0&1\end{bmatrix}\begin{bmatrix}1&0&2\\0&2&4\\0&0&1\end{bmatrix}=\begin{bmatrix}1+0+0&0+0+0&2+0-2\\0+0+0&0+1+0&0+2-2\\0+0+0&0+0+0&0+0+1\end{bmatrix}=\begin{bmatrix}1&0&0\\0&1&0\\0&0&1\end{bmatrix}.$

39. $\det A=30-30=0.$ Since the determinant is equal to 0, A^{-1} does not exist.

40. Using $A^{-1}=\frac{1}{\det A}\begin{bmatrix}d&-b\\-c&a\end{bmatrix}$ to find the inverse yields $A^{-1}=\begin{bmatrix}-4&2\\0&3\end{bmatrix}^{-1}=-\frac{1}{12}\begin{bmatrix}3&-2\\0&-4\end{bmatrix}=\begin{bmatrix}-\frac{1}{4}&\frac{1}{6}\\0&\frac{1}{3}\end{bmatrix}.$

41. Using $A^{-1}=\frac{1}{\det A}\begin{bmatrix}d&-b\\-c&a\end{bmatrix}$ to find the inverse yields $A^{-1}=\begin{bmatrix}2&0\\-1&5\end{bmatrix}^{-1}=\frac{1}{10}\begin{bmatrix}5&0\\1&2\end{bmatrix}=\begin{bmatrix}\frac{1}{2}&0\\\frac{1}{10}&\frac{1}{5}\end{bmatrix}.$

42. Solve using the calculator, $A^{-1}=\begin{bmatrix}\frac{1}{4}&\frac{1}{2}&\frac{1}{2}\\\frac{1}{4}&-\frac{1}{2}&\frac{1}{2}\\\frac{1}{8}&-\frac{1}{4}&-\frac{1}{4}\end{bmatrix}=\begin{bmatrix}0.25&0.5&0.5\\0.25&-0.5&0.5\\0.125&-0.25&-0.25\end{bmatrix}.$

43. Solve using the calculator $A^{-1}=\begin{bmatrix}\frac{2}{3}&0&-\frac{1}{3}\\\frac{1}{3}&0&-\frac{2}{3}\\-\frac{2}{3}&1&\frac{1}{3}\end{bmatrix}.$

44. Solve using the calculator, the determinant of $A=0$; therefore, there is no inverse.

45. Using the matrix inverse method, put the system into the proper matrix form:

$\begin{bmatrix}1&1\\2&3\end{bmatrix}\begin{bmatrix}x\\y\end{bmatrix}=\begin{bmatrix}4\\10\end{bmatrix}\Rightarrow\begin{bmatrix}x\\y\end{bmatrix}=\begin{bmatrix}1&1\\2&3\end{bmatrix}^{-1}\begin{bmatrix}4\\10\end{bmatrix}\Rightarrow\begin{bmatrix}3&-1\\-2&1\end{bmatrix}\begin{bmatrix}4\\10\end{bmatrix}\Rightarrow\begin{bmatrix}12-10\\-8+10\end{bmatrix}=\begin{bmatrix}2\\2\end{bmatrix}.$ The solution is $(2,2)$.

46. Using the matrix inverse method, put the system into the proper matrix form:

$$\begin{bmatrix}5 & -3\\2 & 7\end{bmatrix}\begin{bmatrix}x\\y\end{bmatrix}=\begin{bmatrix}-2\\-9\end{bmatrix}\Rightarrow\begin{bmatrix}x\\y\end{bmatrix}=\begin{bmatrix}5 & -3\\2 & 7\end{bmatrix}^{-1}\begin{bmatrix}-2\\-9\end{bmatrix}\Rightarrow\begin{bmatrix}\frac{7}{41} & \frac{3}{41}\\-\frac{2}{41} & \frac{5}{41}\end{bmatrix}=\begin{bmatrix}-2\\-9\end{bmatrix}\Rightarrow\begin{bmatrix}-\frac{14}{41}-\frac{27}{41}\\\frac{4}{41}-\frac{45}{41}\end{bmatrix}=\begin{bmatrix}-1\\-1\end{bmatrix}.$$

The solution is $(-1,-1)$.

47. Using the matrix inverse method, put the system into the proper matrix form:

$$\begin{bmatrix}2 & 1\\3 & -2\end{bmatrix}\begin{bmatrix}x\\y\end{bmatrix}=\begin{bmatrix}5\\4\end{bmatrix}\Rightarrow\begin{bmatrix}x\\y\end{bmatrix}=\begin{bmatrix}2 & 1\\3 & -2\end{bmatrix}^{-1}\begin{bmatrix}5\\4\end{bmatrix}\Rightarrow\begin{bmatrix}\frac{2}{7} & \frac{1}{7}\\\frac{3}{7} & -\frac{2}{7}\end{bmatrix}\begin{bmatrix}5\\4\end{bmatrix}\Rightarrow\begin{bmatrix}\frac{10}{7}+\frac{4}{7}\\\frac{15}{7}-\frac{8}{7}\end{bmatrix}=\begin{bmatrix}2\\1\end{bmatrix}.$$ The solution is $(2,1)$.

48. Using the matrix inverse method, put the system into the proper matrix form:

$$\begin{bmatrix}1 & -2\\3 & 1\end{bmatrix}\begin{bmatrix}x\\y\end{bmatrix}=\begin{bmatrix}7\\7\end{bmatrix}\Rightarrow\begin{bmatrix}x\\y\end{bmatrix}=\begin{bmatrix}1 & -2\\3 & 1\end{bmatrix}^{-1}\begin{bmatrix}7\\7\end{bmatrix}\Rightarrow\begin{bmatrix}\frac{1}{7} & \frac{2}{7}\\-\frac{3}{7} & \frac{1}{7}\end{bmatrix}\begin{bmatrix}7\\7\end{bmatrix}\Rightarrow\begin{bmatrix}\frac{7}{7}+\frac{14}{7}\\-\frac{21}{7}+\frac{7}{7}\end{bmatrix}=\begin{bmatrix}3\\-2\end{bmatrix}.$$ The solution is $(3,-2)$.

49. Solve on the calculator using the matrix inverse method:

$$\begin{bmatrix}x\\y\\z\end{bmatrix}=\begin{bmatrix}1 & 2 & 0\\0 & 3 & -1\\1 & 2 & -1\end{bmatrix}^{-1}\cdot\begin{bmatrix}-1\\-5\\-3\end{bmatrix}=\begin{bmatrix}1\\-1\\2\end{bmatrix};$$ therefore, the solution is $(1,-1,2)$.

50. The determinant of the coefficient matrix is $\det\begin{bmatrix}3 & -2 & 4\\4 & 1 & -5\\-6 & 4 & -8\end{bmatrix}=0$; therefore, infinitely many solutions or no solutions. Now use Row Echelon Method on the augmented matrix:

$$\left[\begin{array}{ccc|c}3 & -2 & 4 & 1\\4 & 1 & -5 & 2\\-6 & 4 & -8 & -2\end{array}\right]\Rightarrow 2R_1+R_3\rightarrow\left[\begin{array}{ccc|c}3 & -2 & 4 & 1\\4 & 1 & -5 & 2\\0 & 0 & 0 & 0\end{array}\right].$$ Since $0=0$, the equation has an infinite number of solutions. Continue to use Row Echelon Method to find these solutions: $R_1+2R_2\rightarrow\left[\begin{array}{ccc|c}3 & -2 & 4 & 1\\11 & 0 & -6 & 5\\0 & 0 & 0 & 0\end{array}\right].$

Now solve for R_2 for x: $11x-6z=5\Rightarrow 11x=6z+5\Rightarrow x=\frac{6z+5}{11}$. Now substitute $x=\frac{6z+5}{11}$ into R_1 and solve for y:

$$3\left(\frac{6z+5}{11}\right)-2y+4z=1\Rightarrow 18z+15-22y+44z=11\Rightarrow -22y=-62z-4\Rightarrow y=\frac{62z+4}{22}=\frac{31z+2}{11}.$$

The solution is $\left(\frac{6z+5}{11},\frac{31z+2}{11},z\right)$. Other forms are possible.

51. Solve on the calculator using the matrix inverse method:

$$\begin{bmatrix}x\\y\\z\end{bmatrix}=\begin{bmatrix}1 & 1 & 1\\2 & -1 & 0\\0 & 3 & 1\end{bmatrix}^{-1}\cdot\begin{bmatrix}1\\-2\\2\end{bmatrix}=\begin{bmatrix}-1\\0\\2\end{bmatrix};$$ therefore, the solution is $(-1,0,2)$.

52. Solve on the calculator using the matrix inverse method:

$\begin{bmatrix} x \\ y \\ z \end{bmatrix} = \begin{bmatrix} 1 & 0 & 0 \\ 0 & 1 & 1 \\ 2 & 0 & -3 \end{bmatrix}^{-1} \cdot \begin{bmatrix} -3 \\ 6 \\ -9 \end{bmatrix} = \begin{bmatrix} -3 \\ 5 \\ 1 \end{bmatrix}$; therefore, the solution is $(-3,5,1)$.

53. Solve on the calculator using the matrix inverse method:

$\begin{bmatrix} x \\ y \\ z \end{bmatrix} = \begin{bmatrix} 2 & -4 & 4 \\ 1 & -3 & 2 \\ 1 & -1 & 2 \end{bmatrix}^{-1} \cdot \begin{bmatrix} 0 \\ -3 \\ 1 \end{bmatrix} \Rightarrow \varnothing$. Since the inverse of the 3×3 matrix does not exist this system is inconsistent.

54. One solution to the solution set $\{(4-y, y)\}$ is $\{(4-1,1)\} \Rightarrow \{(3,1)\}$. Answers may vary.

55. $\det \begin{bmatrix} -1 & 8 \\ 2 & 9 \end{bmatrix} = -9-16 = -25.$

56. $\det \begin{bmatrix} -2 & 4 \\ 0 & 3 \end{bmatrix} = -6-0 = -6.$

57. Evaluate, expand by the second column. Therefore, $\det = (-)(a_{12})(M_{12}) + (a_{22})(M_{22}) - (a_{32})(M_{32})$:

$-4\left(\det \begin{bmatrix} 3 & 2 \\ -1 & 3 \end{bmatrix}\right) + 0 - 0 = -4(9+2) = -4(11) = -44.$

58. Evaluate, expand by the second column. Therefore, $\det = (-)(a_{12})(M_{12}) + (a_{22})(M_{22}) - (a_{32})(M_{32})$:

$-2\left(\det \begin{bmatrix} 4 & 3 \\ 5 & 2 \end{bmatrix}\right) + 0 - (-1)\left(\det \begin{bmatrix} -1 & 3 \\ 4 & 3 \end{bmatrix}\right) = -2(8-15) + 1(-3-12) = -2(-7) + 1(-15) = 14 - 15 = -1.$

59. If $\det \begin{bmatrix} -3 & 2 \\ 1 & x \end{bmatrix} = 5$, then $-3x - 2 = 5 \Rightarrow -3x = 7 \Rightarrow x = \frac{7}{3}$. The solution set is $\left\{-\frac{7}{3}\right\}$.

60. If $\det \begin{bmatrix} 3x & 7 \\ -x & 4 \end{bmatrix} = 8$, then $12x + 7x = 8 \Rightarrow 19x = 8 \Rightarrow x = \frac{8}{19}$. The solution set is $\left\{\frac{8}{19}\right\}$.

61. Evaluate, expand by the third column. Therefore, $\det = (a_{13})(M_{13}) - (a_{23})(M_{23}) + (a_{33})(M_{33})$:

$0 - (-1)\left(\det \begin{bmatrix} 2 & 5 \\ 0 & 2 \end{bmatrix}\right) + 0 = 4 \Rightarrow 4 = 4$. Since this is always true, all real numbers can be input for x.

62. Evaluate, expand by the first row. Therefore, $\det = (a_{11})(M_{11}) - (a_{12})(M_{12}) + (a_{13})(M_{13})$:

$6x\left(\det \begin{bmatrix} 5 & 3 \\ 2 & -1 \end{bmatrix}\right) - 2\left(\det \begin{bmatrix} 1 & 3 \\ x & -1 \end{bmatrix}\right) + 0 = 2x \Rightarrow 6x(-5-6) - 2(-1-3x) = 2x \Rightarrow$

$-66x + 2 + 6x = 2x \Rightarrow -60x + 2 = 2x \Rightarrow -62x = -2 \Rightarrow x = \frac{-2}{-62}$. The solution set is $\left\{\frac{1}{31}\right\}$.

63. (a) $D = \det \begin{bmatrix} 3 & -1 \\ 2 & 1 \end{bmatrix} = 5.$

(b) $D_x = \det \begin{bmatrix} 28 & -1 \\ 2 & 1 \end{bmatrix} = 30.$

(c) $D_y = \det\begin{bmatrix} 3 & 28 \\ 2 & 2 \end{bmatrix} = -50.$

(d) $x = \frac{D_x}{D} = \frac{30}{5} = 6;\ y = \frac{D_y}{D} = \frac{-50}{5} = -10.$ The solution is $(6, -10)$.

64. (a) $A =$ the coefficient matrix: $A = \begin{bmatrix} 3 & -1 \\ 2 & 1 \end{bmatrix}.$

(b) $B =$ the answer matrix: $B = \begin{bmatrix} 28 \\ 2 \end{bmatrix}.$

(c) To solve for x and y, multiply A^{-1} by B: $\begin{bmatrix} x \\ y \end{bmatrix} = A^{-1}B = \begin{bmatrix} 6 \\ -10 \end{bmatrix}$; therefore, the solution is $(6, -10)$.

65. If $D = 0$, there would be division by 0, which is undefined. The system will have no solutions or infinitely many solutions.

66. Find the determinants $D = \det\begin{bmatrix} 3 & 1 \\ 5 & 4 \end{bmatrix} = 7$, $D_x = \det\begin{bmatrix} -1 & 1 \\ 10 & 4 \end{bmatrix} = -14$, and $D_y = \det\begin{bmatrix} 3 & -1 \\ 5 & 10 \end{bmatrix} = 35$. Then

$x = \frac{D_x}{D} = \frac{-14}{7} = -2$ and $y = \frac{D_y}{D} = \frac{35}{7} = 5$. The solution is $(-2, 5)$.

67. Find the determinants $D = \det\begin{bmatrix} 3 & 7 \\ 5 & -1 \end{bmatrix} = -38$, $D_x = \det\begin{bmatrix} 2 & 7 \\ -22 & -1 \end{bmatrix} = 152$, and $D_y = \det\begin{bmatrix} 3 & 2 \\ 5 & -22 \end{bmatrix} = -76$.

Then $x = \frac{D_x}{D} = \frac{152}{-38} = -4$ and $y = \frac{D_y}{D} = \frac{-76}{-38} = 2$. The solution is $(-4, 2)$.

68. Find the determinants $D = \det\begin{bmatrix} 2 & -5 \\ 3 & 4 \end{bmatrix} = 23$, $D_x = \det\begin{bmatrix} 8 & -5 \\ 10 & 4 \end{bmatrix} = 82$, and $D_y = \det\begin{bmatrix} 2 & 8 \\ 3 & 10 \end{bmatrix} = -4$. Then

$x = \frac{D_x}{D} = \frac{82}{23}$ and $y = \frac{D_y}{D} = \frac{-4}{23}$. The solution is $\left(\frac{82}{23}, -\frac{4}{23}\right)$.

69. Using your calculator, find the following determinants:

$$D = \det\begin{bmatrix} 3 & 2 & 1 \\ 4 & -1 & 3 \\ 1 & 3 & -1 \end{bmatrix} = 3,\ D_x = \det\begin{bmatrix} 2 & 2 & 1 \\ -16 & -1 & 3 \\ 12 & 3 & -1 \end{bmatrix} = -12,\ D_y = \det\begin{bmatrix} 3 & 2 & 1 \\ 4 & -16 & 3 \\ 1 & 12 & -1 \end{bmatrix} = 18, \text{ and}$$

$$D_z = \det\begin{bmatrix} 3 & 2 & 2 \\ 4 & -1 & -16 \\ 1 & 3 & 12 \end{bmatrix} = 6 \text{ Then } x = \frac{D_x}{D} = \frac{-12}{3} = -4,\ y = \frac{D_y}{D} = \frac{18}{3} = 6, \text{ and } z = \frac{D_z}{D} = \frac{6}{3} = 2.$$

The solution is $(-4, 6, 2)$.

70. The determinant of the coefficient matrix is $\det\begin{bmatrix} 5 & -2 & -1 \\ -5 & 2 & 1 \\ 1 & -4 & -2 \end{bmatrix} = 0$; therefore, infinitely many or no

solutions. Now use Row Echelon Method on the augmented matrix: $\left.\begin{matrix} R_3 \to R_1 \\ R_2 \to R_3 \end{matrix}\right\} \to \left[\begin{array}{ccc|c} 1 & -4 & -2 & 0 \\ 5 & -2 & -1 & 8 \\ -5 & 2 & 1 & -8 \end{array}\right] \Rightarrow$

Since $R_2 + R_3$ produces $0 = 0$, there are infinitely many solutions and they are dependent solutions.

Continue to use Row Echelon Method to find these solutions: $R_1 + 2R_3 \to \left[\begin{array}{ccc|c} 1 & -4 & -2 & 0 \\ 5 & -2 & -1 & 8 \\ -9 & 0 & 0 & -16 \end{array}\right]$.

Solve for x using R_3 : $-9x = -16 \Rightarrow x = \dfrac{16}{9}$. Now substitute $x = \dfrac{16}{9}$ into R_1 and solve for y:

$$\frac{16}{9} - 4y - 2z = 0 \Rightarrow 16 - 36y - 18z = 0 \Rightarrow -36y = 18z - 16 \Rightarrow y = \frac{18z - 16}{-36} = \frac{8 - 9z}{18}.$$

The solution is $\left(\dfrac{16}{9}, \dfrac{8 - 9z}{18}, z\right)$.

71. Using your calculator, find the following determinants:

$$D = \det\begin{bmatrix} -1 & 3 & -4 \\ 2 & 4 & 1 \\ 3 & 0 & -1 \end{bmatrix} = 67;\ D_x = \det\begin{bmatrix} 2 & 3 & -4 \\ 3 & 4 & 1 \\ 9 & 0 & -1 \end{bmatrix} = 172;\ D_y = \det\begin{bmatrix} -1 & 2 & -4 \\ 2 & 3 & 1 \\ 3 & 9 & -1 \end{bmatrix} = -14;$$

$$D_z = \det\begin{bmatrix} -1 & 3 & 2 \\ 2 & 4 & 3 \\ 3 & 0 & 9 \end{bmatrix} = -87 \text{ Then } x = \frac{D_x}{D} = \frac{172}{67},\ y = \frac{D_y}{D} = \frac{-14}{67}, \text{ and } z = \frac{D_z}{D} = \frac{-87}{67}.$$

The solution is $\left(\dfrac{172}{67}, -\dfrac{14}{67}, -\dfrac{87}{67}\right)$.

72. Let x = the amount of rice in cups and y = the amount of soybeans in cups. Then from the information the system of equations is $15x + 22.5y = 9.5$ [equation 1] and $810x + 270y = 324$ [equation 2]. Multiply [equation 1] by -12 and add [equation 2] to eliminate y:

$$\begin{array}{rl} -180x - 270y = & -114 \\ 810x + 270y = & 324 \\ \hline 630x \qquad\quad = & 210 \end{array}$$

$\Rightarrow x = \dfrac{210}{630} = \dfrac{1}{3}$. Now substitute $x = \dfrac{1}{3}$ into [equation 1] and solve for y:

$$15\left(\frac{1}{3}\right) + 22.5y = 9.5 \Rightarrow 5 + 22.5y = 9.5 \Rightarrow 2.5y = 4.5 \Rightarrow y = 0.20 \text{ or } \frac{1}{5}.$$

The meal should include $\dfrac{1}{3}$ cup of rice and $\dfrac{1}{5}$ cup of soybeans.

73. Let x = the number of CD's and y = the number of plastic holders. From the information the system of equations is obtained: $x + y = 100$ [equation 1] and $0.40x + 0.30y = 38.00$ [equation 2]. Multiply [equation 1] by -30 and add [equation 2] multiplied by 100 to eliminate y:

$$\begin{array}{rl} -30x - 30y = & -3000 \\ 40x + 30y = & 3800 \\ \hline 10x \qquad\quad = & 800 \end{array}$$

$\Rightarrow x = 80$. Now substitute $x = 80$ into [equation 1] and solve for y:

$80 + y = 100 \Rightarrow y = 20$. They should send 80 CD's and 20 holders.

74. Let x = the number pounds of $4.60 tea, y = the number of pounds of $5.75 tea, and z = the number of pounds of $6.50 tea. From the information, the system of equations is obtained: $x + y + z = 20$ [equation 1], $4.6x + 5.75y + 6.5z = 20(5.25) = 105$ and $x = y + z \Rightarrow x - y - z = 0$. Now create the coefficient matrix and solve on the calculator using the matrix inverse method:

$$A = \begin{bmatrix} 1 & 1 & 1 \\ 4.6 & 5.75 & 6.5 \\ 1 & -1 & -1 \end{bmatrix}. \text{ Therefore, } \begin{bmatrix} x \\ y \\ z \end{bmatrix} = \begin{bmatrix} 1 & 1 & 1 \\ 4.6 & 5.75 & 6.5 \\ 1 & -1 & -1 \end{bmatrix}^{-1} \cdot \begin{bmatrix} 20 \\ 105 \\ 0 \end{bmatrix} = \begin{bmatrix} 10 \\ 8 \\ 2 \end{bmatrix}.$$

They should use 10 pounds of $4.60 tea, 8 pounds of $5.75 tea, and 2 pounds of $6.50 tea.

75. Let x = the amount of 5% solution (ml). y = the amount of 15% solution (ml), and z = the amount of 10% solution (ml). Then from the information, the system of equations is $x + y + z = 20$ [equation 1], $0.05x + 0.15y + 0.10z = 0.08(20) = 1.6$, and $x = y + z + 2 \Rightarrow x - y - z = 2$. Now create the coefficient matrix and solve on the calculator using the matrix inverse method:

$$A = \begin{bmatrix} 1 & 1 & 1 \\ 5 & 15 & 10 \\ 1 & -1 & -1 \end{bmatrix}; \text{ therefore, } \begin{bmatrix} x \\ y \\ z \end{bmatrix} = \begin{bmatrix} 1 & 1 & 1 \\ 5 & 15 & 10 \\ 1 & -1 & -1 \end{bmatrix}^{-1} \cdot \begin{bmatrix} 20 \\ 160 \\ 2 \end{bmatrix} = \begin{bmatrix} 11 \\ 3 \\ 6 \end{bmatrix}.$$

They should use 11 ml of 5% solution, 3 ml of 15% solution, and 6 ml of 10% solution.

76. (a) Using $P = a + bA + cW$ and the data from the table yields,

$$a + 39b + 142c = 113$$
$$a + 53b + 181c = 138$$
$$a + 65b + 191c = 152$$

Now using the capabilities of the calculator and the Reduced Row Echelon Method:

$$\left[\begin{array}{ccc|c} 1 & 39 & 142 & 113 \\ 1 & 53 & 181 & 138 \\ 1 & 65 & 191 & 152 \end{array}\right] = \left[\begin{array}{ccc|c} 1 & 0 & 0 & 32.780488 \\ 0 & 1 & 0 & 0.9024390 \\ 0 & 0 & 1 & 0.3170732 \end{array}\right].$$

Therefore, the equation is $P \approx 32.78 + 0.9024A + 0.3171W$.

(b) Using $A = 55$ and $W = 175$ yields $P \approx 32.78 + 0.9024(55) + 0.3171(175) \Rightarrow P \approx 138$.

77. Using the equation for a polynomial of degree 3 and given points yields:

$(-2,1)$: $-8a + 4b - 2c + d = 1$

$(-1,6)$: $-a + b - c + d = 6$

$(2,9)$: $8a + 4b + 2c + d = 9$

$(3,26)$: $27a + 9b + 3c + d = 26$

The equations can be represented by $\left[\begin{array}{cccc|c} -8 & 4 & -2 & 1 & 1 \\ -1 & 1 & -1 & 1 & 6 \\ 8 & 4 & 2 & 1 & 9 \\ 27 & 9 & 3 & 1 & 26 \end{array}\right]$.

Using the capabilities of the calculator, the Reduced Row Echelon Method produces the solutions:

$a = 1,\ b = 0,\ c = -2,\ d = 5$. The equation is $P(x) = x^3 - 2x + 5$.

78. Using $f(x) = ax^2 + bx + c$ and the data from the graph yields:

From the point $(-6,4)$: $36a - 6b + c = 4$.

From the point $(-4,-2)$: $16a - 4b + c = -2$.

From the point $(2,4)$: $4a + 2b + c = 4$.

Now using your calculator, find the following determinants:

$D = \det\begin{bmatrix} 36 & -6 & 1 \\ 16 & -4 & 1 \\ 4 & 2 & 1 \end{bmatrix} = -96; D_a = \det\begin{bmatrix} 4 & -6 & 1 \\ -2 & -4 & 1 \\ 4 & 2 & 1 \end{bmatrix} = -48;\ D_b = \det\begin{bmatrix} 36 & 4 & 1 \\ 16 & -2 & 1 \\ 4 & 4 & 1 \end{bmatrix} = -192;$

$D_c = \det\begin{bmatrix} 36 & -6 & 4 \\ 16 & -4 & -2 \\ 4 & 2 & 4 \end{bmatrix} = 192$ Then $a = \dfrac{D_a}{D} = \dfrac{-48}{-96} = \dfrac{1}{2},\ b = \dfrac{D_b}{D} = \dfrac{-192}{-96} = 2$, and

$c = \dfrac{D_c}{D} = \dfrac{192}{-96} = -2$. The equation is $P(x) = \dfrac{1}{2}x^2 + 2x - 2$.

79. See Figure 79.

80. See Figure 80.

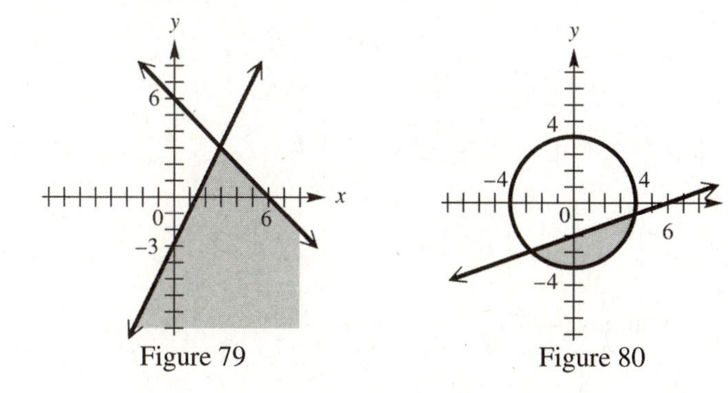

Figure 79 Figure 80

81. Graphing $x \geq 0,\ y \geq 0,\ 3x + 2y \leq 12$ and $5x + y \geq 5$ in the same window yields four vertices. Using the given expression and the ordered pairs of the vertices yields the following solutions:

$(0,5)$: $2(0) + 4(5) = 20$

$(0,6)$: $2(0) + 4(6) = 24$

$(1,0)$: $2(1) + 4(0) = 2$

$(4,0)$: $2(4) + 4(0) = 8$

Therefore, the maximum value is 24 at (0,6).

82. Graphing $x \geq 0,\ y \geq 0,\ x+y \leq 50,\ 2x+y \geq 20$ and $x+2y \geq 30$ in the same window yields five vertices.

Using the given expression and the ordered pairs of the vertices yields the following solutions:

$(0,50)$: $4(0)+2(50)=100$

$(0,20)$: $4(0)+2(20)=40$

$\left(\frac{10}{3},\frac{40}{3}\right)$: $4\left(\frac{10}{3}\right)+2\left(\frac{40}{3}\right)=40$

(30.0): $4(30)+2(0)=120$

$(50,0)$: $4(50)+2(0)=200$

Therefore, the minimum value is 40 at (0,20) and $\left(\frac{10}{3},\frac{40}{3}\right)$

Note: values on the line $2x+y=20$ also give minimums of 40 between the two points.

83. Let x = the number radios produced daily and let y = the number of DVD players produced daily.

The objective function to find the maximum profit is $3x+2y$.

The constraints are

$5 \leq x \leq 12$ (radio production restrictions)

$0 \leq x \leq 30$ (DVD player maximum production)

$x \leq y$ (radio production less than or equal to DVD player production)

Graphing the constraints on the calculator yields four vertices. Inputting these ordered pairs into the objective function yields (5,30): 1125, (5,5): 250, (25,30): 1425, and (25,25): 1250.

The maximum profit is \$1425, this happens when 25 radios and 30 DVD players are manufactured.

84. Factoring $\frac{5x-2}{x^2-4}$ results in $\frac{5x-2}{(x-2)(x+2)}$. Multiply $\frac{5x-2}{(x-2)(x+2)}=\frac{A}{x-2}+\frac{B}{x+2}$ by

$(x-2)(x+2) \Rightarrow 5x-2=A(x+2)+B(x-2)$. Let $x=-2 \Rightarrow -12=B(-4) \Rightarrow B=3$.

Let $x=2 \Rightarrow 8=A(4) \Rightarrow A=2$. The expression can be written $\frac{2}{x-2}+\frac{3}{x+2}$.

85. Factoring $\frac{x+2}{x^3+2x^2+x}$ results in $\frac{x+2}{x(x+1)^2}$. Multiply $\frac{x+2}{x^3-x^2+4x}=\frac{A}{x}+\frac{B}{x+1}+\frac{C}{(x+1)^2}$ by

$x(x+1)^2 \Rightarrow x+2=A(x+1)^2+Bx(x+1)+Cx$. Let $x=0 \Rightarrow 2=A(1) \Rightarrow A=2$. Let

$x=-1 \Rightarrow 1=C(-1) \Rightarrow C=-1$. Let $x=1,\ A=2,\ C=-1 \Rightarrow$

$3=2(4)+B(1)(2)-1 \Rightarrow 3=8=2B-1 \Rightarrow B=-2$

The expression can be written $\frac{2}{x}+\frac{-2}{x+1}+\frac{-1}{(x+1)^2}$ or $\frac{2}{x}-\frac{2}{x+1}-\frac{1}{(x+1)^2}$.

86. Factoring $\dfrac{x+2}{x^3-x^2+4x}$ results in $\dfrac{x+2}{x\left(x^2-x+4\right)}$. Multiply $\dfrac{x+2}{x(x^2-x+4)}=\dfrac{A}{x}+\dfrac{Bx+C}{x^2-x+4}$ by

$x\left(x^2-x+4\right)\Rightarrow\ x+2=A\left(x^2-x+4\right)+(Bx+C)(x)\Rightarrow x+2=(A+B)x^2+(C-A)x+4A.$

By equating coefficients $A+B=0,\ C-A=1,$ and $4A=2\Rightarrow A=\frac{1}{2}$. Since $A+B=0$ and

$A=\frac{1}{2},\ B=-\frac{1}{2}$. Since $C-A=1$ and $A=\frac{1}{2},\ C=\frac{3}{2}$.

The expression can be written $\dfrac{\frac{1}{2}}{x}+\dfrac{-\frac{1}{2}x+\frac{3}{2}}{x^2-x+4}=\dfrac{1}{2x}+\dfrac{-x+3}{2\left(x^2-x+4\right)}$.

87. Factoring $\dfrac{6x^2-x-3}{x^3-x}=\dfrac{6x^2-x-3}{x\left(x^2-1\right)}=\dfrac{6x^2-x-3}{x(x-1)(x+1)}$. Multiply $\dfrac{6x^2-x-3}{x(x-1)(x+1)}=\dfrac{A}{x}+\dfrac{B}{x-1}+\dfrac{C}{x+1}$ by

$x(x-1)(x+1)\Rightarrow\ 6x^2-x-3=A(x-1)(x+1)+(B)(x)(x+1)+(C)x(x-1).$

Let $x=0\Rightarrow -3=-A\Rightarrow A=3$. Let $x=1\Rightarrow 2=2B\Rightarrow B=1$. Let $x=-1\Rightarrow 4=2C\Rightarrow C=2.$

The expression can be written $\dfrac{3}{x}+\dfrac{1}{x-1}+\dfrac{2}{x+1}$.

Chapter 7 Test

1. (a) The first is the equation of a hyperbola; the second is the equation of a line.
   (b) A hyperbola and a line could intersect in 0, 1, or 2 points.
   (c) First solve the second equation for y: $3x+y=1\Rightarrow y=1-3x$. Now substitute $y=1-3x$ into

   Equation 1: $x^2-4(1-3x)^2=-15\Rightarrow x^2-4\left(1-6x+9x^2\right)=-15\Rightarrow$

   $x^2-4+24x-36x^2=-15\Rightarrow -35x^2+24x+11=0\Rightarrow (35x+11)(-x+1)=0\Rightarrow x=\frac{11}{35}$, 1. Now

   substitute these values into $y=1-3x$: $y=1-3(1)\Rightarrow y=-2$ and $y=1-3\left(-\frac{11}{35}\right)\Rightarrow y=\frac{68}{35}$.

   The solutions are $\left(-\frac{11}{35},\frac{68}{35}\right)$ and $(1,-2)$.

   (d) Graph on a calculator shows two points of intersection. See Figure 1.

[-10,10] by [-10,10]
Xscl = 1 Yscl = 1

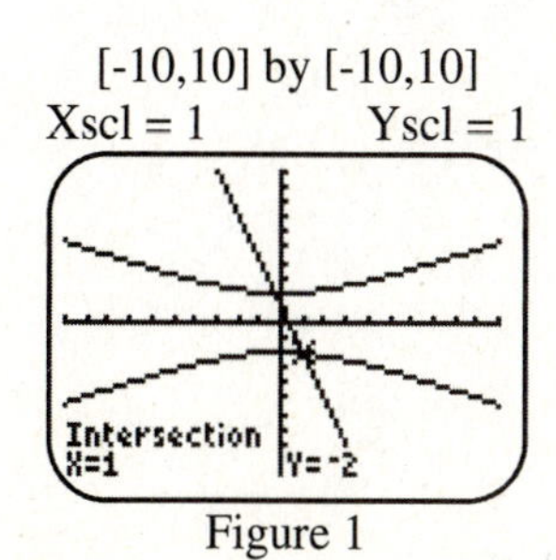

Figure 1

2. Using the Row Echelon Method, the given system of equations yields the matrix,

$\left[\begin{array}{ccc|c} 2 & 1 & 1 & 3 \\ 1 & 2 & -1 & 3 \\ 3 & -1 & 1 & 5 \end{array}\right]$, which by $(R1 \leftrightarrow R2) \rightarrow \left[\begin{array}{ccc|c} 1 & 2 & -1 & 3 \\ 2 & 1 & 1 & 3 \\ 3 & -1 & 1 & 5 \end{array}\right] \Rightarrow (-2R_1 + R_2)$ and

$\Rightarrow (-3R_1 + R_3) \rightarrow \left[\begin{array}{ccc|c} 1 & 2 & -1 & 3 \\ 0 & -3 & 3 & -3 \\ 0 & -7 & 4 & -4 \end{array}\right] \Rightarrow (7R_2 - 3R_3) \rightarrow \left[\begin{array}{ccc|c} 1 & 2 & -1 & 3 \\ 0 & -3 & 3 & -3 \\ 0 & 0 & 9 & -9 \end{array}\right]$. From this matrix, we have

the resulting equation, $9z = -9 \Rightarrow z = -1$. Now use back-substitution: $z = -1$ substituting

into the resulting $x = 2: -12 = C(4) \Rightarrow C = -3.$ yields, $-3y + 3(-1) = -3 \Rightarrow -3y = 0 \Rightarrow y = 0;$

finally substituting $y = 0$ and $z = -1$ into the resulting R_1 yields, $x + 2(0) - (-1) = 3 \Rightarrow x = 2.$

The solution is $\{(2, 0, -1)\}$.

3. (a) $3\begin{bmatrix} 2 & 3 \\ 1 & -4 \\ 5 & 9 \end{bmatrix} - \begin{bmatrix} -2 & 6 \\ 3 & -1 \\ 0 & 8 \end{bmatrix} = \begin{bmatrix} 6 & 9 \\ 3 & -12 \\ 15 & 27 \end{bmatrix} - \begin{bmatrix} -2 & 6 \\ 3 & -1 \\ 0 & 8 \end{bmatrix} = \begin{bmatrix} 6+2 & 9-6 \\ 3-3 & -12+1 \\ 15-0 & 27-8 \end{bmatrix} = \begin{bmatrix} 8 & 3 \\ 0 & -11 \\ 15 & 19 \end{bmatrix}.$

(b) Cannot add $(1\times2)+(1\times2)+(2\times2)$; therefore, the solution set is $\varnothing$.

(c) $\begin{bmatrix} 2 & 1 & -3 \\ 4 & 0 & 5 \end{bmatrix}\begin{bmatrix} 1 & 3 \\ 2 & 4 \\ 3 & -2 \end{bmatrix} = \begin{bmatrix} 2+2-9 & 6+4+6 \\ 4+0+15 & 12+0-10 \end{bmatrix} = \begin{bmatrix} -5 & 16 \\ 19 & 2 \end{bmatrix}.$

4. (a) AB can be found; it will be $n \times n$

(b) BA can be found; it will be $n \times n$

(c) $AB = BA$ is not necessarily true, since matrix multiplication is not commutative.

(d) Since the number of rows of the first matrix is not equal to the number of columns of the second matrix, AC cannot be found.

Since the number of rows of the first matrix is equal to the number of columns of the second matrix, CA can be found; it will be $m \times n.$

5. (a) $\det\begin{bmatrix} 4 & 9 \\ -5 & -11 \end{bmatrix} = -44 + 45 = 1$

(b) Evaluate, expand by the second column. Therefore, $\det = (-)(a_{12})(M_{12}) + (a_{22})(M_{22}) - (a_{32})(M_{32})$:

$0\left(\det\begin{bmatrix} -1 & 9 \\ 12 & -3 \end{bmatrix}\right) + 7\left(\det\begin{bmatrix} 2 & 8 \\ 12 & -3 \end{bmatrix}\right) - 5\left(\det\begin{bmatrix} 2 & 8 \\ -1 & 9 \end{bmatrix}\right) = 0 + 7(-6-96) - 5(18+8) \Rightarrow$

$7(-102) - 5(26) = -844$

6. Find the determinants: $D=\det\begin{bmatrix}2 & -3\\4 & 5\end{bmatrix}=22$, $D_x=\det\begin{bmatrix}-33 & -3\\11 & 5\end{bmatrix}=-132$, and $D_y=\det\begin{bmatrix}2 & -33\\4 & 11\end{bmatrix}=154$.

   Then $x=\dfrac{D_x}{D}=\dfrac{-132}{22}=-6$ and $y=\dfrac{D_y}{D}=\dfrac{154}{22}=7$. The solution set is $\{(-6,7)\}$.

7. (a) For the system: $A=\begin{bmatrix}1 & 1 & -1\\2 & -3 & -1\\1 & 2 & 2\end{bmatrix}$, $X=\begin{bmatrix}x\\y\\z\end{bmatrix}$, and $B=\begin{bmatrix}-4\\5\\3\end{bmatrix}$.

   (b) Solve using the calculator, $A^{-1}=\begin{bmatrix}\frac{1}{4} & \frac{1}{4} & \frac{1}{4}\\ \frac{5}{16} & -\frac{3}{16} & \frac{1}{16}\\ -\frac{7}{16} & \frac{1}{16} & \frac{5}{16}\end{bmatrix}$.

   (c) $\begin{bmatrix}x\\y\\z\end{bmatrix}=\begin{bmatrix}1 & 1 & -1\\2 & -3 & -1\\1 & 2 & 2\end{bmatrix}^{-1}\begin{bmatrix}-4\\5\\3\end{bmatrix}=\begin{bmatrix}1\\-2\\3\end{bmatrix}$. The solution is $\{(1,-2,3)\}$.

   (d) For the new system: $A=\begin{bmatrix}0.5 & 1 & 1\\2 & -3 & -1\\1 & 2 & 2\end{bmatrix}$ and by the calculator, $\det A=0$. Since the determinant equals zero, there is no inverse and the matrix inverse method cannot be used.

8. (a) Using $f(x)=ax^2+bx+c$ and the data from the table yields

   $(0,1.3)$: $0^2a+0b+c=1.3$

   $(100,2.5)$: $100^2a+100b+c=2.5$

   $(200,8.9)$: $200^2a+200b+c=8.9$

   Now using the capabilities of the calculator and the Reduced Row Echelon Method:

   $$\left[\begin{array}{ccc|c}0 & 0 & 1 & 1.3\\100^2 & 100 & 1 & 2.5\\200^2 & 200 & 1 & 8.9\end{array}\right]=\left[\begin{array}{ccc|c}1 & 0 & 0 & 0.000264\\0 & 1 & 0 & -0.014\\0 & 0 & 1 & 1.3\end{array}\right].$$

   Therefore, the equation is $f(x)=0.00026x^2-0.014x+1.3$.

   (b) Using the functions of the calculator we see that that the population will reach 8 billion in about 2040. See Figure 8.

[-5,250] by [0,15]
Xscl = 5 Yscl = 1

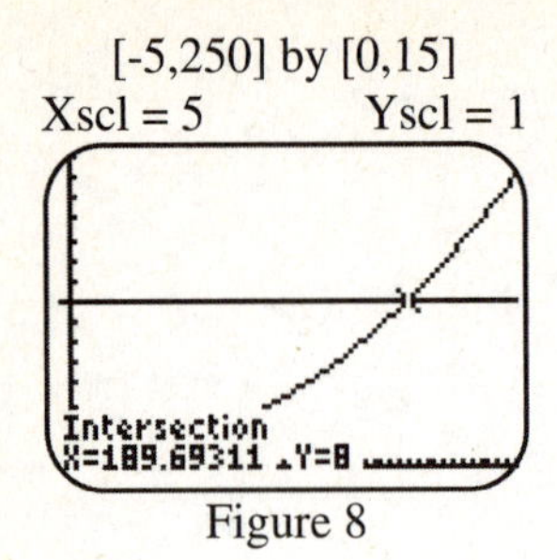

Figure 8

9. B, the graph is shaded above the line; therefore, $y > 2 - x$ and outside the parabola; therefore, $y < x^2 - 5$.

10. Let x = the number type X cabinets and let y = the number of type Y cabinets.

The objective function to find the maximum storage space is $8x + 12y$. The constraints are,

$100x + 200y \le 1400$ (cost of cabinets), $6x + 8y \le 72$ (floor space), and $x \ge 0$ and $y \ge 0$ (cannot have a negative number of cabinets). Graphing the constraints on the calculator yields four vertices. Inputting these ordered pairs into the objective function yields,

$(0,0):\ 8(0) + 12(0) = 0$

$(12,0):\ 8(12) + 12(0) = 96$

$(8,3):\ 8(8) + 12(3) = 100$

$(0.7):\ 8(0) + 12(7) = 84$

The maximum storage is 100 cubic feet, when there are 8 cabinets of type X and 3 cabinets of type Y.

11. Factoring $\dfrac{7x-1}{x^2 - x - 6}$ results in $\dfrac{7x-1}{(x-3)(x+2)}$. Multiply $\dfrac{7x-1}{(x-3)(x+2)} = \dfrac{A}{x-3} + \dfrac{B}{x+2}$ by

$(x-3)(x+2) \Rightarrow 7x - 1 = A(x+2) + B(x-3)$.

Let $x = -2 \Rightarrow -15 = B(-5) \Rightarrow B = 3$. Let $x = 3 \Rightarrow 20 = A(5) \Rightarrow A = 4$.

The expression can be written $\dfrac{4}{x-3} + \dfrac{3}{x+2}$.

12. The expression $\dfrac{x^2 - 11x + 6}{(x+2)(x-2)^2}$ is factored. Multiply $\dfrac{x^2 - 11x + 6}{(x+2)(x-2)^2} = \dfrac{A}{x+2} + \dfrac{B}{x-2} + \dfrac{C}{(x-2)^2}$ by

$(x+2)(x-2)^2 \Rightarrow x^2 - 11x + 6 = A(x-2)^2 + B(x-2)(x+2) + C(x+2)$.

Let $x = -2 \Rightarrow 32 = A(16) \Rightarrow A = 2$. Let $x = 2:\ -12 = C(4) \Rightarrow C = -3$.

Let $x = 0,\ A = 2,\ C = -3:\ 6 = 4(2) - 4B + 2(-3) \Rightarrow$

$6 = 8 - 4B - 6 \Rightarrow 6 = 2 - 4B \Rightarrow 4 = -4B \Rightarrow B = -1$.

The expression can be written $\dfrac{2}{x+2} + \dfrac{-1}{x-2} + \dfrac{-3}{(x-2)^2}$.

Chapter 8: Further Topics in Algebra

8.1: Sequences and Series

1. $a_1 = 4(1)+10 = 14;\ a_2 = 4(2)+10 = 18;\ a_3 = 4(3)+10 = 22;\ a_4 = 4(4)+10 = 26;\ a_5 = 4(5)+10 = 30$

   The first five terms of the sequence are 14, 18, 22, 26, and 30.

3. $a_1 = 2^{1-1} = 2^0 = 1;\ a_2 = 2^{2-1} = 2^1 = 2;\ a_3 = 2^{3-1} = 2^2 = 4;\ a_4 = 2^{4-1} = 2^3 = 8;\ a_5 = 2^{5-1} = 2^4 = 16$

   The first five terms of the sequence are 1, 2, 4, 8, and 16.

5. $a_1 = \left(\frac{1}{3}\right)^1 (1-1) = \left(\frac{1}{3}\right)(0) = 0;\ a_2 = \left(\frac{1}{3}\right)^2 (2-1) = \left(\frac{1}{9}\right)(1) = \frac{1}{9};$

   $a_3 = \left(\frac{1}{3}\right)^3 (3-1) = \left(\frac{1}{27}\right)(2) = \frac{2}{27};\ a_4 = \left(\frac{1}{3}\right)^4 (4-1) = \left(\frac{1}{81}\right)(3) = \frac{1}{27};\ a_5 = \left(\frac{1}{3}\right)^5 (5-1) = \left(\frac{1}{243}\right)(4) = \frac{4}{243}$

   The first five terms of the sequence are 0, $\frac{1}{9}$, $\frac{2}{27}$, $\frac{1}{27}$, and $\frac{4}{243}$.

7. $a_1 = (-1)^1[(2)(1)] = -2;\ a_2 = (-1)^2[(2)(2)] = 4;\ a_3 = (-1)^3[(2)(3)] = -6;$

   $a_4 = (-1)^4[(2)(4)] = 8;\ a_5 = (-1)^5[(2)(5)] = -10$

   The first five terms of the sequence are -2, 4, -6, 8, and -10.

9. $a_1 = \frac{4(1)-1}{1^2+2} = \frac{3}{3} = 1;\ a_2 = \frac{4(2)-1}{2^2+2} = \frac{7}{6};\ a_3 = \frac{4(3)-1}{3^2+2} = \frac{11}{11} = 1;$

   $a_4 = \frac{4(4)-1}{4^2+2} = \frac{15}{3} = \frac{5}{6};\ a_5 = \frac{4(5)-1}{5^2+2} = \frac{19}{27}$ The first five terms of the sequence are 1, $\frac{7}{6}$, 1, $\frac{5}{6}$, and $\frac{19}{27}$.

11. The terms of a sequence are numbered according to their position in the sequence; a_1 is the first term, a_2 is the second term, etc. So a_n will be the nth term in the sequence for any positive integer value of *n*.

13. The sequence has 7 terms; it is finite.

15. The sequence has 4 terms; it is finite.

17. The sequence has no last term; it is infinite.

19. The sequence is defined for integer values of *n* from 1 through 10; it is finite.

21. $a_1 = -2;\ a_2 = a_1 + 3 = -2+3 = 1;\ a_3 = a_2 + 3 = 1+3 = 4;\ a_4 = a_3 + 3 = 4+3 = 7$

    The first four terms of the sequence are -2, 1, 4, and 7.

23. $a_1 = 1;\ a_2 = 1;\ a_3 = a_2 + a_1 = 1+1 = 2;\ a_4 = a_3 + a_2 = 2+1 = 3.$

    The first four terms of the sequence are 1, 1, 2, and 3.

25. $\sum_{i=1}^{5}(2i+1) = 3+5+7+9+11 = 35$

27. $\sum_{j=1}^{4} \frac{1}{j} = \frac{1}{1} + \frac{1}{2} + \frac{1}{3} + \frac{1}{4} = \frac{25}{12}$

29. $\sum_{i=1}^{4} i^i = 1^1 + 2^2 + 3^3 + 4^4 = 288$

31. $\sum_{k=1}^{6} (-1)^k \cdot k = (-1)^1(1) + (-1)^2(2) + (-1)^3(3) + (-1)^4(4) + (-1)^5(5) + (-1)^6(6) = 3$

33. $\sum_{i=2}^{5} (6-3i) = (6-6) + (6-9) + (6-12) + (6-15) = -18$

35. $\sum_{i=-2}^{3} 2(3)^i = 2(3)^{-2} + 2(3)^{-1} + 2(3)^0 + 2(3)^1 + 2(3)^2 + 2(3)^3 = \frac{728}{9}$

37. $\sum_{i=-1}^{5} (i^2 - 2i) = \left[(-1)^2 - 2(-1)\right] + \left[0^2 - 2(0)\right] + \left[1^2 - 2(1)\right] + \left[2^2 - 2(2)\right] + \left[3^2 - 2(3)\right] +$

$\left[4^2 - 2(4)\right] + \left[5^2 - 2(5)\right] = 28$

39. $\sum_{i=1}^{5} (3^i - 4) = (3^1 - 4) + (3^2 - 4) + (3^3 - 4) + (3^4 - 4) + (3^5 - 4) = 343$

41. $\sum_{i=1}^{5} x_i = (-2) + (-1) + (0) + (1) + (2) = -2 - 1 + 0 + 1 + 2$

43. $\sum_{i=1}^{5} (2x_i + 3) = [2(-2) + 3] + [2(-1) + 3] + [2(0) + 3] + [2(1) + 3] + [2(2) + 3] = -1 + 1 + 3 + 5 + 7$

45. $\sum_{i=1}^{3} (3x_i - x_i^2) = \left[3(-2) - (-2)^2\right] + \left[3(-1) - (-1)^2\right] + \left[3(0) - (0)^2\right] = -10 - 4 + 0$

47. $\sum_{i=2}^{5} \frac{x_i + 1}{x_i + 2} = \frac{-1+1}{-1+2} + \frac{0+1}{0+2} + \frac{1+1}{1+2} + \frac{2+1}{2+2} = 0 + \frac{1}{2} + \frac{2}{3} + \frac{3}{4}$

49. $\sum_{i=1}^{4} f(x_i)\Delta x = [4(0) - 7](0.5) + [4(2) - 7](0.5) + [4(4) - 7](0.5) + [4(6) - 7](0.5) = -0.35 + 0.5 + 4.5 + 8.5$

51. $\sum_{i=1}^{4} f(x_i)\Delta x = \left[2(0)^2\right](0.5) + \left[2(2)^2\right](0.5) + \left[2(4)^2\right](0.5) + \left[2(6)^2\right](0.5) = 0 + 4 + 16 + 36$

53. $\sum_{i=1}^{4} f(x_i)\Delta x = \left(\frac{-2}{0+1}\right)(0.5) + \left(\frac{-2}{2+1}\right)(0.5) + \left(\frac{-2}{4+1}\right)(0.5) + \left(\frac{-2}{6+1}\right)(0.5) = -1 - \frac{1}{3} - \frac{1}{5} - \frac{1}{7}$

55. $\sum_{i=1}^{100} 6 = (100)(6) = 600$

57. $\sum_{i=1}^{15} i^2 = \frac{(15)(15+1)[2(15)+1]}{6} = 1240$

59. $\sum_{i=1}^{5} (5i + 3) = 5\left[\frac{(5)(5+1)}{2}\right] + 5(3) = 90$

61. $\sum_{i=1}^{5}(4i^2-2i+6)=4\left[\frac{(5)(5+1)[2(5)+1]}{6}\right]-2\left[\frac{5(5+1)}{2}\right]+5(6)=220$

63. $\sum_{i=1}^{4}(3i^3+2i-4)=3\left[\frac{4^2(4+1)^2}{4}\right]+2\left[\frac{(4)(4+1)}{2}\right]+4(-4)=304$

65. The graphing calculator graph of the first ten terms of this sequence is shown in Figure 65. From this graph, the terms of the sequence appear to converge to the value $\frac{1}{2}$ which, in fact, they do.

67. The graphing calculator graph of the first ten terms of this sequence is shown in Figure 67. From this graph, the terms of the sequence appear to diverge and, in fact, they do.

69. The graphing calculator graph of the first ten terms of this sequence is shown in Figure 69. From this graph, the terms of the sequence appear to converge to the value somewhat greater than 2.5. In fact, they converge to $e \approx 2.71828$.

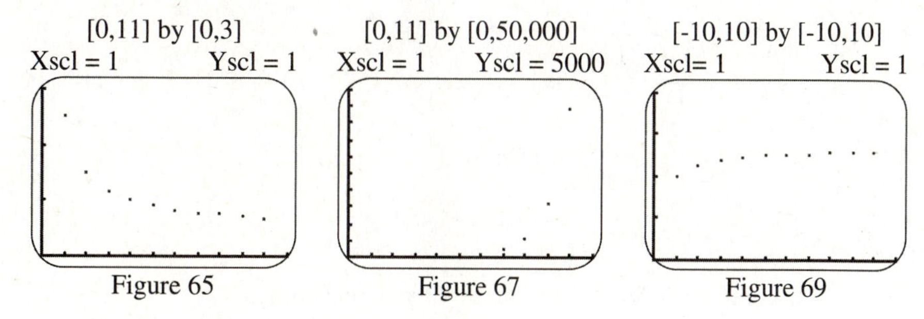

Figure 65 Figure 67 Figure 69

71. The sum of the first six terms of the series is

$$\frac{1}{1^4}+\frac{1}{2^4}+\frac{1}{3^4}+\frac{1}{4^4}+\frac{1}{5^4}+\frac{1}{6^4}=1+\frac{1}{16}+\frac{1}{81}+\frac{1}{256}+\frac{1}{625}+\frac{1}{1296}\approx 1.081123534 \Rightarrow$$

$$\frac{\pi^4}{90}\approx 1.081123534 \Rightarrow \pi^4 \approx 97.30111806 \Rightarrow \pi \approx 3.140721718.$$

This approximation of π, is accurate to three decimal places when rounded to 3.141.

73. (a) The number of bacteria doubles every 40 minutes, so $N_{j+1}=2N_j$ for $j \ge 1$.

(b) Two hours = 120 minutes $\Rightarrow 120=40(j-1)\Rightarrow 3=j-1\Rightarrow j=4$.

$N_1=230; N_2=2(230)=460; N_3=2(460)=920; N_4=2(920)=1840$. If there are initially 230 bacteria, then there will be 1840 bacteria after two hours.

(c) The graph of the first seven terms of this sequence is shown in Figure 73.

(d) From the graph, the growth rate is seen to be very rapid. Doubling the number of bacteria at equal intervals of time produces an exponential growth rate.

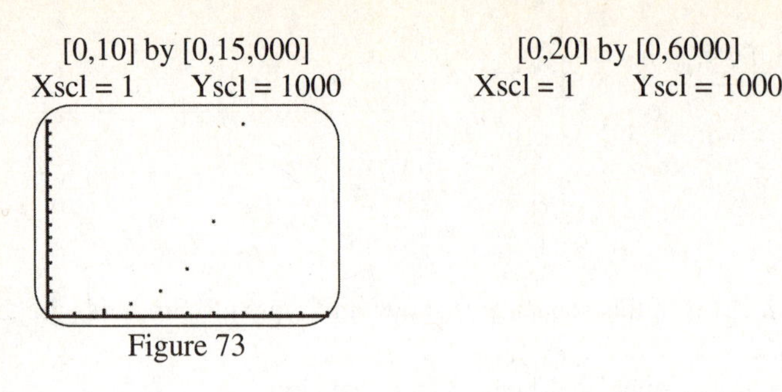

[0,10] by [0,15,000]
Xscl = 1 Yscl = 1000

[0,20] by [0,6000]
Xscl = 1 Yscl = 1000

Figure 73

75. (a) $1+1+\frac{1^2}{2!}+\frac{1^3}{3!}+\frac{1^4}{4!}+\frac{1^5}{5!}+\frac{1^6}{6!}+\frac{1^7}{7!}\approx 2.718254;\ e\approx 2.718282$. Eight terms of the series gives an estimate that matches to four decimal places.

(b) $1+(-1)+\frac{(-1)^2}{2!}+\frac{(-1)^3}{3!}+\frac{(-1)^4}{4!}+\frac{(-1)^5}{5!}+\frac{(-1)^6}{6!}+\frac{(-1)^7}{7!}\approx 0.367857;\ e\approx 0.367879$. Eight terms of the series gives an estimate that matches to four decimal places.

(c) $1+\frac{1}{2}+\frac{\left(\frac{1}{2}\right)^2}{2!}+\frac{\left(\frac{1}{2}\right)^3}{3!}+\frac{\left(\frac{1}{2}\right)^4}{4!}+\frac{\left(\frac{1}{2}\right)^5}{5!}+\frac{\left(\frac{1}{2}\right)^6}{6!}+\frac{\left(\frac{1}{2}\right)^7}{7!}\approx 1.648721;\ \sqrt{e}=e^{1/2}\approx 1.648721$. Eight terms of the series gives an estimate that matches to six decimal places.

8.2: Arithmetic Sequences and Series

1. $5-2=3;\ 8-5=3;\ 11-8=3$. The common difference for this arithmetic sequence is 3.

3. $-2-3=-5;\ -7-(-2)=-5;\ -12-(-7)=-5$. The common difference for this arithmetic sequence is -5.

5. $(2x+5y)-(x+3y)=x+2y;(3x+7y)-(2x+5y)=x+2y$. The common difference for this arithmetic sequence is $x+2y$.

7. $8;\ 8+6=14;\ 14+6=20;\ 20+6=26;\ 26+6=32$. The first five terms of this arithmetic sequence are 8, 14, 20, 26, and 32.

9. $5;\ 5+(-2)=3;\ 3+(-2)=1;\ 1+(-2)=-1;\ (-1)+(-2)=-3$. The first five terms of this arithmetic sequence 5, 3, 1, -1, and -3.

11. $a_3=10;\ a_1=10-2(-2)=14;\ a_2=10-(-2)=12;\ a_4=10+(-2)=8;\ a_5=8+(-2)=6$. The first five terms of this arithmetic sequence are 14, 12, 10, 8, and 6.

13. $a_8=5+(8-1)(2)=19;\ a_n=5+(n-1)(2)=2n+3$. The eighth term of the sequence is 19. The nth term of the sequence is $2n+3$.

15. $a_3=2;\ d=1\Rightarrow a_1=2-2(1)=0;\ a_8=0+(8-1)(1)=7;\ a_n=0+(n-1)(1)$. The eighth term of the sequence is 7. The nth term of the sequence is $n-1$.

17. $a_1 = 8;\ a_6 = 6 \Rightarrow d = 6 - 8 = -2;\ a_8 = 8 + (8-1)(-2) = -6;\ a_n = 8 + (n-1)(-2) = -2n + 10$. The eighth term of the sequence is -6. The nth term of the sequence is $-2n+10$.

19. $a_{10} = 6;\ a_{12} = 15 \Rightarrow 2d = 15 - 6 = 9;\ d = 4.5;\ a_1 = 6 - (9)(4.5) = -3.45;$

$a_8 = -34.5 + (8-1)(4.5) = -3;\ a_n = -34.5 + (n-1)(4.5) = 4.5n - 39$. The eighth term of the sequence is -3 The nth term of the sequence is $4.5n - 39$.

21. $a_1 = x;\ a_2 = x + 3 \Rightarrow d = (x+3) - x = 3;\ a_8 = x + (8-1)(3) = x + 21;\ a_n = x + (n-1)(3) = x + 3n - 3$.

The eighth term of the sequence is $x + 21$. The nth term of the sequence is $x + 3n - 3$.

23. $a_3 = \pi + 2\sqrt{e}; a_4 = \pi + 3\sqrt{e} \Rightarrow d = \left(\pi + 3\sqrt{3}\right) - \left(\pi + 2\sqrt{e}\right) \Rightarrow a_1 = \pi.\ a_8 = \pi + 7\sqrt{e}\ \ a_n = \pi + (n-1)\sqrt{e}$

The eighth term is $\pi + 7\sqrt{e}$ and the nth term is $\pi + (n-1)\sqrt{e}$.

25. $a_5 = 27;\ a_{15} = 87 \Rightarrow 87 = 27 + 10d \Rightarrow 10d = 60 \Rightarrow d = 6;\ 27 = a_1 + (5-1)(6) \Rightarrow\ a_1 = 27 - 24 = 3$.

The first term of the sequence is 3.

27. $S_{16} = -160;\ a_{16} = -25 \Rightarrow -160 = \dfrac{16}{2}[a_1 + (-25)] \Rightarrow 8a_1 = -160 + 200 \Rightarrow a_1 = 5$. The first term of the sequence is 5.

29. $a_8 = -160;\ d = 3 \Rightarrow a_{10} = 8 + (10-1)(3) = 35 \Rightarrow S_{10} = \dfrac{10}{2}\left(8 + 35\right) = 215$. The sum of the first ten terms of the sequence is 215.

31. $a_3 = 5;\ a_4 = 8 \Rightarrow d = 8 - 5 = 3 \Rightarrow 5 = a_1 + (3-1)(3) \Rightarrow a_1 = -1 \Rightarrow a_{10} = -1 + (10-1)(3) = 26 \Rightarrow$

$S_{10} = \dfrac{10}{2}(-1 + 26) = 125$. The sum of the first ten terms of the sequence is 125.

33. $a_1 = 5;\ d = 9 - 5 = 4 \Rightarrow a_{10} = 5 + (10-1)(4) = 41 \Rightarrow S_{10} = \dfrac{10}{2}(5 + 41) = 230$. The sum of the first ten terms of the sequence is 230.

35. $a_1 = 10;\ a_{10} = 5.5 \Rightarrow S_{10} = \dfrac{10}{2}(10 + 55) = 77.5$. The sum of the first ten terms of the sequence is 77.5.

37. $S_{20} = 1090;\ a_{20} = 102 \Rightarrow 1090 = \dfrac{20}{2}(a_1 + 102) \Rightarrow 109 = a_1 + 102 \Rightarrow a_1 = 7 \Rightarrow 102 = 7 + (20-1)d \Rightarrow$

$19d = 95 \Rightarrow d = 5$. The first term of the sequence is 7 and the common difference is 5.

39. $S_{12} = -108;\ a_{12} = -19 \Rightarrow -108 = \dfrac{12}{2}[a_1 + (-19)] \Rightarrow 6a_1 = -108 + 114 = 6 \Rightarrow$

$a_1 = 1 \Rightarrow -19 = 1 + (12-1)d \Rightarrow 11d = -20 \Rightarrow d = -\dfrac{20}{11}$. The first term of the sequence is 1 and the common difference is $-\dfrac{20}{11}$.

41. From the graph, $a_1=-2, a_2=-1 \Rightarrow d=-1-(-2)=1 \Rightarrow a_n=-2+(n-1)=n-3$. Also from the graph, $D:\{1,2,3,4,5,6\}$; $R:\{-2,-1,0,1,2,3\}$. The nth term of the sequence is $n-3$. The domain of the sequence is the set $\{1,2,3,4,5,6\}$ and the range of the sequence is set $\{-2,-1,0,1,2,3\}$.

43. From the graph, $a_1=2.5,\ a_2=2 \Rightarrow d=2-2.5=-0.5 \Rightarrow a_n=2.5+(n-1)(-0.5)=-0.5n+3$. Also from the graph, $D:\{1,2,3,4,5,6\}$; $R:\{0.5,1,1.5,2,2.5\}$. The nth term of the sequence is $-0.5n+3$. The domain of the sequence is the set $\{1,2,3,4,5,6\}$ and the range of the sequence is set $\{0.5,1,1.5,2,2.5\}$.

45. From the graph, $a_1=10,\ a_2=-10 \Rightarrow d=-10-10=-20 \Rightarrow a_n=10+(n-1)(-20)=-20n+30$. Also from the graph, $D:\{1,2,3,4,5\}$; $R:\{-70,-50,-30,-10,10\}$. The nth term of the sequence is $-20n+30$. The domain of the sequence is the set $\{1,2,3,4,5\}$ and the range of the sequence is set $\{-70,-50,-30,-10,10\}$.

47. $a_1=3,\ a_8=17;\ 3+5+7+9+11+13+15+17=S_8=\frac{8}{2}(3+17)=80$. The sum of the series is 80.

49. $a_1=1;\ a_{50}=50;\ 1+2+3+4+\cdots+50=S_{50}=\frac{50}{2}(1+50)=1275$. The sum of the series is 1275.

51. $a_1=-7;\ d=-4-(-7)=3;\ 101=-7+(n-1)(3) \Rightarrow 3n=111 \Rightarrow n=37;$

$-7+(-4)+(-1)+2+5+\ldots+98+101=S_{37}=\frac{37}{2}(-7+101)=1739$. The sum of the series is 1739.

53. $a_1=5(1)=5;\ a_{40}=5(40)=200;\ S_{40}=\frac{40}{20}(5+200)=4100$. The sum of the series is 4100.

55. $a_1=1+4=5;\ a_3=3+4=7;\ \sum_{i=1}^{3}(i+4)=S_3=\frac{3}{2}(5+7)=18$. The sum is 18.

57. $a_1=2(1)+3=5; a_{10}=2(10)+3=23;\ \sum_{j=1}^{10}(2j+3)=S_{10}=\frac{10}{2}(5+23)=140$. The sum is 140.

59. $a_1=-5-8(1)=-13;\ a_{12}=-5-8(12)=-101;\ \sum_{i=1}^{12}(-5-8i)=S_{12}=\frac{12}{2}[-13+(-101)]=-684$.

The sum is -684.

61. $a_1=1;\ a_{1000}=1000;\ \sum_{i=1}^{1000} i=S_{1000}=\frac{1000}{2}(1+1000)=500{,}500$. The sum is 500,500.

63. $a_n=4.2n+9.73$. Using the sequence feature of a graphing calculator, we obtain $s_{10}=328.3$.

65. $a_n=\sqrt{8}n+\sqrt{3}$. Using the sequence feature of a graphing calculator, we obtain $s_{10}=172.884$.

67. $a_1=51;\ d=1;\ 71=51+(n-1)(1) \Rightarrow n=71-50=21;\ \sum_{i=51}^{71} i=s_{21}=\frac{21}{2}(51+71)=1281$. The sum of all the integers from 51 to 71 is 1281.

69. $a_1 = 1$; $a_{12} = 12$; $s_{12} = \frac{12}{2}(1+12) = 78$; chimes per 24 hours $= 2(78) = 156$. Chimes per 30 days $= (30)(156) = 4680$. The clock will chime 4680 times in a month of 30 days.

71. $a_1 = 49{,}000$, $d = 580$, $n = 11$; $a_{11} = 49{,}000 + (11-1)(580) = 54{,}800$. The population five years from now will be 54,800.

73. $a_1 = 18$; $a_{31} = 28$; $s_{31} = \frac{31}{2}(18+28) = 713$. A total of 713 inches of material will be needed.

(b) $a_6 = 98.2 + (6-1)(5.8) = 127.2$. We would expect the child's height to be 127.2 centimeters at age 8.

75. $a_1, a_2, a_3, \ldots$ is an arithmetic sequence $\Rightarrow a_2 = a_1 + d, a_3 = a_2 + d$. For $a_1^2, a_2^2, a_3^2, \ldots$ to also be an arithmetic sequence, $a_2^2 - a_1^2 = a_3^2 - a_2^2 \Rightarrow (a_1+d)^2 - a_1^2 = (a_2+d)^2 - a_2^2 \Rightarrow$

$a_1^2 + 2a_1d + d^2 - a_1^2 = a_2^2 + 2a_2d + d^2 - a_2^2 \Rightarrow 2a_1d + d^2 = 2a_2d + d^2 \Rightarrow a_1d = a_2d \Rightarrow a_1 = a_2 \Rightarrow d = 0$.

For both $a_1, a_2, a_3, \ldots$ and $a_1^2, a_2^2, a_3^2, \ldots$ to be arithmetic sequences, the terms of each are constant. That is, $a_1, a_2, a_3, \ldots = k, k, k, \ldots$ and $a_1^2, a_2^2, a_3^2, \ldots = k^2, k^2, k^2, \ldots$

8.3: Geometric Sequences and Series

1. $a_1 = \frac{5}{3}$; $a_2 = \left(\frac{5}{3}\right)(3) = 5$; $a_3 = (5)(3) = 15$; $a_4 = (15)(3) = 45$. The first four terms of the sequence are $\frac{5}{3}, 5, 15, 45$.

3. $a_4 = 5, a_5 = 10 \Rightarrow r = \frac{10}{5} = 2$; $a_3 = \frac{5}{2}$; $a_2 = \frac{\frac{5}{2}}{2} = \frac{5}{4}$; $a_1 = \frac{\frac{5}{4}}{2} = \frac{5}{8}$. The first five terms of the sequence are $\frac{5}{8}, \frac{5}{4}, \frac{5}{2}, 5, 10$.

5. $a_5 = 5(-2)^{5-1} = (5)(16) = 80$; $a_n = 5(-2)^{n-1}$. The fifth term of the sequence is 80 and the nth term of the sequence is $5(-2)^{n-1}$.

7. $a_2 = -4, r = 3 \Rightarrow a_1 = -\frac{4}{3}$; $a_5 = \left(-\frac{4}{3}\right)(3)^{5-1} = -108$; $a_n = \left(-\frac{4}{3}\right)(3)^{n-1}$. The fifth term of the sequence is -108 and the nth term of the sequence is $\left(-\frac{4}{3}\right)(3)^{n-1}$.

9. $a_4 = 243, r = -3 \Rightarrow a_1 = \frac{243}{(-3)^{4-1}} = -9$; $a_5 = (-9)(-3)^{5-1} = -729$; $a_n = (-9)(-3)^{n-1}$. The fifth term of the sequence is -729 and the nth term of the sequence is $(-9)(-3)^{n-1}$.

11. $r=\frac{-12}{-4}=\frac{-36}{-12}=\frac{-108}{-36}=3;\ a_5=(-4)(3)^{5-1}=-324;\ a_n=(-4)(3)^{n-1}$. The fifth term of the sequence is -324 and the nth term of the sequence is $(-4)(3)^{n-1}$.

13. $r=\frac{2}{\frac{4}{5}}=\frac{5}{2}=\frac{\frac{25}{2}}{5}=\frac{5}{2};\ a_5=\left(\frac{4}{5}\right)\left(\frac{5}{2}\right)^{5-1}=\frac{125}{4};\ a_n=\left(\frac{4}{5}\right)\left(\frac{5}{2}\right)^{n-1}$. The fifth term of the sequence is $\frac{125}{4}$ and the nth term of the sequence is $\left(\frac{4}{5}\right)\left(\frac{5}{2}\right)^{n-1}$.

15. $r=\frac{-5}{10}=\frac{\frac{5}{2}}{-5}=\frac{-\frac{5}{4}}{\frac{5}{2}}=-\frac{1}{2};\ a_5=(10)\left(-\frac{1}{2}\right)^{5-1}=\frac{5}{8};\ a_n=10\left(-\frac{1}{2}\right)^{n-1}$. The fifth term of the sequence is $\frac{5}{8}$ and the nth term of the sequence is $10\left(-\frac{1}{2}\right)^{n-1}$.

17. $a_3=5,\ a_8=\frac{1}{625}\Rightarrow r^{7-2}=\frac{\frac{1}{625}}{5}=\frac{1}{3125}\Rightarrow r=\frac{1}{5};\ a_1=\frac{5}{\left(\frac{1}{5}\right)^2}=125$. The first term of the sequence is 125 and the common ratio is $\frac{1}{5}$.

19. $a_4=-\frac{1}{4},\ a_9=-\frac{1}{128}\Rightarrow r^{8-3}=\frac{-\frac{1}{128}}{-\frac{1}{4}}=\frac{1}{32}\Rightarrow r=\frac{1}{2};\ a_1=\frac{-\frac{1}{4}}{\left(\frac{1}{2}\right)^3}=-2$. The first term of the sequence is -2 and the common ratio is $\frac{1}{2}$.

21. $r=\frac{8}{2}=\frac{32}{8}=\frac{128}{32}=4;\ S_5=\frac{2(1-4^5)}{1-4}=682$. The sum of the first five terms of the sequence is 682.

23. $r=\frac{-9}{18}=\frac{\frac{9}{2}}{-9}=\frac{-\frac{9}{4}}{\frac{9}{2}}=-\frac{1}{2};\ S_5=\frac{18\left[1-\left(-\frac{1}{2}\right)^5\right]}{1-\left(-\frac{1}{2}\right)}=\frac{99}{8}$. The sum of the first five terms of the sequence is $\frac{99}{8}$.

25. $S_5=\frac{(8.423)\left[1-(2.859)^5\right]}{1-2.859}\approx 860.95$. Rounded to the nearest hundredth, the sum of the first five terms of the sequence is 860.95.

27. $a_1=3;\ r=3;\ S_5=\frac{3(1-3^5)}{1-3}=363$. The sum is 363.

29. $a_1 = (48)\left(\frac{1}{2}\right) = 24;\ r = \frac{1}{2};\ S_6 = \frac{24\left[1-\left(\frac{1}{2}\right)^6\right]}{1-\frac{1}{2}} = \frac{189}{4}$. The sum is $\frac{189}{4}$.

31. $a_1 = 2^4 = 16;\ r = 2;\ n = 10-3 = 7;\ S_7 = \frac{16\left(1-2^7\right)}{1-2} = 2032$. The sum is 2032.

33. The sum of the terms of an infinite geometric sequence exists if the absolute value of the common ratio is less than 1.

35. $a_1 = 0.8,\ r = 0.1 \Rightarrow S_\infty = \frac{0.8}{1-0.1} = \frac{8}{9}$. The sum of the geometric series is $\frac{8}{9}$.

37. $a_1 = 0.45,\ r = 0.01 \Rightarrow S_\infty = \frac{0.45}{1-0.01} = \frac{5}{11}$. The sum of the geometric series is $\frac{5}{11}$.

39. $r = \frac{24}{12} = \frac{48}{24} = \frac{96}{48} = 2;\ |r| > 1 \Rightarrow$ the sum will not converge. Because the common ratio, 2, is greater than 1, the sum of the terms of the geometric sequence does not converge.

41. $r = \frac{-24}{-48} = \frac{-12}{-24} = \frac{-6}{-12} = \frac{1}{2};\ |r| < 1 \Rightarrow$ the sum would converge. The common ratio is $\frac{1}{2}$ and the sum of the terms of the geometric sequence would converge.

43. $r = \frac{2}{16} = \frac{\frac{1}{4}}{2} = \frac{\frac{1}{32}}{\frac{1}{4}} = \frac{1}{8};\ S_\infty = \frac{16}{1-\frac{1}{8}} = \frac{128}{7}$. The sum is $\frac{128}{7}$.

45. $r = \frac{10}{100} = \frac{1}{10};\ S_\infty = \frac{100}{1-\frac{1}{10}} = \frac{1000}{9}$. The sum is $\frac{1000}{9}$.

47. $r = \frac{\frac{2}{3}}{\frac{4}{3}} = \frac{\frac{1}{3}}{\frac{2}{3}} = \frac{1}{2};\ S_\infty = \frac{\frac{4}{3}}{1-\frac{1}{2}} = \frac{8}{3}$. The sum is $\frac{8}{3}$.

49. $a_1 = 3,\ r = \frac{1}{4} \Rightarrow S_\infty = \frac{3}{1-\frac{1}{4}} = 4$. The sum is 4.

51. $a_1 = 0.3,\ r = 0.3 \Rightarrow S_\infty = \frac{0.3}{1-0.3} = \frac{3}{7}$. The sum is $\frac{3}{7}$.

53. See Figure 53. Rounded to the nearest thousandth, the sum is 97.739.

55. See Figure 55. Rounded to the nearest thousandth, the sum is 0.212.

```
sum(seq(1.4^n,n,
1,10))
        97.73912924
```

Figure 53

```
sum(seq(2(.4)^n,
n,3,8))
         .21245952
```

Figure 55

57. $S = 1000\left[\frac{(1+0.04)^9 - 1}{0.04}\right] \approx 10{,}582.80$. Rounded to the nearest penny, the future value of the annuity is $\$10{,}582.80$.

59. $S = 2430\left[\frac{(1+0.025)^{10}}{0.025}\right] \approx 27{,}244.22$. Rounded to the nearest penny, the future value of the annuity is $\$27{,}224.22$.

61. (a) $a_1 = 1276(0.916)^1 \approx 1169;\ r = 0.916$. Rounded to the nearest whole number, the first term of the sequence is 1169. The common ratio is 0.916.

(b) $a_{10} = 1276(0.916)^{10} \approx 531;\ a_{20} = 1276(0.916)^{20} \approx 221$. Rounded to the nearest whole number, the tenth term of the sequence is 531 and the twentieth term of the sequence is 221. This means that a person 10 years from retirement should have saved 531% of his or her annual income and a person 20 years from retirement should have saved 221% of his or her annual income.

63. (a) $a_2 = 2a_1;\ a_3 = 2a_2 = 4a_1;\ \ldots a_n = a_1(20)^{n-1}$. The nth term of the geometric series is $a_1(2)^{n-1}$.

(b) $a_1 = 100;\ a_n > 1{,}000{,}000 \Rightarrow (2)^{n-1} > 10{,}000 \Rightarrow n-1 > \frac{4}{\log(2)} \Rightarrow n > 14.25 = 15.$

(c) $n = 4$ time = $(15-1)$ 40 minutes $= 560$ minutes $= 9$ hours 20 minutes. After 9 hours and 20 minutes, the number of bacteria will exceed one million.

65. $a_1 = 0.8,\ r = 0.8 \Rightarrow a_9 = (0.8)(0.8)^{9-1} \approx 0.134 = 13.4\%$. After nine drainings and replacements with water, the strength of the mixture will be approximately 13.4%.

67. $a_1 = (100{,}000)(1-0.2) = 80{,}000;\ r = 0.8 \Rightarrow a_6 = 80{,}000(0.8)^{6-1} = 26{,}214.40\ldots$ The value of the machine at the end of six years will be \$26,214.40.

69. $a_1 = 40,\ r = 0.8 \Rightarrow S_\infty = \frac{40}{1-0.8} = 200$. The total length of arc through which the pendulum will swing is 200 centimeters.

71. $a_1 = 2,\ r = 2 \Rightarrow S_5 = \frac{2(1-2^5)}{1-2} = 62;\ S_{10} = \frac{2(1-2^{10})}{1-2} = 2046$. Going back 5 generations, a person has 62 ancestors and going back 10 generations, a person has 2046 ancestors

73. Option 1: Arithmetic sequences with $a_1 = 5000$ and $d = 10{,}000 \Rightarrow$

$$S_{30} = \frac{30}{2}\left[2(5000) + (30-1)(10{,}000)\right] = 4{,}500{,}000.$$

Option 2: Geometric sequences with $a_1 = 0.01$ and $r = 2 \Rightarrow S_{30} = \frac{0.01(1-2^{30})}{1-2} = 10{,}737{,}418.23$.

Option 1 pays a total of \$4,500,000 while Option 2 pays a total of \$10,737,418.23. Choose Option 2.

75. $a_1 = 2, r = \frac{1}{2} \Rightarrow a_8 = 2\left(\frac{1}{2}\right)^{8-1} = \frac{1}{64}$. The eighth such triangle will have sides of length $\frac{1}{64}$.

Reviewing Basic Concepts (Sections 8.1—8.3)

1. $a_1 = (-1)^{1-1}(4\cdot 1) = 4$; $a_2 = (-1)^{2-1}(4\cdot 2) = -8$; $a_3 = (-1)^{3-1}(4\cdot 3) = 12$; $a_4 = (-1)^{4-1}(4\cdot 4)$;

   $a_5 = (-1)^{5-1}(4\cdot 5) = 20$. The first five terms of the sequence are $4, -8, 12, -16, 20$.

2. $a_1 = 3\cdot 1 + 1 = 4$; $a_5 = 3\cdot 5 + 1 = 16$; $s_5 = \frac{5}{2}(4+16) = 50$. The series sum is 50.

3. From Figure 3, it can be seen that as n increases, the terms of the sequence converge to 1.

[0,11] by [-2,2]
Xscl = 1 Yscl = 1

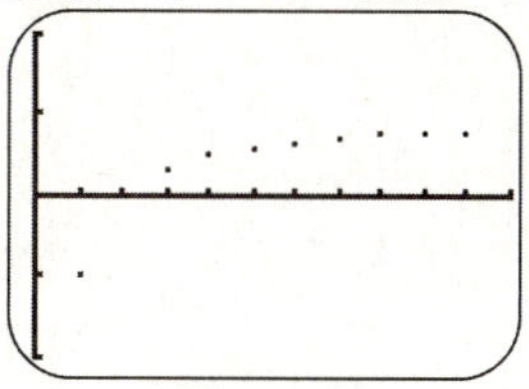

Figure 3

4. $a_1 = 8, d = -2 \Rightarrow a_2 = 8 - 2 = -6, a_3 = 6 - 2 = 4, a_4 = 4 - 2 = 2, a_5 = 2 - 2 = 0$. The first five terms of the sequence are 8, 6, 4, 2, 0.

5. $a_5 = 5, a_8 = 17 \Rightarrow 3d = 17 - 5 \Rightarrow d = 4 \Rightarrow a_1 = 5 - 4\cdot 4 = -11$. The first term of the sequence is -11.

6. $a_{10} = 2 + 5(10-1) = 47$; $S_{10} = \frac{10}{2}(2+47) = 245$. The sum of the first ten terms of the sequence is 245.

7. $a_3 = (-2)(-3)^{3-1} = -18$; $a_n = (-2)(-3)^{n-1}$. .The third term of the sequence is -18 and the nth term of the sequence is $(-2)(-3)^{n-1}$.

8. $r = \frac{3}{5} = \frac{\frac{9}{5}}{3} = \frac{3}{5}$; $\frac{244}{625} = 5\left(\frac{3}{5}\right)^{n-1} \Rightarrow \left(\frac{3}{5}\right)^{n-1} = \frac{243}{3125} \Rightarrow n - 1 = 5 \Rightarrow n = 6$.

   $S_6 = \frac{5\left[1 - \left(\frac{3}{5}\right)^6\right]}{1 - \frac{3}{5}} = \frac{7448}{625}$. The series is geometric with common ratio $\frac{3}{5}$. The series sum is $\frac{7448}{625}$.

9. $a_1 = 3\left(\frac{2}{3}\right) = 2$. $S_\infty = \frac{2}{1 - \frac{2}{3}} = 6$. The series sum is 6.

10. $S = 500\left[\frac{(1+0.035)^{13} - 1}{0.035}\right] \approx 8056.52$. Rounded to the nearest penny, the future value of the annuity is $8,056.52.

8.4: Counting Theory

1. $4! = 4 \cdot 3 \cdot 2 \cdot 1 = 24$

3. $(4-2)! = 2! = 2 \cdot 1 = 2$

5. $\frac{6!}{5!} = \frac{6 \cdot 5 \cdot 4 \cdot 3 \cdot 2 \cdot 1}{5 \cdot 4 \cdot 3 \cdot 2 \cdot 1} = 6$

7. $\frac{8!}{6!} = \frac{8 \cdot 7 \cdot 6 \cdot 5 \cdot 4 \cdot 3 \cdot 2 \cdot 1}{6 \cdot 5 \cdot 4 \cdot 3 \cdot 2 \cdot 1} = 8 \cdot 7 = 56$

9. $3! \cdot 4 = 3 \cdot 2 \cdot 1 \cdot (4) = 24$

11. Divide the common factor $6 \cdot 5 \cdot 4 \cdot 3 \cdot 2 \cdot 1$ from both the numerator and denominator to begin. Then multiply the factors 7 and 8 which remain in the denominator to obtain 56.

13. $P(7,7) = \frac{7!}{(7-7)!} = \frac{7!}{0!} = 7 \cdot 6 \cdot 5 \cdot 4 \cdot 3 \cdot 2 \cdot 1 = 5040$. The value of $P(7,7)$ is 50,400

15. $P(9,2) = \frac{9!}{(9-2)!} = \frac{9!}{7!} = 9 \cdot 8 = 72$. The value of $P(9,2)$ is 72.

17. $P(5,1) = \frac{5!}{(5-1)!} = \frac{5!}{4!} = 5$. The value of $P(5,4)$ is 5.

19. $C(4,2) = \frac{4!}{(4-2)!2!} = \frac{4!}{2!2!} = \frac{4 \cdot 3}{2 \cdot 1} = 6$. The value of $C(4,2)$ is 6.

21. $C(6,0) = \frac{6!}{(6-0)!0!} = \frac{6!}{6!} = 1$. The value of $C(6,0)$ is 1.

23. $C(12,4) = \frac{12!}{(12-4)!4!} = \frac{12!}{8!4!} = \frac{12 \cdot 11 \cdot 10 \cdot 9}{4 \cdot 3 \cdot 2 \cdot 1} = 495$. The value of $C(12,4)$ is 495.

25. ${}_{20}P_5 = 20\ {}_nP_r\ 5 = 1{,}860{,}480$ The value of ${}_{20}P_5$ is $1{,}860{,}480$

27. ${}_{15}P_8 = 15_n P_r 8 = 259{,}459{,}200$. The value of ${}_{15}P_8$ is $259{,}459{,}200$.

29. ${}_{20}C_5 = 20_n C_r 5 = 15{,}504$. The value of ${}_{20}C_5$ is $15{,}504$.

31. ${}_{15}C_8 = 15_n C_r 8 = 6435$. The value of ${}_{15}C_8$ is 6435.

33. (a) A telephone number involves a permutation of digits because order matters.
    (b) A social security number involves a permutation of digits because order matters.
    (c) A hand of cards in poker involves a combination of cards because order does not matter.
    (d) A committee of politicians involves a combination of persons because order does not matter.
    (e) The "combination" on a combination lock involves a permutation of numbers because order matters.
    (f) A lottery choice of six numbers where the order does not matter involves a combination of numbers.
    (g) An automobile license plate involves a permutation of characters because order matters.

35. $5 \times 3 \times 2 = 30$. 30 different types of homes are available.

37. (a) $2 \times 25 \times 24 \times 23 = 27{,}600$. 27,600 radio station calls can be made.

(b) $2\times26\times26\times26=35{,}152$. 35,152 radio station calls can be made.

(c) $2\times24\times24\times1=1104$. 1104 radio station calls can be made.

39. $3\times5=15$. 15 first and middle name combinations are possible.

41. (a) $26\times26\times26\times10\times10\times10=17{,}576{,}000$. 17,576,000 different license plates are possible.

(b) $10\times10\times10\times26\times26\times26=17{,}576{,}000$. 17,576,000 additional plates are possible.

(c) $26\times10\times10\times10\times26\times26\times26=456{,}976{,}000$. The new scheme provides 456,976,000 plates..

43. $6_n P_r 6=720$. 720 arrangements of the people are possible.

45. $6_n P_r 3=120$. 120 course schedules are possible.

47. $15_n P_r 3=2730$. 2730 slates of 3 officers are possible.

49. $5_n P_r 5=120$; $10_n P_r 5=30{,}240$. With 5 players, 120 assignments are possible and with 10 players, 30,240 assignments are possible.

51. $30_n P_r 4=27{,}405$. 27,405 different groups are possible.

53. $6_n C_r 3=20$. 20 different garnished hamburgers are possible.

55. $15_n C_r 2=105$; $15_n C_r 4=1365$. 105 samples of 2 may be drawn and 1365 samples of 4 may be drawn.

57. $8_n C_r 2=28$. 28 samples of 2 may be drawn where both marbles are blue.

59. (a) $9_n C_r 3=84$. 84 delegations are possible.

(b) $5_n C_r 3=10$. 10 delegations could have all liberals.

(c) $(5_n C_r 2)(4_n C_r 1)=(10)(4)=40$. 40 delegations could have 2 liberals and 1 conservative.

(d) $8_n C_r 2=28$. 28 delegations are possible that include the mayor.

61. $8_n P_r 4=1680$. 1680 course schedules are possible.

63. $6_n C_r 4=15$. 15different soups can be made.

65. $12_n P_r 11=479{,}001{,}600$. 479,001,600 different seatings are possible.

67. (a) $8_n C_r 5=56$. 56 committees of all men may be chosen.

(b) $11_n C_r 5=462$. 462 committees of all women may be chosen.

(c) $(8_n C_r 3)(11_n C_r 2)=(56)(55)=3080$. 3080 committees of 3 men and 2 women may be chosen.

(d) $(8_n C_r 5)(11_n C_r 0)+(8_n C_r 4)(11_n C_r 1)+(8_n C_r 3)(11_n C_r 2)+(8_n C_r 2)(11_n C_r 3)=$

$(56)(1)+(70)(11)+(56)(55)+(28)(165)=8526$. 8526 committees with no more than 3 women are possible.

69. $2^{12}=4096$. 4096 codes are possible.

71. $(10\times10\times10)(10\times10\times10)=1{,}000{,}000$. 1,000,000 different combinations are possible.

73. Circular ring $\Rightarrow 1$ key to start; then, $3_n P_r 3=6$ ways to add the remaining 3. The keys can be put on in 6 distinguishable ways.

75. $P(n,n-1)=\frac{n!}{[n-(n-1)]!}=\frac{n!}{1!}=n!$; $P(n,n)=\frac{n!}{(n-n)!}=\frac{n!}{0!}=n!$. Thus, $P(n,n-1)=P(n,n)$.

77. $P(n,0)=\frac{n!}{(n-0)!}=\frac{n!}{n!}=1$. Thus, $P(n,0)=1$.

79. $C(n,0)=\frac{n!}{(n-0)!0!}=\frac{n!}{n!}=1$. Thus, $C(n,0)=1$.

81. $C(n,n-1)=\frac{n!}{[n-(n-1)]!(n-1)!}=\frac{n!}{1!(n-1)!}=\frac{n(n-1)!}{(n-1)!}=n$. Thus, $C(n,n-1)=n$.

8.5: The Binomial Theorem

1. $\frac{6!}{3!3!}=\frac{6\cdot5\cdot4\cdot3\cdot2\cdot1}{3\cdot2\cdot1\cdot3\cdot2\cdot1}=20$. The expression is equal to 20.

3. $\frac{7!}{3!4!}=\frac{7\cdot6\cdot5\cdot4\cdot3\cdot2\cdot1}{3\cdot2\cdot1\cdot4\cdot3\cdot2\cdot1}=35$. The expression is equal to 35.

5. $\binom{8}{3}=C(8,3)=\frac{8!}{(8-3)!3!}=\frac{8!}{5!3!}=56$. The expression is equal to 56.

7. $\binom{10}{8}=C(10,8)=\frac{10!}{(10-8)!10!}=\frac{10!}{2!10!}=45$. The expression is equal to 45.

9. $\binom{13}{13}=C(13,13)=\frac{13!}{(13-13)!13!}=\frac{13!}{13!}=1$. The expression is equal to 1.

11. $\binom{8}{3}=C(8,3)=\frac{8!}{(8-3)!3!}=\frac{8!}{5!3!}=56$. The expression is equal to 56.

13. $\binom{100}{2}=C(100,2)=\frac{100!}{(100-2)!2!}=\frac{100!}{98!2!}=4950$. The expression is equal to 4950.

15. $\binom{5}{0}=C(5,0)=\frac{5!}{(5-0)!0!}=\frac{5!}{5!}=1$. The expression is equal to 1.

17. $\binom{n}{n-1}=\frac{n!}{(n-1)!(n-(n-1))!}=\frac{n!}{(n-1)!(1)!}=\frac{n(n-1)!}{(n-1)!}=n$. The expression is equal to n.

19. The expansion of $(x+y)^8$ has a term where x is raised to each of the following powers: 8, 7, 6, 5, 4, 3, 2, 1, and 0. Thus, there are 9 terms.

21. $(2x)^4=16x^4; (3y)^4=81y^4$. In the expansion of $(2x)^4$, $16x^4$ is the first term and $81y^4$ is the last term.

23. The binomial expansion for $(x+y)^6$ is given by:

$$(x+y)^6=\binom{6}{0}x^6+\binom{6}{1}x^5y+\binom{6}{2}x^4y^2+\binom{6}{3}x^3y^3+\binom{6}{4}x^2y^4+\binom{6}{5}xy^5+\binom{6}{6}y^6=$$

$x^6+6x^5y+15x^4y^2+20x^3y^3+15x^2y^4+6xy^5+y^6$.

25. The binomial expansion for $(p-q)^5$ is given by:

$$(p-q)^5=\binom{5}{0}p^5+\binom{5}{1}p^4(-q)+\binom{5}{2}p^3(-q)^2+\binom{5}{3}p^2(-q)^3+\binom{5}{4}p(-q)^4+\binom{5}{5}(-q)^5=$$

$$p^5-5p^4q+10p^3q^2-10p^2q^3+5pq^4-q^5.$$

27. The binomial expansion for $\left(r^2+s\right)^5$ is given by:

$$\left(r^2+s\right)^5=\binom{5}{0}\left(r^2\right)^5+\binom{5}{1}\left(r^2\right)^4s+\binom{5}{2}\left(r^2\right)^3s^2+\binom{5}{3}\left(r^2\right)^2s^3+\binom{5}{4}\left(r^2\right)s^4+\binom{5}{5}s^5=$$

$$r^{10}+5r^8s+10r^6s^2+10r^4s^3+5r^2s^4+s^5.$$

29. The binomial expansion for $(p+2q)^4$ is given by:

$$(p+2q)^4=\binom{4}{0}p^4+\binom{4}{1}p^3(2q)+\binom{4}{2}p^2(2q)^2+\binom{4}{3}p(2q)^3+\binom{4}{4}(2q)^4=$$

$$p^4+8p^3q+24p^2q^2+32pq^3+16q^4.$$

31. The binomial expansion for $(7p+2q)^4$ is given by:

$$(7p+2q)^4=\binom{4}{0}(7p)^4+\binom{4}{1}(7p)^3(2q)+\binom{4}{2}(7p)^2(2q)^2+\binom{4}{3}(7p)(2q)^3+\binom{4}{4}(2q)^4=$$

$$2401p^4+2744p^3q+1176p^2q^2+224pq^3+16q^4.$$

33. The binomial expansion for $(3x-2y)^6$ is given by:

$$(3x-2y)^6=\binom{6}{0}(3x)^6+\binom{6}{1}(3x)^5(-2y)+\binom{6}{2}(3x)^4(-2y)^2+\binom{6}{3}(3x)^3(-2y)^3+\binom{6}{4}(3x)^2(-2y)^4+$$

$$\binom{6}{5}(3x)(-2y)^5+\binom{6}{6}(-2y)^6=729x^6-2916x^5y+4860x^4y^2-4320x^3y^3+2160x^2y^4-576xy^5+64y^6.$$

35. The binomial expansion for $\left(\frac{m}{2}-1\right)^6$ is given by:

$$\left(\frac{m}{2}-1\right)^6=\binom{6}{0}\left(\frac{m}{2}\right)^6+\binom{6}{1}\left(\frac{m}{2}\right)^5(-1)+\binom{6}{2}\left(\frac{m}{2}\right)^4(-1)^2+\binom{6}{3}\left(\frac{m}{2}\right)^3(-1)^3+\binom{6}{4}\left(\frac{m}{2}\right)^2(-1)^4+$$

$$\binom{6}{5}\left(\frac{m}{2}\right)(-1)^5+\binom{6}{6}(-1)^6=\frac{m^6}{64}-\frac{3m^5}{16}+\frac{15m^4}{16}-\frac{5m^3}{2}+\frac{15m^2}{4}-3m+1.$$

37. The binomial expansion for $\left(\sqrt{2}r+\frac{1}{m}\right)^4$ is given by:

$$\left(\sqrt{2r}+\frac{1}{m}\right)^4=\binom{4}{0}\left(\sqrt{2r}\right)^4+\binom{4}{1}\left(\sqrt{2r}\right)^3\left(\frac{1}{m}\right)+\binom{4}{2}\left(\sqrt{2r}\right)^2\left(\frac{1}{m}\right)^2+\binom{4}{3}\left(\sqrt{2r}\right)\left(\frac{1}{m}\right)^3+\binom{4}{4}\left(\frac{1}{m}\right)^4=$$

$$4r^4+\frac{8\sqrt{2}r^3}{m}+\frac{4\sqrt{2}r^2}{m^2}+\frac{4\sqrt{2}r}{m^3}+\frac{1}{m^4}.$$

39. $\binom{8}{5}(4h)^3(-j)^5=-3584h^3j^5$ The sixth term of the binomial expansion of $(4h-j)^8$ is $-3584n^3j^5$.

41. $\binom{22}{14}\left(a^2\right)\left(b^{14}\right)=319,770a^{16}b^{14}$ The fifteenth term of the binomial expansion of $\left(a^2+b\right)^{22}$ is $319,770a^{16}b^{14}$.

43. $\binom{20}{14}(x)^6\left(-y^3\right)^{14}=38,760x^6y^{42}$ The fifteenth term of the binomial expansion of $\left(x-y^3\right)^{20}$ is $38,760x^6y^{42}$.

45. $\binom{8}{4}\left(3x^7\right)\left(2y^3\right)^4=90,720x^{28}y^{12}$. The middle term of the binomial expansion of $\left(3x^7+2y^3\right)^8$ is the fifth term which is $90,720x^{28}y^{12}$.

47. $n-4=7$ and $n-7=4\Rightarrow n=11$. The coefficients of the fifth and eighth terms in the expansion of $(x+y)^n$ are the same for n=11.

49. $10!=3,628,800$; $\sqrt{2\pi(10)}\left(10^{10}\right)\left(e^{-10}\right)\approx 3,598,695.619$. The exact value of 10! is 3,628,800 and the Stirling's formula approximation of 10! is about 3,598,695.619.

50. $\dfrac{3,628,800-3,598,695.619}{3,628,800}\approx 0.00830=0.830\%$. The percentage error in the Stirling's formula approximation of 10! is about 0.830%.

51. $12!=479,001,600$; $\sqrt{2\pi(12)}\left(12^{12}\right)\left(e^{-12}\right)\approx 475,687,486.5$;

$\dfrac{479,001,600-475,687,486.5}{479,001,600}\approx 0.00692=0.692\%$. The exact value of 12! is 479,001,600 and the Stirling's formula approximation of 12! is about 475,687,486.5 which has a percent error of about 0.692%.

52. $13!=6,227,020,800$; $\sqrt{2\pi(13)}\left(13^{13}\right)\left(e^{-13}\right)\approx 6,187,239,475$;

$\dfrac{6,227,020,800-6,187,239,475}{6,227,020,800}\approx 0.00639=0.639\%$. The exact value of 13! Is 6,227,020,800 and the Stirling's formula approximation of 13! is about 6,187,239,475 which has a percent error of about 0.639%. Based on this series of exercises, it appears that the percent error in the Stirling's formula approximation of *n!* decreases as *n* increases.

53. $(1.02)^{-3}=(1+0.02)^{-3}\approx 1+(-3)(0.02)+\dfrac{(-3)(-3-1)}{2!}(0.02)^2+$

$\dfrac{(-3)(-3-1)(-3-2)}{3!}(0.02)^3\approx 0.942$. To the nearest thousandth, $(1.02)^{-3}$ is 0.942.

55. $(1.01)^{3/2}=(1+0.01)^{3/2}\approx 1+\left(\dfrac{3}{2}\right)(0.01)+\dfrac{\left(\frac{3}{2}\right)\left(\frac{3}{2}-1\right)}{2!}(0.01)^2+\dfrac{\left(\frac{3}{2}\right)\left(\frac{3}{2}-2\right)}{3!}(0.01)^3\approx 1.015$. To the nearest thousandth, $(1.01)^{3/2}$ is 1.015.

Reviewing Basic Concepts (Sections 8.4 and 8.5)

1. $4!=24$. There are 24 different arrangements.
2. $P(7,3)=\dfrac{7!}{(7-3)!}=\dfrac{7!}{4!}=210$.
3. $C(10,4)=\dfrac{10!}{(10-4)!4!}=\dfrac{10!}{6!4!}=210$
4. ${}_6C_2\cdot{}_5C_2\cdot{}_3C_1=450$. There are 450 different arrangements for this basketball team.
5. $9\cdot 4\cdot 2=72$. There are 72 different homes available.
6. To expand $(x+y)^n$ use row *n+1* of Pascal's Triangle.
7. Given $(a+2b)^4$. We will let $x=a$ and $y=2b$ in the binomial theorem. Using the 5th row of Pascal's Triangle we find the coefficients 1,4,6,4,1

   $\Rightarrow (a+2b)^4=(a)^4+4(a)^3(2b)+6(a)^2(2b)^2+4(a)(2b)^3+(2b)^4$
8. $\dbinom{6}{2}(x)^4(-2y)^2=60x^4y^2$. The third term of the binomial $(x-2y)^6$ is $60x^4y^2$.

8.6: Mathematical Induction

1. The domain of the variable must be all positive integers (natural numbers).

3. Since $2^1=2(1)$ and $2^2=2(2)$, the statement is not true for $n=1$ and $n=2$.

5. $3+6+9+\cdots+3n=\dfrac{3n(n+1)}{2}$

   (i) Show that the statement is true for $n=1$: $3(1)=\dfrac{3(1)(2)}{2}\Rightarrow 3=3$.

   (ii) Assume that S_k is true: $3+6+9+\cdots+3k=\dfrac{3k(k+1)}{2}$. Show that S_{k+1} is true: $3+6+\cdots+3(k+1)=\dfrac{3(k+1)(k+2)}{2}$. Add $3(k+1)$ to each side of

$S_k: 3+6+9+\cdots+3k+3(k+1)=\frac{3k(k+1)}{2}+3(k+1)=\frac{3k(k+1)+6(k+1)}{2}=\frac{(k+1)(3k+6)}{2}=$

$\frac{3(k+1)(k+2)}{2}.$

Since S_k implies S_{k+1}, the statement is true for every positive integer n.

7. $5+10+15+\cdots+5n=\frac{5n(n+1)}{2}$

(i) Show that the statement is true for $n=1$: $5(1)=\frac{5(1)(2)}{2}\Rightarrow 5=5.$

(ii) Assume that S_k is true: $5+10+15+\cdots+5k=\frac{5k(k+1)}{2}$. Show that S_{k+1} is true:

$5+10+\cdots+5(k+1)=\frac{5(k+1)(k+2)}{2}$. Add $5(k+1)$ to each side of S_k:

$5+10+15+\cdots+5k+5(k+1)=\frac{5k(k+1)}{2}+5(k+1)=$

$\frac{5k(k+1)+10(k+1)}{2}=\frac{(k+1)(5k+10)}{2}=\frac{5(k+1)(k+2)}{2}.$

Since S_k implies S_{k+1}, the statement is true for every positive integer n.

9. $3+3^2+3^3+\cdots+3^n=\frac{3(3^n-1)}{2}$

(i) Show that the statement is true for $n=1$: $3^1=\frac{3(3^1-1)}{2}\Rightarrow 3=3.$

(ii) Assume that S_k is true: $3+3^2+3^3+\cdots+3^k=\frac{3(3^k-1)}{2}$. Show that S_{k+1} is true:

$3+3^2+3^3+\cdots+3^{k+1}=\frac{3(3^{k+1}-1)}{2}$ Add 3^{k+1} to each side of S_k:

$3+3^2+3^3+\cdots 3^k+3^{k+1}=\frac{3\left(3^k-1\right)}{2}+3^{k+1}=\frac{3\left(3^k-1\right)+2\left(3^{k+1}\right)}{2}=$

$\frac{3^{k+1}-3+2(3^{k+1})}{2}=\frac{3(3^{k+1})-3}{2}=\frac{3(3^{k+1}-1)}{2}$

Since S_k implies S_{k+1}, the statement is true for every positive integer n.

11. $1^3+2^3+3^3+\cdots+n^3=\frac{n^2(n+1)^2}{4}$

(i) Show that the statement is true for $n=1$: $1^3=\frac{1^2(1+1)^2}{4}\Rightarrow 1=1.$

(ii) Assume that S_k is true: $1^3+2^3+3^3+\cdots+k^3=\dfrac{k^2(k+1)^2}{4}$. Show that S_{k+1} is true:

$1^3+2^3+3^3+\cdots+(k+1)^3=\dfrac{(k+1)^2(k+2)^2}{4}$. Add $(k+1)^3$ to each side of

$S_k:\ 1^3+2^3+3^3+\cdots+k^3+(k+1)^3=\dfrac{k^2(k+1)^2}{4}+(k+1)^3=$

$$\frac{k^2(k+1)^2+4(k+1)^3}{4}=\frac{(k+1)^2(k^2+4k+4)}{4}=\frac{(k+1)^2(k+2)^2}{4}$$

Since S_k implies S_{k+1}, the statement is true for every positive integer n.

13. $\dfrac{1}{1\cdot 2}+\dfrac{1}{2\cdot 3}+\cdots+\dfrac{1}{n(n+1)}=\dfrac{n}{n+1}$

(i) Show that the statement is true for $n=1$: $\dfrac{1}{1(1+1)}=\dfrac{1}{1+1}\Rightarrow\dfrac{1}{2}=\dfrac{1}{2}$.

(ii) Assume that S_k is true: $\dfrac{1}{1\cdot 2}+\dfrac{1}{2\cdot 3}+\cdots+\dfrac{1}{k(k+1)}=\dfrac{k}{k+1}$. Show that S_{k+1} is true:

$\dfrac{1}{1\cdot 2}+\dfrac{1}{2\cdot 3}+\cdots+\dfrac{1}{(k+1)(k+2)}=\dfrac{k+1}{k+2}$. Add $\dfrac{1}{(k+1)(k+2)}$ to each side of

$S_k:\ \dfrac{1}{1\cdot 2}+\dfrac{1}{2\cdot 3}+\cdots+\dfrac{1}{(k+1)(k+2)}=\dfrac{k}{k+1}+\dfrac{1}{(k+1)(k+2)}=$

$$\frac{k(k+2)+1}{(k+1)(k+2)}=\frac{k^2+2k+1}{(k+1)(k+2)}=\frac{(k+1)(k+1)}{(k+1)(k+2)}=\frac{k+1}{(k+2)}.$$

Since S_k implies S_{k+1}, the statement is true for every positive integer n.

15. $\dfrac{4}{5}+\dfrac{4}{5^2}+\dfrac{4}{5^3}+\cdots+\dfrac{4}{5^n}=1-\dfrac{1}{5^n}$

(i) Show that the statement is true for $n=1$: $\dfrac{4}{5^1}=1-\dfrac{1}{5^1}\Rightarrow\dfrac{4}{5}=\dfrac{4}{5}$.

(ii) Assume that S_k is true: $\dfrac{4}{5}+\dfrac{4}{5^2}+\dfrac{4}{5^3}+\cdots+\dfrac{4}{5^k}=1-\dfrac{1}{5^k}$.

Show that S_{k+1} is true: $\dfrac{4}{5}+\dfrac{4}{5^2}+\cdots+\dfrac{4}{5^{k+1}}=1-\dfrac{1}{5^{k+1}}$. Add $\dfrac{4}{5^{k+1}}$ to each side of S_k:

$$\frac{4}{5}+\frac{4}{5^2}+\frac{4}{5^3}+\cdots+\frac{4}{5^k}+\frac{4}{5^{k+1}}=1-\frac{1}{5^k}+\frac{4}{5^{k+1}}=1-\frac{1}{5^k}\cdot\frac{5}{5}+\frac{4}{5^{k+1}}=1-\frac{5}{5^{k+1}}+\frac{4}{5^{k+1}}=1-\frac{1}{5^{k+1}}.$$

Since S_k implies S_{k+1}, the statement is true for every positive integer n.

17. $\dfrac{1}{1\cdot 4}+\dfrac{1}{4\cdot 7}+\cdots+\dfrac{1}{(3n-2)(3n+1)}=\dfrac{n}{3n+1}$

(i) Show that the statement is true for $n=1$: $\frac{1}{1\cdot 4}=\frac{1}{3(1)+1}\Rightarrow\frac{1}{4}=\frac{1}{4}$.

(ii) Assume that S_k is true: $\frac{1}{1\cdot 4}+\cdots+\frac{1}{(3k-2)(3k+1)}=\frac{k}{3k+1}$.

Show that S_{k+1} is true: $\frac{1}{1\cdot 4}+\cdots+\frac{1}{[3(k+1)-2][3(k+1)+1]}=\frac{k+1}{3(k+1)+1}$. Add

$\frac{1}{[3(k+1)-2][3(k+1)+1]}$ to each side of S_k: $\frac{1}{1\cdot 4}+\cdots+\frac{1}{[3(k+1)-2][3(k+1)+1]}=$

$\frac{k}{3k+1}+\frac{1}{[3(k+1)-2][3(k+1)+1]}=\frac{k}{3k+1}+\frac{1}{(3k+1)(3k+4)}=\frac{k(3k+4)+1}{(3k+1)(3k+4)}=$

$\frac{3k^2+4k+1}{(3k+1)(3k+4)}=\frac{(3k+1)(k+1)}{(3k+1)(3k+4)}=\frac{k+1}{3k+4}=\frac{k+1}{3(k+1)+1}$.

Since S_k implies S_{k+1}, the statement is true for every positive integer n.

19. When $n=1$, $3^1<6(1)\Rightarrow 3<6$. When $n=2$, $3^2<6(2)\Rightarrow 9<12$. When $n=3$, $3^3>6(3)\Rightarrow 27>18$. For all $n\geq 3$, $3^n>6n$. The only values are 1 and 2.

21. When $n=1$, $2^1>1^2\Rightarrow 2>1$. When $n=2$, $2^2=2^2\Rightarrow 4=4$. When $n=3$, $2^3<3^2\Rightarrow 8<9$. When $n=4$, $2^4=4^2\Rightarrow 16=16$. For all $n\geq 5$, $2^n>n^2$. The only values are 2, 3, and 4..

23. $(a^m)^n=a^{mn}$

(i) Show that the statement is true for $n=1$: $(a^m)^1=a^{m\cdot 1}\Rightarrow a^m=a^m$.

(ii) Assume that S_k is true: $(a^m)^k=a^{mk}$.

Show that S_{k+1} is true: $(a^m)^{k+1}=a^{m(k+1)}$.

Multiply each side of S_k by a^m: $(a^m)^k\cdot(a^m)^1=a^{mk}\cdot a^m\Rightarrow(a^m)^{k+1}=a^{mk+m}\Rightarrow$

$(a^m)^{k+1}=a^{m(k+1)}$.

Since S_k implies S_{k+1}, the statement is true for every positive integer n.

25. $2^n>2n$, if $n\geq 3$

(i) Show that the statement is true for $n=3$: $2^3>2(3)\Rightarrow 8>6$.

(ii) Assume that S_k is true: $2^k>2k$.

Show that S_{k+1} is true: $2^{k+1}>2(k+1)$. Multiply each side of S_k by 2:

$2^k\cdot 2>2k\cdot 2\Rightarrow 2^{k+1}>2(2k)$. Because $2k>k+1$ for all $k>1$, we have $2(2k)>2(k+1)$; therefore,

$2^{k+1}>2(k+1)$.

Since S_k implies S_{k+1}, the statement is true for every positive integer $n \geq 3$.

27. $a^n > 1$, if $a > 1$

(i) Show that the statement is true for $n = 1$: $a^1 > 1 \Rightarrow a > 1$, which is true by the given restriction.

(ii) Assume that S_k is true: $a^k > 1$. Show that S_{k+1} is true: $a^{k+1} > 1$.

Multiply each side of S_k by a: $a^k \cdot a > 1 \cdot a \Rightarrow a^{k+1} > a$. Because $a > 1$, we may substitute 1 for a in the expression. That is, $a^{k+1} > 1$.

Since S_k implies S_{k+1}, the statement is true for every positive integer n.

29. $a^n < a^{n-1}$, if $0 < a < 1$

(i) Show that the statement is true for $n = 1$: $a^1 < a^0 \Rightarrow a < 1$, which is true by the given restriction.

(ii) Assume that S_k is true: $a^k < a^{k-1}$. Show that S_{k+1} is true: $a^{k+1} < a^k$.

Multiply each side of S_k by a: $a^k \cdot a < a^{k-1} \cdot a \Rightarrow a^{k+1} < a^k$.

Since S_k implies S_{k+1}, the statement is true for every positive integer n.

31. $n! > 2^n$, if $n \geq 4$

(i) Show that the statement is true for $n = 4$: $4! > 2^4 \Rightarrow 24 > 16$.

(ii) Assume that S_k is true: $k! > 2^k$. Show that S_{k+1} is true: $(k+1)! > 2^{k+1}$.

Multiply each side of S_k by $k+1$: $(k+1)k! > 2^k(k+1) \Rightarrow (k+1)! > 2^k(k+1)$.

Because $(k+1) > 2$ for all $k \geq 4$, we may substitute 2 for $(k+1)$ in the expression.

That is, $(k+1)! > 2^k(2)$ or $(k+1)! > 2^{k+1}$.

Since S_k implies S_{k+1}, the statement is true for every positive integer $n \geq 4$.

33. The number of handshakes is $\dfrac{n^2 - n}{2}$ if $n \geq 2$.

(i) Show that the statement is true for $n = 2$. The number of handshakes for 2 people is $\dfrac{2^2 - 2}{2} = \dfrac{2}{2} = 1$, which is true.

(ii) Assume that S_k is true: the number of handshakes for k people is $\dfrac{k^2 - k}{2}$.

Show that S_{k+1} is true: the number of handshakes for $k+1$ people is

$\dfrac{(k+1)^2 - (k+1)}{2} = \dfrac{k^2 + 2k + 1 - k - 1}{2} = \dfrac{k^2 + k}{2}$. When a person joins a group of k people, each person must shake hands with the new person. Since there are a total of k people that will shake hands with the new person, the total number of handshakes for $k+1$ people is

$$\frac{k^2-k}{2}+k=\frac{k^2-k+2k}{2}=\frac{k^2+k}{2}.$$

Since S_k, implies $S_{k+1,}$ the statement is true for every positive integer $n \geq 2$.

35. The first figure has perimeter $P=3$. When a new figure is generated, each side of the previous figure increases in length by a factor of $\frac{4}{3}$. Thus, the second figure has perimeter $P=3\left(\frac{4}{3}\right)$, the third figure has perimeter $P=3\left(\frac{4}{3}\right)^2$, and so on. In general, the nth figure has perimeter $P=3\left(\frac{4}{3}\right)^{n-1}$.

37. With 1 ring, 1 move is required. With 2 rings, 3 moves are required. Note that $3=2+1$. With 3 rings, 7 moves are required. Note that $7=2^2+2+1$. With *n* rings, $2^{n-1}+2^{n-2}+\cdots+2^1+1=2^n-1$ moves are required.

    (i) Show that the statement is true for $n-1$. The number of moves for 1 ring is $2^1-1=1$, which is true.

    (ii) Assume that S_k is true: the number of moves for *k* rings is 2^k-1. Show that S_{k+1} is true: the number of moves for $k+1$ rings is $2^{k+1}-1$. Assume $k+1$ rings are on the first peg. Since S_k is true, the top *k* rings can be moved to the second peg in 2^k-1 moves. Now move the bottom ring to the third peg. Since S_k is true, move the *k* rings from the second peg on top of the ring on the third peg in 2^k-1 moves. The total number of moves is $\left(2^k-1\right)+1+\left(2^k-1\right)=2\cdot 2^k-1=2^{k+1}-1$.

    Since S_k implies S_{k+1}, the statement is true for every positive integer *n*.

8.7: Probability

1. Since the coin has a head on each side, the sample space is $S=\{H\}$.

3. Since each of the three coins can be either heads or tails, the sample space is
$S=\{(H,H,H),(H,H,T),(H,T,H),(H,T,T),(T,H,H),(T,H,T),(T,T,H),(T,T,T)\}$.

5. On each spin, the spinner may land on 1, 2, or 3. The sample space is
$S=\{(1,1),(1,2),(1,3),(2,1),(2,2),(2,3),(3,1),(3,2),(3,3)\}$.

7. (a) The event $E_1=\{H\}$. The probability of the event, $P(E_1)=1$.

    (b) The event $E_2=\varnothing$. The probability of the event, $P(E_2)=0$.

9. (a) The event $E_1=\{(1,1),(2,2),(3,3)\}$. The probability of the event, $P(E_1)=\frac{3}{9}=\frac{1}{3}$.

(b) The event $E_2 = \{(1,1),(1,3),(2,1),(2,3),(3,1)(3,3)\}$. The probability of the event, $P(E_2) = \frac{6}{9} = \frac{2}{3}$.

(c) The event $E_3 = \{(2,1),(2,3)\}$. The probability of the event, $P(E_3) = \frac{2}{9}$.

11. All probability values must be greater than or equal to 0 and less than or equal to 1. Since $\frac{6}{5} > 1$, it cannot be a probability.

13. (a) The probability of drawing a yellow marble is $P(\text{yellow}) = \frac{3}{15} = \frac{1}{5}$.

(b) The probability of drawing a black marble is $P(\text{black}) = \frac{0}{15} = 0$.

(c) The probability of drawing a yellow or white marble is $P(\text{yellow} \cup \text{white}) = \frac{3+4}{15} = \frac{7}{15}$.

(d) $P(\text{yellow}) = \frac{1}{5}$; $P(\text{not yellow}) = \frac{4}{5}$. The odds in favor of drawing a yellow marble are $\frac{\frac{1}{5}}{\frac{4}{5}} = \frac{1}{4}$ or 1 to 4.

(e) $P(\text{blue}) = \frac{8}{15}$; $P(\text{not blue}) = \frac{7}{15}$. The odds against drawing a blue marble are $\frac{\frac{7}{15}}{\frac{8}{15}} = \frac{7}{8}$ or 7 to 8.

15. $P(\text{sum is } 5) = \frac{2}{10} = \frac{1}{5}$; $P(\text{sum is not } 5) = \frac{8}{10} = \frac{4}{5}$. The odds in favor of the sum being 5 are $\frac{\frac{1}{5}}{\frac{4}{5}} = \frac{1}{4}$ or 1 to 4.

17. Odds in favor of candidate are 3 to 2 $\Rightarrow P(\text{lose}) = \frac{2}{3+2} = \frac{2}{5}$. The probability the candidate will lose is $\frac{2}{5}$.

19. (a) The probability of an uncle or brother arriving first is $P(\text{uncle} \cup \text{brother}) = \frac{2+3}{10} = \frac{5}{10} = \frac{1}{2}$.

(b) The probability of a brother or cousin arriving first is $P(\text{brother} \cup \text{cousin}) = \frac{3+4}{10} = \frac{7}{10}$.

(c) The probability of a brother or her mother arriving first is $P(\text{brother} \cup \text{mother}) = \frac{3+1}{10} = \frac{4}{10} = \frac{2}{5}$.

21. (a) $P(E) = -0.1$ matches with statement F, the event is impossible, because a probability value cannot be negative.

(b) $P(E) = 0.01$ matches with statement D, the event is very unlikely to occur, because the probability value is relatively low.

(c) $P(E) = 1$ matches with statement A, the event is certain to occur.

(d) $P(E) = 2$ matches with statement F, the probability cannot occur, because a probability value cannot be greater than 1.

(e) $P(E) = 0.99$ matches with statement C, the event is very likely to occur, because the probability value is relatively high.

(f) $P(E) = 0$ matches with statement B, the event is impossible.

(g) $P(E) = 0.5$ matches with statement E, the event is just as likely to occur as not occur.

23. (a) The probability that a randomly selected patient will need a kidney or a heart transplant is

$$P(\text{kidney} \cup \text{heart}) = \frac{35{,}025 + 3774}{51{,}277} \approx 0.76.$$

(b) The probability that a randomly selected patient will need neither a kidney nor a heart transplant is $P(\text{not kidney} \cap \text{not heart}) = P'(\text{kidney} \cup \text{heart}) \approx 1 - 0.76 \approx 0.24.$

25. (a) $P(\text{less than } \$20) = 0.25 + 0.37 = 0.62$ .The probability of a purchase that is less then \$20 is 0.62.

(b) $P(\$40 \text{ or more}) = 0.09 + 0.07 + 0.08 + 0.03 = 0.27.$ The probability of a purchase that is \$40 or more is 0.27.

(c) $P(\text{more than } \$99.99) = 0.08 + 0.03 = 0.11.$ The probability of a purchase that is more than \$99.99 is 0.11.

(d) $P(\text{less than } \$100) = 1 - 0.11 = 0.89.$ The probability of a purchase that is less than \$100 is 0.89.

27. Number of picks with all cards correct except the heart $= 12$.
Number of picks with all cards correct except the club $= 12$.
Number of picks with all cards correct except the diamond $= 12$.
Number of picks with all cards correct except the spade $= 12$. Total number of picks with all cards correct but one = 48. $P(\text{all cards correct but one}) = \frac{48}{28{,}561} \approx 0.001681.$ The probability of getting three of the four selections correct is $\frac{48}{28{,}561}$ or approximately $0.001681.$

29. (a) P (male selected) $= 1 - 0.28 = 0.72.$ The probability that a male worker is selected is 0.72.

(b) P (5 years or less) $= 1 - 0.3 = 0.7.$ The probability that a worker is selected who has worked for the company 5 years or less is 0.7.

(c) P (contribute $\cup$ female) $=$ P (contribute) $+$ P (female) $-$ P (contribute $\cap$ female)

$= 0.65 + 0.28 - \frac{0.28}{2} = 0.79.$ The probability that a worker is selected who contributes to the retirement plan or is female is 0.79.

31. There are $5^2 = 32$ possible outcomes, each with probability $\frac{1}{32}$. $\binom{5}{2} = 10$ outcomes with 2 girls (and 3 boys)

$\Rightarrow P(\text{2 girls and 3 boys}) = 10\left(\frac{1}{32}\right) = \frac{5}{16} = 0.3125$. The probability of having exactly 2 girls and 3 boys is

$\frac{5}{16} = 0.3125$.

33. $\binom{5}{0} = 1$ outcome has no girls $\Rightarrow P(\text{no girls}) = 1\left(\frac{1}{32}\right) = \frac{1}{32} = 0.03125$.

The probability of having no girls is $\frac{1}{32} = 0.03125$.

35. $\binom{5}{5} + \binom{5}{4} + \binom{5}{3} = 1 + 5 + 10 = 16$ outcomes have at least 3 boys $\Rightarrow P(\text{at least 3 boys}) = 16\left(\frac{1}{32}\right) = \frac{1}{2} = 0.5$.

The probability of having at least 3 boys is $\frac{1}{2} = 0.5$.

37. $P(\text{1 student smokes less than 10 per day}) = 0.45 + 0.24 = 0.69$. $P(\text{4 of 10 smoke less than 10 per day}) =$

$\binom{10}{4}(0.69)^4(1-0.69)^6 \approx 0.042246$. The probability that 4 of 10 students selected at random smoked less than 10 cigarettes per day is approximately 0.042246.

39. $P(\text{1 student smokes between 1 and 19 per day}) = 0.023831$.

$P(\text{fewer than 2 smoke between 1 and 19 per day}) =$

$P(\text{0 smoke between 1 and 19 per day}) + P(\text{ 1 smokes between 1 and 19 per day}) =$

$= \binom{10}{0}(0.44)^0(1-0.44)^{10} + \binom{10}{1}(0.44)^1(1-0.44)^9 \approx 0.026864$. The probability that fewer than 2 of 10 students selected at random smoked between 1 and 19 cigarettes per day is approximately 0.026864.

41. $P(\text{exactly 12 ones}) = \binom{12}{12}\left(\frac{1}{6}\right)^{12}\left(\frac{5}{6}\right)^0 = \frac{1}{6^{12}} \approx 4.6 \times 10^{-10}$. The probability of rolling exactly 12 ones is

$\frac{1}{6^{12}} \approx 4.6 \times 10^{-10}$.

43. $P(\text{no more than 3 ones}) = P(\text{0 ones}) + P(\text{1 ones}) + P(\text{2 ones}) + P(\text{3 ones}) =$

$\binom{12}{0}\left(\frac{1}{6}\right)^0\left(\frac{5}{6}\right)^{12} + \binom{12}{1}\left(\frac{1}{6}\right)^1\left(\frac{5}{6}\right)^{11} + \binom{12}{2}\left(\frac{1}{6}\right)^2\left(\frac{5}{6}\right)^{10} + \binom{12}{3}\left(\frac{1}{6}\right)^3\left(\frac{5}{6}\right)^9 \approx 0.875$. The probability of rolling no more than 3 ones is approximately 0.875.

45. $P(\text{fewer than 4}) = P(1) + P(\text{2 or 3}) = 0.2 + 0.29 = 0.49$. The probability that a randomly selected student applied to fewer than 4 colleges is 0.49.

47. $P(\text{more than } 3) = P(4\text{-}6) + P(7 \text{ or more}) = 0.37 + 0.14 = 0.51$. The probability that a randomly selected student applied to more than 3 colleges is 0.51.

49. (a) $P(5 \text{ of } 53 \text{ are color blind}) = \binom{53}{5}(0.042)^5(1-0.042)^{48} \approx 0.047822$. The probability that exactly 5 of 53 men are color blind is approximately 0.047822.

(b) $P(\text{no more than } 5) = P(0) + P(1) + P(2) + P(3) + P(4) + P(5) =$

$$\binom{53}{0}(0.042)^0(1-0.42)^{53} + \binom{53}{1}(0.042)^1(1-0.042)^{52} + \binom{53}{2}(0.042)^2(1-0.042)^{51} +$$

$$\binom{53}{3}(0.042)^3(1-0.042)^{50} + \binom{53}{4}(0.042)^4(1-0.042)^{49} + \binom{53}{5}(0.042)^5(1-0.042)^{48} \approx 0.976710.$$

The probability that no more than 5 of 53 men are color blind is approximately 0.976710.

(c) $P(\text{none are color blind}) = \binom{53}{0}(0.042)^0(1-0.042)^{53} \approx 0.102890$.

$P(\text{at least 1 color blind}) = 1 - P(\text{none are color blind}) \approx 0.897110$.

The probability that at least 1 man of 53 men is color blind is approximately 0.897110.

51. (a) $I = 2, S = 4, p = 0.1 \Rightarrow q(1-p)^I = (1-0.1)^2 = 0.81.$.

$P(3 \text{ not becoming infected}) = \binom{4}{3}(0.81)^3(1-0.81)^{4-3} \approx 0.404 = 40.4\%$. The probability of 3 family members not becoming infected is approximately 40.4%.

(b) $I = 2, S = 4, p = 0.5 \Rightarrow q = (1-p)^I = (1-0.5)^2 = 0.25$.

$P(3 \text{ not becoming infected}) = \binom{4}{3}(0.25)^3(1-0.25)^{4-3} \approx 4.7\%$. The probability of 3 family members not becoming infected is approximately 4.7%.

(c) $I = 1, S = 9, p = 0.5 \Rightarrow q = (1-p)^I = (1-0.5)^1 = 0.5$.

$P(0 \text{ not becoming infected}) = \binom{9}{0}(0.5)^0(1-0.5)^{9-0} \approx 0.002$

The probability that all of the other family members becoming sick is approximately 0.2%. Thus, it is unlikely that everyone in a large family would become sick even though the disease is highly infectious.

Reviewing Basic Concepts (Sections 8.6 and 8.7)

1. $4 + 8 + 12 + 16 + \cdots + 4n = 2n(n+1)$.

(i) Show that the statement is true for $n = 1$: $4(1) = 2(1)(1+1) \Rightarrow 4 = 4$.

(ii) Assume that S_k is true: $4+8+12+16+\cdots+4k=2k(k+1)$. Show that S_{k+1} is true:

$4+8+12+16+\cdots+4(k+1)=2(k+1)(k+2)$. Add $4(k+1)$ to each side of S_k :

$4+8+12+16+\cdots+4k+4(k+1)=2k(k+1)+4(k+1)$.

$=2k^2+2k+4k+4=2k^2+6k+4=2(k^2+3k+2)=2(k+1)(k+2)$.

Since S_k implies S_{k+1}, the statement is true for every positive integer n.

2. $n^2 \le 2^n$, if $n \ge 4$

(i) Show that the statement is true for $n=4$: $4^2 \le 2^4 \Rightarrow 16 \le 16$.

(ii) Assume that S_k is true: $k^2 \le 2^k$. Show that S_{k+1} is true: $(k+1)^2 \le 2^{(k+1)}$.

Multiply each side of S_k by 2: $k^2 \cdot 2 \le 2^k \cdot 2 \Rightarrow 2k^2 \le 2^{k+1}$.

Because $(k+1)^2 < 2k^2$ for all $k \ge 4$, we may substitute $(k+1)^2$ for $2k^2$ in the expression.

That is, $(k+1)^2 \le 2^{k+1}$.

Since S_k implies S_{k+1}, the statement is true for every positive integer $n \ge 4$.

3. The sample space S for tossing a coin twice is $S=\{(H,H),(H,T),(T,H),(T,T)\}$.

4. 36 possible outcomes, each with probability $\frac{1}{36}$. 2 outcomes yield 11, $\{(5,6),(6,5)\}$.

Therefore, $1^3=1; 1^2(2(1)^2-1)=1(2-1)=1$; The probability of rolling a sum of 11 is $\frac{2}{36}=\frac{1}{18}$.

5. $\binom{4}{4}=1$ way of drawing 4 aces; $\binom{4}{1}=4$ ways of drawing 1 queen; $\binom{52}{5}=2{,}598{,}960$ ways of drawing 5

cards. P (4 aces and 1 queen) $=\frac{1\times 4}{2{,}598{,}960}\approx 0.0000015$. The probability of drawing 4 aces and 1 queen is

approximately 0.0000015.

6. From number 5, the probability of drawing 4 aces and 1 queen is given by $\frac{4}{2{,}598{,}960}$. Therefore, the

probability of not drawing 4 aces and 1 queen is given by $1-\frac{4}{2{,}598{,}960}=\frac{2{,}598{,}956}{2{,}598{,}960}\approx 0.9999985$.

7. P (rain) $=\frac{3}{3+7}=\frac{3}{10}$. The probability of rain is $\frac{3}{10}$.

8. P (female) $=\frac{2.81-1.45}{2.81}\approx 0.484$. The probability that a randomly selected graduate is female is

approximately 0.484.

Chapter 8 Review Exercises

1. $a_1 = \frac{1}{1+1} = \frac{1}{2}; a_2 = \frac{2}{2+1} = \frac{2}{3}; a_3 = \frac{3}{3+1} = \frac{3}{4}; a_4 = \frac{4}{4+1} = \frac{4}{5}; a_5 = \frac{5}{5+1} = \frac{5}{6}$. The first five terms of the sequence are $\frac{1}{2}, \frac{2}{3}, \frac{3}{4}, \frac{4}{5}$, and $\frac{5}{6}$. The sequence is neither arithmetic nor geometric.

2. $a_1 = (-2)^1 = -2; a_2 = (-2)^2 = 4; a_3 = (-2)^3 = -8; a_4 = (-2)^4 = 16; a_5 = (-2)^5 = -32$. The first five terms of the sequence are –2, 4, –8, 16, and –32. The sequence is geometric with common ratio –2.

3. $a_1 = 2(1+3) = 8; a_2 = 2(2+3) = 10; a_3 = 2(3+3) = 12; a_4 = 2(4+3) = 14; a_5 = 2(5+3) = 16$. The first five terms of the sequence are 8, 10, 12, 14, and 16. The sequence is arithmetic with common difference 2.

4. $a_1 = 1(1+1) = 2; a_2 = 2(2+1) = 6; a_3 = 3(3+1) = 12; a_4 = 4(4+1) = 20; a_5 = 5(5+1) = 30$. The first five terms of the sequence are 2, 6, 12, 20, and 30. The sequence is neither arithmetic nor geometric.

5. $a_1 = 5; a_2 = 5-3; a_3 = 2-3; a_4 = -1-3 = -4; a_5 = -4-3 = -7$. The first five terms of the sequence are 5, 2, –1, –4, and –7. The sequence is arithmetic with common difference –3.

6. $a_2 = 10, d = -2 \Rightarrow a_1 = 10+2 = 12; a_3 = 10-2 = 8; a_4 = 8-2 = 6; a_5 = 6-2 = 4$. The first five terms of the sequence are 12, 10, 8, 6, and 4.

7. $a_3 = \pi, a_4 = 1 \Rightarrow d = 1-\pi \Rightarrow a_2 = \pi-(1-\pi) = 2\pi-1; a_1 = 2\pi-1-(1-\pi) = 3\pi-2;$
$a_5 = 1+(1-\pi) = -\pi+2$. The first five terms of the sequence are $3\pi-2, 2\pi-1, \pi, 1,$ and $-\pi+2$.

8. $a_1 = 6, r = 2 \Rightarrow a_2 = 6\cdot 2 = 12; a_3 = 12\cdot 2 = 24; a_4 = 24\cdot 2 = 48; a_5 = 48\cdot 2 = 96$. The first five terms of the sequence are 6, 12, 24, 48, and 96.

9. $a_1 = -5, a_2 = -1 \Rightarrow r = \frac{1}{5} \Rightarrow a_3 = (-1)\left(\frac{1}{5}\right) = -\frac{1}{5}; a_4 = \left(\frac{-1}{5}\right)\left(\frac{1}{5}\right) = -\frac{1}{25};\ a_5 = \left(-\frac{1}{25}\right)\left(\frac{1}{5}\right) = -\frac{1}{125}$. The first five terms of the sequence are $-5, -1, -\frac{1}{5}, -\frac{1}{25}, -\frac{1}{125}$.

10. $a_5 = -3, a_{15} = 17 \Rightarrow 17 = -3+10d \Rightarrow d = 2 \Rightarrow -3 = a_1 + 4\cdot 2 \Rightarrow a_1 = -11 \Rightarrow$
$a_n = -11+2(n-1) = 2n-13$. The first term of the sequence is –11 and the nth term of the sequence is given by $a_n = 2n-13$.

11. $a_1 = -8, a_7 = -\frac{1}{8} \Rightarrow -\frac{1}{8} = (-8)r^6 \Rightarrow r^6 = \frac{1}{64} \Rightarrow r = \pm\frac{1}{2}$. Therefore, $a_4 = (-8)\left(\frac{1}{2}\right)^3 = -1$ and
$a_n = (-8)\left(\frac{1}{2}\right)^{n-1} = -\left(\frac{1}{2}\right)^{n-4}$ or $a_4 = (-8)\left(-\frac{1}{2}\right)^3 = 1$ and $a_n = (-8)\left(-\frac{1}{2}\right)^{n-1} = \left(-\frac{1}{2}\right)^{n-4}$. Either the common ratio is $\frac{1}{2}$ and the fourth term of the sequence is –1 and the nth term of the sequence is given by

$a_n = (-8)\left(\frac{1}{2}\right)^{n-1} = -\left(\frac{1}{2}\right)^{n-4}$ or the common ratio is $-\frac{1}{2}$ and the fourth term of the sequence is 1 and the

n^{th} term of the sequence is given by $a_n = (-8)\left(-\frac{1}{2}\right)^{n-1} = \left(-\frac{1}{2}\right)^{n-4}$.

12. $a_1 = 6, d = 2 \Rightarrow a_8 = 6 + (8-1)(2) = 20.$ The eighth term of the sequence is 20.

13. $a_1 = 6x - 9, a_2 = 5x + 1 \Rightarrow d = (5x+1) - (6x-9) = -x + 10 \Rightarrow a_8 = (6x-9) + (8-1)(-x+10) = -x + 61.$

The eighth term of the sequence is –x+61.

14. $a_1 = 2, d = 3 \Rightarrow a_{12} = 2 + (12-1)(3) = 35 \Rightarrow S_{12} = \frac{12}{2}(2 + 35) = 222.$ The sum of the first twelve terms of the sequence is 222.

15. $a_2 = 6, d = 10 \Rightarrow a_1 = 6 - 10 = -4 \Rightarrow a_{12} = -4 + (12-1)(10) = 106 \Rightarrow S_{12} = \frac{12}{2}(-4 + 106) = 612.$ The sum of the first twelve terms of the sequence is 612.

16. $a_1 = -2, r = 3 \Rightarrow a_5 = (-2)(3)^{5-1} = -162.$ The fifth term of the sequence is -162.

17. $a_3 = 4, r = \frac{1}{5} \Rightarrow a_1 = \frac{4}{\left(\frac{1}{5}\right)^2} = 100 \Rightarrow a_5 = (100)\left(\frac{1}{5}\right)^{5-1} = \frac{100}{625} = \frac{4}{25}.$ The fifth term of the sequence is $\frac{4}{25}$.

18. $a_1 = 3, r = 2 \Rightarrow S_4 = \frac{3(1-2^4)}{1-2} = 45.$ The sum of the first four terms of the sequence is 45.

19. $a_1 = \frac{3}{4}, a_2 = -\frac{1}{2}, a_3 = \frac{1}{3} \Rightarrow r = -\frac{2}{3} \Rightarrow S_4 = \frac{\frac{3}{4}(1-(-\frac{2}{3})^4)}{1-(-\frac{2}{3})} = \frac{13}{36}.$ The sum of the first four terms is $\frac{13}{36}$.

20. $S = 2000\left[\frac{(1+0.03)^5 - 1}{0.03}\right] = 10{,}618.27.$ The future value of the annuity is \$10,618.27.

21. $a_1 = (-1)^{1-1} = 1; r = -1; \sum_{i=1}^{7}(-1)^{i-1} = S_7 = \frac{1\left[1-(-1)^7\right]}{1-(-1)} = \frac{1(2)}{2} = 1.$ The value of the sum is 1.

22. $\sum_{i=1}^{5}(i^2 + i) = \sum_{i=1}^{5} i^2 + \sum_{i=1}^{5} i = \frac{5(5+1)(2\cdot 5+1)}{6} + \frac{5(5+1)}{2} = 55 + 15 = 70.$ The value of the sum is 70.

23. $\sum_{i=1}^{4}\frac{i+1}{i} = \frac{2}{1} + \frac{3}{2} + \frac{4}{3} + \frac{5}{4} = \frac{73}{12}.$ The value of the sum is $\frac{73}{12}$.

24. $a_1 = 3(1) - 4 = -1; a_{10} = 3(10) - 4 = 26; \sum_{j=1}^{10}(3j - 4) = S_{10} = \frac{10}{2}(-1 + 26) = 125.$

The value of the sum is 125.

25. $\sum_{j=1}^{2500} j = \frac{2500(2500+1)}{2} = 3{,}126{,}250.$ The value of the sum is 3,126,250.

26. $a_1 = 4 \cdot 2^1 = 8$; $r = 2$; $\sum_{i=1}^{5} 4 \cdot 2^i = S_5 = \dfrac{8(1-2^5)}{1-2} = 248$. The value of the sum is 248.

27. $a_1 = \left(\dfrac{4}{7}\right)^1 = \dfrac{4}{7}$; $r = \dfrac{4}{7}$; $\sum_{i=1}^{\infty}\left(\dfrac{4}{7}\right)^i = S_\infty = \dfrac{\frac{4}{7}}{1-\frac{4}{7}} = \dfrac{4}{3}$. The value of the sum is $\dfrac{4}{3}$.

28. $r = \dfrac{6}{5} \Rightarrow |r| \geq 1 \Rightarrow \sum_{i=1}^{\infty} -2\left(\dfrac{6}{5}\right)^i$ does not exist.

29. $a_1 = 24$, $a_2 = 8$, $a_3 = \dfrac{8}{3}$, $a_4 = \dfrac{8}{9} \Rightarrow r = \dfrac{1}{3} \Rightarrow S_\infty = \dfrac{24}{1-\frac{1}{3}} = 36$. The value of the sum is 36.

30. $a_1 = -\dfrac{3}{4}$, $a_2 = \dfrac{1}{2}$, $a_3 = -\dfrac{1}{3}$, $a_4 = \dfrac{2}{9} \Rightarrow r = -\dfrac{2}{3} \Rightarrow S_\infty = \dfrac{-\frac{3}{4}}{1-\left(-\frac{2}{3}\right)} = -\dfrac{9}{20}$. The value of the sum is $-\dfrac{9}{20}$.

31. $a_1 = \dfrac{1}{12}$, $a_2 = \dfrac{1}{6}$, $a_3 = \dfrac{1}{3}$, $a_4 = \dfrac{2}{3} \Rightarrow r = 2 \Rightarrow |r| \geq 1 \Rightarrow S_\infty$ diverges. The value of the sum diverges and does not exist.

32. $a_1 = 0.9$, $a_2 = 0.09$, $a_3 = 0.009$, $a_4 = 0.0009 \Rightarrow r = \dfrac{1}{10} \Rightarrow S_\infty = \dfrac{0.9}{1-\frac{1}{10}} = 1$. The value of the sum is 1.

33. $\sum_{i=1}^{4}(x_i^2 - 6) = \sum_{i=1}^{4} x_i^2 - \sum_{i=1}^{4} 6 = \sum_{i=1}^{4}(x_i+1)^2 - 24 = \dfrac{3(3+1)(2\cdot 3+1)}{6} - 24 = -10$. The value of the sum is –10.

34. $\sum_{i=1}^{6} f(x_i)\Delta x = (0-2)^3(0.1) + (1-2)^3(0.1) + (2-2)^3(0.1) + (3-2)^3(0.1) +$

$(4-2)^3(0.1) + (5-2)^3(0.1) = 2.7$. The value of the sum is 2.7.

35. $a_1 = 4$, $a^2 = -1$, $a_3 = -6 \Rightarrow d = -5 \Rightarrow a_n = 4 - 5(n-1) = -5n + 9$.

$a_n = -66 = -5n + 9 \Rightarrow n = 15 \Rightarrow 4 - 1 - 6 - \ldots - 66 = \sum_{i=1}^{15}(-5i+9)$.

The sum may be written as $\sum_{i=1}^{15}(-5i+9)$.

36. $a_1 = 10$, $a_2 = 14$, $a_3 = 18 \Rightarrow d = 4 \Rightarrow a_n = 10 + 4(n-1) = 4n + 6$..

$a_n = 86 = 4n + 6 \Rightarrow n = 20 \Rightarrow 10 + 14 + 18 + \ldots + 86 = \sum_{i=1}^{20}(4i+6)$. The sum may be written as $\sum_{i=1}^{20}(4i+6)$.

37. $a_1 = 4$, $a_2 = 12$, $a_3 = 36 \Rightarrow r = 3 \Rightarrow a_n 4(3)^{n-1}$. $a_n = 972 = 4(3)^{n-1} \Rightarrow$

$n - 1 = 5 \Rightarrow n = 6 \Rightarrow 4 + 12 + 36 + \ldots + 972 = \sum_{i=1}^{6} 4(3)^{n-1}$. The sum may be written as $\sum_{i=1}^{6} 4(3)^{n-1}$.

38. $a_n = \dfrac{n}{n+1}$ for $5 \leq n \leq 12$; $\dfrac{5}{6} + \dfrac{6}{7} + \dfrac{7}{8} + \ldots + \dfrac{12}{13} = \sum_{i=5}^{12}\dfrac{i}{i+1}$. The sum may be written as $\sum_{i=5}^{12}\dfrac{i}{i+1}$.

39. $9! = 362{,}880$. The value of $9!$ is 362,880.

40. $P(9,2) = 9_n P_r 2 = 72.$ The value of $P(9,2)$ is 72.

41. $P(6,0) = 6_n P_r 0 = 1.$ The value of $P(6,0)$ is 1.

42. $C(10,5) = 10_n C_r 5 = 252.$ The value of $C(10,5)$ is 252.

43. $\binom{8}{3} = 8_n C_r 3 = 56.$ The value of $\binom{8}{3}$ is 56.

44. Order is significant in the digits of a student identification number. Thus, such a number is an example of a permutation.

45. $(x+2y)^4 = \binom{4}{0}x^4(2y)^0 + \binom{4}{1}x^3(2y)^1 + \binom{4}{2}x^2(2y)^2 + \binom{4}{3}x^1(2y)^3 + \binom{4}{4}x^0(2y)^4 =$

$x^4 + 8x^3y + 24x^2y^2 + 32xy^3 + 16y^4$. The binomial expansion of $(x+2y)^4$ is

$x^4 + 8x^3y + 24x^2y^2 + 32xy^3 + 16y^4$.

46. $(3z-5w)^3 = \binom{3}{0}(3z)^3(-5w)^0 + \binom{3}{1}(3z)^2(-5w)^1 + \binom{3}{2}(3z)^1(-5w)^2 + \binom{3}{3}(3z)^0(-5w)^3 = \backslash$

$27z^3 - 135z^2w + 225zw^2 - 125w^3$. The binomial expansion of $(3z-5w)^3$ is

$27z^3 - 135z^2w + 225zw^2 - 125w^3$.

47. $\left(3\sqrt{x} - \frac{1}{\sqrt{x}}\right)^5 = \binom{5}{0}(3x^{1/2})^5(-x^{-1/2})^0 + \binom{5}{1}(3x^{1/2})^4(-x^{-1/2})^1 + \binom{5}{2}(3x^{1/2})^3(-x^{-1/2})^2 +$

$\binom{5}{3}(3x^{1/2})^2(-x^{-1/2})^3 + \binom{5}{4}(3x^{1/2})^1(-x^{-1/2})^4 + \binom{5}{5}(3x^{1/2})^0(-x^{-1/2})^5 =$

$243x^{5/2} - 405x^{3/2} + 270x^{1/2} - 90x^{-1/2} + 15x^{-3/2} - x^{-5/2}$. The binomial expansion of $\left(3\sqrt{x} - \frac{1}{\sqrt{x}}\right)^5$ is

$243x^{5/2} - 405x^{3/2} + 270x^{1/2} - 90x^{-1/2} + 15x^{-3/2} - x^{-5/2}$.

48. $(m^3 - m^{-2})^4 = \binom{4}{0}(m^3)^4(-m^{-2})^0 + \binom{4}{1}(m^3)^3(-m^{-2})^1 + \binom{4}{2}(m^3)^2(-m^{-2})^2 +$

$\binom{4}{3}(m^3)^1(-m^{-2})^3 + \binom{4}{4}(m^3)^0(-m^{-2})^4 = m^{12} - 4m^7 + 6m^2 - 4m^{-3} + m^{-8}$. The binomial expansion of

$(m^3 - m^{-2})^4$ is $m^{12} - 4m^7 + 6m^2 - 4m^{-3} + m^{-8}$.

49. $\binom{8}{5}(4x)^3(-y)^5 = -3584x^3y^5$. The sixth term of $(4x-y)^8$ is $-3584x^3y^5$.

50. $\binom{14}{6}(m)^8(-3n)^6 = 2{,}189{,}187m^8n^6$. The seventh term of $(m-3n)^{14}$ is $2{,}189{,}187\,m^8n^6$.

51. $(x+2)^{12} = \binom{12}{0}(x)^{12}(2)^0 + \binom{12}{1}(x)^{11}(2)^1 + \binom{12}{2}(x)^{10}(2)^2 +$

$\binom{12}{3}(x)^9(2)^3 + \ldots = x^{12} + 24x^{11} + 264x^{10} + 1760x^9 + \ldots$. The first four terms of $(x+12)^{12}$ are

$x^{12} + 24x^{11} + 264x^{10} + 1760x^9$.

52. $(2a+5b)^{16} = \ldots \binom{16}{14}(2a)^2(5b)^{14} + \binom{16}{15}(2a)^1(5b)^{15} + \binom{16}{16}(2a)^0(5b)^{16} =$

$\ldots + 480 \cdot 5^{14} a^2 b^{14} + 32 \cdot 5^{15} ab^{15} + 5^{16} b^{16}$. The last three terms of $(2a+5b)^{16}$ are

$480 \cdot 5^{14} a^2 b^{14} + 32 \cdot 5^{15} ab^{15} + 5^{16} b^{16}$.

53. Statements which are defined on the natural numbers; that is, statements which have the natural numbers as their domain, are proved by mathematical induction. An example is $\sum_{i=1}^{n} i = \frac{n(n+1)}{2}$.

54. A proof by mathematical induction consists of two steps. First, show that the statement is true for $n = 1$. Then show that if the statement is true for $n = k$, it must follow that the statement is also true for $n = k + 1$.

55. Prove $1+3+5+7+\ldots+(2n-1) = n^2$.

Step 1: $1 = 1^2 \Rightarrow 1+3+5+7+\ldots+(2n-1) = n^2$ for $n = 1$.

Step 2: Assume $1+3+5+7+\ldots+(2k-1) = k^2$, then

$1+3+5+7+\ldots+(2k-1)+(2k+1) = k^2+2k+1 \Rightarrow 1+3+5+7+\ldots+(2k-1)+(2(k+1)-1) = (k+1)^2$.

Thus, if $1+3+5+7+\ldots+(2n-1) = n^2$ for $n = k$, then it must follow that

$1+3+5+7+\ldots+(2n-1) = n^2$ for $n = k+1$ also.

56. Prove $2+6+10+14+\ldots+(4n-2) = 2n^2$.

Step 1: $2 = 2(1)^2 \Rightarrow 2+6+10+14+\ldots+(4n-2) = 2n^2$ for n = 1.

Step 2: Assume $2+6+10+14+\ldots+(4k-2) = 2k^2$, then

$2+6+10+14+\ldots+(4k-2)+(4k+2) = 2k^2+4k+2 \Rightarrow$

$2+6+10+14+\ldots+(4k-2)+(4(k+1)-1) = 2(k+1)^2$. Thus if $2+6+10+14+\ldots+(4n-2) = 2n^2$ for $n = k$, then it must follow that $2+6+10+14+\ldots+(4n-2) = 2n^2$ for $n = k+1$ also.

57. Prove $2+2^2+2^3+\ldots+2^n = 2(2^n-1)$.

Step 1: $2 = 2(2^1-1) \Rightarrow 2+2^2+2^3+\ldots+2^n = 2(2^n-1)$ for $n = 1$.

Step 2: Assume $2+2^2+2^3+\ldots+2^k = 2(2^k-1)$, then

$2+2^2+2^3+\ldots+2^k+2^{k+1} = 2(2^k-1)+2^{k+1} = 2^{k+1}-2+2^{k+1} = 2(2^{k+1}-1)$ Thus, if

$2+2^2+2^3+...+2^n=2(2^n-1)$ for $n=k$,then it must follow that $2+2^2+2^3+...+2^n=2(2^n-1)$ for $n=k+1$ also.

58. Prove $1^3+3^3+5^3+...+(2n-1)^2=n^2(2n^2-1)$.

Step 1: $1^3=1; 1^2(2(1)^2-1)=1(2-1)=1$, so $1^3+3^3+5^3+...+(2n-1)^3=n^2(2n^2-1)$ for n = 1.

Step 2: Assume $1^3+3^3+5^3+...+(2k-1)^3=k^2(2k^2-1)$. Then,

$1^3+3^3+5^3+...+(2k-1)^3+(2k+1)^3=k^2(2k^2-1)+(2k=1)^3$. Then

$1^3+3^3+5^3+...+(2k-1)^3+(2(k+1)-1)^3=2k^4+8k^3+11k^2+6k+1=$

$(k^2+2k+1)(2k^2+4k+1)=(k+1)^2(2(k+1)^2-1)$. Thus, if $1^3+3^3+5^3+...+(2n-1)^3=n^2(2n^2-1)$ for $n=k$, then it must follow that $1^3+3^3+5^3+...+(2n-1)^3=n^2(2n^2-1)$ for $n=k+1$ also.

59. $2\times4\times3\times2=48$. The number of wedding arrangements possible is 48.

60. $5\times3\times6=90$. The number of available couches is 90.

61. $4{}_nP_r4=24$. The jobs can be assigned in 24 different ways.

62. (a) $\begin{pmatrix}6\\3\end{pmatrix}=20$. The number of different delegations possible is 20.

(b) $\begin{pmatrix}5\\2\end{pmatrix}=10$. If the president must attend, 10 different delegations are possible.

63. $9{}_nP_r3=504$. There are 504 different ways the winners could be determined.

64. $26\times10\times10\times10\times26\times26\times26=456{,}976{,}000$. The number of possible different license plates is 456,976,000.

65. (a) P(green) = $\frac{4}{15}$. The probability of drawing a green marble is $\frac{4}{15}$.

(b) P(not black) = $\frac{4+6}{15}=\frac{10}{15}=\frac{2}{3}$. The probability of drawing a marble that is not black is $\frac{2}{3}$.

(c) P(blue) = $\frac{0}{15}=0$. The probability of drawing a green blue is 0.

66. (a) P(green) = $\frac{4}{15}\Rightarrow\frac{4}{15-4}=\frac{4}{11}$. The odds in favor of drawing a green marble are 4 to 11.

(b) P(not white) = $\frac{4+5}{15}=\frac{9}{15}=\frac{3}{5}\Rightarrow\frac{3}{5-3}=\frac{3}{2}$. The odds against drawing a white marble are 3 to 2.

(c) From (b), the odds in favor of drawing a marble that is not white are 3 to 2.

67. P(black king) = $\frac{2}{52}=\frac{1}{26}$. The probability of drawing a black king is $\frac{1}{26}$.

68. P(face or ace) = $\frac{16}{52}=\frac{4}{13}$. The probability of drawing a face card or an ace is $\frac{4}{13}$.

69. $P(\text{ace} \cup \text{diamond}) = P(\text{ace}) + P(\text{diamond}) - P(\text{ace} \cap \text{diamond}) = \frac{4}{52} + \frac{13}{52} - \frac{1}{52} = \frac{16}{52} = \frac{4}{13}$. The probability of drawing an ace or a diamond is $\frac{4}{13}$.

70. $P(\text{not a diamond}) = 1 - P(\text{diamond}) = 1 - \frac{1}{4} = \frac{3}{4}$.

The probability of drawing a card that is not a diamond is $\frac{3}{4}$.

71. $P(\text{no more than 3}) = 1 - [P(4) + P(5)] = 1 - (0.08 + 0.06) = 0.86$. The probability that no more than 3 filters are defective is 0.86.

72. $P(\text{at least 2}) = 1 - [P(0) + P(1)] = 1 - (0.31 + 0.25) = 0.44$. The probability that at least 2 filters are defective is 0.44.

73. $P(\text{more than 5}) = 0$. Only 5 filters were selected, so it is not possible for more than 5 to be defective. Note that, from the probabilities in the table, $P(0) + P(1) + P(2) + P(3) + P(4) + P(5) = 1$.

74. $\binom{12}{2}\left(\frac{1}{6}\right)^2\left(\frac{5}{6}\right)^{10} \approx 0.296$. The probability that exactly 2 of 12 rolls results in a 5 is approximately 0.296.

75. $\binom{10}{4}\left(\frac{1}{2}\right)^4\left(\frac{1}{2}\right)^6 \approx 0.205$. The probability that exactly 4 of 10 coins tossed result in a tail is approximately 0.205.

76. (a) $P(\text{conservation}) = \frac{56.51}{282.2} \approx 0.2002$. The probability that a randomly selected student is in the conservative group is approximately 0.2002.

(b) $P(\text{far left or far right}) = \frac{7.06 + 3.673}{282.2} \approx 0.0380$. The probability that a randomly selected student is on the far left or the far right is approximately 0.0380.

(c) $P(\text{not middle of road}) = 1 - P(\text{middle of road}) = 1 - \frac{143.5}{282.2} \approx 0.4915$. The probability that a randomly selected student is not middle of the road is approximately 0.4915.

Chapter 8 Test

1. (a) $a_1 = (-1)^1(1+2) = -3$; $a_2 = (-1)^2(2+2) = 4$; $a_3 = (-1)^3(3+2) = -5$; $a_4 = (-1)^4(4+2) = 6$; $a_5 = (-1)^5(5+2) = -7$. The first five terms of the sequence are $-3, 4, -5, 6$, and -7. The sequence is neither arithmetic nor geometric.

(b) $a_1=(-3)\left(\frac{1}{2}\right)^1=-\frac{3}{2}$; $a_2=(-3)\left(\frac{1}{2}\right)^2=-\frac{3}{4}$; $a_3=(-3)\left(\frac{1}{2}\right)^3=-\frac{3}{8}$;

$a_4=(-3)\left(\frac{1}{2}\right)^4=-\frac{3}{16}$; $a_5=(-3)\left(\frac{1}{2}\right)^5=-\frac{3}{32}$. The first five terms of the sequence are

$-\frac{3}{2}, -\frac{3}{4}, -\frac{3}{8}, -\frac{3}{16}$, and $-\frac{3}{32}$. The sequence is geometric with common ratio $\frac{1}{2}$.

(c) $a_1=2$; $a_2=3$; $a_3=3+2(2)=7$; $a_4=7+2(3)=13$; $a_5=13+2(7)=27$. The first five terms of the sequence are $2, 3, 7, 13$, and 27. The sequence is neither arithmetic nor geometric.

2. (a) $a_1=1$, $a_3=25 \Rightarrow 25=1+(3-1)d \Rightarrow 2d=24 \Rightarrow d=12 \Rightarrow a_5=1+(5-1)(12)=49$. The fifth term of the sequence is 49.

(b) $a_1=81$, $r=-\frac{2}{3} \Rightarrow a_5=81\left(-\frac{2}{3}\right)^{5-1}=16$. The fifth term of the sequence is 16.

3. (a) $a_1=-43$, $d=12 \Rightarrow a_{10}=-43+12(10-1)=65 \Rightarrow S_{10}=\frac{10}{2}(-43+65)=110$. The sum of the first ten terms of the sequence is 110.

(b) $a_1=5$, $r=-2 \Rightarrow S_{10}=\dfrac{5(1-(-2)^{10})}{1-(-2)}=-1705$.

The sum of the first ten terms of the sequence is -1705.

4. (a) $a_1=5(1)+2=7$; $a_{30}=5(30)+2=152$; $d=5 \Rightarrow \sum_{i=1}^{30}(5i+2)=S_{30}=\frac{30}{2}(7+152)=2385$. The value of the sum is 2385.

(b) $a_1=(-3)(2)^1=-6$; $r=2 \Rightarrow \sum_{i=1}^{5}(-3\cdot 2^i)=\dfrac{-6(1-(2)^5)}{1-2}=-186$. The value of the sum is -186.

(c) $r=2 \Rightarrow |r| \geq 1 \Rightarrow \sum_{i=1}^{\infty}(2^i)\cdot 4$ does not exist. The value of the sum diverges and does not exist.

(d) $a_1=54\left(\frac{2}{9}\right)^1=12$; $r=\frac{2}{9} \Rightarrow \sum_{i=1}^{\infty}54\left(\frac{2}{9}\right)^i=S_\infty=\dfrac{12}{1-\frac{2}{9}}=\frac{108}{7}$. The value of the sum is $\frac{108}{7}$.

5. (a) $10_n C_r\, 2=45$. The value of $10_n C_r\, 2=45$.

(b) $\binom{7}{3}=7_n C_r\, 3=35$. The value of $\binom{7}{3}$ is 35.

(c) $7!=5040$. The value of $7!$ is 5040.

(d) $P(11,\ 3)=11_n P_r\, 3=990$. The value of $P(11,\ 3)$ is 990.

6. (a) $(2x-3y)^4=\binom{4}{0}(2x)^4(-3y)^0+\binom{4}{1}(2x)^3(-3y)^1+\binom{4}{2}(2x)^2(-3y)^2+$

$$\binom{4}{3}(2x)^1(-3y)^3+\binom{4}{4}(2x)^0(-3y)^4 = 16x^4-96x^3y+216x^2y^2-216xy^3+81y^4.$$

The binomial expansion of $(2x-3y)^4$ is $16x^4-96x^3y+216x^2y^2-216xy^3+81y^4$.

(b) $\binom{6}{2}(w)^4(-2y)^2=60w^4y^2$. The third term of the expansion of $(w-2y)^6=60w^4y^2$.

7. Prove $8+14+20+26+...+(6n+2)=3n^2+5n$.

Step 1: $3(1)^2+5(1)=8 \Rightarrow 8+14+20+26+...+(6n+2)=3n^2+5n$ for $n=1$.

Step 2: Assume $8+14+20+26+...+(6k+2)=3k^2+5k$, then

$8+14+20+26+...+(6k+2)+(6k+8)=3k^2+5k+6k+8 \Rightarrow$

$8+14+20+26+...+(6k+2)+[6(k+1)+2]=3k^2+11k+8=$

$3k^2+6k+3+5k+5=3(k+1)^2+5(k+1)$. Thus, if $8+14+20+26+...+(6n+2)=3n^2+5n$ for $n=k$, then it must follow that $8+14+20+26+...+(6n+2)=3n^2+5n$ for $n=k+1$ also.

8. $4\times3\times2=24$. 24 different types of shoes can be made.

9. $20_n P_r 3=6840$. The three offices can be filled in 6840 different ways.

10. $\binom{8}{2}\binom{12}{3}=28\cdot220=6160$. Two men and three women can be chosen in 6160 different ways.

11. (a) $P(\text{red } 3)=\frac{2}{52}=\frac{1}{26}$. The probability of drawing a red three is $\frac{1}{26}$.

(b) $4\times3=12$ cards are face cards $\Rightarrow 52-12=40$ cards are not face cards $\Rightarrow P(\text{not a face card})=$

$\frac{40}{52}=\frac{10}{13}$. The probability of drawing a card that is not a face card is $\frac{10}{13}$.

(c) $P(\text{king} \cup \text{spade})=P(\text{king})+P(\text{spade})-P(\text{king} \cap \text{spade})=\frac{4}{52}+\frac{13}{52}-\frac{1}{52}=\frac{16}{52}=\frac{4}{13}$.

(d) $P(\text{facecard})=\frac{12}{52}=\frac{3}{13}\Rightarrow\frac{3}{13-3}=\frac{3}{10}$. The odds in favor of drawing a face card are 3 to 10.

12. (a) $\binom{8}{3}\left(\frac{1}{6}\right)^3\left(\frac{5}{6}\right)^5\approx0.104$. The probability that exactly 3 of 8 rolls will result in a 4 is approximately 0.104.

(b) $\binom{8}{8}\left(\frac{1}{6}\right)^8\left(\frac{5}{6}\right)^0\approx0.000000595$. The probability that all 8 of 8 rolls will result in a 6 is approximately 0.000000595.

Chapter R: Basic Algebraic Concepts

R.1: Exponents and Polynomials

1. $(-4)^3 \cdot (-4)^2 = (-4)^{3+2} = (-4)^5$

3. $2^0 = 1$

5. $(5m)^0 = 1$, if $m \neq 0$

7. $(2^2)^5 = 2^{2\cdot 5} = 2^{10}$

9. $(2x^5y^4)^3 = 2^3(x^5)^3(y^4)^3 = 2^3x^{15}y^{12}$ or $8x^{15}y^{12}$

11. $-\left(\frac{p^4}{q}\right)^2 = -\left(\frac{p^{4\cdot 2}}{q^2}\right) = -\frac{p^8}{q^2}$

13. $-5x^{11}$ is a polynomial. It is a monomial since it has one term. It has degree 11 since 11 is the highest exponent.

15. $18p^5q + 6pq$ is a polynomial. It is a binomial since it has two terms. It has degree 6 since 6 is the sum of the exponents in the term $18p^5q$. (The term $6pq$ has degree 2.)

17. $\sqrt{2}x^2 + \sqrt{3}x^6$ is a polynomial. It is a binomial since it has two terms. It has degree 6 since 6 is the highest exponent.

19. $\frac{1}{3}r^2s^2 - \frac{3}{5}r^4s^2 + rs^3$ is a polynomial. It is a trinomial since it has three terms. It has degree 6 since 6 is the sum of the exponents in the term $-\frac{3}{5}r^4s^2$. (The other terms have degree 4.)

21. $-5\sqrt{z} + 2\sqrt{z^3} - 5\sqrt{z^5} = -5z^{1/2} + 2z^{3/2} - 5z^{5/2}$ is not a polynomial since the exponents are not integers.

23. $(4m^3 - 3m^2 + 5) + (-3m^3 - m^2 + 5) = (4m^3 - 3m^3) + (-3m^2 - m^2) + (5 + 5) = m^3 - 4m^2 + 10$

25. $(8p^2 - 5p) - (3p^2 - 2p + 4) = 8p^2 - 5p - 3p^2 + 2p - 4 = 5p^2 - 3p - 4$

27. $-(8x^3 + x - 3) + (2x^3 + x^2) - (4x^2 + 3x - 1) = -8x^3 - x + 3 + 2x^3 + x^2 - 4x^2 - 3x + 1 = -6x^3 - 3x^2 - 4x + 4$

29. $(5m - 6)(3m + 4) = 5m \cdot 3m + 5m \cdot 4 - 6 \cdot 3m - 6 \cdot 4 = 15m^2 + 20m - 18m - 24 = 15m^2 + 2m - 24$

31. $\left(2m - \frac{1}{4}\right)\left(3m + \frac{1}{2}\right) = 2m \cdot 3m + 2m \cdot \frac{1}{2} - \frac{1}{4} \cdot 3m - \frac{1}{4} \cdot \frac{1}{2} = 6m^2 + m - \frac{3}{4}m - \frac{1}{8} = 6m^2 + \frac{1}{4}m - \frac{1}{8}$

33. $2b^3(b^2 - 4b + 3) = 2b^3 \cdot b^2 + 2b^3 \cdot (-4b) + 2b^3 \cdot 3 = 2b^5 - 8b^4 + 6b^3$

35. $(m - n + k)(m + 2n - 3k) = m^2 + 2mn - 3km - mn - 2n^2 + 3kn + km + 2kn - 3k^2 =$ $m^2 + mn - 2km - 2n^2 + 5kn - 3k^2$

37. To find the square of a binomial, find the sum of the square of the first term, twice the product of the two terms, and the square of the last term.

39. $(2m+3)(2m-3)=(2m)^2-(3)^2=4m^2-9$

41. $(4m+2n)^2=(4m)^2+2(4m)(2n)+(2n)^2=16m^2+16mn+4n^2$

43. $(5r+3t^2)^2=(5r)^2+2(5r)(3t^2)+(3t^2)^2=25r^2+30rt^2+9t^4$

45. $[(2p-3)+q]^2=(2p-3)^2+2(2p-3)(q)+(q)^2=[(2p)^2-2(2p)(3)+3^2]+4pq-6q+q^2=$

$4p^2-12p+9+4pq-6q+q^2$

47. $[(3q+5)-p][(3q+5)+p]=(3q+5)^2-(p)^2=[(3q)^2+2(3q)(5)+(5)^2]-p^2=9q^2+30q+25-p^2$

49. $[(3a+b)-1]^2=(3a+b)^2-2(3a+b)(1)+(-1)^2=[(3a)^2+2(3a)(b)+(b)^2]-6a-2b+1=$

$9a^2+6ab+b^2-6a-2b+1$

51. $(6p+5q)(3p-7q)=(6p)(3p)+(6p)(-7q)+(5q)(3p)+(5q)(-7q)=18p^2-42pq+15pq-35q^2=$

$18p^2-27pq-35q^2$

53. $(p^3-4p^2+p)-(3p^2+2p+7)=p^3-4p^2+p-3p^2-2p-7=p^3-7p^2-p-7$

55. $y(4x+3y)(4x-3y)=y[(4x)^2-(3y)^2]=y[16x^2-9y^2]=16x^2y-9y^3$

57. $(2z+y)(3z-4y)=(2z)(3z)+(2z)(-4y)+(y)(3z)+(y)(-4y)=6z^2-8yz+3yz-4y^2=6x^2-5yz-4y^2$

59. $(3p+5)^2=(3p)^2+2(3p)(5)+5^2=9p^2+30p+25$

61. $p(4p-6)+2(3p-8)=4p^2-6p+6p-16=4p^2-16$

63. $-y(y^2-4)+6y^2(2y-3)=-y^3+4y+12y^3-18y^2=11y^3-18y^2+4y$

R.2: Factoring

1. (a) $(x+5y)^2=x^2+10xy+25y^2$; B

   (b) $(x-5y)^2=x^2-10xy+25y^2$; C

   (c) $(x+5y)(x-5y)=x^2-25y^2$; A

   (d) $(5y+x)(5y-x)=25y^2-x^2$; D

3. $4k^2m^3+8k^4m^3-12k^2m^4=4k^2m^3(1+2k^2-3m)$

5. $2(a+b)+4m(a+b)=(a+b)(2+4m)=2(a+b)(1+2m)$

7. $(2y-3)(y+2)+(y+5)(y+2)=(y+2)(2y-3+y+5)=(y+2)(3y+2)$

9. $(5r-6)(r+3)-(2r-1)(r+3)=(r+3)(5r-6-(2r-1))=(r+3)(3r-5)$

11. $2(m-1)-3(m-1)^2+2(m-1)^3=(m-1)(2-3(m-1)+2(m-1)^2)=$

$(m-1)(2-3m+3+2(m^2-2m+1))=(m-1)(5-3m+2m^2-4m+2)=(m-1)(2m^2-7m+7)$

13. $6st+9t-10s-15=3t(2s+3)-5(2s+3)=(2s+3)(3t-5)$

15. $10x^2-12y+15x-8xy=10x^2+15x-8xy-12y=5x(2x+3)-4y(2x+3)=(2x+3)(5x-4y)$

17. $t^3+2t^2-3t-6=t^2(t+2)-3(t+2)=(t+2)(t^2-3)$

19. $(8a-3)(2a-5)=[-1(-8a+3)][-1(-2a+5)]=(3-8a)(5-2a)$; Both are correct.

21. $8h^2-24h-320=8(h^2-3h-40)=8(h-8)(h+5)$

23. $9y^4-54y^3+45y^2=9y^2(y^2-6y+5)=9y^2(y-1)(y-5)$

25. $14m^2+11mr-15r^2=(7m-5r)(2m+3r)$

27. $12s^2+11st-5t^2=(3s-t)(4s+5t)$

29. $30a^2+am-m^2=(5a+m)(6a-m)$

31. $18x^5+15x^4z-75^3z^2=3x^3(6x^2+5xz-25z^2)=3x^3(2x+5z)(3x-5z)$

33. $16p^2-40p+25=(4p-5)^2$

35. $20p^2-100pq+125q^2=5(4p^2-20pq+25q^2)=5(2p-5q)^2$

37. $9m^2n^2-12mn+4=(3mn-2)^2$

39. $(2p+q)^2-10(2p+q)+25=((2p+q)-5)((2p+q)-5)=(2p+q-5)^2$

41. $9a^2-16=(3a+4)(3a-4)$

43. $25s^4-9t^2=(5s^2+3t)(5s^2-3t)$

45. $(a+b)^2-16=(a+b+4)(a+b-4)$

47. $p^4-625=(p^2+25)(p^2-25)=(p^2+25)(p+5)(p-5)$

49. $x^4-1=(x^2+1)(x^2-1)=(x^2+1)(x+1)(x-1)$; B

51. $8-a^3=2^3-a^3=(2-a)(4+2a+a^2)$

53. $125x^3-27=(5x)^3-3^3=(5x-3)(25x^2+15x+9)$

55. $27y^9+125z^6=(3y^3)^3+(5z^2)^3=(3y^3+5z^2)(9y^6-15y^3z^2+25z^4)$

57. $(r+6)^3-216=(r+6)^3-6^3=(r+6-6)((r+6)^2+6(r+6)+36)=$

$r(r^2+12r+36+6r+36+36)=r(r^2+18r+108)$

59. $27-(m+2n)^3=3^3-(m+2n)^3=(3-(m+2n))(9+3(m+2n)+(m+2n)^2)$

$(3-m-2n)(9+3m+6n+m^2+4mn+4n^2)$

61. $3a^4+14a^2-5$; Let $u=a^2 \Rightarrow 3u^2+14u-5=(3u-1)(u+5)$. Then $a^2=u \Rightarrow (3a^2-1)(a^2+5)$ was not substituted back in for u.

63. a^4-2a^2-48; Let $u=a^2 \Rightarrow u^2-2u-48=(u-8)(u+6)$. Then $a^2=u \Rightarrow (a^2-8)(a^2+6)$.

65. $6(4z-3)^2+7(4z-3)-3$; Let $u=4z-3 \Rightarrow 6u^2+7u-3=(3u-1)(2u+3)$. Then

$4z-3=u \Rightarrow (3(4z-3)-1)(2(4z-3)+3)=(12z-10)(8z-3)=2(6z-5)(8z-3)$.

67. $20(4-p)^2-3(4-p)-2$; Let $u=4-p \Rightarrow 20u^2-3u-2=(5u-2)(4u+1)$. Then

$4-p=u \Rightarrow (5(4-p)-2)(4(4-p)+1)=(18-5p)(17-4p)$.

69. $4b^2+4bc+c^2-16=(4b^2+4b+c^2)-16=(2b+c)^2-16=(2b+c+4)(2b+c-4)$.

71. $x^2+xy-5y-5y=x(x+y)-5(x+y)=(x+y)(x-5)$

73. $p^4(m-2n)+q(m-2n)=(m-2n)(p^4+q)$

75. $4z^2+28z+49=(2z+7)^2$

77. $1000x^3+343y^3=(10x)^3+7(y)^3=(10x+7y)(100x^2-70xy+49y^2)$

79. $125m^6-216=(5m^2)^3-6^3=(5m^2-6)(25m^4+30m^2+36)$

81. $12m^2+16mn-35n^2=(6m-7n)(2m+5n)$

83. $4p^2+3p-1=(4p-1)(p+1)$

85. $144z^2+121$ does not factor, it is prime.

87. $(4t+5)^2+16(4t+5)+64$; Let $u=4t+5 \Rightarrow u^2+16u+64=(u+8)^2$ Then

$4t+5=u \Rightarrow (4t+5+8)^2=(4t+13)^2$

R.3: Rational Expressions

1. $\dfrac{x-2}{x+6}$; $x+6=0, \{x \mid x \neq -6\}$

3. $\dfrac{2x}{5x-3}$; $5x-3=0, \left\{x \mid x \neq \dfrac{3}{5}\right\}$

5. $\dfrac{-8}{x^2+1}$; No restrictions since $x^2+1>0$. Domain: $(-\infty, \infty)$

7. $\frac{3x+7}{(4x+2)(x-1)}$; $4x+2=0 \Rightarrow 4x=-2 \Rightarrow x=-\frac{2}{4}=-\frac{1}{2}$ and $x-1=0 \Rightarrow x=1$; $\left\{x \mid x \neq -\frac{1}{2}, 1\right\}$

9. $\frac{25p^3}{10p^2}=\frac{5p^2(5p)}{5p^2(2)}=\frac{5p}{2}$

11. $\frac{8k+16}{9k+18}=\frac{8(k+2)}{9(k+2)}=\frac{8}{9}$

13. $\frac{3(t+5)}{(t+5)(t-3)}=\frac{3}{t-3}$

15. $\frac{8x^2+16x}{4x^2}=\frac{8x(x+2)}{4x^2}=\frac{2(x+2)}{x}=\frac{2x+4}{x}$

17. $\frac{m^2-4m+4}{m^2+m-6}=\frac{(m-2)^2}{(m-2)(m+3)}=\frac{m-2}{m+3}$

19. $\frac{8m^2+6m-9}{16m^2-9}=\frac{(4m-3)(2m+3)}{(4m-3)(4m+3)}=\frac{2m+3}{4m+3}$

21. $\frac{15p^3}{9p^2}\div\frac{6p}{10p^2}=\frac{15p^3}{9p^2}\cdot\frac{10p^2}{6p}=\frac{25p^2}{9}$

23. $\frac{2k+8}{6}\div\frac{3k+12}{2}=\frac{2(k+4)}{6}\times\frac{2}{3(k+4)}=\frac{2}{9}$

25. $\frac{x^2+x}{5}\cdot\frac{25}{xy+y}=\frac{x(x+1)}{5}\cdot\frac{25}{y(x+1)}=\frac{5x}{y}$

27. $\frac{4a+12}{2a-10}\div\frac{a^2-9}{a^2-a-20}=\frac{4(a+3)}{2(a-5)}\cdot\frac{(a+4)(a-5)}{(a-3)(a+3)}=\frac{2(a+4)}{a-3}$

29. $\frac{p^2-p-12}{p^2-2p-15}\cdot\frac{p^2-9p+20}{p^2-8p+16}=\frac{(p-4)(p+3)}{(p+3)(p-5)}\cdot\frac{(p-5)(p-4)}{(p-4)(p-4)}=1$

31. $\frac{m^2+3m+2}{m^2+5m+4}\div\frac{m^2+5m+6}{m^2+10m+24}=\frac{(m+2)(m+1)}{(m+4)(m+1)}\cdot\frac{(m+4)(m+6)}{(m+3)(m+2)}=\frac{m+6}{m+3}$

33. $\frac{2m^2-5m-12}{m^2-10m+24}\div\frac{4m^2-9}{m^2-9m+18}=\frac{(2m+3)(m-4)}{(m-4)(m-6)}\cdot\frac{(m-3)(m-6)}{(2m+3)(2m-3)}=\frac{m-3}{2m-3}$

35. $\frac{x^3+y^3}{x^2-y^2}\cdot\frac{x+y}{x^2-xy+y^2}=\frac{(x+y)(x^2-xy+y^2)}{(x+y)(x-y)}\cdot\frac{x+y}{x^2-xy+y^2}=\frac{x+y}{x-y}$

37. $\frac{x^3+y^3}{x^3-y^3}\cdot\frac{x^2-y^2}{x^2+2xy+y^2}=\frac{(x+y)(x^2-xy+y^2)}{(x-y)(x^2+xy+y^2)}\cdot\frac{(x-y)(x+y)}{(x+y)(x+y)}=\frac{x^2-xy+y^2}{x^2+xy+y^2}$

39. A: $\frac{x-4}{x+4}\neq -1$, B: $\frac{-x-4}{x+4}=\frac{-1(x+4)}{x+4}=-1$, C: $\frac{x-4}{4-x}=\frac{x-4}{-1(x-4)}=-1$, D: $\frac{x-4}{-x-4}=\frac{x-4}{-1(x+4)}\neq -1$

41. $\frac{3}{2k}+\frac{5}{3k}=\frac{9}{6k}+\frac{10}{6k}=\frac{19}{6k}$

43. $\frac{a+1}{2}-\frac{a-1}{2}=\frac{a+1-(a-1)}{2}=\frac{2}{2}=1$

45. $\frac{3}{p}+\frac{1}{2}=\frac{6}{2p}+\frac{p}{2p}=\frac{6+p}{2p}$

47. $\frac{1}{6m}+\frac{2}{5m}+\frac{4}{m}=\frac{5(1)+6(2)+30(4)}{30m}=\frac{5+12+120}{30m}=\frac{137}{30m}$

49. $\frac{1}{a+1}-\frac{1}{a-1}=\frac{a-1-(a+1)}{(a+1)(a-1)}=\frac{-2}{(a+1)(a-1)}$

51. $\frac{m+1}{m-1}+\frac{m-1}{m+1}=\frac{(m+1)^2+(m-1)^2}{(m-1)(m+1)}=\frac{m^2+2m+1+m^2-2m+1}{(m-1)(m+1)}=\frac{2m^2+2}{(m-1)(m+1)}$

53. $\frac{3}{a-2}-\frac{1}{2-a}=\frac{3}{a-2}-\frac{-1}{a-2}=\frac{3+1}{a-2}=\frac{4}{a-2}$ or $\frac{-4}{2-a}$

55. $\frac{x+y}{2x-y}-\frac{2x}{y-2x}=\frac{x+y}{2x-y}-\frac{-2x}{2x-y}=\frac{x+y+2x}{2x-y}=\frac{3x+y}{2x-y}$ or $\frac{-3x-y}{y-2x}$

57. $\frac{1}{a^2-5a+6}-\frac{1}{a^2-4}=\frac{1}{(a-2)(a-3)}-\frac{1}{(a-2)(a+2)}=\frac{1(a+2)-1(a-3)}{(a-2)(a-3)(a+2)}=$

$\frac{a+2-a+3}{(a-2)(a-3)(a+2)}=\frac{5}{(a-2)(a-3)(a+2)}$

59. $\frac{1}{x^2+x-12}-\frac{1}{x^2-7x+12}+\frac{1}{x^2-16}=\frac{1}{(x+4)(x-3)}-\frac{1}{(x-3)(x-4)}+\frac{1}{(x-4)(x+4)}=$

$\frac{1(x-4)-1(x+4)+1(x-3)}{(x+4)(x-3)(x-4)}=\frac{x-4-x-4+x-3}{(x+4)(x-3)(x-4)}=\frac{x-11}{(x+4)(x-3)(x-4)}$

61. $\frac{3a}{a^2+5a-6}-\frac{2a}{a^2+7a+6}=\frac{3a}{(a+6)(a-1)}-\frac{2a}{(a+6)(a+1)}=\frac{3a(a+1)-2a(a-1)}{(a+6)(a-1)(a+1)}=$

$\frac{3a^2+3a-2a^2+2a}{(a+6)(a-1)(a+1)}=\frac{a^2+5a}{(a+6)(a-1)(a+1)}$

63. $\frac{1+\frac{1}{x}}{1-\frac{1}{x}}=\frac{x(1+\frac{1}{x})}{x(1-\frac{1}{x})}=\frac{x+1}{x-1}$

65. $\frac{\frac{1}{x+1}-\frac{1}{x}}{\frac{1}{x}}=\frac{x(x+1)(\frac{1}{x+1}-\frac{1}{x})}{x(x+1)(\frac{1}{x})}=\frac{x-(x+1)}{x+1}=\frac{-1}{x+1}$

67. $\frac{1+\frac{1}{1-b}}{1-\frac{1}{1+b}}=\frac{(1-b)(1+b)(1+\frac{1}{1-b})}{(1-b)(1+b)(1-\frac{1}{1+b})}=\frac{(1-b)(1+b)+(1+b)}{(1-b)(1-b)-(1-b)}=\frac{1-b^2+1+b}{1-b^2-1+b}=\frac{-b^2+b+2}{-b^2+b}=\frac{(2-b)(1+b)}{b(1-b)}$

69. $\dfrac{m-\frac{1}{m^2-4}}{\frac{1}{m+2}}=\dfrac{(m^2-4)(m-\frac{1}{m^2-4})}{(m^2-4)(\frac{1}{m+2})}=\dfrac{m(m^2-4)-1}{m-2}=\dfrac{m^3-4m-1}{m-2}$

R.4: Negative and Rational Exponents

1. $\left(\dfrac{4}{9}\right)^{3/2}=\left[\left(\dfrac{4}{9}\right)^{1/2}\right]^3=\left(\dfrac{2}{3}\right)^3=\dfrac{2^3}{3^3}=\dfrac{8}{27}$; E

3. $-\left(\dfrac{9}{4}\right)^{3/2}=\left[\left(\dfrac{9}{4}\right)^{1/2}\right]^3=-\left(\dfrac{3}{2}\right)^3=-\dfrac{3^3}{2^3}=-\dfrac{27}{8}$; F

5. $\left(\dfrac{8}{27}\right)^{2/3}=\left[\left(\dfrac{8}{27}\right)^{1/3}\right]^2=\left(\dfrac{2}{3}\right)^2=\dfrac{2^2}{3^2}=\dfrac{4}{9}$; D

7. $-\left(\dfrac{27}{8}\right)^{2/3}=\left[\left(\dfrac{27}{8}\right)^{1/3}\right]^2=-\left(\dfrac{3}{2}\right)^2=-\dfrac{3^2}{2^2}=-\dfrac{9}{4}$; B

9. $(-4)^{-3}=\left(-\dfrac{1}{4}\right)^3=-\dfrac{1}{64}$

11. $\left(\dfrac{1}{2}\right)^{-3}=2^3=8$

13. $-4^{1/2}=-(4^{1/2})=-2$

15. $8^{2/3}=(8^{1/3})^2=2^2=4$

17. $27^{-2/3}=\left[\left(\dfrac{1}{27}\right)^{1/3}\right]^2=\left(\dfrac{1}{3}\right)^2=\dfrac{1}{9}$

19. $\left(\dfrac{27}{64}\right)^{-4/3}=\left[\left(\dfrac{64}{27}\right)^{1/3}\right]^4=\left(\dfrac{4}{3}\right)^4=\dfrac{256}{81}$

21. $(16p^4)^{1/2}=4p^2$

23. $(27x^6)^{2/3}=[(27x^6)^{1/3}]^2=(3x^2)^2=9x^4$

25. $2^{-3}\cdot 2^{-4}=2^{-7}=\dfrac{1}{2^7}$

27. $27^{-2}\cdot 27^{-1}=27^{-3}=\dfrac{1}{27^3}$

29. $\dfrac{4^{-2}\cdot 4^{-1}}{4^{-3}}=4^{(-2+(-1)-(-3))}=4^0=1$

31. $(m^{2/3})(m^{5/3})=m^{(2/3+5/3)}=m^{7/3}$

33. $(1+n)^{1/2}(1+n)^{3/4}=(1+n)^{(2/4+3/4)}=(1+n)^{5/4}$

35. $(2y^{3/4}z)(3y^{-2}z^{-1/3}) = 6y^{(3/4-8/4)}z^{(1-1/3)} = 6y^{-5/4}z^{2/3} = \dfrac{6z^{2/3}}{y^{5/4}}$

37. $(4a^{-2}b^{7})^{1/2}\cdot(2a^{1/4}b^{3})^{5} = (2a^{-1}b^{7/2})(2^{5}a^{5/4}b^{15}) = 2^{6}a(^{-1+5/4})b^{(7/2+30/2)} = 2^{6}a^{1/4}b^{37/2}$

39. $\left(\dfrac{r^{-2}}{s^{-5}}\right)^{-3} = \dfrac{r^{6}}{s^{15}}$

41. $\left(\dfrac{-a}{b^{-3}}\right)^{-1} = \left(\dfrac{b^{-3}}{-a}\right) = \dfrac{-1}{ab^{3}}$

43. $\dfrac{12^{5/4}\cdot y^{-2}}{12^{-1}y^{-3}} = 12^{(5/3-(-1))}y^{(-2-(-3))} = 12^{9/4}y$

45. $\dfrac{8p^{-3}(4p^{2})^{-2}}{p^{-5}} = \dfrac{8p^{5}}{p^{3}(4p^{2})^{2}} = \dfrac{8p^{5}}{16p^{7}} = \dfrac{1}{2p^{2}}$

47. $\dfrac{m^{7/3}n^{-2/5}p^{3/8}}{m^{-2/3}n^{3/5}p^{-5/8}} = m^{(7/3+2/3)}n^{(-2/5-3/5)}p^{(3/8+5/8)} = m^{3}n^{-1}p^{1} = \dfrac{m^{3}p}{n}$

49. $\dfrac{-4a^{-1}a^{2/3}}{a^{-2}} = -4a^{(-3/3+2/3+6/3)} = -4a^{5/3}$

51. $\dfrac{(k+5)^{1/2}(k+5)^{-1/4}}{(k+5)^{3/4}} = (k+5)^{(1/2-1/4-3/4)} = (k+5)^{-1/2} = \dfrac{1}{(k+5)^{1/2}}$

53. $y^{5/8}(y^{3/8}-10y^{11/8}) = y^{(5/8+3/8)} - 10y^{(5/8+11/8)} = y - 10y^{2}$

55. $-4k(k^{7/3}-6k^{1/3}) = -4k^{(1+7/3)} + 24k^{(1+1/3)} = -4k^{10/3} + 24k^{4/3}$

57. $(x+x^{1/2})(x-x^{1/2}) = x^{2} - x^{(1+1/2)} + x^{(1+1/2)} - x^{(1/2+1/2)} = x^{2} - x$

59. $(r^{1/2}-r^{-1/2})^{2} = (r^{1/2})^{2} - 2(r^{1/2})(r^{-1/2}) + (r^{-1/2})^{2} = r - 2r^{0} + r^{-1} = r - 2 + \dfrac{1}{r}$

61. $4k^{-1} + k^{-2} = 4k^{1}\cdot k^{-2} + k^{-2} = k^{-2}(4k+1)$

63. $9z^{-1/2} + 2z^{1/2} = 9z^{-1/2} + 2z^{-1/2}\cdot z^{2/2} = z^{-1/2}(9+2z)$

65. $p^{-3/4} - 2p^{-7/4} = p^{-7/4}\cdot p^{4/4} - 2p^{-7/4} = p^{-7/4}(p-2)$

67. $(p+4)^{-3/2} + (p+4)^{-1/2} + (p+4)^{1/2} = (p+4)^{-3/2}(1+(p+4)+(p+4)^{2}) =$

$(p+4)^{-3/2}(1+p+4+p^{2}+8p+16) = (p+4)^{-3/2}(p^{2}+9p+21)$

R.5: Radicals

1. $(-3x)^{1/3} = \sqrt[3]{-3x}$; F

3. $(-3x)^{-1/3} = \dfrac{1}{(-3x)^{1/3}} = \dfrac{1}{\sqrt[3]{-3x}}$; H

5. $(3x)^{1/3} = \sqrt[3]{3x}$; G

7. $(3x)^{-1/3} = \dfrac{1}{(3x)^{1/3}} = \dfrac{1}{\sqrt[3]{3x}}$; C

9. $(-m)^{2/3} = \sqrt[3]{(-m)^2}$ or $(\sqrt[3]{-m})^2$

11. $(2m+p)^{2/3} = \sqrt[3]{(2m+p)^2}$ or $(\sqrt[3]{(2m+p)})^2$

13. $\sqrt[5]{k^2} = k^{2/5}$

15. $-3\sqrt{5p^3} = -3(5p^3)^{1/2} = -3 \cdot 5^{1/2} p^{3/2}$

17. A: $\sqrt{ab} = \sqrt{a} \cdot \sqrt{b}$ is true for $a > 0, b > 0$.

19. $\sqrt{9ax^2} = 3x\sqrt{a}$ is true for all $x \ge 0$.

21. $\sqrt[3]{125} = \sqrt[3]{5^3} = 5$

23. $\sqrt[5]{-3125} = \sqrt[5]{(-5)^5} = -5$

25. $\sqrt{50} = \sqrt{25 \cdot 2} = 5\sqrt{2}$

27. $\sqrt[3]{81} = \sqrt[3]{3^4} = 3\sqrt[3]{3}$

29. $-\sqrt[4]{32} = -\sqrt[4]{2^5} = -2\sqrt[4]{2}$

31. $-\sqrt{\dfrac{9}{5}} = -\dfrac{\sqrt{9}}{\sqrt{5}} \cdot \dfrac{\sqrt{5}}{\sqrt{5}} = -\dfrac{3\sqrt{5}}{5}$

33. $-\sqrt[3]{\dfrac{4}{5}} = -\dfrac{\sqrt[3]{4}}{\sqrt[3]{5}} \cdot \dfrac{\sqrt[3]{5^2}}{\sqrt[3]{5^2}} = -\dfrac{\sqrt[3]{100}}{5}$

35. $\sqrt[3]{16(-2)^4(2)^8} = \sqrt[3]{(2)^4(-2)^4(2)^8} = \sqrt[3]{(2)^{12}(-2)^4} = 2^4(-2)\sqrt[3]{-2} = -32\sqrt[3]{-2}$ or $32\sqrt[3]{2}$

37. $\sqrt{8x^5z^8} = \sqrt{2^3x^5z^8} = 2x^2z^4\sqrt{2x}$

39. $\sqrt[3]{16z^5x^8y^4} = \sqrt[3]{2^4z^5x^8y^4} = 2zx^2y\sqrt[3]{2z^2x^2y}$

41. $\sqrt[4]{m^2n^7p^8} = np^2\sqrt[4]{m^2n^3}$

43. $\sqrt[4]{x^4 + y^4}$ cannot be simplified.

45. $\sqrt{\dfrac{2}{3x}} = \dfrac{\sqrt{2}}{\sqrt{3x}} \cdot \dfrac{\sqrt{3x}}{\sqrt{3x}} = \dfrac{\sqrt{6x}}{3x}$

47. $\sqrt{\dfrac{x^5y^3}{z^2}} = \dfrac{\sqrt{x^5y^3}}{\sqrt{z^2}} = \dfrac{x^2y\sqrt{xy}}{z}$

49. $\sqrt[3]{\dfrac{8}{x^2}} = \dfrac{\sqrt[3]{2^3}}{\sqrt[3]{x^2}} \cdot \dfrac{\sqrt[3]{x}}{\sqrt[3]{x}} = \dfrac{2\sqrt[3]{x}}{x}$

51. $\sqrt[4]{\dfrac{g^3h^5}{9r^6}} = \dfrac{\sqrt[4]{g^3h^5}}{\sqrt[4]{3^2r^6}} = \dfrac{h\sqrt[4]{g^3h}}{r\sqrt[4]{3^2r^2}} \cdot \dfrac{\sqrt[4]{3^2r^2}}{\sqrt[4]{3^2r^2}} = \dfrac{h\sqrt[4]{9g^3hr^2}}{3r^2}$

53. $\dfrac{\sqrt[3]{mn} \cdot \sqrt[3]{m^2}}{\sqrt[3]{n^2}} \cdot \dfrac{\sqrt[3]{n}}{\sqrt[3]{n}} = \dfrac{\sqrt[3]{m^3n^2}}{n} = \dfrac{m\sqrt[3]{n^2}}{n}$

55. $\dfrac{\sqrt[4]{32x^5y} \cdot \sqrt[4]{2xy^4}}{\sqrt[4]{4x^3y^2}} = \dfrac{\sqrt[4]{2^6x^6y^5}}{\sqrt[4]{2^2x^3y^2}} \cdot \dfrac{\sqrt[4]{2^2xy^2}}{\sqrt[4]{2^2xy^2}} = \dfrac{\sqrt[4]{2^8x^7y^7}}{2xy} = \dfrac{4x\sqrt[4]{x^3y^3}}{2xy} = 2\sqrt[4]{x^3y^3}$

57. $\sqrt[3]{\sqrt{4}} = \sqrt[3]{2}$

59. $\sqrt[6]{\sqrt[3]{x}} = (x^{1/3})^{1/6} = x^{1/18} = \sqrt[18]{x}$

61. $4\sqrt{3} - 5\sqrt{12} + 3\sqrt{75} = 4\sqrt{3} - 5(2\sqrt{3}) + 3(5\sqrt{3}) = 4\sqrt{3} - 10\sqrt{3} + 15\sqrt{3} = 9\sqrt{3}$

63. $3\sqrt{28p} - 4\sqrt{63p} + \sqrt{112p} = 3(2\sqrt{7p}) - 4(3\sqrt{7p}) + 4\sqrt{7p} = 6\sqrt{7p} - 12\sqrt{7p} + 4\sqrt{7p} = -2\sqrt{7p}$

65. $2\sqrt[3]{3} + 4\sqrt[3]{24} - \sqrt[3]{81} = 2\sqrt[3]{3} + 4(2\sqrt[3]{3}) - 3\sqrt[3]{3} = 2\sqrt[3]{3} + 8\sqrt[3]{3} - 3\sqrt[3]{3} = 7\sqrt[3]{3}$

67. $\dfrac{1}{\sqrt{3}} - \dfrac{2}{\sqrt{12}} + 2\sqrt{3} = \dfrac{1}{\sqrt{3}} \cdot \dfrac{\sqrt{3}}{\sqrt{3}} - \dfrac{2}{2\sqrt{3}} \cdot \dfrac{\sqrt{3}}{\sqrt{3}} + 2\sqrt{3} = \dfrac{\sqrt{3}}{3} - \dfrac{2\sqrt{3}}{6} + 2\sqrt{3} = \dfrac{2\sqrt{3} - 2\sqrt{3} + 12\sqrt{3}}{6} = \dfrac{12\sqrt{3}}{6} = 2\sqrt{3}$

69. $\dfrac{5}{\sqrt[3]{2}} - \dfrac{2}{\sqrt[3]{16}} + \dfrac{1}{\sqrt[3]{54}} = \dfrac{5}{\sqrt[3]{2}} \cdot \dfrac{\sqrt[3]{4}}{\sqrt[3]{4}} - \dfrac{2}{\sqrt[3]{16}} \cdot \dfrac{\sqrt[3]{4}}{\sqrt[3]{4}} + \dfrac{1}{\sqrt[3]{54}} \cdot \dfrac{\sqrt[3]{4}}{\sqrt[3]{4}} = \dfrac{5\sqrt[3]{4}}{2} - \dfrac{\sqrt[3]{4}}{2} + \dfrac{\sqrt[3]{4}}{6} = \dfrac{15\sqrt[3]{4} - 3\sqrt[3]{4} + \sqrt[3]{4}}{6} = \dfrac{13\sqrt[3]{4}}{6}$

71. $(\sqrt{2} + 3)(\sqrt{2} - 3) = (\sqrt{2})^2 - 3^2 = 2 - 9 = -7$

73. $(\sqrt[3]{11} - 1)(\sqrt[3]{11^2} + \sqrt[3]{11} + 1) = \sqrt[3]{11} \cdot \sqrt[3]{11^2} + \sqrt[3]{11} \cdot \sqrt[3]{11} + \sqrt[3]{11} \cdot 1 - \sqrt[3]{11^2} - \sqrt[3]{11} - 1 =$

$11 + \sqrt[3]{11^2} + \sqrt[3]{11} - \sqrt[3]{11^2} - \sqrt[3]{11} - 1 = 11 - 1 = 10$

75. $(\sqrt{3} + \sqrt{8})^2 = (\sqrt{3})^2 + 2\sqrt{3} \cdot \sqrt{8} + (\sqrt{8})^2 = 3 + 2\sqrt{24} + 8 = 11 + 2(2\sqrt{6}) = 11 + 4\sqrt{6}$

77. $(3\sqrt{2} + \sqrt{3})(2\sqrt{3} - \sqrt{2}) = 3\sqrt{2} \cdot 2\sqrt{3} + 3\sqrt{2}(-\sqrt{2}) + \sqrt{3} \cdot 2\sqrt{3} + \sqrt{3}(-\sqrt{2}) = 6\sqrt{6} - 6 + 6 - \sqrt{6} = 5\sqrt{6}$

79. $\dfrac{8}{\sqrt{5}} = \dfrac{8}{\sqrt{5}} \cdot \dfrac{\sqrt{5}}{\sqrt{5}} = \dfrac{8\sqrt{5}}{5}$

81. $\dfrac{6}{\sqrt[3]{x^2}} = \dfrac{6}{\sqrt[3]{x^2}} \cdot \dfrac{\sqrt[3]{x}}{\sqrt[3]{x}} = \dfrac{6\sqrt[3]{x}}{x}$

83. $\dfrac{\sqrt{3}}{\sqrt{5} + \sqrt{3}} \cdot \dfrac{\sqrt{5} - \sqrt{3}}{\sqrt{5} - \sqrt{3}} = \dfrac{\sqrt{3} \cdot \sqrt{5} - \sqrt{3} \cdot \sqrt{3}}{(\sqrt{5})^2 - (\sqrt{3})^2} = \dfrac{\sqrt{15} - 3}{5 - 3} = \dfrac{\sqrt{15} - 3}{2}$

85. $\dfrac{1 + \sqrt{3}}{3\sqrt{5} + 2\sqrt{3}} \cdot \dfrac{3\sqrt{5} - 2\sqrt{3}}{3\sqrt{5} - 2\sqrt{3}} = \dfrac{3\sqrt{5} - 2\sqrt{3} + \sqrt{3} \cdot 3\sqrt{5} - 2(\sqrt{3})^2}{(3\sqrt{5})^2 - (2\sqrt{3})^2} = \dfrac{3\sqrt{5} - 2\sqrt{3} + 3\sqrt{15} - 6}{45 - 12} = \dfrac{3\sqrt{5} - 2\sqrt{3} + 3\sqrt{15} - 6}{33}$

87. $\dfrac{p}{\sqrt{p} + 2} \cdot \dfrac{\sqrt{p} - 2}{\sqrt{p} - 2} = \dfrac{p\sqrt{p} - 2p}{(\sqrt{p})^2 - (2)^2} = \dfrac{p(\sqrt{p} - 2)}{p - 4}$

89. $\dfrac{a}{\sqrt{a+b}-1}\cdot\dfrac{\sqrt{a+b}+1}{\sqrt{a+b}+1}=\dfrac{a(\sqrt{a+b}+1)}{(\sqrt{a+b})^2-(1)^2}=\dfrac{a(\sqrt{a+b}+1)}{a+b-1}$

Chapter R-Test

1. (a) $(-3)^4\cdot(-3)^5=(-3)^{4+5}=(-3)^9$

(b) $\left(2x^3y\right)^2=2^2\left(x^3\right)^2y^2=4x^6y^2$

(c) $(-5z)^0=1$

(d) $\left(-\dfrac{4}{5}\right)^2=\dfrac{4^2}{5^2}$

(e) $-\left(\dfrac{m^2}{p}\right)^3=-\dfrac{m^6}{p^3}$

2. (a) $\left(x^2-3x+2\right)-\left(x-4x^2\right)+3x(2x+1)=\left(x^2-3x+2\right)-\left(x-4x^2\right)+6x^2+3x=$

$x^2-3x+2-x+4x^2+6x^2+3x=11x^2-x+2$

(b) $(6r-5)^2=(6r-5)(6r-5)=36r^2-60r+25$

(c) $(u+2)\left(3u^2-u+4\right)=3u^3-u^2+4u+6u^2-2u+8=3u^3+5u^2+2u+8$

(d) $(4x-5)(4x+5)=16x^2-25$

(e) $[(5p-1)+4]^2=[5p+3]^2=(5p+3)(5p+3)=25p^2+30p+9$

3. (a) $6x^2-17x+7=(3x-7)(2x-1)$

(b) $x^4-16=\left(x^2\right)^2-4^2=\left(x^2-4\right)\left(x^2+4\right)=(x-2)(x+2)\left(x^2+4\right)$

(c) $z^2-6zk-16k^2=(z+2k)(z-8k)$

(d) $x^3y^2-9x^3-8y^2+72=x^3\left(y^2-9\right)-8\left(y^2-9\right)=\left(y^2-9\right)\left(x^3-8\right)=$

$(x-2)\left(x^2+2x+4\right)(y+3)(y-3)$

4. (a) $\dfrac{16x^3}{4x^5}=\dfrac{4}{x^2};\{x\mid x\neq 0\}$

(b) $\dfrac{1+k}{k^2-1}=\dfrac{1+k}{(k+1)(k-1)}=\dfrac{1}{k-1};\{k\mid k\neq -1,1\}$

(c) $\dfrac{x^2+x-2}{x^2+5x+6}=\dfrac{(x+2)(x-1)}{(x+2)(x+3)}=\dfrac{x-1}{x+3};\{x\,|\,x\neq -3,-2\}$

5. (a) $\dfrac{5x^2-9x-2}{30x^3+6x^2}\cdot\dfrac{2x^8+6x^7+4x^6}{x^4-3x^2-4}=\dfrac{(5x+1)(x-2)}{6x^2(5x+1)}\cdot\dfrac{2x^6(x+2)(x+1)}{(x+2)(x-2)\left(x^2+1\right)}=\dfrac{x^4(x+1)}{3\left(x^2+1\right)}$

(b) $\dfrac{x}{x^2+3x+2}+\dfrac{2x}{2x^2-x-3}=\dfrac{x}{(x+2)(x+1)}+\dfrac{2x}{(2x-3)(x+1)}=\dfrac{x(2x-3)+2x(x+2)}{(2x-3)(x+2)(x+1)}=$

$\dfrac{4x^2+x}{(2x-3)(x+2)(x+1)}=\dfrac{x(4x+1)}{(2x-3)(x+2)(x+1)}$

(c) $\dfrac{a+b}{2a-3}-\dfrac{a-b}{3-2a}=\dfrac{a+b}{2a-3}+\dfrac{a-b}{2a-3}=\dfrac{2a}{2a-3}=\dfrac{-2a}{3-2a}$

(d) $\dfrac{g-\frac{2}{g}}{g-\frac{4}{g}}=\dfrac{g^2-2}{g^2-4}=\dfrac{g^2-2}{(g-2)(g+2)}$

6. (a) $(-7)^{-2}=\dfrac{1}{(-7)^2}=\dfrac{1}{49}$

(b) $\left(16x^8\right)^{3/4}=16^{3/4}\left(x^8\right)^{3/4}=8x^6$

(c) $\left(\dfrac{64}{27}\right)^{-2/3}=\left(\dfrac{27}{64}\right)^{2/3}=\dfrac{9}{16}$

7. (a) $\dfrac{5^{-3}\cdot 5^{-1}}{5^{-2}}=\dfrac{5^{-4}}{5^{-2}}=5^{-2}=\dfrac{1}{5^2}$

(b) $(x+2)^{-1/5}(x+2)^{-7/10}=(x+2)^{-9/10}=\dfrac{1}{(x+2)^{9/10}}$

(c) $\left(m^{-1}n^{1/2}\right)^4\left(4m^3n^{-2}\right)^{-1/2}=\left(m^{-4}n^2\,\dfrac{1}{2}m^{-3/2}n\right)=\dfrac{n^3}{2m^{11/2}}$

8. (a) $\sqrt[4]{16}=2$

(b) $\sqrt[3]{\dfrac{2}{3}}=\dfrac{\sqrt[3]{2}}{\sqrt[3]{3}}=\dfrac{\sqrt[3]{2\cdot 9}}{\sqrt[3]{3\cdot 9}}=\dfrac{\sqrt[3]{18}}{3}$

(c) $\sqrt{18x^5y^8}=\sqrt{9x^4y^8}\sqrt{2x}=3x^2y^4\sqrt{2x}$

(d) $\dfrac{\sqrt[4]{pq}\cdot\sqrt[4]{q^2}}{\sqrt[4]{p^3}}=\sqrt[4]{\dfrac{pq^3}{p^3}}=\sqrt[4]{\dfrac{q^3}{p^2}}=\dfrac{\sqrt[4]{p^2q^3}}{p}$

9. (a) $\sqrt{32x}+\sqrt{2x}-\sqrt{18x}=4\sqrt{2x}+\sqrt{2x}-3\sqrt{2x}=2\sqrt{2x}$

(b) $\left(\sqrt[3]{2}+4\right)\left(\sqrt[3]{4}+1\right)=\sqrt[3]{8}+\sqrt[3]{2}+4\sqrt[3]{4}+4=6+\sqrt[3]{2}+4\sqrt[3]{4}$

(c) $\left(\sqrt{x}-\sqrt{y}\right)\left(\sqrt{x}+\sqrt{y}\right)=x-\sqrt{xy}+\sqrt{xy}-y=x-y$

10. $\dfrac{14}{\sqrt{11}-\sqrt{7}}=\dfrac{14}{\sqrt{11}-\sqrt{7}}\cdot\dfrac{\sqrt{11}+\sqrt{7}}{\sqrt{11}+\sqrt{7}}=\dfrac{14\left(\sqrt{11}+\sqrt{7}\right)}{11-7}=\dfrac{7\left(\sqrt{11}+\sqrt{7}\right)}{2}$